AMERICAN REGISTRY OF PATHOLOGY

Publication & Education

ARP is pleased to inform you that when purchased from an approved seller, this fascicle comes with a digital version.

To access the online, searchable version:

- Go to www.arppress.org.
- Register your email.

For assistance, email ARP at admin@arppress.org.

Tumors of the Adrenal Glands and Extra-Adrenal Paraganglia

AFIP Atlases of Tumor and Non-Tumor Pathology

AMERICAN REGISTRY OF PATHOLOGY
Publication & Education

ARP PRESS

Arlington, Virginia

Director of Publications: Amy Goldenberg, PhD
Production Editor: Dian S. Thomas
Technical Editor: Elizabeth Tomlinson
Director of Operations, Sales, and Marketing: Heidi Guerrero
Copyeditor: Audrey Kahn

American Registry of Pathology
Arlington, Virginia 22203
www.arppress.org
ISBN 1-933477-47-4
978-1-933477-47-3

Copyright © 2025 The American Registry of Pathology

All rights reserved. No part of this publication may be reproduced or transmitted in any form or by any means without the written permission of the publisher. This includes electronic, mechanical, photocopy, recording, or any other information storage and retrieval system.

AFIP ATLASES OF TUMOR AND NON-TUMOR PATHOLOGY

Fifth Series
Fascicle 21

TUMORS OF THE ADRENAL GLANDS AND EXTRA-ADRENAL PARAGANGLIA

by

Ronald R. de Krijger, MD, PhD
Professor, Department of Pathology
University Medical Center and
Princess Máxima Center for Pediatric Oncology
Utrecht, The Netherlands

Arthur S. Tischler, MD
Tufts University School of Medicine, Tufts Medical Center
Department of Pathology and Laboratory Medicine
Boston, Massachusetts

Sylvia L. Asa, MD, PhD
Consultant in Endocrine Pathology, Department of Pathology
University Hospitals Cleveland Medical Center, and
University Health Network, Toronto
Professor, Department of Pathology
Case Western Reserve University
Cleveland, Ohio

Ernest E. Lack, MD
Senior Consulting Pathologist Endocrine Pathology
Joint Pathology Center
Silver Spring, Maryland

Marco Volante, MD, PhD
Department of Oncology
University of Turin
Orbassano, Turin, Italy

Published by the
American Registry of Pathology
Arlington, Virginia
2025

AFIP ATLASES OF TUMOR AND NON-TUMOR PATHOLOGY

EDITOR
Jason L. Hornick, MD, PhD
Professor of Pathology, Harvard Medical School
Director of Surgical Pathology and Immunohistochemistry
Brigham and Women's Hospital
Boston, Massachusetts

EDITORIAL ADVISORY BOARD

Laura C. Collins, MBBS	Beth Israel Deaconess Medical Center Boston, Massachusetts
Toby C. Cornish, MD, PhD	University of Colorado School of Medicine Aurora, Colorado
Cristina Magi-Galluzzi, MD, PhD	University of Alabama Birmingham, Alabama
Robert P. Hasserjian, MD	Massachusetts General Hospital Boston, Massachusetts
Arie Perry, MD	University of California San Francisco, California
Lynette M. Sholl, MD	Brigham and Women's Hospital Boston, Massachusetts
Tomas Slavik, MBChB, FCPath(SA), MMed	Ampath Pathology Laboratories Pretoria, South Africa
Rhonda K. Yantiss, MD	University of Miami Miller School of Medicine Miami, Florida
Matthew M. Yeh, MD, PhD	University of Washington Seattle, Washington

Manuscript reviewed by:
Justine A. Barletta, MD
C. Christofer Juhlin, MD, PhD

DEDICATIONS

This book is dedicated to the many patients who have suffered from adrenal and paraganglionic diseases and who have taught us the importance of accurate diagnosis.

EDITOR'S NOTE

The Atlases of Tumor Pathology have a long and distinguished history. They were first conceived at a cancer research meeting held in St. Louis in September 1947, as an attempt to standardize the nomenclature of neoplastic diseases. The first series was sponsored by the National Academy of Sciences-National Research Council. The organization of this formidable effort was entrusted to the Subcommittee on Oncology of the Committee on Pathology, and Dr. Arthur Purdy Stout was the first editor-in-chief. Many of the illustrations were provided by the Medical Illustration Service of the Armed Forces Institute of Pathology (AFIP), the type was set by the Government Printing Office, and the final printing was done at the Armed Forces Institute of Pathology. The American Registry of Pathology (ARP) purchased the Fascicles from the Government Printing Office and sold them virtually at cost. Over a period of 20 years, approximately 15,000 copies each of nearly 40 Fascicles were produced. The worldwide impact of these publications over the years has largely surpassed the original goal. They quickly became among the most influential publications on tumor pathology, primarily because of their overall high quality, but also because their low cost made them easily accessible the world over to pathologists and other students of oncology.

Upon completion of the first series, the National Academy of Sciences-National Research Council handed further pursuit of the project over to the newly created Universities Associated for Research and Education in Pathology (UAREP). The Second Series was started, generously supported by grants from the AFIP, the National Cancer Institute, and the American Cancer Society. Dr. Harlan I. Firminger became the editor-in-chief and was succeeded by Dr. William H. Hartmann. The Second Series Fascicles were produced as bound volumes instead of loose leaflets. They featured a more comprehensive coverage of the subjects, to the extent that the Fascicles could no longer be regarded as "atlases" but rather as monographs describing and illustrating in detail the tumors and tumor-like conditions of the various organs and systems.

Dr. Juan Rosai was appointed as editor-in-chief of the Third Series, and Dr. Leslie Sobin became associate editor. A distinguished Editorial Advisory Board was also convened, and these outstanding pathologists and educators played a major role in the success of this series, the first publication of which appeared in 1991 and the last (number 32) in 2003.

The same organizational framework applied to the Fourth Series, meticulously edited by Dr. Steven Silverberg with Dr. Ronald DeLellis as the associate editor. With UAREP and AFIP no longer functioning, ARP remained the responsible organization. The Fourth Series volumes were hardbound with illustrations almost exclusively in color. There was also an increased emphasis on the cytopathologic (intraoperative, exfoliative, or fine needle aspiration) and molecular features that are important in diagnosis and prognosis. At the time of the Fourth Series, ARP also produced

Atlases of Non-Tumor Pathology; these volumes were numbered separately from the tumor volumes.

As in the prior series, the goal of the Fifth Series includes a continuous attempt to correlate, whenever possible, the nomenclature used in the Fascicles with that proposed by the World Health Organization Classification of Tumors, as well as to ensure a consistency of style. Including molecular diagnostics is more important than ever, as is the availability of an online component, a more nimble website (www.arppress.org), and a social media presence. Now in this series, the tumor and non-tumor volumes are combined and consecutively numbered as the Atlases of Tumor and Non-Tumor Pathology. Close cooperation between the various authors and their respective liaisons from the Editorial Board will continue to be emphasized in order to minimize unnecessary repetition and discrepancies in the text and illustrations.

Particular thanks are due to the members of the Editorial Advisory Board, our reviewers, the editorial and production staff, and the individual Fascicle authors for their ongoing efforts to ensure that this series is a worthy successor to the previous four.

Jason L. Hornick, MD, PhD

PREFACE AND ACKNOWLEDGEMENTS

This fifth series ARP Fascicle Tumors of the Adrenal Glands and Extra-adrenal Paraganglia follows the fourth series volume published under the aegis of the American Registry of Pathology in 2007 by Dr. Ernest E. Lack. In the long intervening period, two versions of the WHO Endocrine blue book have appeared, in 2017 and 2022. Several authors of this Fascicle were largely involved in those two books as well, benefitting the shaping of the current work, albeit with a different scope, on the fascinating anatomy and pathology of these organs.

The adrenal gland is a complex endocrine organ composed of two distinct cell types, steroidogenic cells and paraneuronal neuroendocrine cells, the latter also forming the extra-adrenal paraganglia. These different cell types produce hormones of distinct families; the paraneurons produce catecholamines and the steroidogenic cells produce glucocorticoids, mineralocorticoids and/or sex steroids. The tumors that arise in the adrenals and extra-adrenal paraganglia are not only structurally and hormonally complex, they are also intriguing at the molecular level where many are recognized as components and harbingers of familial genetic tumor syndromes encompassing a variety of mutations affecting diverse signaling pathways. In this book, we have tried to summarize and illustrate the large body of knowledge in this field to provide a comprehensive reference for the clinical, biochemical, radiologic, histologic, immunohistochemical, and molecular features of the various tumors and tumor-like conditions that affect the adrenals and extra-adrenal paraganglia.

This work would not have been possible without the contributions of many individuals. Our experience has been based on consultations received from Pathologists, Endocrinologists, Surgeons, and Radiologists who see and treat patients with these tumors, and we offer our sincerest thanks for the education we have gained from these cases. We also thank the students, residents, fellows, and other trainees who have worked with us over the years; their questions and challenges have ensured continued investigation into adrenal and paraganglionic pathology and pathophysiology. Finally, we acknowledge the contributions of early anatomists who laid the foundation for understanding the histopathology of the adrenal gland and paraganglia and whose contributions in some cases were forgotten, only to be rediscovered as something new.

We owe a debt of gratitude to our mentors who were giants in the field of endocrine pathology: Gianni Bussolati, Richard Cohen, Ronald DeLellis, Philip Heitz, Paul Komminoth, Kalman Kovacs, Riccardo Lloyd, and Mauro Papotti. We also thank the reviewers of this work, Christofer Juhlin, and Justine Barletta, for their guidance and assistance that have improved the book.

None of this would have been possible without the support of our partners, Annet de Krijger-Wiebes, Judi Tischler, Erika Volante, and families whose patience has been unwavering during the long hours at work required to generate this book. Dr. Shereen Ezzat, Sylvia's partner and best friend, has a depth of knowledge about endocrine tumors that is unsurpassed, and his intellectual and emotional contributions were critical to this endeavor. Marco Volante will be always thankful to his father, Giuseppe Volante, for his long-life dedication to the pathology discipline and for having been an extraordinary and endless example of devotion to others (Nec morti esse locum, sed viva volare sideris in numerum atque alto succedere caelo; Publius Virgilius Maro Georgica. IV, 226-7).

Ronald R. de Krijger, MD, PhD
Arthur S. Tischler, MD
Sylvia L. Asa, MD, PhD
Ernest E. Lack, MD
Marco Volante, MD, PhD

Permissions to use copyrighted material has been granted by:

Elsevier

Endocrine Practice 2020;26:1366-83. For table 6-1.
Human Pathology 2021;110:50-61. For table 6-10.
Netter™ Images. The CIBA Collection Of Medical Illustrations;1965. For figure 3-1.

Lippincott Williams and Wilkins

Principles and Practice of Pediatric Oncology 1997:761-97. For figure 11-2.

Wolters Kluwer Health, Inc.

Journal of Clinical Oncology 1993;11:1466-77. For table 11-2.
Journal of Clinical Oncology 2009;27:289-97. For table 11-3.

CONTENTS

1 PHYSIOLOGY, DEVELOPMENT, AND ANATOMY OF THE ADRENAL GLANDS

The adrenal glands are composite endocrine organs consisting of the mesodermal-derived cortex and neural crest-derived medulla. These components interact with each other, with peripheral endocrine and nonendocrine tissues, and with the central nervous system (CNS) in order to develop normally during embryogenesis and to function normally throughout life.

A fundamental knowledge of normal adrenal function, development, and structure is essential to interpret the pathology and pathobiology of these complex organs. Familiarity with normal adrenal development from an anatomic standpoint is important both in helping to discriminate pathologic findings from normal anatomic variants (1) and for increasing awareness of variants such as adrenal cortical ectopia and fusions that can be diagnostic pitfalls, occasionally mimicking neoplasms. Molecular identification of factors that regulate adrenal development has provided new biomarkers for the diagnosis and prognostication of cortical and medullary tumors, while molecular studies of tumors have provided new insights into normal development. For example, the transcription factor steroidogenic factor 1 (SF1), which is needed for the development of steroid-producing cells (2), serves as a marker for steroid-producing tumors, and interacts in normal development with beta-catenin, which is frequently dysregulated in adrenal cortical carcinomas (3).

PHYSIOLOGY

Adrenal Cortex

The adrenal cortex plays vital roles via the actions of three major classes of steroid hormones: glucocorticoids, mineralocorticoids, and sex hormones (4,5). Their respective biosynthetic pathways are illustrated in figure 1-1. In the adult cortex, the hormone produced in the largest quantity is cortisol. In contrast, the predominant products of the fetal cortex are dehydroepiandrosterone (DHEA) and DHEA-sulfate (DHEAS).

Glucocorticoids regulate the intermediary metabolism of carbohydrates, fat, and proteins, and affect processes including wound healing, growth, inflammation, and functions of the immune system. In addition to these systemic effects, they contribute to survival of adrenal medullary endocrine cells, which are called chromaffin cells, during embryogenesis, and are needed for the development of epinephrine-producing chromaffin cells (6). They also modulate hormonal function in adult life by inducing catecholamine-synthesizing enzymes, tyrosine hydroxylase and phenylethanolamine N-methyltransferase (6). Catecholamines regulate steroidogenesis by several mechanisms (7). In humans and other mammals, corticomedullary interactions are facilitated by the complete envelopment of the medulla by the cortex. In contrast, birds and reptiles achieve the same effect by the intermingling of cell types.

The principal roles of mineralocorticoids are the regulation of water and electrolyte balance and the maintenance of intravascular volume through retention of sodium and elimination of potassium. There are still poorly understood additional functions that may include CNS effects on behavior and cognition, and interactions with glucocorticoid receptors (8).

Adrenal-derived sex hormones are particularly important in utero, where the fetal cortex produces testosterone and androstenedione to modulate the development of the fetal genitalia (9), and DHEA and DHEAS to serve as estrogen precursors (10). DHEAS is metabolized in the fetal liver to 16-hydroxy DHEAS, and both of these steroids can be hydrolyzed in the placenta, with subsequent aromatization to form estrone, estradiol, and estriol. The adrenal gland thereby

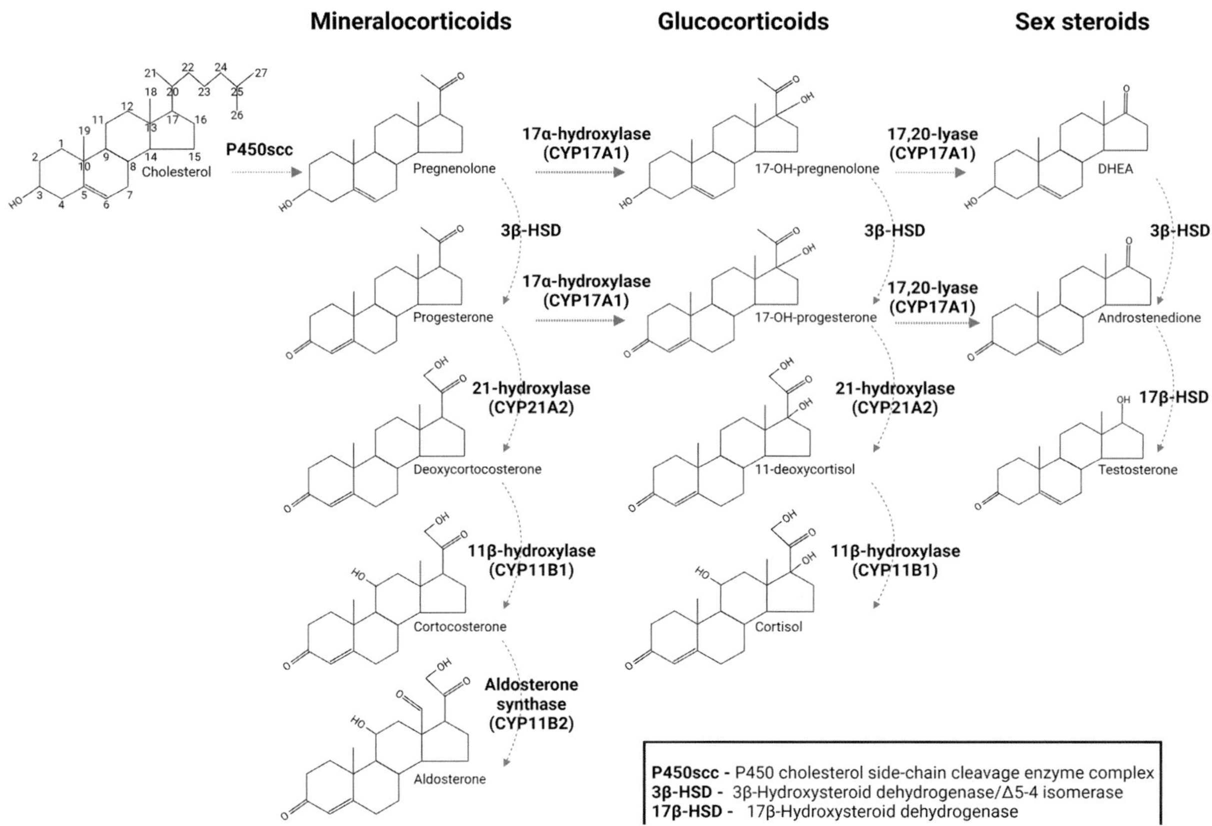

Figure 1-1

MAJOR BIOSYNTHETIC PATHWAYS FOR STEROIDOGENESIS

Both adrenal gland and gonadal pathways are shown. Functions distinguishing the adrenal cortical zones from each other and from the gonads are determined by differential expression of steroidogenic enzymes. CYP450c17 is not expressed in the adrenal zona glomerulosa, allowing progesterone to be 21-hydroxylated by P450c21 for aldosterone synthesis. A critical distinction between the adrenal and gonadal pathways is that the predominant "androgenic" form of 17β-hydroxy-steroid dehydrogenase (17β-HSD type 3) is expressed in the testis but not in the adrenal glands, which therefore have minimal ability to convert the weak androgens DHEA and androstenedione into testosterone. "17β-HSD" encompasses a large family of isoenzymes encoded by separate genes and differentially expressed in a wide variety of tissues. A small amount of testosterone may be produced in the adrenal zona reticularis by 17β-HSD type 5 (also known as aldo-keto reductase family 1 member C3 [AKR1C3]). (Fig. 5 from Asa SL, Uccella S, Tischler A. The unique importance of differentiation and function in endocrine neoplasia. Endocrine Pathol 2023;34:382-92.)

serves indirectly as a major source of estrogens in pregnancy while the placenta, which is the direct source, uses the adrenal glands as a supplier of the necessary precursors (10a). However, the precise roles of adrenal hormones in pregnancy remain unclear, as fetal adrenal steroidogenesis may not be essential for fetal development or survival (4). In adults, the adrenal cortex provides about 1 percent of biologically important androgens in men (most are of testicular origin), and roughly half of all androgens in women (11).

Steroid biosynthesis in the adrenal gland is regulated predominantly by pituitary-derived adrenocorticotropic hormone (ACTH) for glucocorticoids and the renin-angiotensin system for mineralocorticoids. In addition, intra-adrenal production of ACTH by a subset of adrenal medullary cells may provide a local paracrine regulatory mechanism (11a).

Adrenal Medulla

The biosynthetic pathways for catecholamines, the major hormones of the adrenal medulla, are shown in figure 1-2. The initial and rate-limiting enzyme is tyrosine hydroxylase (TH), which converts the amino acid tyrosine

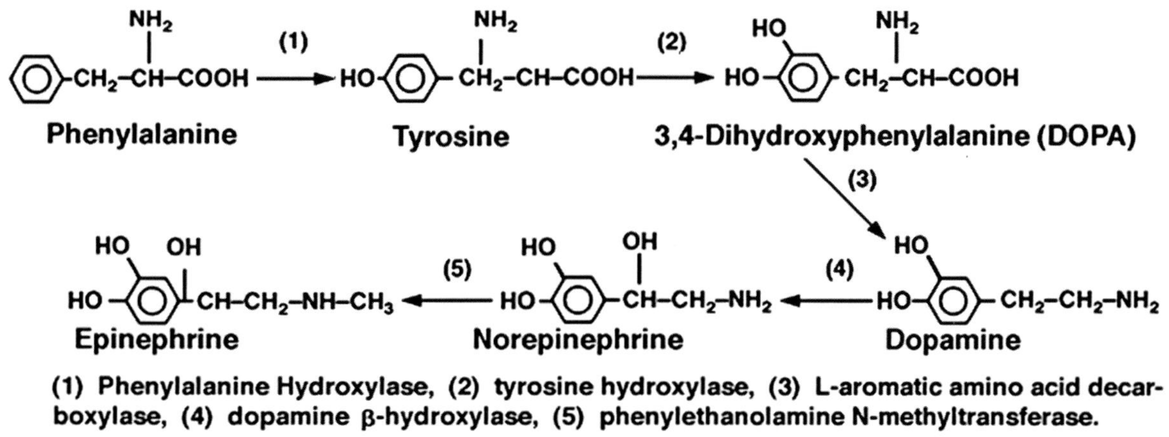

Phenylalanine —(1)→ Tyrosine —(2)→ 3,4-Dihydroxyphenylalanine (DOPA) —(3)→ Dopamine —(4)→ Norepinephrine —(5)→ Epinephrine

(1) Phenylalanine Hydroxylase, (2) tyrosine hydroxylase, (3) L-aromatic amino acid decarboxylase, (4) dopamine β-hydroxylase, (5) phenylethanolamine N-methyltransferase.

Figure 1-2

BIOSYNTHETIC PATHWAYS FOR CATECHOLAMINES

Tyrosine hydroxylase is required for essentially all catecholamine synthesis, while phenylethanolamine N-methyltransferase (PNMT) identifies cells capable of producing epinephrine.

into 3, 4-dihydroxyphenylalanine (DOPA); the cytosolic enzyme L-aromatic amino acid decarboxylase (AADC) then converts DOPA into dopamine, which is transported into the storage vesicle where the membrane-bound enzyme dopamine beta-hydroxylase (DBH) mediates the formation of norepinephrine. The terminal enzyme in catecholamine synthesis is phenylethanolamine N-methyltransferase (PNMT) (12), which is present in the cytosol. This enzyme methylates norepinephrine to form epinephrine, which is taken up again into the vesicle for storage.

Corticosteroids work in conjunction with other transcriptional regulators to control the PNMT gene by increasing PNMT transcriptional activity and ensuring appropriate processing of PNMT messenger RNA (13). The adrenal gland is the body's almost exclusive source of circulating epinephrine, also known as adrenaline, although small quantities are reported to be produced in other cell types for local use (14). The ratio of epinephrine to norepinephrine in the normal adult adrenal medulla is about 4 to 1 (12).

The classic role of the adrenal medulla is reacting to stress and changes in the environment that mandate rapid adaptations: the "fight or flight response." This homeostatic mechanism is mediated by catecholamines, mainly epinephrine. The rapidity of the response is enabled by the ability of the neuroendocrine cells comprising the medulla to respond to trans-synaptic signals from preganglionic sympathetic nerves by exocytosis of presynthesized pools of catecholamines stored in secretory granules (15). The clearance of catecholamines within the blood is also rapid, and epinephrine and norepinephrine have a half-life of only 1 to 2 minutes. In a complementary role to the adrenal medulla, the cortex also reacts to stress but does so more slowly, with de novo synthesis of cortisol from intracellular cholesterol-containing lipid in response to ACTH (4). In addition, sustained exposure to ACTH under conditions of chronic stress increases the adrenal glands' capacity to produce cortisol by exerting a trophic effect on the zona fasciculata.

The fight or flight response is a function of adult adrenal glands, in which the chromaffin cell acts essentially as a modified postganglionic sympathetic neuron. During much of fetal life, the adrenal gland contains very few chromaffin cells, and most of the chromaffin tissue resides in extra-adrenal sites, particularly the organ(s) of Zuckerkandl, where norepinephrine is the predominant catecholamine. Catecholamines detectable in the early fetal adrenal gland also contain a high proportion of norepinephrine, suggestive of functional immaturity. Fetal chromaffin tissue may be involved in the homeostatic maintenance of vascular tone and blood pressure in utero. An important stimulus of catecholamine secretion, both from extra-adrenal sites and from adrenal chromaffin cells before the onset of functional innervation, is hypoxia (16,17).

In addition to catecholamines, chromaffin cells produce a variety of regulatory peptides. The adrenal medulla is a prototype for amine- and peptide hormone-producing neuroendocrine cells throughout the body, and much of what is now known about those cells is derived from chromaffin cell biology. As a prime example, chromogranin A and other granin proteins were originally isolated from bovine chromaffin cells (18).

As with other neuroendocrine cells, normal chromaffin cells can produce more than one hormone. The major peptide hormones in the normal human adrenal medulla are endogenous opiate peptides called enkephalins (19,20), which may participate in stress responses by exerting antinociceptive effects in specific areas of the brain. Others include somatostatin, substance P (19), and ACTH (11a). Neuropeptides co-secreted with catecholamines may contribute to a coordinated stress response affecting both peripheral targets and the adrenal cortex (21).

The concept popular in the 1970s of a shared neural crest origin of all peptide-producing neuroendocrine cells has long since been disproved, except for the adrenal medulla and other paraganglia (see chapter 8). Nonetheless, because of the close functional similarities of all neuroendocrine cells, adrenal medullary tumors sometimes ectopically produce hormones characteristically associated with other locations.

EMBRYOLOGY AND DEVELOPMENTAL ANATOMY

Adrenal Cortex

The adrenal glands undergo a series of migrations and remodelings during development and postnatally, up to and beyond puberty (2). Although the molecular mechanisms regulating the development and functions of the cortex and medulla are in many respects still poorly understood, great strides have resulted from technologic advances in embryology and molecular endocrinology in the past decade. It is known that the transcription factors SF1 (encoded by *NR5A1* gene) and dosage-sensitive sex reversal DAX1 (encoded by *NROB1* gene) play key roles in early human adrenal cortical and reproductive development, but loss of SF1 function only rarely causes adrenal insufficiency, most often in combination with testicular dysgenesis, while X-linked adrenal hypoplasia congenita is associated with loss of DAX1 function (2,13,22). Wnt/β-catenin signaling is later involved in maintaining cortical zonation and cell replenishment (2,13,22). Signaling by numerous other ligands, including corticotrophin-releasing hormone (CRH), ACTH, angiotensin 2, fibroblast growth factor (FGF), and insulin-like growth factor (IGF) 1 and 2 is involved at various stages of cortical zonation (2,13,22). FGF signaling occurs through FGFR2 (isoform IIIb), which binds multiple FGFs; FGF1 is thought to be the main ligand (23).

Development of the adrenal medulla starts with migrating neural crest cells that express familiar transcription factors including SOX10 and FOXD3, followed by PHOX2B, ASCL1, and others. While these and other factors are expressed during both sympathetic neuronal and chromaffin cell development (24), many additional markers demonstrate different overall signatures for most developing chromaffin cells and neurons (25). Bridge cells and fate convergence add great complexity to developmental schema (26).

The historic and also most precise vantage point for assessing early human embryology at the anatomic level is the Carnegie International Staging System, named after the Carnegie Institute in Washington, DC. Based on studies beginning in the early 20th century, the Institute divided the first 8 weeks postovulation into 23 stages on the basis of morphologic development. While the system is not directly dependent on embryonic age or size, it correlates roughly with age, crown rump length, and main external characteristics, and allows for standardized assessments of growth and development between different investigators. Digitized images of the original Carnegie Institute slides are now available online (27), together with newly acquired molecular and experimental data from multiple databases. However, from a practical pathology standpoint and for later periods in gestation, more approximate dating based on days post-conception is usually employed.

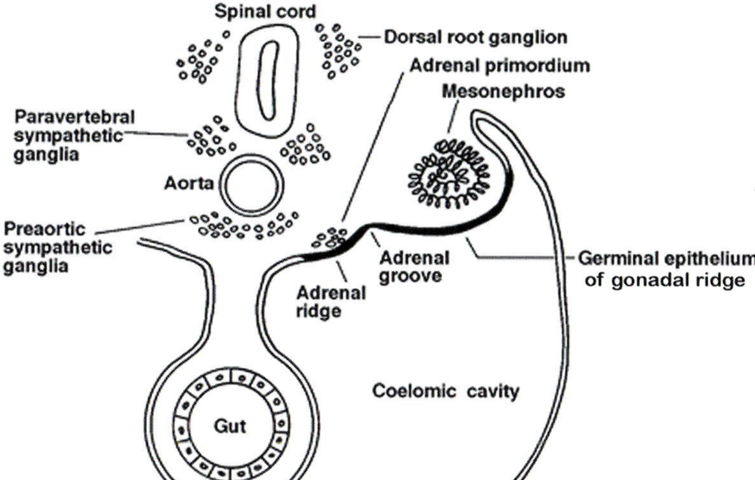

Figure 1-3

PRIMORDIA OF ADRENAL GLAND

Schematic cross section of human embryo shows primordia of adrenal glands, which are located lateral to the base of the dorsal mesentery, separated from the gonadal primordium by the adrenal groove. (Fig. 1-3 from Lack EE, Kozakewich HP. Embryology, developmental anatomy and selected aspects of non-neoplastic pathology. Pathology of the adrenal glands. New York: Churchill Livingstone; 1990:1-74.)

The adrenal glands begin to form as adreno-gonadal primordia (AGP) at stage 14 (about 20 to 30 days), from mesoderm just lateral to the dorsal mesentery near the cephalic end of the mesonephros in an area called the urogenital ridge. The AGP then separates, with a subset of cells migrating dorsomedially to form adrenal primordia (2), while most cells migrate dorsolaterally to form the primordial gonads. The two primordia are separated by the adrenal grooves (fig. 1-3).

From stages 15 to 18 (about 33 to 37 days), the adrenal glands enlarge and elongate into cigar-shaped structures extending from vertebral segments T6 to L1, lateral to the aorta and mesogastrium (3). This growth occurs as new cells proliferate in the subcapsular area, and migrate or expand centripetally to form the inner part of the gland, These areas start to become cytologically distinct at about 8 weeks, with the small proliferative subcapsular cells forming a densely packed band called the definitive zone, and a central area composed of large eosinophilic cells and called the fetal or provisional zone (2). By about 4 months' gestation, the adrenal glands are slightly larger than the kidneys and composed mostly of fetal/ provisional cortical cells arranged as irregular cords with intervening sinusoids (figs. 1-4, 1-5). The fetal adrenal gland is one of the largest organs in the abdomen, consisting of about 0.4 percent of total body weight from months 3 to 6 (fig. 1-6) (22).

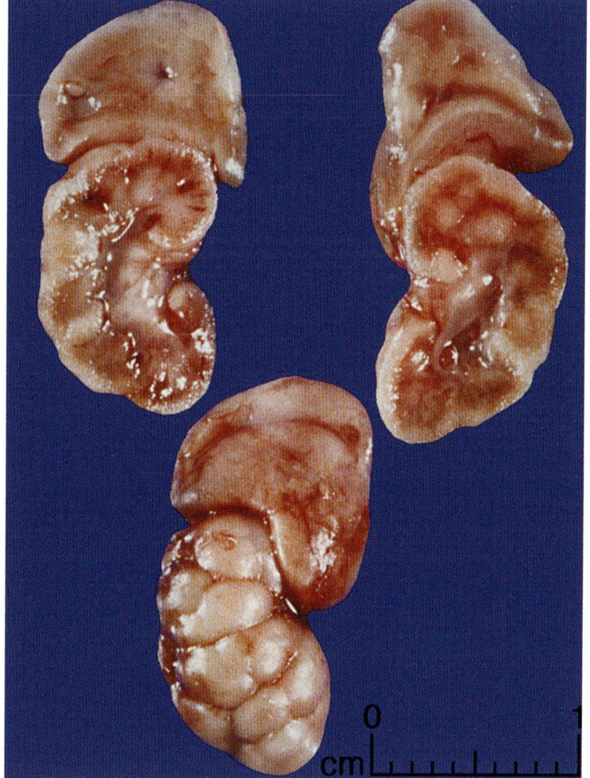

Figure 1-4

FETAL ADRENAL GLAND AND KIDNEY

Coronal sections (left and right) of adrenal gland and kidney from a fetus of about 18 weeks' gestational age. External aspect of one specimen is at bottom. Medullary tissue is not evident macroscopically at this early stage of development.

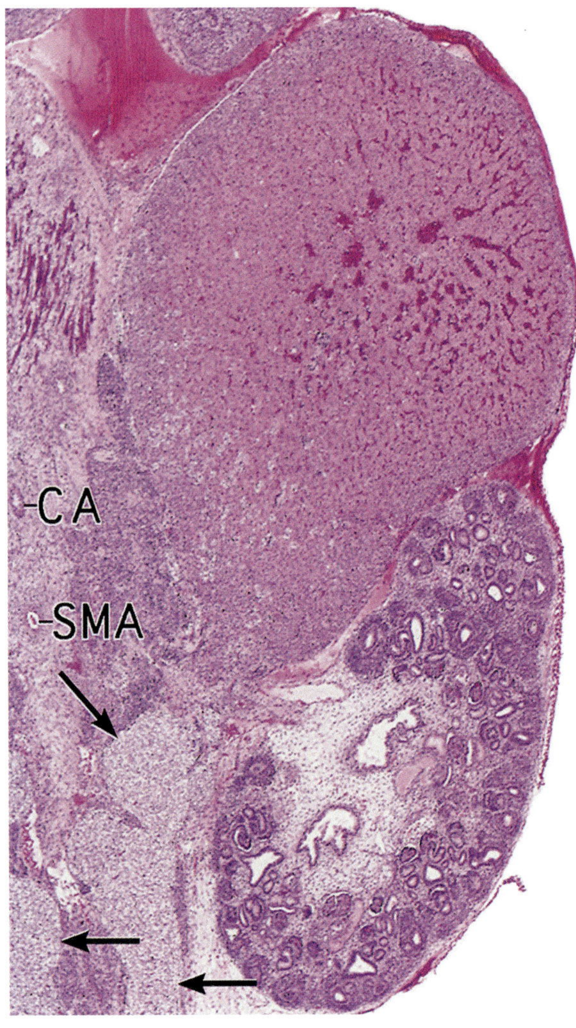

Figure 1-5

FETAL ADRENAL GLAND AND KIDNEY

Coronal section of adrenal kidney and adjacent tissue from a fetus at approximately 11 weeks' gestational age. The developing para-aortic sympathetic plexus has a close relationship with the medial and inferior aspects of the adrenal gland. Small pale-staining clusters of medullary progenitors separate cords of fetal or provisional cortical cells in the inferomedial aspect of the gland. Arrows point to extra-adrenal paraganglia, which are fully developed before the adrenal medulla exists and supply most of the circulating catecholamines in fetal life (see chapter 9). Celiac arteries (CA) and superior mesenteric arteries (SMA) are also present.

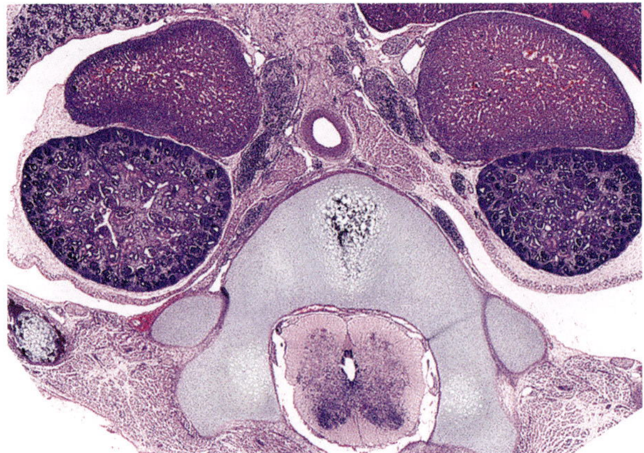

Figure 1-6

ANATOMIC RELATIONSHIPS OF FETAL ADRENAL GLANDS

Transverse section of upper abdomen in a fetus of about 10 weeks' gestational age. Liver is at the top left side of the image, pancreas at top right. The developing para-aortic and preaortic neural plexus and ganglia are medial to the adrenal glands. Both adrenal glands in this plane of section lie anterior and medial to the kidneys, and at this stage most of the adrenal glands are composed of fetal or provisional cells. The developing prevertebral sympathetic plexus is located on either side of the aorta.

During the late second trimester the definitive zone begins to form arched columns of cells that extend to the outer fetal zone. During the third trimester, these may elongate and begin to resemble the zona fasciculata of the adult cortex (10). Near term, the adrenal gland consists of about 75 to 85 percent fetal cortex (2) surrounded by a thin rim of definitive cortex (fig. 1-7).

Expression of steroidogenic enzymes and steroid production correlate with developmental changes in adrenal morphology (2). Enzymes required for steroidogenesis appear at about 50 to 52 days postconception (4). Patterns of expression undergo dynamic remodeling throughout gestation to allow for specific developmental needs (10). These changes include an early peak in cortisol synthesis followed by suppressed synthesis until late in pregnancy (10).

The fetal zone rapidly degenerates postnatally, resulting in a marked decrease in adrenal gland volume and weight, and a marked decline in adrenal androgens (see Adrenal Gland Weight). Parturition is believed to be the triggering event but the mechanism is unclear (4). Concurrently with fetal cortex involution, the zona glomerulosa and zona fasciculata develop adult-type morphology under the trophic influence of angiotensin 2 and ACTH, respectively (28). The lost synthesis of adrenal androgens that accompanies postnatal degeneration of the fetal cortex is eventually recovered, along with regained adrenal weight, but adrenal gland development is not completed until puberty, with the establishment of the zona reticularis, which begins to form at ages 6 to 8 years in girls and 7 to 9 years in boys (2). This milestone is called "adrenarche" (2,10).

Because of the anatomic relationships during adrenal gland development (see fig. 1-3), adrenal cortical tissue can be found later in life anywhere in the vicinity of the fetal adrenal primordium or along the line of descent of the gonads in males and females, including gonadal adnexal structures and, rarely, within the substance of the gonad or kidney. These ectopic rests can be a source of confusion and be interpreted clinically or microscopically as metastases from adrenal cortical or other tumors. Ectopic adrenal cortical rests frequently recapitulate the architecture of normal adrenal cortex but lack medulla, particularly at increasing distances from the native adrenal gland.

Adrenal Medulla

The adrenal medulla does not begin to form until Carnegie stages 16 to 19 (about 6 weeks) as the cortical primordia start to be colonized by immature medullary precursors migrating from the neural crest. This occurs concurrently with the formation and enlargement of the paravertebral sympathetic ganglia. The longstanding concept of a single sympathoadrenal precursor that gives rise to either chromaffin cells or sympathetic neurons, depending on its environment, has largely been overturned by new lineage-tracing embryologic techniques showing progenitors with predetermined fates and separate routes of migration contributing to the development of these two closely related cell types.

While cells destined for sympathetic ganglia predominantly migrate to their destinations directly from the neural crest, the adrenal medulla is formed in large part from neural crest-derived cells that migrate from the dorsal root ganglia (DRG). These cells have been named Schwann cell precursors (SCPs) (25), somewhat confusingly, since they also give rise to other cell types (24). Derivatives of SCPs migrate along the DRG axons to join with preganglionic sympathetic axons emerging from the spinal cord, then travel laterally along those axons to enter the adrenal gland (25). "Bridge cells," indicating transitions between neuroblast and chromaffin cell lineages, also contribute to adrenal development (fig. 1-8) (29). Comparable axonally guided migration of neural crest progenitors along nerves near the superior cervical ganglion has been proposed for development of carotid body chief cells (30).

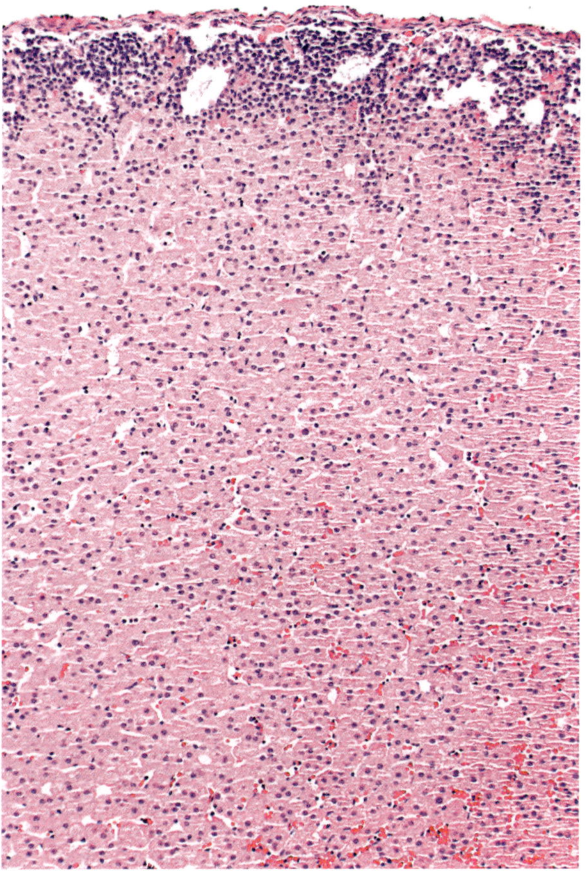

Figure 1-7

LATE GESTATION ADRENAL GLAND FROM A PREMATURE STILLBORN

Most of this adrenal gland is composed of provisional cortical cells with abundant eosinophilic cytoplasm. There is a vague suggestion of columns resembling zona fasciculata, topped by a thin band of darker definitive cortex, with small cysts forming arches.

This model may help to account for the small numbers of neurons present in normal adrenal medulla and chromaffin cells within sympathetic ganglia (31,32) (see chapter 8). It is possible that precursors of those cells correspond to distinctive subsets of cells identified in Carnegie

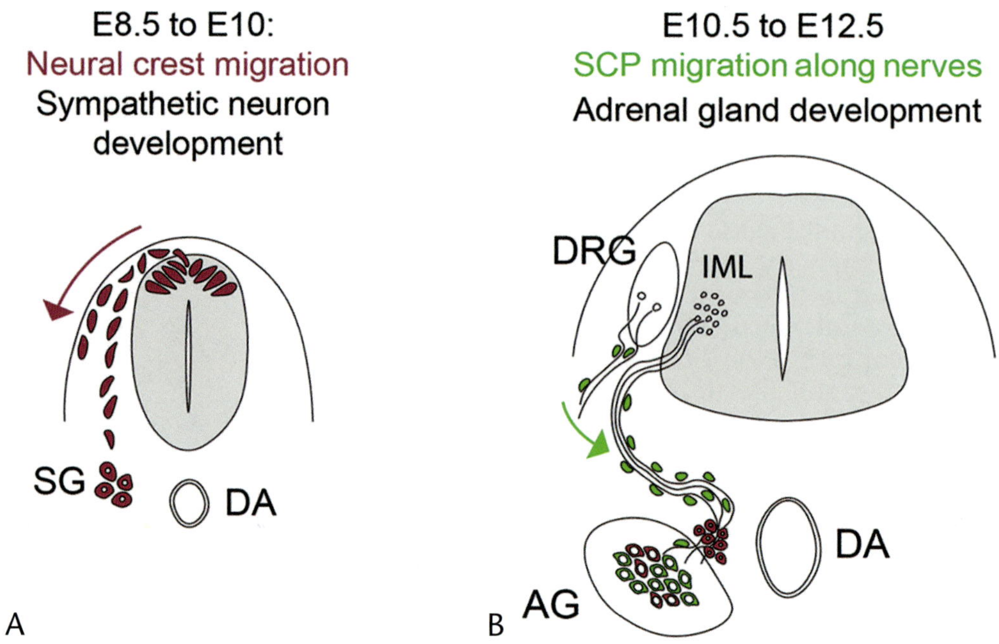

Figure 1-8

CURRENT CONCEPT OF NEURAL CREST CELL MIGRATION

A: Neural crest cells directly migrate laterally and ventrally to give rise to cutaneous melanocytes and sympathetic ganglia (SG). B: Slightly later in embryogenesis a different group of neural crest-derived cells migrates from the dorsal route ganglia (DRG) along sensory axons to reach preganglionic sympathetic axons emerging from the intermediolateral column (IML) of the spinal cord. They then travel to the developing adrenal glands (AG) guided by those axons, which will also provide innervation of the medulla (DA = dorsal aorta). (Courtesy of Drs. M. Kastriti and I. Adameyko, Vienna, Austria)

Institute studies on the basis of size and shape that apparently reach the adrenal glands from within sympathetic ganglia (33), and with bridge cells in molecular lineage tracing (24). While the model was initially based on mouse studies, single-cell transcriptomics and lineage tracing in early human embryos confirm that intra-adrenal sympathoblasts are derived from nerve-associated SCPs, similarly to chromaffin cells, while extra-adrenal sympathoblasts arise mostly from direct neural crest cell migration (29).

Adrenal medullary precursors enter the dorsal and medial aspect of the adrenal glands (see figs. 1-5 and 1-9), while the ventrolateral part remains relatively intact. As the medulla enlarges, small collections of cells separate the cords of cortical cells. At Carnegie stage 19, there are small nests of immature cells scattered throughout most of the cortex, and at stage 23, there is cytologic evidence of a cell population maturing into chromaffin cells. This is apparent as cells with an increased amount of pale-staining cytoplasm and slightly enlarged nuclei with dispersed chromatin (fig. 1-10). At the same time, smaller immature cells continue to be present, both admixed with clear cells or in separate aggregates. Some of the latter include pseudorosettes (Homer Wright rosettes) (fig. 1-10), which can also be found in the developing sympathetic chain (see fig. 1-9). With postnatal regression of the fetal cortex, the developing medullary cells aggregate to the area of the central vein. However, the adrenal medulla achieves most of its volume during the first years after birth. During that time, it exists concurrently with the organ(s) of Zuckerkandl, which involutes at about age 3 years (34).

A consistent histologic finding in the normal development of the adrenal glands is the presence of nodular aggregates of primitive sympathetic cells ("neuroblastic nodules") (fig. 1-11). The striking resemblance of these to miniature developing sympathetic ganglia has been noted in embryologic studies (29). They

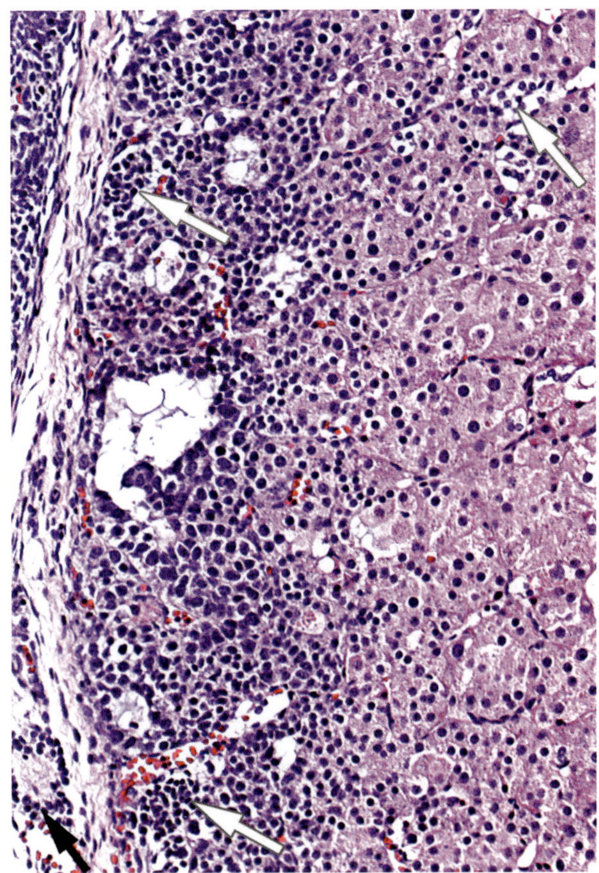

Figure 1-9

MEDIAL ASPECT OF DEVELOPING ADRENAL GLAND IN HUMAN FETUS

Coronal section of fetal adrenal gland shows close association with developing sympathetic plexus (on far left). Small nests of primitive sympathetic cells (white arrows) are present in the definitive cortex beneath the adrenal capsule, percolating through the definitive and provisional cortex where some enlarge and acquire pale cytoplasm. They are most likely future chromaffin cells. A pseudorosette (Homer Wright rosette), which can be a normal developmental finding, is present in the developing sympathetic plexus (black arrow).

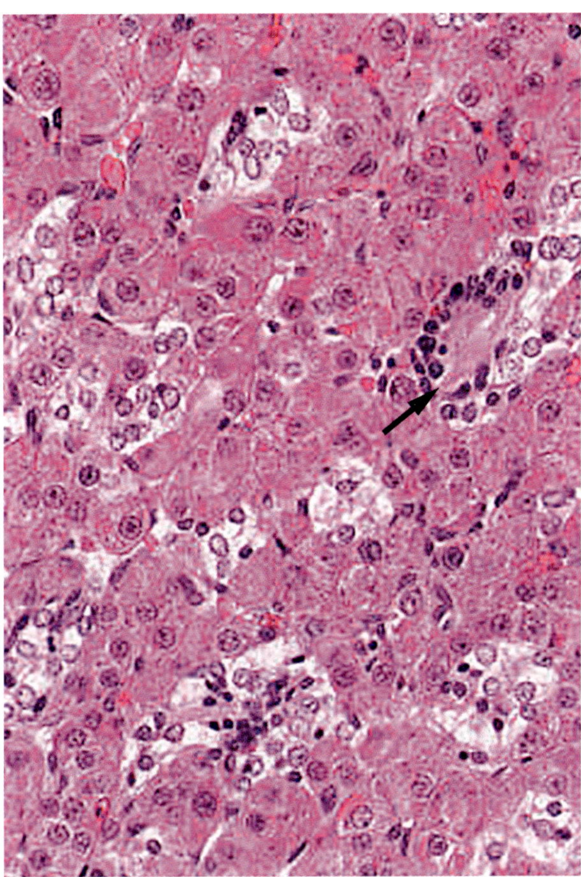

Figure 1-10

PROVISIONAL ADRENAL CORTEX WITH MEDULLARY PROGENITORS

Aggregates of presumptive chromaffin cells with clear cytoplasm coexist with primitive sympathetic cells, some of which are arranged in a pseudorosette with central pale fibrillary matrix (neuropil) (arrow).

may be superficially located near the adrenal capsule early in gestation (note small subcapsular aggregates in figure 1-9), but in older fetuses are located more centrally as a result of inward migration or relatively greater growth of the cortex. Some neuroblastic nodules become cystic, but usually not until the 16th week of gestation (fig. 1-12).

The number and size of neuroblastic nodules increase with age, peaking at 17 to 20 weeks' gestational age, and there is a tendency to regress in older fetuses. However, some may linger until birth or early infancy (35). In all age groups combined, neuroblastic nodules range from 60 x 60 µm to a maximum size of about 200 x 400 µm (35). Immunohistochemical stains for tyrosine hydroxylase are positive in maturing chromaffin cells (fig. 1-13), some of which also stain for chromogranin A. Most neuroblastic nodules immunostain for S-100 protein, showing occasional cells consistent with SCPs, early sustentacular cells, or Schwann cells (fig. 1-14) (36).

Hypothetically, neuroblastic nodules may give rise to normal adrenal medullary cell types, undergo regression or apoptosis, or become the

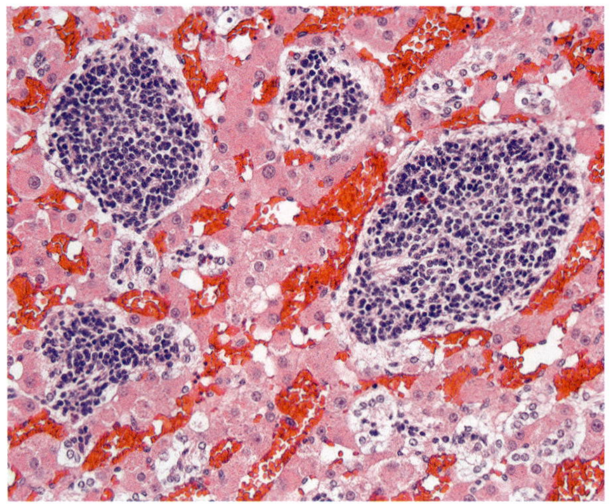

Figure 1-11

NEUROBLASTIC NODULES AND MATURING MEDULLARY PROGENITORS IN DEVELOPING FETAL ADRENAL GLAND

Fetal cortex with interspersed aggregates of immature cells with varied sizes and cytologic features. These include presumptive chromaffin cells with pale staining cytoplasm and small nodules of primitive neuroblastic cells. Some neuroblastic nodules contain cells arranged in pseudorosettes.

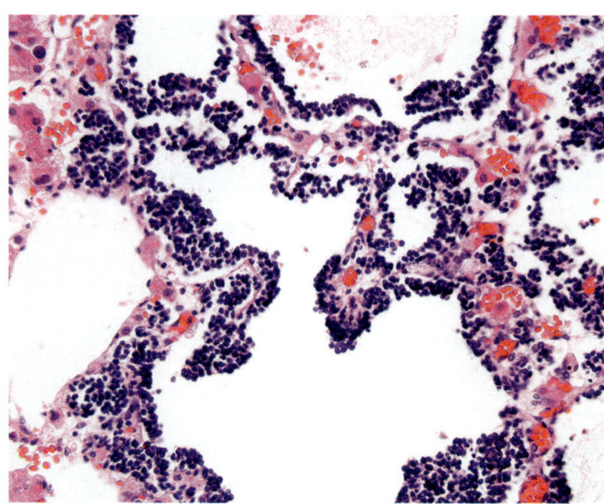

Figure 1-12

CYSTIC DEGENERATION OF NEUROBLASTIC NODULES

A few nodules of primitive neuroblastic cells in this adrenal gland of about 16 weeks' gestation show prominent cystic change, which may be a step in eventual involution.

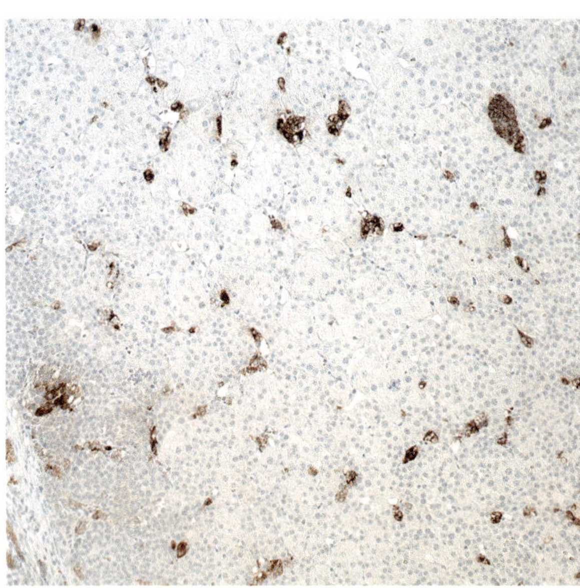

Figure 1-13

TYROSINE HYDROXYLASE IN DEVELOPING FETAL ADRENAL GLAND

Most of the medullary progenitor cells in this image express TH. Because TH is a cytosolic enzyme, it can often detect ability to produce catecholamines in immature or poorly differentiated cells with sparse secretory granules that are negative for granin proteins.

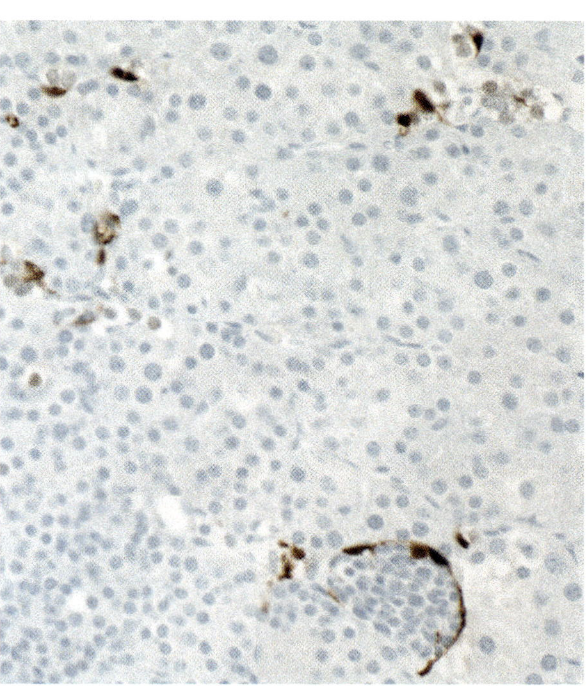

Figure 1-14

S-100 PROTEIN IN DEVELOPING FETAL ADRENAL GLAND

Some S-100 protein-positive cells are present at the periphery of a small neuroblastic aggregate (right), while others are associated with scattered more mature-appearing cells with clear cytoplasm (left).

nidus for in situ or clinically overt neuroblastoma or a related tumor (see chapter 10). They can become a significant pitfall in the differential diagnosis of in situ neuroblastoma (1).

Immunohistochemical staining for proliferation markers shows that from early stages of adrenal gland development expansion of the adrenal cortex is driven mainly by proliferation of cells in the definitive zone, with little proliferation in the centrally located provisional cortex (fig. 1-15) (2). In contrast, expansion of the medulla, at least early on, appears to simultaneously involve immigration, proliferation, and centripetal migration of morphologically diverse medullary progenitors (fig. 1-15).

WEIGHT AND GROSS ANATOMY

Adrenal Gland Weight

Weighing of the adrenal glands requires careful removal of attached connective tissue and fat in order to obtain reliable specimen weight. Biologic factors affecting adrenal weight include age, abnormalities in development, and the presence and chronicity of various diseases (37). Physical data regarding adrenal gland weights in infancy and childhood are based per force on postmortem studies (38). Figure 1-16 shows the average combined weights of meticulously dissected adrenal glands from 226 individuals ranging from 30 weeks' gestational age to 35 years (38).

The adrenal glands grow rapidly toward the end of pregnancy but there is a dramatic decrease in weight in the first few weeks postnatally caused almost entirely by regression of the fetal cortex. These data are corroborated and supplemented by volumetric fetal ultrasound studies, which also show disease-related effects on the time course for regression of the fetal cortex in living patients (39). In a study of adrenal glands obtained surgically from adult females undergoing staged bilateral adrenalectomy for advanced breast cancer, the average weight of individual glands was 4.0 g (1 SD = 0.8 g). Adrenal glands obtained at autopsy were heavier (average, 6.0 g), probably due to trophic stimulation by ACTH resulting from illness-related stress (40). Adrenal glands from patients given a crude extract of ACTH prior to surgery underwent a nearly two-fold enlargement (average weight, 8.1 g), and showed conversion of the pale, lipid-rich cells of the zona fasciculata to cells with compact, eosinophilic cytoplasm (40).

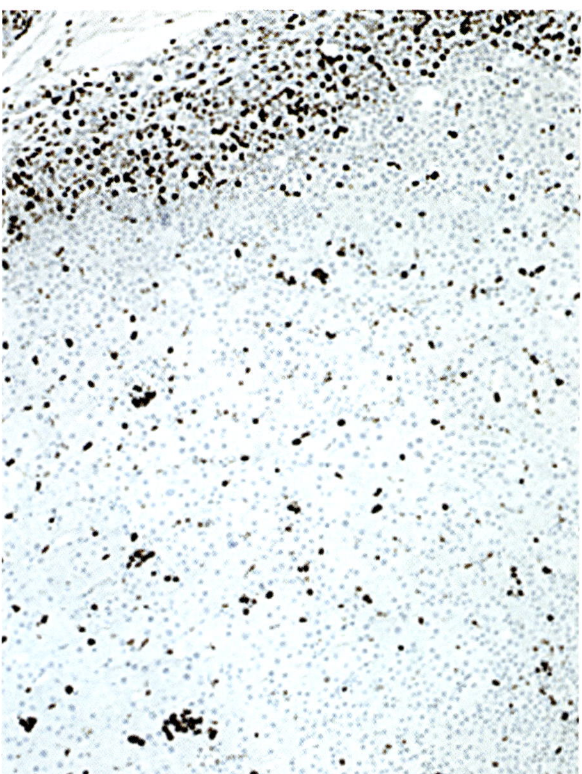

Figure 1-15

KI-67 IN FETAL ADRENAL GLAND

Immunoreactivity for Ki-67 is abundant in the definitive zone, which is the major site of cortical cell generation. Scattered individual stained cortical cells are also present in the provisional zone, along with aggregates of labeled medullary progenitors.

In children and adults in the United States, there are no significant differences in adrenal weights with regard to sex or laterality. Since 98 percent of apparently normal glands removed surgically weigh less than 6 g, adrenal glands heavier than this may be abnormal, but this assessment must be based upon weights of meticulously dissected glands and correlation with clinical and endocrinologic data. Gross dissection and weighing of the adrenal medulla has shown that this component accounts for about 10 percent of the weight of the entire gland (0.431 g left, 0.448 g right) (41). In a study of a large Chinese population (n=333), patients with

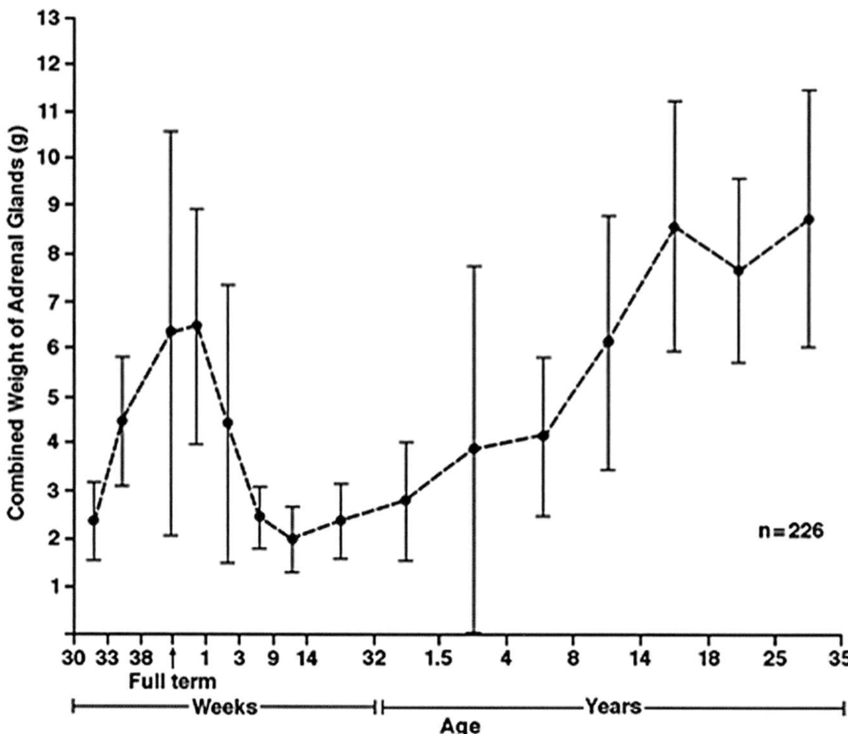

Figure 1-16

AVERAGE COMBINED WEIGHT OF ADRENAL GLANDS FROM 226 PATIENTS AT AUTOPSY

The patients ranged from 30 weeks' gestational age to 35 years. Vertical bars represent one standard deviation. There is a marked decrease in combined weight during the first few weeks of life. (Fig. 1-11 from Lack EE, Kozakewich HP. Embryology, developmental anatomy and selected aspects of non-neoplastic pathology. Pathology of the adrenal glands. New York: Churchill Livingstone; 1990:1-74.)

hypertension or lung cancer had significantly heavier adrenal glands; for those with lung cancer, ectopic ACTH production resulting in adrenal cortical hyperplasia was considered to be a possible factor for the increased weight (37).

Adrenal Gland Gross Anatomy

In newborns, the adrenal glands have a relatively smooth capsular surface (fig. 1-17). The gland may appear dark red to brown on transverse section due to a combination of regression of fetal cortex, congestion, and increased density of vascular sinusoids (fig. 1-18). On gross inspection, this appearance may give the erroneous impression of adrenal hemorrhage or infarction. In addition, the glands may be soft. Medullary tissue is not grossly identifiable in the newborn gland under normal circumstances since it constitutes less than 1 percent of the total gland volume.

The size and configuration of the gland can be altered by some congenital abnormalities, such as an extremely small size in congenital adrenal hypoplasia and a more rounded contour with ipsilateral renal agenesis. The adrenal glands in anencephalic newborns are small, with an average combined weight that can be less than 1.0 g (38). In anencephaly, the fetal cortex is markedly thin, although it is often normal in size and structure until approximately 20 weeks' gestation (38); by birth, the fetal cortex has undergone marked regression, leaving the definitive/adult cortex relatively prominent (38).

In adults, the right adrenal gland is roughly pyramidal while the left is elongated or crescentic (fig. 1-19). The posterior or dorsal surface of the gland is convex, with a ridge (or crista) that is often more prominent in the tail of the gland on the superior and lateral aspect (fig. 1-19A).

The anterior (or ventral) surfaces are relatively smooth and there may be a shallow groove, usually on the left side, which contains a segment of the adrenal vein (fig. 1-19B). The crista is flanked by flattened projections or alae (wings); these medial and lateral limbs are evident with computerized tomography (CT) or magnetic resonance imaging (MRI) (fig. 1-20).

The arterial supply to the adrenal gland has a triple origin from the inferior phrenic artery, the aorta, and the renal artery (fig. 1-21). The adrenal vein is much longer on the left side and

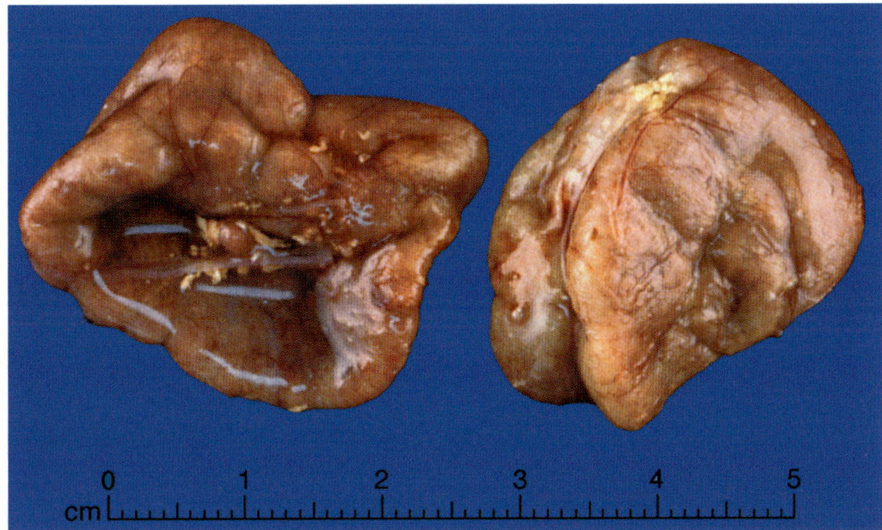

Figure 1-17

DORSAL ASPECT OF FETAL ADRENAL GLANDS

The dorsal surface of both adrenal glands from a 35-week premature infant shows a ridge (or crista) and a relatively smooth cortical surface. Each gland weighed 3 g.

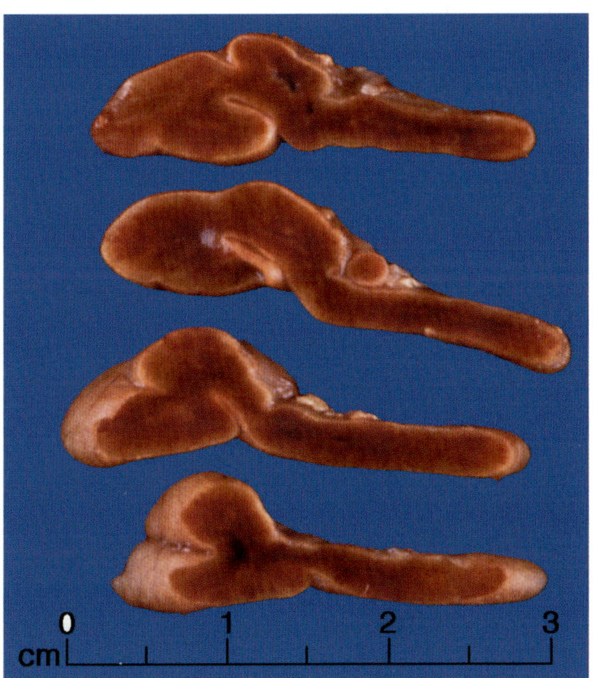

Figure 1-18

ADRENAL GLAND FROM A 35-WEEK PREMATURE INFANT

In this transverse section, much of the fetal or provisional cortex has a dark appearance. This may be a normal feature and should not be confused with adrenal hemorrhage.

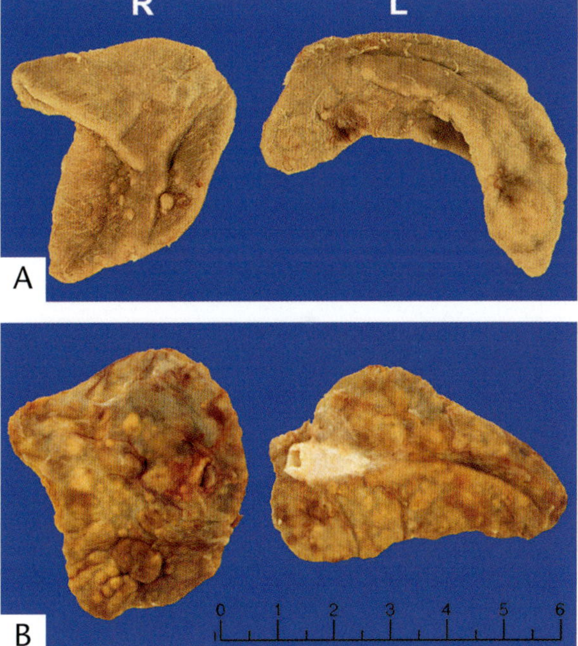

Figure 1-19

DORSAL AND VENTRAL SURFACES OF ADULT ADRENAL GLAND

A: In the adult, the dorsal surface of the right adrenal gland is pyramidal, while the opposite gland is more elongated. A few cortical extrusions are present.

B: Central adrenal vein exits as a single vein from the relatively flat ventral surface of the adrenal gland near the junction of the head and body. The adrenal vein is most evident in the elongated left gland and is associated with a well-developed groove. The pyramidal-shaped right adrenal gland has a shorter vein.

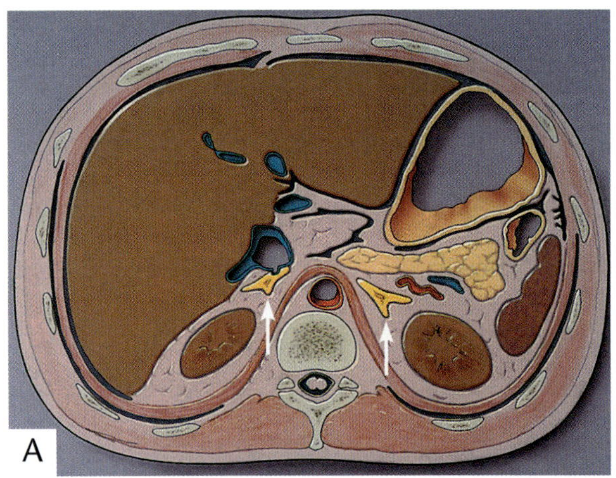

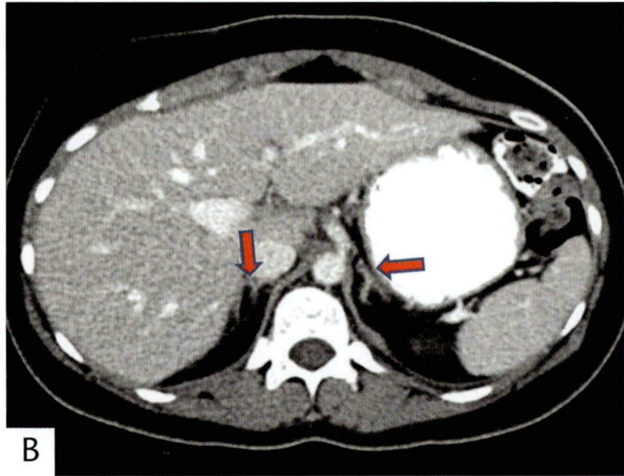

Figure 1-20

ADRENAL GLANDS IN TRANSVERSE PLANE THROUGH ABDOMEN

A: The adrenal ridge (or crista) is located on the dorsal aspect (arrows) and is flanked by medial and lateral wings (or alae), creating a Y-shaped appearance. The right adrenal gland has a relatively short adrenal vein which drains directly into the inferior vena cava.

B: Computerized tomography (CT) scan corresponding to A. The adrenal glands are indicated by arrows.

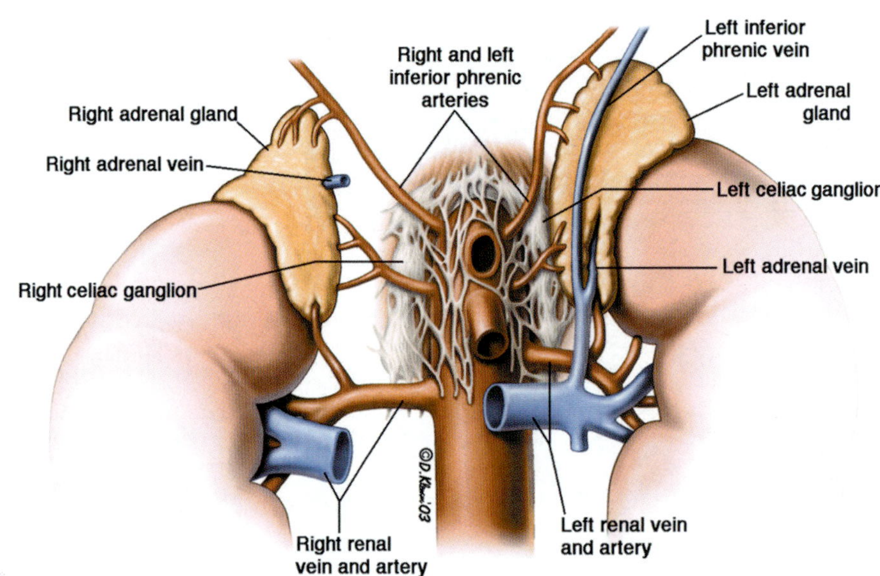

Figure 1-21

BLOOD SUPPLY OF THE ADRENAL GLANDS

In the coronal plane, the adrenal glands have a relatively flat ventral or anterior surface. Arterial supply is from the inferior phrenic artery, the aorta, and the renal artery. The left adrenal vein drains into the renal vein while the right adrenal vein is much shorter and drains into the inferior vena cava.

normally drains into the renal vein; the right adrenal vein is short and empties directly into the inferior vena cava (figs. 1-20A, 1-21).

In transverse sections perpendicular to the long axis, the gland can arbitrarily be divided into head (inferomedial one third), body (middle one third), and tail (superolateral one third) regions. The adrenal medulla is concentrated in the head and body of the gland, often with some extension into proximal parts of one of the ala. Extension into distal parts of alae is usually indicative of adrenal medullary hyperplasia. The medulla has an ellipsoid shape near the extremity of the head and an omega or sickle shape near the body (fig. 1-22). Using morphometric techniques, the ratio of cortex to medulla is approximately 10 to 1, consistent with gross medullary weight (41).

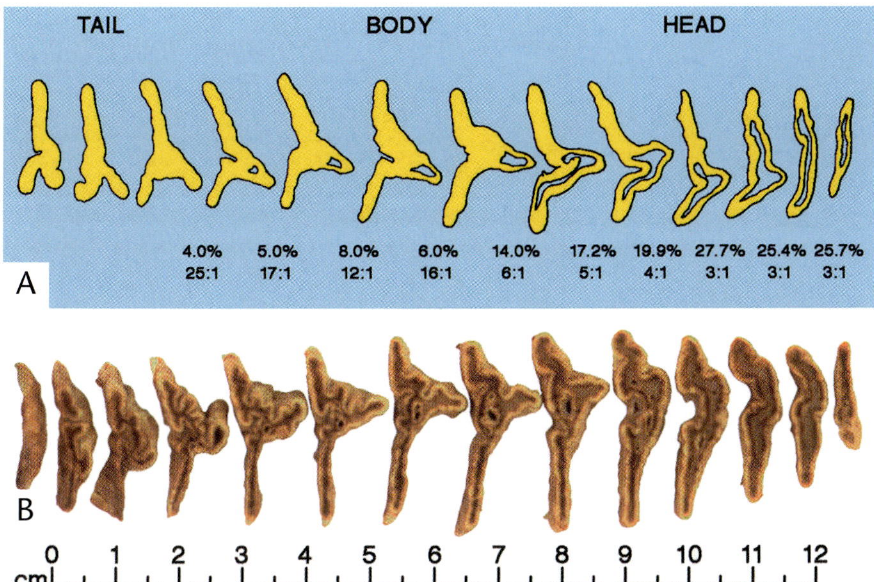

Figure 1-22

TOPOGRAPHY OF THE ADULT ADRENAL GLAND

A: Serial transverse sections show three regions of the gland: the head (medial and inferior in situ), body, and tail (superior and lateral in situ). Most of the chromaffin tissue is concentrated in the body and head of the gland; none is present in the tail. The percentage of cross-sectional area occupied by medulla is indicated at the top along with the ratio of area occupied by cortex versus medulla (overall ratio about 10 to 1).

B: In another adrenal gland, the medullary compartment is gray-white and is most prominent in the body and head of the gland.

The normal adult adrenal cortex is a fairly uniform rim, approximately 1-mm thick, beneath the adrenal capsule. The outer zones are bright yellow due to the combined presence of lipid stores and cytochromes (fig. 1-23) (42), with occasional areas of darker mottling. The thin zona reticularis is a darker yellow-brown caused by the combination of less stored lipid and the accumulation of lipofuscin, a wear and tear pigment that accumulates with age and oxidative damage. The medulla is a gray tissue concentrated in the head and body of the gland (see fig. 1-22) and should not be confused with the brown-yellow inner cortex. Where chromaffin cells are absent (e.g., tail of the gland), the opposing layers of adrenal cortex are demarcated by the interalar raphe, which may appear as a wider pigmented zone because of the juxtaposition of zona reticularis on either side. There may be variation in cortical thickness from gland to gland, or in different areas of the same gland. This variation is particularly evident in elderly individuals due to an increased prevalence of cortical nodularity (fig. 1-24).

A partial or complete cuff of lipid-containing cortical cells is present about the intraglandular portion of the central vein or its tributaries (see fig. 1-23). It is also common to see small (usually 1 to 2 mm), intramedullary foci of cortical tissue, with or without apparent connection to overlying cortex (see fig. 1-23). In adrenal glands

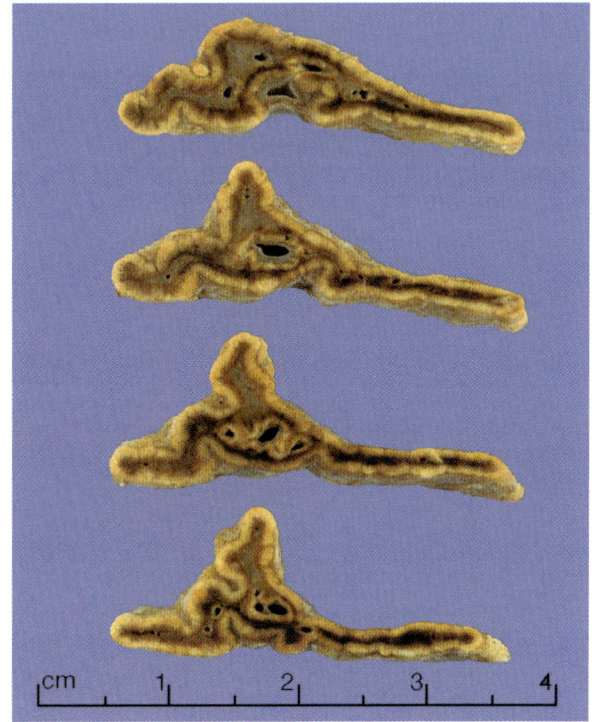

Figure 1-23

BODY AND HEAD OF NORMAL ADRENAL GLAND

In this transverse section, the medullary compartment appears dull gray in contrast to the bright yellow cortex. Partial to complete cuffs of cortex are present around the central adrenal vein and its tributaries. Note also the brown zona reticularis at the inner cortex. There is a small cortical extrusion on the left side of the top section.

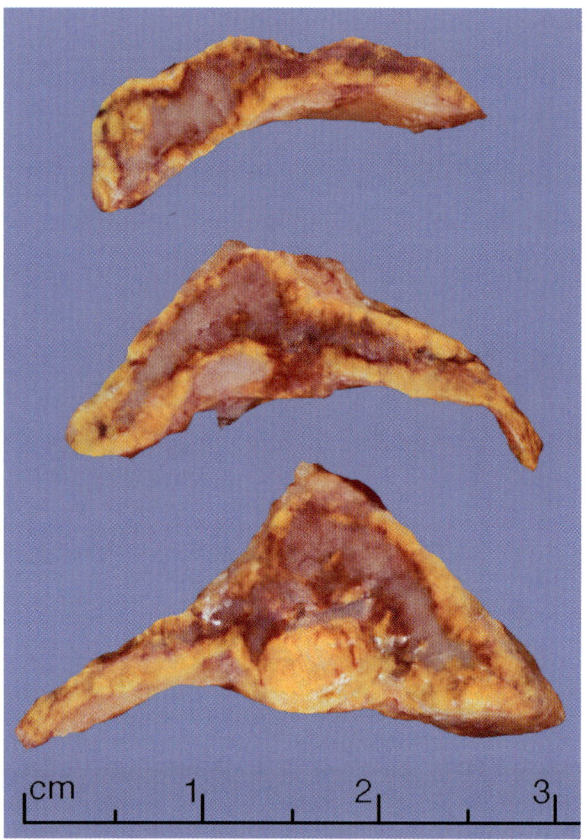

Figure 1-24

BODY AND HEAD OF A NODULAR ADRENAL GLAND

In this architecturally distorted gland in transverse section, the adrenal cortex is bright yellow, with focal nodularity. Medullary tissue is prominent and extends into both wings (alae) of the gland. This degree of medullary prominence is within the range of normal. The adrenal gland was obtained during radical nephrectomy from a patient with no known endocrine abnormalities.

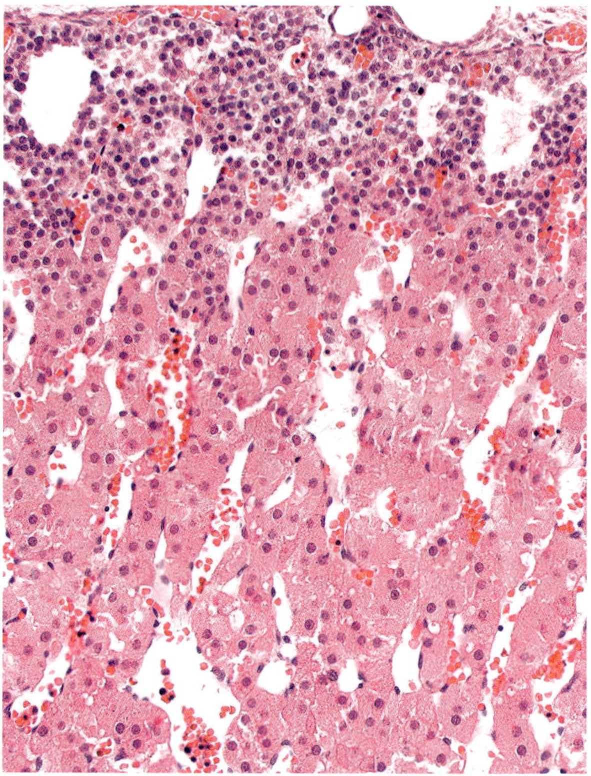

Figure 1-25

PERINATAL ADRENAL GLAND SHOWING DEFINITIVE AND PROVISIONAL ZONES BEFORE INVOLUTION

Definitive cortex forms a thin rim of subcapsular cells, which have a high nuclear to cytoplasmic ratio. Most of the cortex is composed of fetal, or provisional, cortical cells.

with cortical nodularity or hyperplasia there can be marked distortion of corticomedullary distribution, including discontinuous islands of medulla, interalar or intercristal bridges of cortex to cortex, and cortical micronodules (fig. 1-24).

Atrophy of the adrenal cortex may be caused by feedback inhibition of pituitary ACTH secretion in patients with high levels of circulating cortisol, whether iatrogenic or derived from an autonomously functioning contralateral cortical neoplasm. A diagnosis of adrenal medullary hyperplasia should generally not be made in cases with cortical atrophy.

MICROSCOPIC ANATOMY

Postnatal Adrenal Gland

Immediately prior to birth, the adrenal cortex still has a biphasic structure that consists of a thin subcapsular adult or definitive zone and a wide inner zone composed of fetal, or provisional, cortical cells (fig. 1-25). Cells of the definitive cortex are relatively small (10 to 20 um), with dark-staining nuclei and scant cytoplasm, while the fetal cortical cells are larger (20 to 50 um), with voluminous, granular and eosinophilic cytoplasm and more vesicular nuclei, often with a small nucleolus. The initial mode of cell death in the involuting fetal cortex is believed to be apoptosis (10), but because of the rapidity of cortical involution, advanced features of regression are usually present in adrenal glands

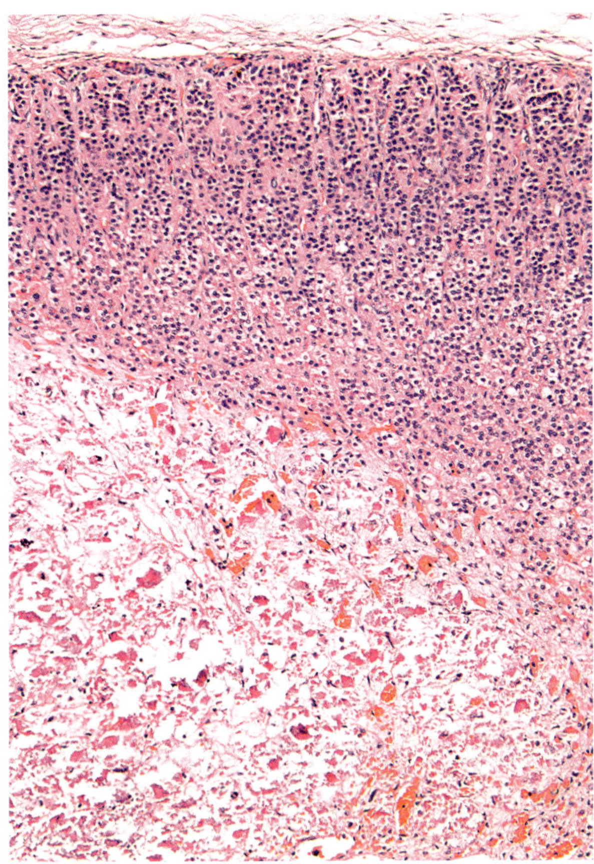

Figure 1-26

POSTNATAL REGRESSION OF PROVISIONAL ADRENAL CORTEX

This 3-week postnatal adrenal gland shows necrosis and debris in the central part of the gland and a small amount of residual provisional cortex beneath the developing zona fasciculata.

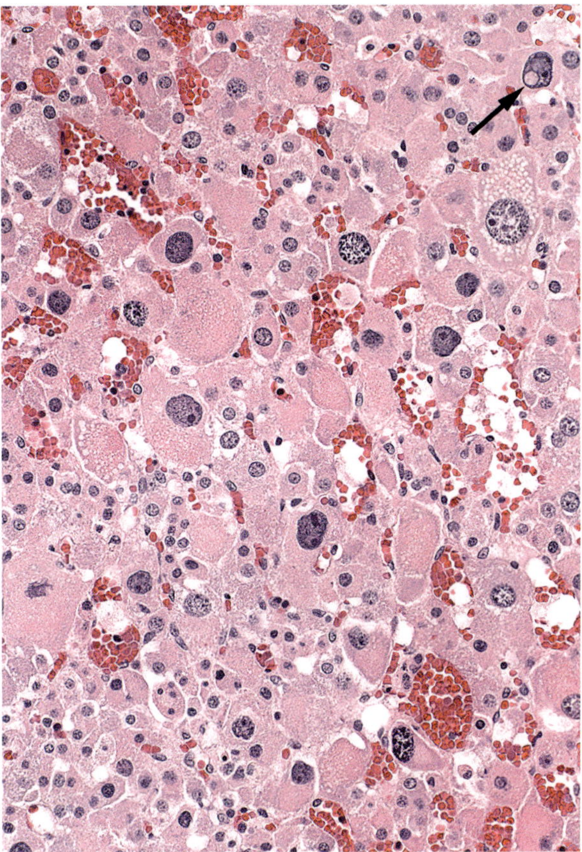

Figure 1-27

CYTOMEGALY IN REGRESSING FETAL CORTEX FROM A STILLBORN

The nuclei are extremely large and occasional nuclear pseudoinclusions are formed by invagination of cytoplasm (arrow).

seen in the postnatal period. These include coagulative necrosis, cytolysis with nuclear pyknosis and karyorrhexis, and evidence of recent and old hemorrhage (fig. 1-26).

Dystrophic calcification may be present, and if detected incidentally on imaging studies, may potentially arouse suspicion for a neuroblastic tumor. Prominent vascular sinusoids can simulate a hemangioma or lymphangioma. Microcysts are often observed in the definitive cortex of premature stillborns and newborn infants (see fig. 1-25) and are generally considered to be a degenerative change associated with in utero stress (see figs. 1-7, 1-9). Microcysts have been correlated with a shorter gestational period and shorter survival time following birth (43,44).

Other microscopic findings variably present in the newborn cortex and ascribed to several types of intrauterine stress include vacuolar change of cells in the outer fetal cortex, extramedullary hematopoiesis, and cytomegaly (38).

Adrenal Cytomegaly. Adrenal cytomegaly has been reported in 0.8 percent of pediatric autopsies (45) and occasionally in older patients (45). It is usually an incidental finding in glands that are otherwise grossly normal, and may be focal or diffuse (fig. 1-27). Affected cells are limited to the fetal cortex. They may be as large as 120 µm in diameter and show marked nucleomegaly with pleomorphism and hyperchromasia. Nuclei of megalic cells may contain over 25 times the normal amount of DNA (45). Characteristically, there are no identifiable mitotic figures.

Adult Adrenal Gland

Normal Capsule and Cortex. The adrenal capsule is composed of hypocellular fibrous tissue with coarse hyalinized collagen and elastic fibers. The capsule is usually thin, but may vary in thickness from gland to gland or in different areas of the same gland; occasionally, the capsule appears focally absent. Capsular arterioles are often present in random sections just outside the capsule. Small myelinated nerves penetrate the adrenal capsule, primarily on the posterior or dorsal surface. Extrusions of adrenal cortical tissue through the capsule and accessory cortical nodules are common in adrenal glands from adults and elderly persons, and can be a source of confusion when assessing invasiveness of cortical neoplasms (fig. 1-28).

There are three well-defined zones of cortex (figs. 1-29–1-31): the zona glomerulosa (from Latin glomus, ball), zona fasciculata (from Latin fasciculus, a little bundle), and zona reticularis (from Latin reticulum, a little net). The zona glomerulosa is a thin, usually discontinuous layer beneath the capsule with a ball-like arrangement of cells; it comprises about 5 to 10 percent of the cortex in some areas of the gland. The zona fasciculata comprises about 75 percent of the thickness of the cortex and consists of radial columns or cords of cells that have ample, lipid-rich cytoplasm, contrasting with the relatively sparse cytoplasm of the zona glomerulosa. Close scrutiny of these cells with optimal preservation/fixation reveals fine lattice-like vacuolization of cytoplasm, which is pale-staining due to the accumulation of lipid droplets. The orderly appearance of the zone fasciculata is frequently altered by patchy lipid depletion in response to physiologic stress (fig. 1-30). The inner zona fasciculata merges often imperceptibly with the zona reticularis, which was so-named because of its diffuse net-like appearance caused by irregularly arranged cells with interspersed vascular spaces. The cells in the zona reticularis have compact eosinophilic cytoplasm frequently containing prominent lipochrome pigment (fig. 1-30). The corticomedullary interface is often smooth or delicately undulating, but can be irregular, with intermingling of cortical cells and chromaffin cells. The architectural arrangements of cells in each zone of the adrenal gland are accentuated by staining for reticulin (fig. 1-31).

Immunohistochemical studies of the distribution of steroidogenic enzymes have been used to determine the functional correlates of anatomic cortical zonation (see fig. 1-1). Cytochrome CYP450C21 (C21 hydroxylation) is present in all three zones of the adrenal cortex, reflecting its role in biosynthesis of both mineralocorticoids and glucocorticoids, while cytochrome CYP45017α is not present in the zona glomerulosa, thus confirming the exclusive localization of glucocorticoid and androgen biosynthesis in the zona fasciculata and zona reticularis. CYP11B2, leading to the synthesis of aldosterone, is normally only expressed in the zona glomerulosa (46), and recently developed antibodies have been valuable in documenting a variety of abnormalities in the adrenal glands of patients with hyperaldosteronism (47).

In the normal adult adrenal cortex, immunohistochemical staining for Ki-67 shows the greatest amount of proliferative activity in the outer part of the zona fasciculata, with lesser amounts in the reticularis and glomerulosa (fig. 1-32) (10,48). Unequivocal mitotic figures are rarely seen by routine light microscopy.

Stress-Related Cortical Changes. Several morphologic changes can occur during normal physiologic adrenal cortical responses to illness or other stress. The earliest and most common is lipid depletion. In this situation, pale-staining, lipid-rich cells of the zona fasciculata convert to cells with compact, eosinophilic cytoplasm, often in a patchy pattern, as they respond to ACTH by increasing cortisol production (see fig. 1-30). Longer-term stimulation by ACTH results in irregular cortical thickening, with a predominance of lipid-depleted cells. In patients with extremely high levels of ACTH, as ectopically produced by neuroendocrine tumors, the macroscopic appearance may be striking, with uniform lipid depletion and medium brown coloration. Almost the entire thickness of the cortex may be uniformly involved (see chapter 3). Other changes less frequently observed include tubular degeneration, in which normally solid cords of cells in the outer cortex convert to tubular structures, sometimes containing proteinaceous material or debris (fig. 1-33) (49).

Miscellaneous Cortical Findings. *"Fissure" Formation.* Fissures or cyst-like cavities can form in adrenal gland specimens consequent to the

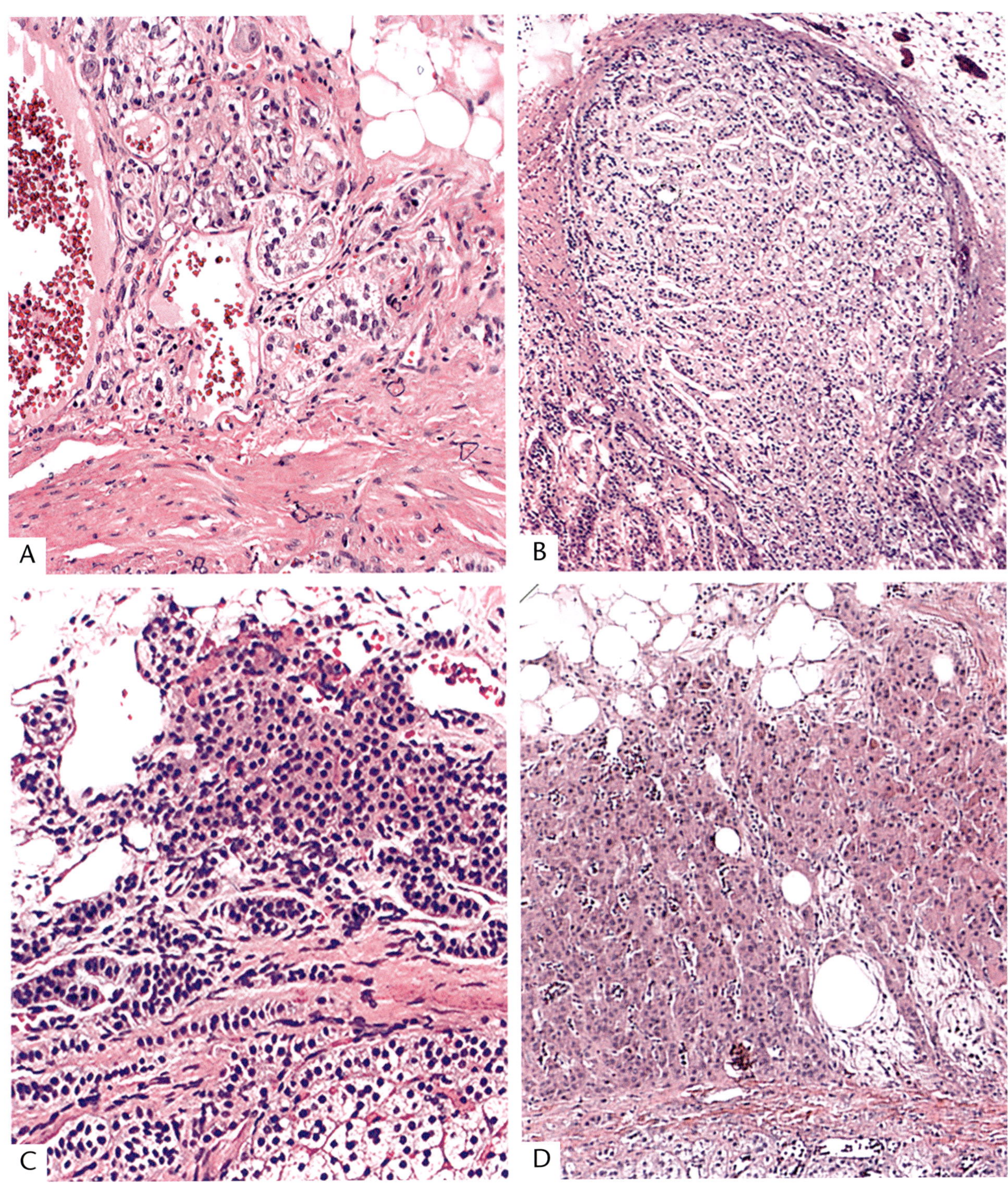

Figure 1-28

EXTRUSION OF CORTICAL TISSUE THROUGH THE CAPSULE OF NORMAL ADRENAL GLAND

A: Extracapsular perivascular extrusion of cortical cells.
B: Nodular protrusion of cortical cells still partly covered by attenuated capsule.
C: Unencapsulated cortical cells in periadrenal fat.
D: Florid proliferation of extracapsular cortical cells associated with adrenal cortical hyperplasia.

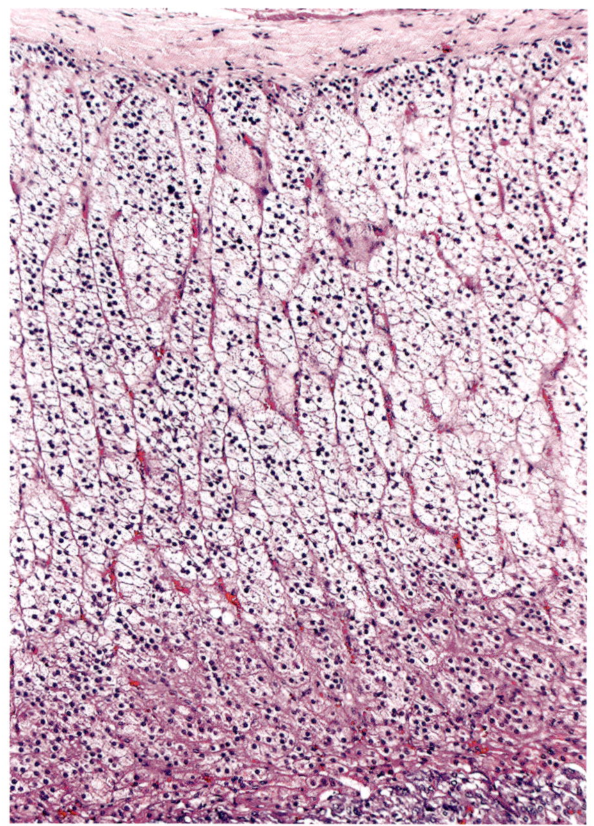

Figure 1-29

ZONATION OF ADULT ADRENAL GLAND

An indistinct, discontinuous zona glomerulosa beneath the adrenal capsule has a ball-like arrangement of cells, with less abundant lipid-rich cytoplasm than cells of the zona fasciculata. Radial cords of zona fasciculata cells with abundant lipid merge with zona reticularis, where most of the cells are lipid poor and have compact eosinophilic cytoplasm. A few cells in the zona reticularis show faint brown lipochrome pigmentation. A sharp corticomedullary junction and a portion of adrenal medulla are at the bottom of the field.

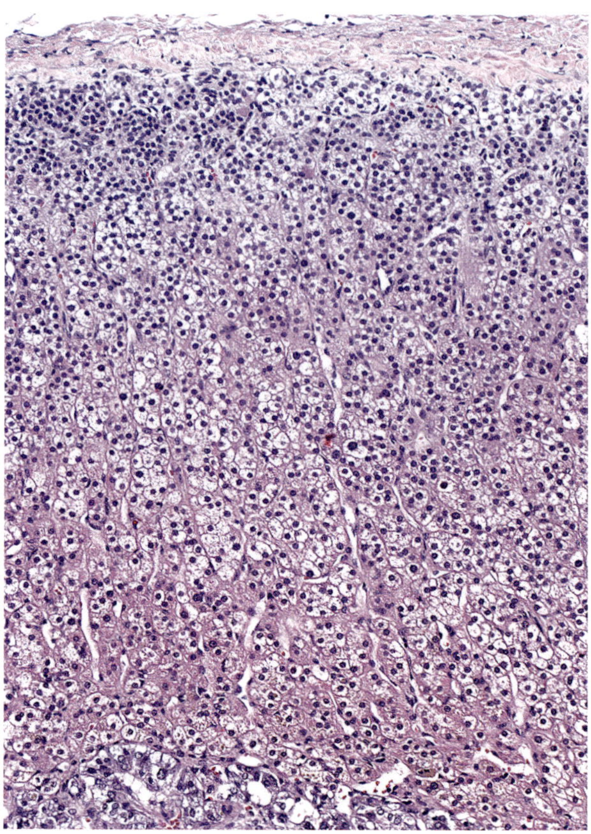

Figure 1-30

LIPID DEPLETION IN CORTICAL CELLS NORMALLY RICH IN CYTOPLASMIC LIPID

There is distortion of the zona fasciculata caused by patchy lipid depletion. In some areas, lipid is absent in the outer cortex but prominent in the inner zona fasciculata. This pattern has been called as "lipid reversion" and suggests recovery from stress (12).

combined effects of autolysis and physical manipulation (fig. 1-34). The plane of separation is often near or at the corticomedullary junction, and there may be almost complete separation of medulla from cortex. There is usually no significant inflammatory reaction. This artifactual change undoubtedly contributed to the misconception of the adrenal glands as hollow structures (hence the term "suprarenal capsules") (50).

Fat Cell Metaplasia. In some cases the zona fasciculata contains small clear spaces caused by confluent lipid vacuoles, or areas of overt fat cell metaplasia (lipomatous foci) (fig. 1-35), which may be accompanied by lymphocytes and/or myeloid metaplasia. Fat cell metaplasia is observed in 5 percent of adult adrenal glands, with equal sex distribution, and does not appear to correlate with generalized adiposity; the incidence increases with age (51).

Focal Lymphocytic Infiltration. Small interstitial or perivenous foci of lymphocytes and occasional plasma cells are a frequent incidental finding in the adrenal cortex and medulla. They are most common in elderly patients of both sexes. Although usually not clinically significant, they may be associated with autoimmune inflammation in other organs (52). Lymphocytic adrenal medullitis has been reported in patients with rabies and in occasional patients with paroxysmal hypertensive episodes.

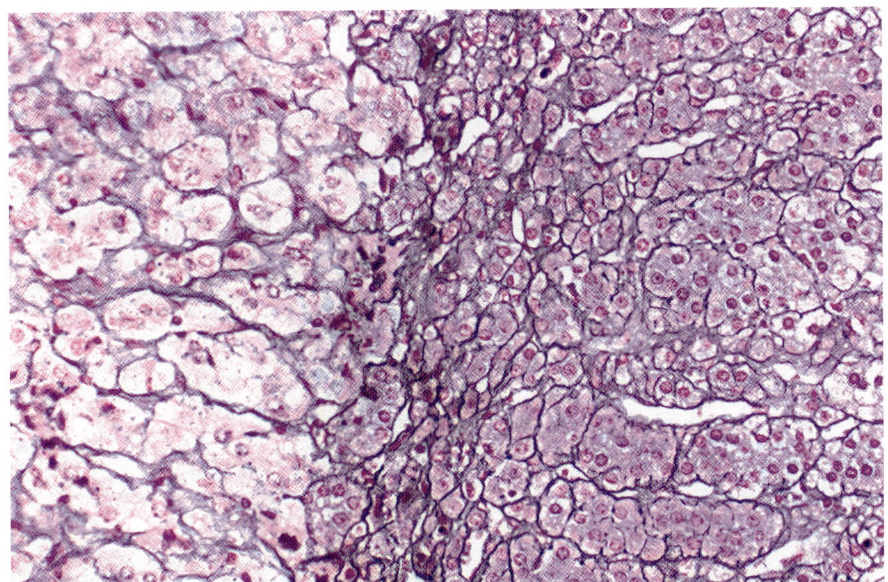

Figure 1-31

RETICULIN STAIN

Reticulin stain highlights the columns of the zona fasciculata (right), net-like arrangement of the zona reticularis, and nests ("Zellballen") of chromaffin cells in the adrenal medulla.

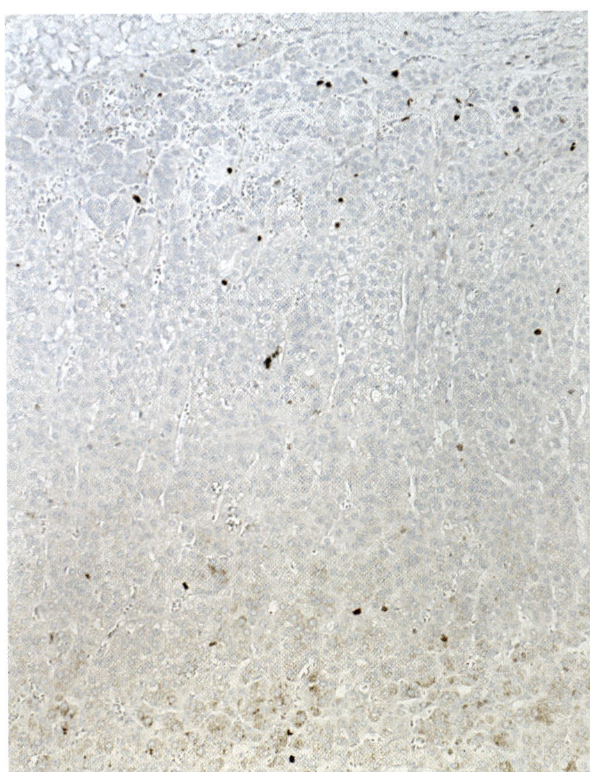

Figure 1-32

Ki-67 STAIN

Ki-67 staining is relatively sparse in the unstressed normal adult adrenal gland, with the greatest amount of staining in the outer part of the zona fasciculata. Rare stained cells are present in the zona reticularis, which is highlighted by the intrinsic brown pigment.

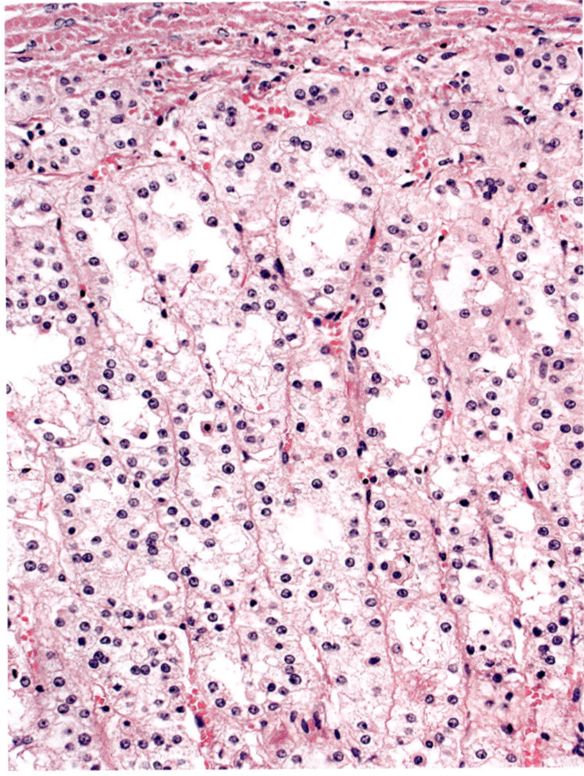

Figure 1-33

TUBULAR DEGENERATION OF OUTER ZONA FASCICULATA

Solid columns and cords of lipid-depleted cells are converted into hollow tubules containing occasional degenerating cortical cells.

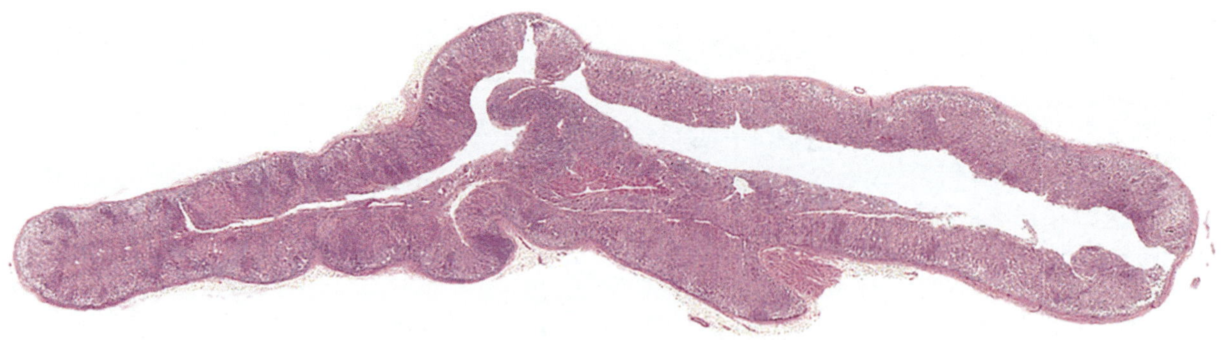

Figure 1-34

ARTIFACTUAL FISSURE FORMATION IN A NORMAL ADRENAL GLAND

Autolysis at a plane roughly between cortex and medulla can create the appearance of an empty sac.

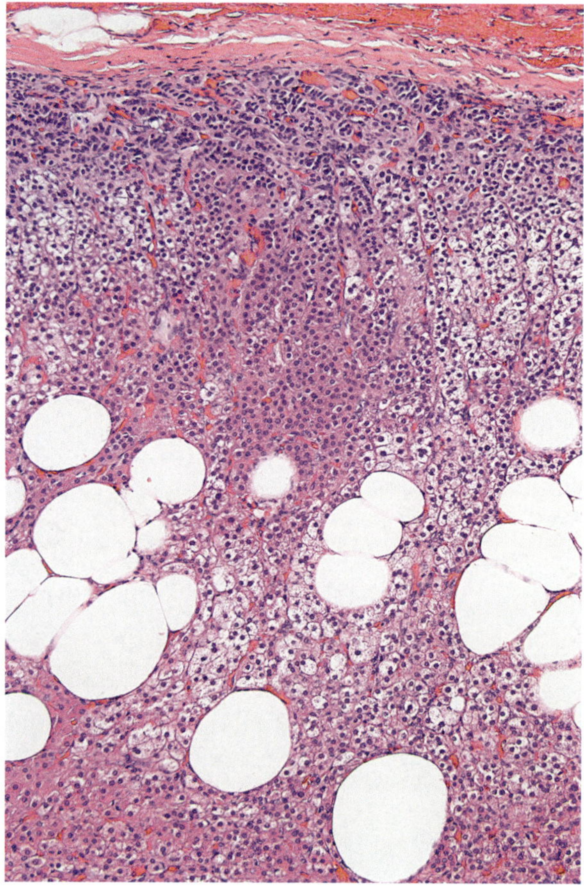

Figure 1-35

FAT CELL METAPLASIA

Normal adrenal cortex in an adult shows patchy areas with confluent lipid vacuoles and some individual cells resembling mature adipocytes. This has been referred to as lipomatous metaplasia.

"Ovarian Thecal Metaplasia." This term is most likely a misnomer for small, partially hyalinized spindle cell proliferations in the adrenal cortex that resemble mesenchymal cells of ovarian stroma, thus prompting the designation ovarian thecal metaplasia. The lesions are often wedge-shaped, attached to the adrenal capsule, and may extend between or surround small nests of cortical cells (fig. 1-36).

These proliferations have a maximum size of about 2 mm and are multiple in about half of the cases and bilateral in roughly a third. They are reported in 4.3 percent of women, who are usually postmenopausal, and occasionally in men. Some foci are quite hyalinized while others are more cellular. Dystrophic calcification is present occasionally. The histogenesis is not entirely settled; however, the morphology and immunohistochemical profile suggest origin from myofibroblasts in some cases (53).

Hyaline Globules. Rarely, intracytoplasmic hyaline globules are reported within the cells of the zona fasciculata in humans dying of various conditions including streptococcal meningitis, chronic renal failure, pneumonia, and overexposure to cold. The example seen in figure 1-37 was an incidental finding in the cortex attached to an aldosterone-secreting cortical adenoma.

Leydig Cells. Leydig-like cells containing Reinke crystals have been observed by electron microscopy in a single study of normal adrenal cortex (54). An additional incidental example is shown in figure 1-38. While such cells may be the precursors of rare cortical adenomas with

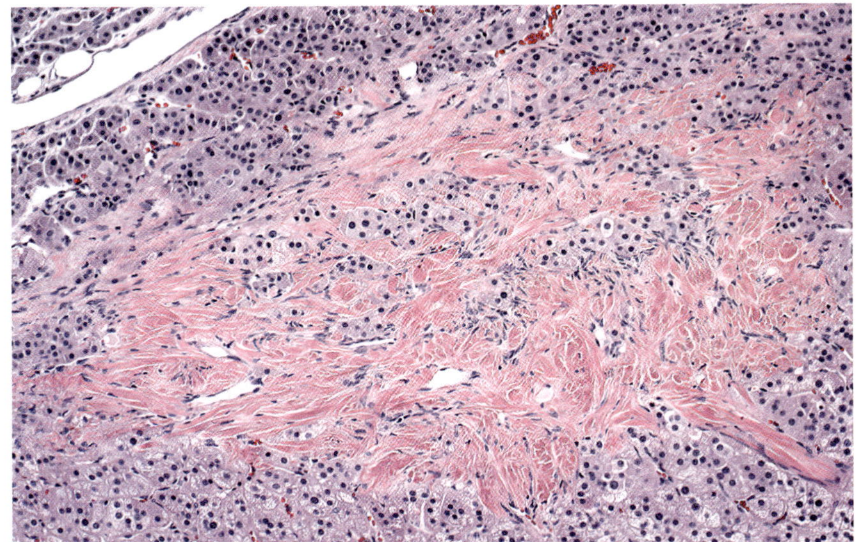

Figure 1-36

OVARIAN THECAL METAPLASIA

Ovarian thecal metaplasia appears as a broad zone of bland spindle cells. Nests of pale-staining cortical cells are present within the spindle cell component.

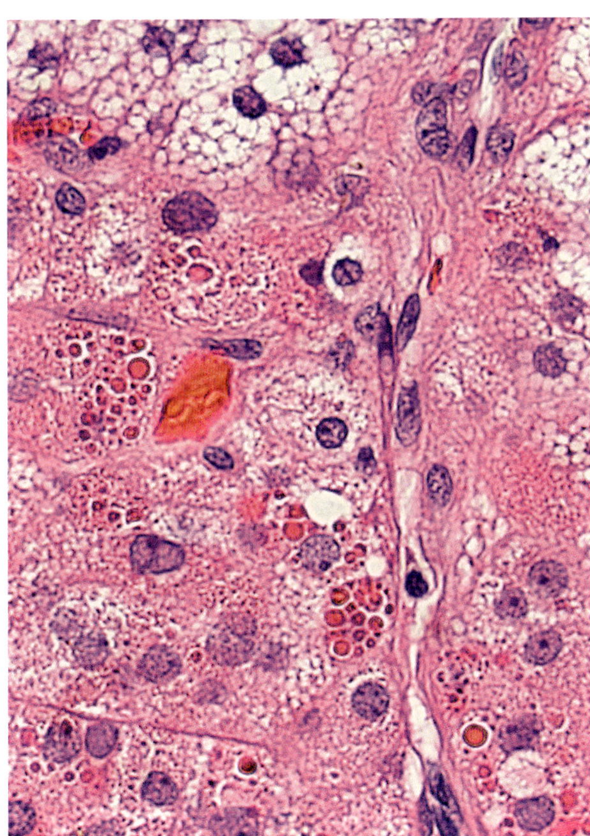

Figure 1-37

HYALINE GLOBULES IN ADRENAL CORTEX

These hyaline globules were noted incidentally in the normal part of an adrenal gland removed because of an aldosterone-secreting cortical adenoma. There were no spironolactone bodies in any part of the specimen, and the patient had not received spironolactone.

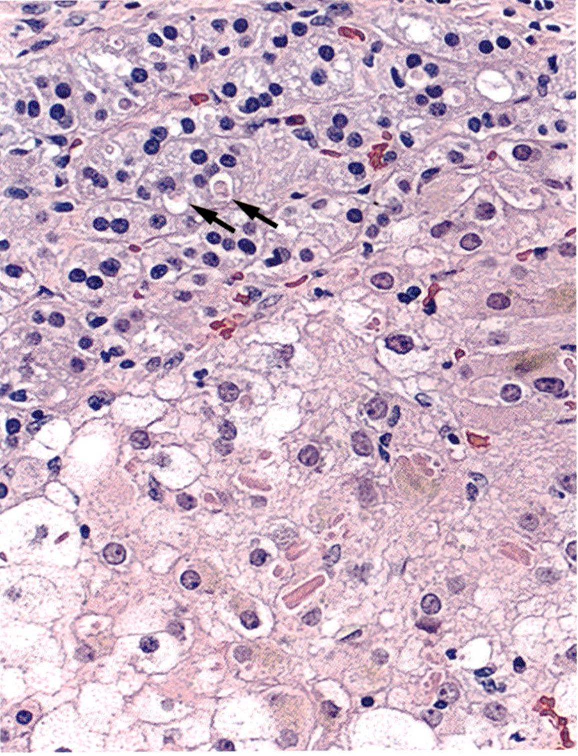

Figure 1-38

INTRACORTICAL LEYDIG CELLS

A solitary cluster of Leydig (or hilus) cells with numerous Reinke crystalloids were noted incidentally in the normal part of an adrenal gland removed because of an aldosterone-secreting cortical adenoma. Most of the Leydig cells also contain granular brown lipofuscin pigment. The patient had been treated with spironolactone for primary hyperaldosteronism, and some cells of the adjacent zona glomerulosa contain spironolactone bodies (arrows).

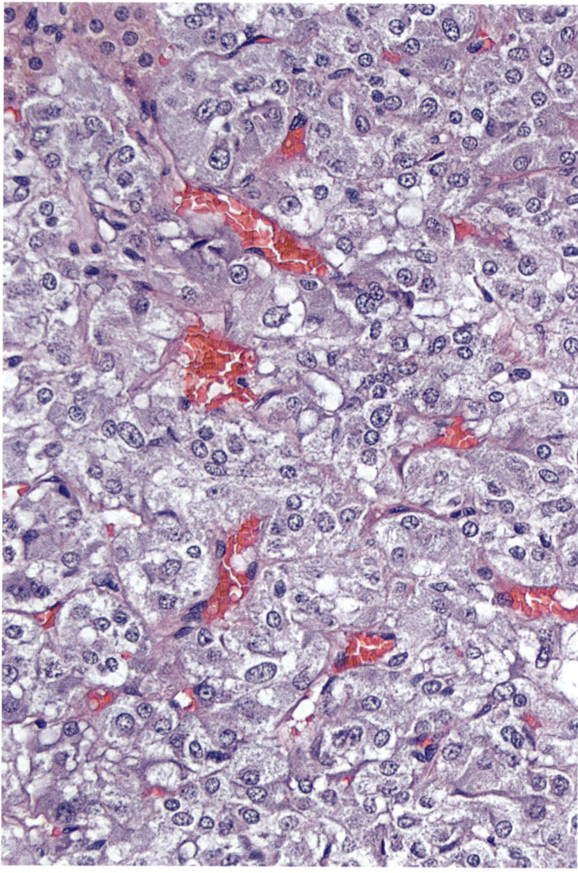

Figure 1-39

NORMAL ADRENAL MEDULLA IN AN ADULT

Well-preserved chromaffin cells are arranged in nests and anastomosing short cords. The cells vary in size, with some larger or smaller than the zona reticularis cells at top left. A few cells have moderately conspicuous nucleoli.

Leydig cell characteristics (55), most normal adrenal glands lack cells with Reinke crystals and are immunohistochemically negative for 17β-hydroxysteroid dehydrogenase, which identifies normal testicular Leydig cells and Leydig cell tumors (see fig. 1-1) (56).

Adrenal Medulla. The catecholamine producing neuroendocrine cells of the adrenal medulla and related tissues were named chromaffin cells in the late 19th century when they were found to develop a reddish brown color in a solution containing chromate salts. At the time, the color change was believed to result from affinity for chromium, and hence called the chromaffin reaction (57), although it was subsequently found to result from oxidation of stored catecholamines (50). The chromaffin reaction is no longer used in diagnosis but is occasionally revisited out of historic interest (58).

Chromaffin cells in the adrenal medulla are arranged in discrete nests or balls, still often referred to as "Zellballen" as in the original German description. There may also be short anastomosing cords or an admixture of architectural patterns (fig. 1-39). Staining for reticulin highlights the organoid arrangement of the cells (see fig. 1-31).

Chromaffin cells have amphophilic to basophilic cytoplasm, indistinct borders, and usually a single nucleus that is slightly eccentric in location. Pinpoint basophilic cytoplasmic granules are seen in well-preserved specimens. The nuclei are often round or oval, with smooth nuclear contours, finely stippled chromatin, and inconspicuous nucleoli, but in some glands there is considerable nuclear pleomorphism, with enlargement, hyperchromasia, and occasional prominent nucleoli. Mitotic figures are absent in the normal medulla and immunohistochemical staining for Ki-67, if present at all, is seen only in extremely rare cells. The cytologic detail of chromaffin cells is very dependent on the promptness and method of fixation; cytoplasmic vacuolization and blurring of cell membranes often occur if fixation is suboptimal.

Chromaffin cells in normal adult adrenal glands occasionally contain intracytoplasmic hyaline globules. The globules are eosinophilic, range in size from 1 to 25 µm, and are periodic acid–Schiff (PAS) positive and diastase resistant. Although reported in the majority of adrenal glands from adults, they may be difficult to find on casual inspection (fig. 1-40). There appears to be no correlation with sex, race, or age, but in the authors' experience, they are uncommon in the pediatric age group. They have been reported with increased frequency in patients with chronic neurologic disorders such as Parkinson disease (59), and are common with some pheochromocytomas.

Although usually confined to the adrenal medulla, chromaffin cells are also sometimes found in small foci near the adrenal capsule (fig. 1-41A) or in periadrenal connective tissue, sometimes associated with small nerves. These foci may result from aberrant migration of medullary progenitors (see fig. 1-8), and they can be

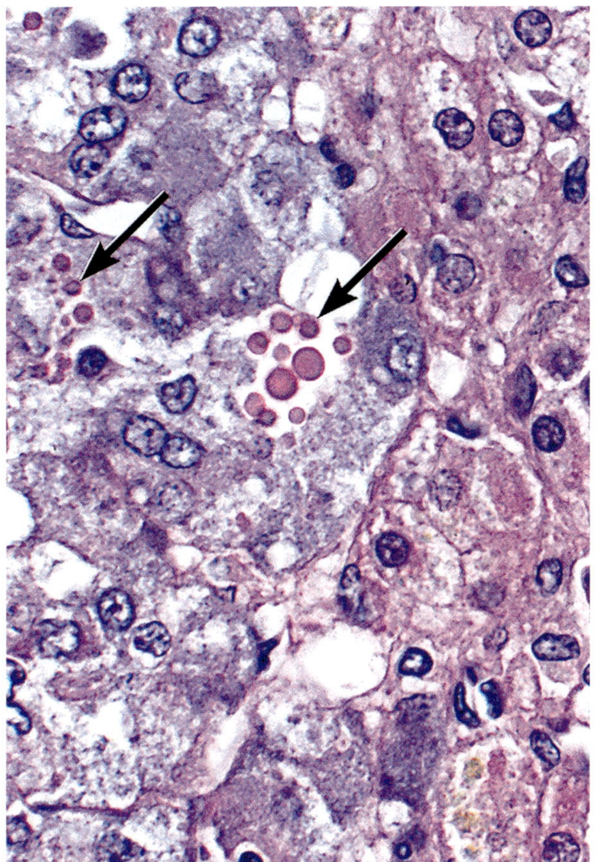

Figure 1-40

INTRACYTOPLASMIC HYALINE GLOBULES IN NORMAL ADRENAL MEDULLA

Globules are present in several chromaffin cells. A few cells contain numerous small globules.

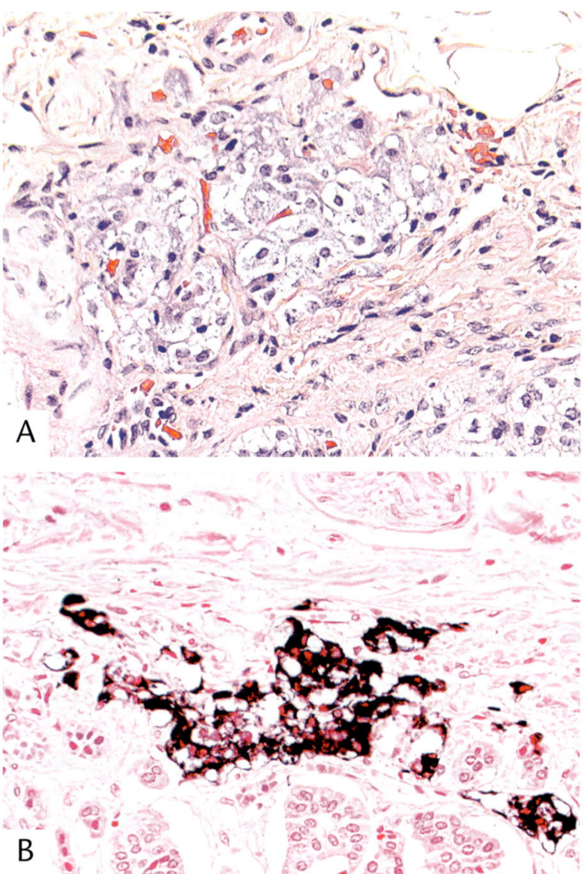

Figure 1-41

JUXTACAPSULAR RESTS OF CHROMAFFIN CELLS

A: Nest of chromaffin cells in the outer adrenal cortex just outside the capsule of the adrenal gland. A capsular arteriole is present at the top of the field.

B: Immunostain for chromogranin A shows a small nest of chromaffin cells in the outer cortex just beneath the capsule of the adrenal gland.

pronounced in experimental studies in which migration is apparently arrested by inactivation of developmentally important molecules (60). They are typically unencapsulated, which may help to distinguish them from extra-adrenal paraganglia. Immunostaining for chromogranin or other neuroendocrine markers may help to identify these nests (fig. 1-41B). Similarly to cortical extrusions and cortical tumors, small collections of chromaffin cells outside the adrenal gland can be a source of confusion in assessing invasiveness or metastasis of pheochromocytomas.

Sustentacular cells form a second distinctive cell population in the adrenal medulla. These enigmatic cells may be a type of peripheral nerve glial cell. They express the same phenotype markers as SCPs, which give rise to subsets of chromaffin cells and neurons during development (25,29), and some sustentacular cells reportedly contribute to the proliferative response of adult carotid body chief cells to hypoxia (61). However, their direct transdifferentiation in any adult organ is questionable (62). Sustentacular cells are vividly demonstrated by immunostaining for S-100 protein (fig. 1-42A) or SOX10 (fig. 1-42B), both of which also highlight satellite cells of intrinsic adrenal neurons (shown for S-100 protein in fig. 1-43).

In addition to chromaffin and sustentacular cells, the adrenal medulla in humans and other

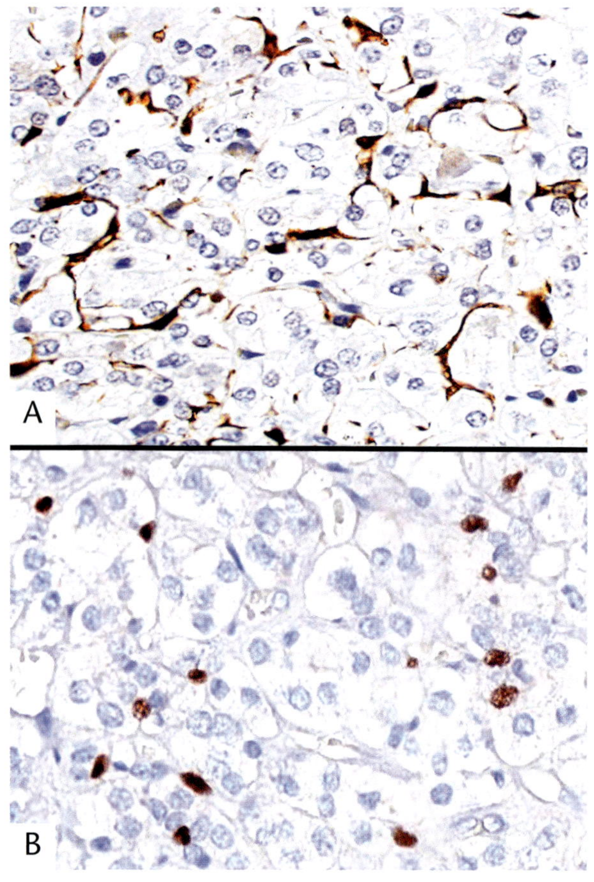

Figure 1-42

SUSTENTACULAR CELLS

Immunohistochemical stains for S-100 protein (A) or SOX10 (B) show numerous sustentacular cells around and within groups of chromaffin cells.

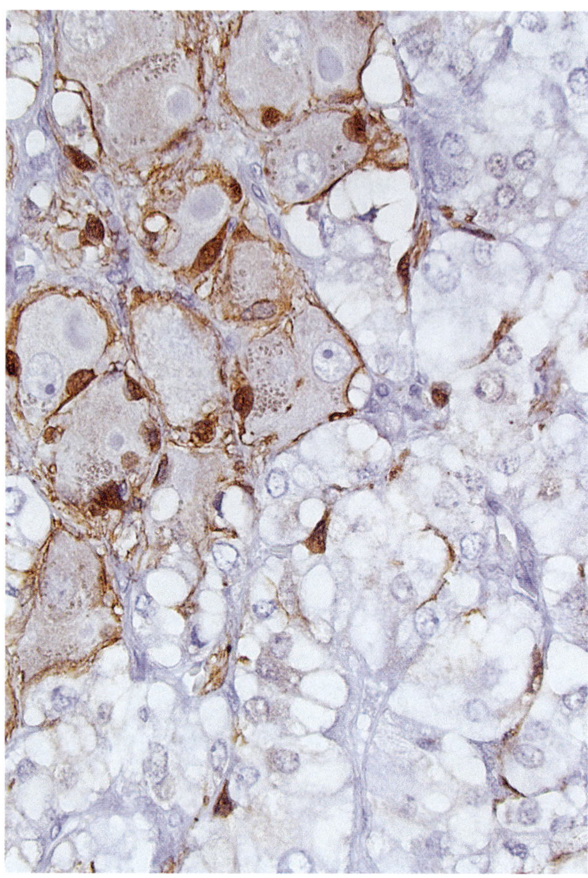

Figure 1-43

SATELLITE CELLS OF ADRENAL MEDULLARY GANGLION

Immunohistochemical stain for S-100 protein highlights both sustentacular cells and satellite cells in a group of medullary ganglion cells.

species contains several morphologically and functionally distinct types of intrinsic neurons ("ganglion cells"). These may be found in aggregates comparable to ganglia (fig. 1-43), or singly (fig. 1-44), sometimes in association with small myelinated nerves. The human adrenal gland is reported to contain about 40 neuronal cell bodies/mm^2 of medulla (63). Some are immunoreactive for tyrosine hydroxylase (TH), similar to sympathetic neurons, while others are TH negative, cholinergic and/or peptidergic, and resemble parasympathetic enteric neurons. The TH-positive and TH-negative subsets are further distinguished, respectively, by expression of the peptide neurotransmitters vasoactive intestinal peptide (VIP) (fig. 1-44) and substance P (63). Processes originating from intrinsic neurons or the rich splanchnic innervation by preganglionic sympathetic fibers can also be stained for neurofilament proteins or other generic neuronal markers. The functions of intrinsic adrenal neurons are poorly understood but may include regulation of vascular tone and modulation of both medullary and cortical hormone secretion.

Adrenal Vasculature. The three arteries supplying the adrenal glands divide repeatedly into as many as 50 branches, which partially invest the capsule as the capsular arterioles (seen in fig. 1-37) (30). There is a single central or main adrenal vein, which typically exits the gland on the ventral or flat surface (see fig. 1-19). A cuff of cortical tissue partially envelops the central vein throughout its length in much of the gland (see figs. 1-23 and 1-45).

Once the arterioles penetrate the cortex, they enter a thin subcapsular plexus that gives rise to capillary channels, which course centrally through the zona fasciculata to join a rich vascular plexus in the zona reticularis. From there, blood flows into venous sinuses in the medulla to enter tributaries of the central vein. In areas of the gland with no medulla, blood from the reticular plexus courses directly into tributaries of the central adrenal vein (64).

The microscopic anatomy of the central vein and its tributaries is remarkable for its distinctive discontinuous medial musculature organized into longitudinal bundles (fig. 1-45). Contraction of the longitudinal muscle bundles is thought to aid in closure, or damming up, of blood in the venous sinuses and reticular plexus; in this way, the arcades of smooth muscle act as "sluice gates" regulating the degree of congestion of the zona reticularis and inner zona fasciculata. With relaxation of muscle bundles and elastic recoil of venous sinuses there may be rapid, intermittent release of hormones into the central adrenal vein.

An important aspect of the adrenal vasculature is that the discontinuous musculature of the central vein and venous tributaries can be a source of confusion when assessing vascular invasion by medullary or cortical neoplasms.

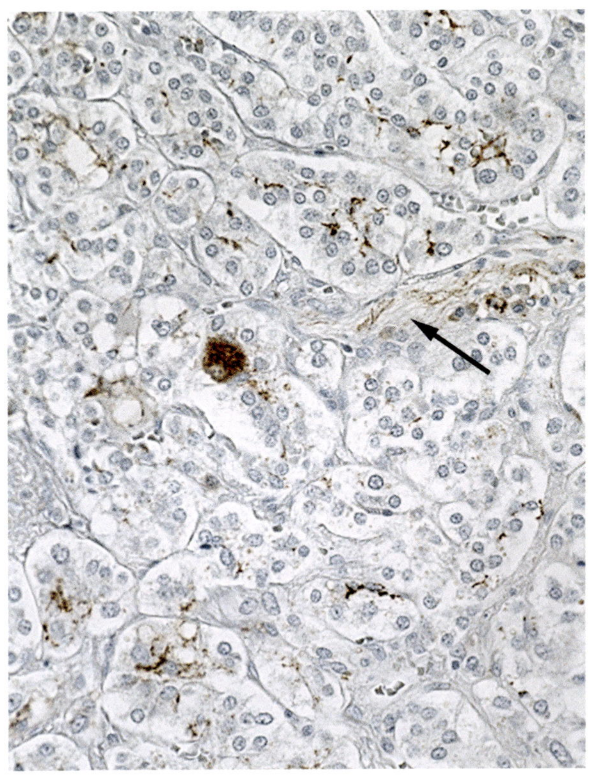

Figure 1-44

VASOACTIVE INTESTINAL PEPTIDE (VIP) IN NORMAL ADRENAL MEDULLA

The figure shows a single VIP-expressing neuron, a myelinated nerve containing VIP-positive axons (arrow), and branching VIP-positive terminal fibers between chromaffin cells. The chromaffin cells are negative for VIP.

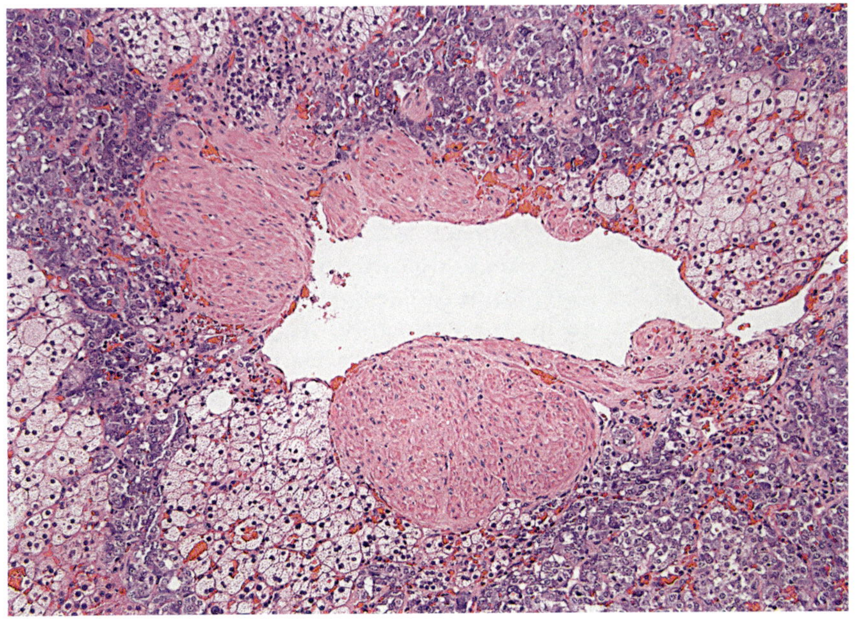

Figure 1-45

DISTINCTIVE ARCHITECTURE OF THE CENTRAL VEIN IN NORMAL ADRENAL GLAND

Distinctive arcades of interrupted smooth muscle form the wall of the central adrenal vein and its tributaries. The vein is surrounded by chromaffin cells and a partial cuff of cortical cells.

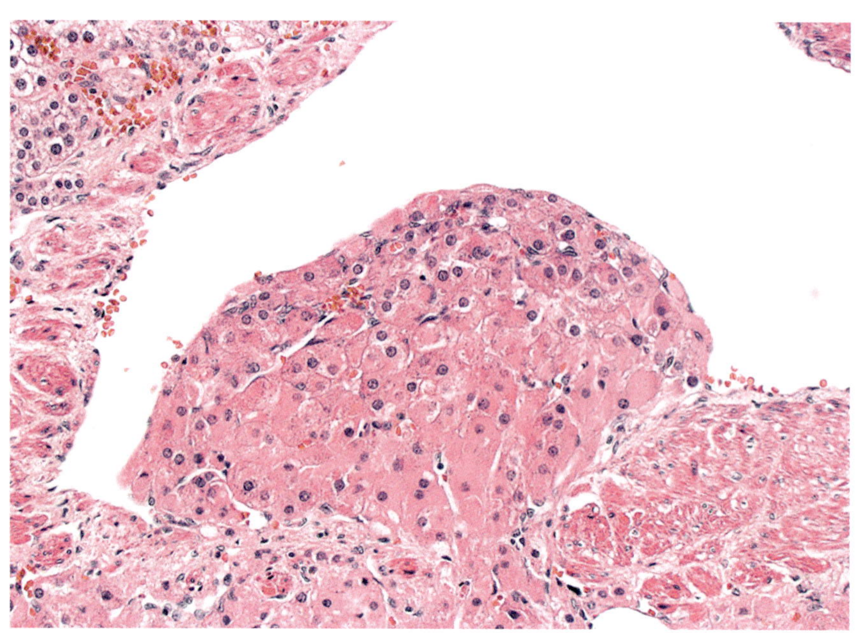

Figure 1-46

PROTRUSION OF NORMAL CORTICAL CELLS INTO THE LUMEN OF THE CENTRAL VEIN

Where smooth muscle is deficient, chromaffin or cortical cells may come into close proximity with the vascular space and can form intravascular protrusions covered by an intact endothelial layer.

Normal adrenal tissue can bulge toward the lumen of the vein, thus creating a false impression of vascular invasion (fig. 1-46). In some neoplasms, the discontinuities in muscular arcades facilitate invasion of vascular spaces.

ELECTRON MICROSCOPY

While electron microscopy is now seldom employed in tumor diagnosis, familiarity with the ultrastructural characteristics of different cell types helps to provide a depth of knowledge important for other purposes. These include building a foundation for understanding the subcellular localization of immunohistochemical markers and interpreting the results of immunohistochemical studies.

The enzymes in steroidogenesis are localized to mitochondria and smooth endoplasmic reticulum (65,66), and these organelles are specialized to reflect their biosynthetic functions. The three zones of the cortex can be distinguished from each other on the basis of mitochondrial architecture and cytoplasmic contents (fig. 1-47) (67). Zonal differences in mitochondria are most pronounced between the zona glomerulosa, where the cristae are lamellar (67) or plate-like (68), and the fasciculata/reticularis where cristae are tubular (67) or vesicular (fig. 1-48) (68). This may help to characterize a tumor as mineralocorticoid or glucocorticoid-/sex steroid-producing if antibodies for the CYP enzymes are unavailable. Mitochondrial morphology, however, is dynamic (69). While smooth endoplasmic reticulum and lipid are present in all cortical cells, lipid is normally most abundant in the fasciculata. Numerous lipofuscin pigment granules are the most distinctive feature of the zona reticularis (67,68). Almost all of the cortical cells are in close proximity to capillary channels lined by an attenuated endothelium with fenestrations, and in most areas, there is a continuous, well-developed basal lamina with scattered pericytes. The perivascular spaces within the adrenal cortex are similar to the space of Disse in the liver.

The dominant ultrastructural feature of adrenal chromaffin cells is the presence of membrane–bound dense core neurosecretory granules, which contain a variable mix of catecholamines, granin proteins, neuropeptides, and other constituents (18). The granules typically range in size from 150 to 250 nm (50); may be round, elongated, crescentic, or "dumbbell" shaped; and vary in number from cell to cell. Under appropriate fixation conditions (70), some granules have small, uniformly dense dark cores and a wide asymmetric halo, while others have a lighter, more granular core and a tight limiting membrane (fig. 1-49). The former are associated with norepinephrine storage, and the latter with epinephrine (70).

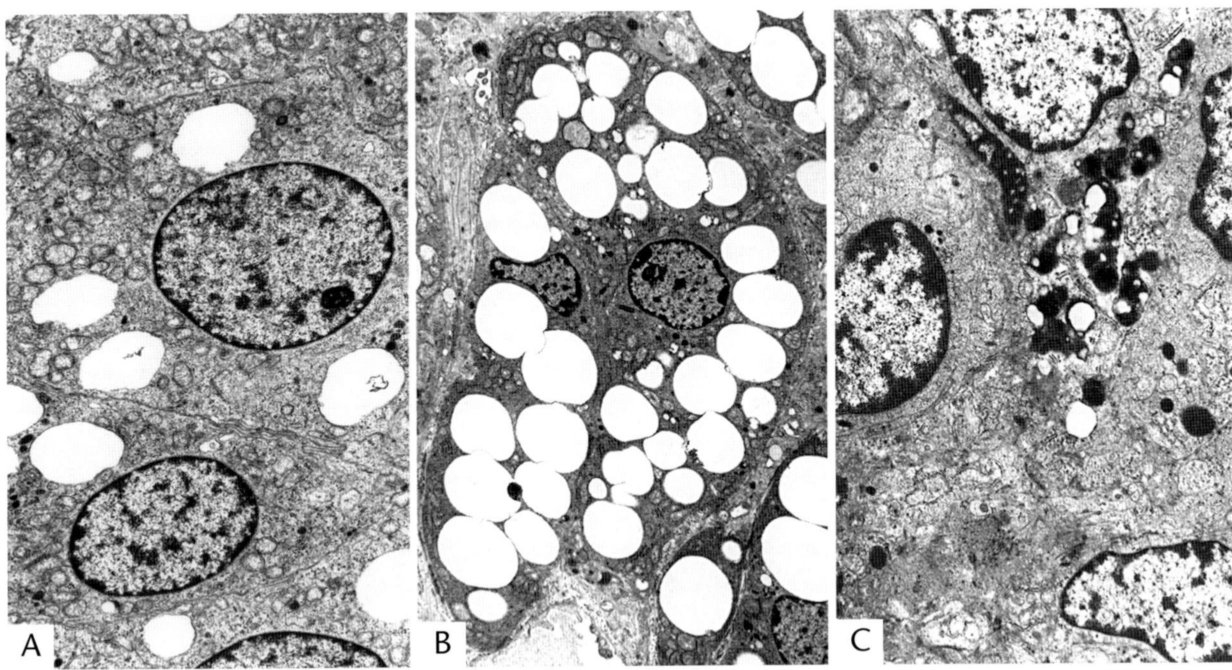

Figure 1-47

ULTRASTRUCTURAL CORRELATES OF CORTICAL ZONATION

At low magnification, cells of the zona glomerulosa (A) contain sparse lipid and have round to oval mitochondria with lamellar cristae. Cells of the zona fasciculata (B) contain many large lipid droplets and abundant smooth endoplasmic reticulum. In the zona reticularis (C), cells are relatively depleted of lipid droplets. Lysosomes and lipofuscin granules are prominent, corresponding to the brown pigment seen in routine light microscopy.

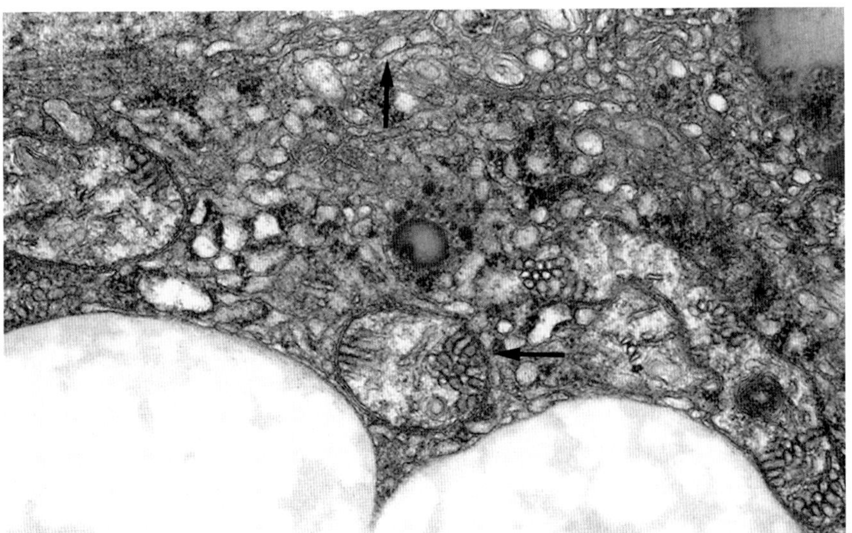

Figure 1-48

ZONA FASCICULATA

Features associated with steroidogenesis include abundant smooth endoplasmic reticulum (top arrow), which may form a complex network of anastomosing tubules, mitochondria with tubular cristae (bottom arrow), and lipid droplets (bottom of field).

In human adrenal glands, most chromaffin cells have heterogeneous granule populations which vary in size and shape (fig. 1-49). A moderate amount of rough endoplasmic reticulum is present in chromaffin cells but smooth endoplasmic reticulum and lipid are sparse. Lipochrome pigment is present in some cells. The vascular network is delicate and lined by endothelial cells with some cytoplasmic fenestrations, similar to those in the cortex.

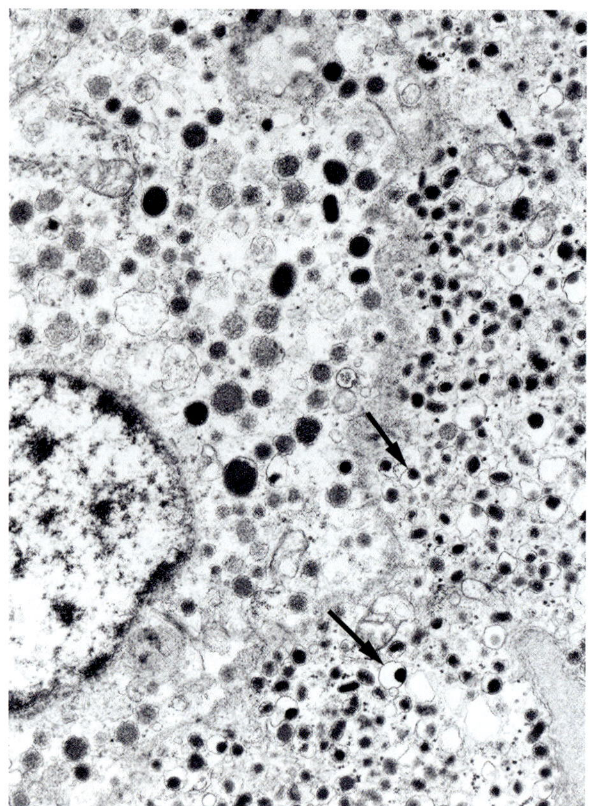

Figure 1-49

CHROMAFFIN CELLS IN ADRENAL MEDULLA

Portions of adjacent chromaffin cells with heterogeneous secretory granules. With fixation in glutaraldehyde and postfixation in osmium (58), norepinephrine-containing granules have small, uniformly dark dense cores, sometimes with a wide asymmetric halo (arrows), while those containing epinephrine have lighter, more granular cores and a tight limiting membrane. A small amount of rough endoplasmic reticulum is at lower left.

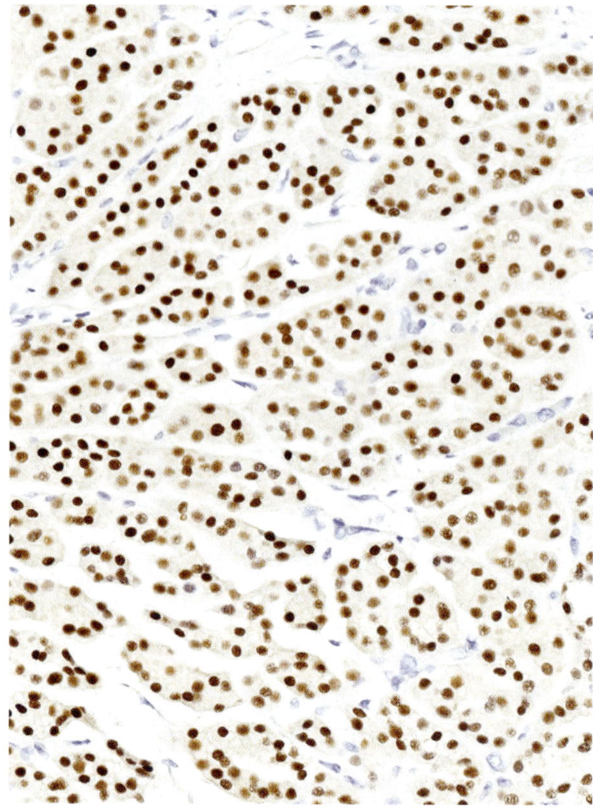

Figure 1-50

STEROIDOGENIC FACTOR 1 (SF1) IN NORMAL ADRENAL CORTEX

SF1 controls the development of steroidogenic cells in the adrenal cortex and gonads, and distinguishes adrenal cortex from medulla.

NORMAL ADRENAL GLAND MARKERS IN TUMOR PATHOLOGY

Developmental and functional markers expressed in the normal adrenal gland play an important role in characterization of adrenal cortical and medullary tumors, as discussed in detail in subsequent chapters. An ever-widening array of reliable antibodies is now available to pathologists for immunohistochemical studies of these tumors. This is occurring in step with a comparably expanding array of applications, including differential diagnosis, functional characterization, risk stratification, identification of drug targets, and screening for hereditary disorders (71). Challenges for pathologists are to determine which immunohistochemical markers are the most informative and most specific for each application, and to discard older, less useful ones along the way.

Relevant markers include transcription factors and growth factors participating in the development of the cortex or medulla, functional markers of steroid- or catecholamine-synthesizing cells, and a variety of cortical markers (in particular) with poorly understood functions. Diagnostically important transcription factors with high specificity in appropriate context are SF1 for adrenal cortex (fig. 1-50) and GATA3 for adrenal medulla and extra-adrenal paraganglia (fig. 1-51). Adrenal medulla is also distinguished from cortex by expression of INSM1, which is expressed in essentially all neuroendocrine cells and is more sensitive than functional markers, particularly for poorly differentiated tumors (72).

Functional markers for the cortex or medulla include steroidogenic or catecholamine-synthesizing enzymes, some granin proteins, which are the major constituent of neuroendocrine secretory granules, and some components of the neuroendocrine secretory apparatus (71,73,74). Additional cortical markers include inhibin (fig. 1-52) and others with poorly understood functions such as Melan-A (MART-1) (fig. 1-53) and calretinin (fig. 1-54). The expression of markers within the normal cortex is highly variable, consistent with functional regulation.

Chosen wisely, normal adrenal markers discriminate between cortical and medullary tumors and help to distinguish them from other tumor types. Caveats are that some markers are expressed normally in both cortex and medulla, or are normally limited to one or the other region but expressed incongruously in tumors. An important example of the former is synaptophysin (fig. 1-55); inhibin is an example of the latter (74,75), being especially associated with pseudohypoxic pheochromocytomas and paragangliomas (76). Calretinin distinguishes cortical cells from chromaffin cells and cortical tumors from pheochromocytomas, but may stain neuronal cells and processes in the medulla. It is also expressed in many nonendocrine tumors (74).

Synaptophysin is a constituent of the membranes enclosing neuroendocrine secretory granules and synaptic vesicles in neurons, where it associates with other vesicle proteins and plays a still incompletely understood role in trafficking of those secretory organelles to and from the cell membrane. However, it is also expressed in poorly differentiated neuroendocrine carcinomas that have scant or no secretory granules. Its role in the adrenal cortex is unknown. Components of the vesicle docking complex, including synaptobrevin 2, a major partner of synaptophysin, are expressed in a variety of non-neural tissues including the adrenal cortex (77) and may play a role in lipid trafficking. In contrast to synaptophysin, granin proteins, including chromogranin A and chromogranin B, are constituents of the secretory granule matrix (18) and are not expressed in the adrenal cortex. While synaptophysin cannot be used to distinguish between cortical and chromaffin cell tumors, chromogranin A is currently the most reliable widely used generic neuroendocrine marker for that purpose (fig. 1-56) but may be absent in sparsely granulated tumors.

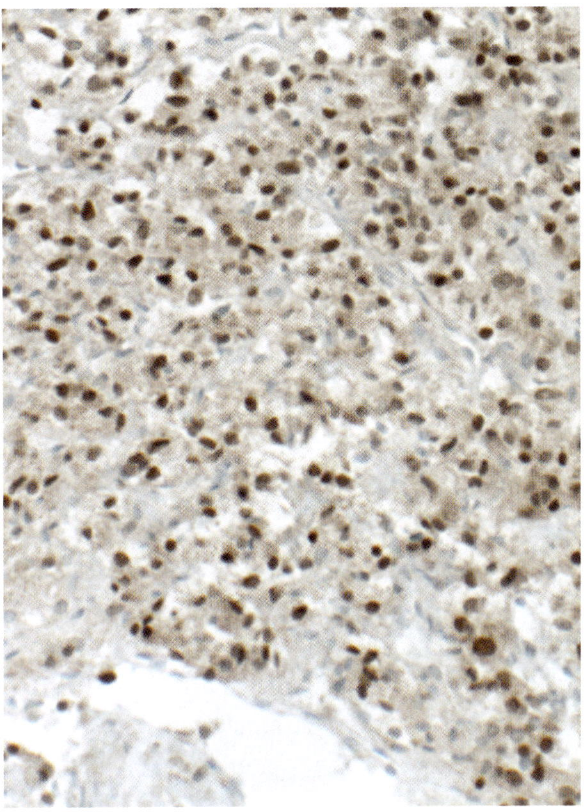

Figure 1-51

GATA3 IN NORMAL ADRENAL MEDULLA

GATA3 is expressed in normal and neoplastic adrenal medulla and extra-adrenal paraganglia but not adrenal cortex. It is a valuable marker for diagnosis but is widely expressed in various tissues and must be considered in appropriate anatomic context together with other positive and negative markers.

The same purpose can usually be served by staining for tyrosine hydroxylase (TH), which provides additional, specific functional information that can be used to distinguish catecholamine-producing tumors from biochemically silent paragangliomas and other neuroendocrine neoplasms. Expression of TH is the *sine qua non* for catecholamine production, and therefore useful for demonstrating catecholamine-synthesizing ability, while expression of dopamine B-hydroxylase (DBH) or phenylethanolamine

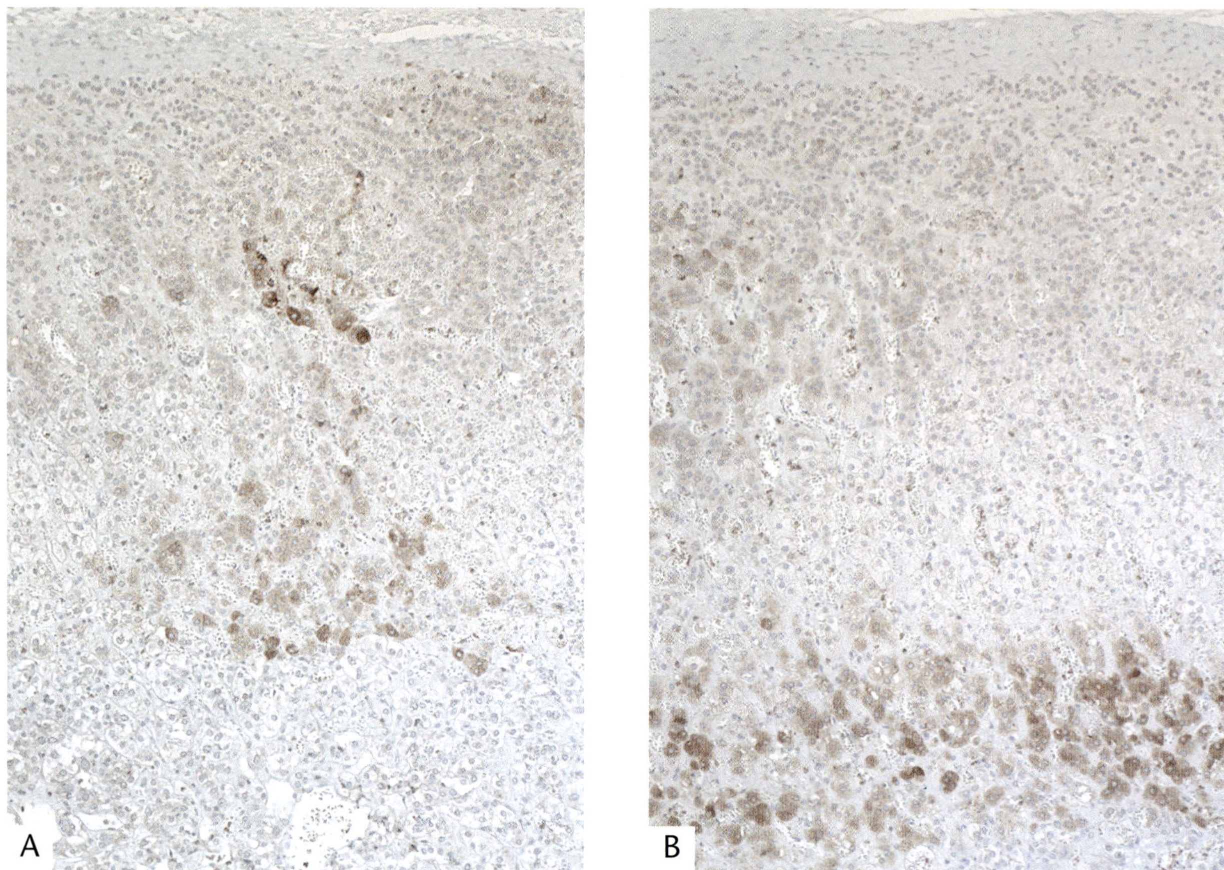

Figure 1-52

INHIBIN EXPRESSION IN NORMAL ADRENAL GLAND

Inhibin is expressed in the adrenal cortex in a variable and often patchy distribution, suggesting an association with different functional states. The greatest expression is in the zona reticularis, consistent with reported preferential expression in androgen-producing tumors (62). A: Patchy distribution in zona fasciculata and reticularis (negative medulla at bottom of field). B: Extensive distribution in zona reticularis.

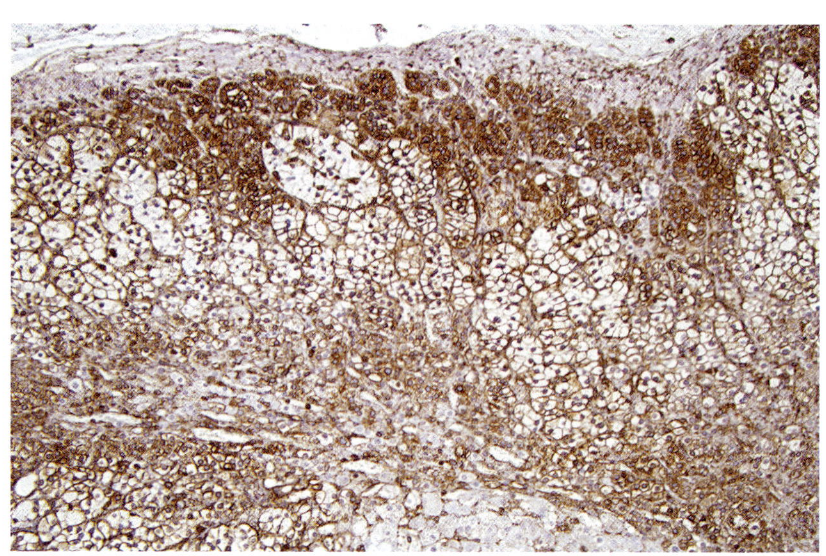

Figure 1-53

MELAN-A IN NORMAL ADRENAL GLAND

Melan-A is expressed throughout the cortex but staining tends to be most intense in the zona glomerulosa (negative medulla at bottom of field).

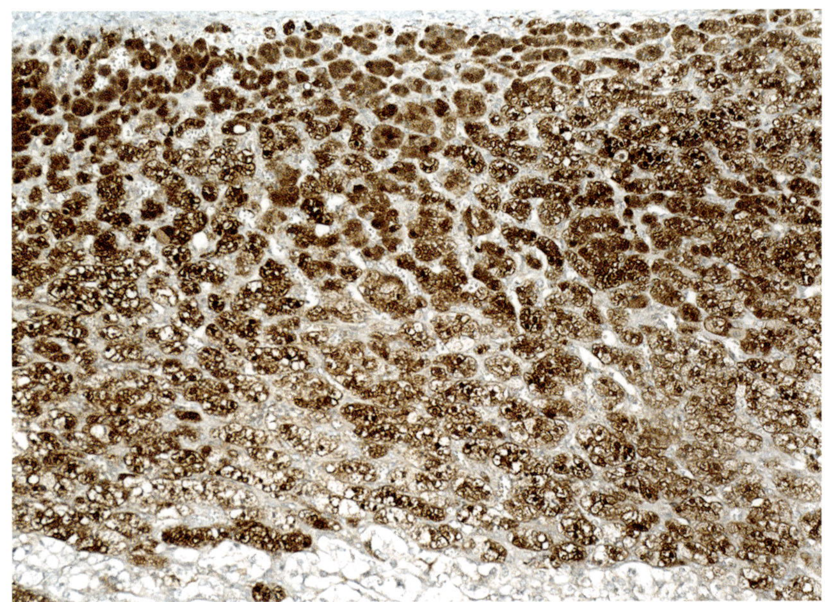

Figure 1-54

CALRETININ IN NORMAL ADRENAL GLAND

Calretinin typically shows diffuse strong expression throughout the cortex (negative medulla at bottom of field).

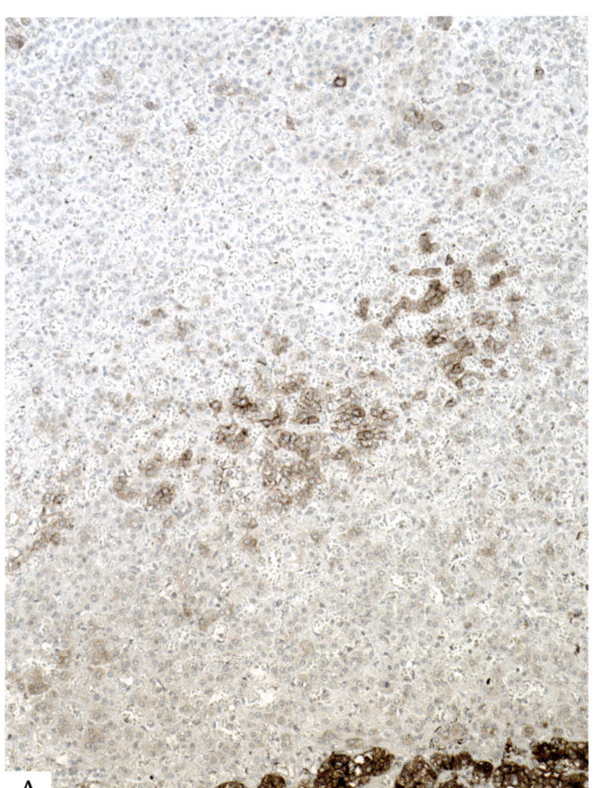

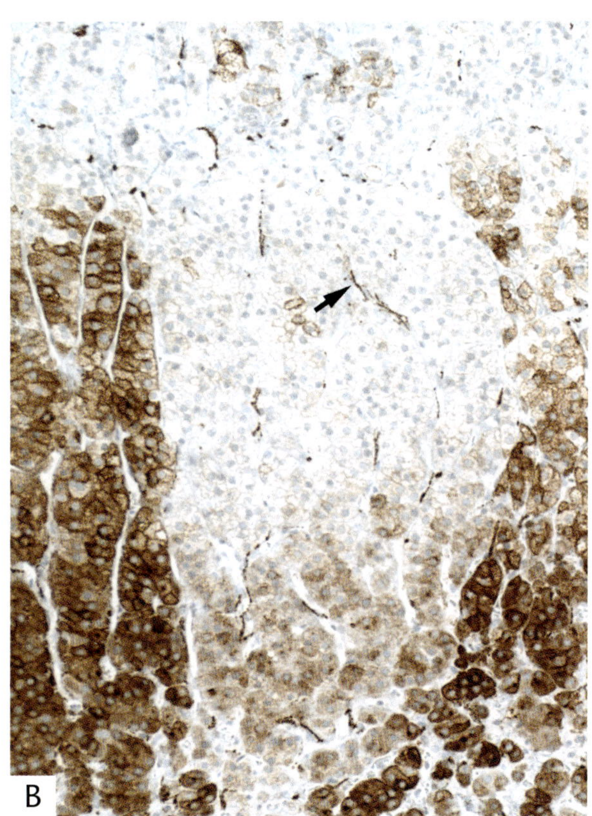

Figure 1-55

SYNAPTOPHYSIN IN NORMAL ADRENAL GLAND

A: Synaptophysin expression is shown in a patchy distribution in the zona fasciculata. This contrasts with diffuse strong staining in the adrenal medulla at bottom of the field.

B: Strong but variable expression of synaptophysin in the zona fasciculata and reticularis. A few immunoreactive nerve fibers traverse the cortex in the unstained area (arrow).

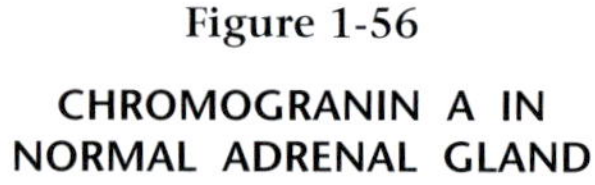

Figure 1-56

CHROMOGRANIN A IN NORMAL ADRENAL GLAND

Chromaffin cells in the adrenal medulla characteristically show diffuse, intense immunoreactivity for chromogranin A. There is no staining in the adrenal cortex or in perivascular cuffs of cortical tissue within the medulla.

N methyltransferase (PNMT) can suggest the specific catecholamines produced (78).

While immunohistochemical studies of the distribution of steroidogenic enzymes have been used to determine the functional correlates of anatomic cortical zonation, applications to pathology have historically been hampered by lack of adequate antibodies. This is due to the multifunctionality of some steroidogenic enzymes and the over 90 percent structural homology between different enzymes in some cases (46). This situation has recently improved with the development of new monoclonal antibodies (71), recent examples being antibodies against CYP11B2 to identify cells and tumors producing aldosterone (46) and CYP11B1 to identify those producing cortisol (47). These two immunohistochemical markers have led to greatly increased understanding of normal and pathologic mineralocorticoid-producing cells. New findings include transition from continuous to patchy distribution of aldosterone-producing cells in the normal zona glomerulosa from childhood to adult life, heterogeneity of cell populations in the normal zona glomerulosa and aldosterone-producing neoplasms, and existence of multiple types of aldosterone-producing lesions in patients with primary hyperaldosteronism.

An additional immunohistochemical marker that has yet to fully find its place in pathology studies is expression of somatostatin receptors. Five different receptors, designated somatostatin receptor types 1-5 (SSTR1-5) are variably expressed in adrenal cortex and medulla and in their corresponding tumors (79). Although their roles in regulating hormone synthesis and secretion are not fully established, they are potential targets for both imaging and treatment (80). Immunohistochemical staining for SSTRs may help to triage patients for these purposes (81).

REFERENCES

1. Ikeda Y, Lister J, Bouton JM, Buyukpamukcu M. Congenital neuroblastoma, neuroblastoma in situ, and the normal fetal development of the adrenal. J Pediatr Surg 1981;16:636-44.
2. Xing Y, Lerario AM, Rainey W, Hammer GD. Development of adrenal cortex zonation. Endocrinol Metab Clin North Am 2015;44:243-74.
3. Kim AC, Reuter AL, Zubair M, et al. Targeted disruption of beta-catenin in SF1-expressing cells impairs development and maintenance of the adrenal cortex. Development 2008;135:2593-602.
4. Miller WL. Steroidogenesis: unanswered questions. Trends Endocrinol Metab 2017;28:771-93.
5. Moeller G, Adamski J. Integrated view on 17beta-hydroxysteroid dehydrogenases. Mol Cell Endocrinol 2009;301:7-19.
6. Unsicker K, Huber K, Schober A, Kalcheim C. Resolved and open issues in chromaffin cell development. Mech Dev 2013;130:324-9.
7. Schinner S, Bornstein SR. Cortical-chromaffin cell interactions in the adrenal gland. Endocr Pathol 2005;16:91-8.
8. Gomez-Sanchez E, Gomez-Sanchez CE. The multifaceted mineralocorticoid receptor. Compr Physiol 2014;4:965-94.
9. Goto M, Hanley KP, Marcos J, et al. In humans, early cortisol biosynthesis provides a mechanism to safeguard female sexual development. J Clin Invest 2006;116:953-60.
10. Ishimoto H, Jaffe RB. Development and function of the human fetal adrenal cortex: a key component in the feto-placental unit. Endocr Rev 2011;32:317-55.

10a. Broderick R, Ramadurai S, Toth K, et al. Cell cycle-dependent mobility of CDC45 determined in vivo by fluorescence correlation spectroscopy. PloS One 2012;7:e35537.

11. Turcu A, Smith JM, Auchus R, Rainey WE. Adrenal androgens and androgen precursors-definition, synthesis, regulation and physiologic actions. Compr Physiol 2014;4:1369-81.

11a. Lefebvre H, Thomas M, Duparc C, Bertherat J, Louiset E. Role of ACTH in the interactive/paracrine regulation of adrenal steroid secretion in physiological and pathophysiological conditions. Front Endocrinol (Lausanne) 2016:7:98.

12. Eisenhofer G, Kopin IJ, Goldstein DS. Catecholamine metabolism: a contemporary view with implications for physiology and medicine. Pharmacol Rev 2004;56:331-49.
13. Wong DL. Why is the adrenal adrenergic? Endocr Pathol 2003;14:25-36.
14. Ziegler MG, Bao X, Kennedy BP, Joyner A, Enns R. Location, development, control, and function of extraadrenal phenylethanolamine N-methyltransferase. Ann N Y Acad Sci 2002;971:76-82.
15. Guerineau NC. Gap junction communication between chromaffin cells: the hidden face of adrenal stimulus-secretion coupling. Pflugers Arch 2018;470:89-96.
16. Nurse CA, Salman S, Scott AL. Hypoxia-regulated catecholamine secretion in chromaffin cells. Cell Tissue Res 2018;372:433-41.
17. Huber K, Janoueix-Lerosey I, Kummer W, Rohrer H, Tischler AS. The sympathetic nervous system: malignancy, disease, and novel functions. Cell Tissue Res 2018;372:163-70.
18. Helle KB, Metz-Boutigue MH, Cerra MC, Angelone T. Chromogranins: from discovery to current times. Pflugers Arch 2018;470:143-54.
19. Lundberg JM, Hamberger B, Schultzberg M, et al. Enkephalin- and somatostatin-like immunoreactivities in human adrenal medulla and pheochromocytoma. Proc Natl Acad Sci U S A 1979;76:4079-83.
20. Whitworth EJ, Kosti O, Renshaw D, Hinson JP. Adrenal neuropeptides: regulation and interaction with ACTH and other adrenal regulators. Microsc Res Tech 2003;61:259-67.
21. Podvin S, Bundey R, Toneff T, Ziegler M, Hook V. Profiles of secreted neuropeptides and catecholamines illustrate similarities and differences in response to stimulation by distinct secretagogues. Mol Cell Neurosci 2015;68:177-85.
22. Pignatti E, Leng S, Carlone DL, Breault DT. Regulation of zonation and homeostasis in the adrenal cortex. Mol Cell Endocrinol 2017;441:146-55.
23. Guasti L, Candy Sze WC, McKay T, Grose R, King PJ. FGF signalling through FGFR2 isoform IIIb regulates adrenal cortex development. Mol Cell Endocrinol 2013;371:182-8.
24. Chan WH, Anderson CR, Gonsalvez DG. From proliferation to target innervation: signaling molecules that direct sympathetic nervous system development. Cell Tissue Res 2018;372:171-93.
25. Furlan A, Dyachuk V, Kastriti ME, et al. Multipotent peripheral glial cells generate neuroendocrine cells of the adrenal medulla. Science 2017; 357:eaal3753.
26. Erickson AG, Kameneva P, Adameyko I. The transcriptional portraits of the neural crest at the individual cell level. Semin Cell Dev Biol 2023; 138:68-80.

27. Gasser RF, Cork RJ, Stillwell BJ, McWilliams DT. Rebirth of human embryology. Dev Dyn 2014;243:621-8.
28. Martucci VL, Emaminia A, del Rivero J, et al. Succinate dehydrogenase gene mutations in cardiac paragangliomas. Am J Cardiol 2015;115:1753-9.
29. Kameneva P, Artemov AV, Kastriti ME, et al. Single-cell transcriptomics of human embryos identifies multiple sympathoblast lineages with potential implications for neuroblastoma origin. Nat Genet 2021;53:694-706.
30. Hockman D, Adameyko I, Kaucka M, et al. Striking parallels between carotid body glomus cell and adrenal chromaffin cell development. Dev Biol 2018;444(Suppl 1):S308-24.
31. Helen P, Alho H, Hervonen A. Ultrastructure and histochemistry of human SIF cells and paraganglia. Adv Biochem Psychopharmacol 1980;25: 149-52.
32. Tischler AS, Asa SL. Paraganglia. In: Mills SE, ed. Histology for pathologists, 5th ed. Philadelphia: Wolters Kluwer; 2020:1274-95.
33. Crowder RE. The development of the adrenal gland in man, with special reference to origin and ultimate location of cell types and evidence in favor of the "cell migration" theory. Contrib Embryol Carneg Instn 1957;36:193-210.
34. Schober A, Parlato R, Huber K, et al. Cell loss and autophagy in the extra-adrenal chromaffin organ of Zuckerkandl are regulated by glucocorticoid signalling. J Neuroendocrinol 2013;25:34-47.
35. Turkel SB, Itabashi HH. The natural history of neuroblastic cells in the fetal adrenal gland. Am J Pathol 1974;76:225-44.
36. Magro G, Grasso S. Immunohistochemical identification and comparison of glial cell lineage in foetal, neonatal, adult and neoplastic human adrenal medulla. Histochem J 1997;29:293-9.
37. Lam KY, Chan AC, Lo CY. Morphological analysis of adrenal glands: a prospective analysis. Endocr Pathol 2001;12:33-8.
38. Lack EE, Kozakewich HP. Embryology, developmental anatomy and selected aspects of non-neoplastic pathology. In: Lack EE, ed. Pathology of the adrenal glands. New York: Churchill Livingstone; 1990:1-74.
39. Helfer TM, Rolo LC, Okasaki NA, et al. Reference ranges of fetal adrenal gland and fetal zone volumes between 24 and 37 + 6 weeks of gestation by three-dimensional ultrasound. J Matern Fetal Neonatal Med 2017;30:568-73.
40. Studzinski GP, Hay DC, Symington T. Observations on the weight of the human adrenal gland and the effect of preparations of corticotropin of different purity on the weight and morphology of the human adrenal gland. J Clin Endocrinol Metab 1963;23:248-54.
41. Kreiner E. Weight and shape of the human adrenal medulla in various age groups. Virchows Arch A Pathol Anat Histol 1982;397:7-15.
42. Pina-Oviedo S, Ortiz-Hidalgo C, Ayala AG. Human colors-the rainbow garden of pathology: what gives normal and pathologic tissues their color? Arch Pathol Lab Med 2017;141:445-62.
43. Jacques SM, Qureshi F. Adrenal histologic stress-related changes in third trimester stillbirth. Pediatr Dev Pathol 2017;20:112-9.
44. Delprado WJ, Baird PJ. The fetal adrenal gland: definitive cortex cystic change, lipid patterns, and their relationship to fetal disease and maturity. Pathology 1984;16:312-7.
45. Favara BE, Steele A, Grant JH, Steele P. Adrenal cytomegaly: quantitative assessment by image analysis. Pediatr Pathol 1991;11:521-36.
46. Gomez-Sanchez CE, Gomez-Sanchez EP, Nishimoto K. Immunohistochemistry of the human adrenal CYP11B2 in normal individuals and in patients with primary aldosteronism. Horm Metab Res 2020;52:421-6.
47. Sun L, Jiang Y, Xie J, et al. Immunohistochemical Analysis of CYP11B2, CYP11B1 and beta-catenin helps subtyping and relates with clinical characteristics of unilateral primary aldosteronism. Front Mol Biosci 2021;8:751770.
48. Sasano H, Imatani A, Shizawa S, Suzuki T, Nagura H. Cell proliferation and apoptosis in normal and pathologic human adrenal. Mod Pathol 1995;8:11-7.
49. Murray D, Mackay AM. Studies on tubular degeneration of the adrenal cortex. J Pathol 1972; 106:Pix.
50. Carmichael SW. The history of the adrenal medulla. Rev Neurosci 1989;2:83-100.
51. Saeger W, Reinhard K. Fat-cell metaplasia in the adrenal cortex: incidence, structure, and correlation to basic diseases in a postmortem series. Endocr Pathol 1998;9:241-7.
52. Hayashi Y, Hiyoshi T, Takemura T, Kurashima C, Hirokawa K. Focal lymphocytic infiltration in the adrenal cortex of the elderly: immunohistological analysis of infiltrating lymphocytes. Clin Exp Immunol 1989;77:101-5.
53. Mete O, Raphael S, Pirzada A, Asa SL. Is adrenal ovarian thecal metaplasia a misnomer? Report of three cases of radial scar-like spindle cell myofibroblastic nodule of the adrenal gland. Endocr Pathol 2011;22:222-5.
54. Magalhàes MC. A new crystal-containing cell in human adrenal cortex. J Cell Biol 1972;55:126-33.
55. Pollock WJ, McConnell CF, Hilton C, Lavine RL. Virilizing leydig cell adenoma of adrenal gland. Am J Surg Pathol 1986;10:816-22.

56. Mesa H, Gilles S, Datta MW, et al. Immunophenotypic differences between neoplastic and non-neoplastic androgen-producing cells containing and lacking Reinke crystals. Virchows Arch 2016;469:679-86.
57. Kohn A. Die chromaffinen Zellen des sympathicus. Anat Anz 1898;15:399-400.
58. Bausch B, Tischler AS, Schmid KW, Leijon H, Eng C, Neumann HP. Max Schottelius: pioneer in pheochromocytoma. J Endocr Soc 2017;1:957-64.
59. Sugie M, Goto J, Kawamura M, Ota H. Increased norepinephrine-associated adrenomedullary inclusions in Parkinson's disease. Pathol Int 2005; 55:130-6.
60. Eckardt L, Prange-Barczynska M, Hodson EJ, et al. Developmental role of PHD2 in the pathogenesis of pseudohypoxic pheochromocytoma. Endocr Relat Cancer 2021;28:757-72.
61. Sobrino V, Platero-Luengo A, Annese V, et al. Neurotransmitter modulation of carotid body germinal niche. Int J Mol Sci 2020;21:8231.
62. Fielding JW, Hodson EJ, Cheng X, et al. PHD2 inactivation in Type I cells drives HIF-2alpha-dependent multilineage hyperplasia and the formation of paraganglioma-like carotid bodies. J Physiol 2018;596:4393-412.
63. Colombo-Benkmann M, Klimaschewski L, Heym C. Immunohistochemical heterogeneity of nerve cells in the human adrenal gland with special reference to substance P. J Histochem Cytochem 1996;44:369-75.
64. Dobbie JW, Symington T. The human adrenal gland with special reference to the vasculature. J Endocrinol 1966;34:479-89.
65. Pelletier G, Li S, Luu-The V, Tremblay Y, Belanger A, Labrie F. Immunoelectron microscopic localization of three key steroidogenic enzymes (cytochrome P450(scc), 3 beta-hydroxysteroid dehydrogenase and cytochrome P450(c17)) in rat adrenal cortex and gonads. J Endocrinol 2001;171:373-83.
66. Ishimura K, Fujita H. Light and electron microscopic immunohistochemistry of the localization of adrenal steroidogenic enzymes. Microsc Res Tech 1997;36:445-53.
67. Long JA, Jones AL. Observations on the fine structure of the adrenal cortex of man. Lab Invest 1967;17:355-70.
68. Belloni AS, Mazzocchi G, Mantero F, Nussdorfer GG. The human adrenal cortex: ultrastructure and base-line morphometric data. J Submicrosc Cytol 1987;19:657-68.
69. Pánek T, Eliáš M, Vancová M, Lukeš J, Hashimi H. Returning to the fold for lessons in mitochondrial crista diversity and evolution. Curr Biol 2020;30:R575-88.
70. Coupland RE, Hopwood D. The mechanism of the differential staining reaction for adrenaline-and noreadrenaline-storing granules in tissues fixed in glutaraldehyde. J Anat 1966;100: 227-43.
71. Mete O, Asa SL, Giordano TJ, Papotti M, Sasano H, Volante M. Immunohistochemical biomarkers of adrenal cortical neoplasms. Endocr Pathol 2018;29:137-49.
72. Maleki Z, Nadella A, Nadella M, Patel G, Patel S, Kholova I. INSM1, a novel biomarker for detection of neuroendocrine neoplasms: cytopathologists' view. Diagnostics (Basel) 2021;11:2172.
73. Sangoi AR, McKenney JK. A tissue microarray-based comparative analysis of novel and traditional immunohistochemical markers in the distinction between adrenal cortical lesions and pheochromocytoma. Am J Surg Pathol 2010; 34:423-32.
74. Zhang PJ, Genega EM, Tomaszewski JE, Pasha TL, LiVolsi VA. The role of calretinin, inhibin, melan-A, BCL-2, and C-kit in differentiating adrenal cortical and medullary tumors: an immunohistochemical study. Mod Pathol 2003;16: 591-7.
75. Arola J, Liu J, Heikkila P, et al. Expression of inhibin alpha in adrenocortical tumours reflects the hormonal status of the neoplasm. J Endocrinol 2000;165:223-9.
76. Mete O, Pakbaz S, Lerario AM, Giordano TJ, Asa SL. Significance of alpha-inhibin expression in pheochromocytomas and paragangliomas. Am J Surg Pathol 2021;45:1264-73.
77. Rossetto O, Gorza L, Schiavo G, Schiavo N, Scheller RH, Montecucco C. VAMP/synaptobrevin isoforms 1 and 2 are widely and differentially expressed in nonneuronal tissues. J Cell Biol 1996;132:167-79.
78. Tischler AS. Triple immunohistochemical staining for bromodeoxyuridine and catecholamine biosynthetic enzymes using microwave antigen retrieval. J Histochem Cytochem 1995;43:1-4.
79. Mariniello B, Finco I, Sartorato P, et al. Somatostatin receptor expression in adrenocortical tumors and effect of a new somatostatin analog SOM230 on hormone secretion in vitro and in ex vivo adrenal cells. J Endocrinol Invest 2011; 34:e131-8.
80. Janssen I, Blanchet EM, Adams K, et al. Superiority of [68Ga]-DOTATATE PET/CT to other functional imaging modalities in the localization of SDHB-associated metastatic pheochromocytoma and paraganglioma. Clin Cancer Res 2015; 21:3888-95.
81. Korner M, Waser B, Schonbrunn A, Perren A, Reubi JC. Somatostatin receptor subtype 2A immunohistochemistry using a new monoclonal antibody selects tumors suitable for in vivo somatostatin receptor targeting. Am J Surg Pathol 2012;36:242-52.

2 CONGENITAL DISORDERS OF THE ADRENAL GLAND

Congenital adrenal gland abnormalities may manifest anatomically, hormonally, or as a combination of the two, and thus present challenges to pathologists and clinicians. A common occurrence is the need to distinguish ectopic cortical tissue from tumor in cancer staging, abdominal surgery, or assessment of donor organs for transplantation. In other situations, an underlying genetic disorder affecting hormone synthesis may alter adrenal anatomy or a functional tumor may arise in ectopic or malformed adrenal tissue.

ADRENAL ADHESION, UNION, AND FUSION

Adrenal adhesion and *adrenal union* are distinguished by the presence or absence, respectively, of a continuous connective tissue capsule (1). In adrenal adhesion, which may involve either kidney or liver, there is an intervening thick or thin capsule (fig. 2-1A), while in adrenal union, there is intermingling of parenchymal cells of the two tissues (fig. 2-1B). In adrenal union, the intermingling may lead to separated, discontinuous nests of adrenal cortical cells which should not be misinterpreted as "invasion" (fig. 2-2). The morphologic features of the discontinuous nests in adrenal union can be identical to those of adrenal heterotopia, which also may involve the liver and kidney. Adrenal-hepatic and adrenal-renal fusion rarely give rise to an intrahepatic or intrarenal adrenal cortical neoplasm (2,3).

Adrenal fusion is a rare anomaly in which the adrenal glands are united in the midline (fig. 2-3), with some medial deviation of the kidneys in most cases. The fused adrenal glands may have a horseshoe or butterfly appearance. This abnormality is sometimes associated with midline congenital defects such as spinal dysraphism and indeterminate visceral situs, and occasionally, the Cornelia de Lange syndrome (mental and growth retardation, synophrys, anteverted nostrils, low-set ears, and spade-like hands with short tapering fingers) (1). It is also associated with major central nervous system (CNS) malformations, renal agenesis, and complex cardiac anomalies (4). *Renal agenesis* is sometimes associated with an adrenal gland of abnormal shape, which may be oval with a smooth contour.

ADRENAL HYPOPLASIA AND AGENESIS

Adrenal hypoplasia congenita (AHC) refers to adrenal cortical underdevelopment, a complex disorder that is divided into a miniature adult form and a cytomegalic form (5). AHC can occur as an X-linked form due to *DAX1* mutation, and is associated with a number of single gene disorders and various chromosomal abnormalities. Bilateral adrenal agenesis is a rare condition that is incompatible with postnatal life without hormone replacement (6).

ECTOPIC ADRENAL TISSUES

Adrenal tissue in ectopic locations, i.e., outside of the adrenal gland, goes by various names including *ectopic, heterotopic,* and *accessory adrenal tissue* or *adrenal rests*. These terms are often used interchangeably. Some authors attempt to refine the definitions, applying heterotopic to locations that cannot be explained by usual embryonic development and accessory to those that can. Ectopic adrenal tissue has several implications with regard to tumor diagnosis. Benign adrenal cortical tissue can mimic a variety of clear cell neoplasms, creating challenges in cancer diagnosis and staging, and in the assessment of donor organ transplant specimens. It can also give rise to adrenal cortical neoplasia in unexpected locations. Familiarity with its distribution and appearance is therefore essential to avoid diagnostic mishaps.

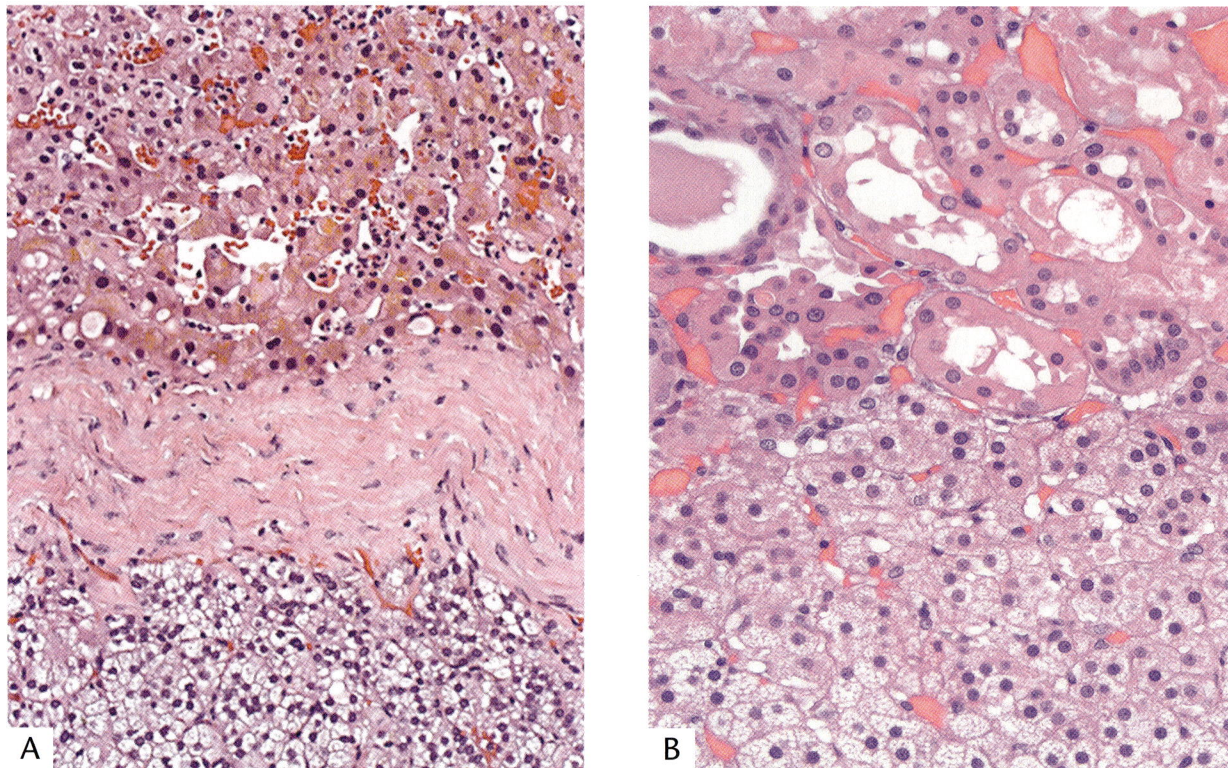

Figure 2-1

ADRENAL-HEPATIC ADHESION AND ADRENAL-RENAL UNION

A: A thick capsule separates adrenal cortex (right) from hepatic parenchyma (left) in this example of adrenal-hepatic adhesion. B: In adrenal-renal union (or fusion), there is no intervening capsule between adrenal cortex and renal parenchyma.

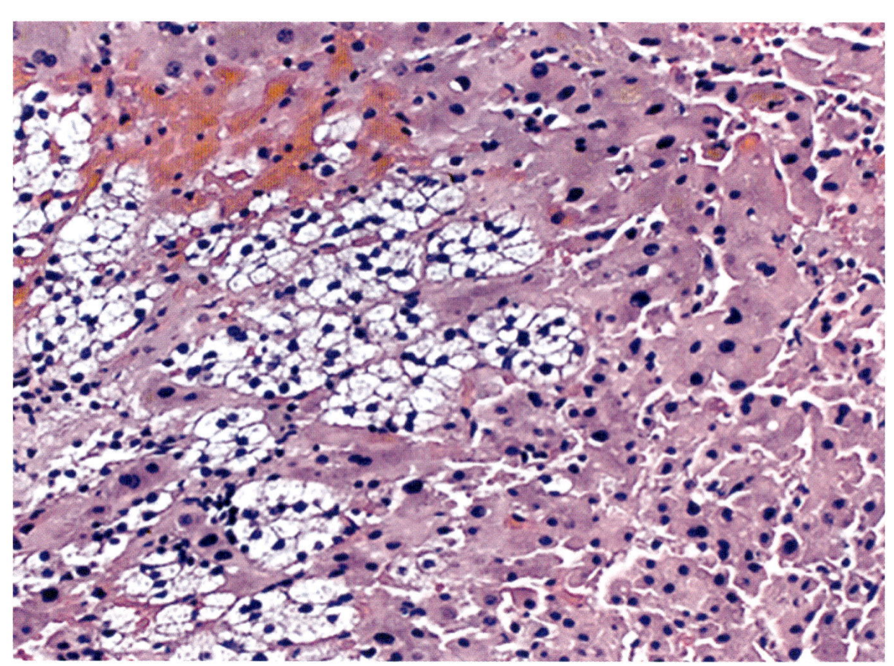

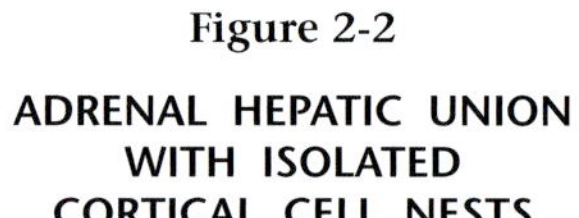

Figure 2-2

ADRENAL HEPATIC UNION WITH ISOLATED CORTICAL CELL NESTS

Heterotopic adrenal cortical tissue is within liver parenchyma, with isolated nests of cortical cells. Intrahepatic adrenal heterotopia was present just beneath the liver capsule on the undersurface of the right lobe. There was also adrenal-hepatic adhesion and union on that side.

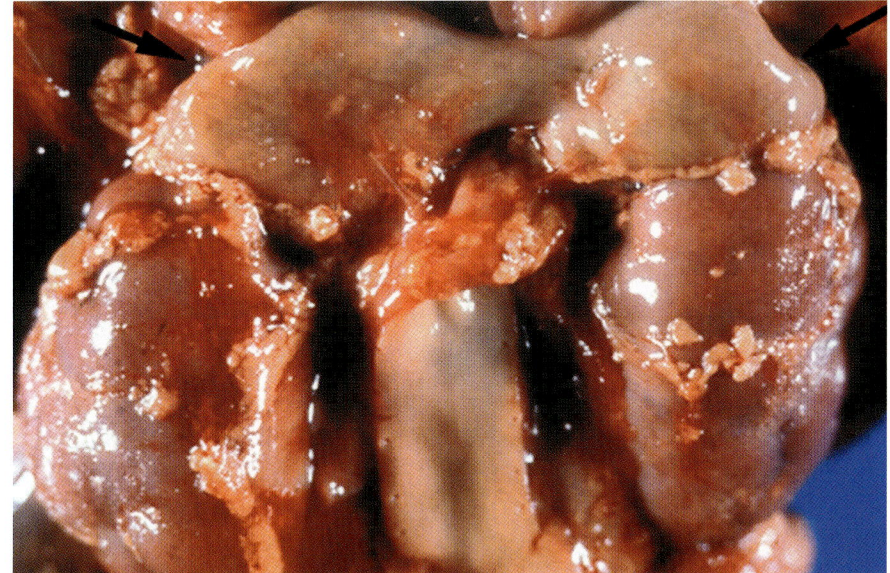

Figure 2-3

FUSION OF ADRENAL GLANDS

Newborn infant at autopsy with adrenal glands (arrows) fused in the midline anterior to the aorta.

Ectopic adrenal tissue is most often found in the upper abdomen or anywhere along the path of descent of the gonads. This anatomic localization is due to the close embryologic spatial relationship between the gonadal ridge and adrenal glands (see chapter 1). In rare cases, adrenal heterotopia occurs in bizarre anatomic sites which defy logical embryologic explanation, e.g., placenta (7), lung (8), pituitary gland (9,10), and intracranially (11). Rare reverse examples of other types of tissues located heterotopically within the adrenal have also been reported. These include intra-adrenal hepatic (12) and thyroid gland (13) tissue. Grossly, accessory adrenal tissue most often is seen as a circumscribed yellow nubbin less than 4 mm in size. One of the most frequent sites is the area of the celiac axis. In a study of 100 consecutive autopsies, these nubbins were identified in the area of the celiac plexus in 32 percent of cases, and in half of these (16 percent), they consisted of both cortex and medulla (14). A medullary component is usually absent in sites distant from the adrenal gland (7) and usually only cortex is present (fig. 2-4).

Adrenal cortical tissue is found in the adult kidney, usually in a subcapsular location in the upper pole, in 0.1 to 6.0 percent of individuals at autopsy (15). Complete removal of the renal capsule with exposure of the entire cortical surface may facilitate recognition of the adrenal tissue. Intrarenal adrenal tissue especially may be confused with renal cell carcinoma, both because of its clear cells and the frequent lack of circumscription. Immunohistochemistry may therefore be required for differentiation. A panel including positive staining for inhibin and steroidogenic factor 1 (SF1) and negative staining for PAX2 and PAX8 identifies adrenal tissue in most cases (fig. 2-5) (16). An unusual example in a dysplastic pediatric kidney is shown in figure 2-6.

Adrenal tissue is frequently encountered in the broad ligament near the ovary in infant girls. In one study, it was identified in 23.3 percent of cases anywhere in the broad ligament from the junction with the mesosalpinx to its lateral attachment, and was bilateral in 6.7 percent of cases (17). In boys, adrenal cortical tissue was found along the spermatic cord in 3.8 percent of children undergoing inguinoscrotal surgery, and in 9.3 percent who were operated on for an undescended testis where a longer segment of spermatic cord was dissected (18). Using a serial blocking technique, adrenal cortical tissue was found in 15 of 200 testicular specimens from 100 male infants less than 1 year of age (7.5 percent), and was bilateral in 4 percent (19). The nodules were circumscribed, round to ovoid, measured 0.5 to 7.0 mm in diameter, and were located in connective tissue of the distal spermatic cord or the area of the hilum of the testis (fig. 2-7). It is rare to see an adrenal cortical rest located within testicular parenchyma (fig. 2-8) (19) or in the substance of the ovary (20).

Figure 2-4

HETEROTOPIC ADRENAL CORTICAL TISSUE

These nubbins of heterotopic adrenal cortical tissue resemble miniature adrenal glands.

A: Typical example in which medulla is absent.

B: An unusual example in which dystrophic calcification is present centrally.

C: A small amount of medulla is identified by immunohistochemical stain for chromogranin A.

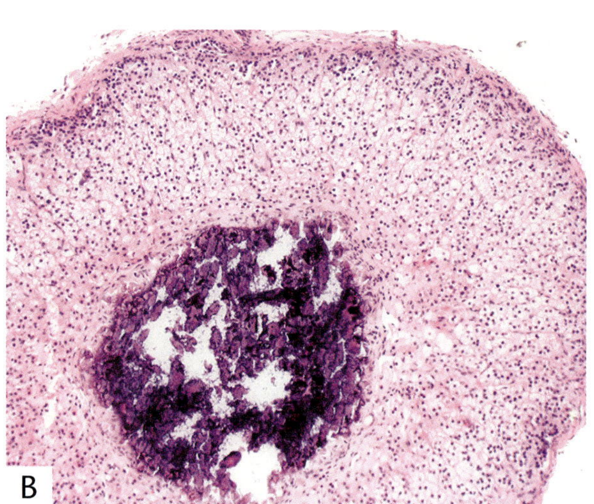

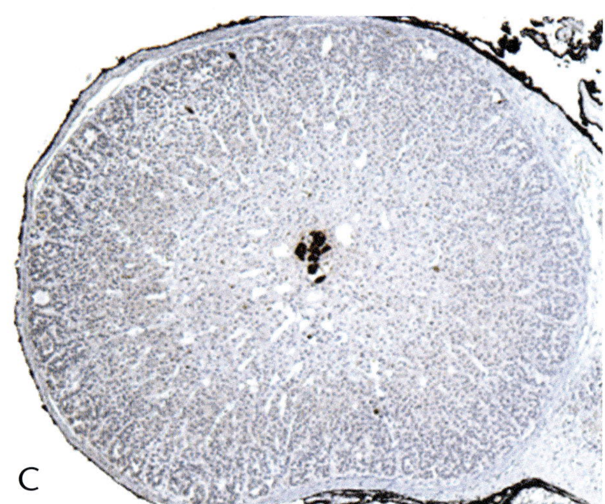

In children and adults, adrenal cortical tissue is occasionally found coincidentally together with a benign proliferative lesion, potentially causing diagnostic confusion (fig. 2-9). Examples of adrenal cortical neoplasms in unusual sites include adrenal cortical carcinoma arising in liver, kidney, scrotum, and spinal cord (21). There is also one report of pheochromocytoma arising in ectopic adrenal tissue (22). These neoplasms can pose severe diagnostic challenges. Although the developmentally predictable distribution of ectopic adrenal tissue helps explain where some of these arise, some cases defy ready embryologic explanation.

CONGENITAL ADRENAL HYPERPLASIA

The first unmistakable and thorough account of *congenital adrenal hyperplasia* (CAH), also known as *adrenogenital syndrome*, was given by the Italian anatomist de Crecchio in 1865 (23). CAH results from a defect in any one of several enzymatic steps involved in steroid synthesis (24). Steroidogenic pathways are diagrammed in chapter 1. CAH is an inborn error of metabolism which has an autosomal recessive mode of inheritance and is the most common cause of ambiguous genitalia in infants. This disorder is the most common cause of primary adrenal insufficiency in the pediatric age group. About 90 to 95 percent of all cases of CAH are due to 21-hydroxylase deficiency, a disorder of cortisol and aldosterone biosynthesis due to mutations in the *CYP21* gene that encodes adrenal 21-hydroxylase P-450c21 (25).

There are several forms of CAH: 1) a "classic" form that has an incidence between 1 in 5,000 and 1 in 15,000 live births in most white populations; 2) a "nonclassic" form, one of the

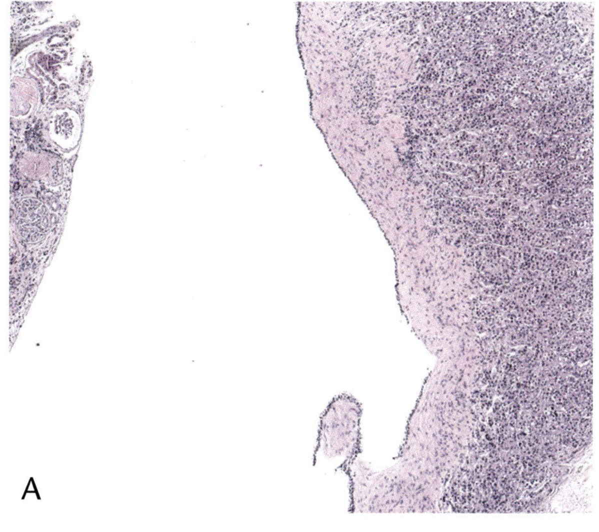

Figure 2-5

ADRENAL CORTICAL TISSUE IN A DONOR KIDNEY BIOPSY

A: Uncircumscribed adrenal cortical tissue was present between the renal capsule and a renal subcortical cyst. A small amount of normal kidney is present at left. The empty space is the cyst lumen.

B: Recognition of the fascicular arrangement of cytologically bland adrenal cells in frozen section allowed the transplant to proceed.

C: The diagnosis was confirmed by immunohistochemical staining for inhibin.

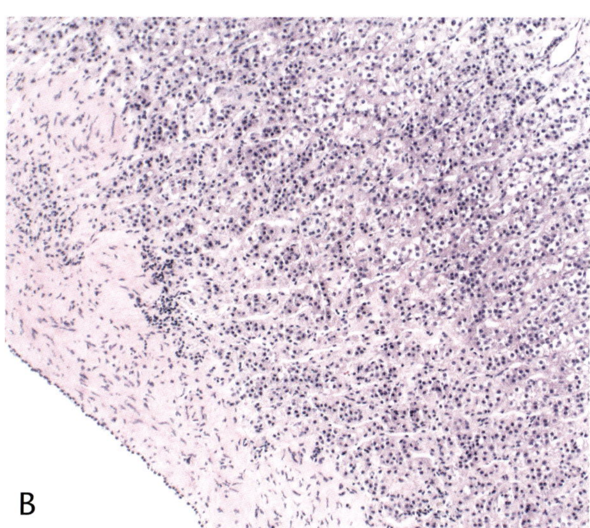

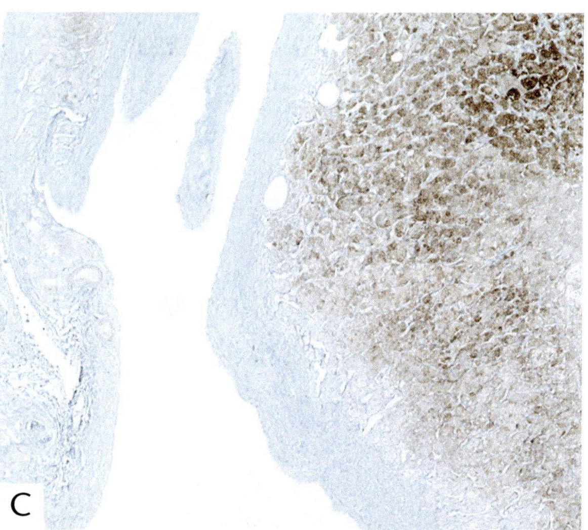

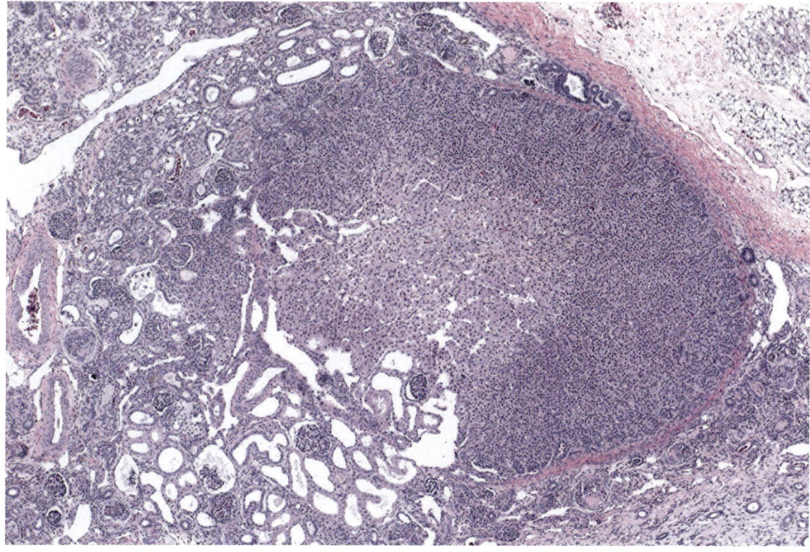

Figure 2-6

ADRENAL CORTICAL TISSUE IN A CONGENITAL DYSPLASTIC KIDNEY

This dysplastic kidney removed from a 5-week-old girl contains a prominent focus of benign immature adrenal cortex.

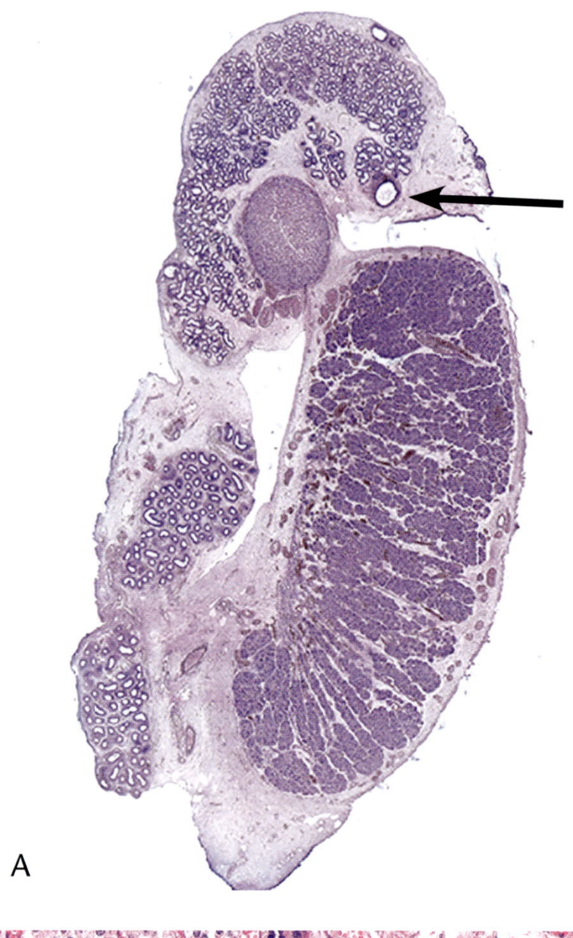

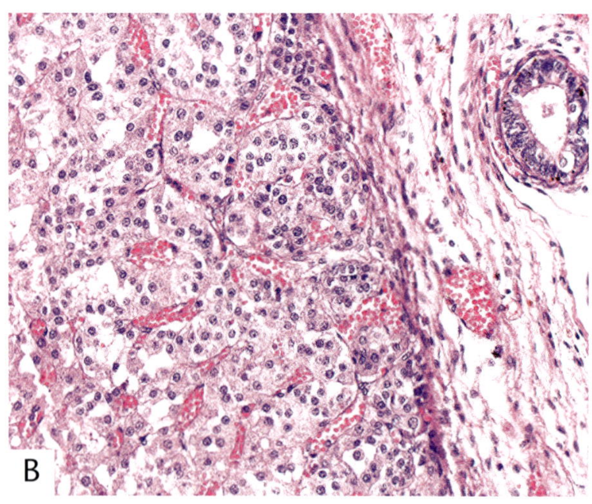

Figure 2-7

ACCESSORY ADRENAL CORTICAL TISSUE

A: Longitudinal section of a newborn testis shows a nodule of heterotopic/accessory adrenal cortical tissue in the hilum of the testis (arrow), adjacent to the head of the epididymis.

B: The adrenal tissue from this case is composed of only cortical cells, without a medullary component.

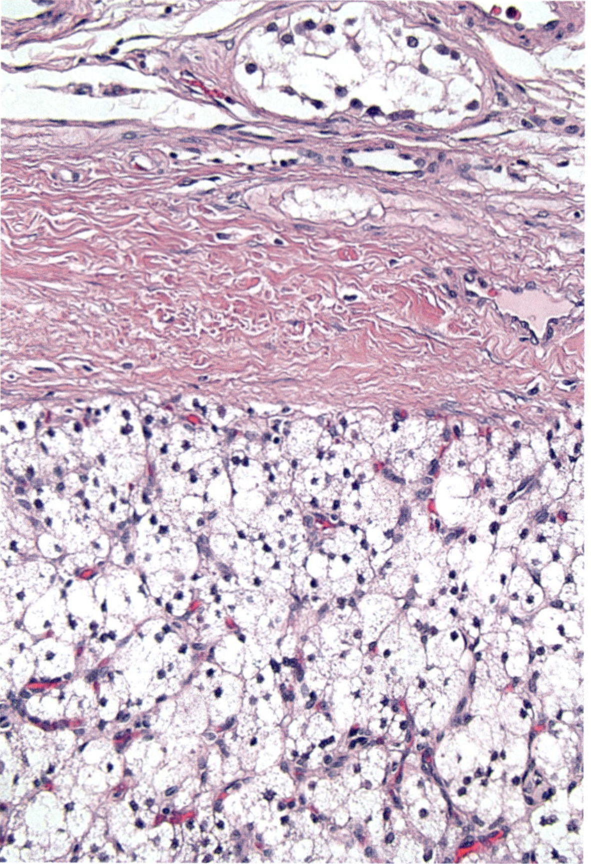

Figure 2-8

HETEROTOPIC/ACCESSORY ADRENAL CORTICAL TISSUE WITHIN THE TESTIS

Heterotopic or accessory adrenal cortical tissue is encapsulated and presents as a small nodule within the testicular parenchyma near the hilum of the testis. Seminiferous tubules are is at the top of field. Inguinal orchiectomy was performed for possible malignancy. The nodule measured 1.4 cm in diameter. It is composed of cells with pale-staining, lipid-rich, finely vacuolated cytoplasm.

most frequent autosomal recessive disorders in the general white population; and 3) a "cryptic" form in which biochemical abnormalities may exist, but the patients are asymptomatic. In two thirds of patients with the classic form of CAH, biosynthesis of aldosterone is blocked, resulting in "salt wasting"; the remaining one third have a simple virilizing form of the disease. Other enzymatic deficiencies causing CAH include: 1) 11-beta-hydroxylase deficiency which causes virilization and often hypertension due to the accumulation of deoxycorticosterone; 2) 3-beta-hydroxysteroid dehydrogenase deficiency

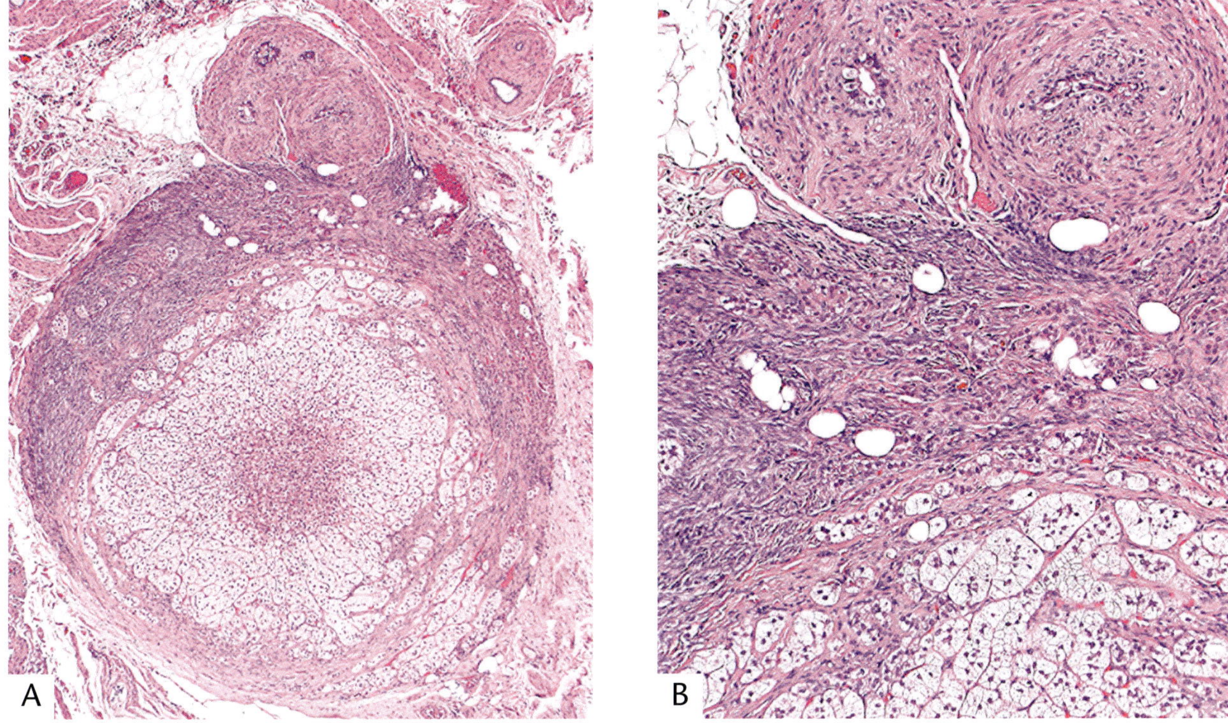

Figure 2-9

ACCESSORY ADRENAL CORTICAL TISSUE WITH OVARIAN THECAL METAPLASIA

A: The heterotopic adrenal cortical rest was located near the broad ligament in an adult woman. Cortical cells are present in the center of the nodule and are surrounded by a mantle of ovarian thecal metaplasia.

B: Scattered adrenal cortical cells, some with lipid-depleted cytoplasm, are present within the spindle cell stroma of ovarian thecal metaplasia.

which results in disorders of sex development and salt-wasting in severe cases; 3) 17-alpha-hydroxylase deficiency which is associated with hypertension, hypokalemia, and incomplete masculinization; and 4) a mutation in *STAR* gene which encodes the steroidogenic acute regulatory protein StAR. This protein plays a critical role in regulating the rate-limiting step in steroid hormone biosynthesis, cholesterol side-chain cleavage, by facilitating the movement of cholesterol to the inner mitochondrial membrane where the reaction takes place. *STAR* mutations cause *congenital lipoid adrenal hyperplasia,* the most severe form of CAH (26). The disorder is usually fatal despite replacement therapy.

The clinical manifestations of CAH are the result of a deficiency of a particular steroid, such as cortisol, and/or the effects of steroids that accumulate proximal to the site of enzymatic deficiency and may be shunted into alternate biosynthetic pathways, particularly androgen synthesis. Males with the salt-losing form of 21-hydroxylase deficiency may have signs and symptoms (e.g., vomiting, dehydration, and hypotension) that resemble an addisonian crisis within a few weeks of birth. If the female fetus is exposed to increased androgens in utero there is a variable degree of virilization of external genitalia, while the internal female organs are relatively normal. The typically ambiguous genitalia in these female infants include clitoromegaly and fusion of labioscrotal folds, which may be bulbous and rugate, thus simulating a scrotum (fig. 2-10). The clitoris may be bound somewhat by a "chordee." In a small number of infants, the degree of virilization is so marked that there is a fully masculinized penile urethra. The masculinized female may be incorrectly classified as a male (fig. 2-10), but the error is recognized when a salt-losing crisis develops at 1 to 4 weeks of age. Masculinization of the external genitalia begins by 8 weeks of gestation. A prenatal diagnosis is

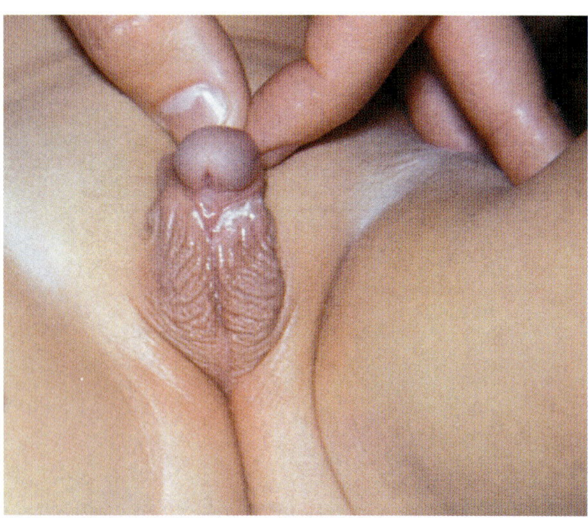

Figure 2-10

CONGENITAL ADRENAL HYPERPLASIA

This 18-month-old female was mistakenly assigned a male gender. The markedly enlarged clitoris resembles a penis. The hypertrophied clitoris had a urethral opening near the base. The rugosity of the partially fused labio-scrotal folds simulates a scrotum.

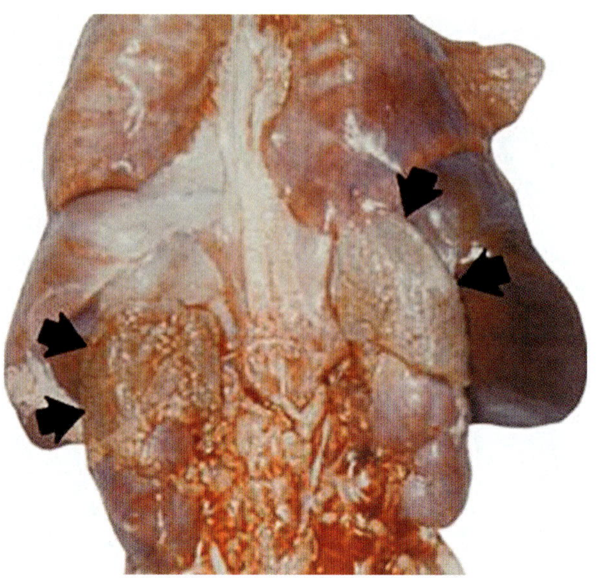

Figure 2-11

CONGENITAL ADRENAL HYPERPLASIA

Autopsy of an infant who died of the severe salt-losing form of 21-hydroxylase deficiency shows enlargement of both adrenal glands (arrows). The glands have a convoluted or cerebriform surface and are darker than normal due to intense persistent stimulation by adrenocorticotropic hormone (ACTH).

possible, and maternal dexamethasone treatment may prevent virilization of the female fetus, thus obviating the need for corrective genital surgery after birth. Treatment failures may be due to early cessation of therapy, late start of treatment, noncompliance, suboptimal dosing, or differences in dexamethasone metabolism (25).

Adrenal Hyperplasia in the Setting of CAH

CAH may be fatal if unrecognized or untreated. At autopsy, the adrenal glands are enlarged, often tan or brown, and have a convoluted or cerebriform surface with redundant folds due to cortical hyperplasia (fig. 2-11). Individual adrenal glands in children weigh 10 to 15 g, while in older patients are 30 to 35 g (1). In most untreated cases of CAH the adrenal glands are darker than normal. In congenital lipoid adrenal hyperplasia, the accumulation of cholesterol and cholesterol esters may give rise to a nodular cortex which is bright yellow or white. Microscopically, there may be cholesterol clefts with a foreign body giant cell reaction and dystrophic calcification.

In CAH, intense persistent trophic stimulation by adrenocorticotropic hormone (ACTH) results in marked hyperplasia of the zona fasciculata with conversion of pale-staining, lipid-rich cells into lipid-depleted cells with compact, eosinophilic cytoplasm, similar to those of the zona reticularis (fig. 2-12); these morphologic findings, however, may vary if the patient is partially treated with exogenous corticosteroids. This lipid depletion contributes, in large measure, to the dark color of the glands. With partial or incomplete steroid replacement, the histologic features may become altered, with columns and cords of lipid-rich cells admixed with some cells having lipid-depleted cytoplasm. Heterotopic or accessory adrenal cortical tissue can also become enlarged and hyperplastic.

Adrenal Cortical Neoplasms in the Setting of CAH

Persistent trophic stimulation of adrenal cortical tissue by ACTH gives rise to diffuse, and at times, slightly nodular hyperplasia, and there have been rare cases of *adrenal cortical neoplasia* occurring in the setting of CAH. The reported tumors have included both adrenal cortical

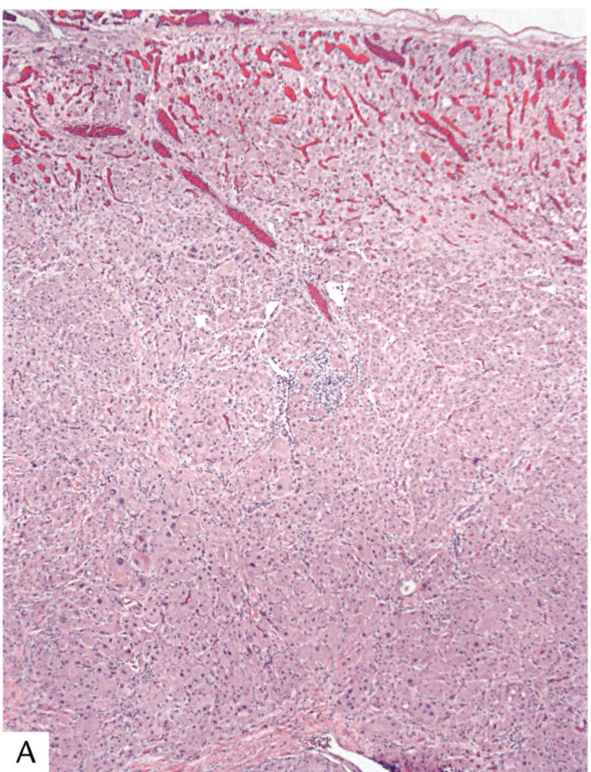

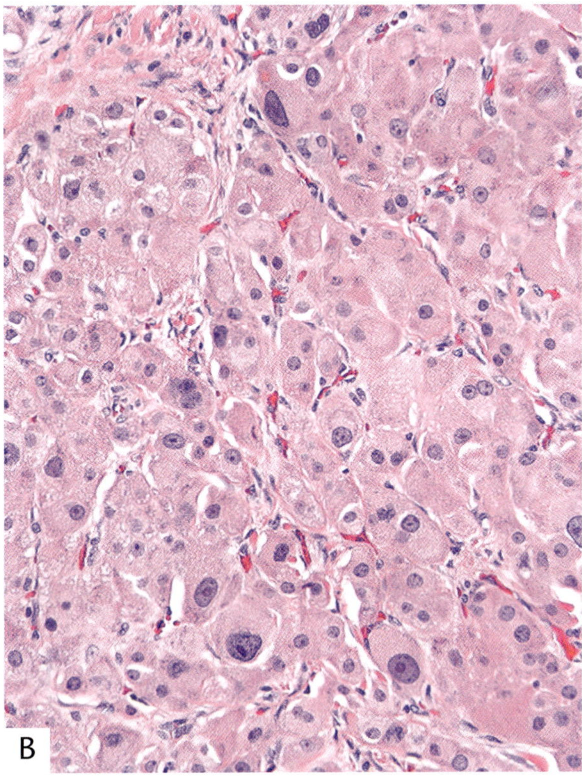

Figure 2-12

CONGENITAL ADRENAL HYPERPLASIA

A: Marked expansion of the zona fasciculata in an older child with congenital adrenal hyperplasia (CAH). Many cells have lipid-depleted cytoplasm due to sustained trophic stimulation by ACTH.

B: Hyperplastic zona fasciculata in CAH is composed of cells with lipid-depleted, compact, eosinophilic cytoplasm. Occasional cells have mildly enlarged hyperchromatic nuclei.

adenomas (27) and carcinomas (28). In two cases, adrenal cortical carcinomas developed in patients with a longstanding (30 years and 36 years) history of virilization with onset in childhood, suggesting untreated CAH (29). An increased incidence of incidental adrenal nodule(s) occurs in patients with homozygous (82 percent of patients; 2 cases were bilateral) and heterozygous (45 percent of patients) traits for CAH; the lesions were assumed to be adrenal cortical adenomas without evidence of excess steroid secretion (30).

In both homozygous and heterozygous CAH patients, tumors range from 0.5 cm to 5.0 cm in diameter. It is uncertain whether these lesions represent a true cortical neoplasm that was not hyperfunctional, or a dominant macronodule arising in a background of cortical hyperplasia (fig. 2-13).

Adrenal Rest Tumors in the Setting of CAH

Adrenal rest tumors are proliferative lesions of uncertain histogenesis that typically arise in or near the testis or ovary in patients with CAH. They are rare in patients without CAH. *Testicular adrenal rest tumors* (TARTs) are more common, arising in approximately 40 percent of male CAH patients, and are a common cause of infertility in these individuals (31). *Ovarian adrenal rest tumors* (OARTs) in contrast are very rare (32). In a study of 13 female patients with CAH, no OARTs were detected by ovarian imaging (33). Extratesticular (34) or extraovarian (35) adrenal rest tumors are also occasionally detected in patients with CAH.

Excess ACTH stimulation is believed to be the major contributor to development of adrenal rest tumors in patients with CAH, and this

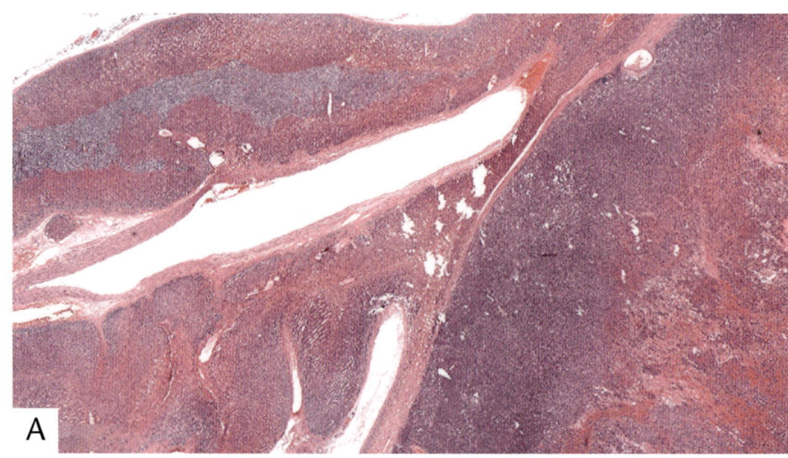

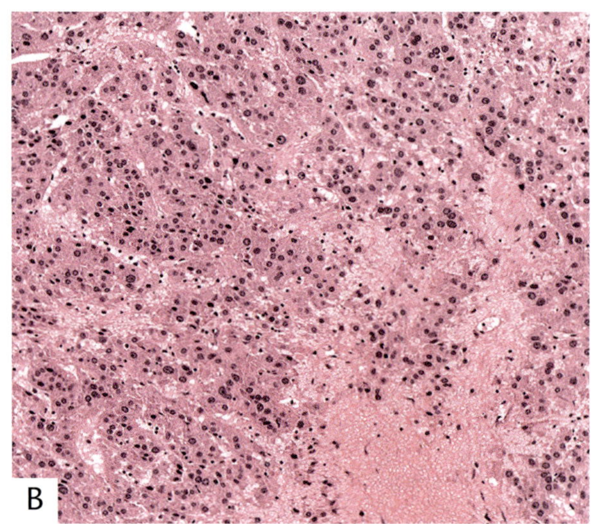

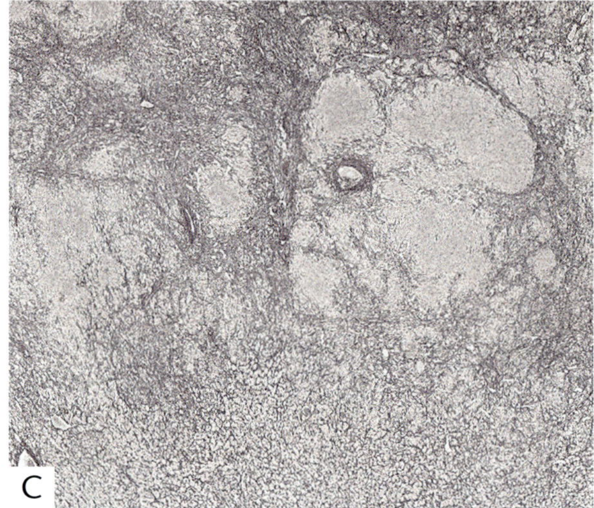

Figure 2-13

ADRENAL CORTICAL NEOPLASM ASSOCIATED WITH CONGENITAL ADRENAL HYPERPLASIA

A: A large adrenal cortical tumor present on the right is seen next to hyperplastic lipid-depleted adrenal cortex and a normal adrenal medulla on the left.

B: Higher magnification of the tumor shows areas of hemorrhage and hemorrhagic necrosis.

C: Reticulin stain in this tumor shows areas of disruption, often associated with carcinoma. However other findings for malignancy were inconclusive.

may be exacerbated by poor medical compliance with replacement steroid therapy in some cases. Conversely, the tumors may shrink in response to glucocorticoid administration. However, factors in addition to ACTH must be involved, as evidenced by the rarity of these lesions in other diseases associated with ACTH, and the male predilection.

One adrenal rest tumor reported in a patient without CAH included bilateral TARTs in a male patient with von Hippel-Lindau disease and Cushing syndrome after bilateral adrenalectomy; bilateral testicular venous sampling revealed elevated glucocorticoids that were responsive to dexamethasone suppression. Conservative management with appropriate steroid therapy and radiographic evaluation was successful in that case (36). TARTs and hyperplasia have also been noted in patients with Nelson syndrome (37) and primary Addison disease (38).

Biologic Behavior. Adrenal rest tumors histologically resemble Leydig cell tumors, but have features in common with hyperplastic adrenal cortical cells, including ACTH as the major trophic factor stimulating their growth. They variably express molecular and immunohistochemical markers of adrenal cortical cells and/or Leydig cells, and may produce a variety of steroid hormones or precursors. In patients with CAH, they typically lack 21-hydroxylase and do not produce cortisol or aldosterone (31). However, in patients with ACTH excess derived from the pituitary gland or an ectopic source, the tumors can produce cortisol and cause persistent Cushing disease/syndrome. Because TARTs have both adrenocortical and Leydig cell characteristics, a pluripotent cell is hypothesized (31).

Several studies have documented a dependence on ACTH, with increased levels of steroids in the testicular venous effluent. Suppression of ACTH with dexamethasone can decrease testicular size to near normal, while trophic stimulation can cause testicular enlargement with recurrence of pain and tenderness (fig. 2-14). Orchiectomy may not be necessary since the tumor may not be truly neoplastic in view of its dependence on ACTH and the lack of malignant behavior in cases reported to date (36,39). Rarely, TARTs and adrenal cortical lesions are mistakenly diagnosed as malignant based upon cytologic atypia and nuclear pleomorphism (40).

Clinical Features. A review of testicular tumors in CAH showed that two thirds were palpable (up to 10 cm), occurrence was usually in early adult life (average age, 22.5 years), smaller tumors (less than 2 cm) were usually seen in children, and 86 percent were located in the hilum of the testis (39). The tumors were bilateral in 83 percent of patients (39) compared to only 2.5 percent bilaterality for Leydig cell tumors (41). Ultrasonography is more sensitive for detecting testicular nodules than manual palpation (42,43) and is the most cost-effective and quickest imaging technique. In cases of partial orchiectomy, magnetic resonance imaging (MRI) has been recommended because it shows lesion margins optimally (33). Using ultrasound, testicular abnormalities are noted in 27 (43) to 94 percent (30) of postpubertal male patients with CAH. Lesions vary from 2 to 40 mm and are typically located adjacent to the hilum of the testis (33).

In a review of 244 patients with CAH, the prevalence of TARTs was 33 percent in boys and 44 percent in adult men (44). Many of these testicular nodules are clinically undetectable. In another study, the nodules were hypoechoic, ranged in size from 0.2 to 2.8 cm (average, 1.6 cm), were bilateral in 6 of 8 patients (75 percent), and multifocal in each case (43). In a study of 36 males with CAH followed over a 30-year period, 3 patients (8.2 percent) developed a testicular mass 1 to 2 cm in size; the nodules were bilateral in 2 patients, and typically occupied the upper half of the testicle near the hilum (note the location of the adrenal cortical rest in figure 2-7) (45).

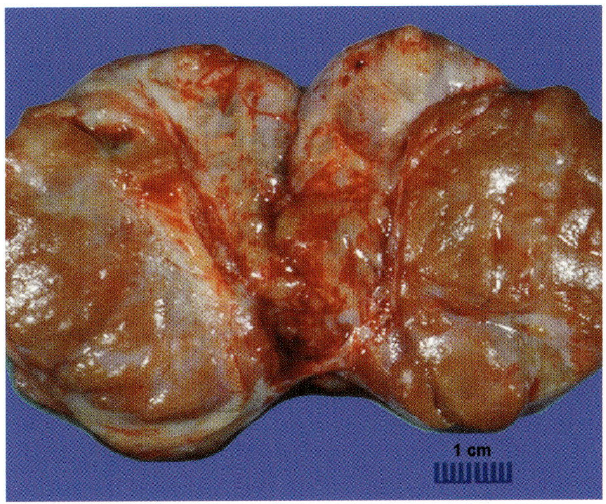

Figure 2-14

BILATERAL ORCHIECTOMY SPECIMENS FROM A 25-YEAR-OLD MALE WITH SALT-LOSING FORM OF 21-HYDROXYLASE DEFICIENCY

Both testes were grossly nearly identical and were almost completely replaced by bulging nodules of tan tumor. Cross section of one of the testicular tumors is shown here. Surgery was done to alleviate severe pain and swelling, which was not adequately controlled with suppressive doses of dexamethasone. The testicular tumor was shown to be ACTH dependent.

Fewer than 20 OARTs have been reported. Similarly to TARTs, they may be unilateral or bilateral. They may be asymptomatic or associated with virilization, and can be difficult to detect in imaging studies (46).

Pathologic Findings. On cross section, the larger tumors are unencapsulated, bulging, multinodular masses separated by prominent bands of fibrous connective tissue (fig. 2-14). They are often light tan-brown, due in part to lipochrome pigment and probably diminished cytoplasmic lipid, particularly in cases with inadequate suppression of ACTH. In several cases there have been multiple extratesticular nodules as large as 1.5 cm along the spermatic cord or adjacent to the epididymis (39).

Microscopically, there are interconnecting sheets and nests of cells with granular pink cytoplasm and relatively distinct cell borders (fig. 2-15A–C). The tumor also has an intimate pattern of reticulin, with isolation of individual or small nests of cells (fig. 2-15D).

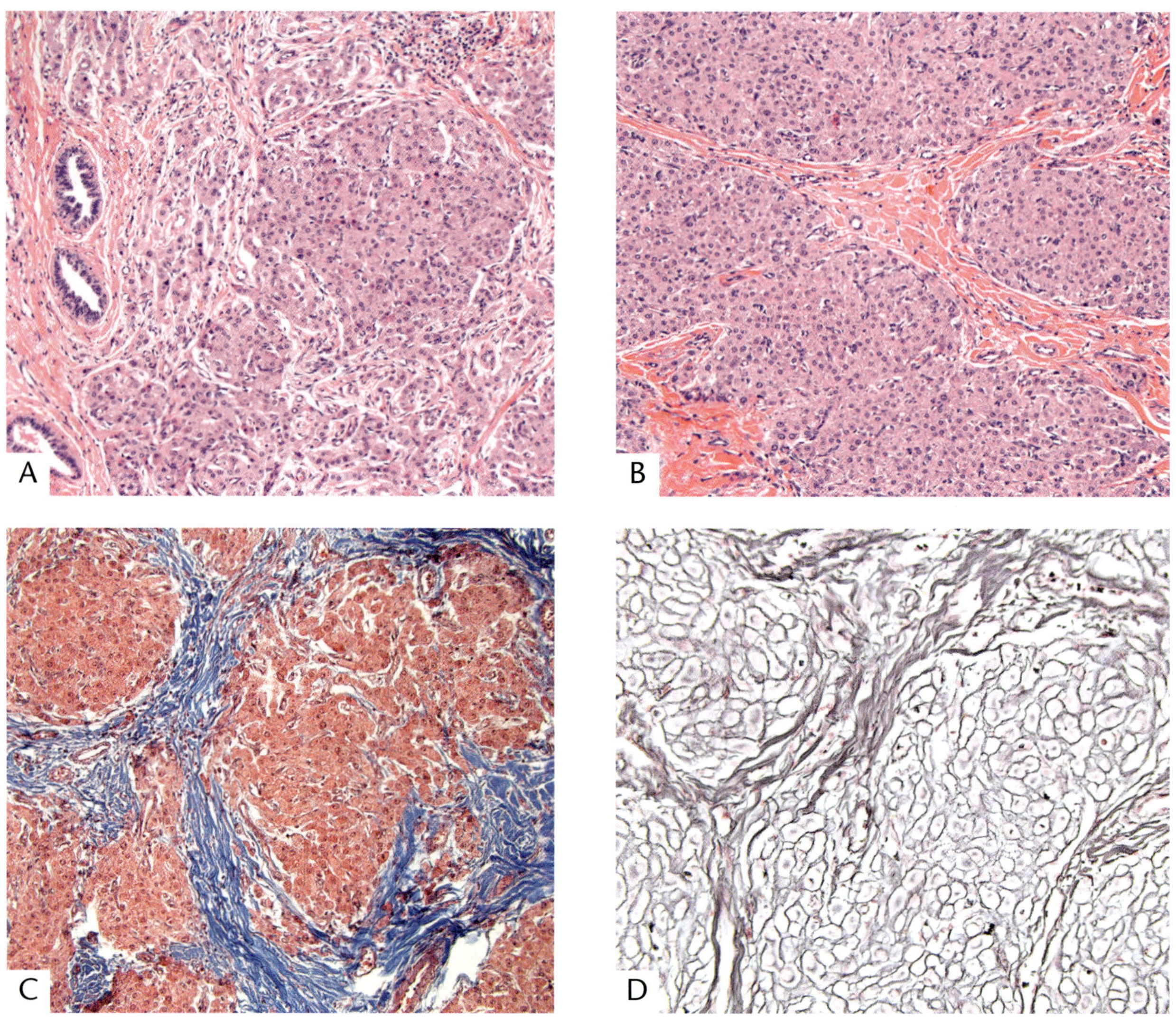

Figure 2-15

TESTICULAR ADRENAL REST TUMOR IN A PATIENT WITH SALT-LOSING FORM OF 21-HYDROXYLASE DEFICIENCY

A: Tumor cells are arranged in lobules and solid sheets, with some intervening fibrous stroma. A small portion of rete testis is present at left.

B: Tumor has prominent intersecting bands of fibrous tissue.

C: Stain for collagen accentuates the prominent fibrous stroma between lobules.

D: The reticulin staining pattern isolates individual and small groups of cells.

This is in contrast to the broad radial cords of cells usually seen in hyperplasia of eutopic or heterotopic adrenal cortex. Occasionally, there are interstitial adipocytes.

Sections taken through the testicular hilum and rete testis may show involvement by tumor (fig. 2-15A). Most tumor cells contain prominent lipochrome pigment (lipofuscin). Nuclei are fairly uniform and round to oval, with one or two small, central to eccentric nucleoli (fig. 2-16); occasionally, there is some nuclear enlargement. Mitotic figures are uncommon. Adjacent testicular parenchyma may appear almost normal or atrophic, with sclerosis and decreased spermatogenesis. There is a great resemblance to Leydig cells, and the main differential diagnosis is Leydig cell tumor.

Distinction between adrenal rest tumor and Leydig cell tumor is important because adrenal rest tumors are invariably benign, while approximately 10 percent percent of Leydig cell tumors are malignant (31). Differential diagnosis requires an integrated morphologic and clinical assessment. Findings that favor adrenal rest tumor include the background of CAH, frequent bilaterality, and absence of Reinke crystalloids (31), which are a pathognomonic marker for Leydig cells, found in about 35 percent of Leydig cell tumors (41). Extensive molecular and immunohistochemical studies show that TARTs express some markers for androgen-producing cells but lack specific markers for Leydig cells. They also variably express adrenal cortical markers, including proteins involved in adrenal steroidogenesis (31).

Other Tumors in the Setting of CAH

Adrenal myelolipoma is the most common adrenal tumor reported in the setting of CAH (47). In some cases, especially when CAH is poorly controlled, myelolipomas are of giant proportions and may also be bilateral (48). In these cases there is marked hyperplasia of adrenal cortical cells admixed with the myelolipoma (fig. 2-17). *Adrenal adenomas* or *carcinomas* have been reported but are rare (47). Several extra-adrenal seemingly unrelated neoplasms have been reported such as *osteosarcoma, Ewing sarcoma, astrocytoma* (49), and *hepatocellular carcinoma* (47).

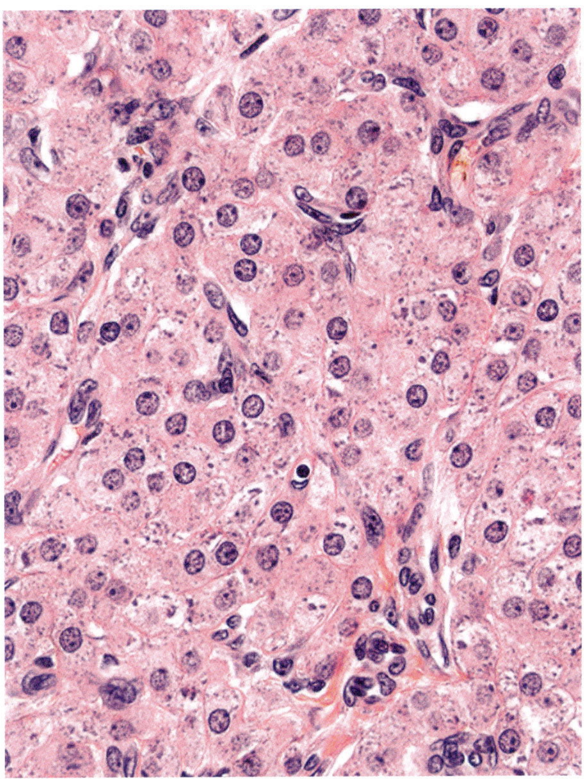

Figure 2-16

TESTICULAR TUMOR IN CONGENITAL ADRENAL HYPERPLASIA

Individual cells have distinct cell borders and rounded nuclei, often with one or two eccentric nucleoli. Many cells contain coarse, granular pigment, which is lipofuscin.

BECKWITH-WIEDEMANN SYNDROME

Beckwith-Wiedemann syndrome (BWS) is a congenital human overgrowth disorder associated with dysregulation of the imprinting of genes in chromosome 11p15.5 (50). BWS was described by Wiedemann in 1964 (51) and by Beckwith in 1969 (52). The major anatomic features are congenital abdominal wall defects, macroglossia, lateralized somatic hypertrophy, and visceral overgrowth, and BWS was therefore originally described as *exomphalos-macroglossia-gigantism (EMG) syndrome*. Additional major clinical manifestations are neonatal hypoglycemia and predisposition to embryonal tumors.

BWS is now considered a spectrum disorder, with a range of clinical characteristics (50). In a population-wide study based on registries from 16 European countries, the estimated frequency of BWS is 1 in 10,340 births, with boys and girls affected equally. Approximately 90 percent of cases are sporadic (53). There is a high frequency of premature births (40.6 percent of males and 33.9 percent of females), and an infant death rate of about 10 percent in the first year of life (53).

The molecular genetic underpinnings of BWS consist of variable alterations in clusters of imprinted genes in the 11p15 region, and BWS patients segregate into subgroups in which the type and severity of disease manifestation correlate with molecular mechanism (50,54). Uniparental disomy and somatic mosaicism may complicate molecular assessment in some cases (50,55). Functional consequences of the epigenetic and genetic alterations that

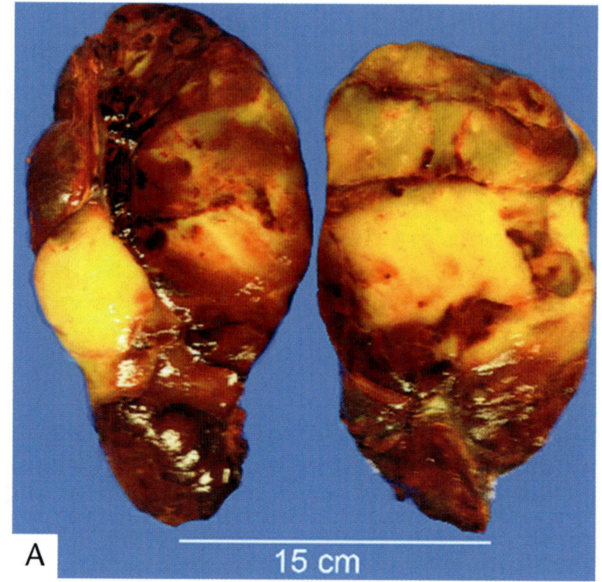

Figure 2-17

ADRENAL MYELOLIPOMA IN CONGENITAL ADRENAL HYPERPLASIA

A: Huge, bilateral adrenal myelolipomas in an adult male with CAH. The variegated yellow and red coloration results from heterogenous admixture of hematopoietic elements and fat within the tumor.

B: In another case, hyperplastic lipid-depleted cortical cells are present at right of the myelolipoma.

C: The hyperplastic adrenal cortical tissue is remarkably similar to that of the testicular tumors of CAH illustrated previously. Cells are lipid depleted and have abundant compact eosinophilic cytoplasm. The granular pigment within some cells is lipofuscin.

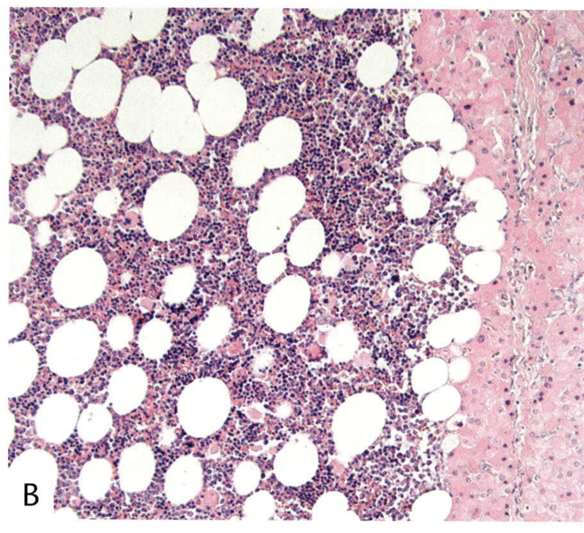

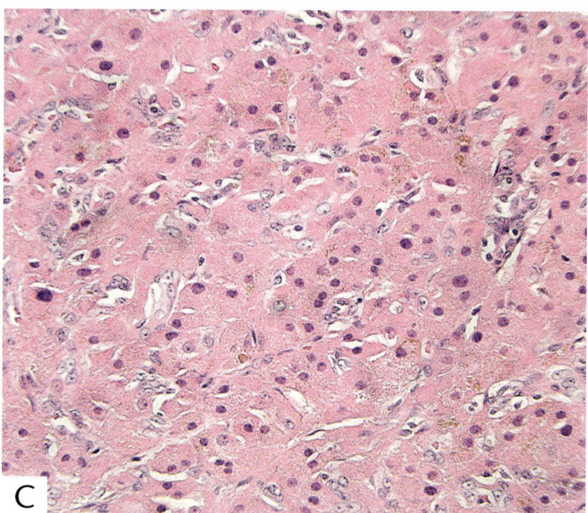

lead to overgrowth include overexpression of IGF2 and/or downregulation of the cyclin-dependent kinase inhibitor 1C (*CDKN1C*) gene (50), which encodes the cell cycle checkpoint protein p57KIP2. In contrast to downregulation, which is seen predominantly in sporadic BWS, loss-of-function point mutations in *CDKN1C* are found in up to 40 percent of familial cases and only a few sporadic ones (50,56).

Causes of hyperinsulinemic hypoglycemia in BWS include mutations of *ABCC8* and *KCNJ11*, which respectively encode the sulfonylurea receptor 1 (SUR-1) and subunits of the ATP-sensitive potassium (KATP) channel in pancreatic B cells. These genes are not imprinted but are located near the BWS region on chromosome 11p (57). In severe cases, hypoglycemia may lead to permanent brain damage, mental deficiency, or death, and some patients require pancreatectomy to avoid these complications. Histologically, the excised pancreatic tissue shows expansion of the pancreatic islets with loss of p57KIP2 (57).

The overall tumor risk in BWS is 8 percent, but varies up to 28 percent in different molecular genetic subgroups. In addition, each subgroup is associated with a predisposition to different types of tumors. The most common are Wilms tumor (52 percent), hepatoblastoma (14 percent), neuroblastoma (10 percent), and

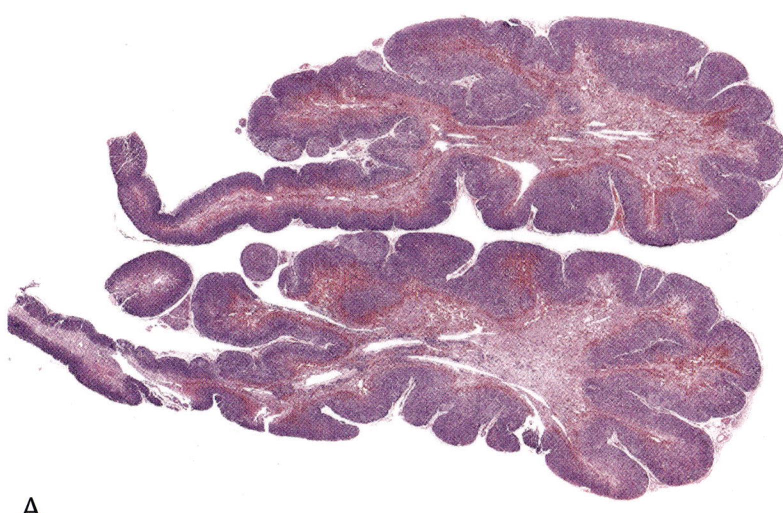

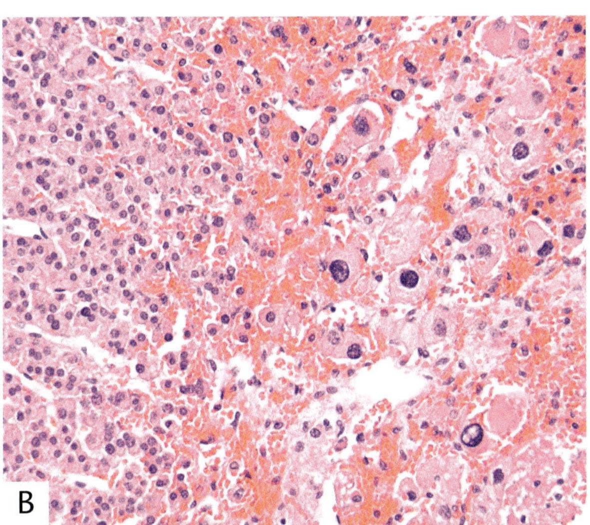

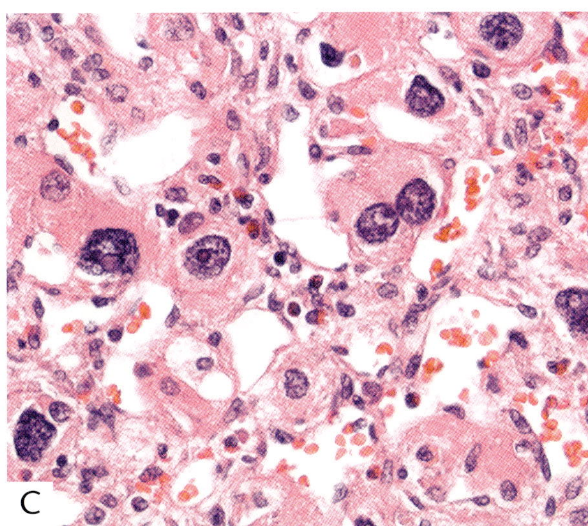

Figure 2-18

BECKWITH-WIEDEMANN SYNDROME

A: Transverse sections of adrenal gland from a 3-week-old infant with Beckwith-Wiedemann syndrome are enlarged and show redundant cerebriform folds of cortex with demarcation between adult or definitive cortex and regressing and hemorrhagic regressing fetal or provisional cortex.

B: Developing adult cortex at left of field, with fetal cortex at right. At this early stage in postnatal life, fetal cortex is undergoing normal regression.

C: Marked adrenal cytomegaly involves cells of the fetal, or provisional, cortex. Nucleomegaly, hyperchromasia, and occasional prominent nucleoli are seen. Other fields had well-developed nuclear pseudoinclusions. Mitotic figures were not identified.

rhabdomyosarcoma (5 percent). Adrenal cortical carcinoma accounts for approximately 3 percent (50,58). Rare associated neoplasms include pancreatoblastoma (55), cardiac tumors (59), and pheochromocytoma (60,61). In patients with pheochromocytoma, physical manifestations of BWS can be subtle, leading to the suggestion that 11p15 paternal uniparental disomy may be an underdetected mechanism of sporadic pheochromocytoma in the general pediatric population (61). Copy number variants involving chromosome 11p are common in pheochromocytomas and may include the region containing the IGF1 receptor, suggesting a possible role of aberrant IGF signaling in tumor development (60).

In addition to cortical carcinomas and pheochromocytomas, the adrenal glands in BWS show a variety of hyperplastic and dysmorphogenetic changes. The adrenal glands are enlarged, with a combined weight often up to 16 g, and due to cortical hyperplasia, may show redundant cortical folds and nodularity that have a cerebriform configuration (fig. 2-18A). Marked adrenal cytomegaly is a characteristic feature, and is bilateral, usually affecting most cells in the fetal cortex (fig. 2-18B,C). In addition to enlarged pleomorphic nuclei,

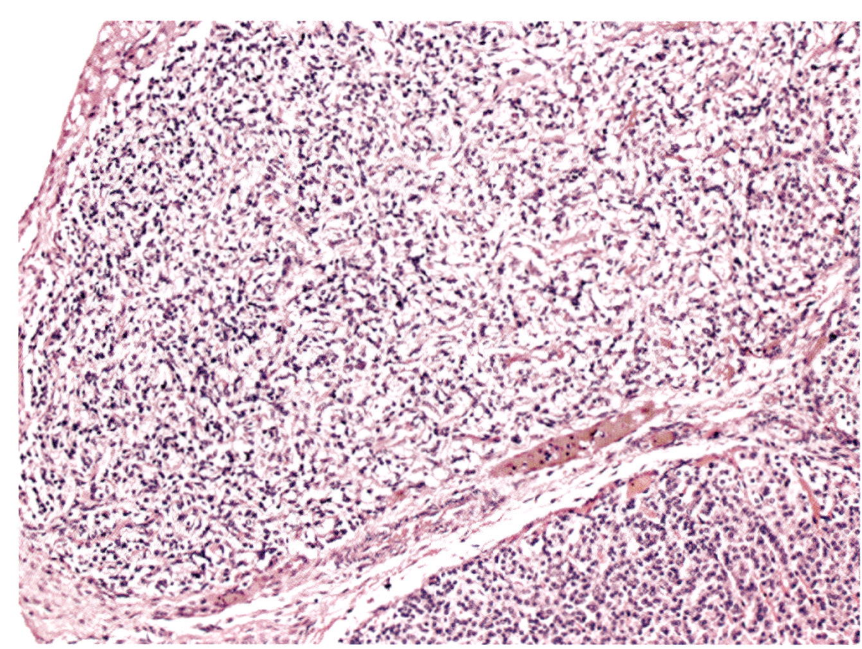

Figure 2-19

BECKWITH-WIEDEMANN SYNDROME

Chromaffin cells forming a nodule protruding from the fetal cortex are inappropriately mature for this stage of development and have clear cytoplasm. There is a morphologic resemblance of the nodule to normal extra-adrenal paraganglia (see chapter 1).

pseudoinclusions are often seen. Cortical microcysts may be present in the adult cortex on histologic examination. Rarely, gross hemorrhagic macrocysts of the adrenal cortex present as an abdominal mass in the fetus and neonate with BWS (62). Chromaffin cells may be strikingly hyperplastic within adrenal (fig. 2-19) and extra-adrenal sites (52), although this is not a consistent finding. In some cases, the adrenal chromaffin cells form organoid nodules that resemble extra-adrenal paraganglia, both architecturally and by the absence of phenylethanolamine N-methyltransferase expression (63), suggesting abnormalities in developmental mechanisms that normally lead to divergence of adrenal and extra-adrenal progenitors.

REFERENCES

1. Lack EE, Kozakewich HP. Embryology, developmental anatomy and selected aspects of non-neoplastic pathology. In: Lack EE, ed. Pathology of the adrenal glands. Churchill Livingstone; 1990:1-74.
2. Angkathunyakul N, Mingphruedhi S, Tannaphai P. Non-functioning adrenal cortical adenoma arising in adrenohepatic fusion, mimicking hepatocellular carcinoma-a case report. Asian Arch Pathol 2014;10:83-8.
3. Patel V, Bejarano PA, Parlade A, Muruve N. Adrenal-renal fusion giving rise to an intrarenal adrenocortical adenoma: a novel case report with review of the literature. Endocrinol Metab Syndr 2016;5:247-51.
4. Klatt EC, Pysher TJ, Pavlova Z. Adrenal fusion. Pediatr Dev Pathol 1998;1:475-9.
5. McCabe ER. Adrenal hypoplasias and aplasias. In: Scriver CR, Beaudet AL, Sly W, Valle D, eds. The metabolic and molecular bases of inherited diseases, 8th ed. New York: McGraw Hill; 2001.
6. D'Arcy C, Pertile M, Goodwin T, Bittinger S. Bilateral congenital adrenal agenesis: a rare disease entity and not a result of poor autopsy technique. Pediatr Dev Pathol 2014;17:308-11.
7. Zhong H, Xu B, Popiolek DA. Growth patterns of placental and paraovarian adrenocortical heterotopias are different. Case Rep Pathol 2013; 2013:205692.
8. Armin A, Castelli M. Congenital adrenal tissue in the lung with adrenal cytomegaly. Case report and review of the literature. Am J Clin Pathol 1984;82:225-8.

9. Oka H, Kameya T, Sasano H, et al. Pituitary choristoma composed of corticotrophs and adrenocortical cells in the sella turcica. Virchows Arch 1996;427:613-7.
10. Mete O, Ng T, Christie-David D, McMaster J, Asa SL. Silent corticotroph adenoma with adrenal cortical choristoma: a rare but distinct morphological entity. Endocr Pathol 2013;24:162-6.
11. Wiener MF, Dallgaard SA. Intracranial adrenal gland; a case report. AMA Arch Pathol 1959;67:228-33.
12. Honoré LH. Intra-adrenal hepatic heterotopia. J Urol 1985;133:652-4.
13. Casadei GP, Bertarelli C, Giorgini E, Cremonini N, de Biase D, Tallini G. Ectopic thyroid tissue in the adrenal gland: report of a case. Int J Surg Pathol 2015;23:170-5.
14. Graham LS. Celiac accessory adrenal glands. Cancer 1953;6:149-52.
15. Ye H, Yoon GS, Epstein JI. Intrarenal ectopic adrenal tissue and renal-adrenal fusion: a report of nine cases. Mod Pathol 2009;22:175-81.
16. Yousif MQ, Salih ZT, DeYoung BR, Qasem SA. Differentiating intrarenal ectopic adrenal tissue from renal cell carcinoma in the kidney. Int J Surg Pathol 2018;26:588-92.
17. Falls JL. Accessory adrenal cortex in the broad ligament: incidence and functional significance. Cancer 1955;8:143-50.
18. Mares AJ, Shkolnik A, Sacks M, Feuchtwanger MM. Aberrant (ectopic) adrenocortical tissue along the spermatic cord. J Pediatr Surg 1980;15:289-92.
19. Roosen-Runge EC, Lund J. Abnormal sex cord formation and an intratesticular adrenal cortical nodule in a human fetus. Anat Rec 1972;173:57-67.
20. Symonds DA, Driscoll SG. An adrenal cortical rest within the fetal ovary: report of a case. Am J Clin Pathol 1973;60:562-4.
21. Cornejo KM, Afari HA, Sadow PM. Adrenocortical carcinoma arising in an adrenal rest: a case report and review of the literature. Endocr Pathol 2017;28:165-70.
22. Ohsugi H, Takizawa N, Kinoshita H, Matsuda T. Pheochromocytoma arising from an ectopic adrenal tissue in multiple endocrine neoplasia type 2A. Endocrinol Diabetes Metab Case Rep 2019;2019:19-0073.
23. de Crecchio L. Sopra un caso di apparenze virili in una donna. Morgagni 1865;7:151-88.
24. El-Maouche D, Arlt W, Merke DP. Congenital adrenal hyperplasia. Lancet 2017;390:2194-210.
25. Merke DP, Auchus RJ. Congenital adrenal hyperplasia due to 21-hydroxylase deficiency. N Engl J Med 2020;383:1248-61.
26. Kim CJ. Congenital lipoid adrenal hyperplasia. Ann Pediatr Endocrinol Metab 2014;19:179-83.
27. Pang S, Becker D, Cotelingam J, Foley TP Jr, Drash AL. Adrenocortical tumor in a patient with congenital adrenal hyperplasia due to 21-hydroxylase deficiency. Pediatrics 1981;68:242-6.
28. Bauman A, Bauman CG. Virilizing adrenocortical carcinoma. Development in a patient with salt-losing congenital adrenal hyperplasia. JAMA 1982;248:3140-1.
29. Nogeire C, Fukushima DK, Hellman L, Boyar RM. Virilizing adrenal cortical carcinoma. Cancer 1977;40:307-13.
30. Jaresch S, Kornely E, Kley HK, Schlaghecke R. Adrenal incidentaloma and patients with homozygous or heterozygous congenital adrenal hyperplasia. J Clin Endocrinol Metab 1992;74:685-9.
31. Engels M, Span PN, van Herwaarden AE, Sweep F, Stikkelbroeck N, Claahsen-van der Grinten HL. Testicular adrenal rest tumors: current insights on prevalence, characteristics, origin, and treatment. Endocr Rev 2019;40:973-87.
32. Bouzidi L, Triki M, Charfi S, et al. Incidental finding of bilateral ovarian adrenal rest tumor in a patient with congenital adrenal hyperplasia: a case report and brief review. Pediatr Dev Pathol 2021;24:137-41.
33. Stikkelbroeck NM, Hermus AR, Schouten D, et al. Prevalence of ovarian adrenal rest tumours and polycystic ovaries in females with congenital adrenal hyperplasia: results of ultrasonography and MR imaging. Eur Radiol 2004;14:1802-6.
34. Claahsen-van der Grinten HL, Duthoi K, Otten BJ, d'Ancona FC, Hulsbergen-vd Kaa CA, Hermus AR. An adrenal rest tumour in the perirenal region in a patient with congenital adrenal hyperplasia due to congenital 3beta-hydroxysteroid dehydrogenase deficiency. Eur J Endocrinol 2008;159:489-91.
35. Sisto JM, Liu FW, Geffner ME, Berman ML. Para-ovarian adrenal rest tumors: gynecologic manifestations of untreated congenital adrenal hyperplasia. Gynecol Endocrinol 2018;34:644-6.
36. Weeks DC, Walther MM, Stratakis CA, Hwang JJ, Linehan WM, Phillips JL. Bilateral testicular adrenal rests after bilateral adrenalectomies in a cushingoid patient with von Hippel-Lindau disease. Urology 2004;63:981-2.
37. Johnson RE, Scheithauer B. Massive hyperplasia of testicular adrenal rests in a patient with Nelson's syndrome. Am J Clin Pathol 1982;77:501-7.
38. Al-Ahmadie HA, Stanek J, Liu J, Mangu PN, Niemann T, Young RH. Ovarian 'tumor' of the adrenogenital syndrome: the first reported case. Am J Surg Pathol 2001;25:1443-50.
39. Rutgers JL, Young RH, Scully RE. The testicular "tumor" of the adrenogenital syndrome. A report of six cases and review of the literature on testicular masses in patients with adrenocortical disorders. Am J Surg Pathol 1988;12:503-13.

40. Marianovsky V, Bogdanova O, Tsvetkov M, Serteva D, Mladenov B. Testicular adrenal rest tumors (TARTS) with unusual histological features in congenital adrenal hyperplasia (CAH). Urol Case Rep 2015;3:126-8.
41. Kim I, Young RH, Scully RE. Leydig cell tumors of the testis. A clinicopathological analysis of 40 cases and review of the literature. Am J Surg Pathol 1985;9:177-92.
42. Seidenwurm D, Smathers RL, Kan P, Hoffman A. Intratesticular adrenal rests diagnosed by ultrasound. Radiology 1985;155:479-81.
43. Vanzulli A, DelMaschio A, Paesano P, et al. Testicular masses in association with adrenogenital syndrome: US findings. Radiology 1992;183:425-9.
44. Finkielstain GP, Kim MS, Sinaii N, et al. Clinical characteristics of a cohort of 244 patients with congenital adrenal hyperplasia. J Clin Endocrinol Metab 2012;97:4429-38.
45. Srikanth MS, West BR, Ishitani M, Isaacs H Jr, Applebaum H, Costin G. Benign testicular tumors in children with congenital adrenal hyperplasia. J Pediatr Surg 1992;27:639-41.
46. Chen HD, Huang LE, Zhong ZH, et al. Ovarian adrenal rest tumors undetected by imaging studies and identified at surgery in three females with congenital adrenal hyperplasia unresponsive to increased hormone therapy dosage. Endocr Pathol 2017;28:146-51.
47. Vergani E, Bruno C, Raimondo S, et al. Recurrent hepatocellular carcinoma and non-classic adreno-genital syndrome. Eur Rev Med Pharmacol Sci 2020;24:4172-9.
48. Al-Bahri S, Tariq A, Lowentritt B, Nasrallah DV. Giant bilateral adrenal myelolipoma with congenital adrenal hyperplasia. Case Rep Surg 2014; 2014:728198.
49. Duck SC. Malignancy associated with congenital adrenal hyperplasia. J Pediatr 1981;99:423-4.
50. Wang KH, Kupa J, Duffy KA, Kalish JM. Diagnosis and management of Beckwith-Wiedemann syndrome. Front Pediatr 2020;7:562.
51. Wiedemann HR. [Familial malformation complex with umbilical hernia and macroglossia—a "new syndrome"?]. J Genet Hum 1964;13:223-32. [French]
52. Beckwith JB. Macroglossia, omphalocele, adrenal cytomegaly, gigantism, and hyperplastic visceromegaly. Birth Defects 1969;5:188-96.
53. Barisic I, Boban L, Akhmedzhanova D, et al. Beckwith Wiedemann syndrome: a population-based study on prevalence, prenatal diagnosis, associated anomalies and survival in Europe. Eur J Med Genet 2018;61:499-507.
54. Brioude F, Lacoste A, Netchine I, et al. Beckwith-Wiedemann syndrome: growth pattern and tumor risk according to molecular mechanism, and guidelines for tumor surveillance. Horm Res Paediatr 2013;80:457-65.
55. Lee CT, Tung YC, Hwu WL, et al. Mosaic paternal haploidy in a patient with pancreatoblastoma and Beckwith-Wiedemann spectrum. Am J Med Genet A 2019;179:1878-83.
56. Lam WW, Hatada I, Ohishi S, et al. Analysis of germline CDKN1C (p57KIP2) mutations in familial and sporadic Beckwith-Wiedemann syndrome (BWS) provides a novel genotype-phenotype correlation. J Med Genet 1999;36:518-23.
57. Kalish JM, Boodhansingh KE, Bhatti TR, et al. Congenital hyperinsulinism in children with paternal 11p uniparental isodisomy and Beckwith-Wiedemann syndrome. J Med Genet 2016;53:53-61.
58. Maas SM, Vansenne F, Kadouch DJ, et al. Phenotype, cancer risk, and surveillance in Beckwith-Wiedemann syndrome depending on molecular genetic subgroups. Am J Med Genet A 2016;170:2248-60.
59. Longardt AC, Nonnenmacher A, Graul-Neumann L, et al. Fetal intracardiac rhabdomyoma in Beckwith-Wiedemann syndrome. J Clin Ultrasound 2014;42:569-73.
60. Caza T, Manwaring J, Riddell J. Recurrent, bilateral, and metastatic pheochromocytoma in a young patient with Beckwith-Wiedemann syndrome: a genetic link? Can Urol Assoc J 2017;11:E240-3.
61. Kalish JM, Conlin LK, Mostoufi-Moab S, et al. Bilateral pheochromocytomas, hemihyperplasia, and subtle somatic mosaicism: the importance of detecting low-level uniparental disomy. Am J Med Genet A 2013;161A:993-1001.
62. Anoop P, Anjay MA. Bilateral benign haemorrhagic adrenal cysts in Beckwith-Wiedemann syndrome: case report. East Afr Med J 2004;81:59-60.
63. Tischler AS, Semple J. Adrenal medullary nodules in Beckwith-Wiedemann syndrome resemble extra-adrenal paraganglia. Endocr Pathol 1996; 7:265-72.

3 CLINICOPATHOLOGIC FEATURES OF THE ADRENAL CORTEX

Because the adrenal cortex is the site of synthesis of several hormones, adrenal cortical tumors can be associated with endocrine syndromes that vary depending on the function of the hormone(s) released. The main hormones produced by the adrenal cortex are glucocorticoids, mineralocorticoids, and sex steroids. Excess glucocorticoids give rise to Cushing syndrome, mineralocorticoid excess causes Conn syndrome, and sex steroid excess results in virilization or more rarely, feminization. A mixed endocrine syndrome can result from a combination of glucocorticoid and sex steroid secretion.

The complexity of hormone regulation can also result in hyperplasia of the adrenal cortex, which is secondary to neoplasms in other body sites. In some clinical scenarios, the adrenal glands are resected to prevent hormone excess even when the primary pathology is elsewhere, and the pathologist must assess the cause of that secondary process, e.g., ectopic adrenocorticotropic hormone (ACTH) syndrome. In other patients, bilateral adrenal diseases cause adrenocortical hypofunction. In this context, pathologists have a great impact in the interpretation of the clinical-pathologic, biochemical, and radiographic features underpinning these conditions, with a critical role in the multidisciplinary team (1).

CUSHING SYNDROME

Cushing syndrome is a systemic disorder originally described by Harvey Cushing in 1932 (2) and caused by excess glucocorticoids that result in alterations in the structure and function of multiple tissues (fig. 3-1). The most obvious changes include weight gain that causes truncal obesity, moon facies, and increased fat around the neck and between the scapulae (so called "buffalo hump"); unlike in generalized weight gain, there is peripheral wasting, with thin extremities and weakness. The skin becomes thin and pigmented striae are prominent. Easy bruising is common. One of the most common presenting features is insulin resistance and type 2 diabetes, often overlooked as a pivotal manifestation of glucocorticoid excess. Patients may develop hypertension, dyslipidemia, and atherosclerosis, which predispose to cardiovascular events that are more common in patients with this disorder. Less obvious changes include osteoporosis and immunosuppression that can cause poor wound healing and may lead to opportunistic infection. Neuropsychiatric disturbances may also be a complication of glucocorticoid excess.

Because of the heterogeneity of clinical presentations, and the common occurrence of many of the individual signs and symptoms, Cushing syndrome may be misdiagnosed or overlooked. Moreover, features of Cushing syndrome may be mild or absent despite increased cortisol levels, a situation known as *"subclinical" Cushing syndrome* or more recently, *autonomous cortisol secretion*, which also has a high risk of mortality (3).

Cushing syndrome is associated with exogenous or endogenous cortisol excess (Table 3-1). The most common cause of Cushing syndrome in the developed world is iatrogenic exogenous administration of steroids for the treatment of numerous conditions, ranging from asthma to inflammatory gastrointestinal and skin disorders. In this setting, the adrenal glands may be atrophic due to suppression of the hypothalamic-pituitary-adrenal axis.

Endogenous Cushing syndrome has an incidence of approximately 2 to 3 per 1,000,000 population per year and a prevalence of almost 40 per million population (3). It is most frequent in females in the 4th to 5th decades, however, sex and age predominance are largely biased by the underlying cause of cortisol oversecretion. Cushing syndrome due to ectopic ACTH/corticotrophin-releasing hormone (CRH) production

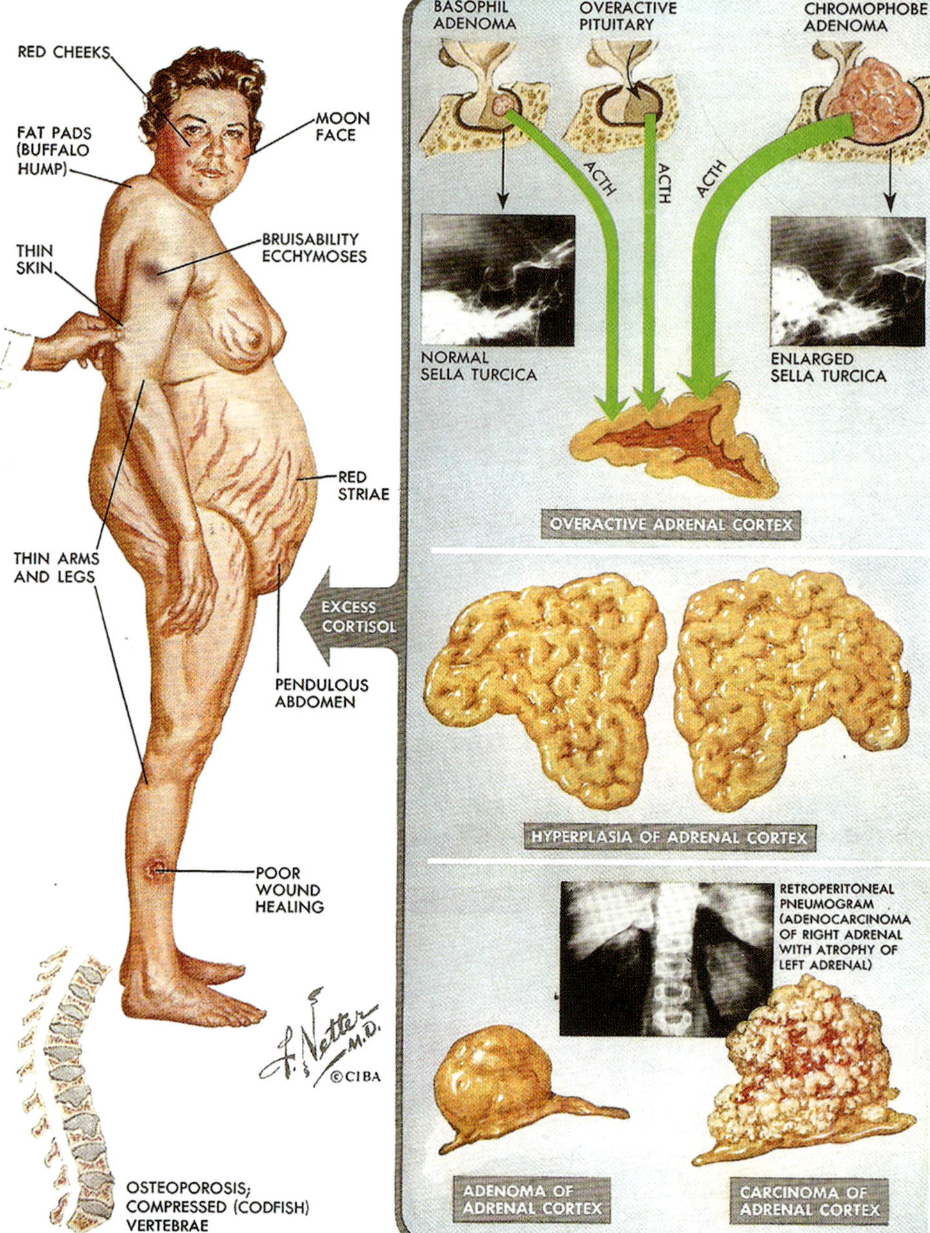

Figure 3-1

CUSHING SYNDROME

The clinical features of this disorder (left) can be caused by pituitary adrenocorticotropic hormone (ACTH) excess (top right), ectopic ACTH causing massive adrenal hyperplasia (right middle), or primary adrenal disease (bottom right). (From Netter FH. The CIBA collection of Medical Illustrations, Volume 4, Endocrine System and Selected Metabolic Diseases, 1965. 2024. Used with permission of Elsevier. All rights reserved.)

Table 3-1

ETIOLOGY OF CUSHING SYNDROME

Exogenous	Endogenous	
	ACTH Dependent (80-85%)	ACTH Independent (15-20%)
Glucocorticoid administration	Pituitary corticotroph tumor (80%)	Adrenocortical adenoma (60%)
Synthetic ACTH administration	Ectopic ACTH-producing tumor (20%)	Adrenocortical carcinoma (up to 30%)
	Corticotropin-releasing hormone producing tumor (<1%)	Primary bilateral macro- and micronodular disease (up to 10%)

is a disease of adult men, while most patients with ACTH-dependent Cushing syndrome related to a pituitary tumor are women of reproductive age. In children under the age of 7 years, an adrenal cortical neoplasm (either adrenal cortical adenoma or carcinoma) appears to be the most common cause of Cushing syndrome.

Overall, the most common cause of endogenous Cushing syndrome is an ACTH-producing pituitary tumor, the disorder known as *Cushing disease*. This accounts for about 70 percent of cases of Cushing syndrome and because of the marked predominance of young women who develop this disorder, it skews the overall prevalence. The second most common cause is ACTH-independent adrenal disease, and in this group, adrenal adenomas are the most common underlying pathology. Ectopic production of ACTH and/or CRH is responsible for only about 10 percent of cases.

An important differential diagnosis is the *pseudo-Cushing syndrome*, which is associated with neuropsychiatric disorders, insulin-resistant obesity, and polycystic ovary syndrome (4). There is mild to moderate ACTH-dependent hypercortisolism that is unrelated to an ACTH-secreting tumor but rather is the result of CRH or arginine vasopressin (AVP) hypothalamic hypersecretion through activation of various neural pathways.

Cushing syndrome is diagnosed by identifying elevated cortisol levels in the serum, saliva, or urine. This is not a simple measurement since the secretion of cortisol is pulsatile and has diurnal variation, with a peak in early morning and a nadir at about 4 pm. To account for these variations in cortisol levels, serum cortisol is measured at 8 am and 4 pm to determine if there is a loss of diurnal variation, which is an early feature of Cushing syndrome. Assays must be established to measure free cortisol, since the amount of cortisol-binding globulin (CBG) can also vary based on other conditions. Because the concentration of cortisol in saliva is in equilibrium with free cortisol in plasma, measurements of salivary cortisol can also be used to screen for this disorder (5). Ultimately, hypercortisolism is established by measuring the 24h-hour urinary free cortisol, which is an integrated measure.

Cortisol is regulated by ACTH, and in primary adrenal disease, the elevated cortisol levels are usually associated with marked suppression of the hypothalamus and pituitary gland, so that ACTH is low. In patients with elevated ACTH or with ACTH levels in the normal range but inappropriately high for the level of glucocorticoids, the dexamethasone suppression test is used to distinguish pituitary ACTH-dependent disease from ectopic disease. This test is based on the fact that even in pathologic conditions, this potent steroid inhibits pituitary ACTH secretion; the response, however, is attenuated in pituitary neoplasia and even more so in most cases of ectopic ACTH/CRH secretion.

Increased cortisol levels are seen in the *generalized glucocorticoid resistance syndrome* caused by mutations in the glucocorticoid receptor gene *NR3C1* (6). Patients with this syndrome do not have features of Cushing syndrome, but the lack of negative feedback of cortisol on the hypothalamic-pituitary-adrenal axis results in increased ACTH levels that induce further cortisol increases that are associated with androgen and mineralocorticoid secretion. Patients develop hypertension and hypokalemia, may have hirsutism, and develop adrenal hyperplasia (7).

Adrenal pathology in endogenous Cushing syndrome includes hyperplasia and neoplasia with various morphologies. In patients with pituitary or ectopic ACTH excess, hyperplasia predominates, but can be admixed with nodules due to the common occurrence of sporadic nodular disease and incidental adrenal adenomas. Primary causes of adrenal glucocorticoid excess include various adrenal cortical nodular diseases (chapter 4), adenomas (chapter 5), and carcinomas (chapter 6). Familial predisposition to these various proliferative disorders are summarized in chapter 7.

ACTH-Dependent Adrenocortical Hyperplasia

In the adrenal gland, ACTH binds to melanocortin-2 receptors, activating adenyl cyclase through the Gs protein alpha-subunit (GNAS), and leading to cAMP production. CAMP binding to the regulatory subunit of protein kinase A releases its catalytic subunit, and allows phosphorylation of downstream targets (8), which in turn results in cortisol synthesis and secretion and, if prolonged, hyperplasia of cortical cells. In ACTH-dependent Cushing syndrome, this is reflected in lipid depletion and bilateral hyperplasia that varies with the severity and duration of disease. In general, pituitary ACTH excess is

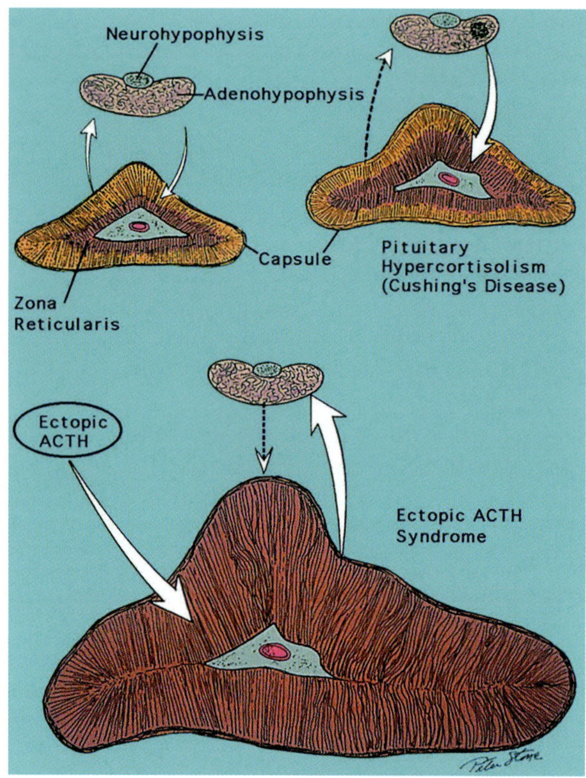

Figure 3-2

ETIOLOGY OF NONIATROGENIC ACTH-DEPENDENT CUSHING SYNDROME

Normal hypothalamic-pituitary-adrenal axis (upper left), pituitary-dependent Cushing disease (upper right), and ectopic ACTH syndrome.

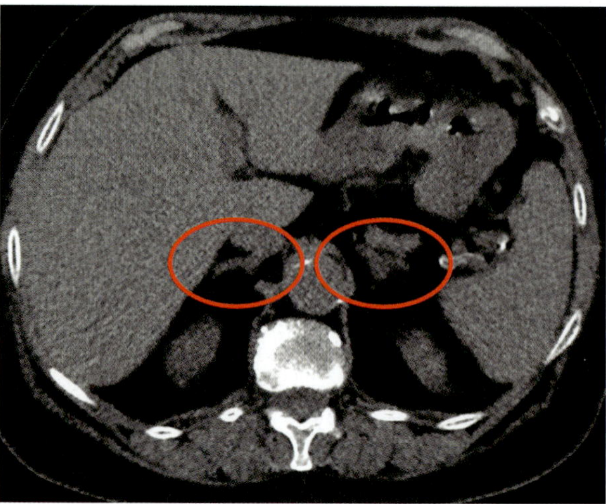

Figure 3-3

CUSHING DISEASE

Bilateral adrenal hyperplasia is seen by computerized tomography (CT) in a patient with an ACTH-dependent, recurrent ACTH-secreting pituitary tumor. (Courtesy of Prof. A. Veltri and Dr. F. Solitro, San Luigi Hospital, Turin, Italy)

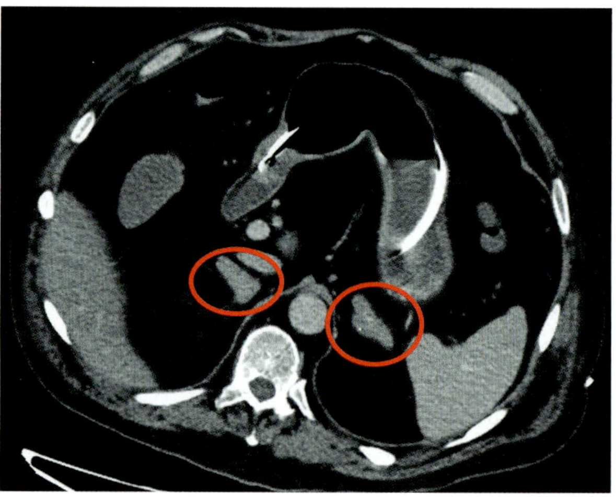

Figure 3-4

CUSHING SYNDROME

Magnetic resonance image (MRI) of bilateral adrenal hyperplasia in a patient with ACTH-dependent Cushing syndrome due to ectopic ACTH secretion.

associated with mild hyperplasia whereas ectopic ACTH excess tends to be much more severe, resulting in massive hyperplasia (fig. 3-2).

Imaging Features. In adults, diffuse adrenal cortical hyperplasia is defined radiologically on computerized tomography (CT) or magnetic resonance imaging (MRI) scans as a body width of greater than 10 mm and adrenal limb thickness of greater than 5 mm, in the absence of measurable adrenal nodules (figs. 3-3, 3-4) (9). In ACTH-dependent Cushing syndrome, imaging and biochemical procedures are designed to detect the source of ACTH production, and include pituitary MRI and bilateral inferior petrosal sinus sampling, to confirm the pituitary gland as the source of ACTH excess. In the absence of pituitary excess of ACTH relative to the systemic circulation, an ectopic source must be considered.

Imaging to localize an ectopic source should include the neck (to detect potential medullary thyroid carcinoma), the thorax (for thymic or pulmonary neuroendocrine tumors), the abdomen (where pancreatic neuroendocrine tumors are the most common source of ectopic

ACTH), and the pelvis. The use of Ga68-dotate positron emission tomography (PET) scan may be required to identify an occult source.

Pathologic Findings. Careful gross dissection of the adrenal glands, with removal of all peri-adrenal connective tissue and fat, is necessary for the accurate determination of the size and weight of the gland. In patients with pituitary ACTH excess, changes may be subtle and up to 30 percent of patients have grossly normal-appearing adrenal glands (10). There may be a slight increase in size and weight, and careful inspection may reveal a faint, somewhat irregular junction between the pale yellow outer aspect of the cortex and the inner darker one (fig. 3-5), due to conversion of vacuolated lipid-rich cells to lipid-depleted cells with more compact, eosinophilic cytoplasm. In patients with severe ectopic ACTH excess, diffuse hyperplasia appears as a generalized thickening of the adrenal cortex (fig. 3-6); the glands are most commonly bilaterally smoothly enlarged, with individual weights from 6 to 12 g, and retain the normal adrenal shape. The cortex becomes completely lipid depleted and loses its yellow color; instead, it develops a brown appearance.

The histologic morphology of diffuse hyperplasia can be subtle. In most patients with pituitary ACTH excess, the glands show preservation of the normal cortical zonation with slight expansion of the zona reticularis. In the early

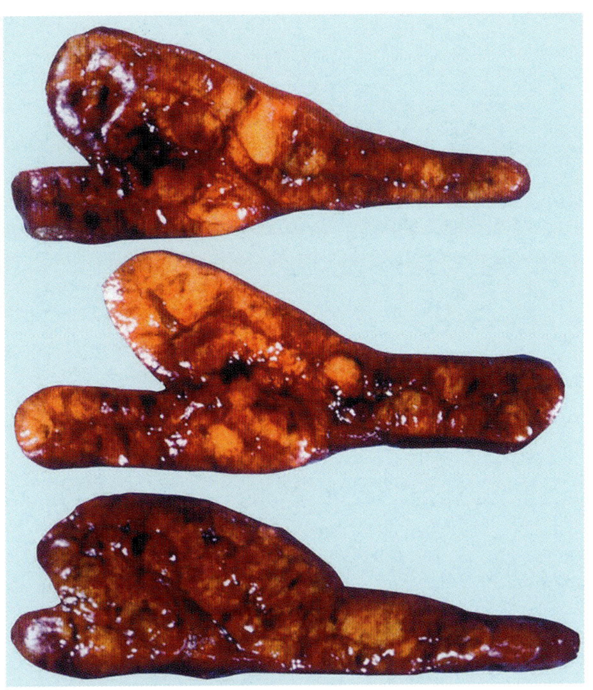

Figure 3-5

CUSHING DISEASE WITH DIFFUSE AND MICRONODULAR HYPERPLASIA

Transverse sections of adrenal gland in a patient with pituitary ACTH-dependent hypercortisolism (Cushing disease). The adrenal gland was only mildly enlarged and shows both diffuse and micronodular hyperplasia. Much of the adrenal cortex is tan to brown although pale yellow areas can be seen.

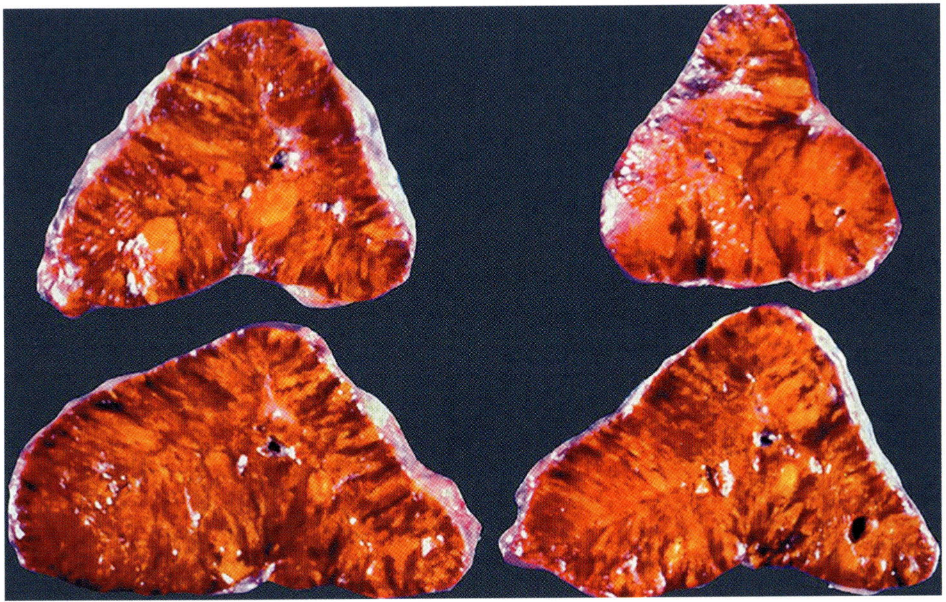

Figure 3-6

ECTOPIC ACTH SYNDROME

Transverse sections of adrenal glands surgically resected from a patient with ectopic ACTH syndrome due to a bronchial neuroendocrine tumor. The tumor had remained occult for several years. The right adrenal gland weighed 12 g and the left 11 g. The dark appearance is due to intense stimulation by ACTH. Vague nodularity is seen in some areas.

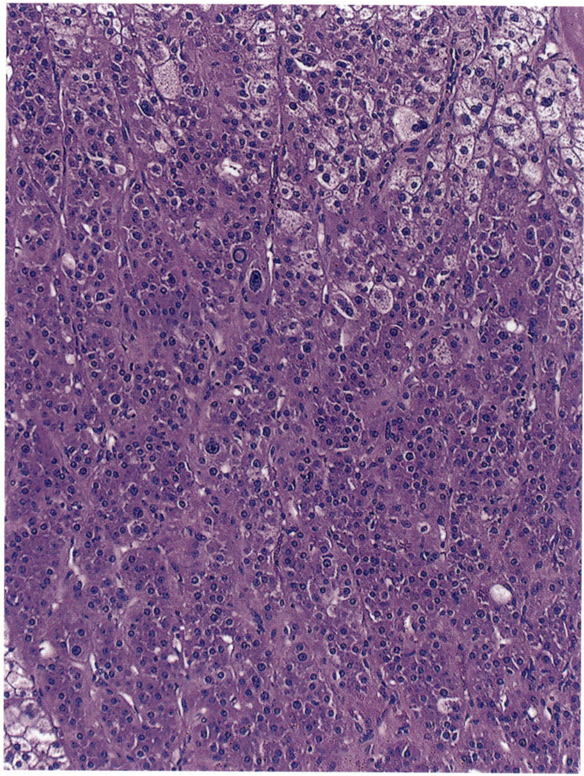

Figure 3-7

CUSHING DISEASE

Under the trophic influence of ACTH, much of the zona fasciculata is converted to cells with lipid-depleted, compact, eosinophilic cytoplasm.

stages of Cushing disease, it may be difficult to recognize adrenal hyperplasia without clinical/endocrinologic correlation, since the extent to which lipid-rich cells of the zona fasciculata are converted to compact lipid-poor cells is variable (figs. 3-7, 3-8). The adrenal medulla appears normal. Ectopic ACTH- or CRH-secreting tumors result in more pronounced diffuse cortical hyperplasia, with loss of normal zonation of the lipid-rich fasciculata layer. The cortex is instead composed of a thick layer of acidophilic compact cells (figs. 3-9–3-11).

Transition from diffuse to nodular architecture in ACTH-dependent hyperplasia may occur in advanced stages of Cushing disease but the mechanisms are still uncertain; nodularity may be a reflection of irregular growth or focal degeneration in a chronically stimulated gland (fig. 3-12). However, because of the common occurrence of incidental adrenal cortical nodules, the presence of nodules may simply be evidence of underlying unrelated sporadic nodular disease, even nodules larger than 1 cm.

An unusual situation arises when the source of ACTH or CRH excess is an adrenal paraganglioma (pheochromocytoma) (fig. 3-13) (11,12). This can cause bilateral cortical hyperplasia but the cortical response is usually asymmetric and more pronounced on the side of the lesion, which is clinically identified as a unilateral adrenal mass. These tumors can be mistaken for adrenal cortical tumors causing Cushing syndrome.

Treatment. The treatment of adrenal cortical hyperplasia in patients with ACTH-dependent Cushing syndrome is usually to address the primary disorder. Removal of the pituitary lesion or the source of ectopic ACTH is the primary goal, but in refractory cases, bilateral adrenalectomy is performed to reduce the complications of cortisol excess. Medical therapy with steroidogenesis inhibitors (13) or targeted G-protein coupled receptor therapy (14) may be also considered before or as an alternative to surgery.

PRIMARY ALDOSTERONISM (CONN SYNDROME)

Primary aldosteronism, also known as *Conn syndrome*, is a heterogeneous group of disorders caused by the autonomous overproduction of aldosterone. Aldosterone, the steroid hormone produced by the zona glomerulosa of the adrenal cortex, is a potent mineralocorticoid hormone regulated by the renal renin–angiotensin II system, which is in turn activated by hyperkalemia, hyponatremia, and reduced circulating blood volume, reflected in reduced renal blood supply. Aldosterone acts mainly on the renal distal convoluted tubule, where it binds the cytosolic mineralocorticoid receptor to modulate sodium and potassium channels that regulate renal tubular sodium reabsorption and urinary potassium excretion.

Patients with primary aldosteronism present with hypertension, hypokalemia, and metabolic alkalosis with suppression of plasma renin activity. Primary aldosteronism is the most common cause of endocrine hypertension and is associated with a significant increase in the risk of cardiovascular complications, including myocardial and vascular endothelial dysfunction, as

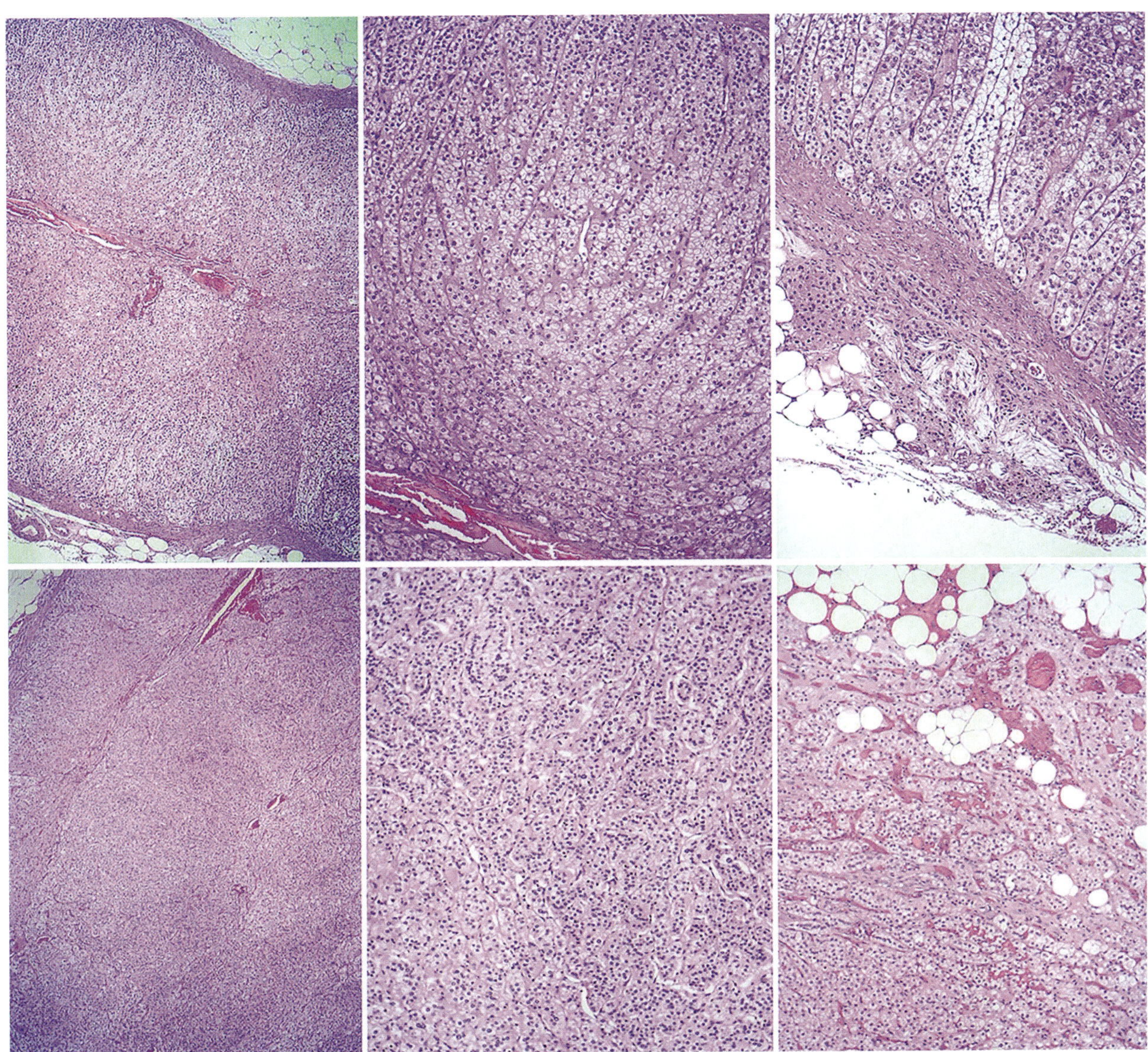

Figure 3-8

CUSHING DISEASE

Histology of left (upper panels) and right (lower panels) adrenal glands from a patient with a recurrent ACTH-secreting pituitary tumor (same case as fig. 3-3). Both glands show a similar morphologic appearance, with a homogeneous increase of cortical thickness with cortical extracapsular extrusions (right panels).

well as nephropathy and osteoporosis. Injuries in other target organs include inflammatory bowel disease and pulmonary lesions. Due to the variety of tissues damaged, it is important to detect and treat primary aldosteronism as early as possible to avoid or reduce its intractable organ injuries (15).

The frequency of primary hyperaldosteronism as a cause of hypertension is influenced by the diagnostic algorithmic approach, but in large population-based studies it represents up to 12 percent of cases, and is the second most frequent cause of secondary hypertension (16). Primary aldosteronism is frequently not correctly identified, since specific and easily identifiable features are lacking, and clinicians may not recognize the disease among the broad group of patients with hypertension.

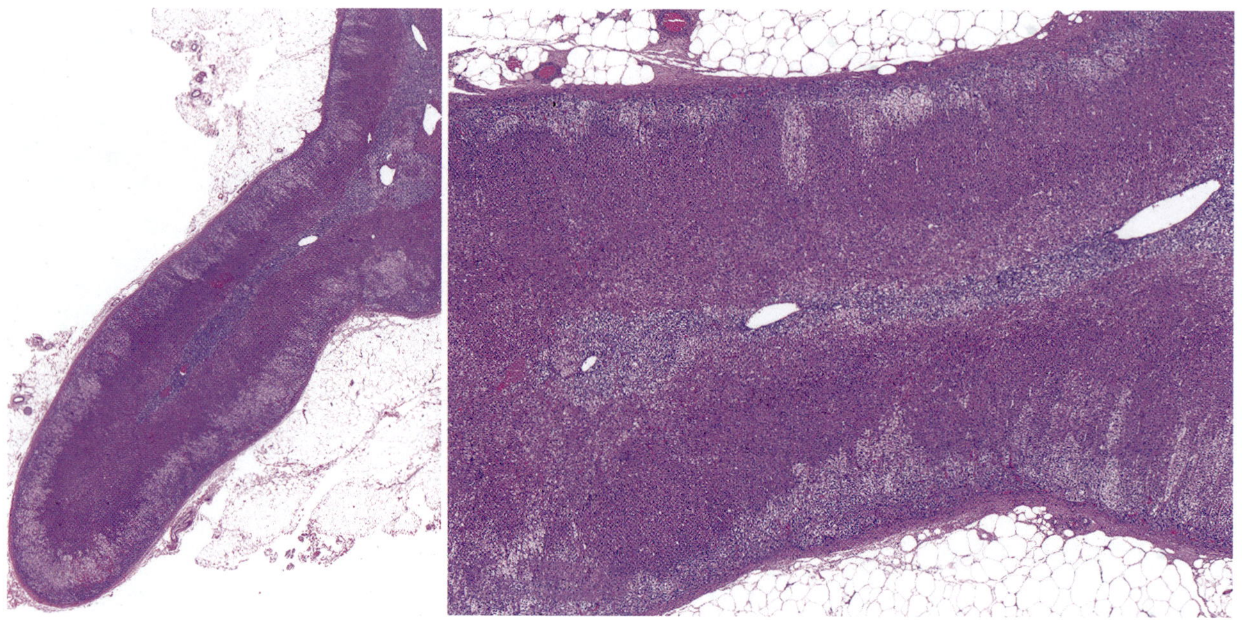

Figure 3-9

ECTOPIC ACTH SYNDROME

Adrenal cortical hyperplasia shows acidophilic cells expanding the zona fasciculata of the adrenal cortex.

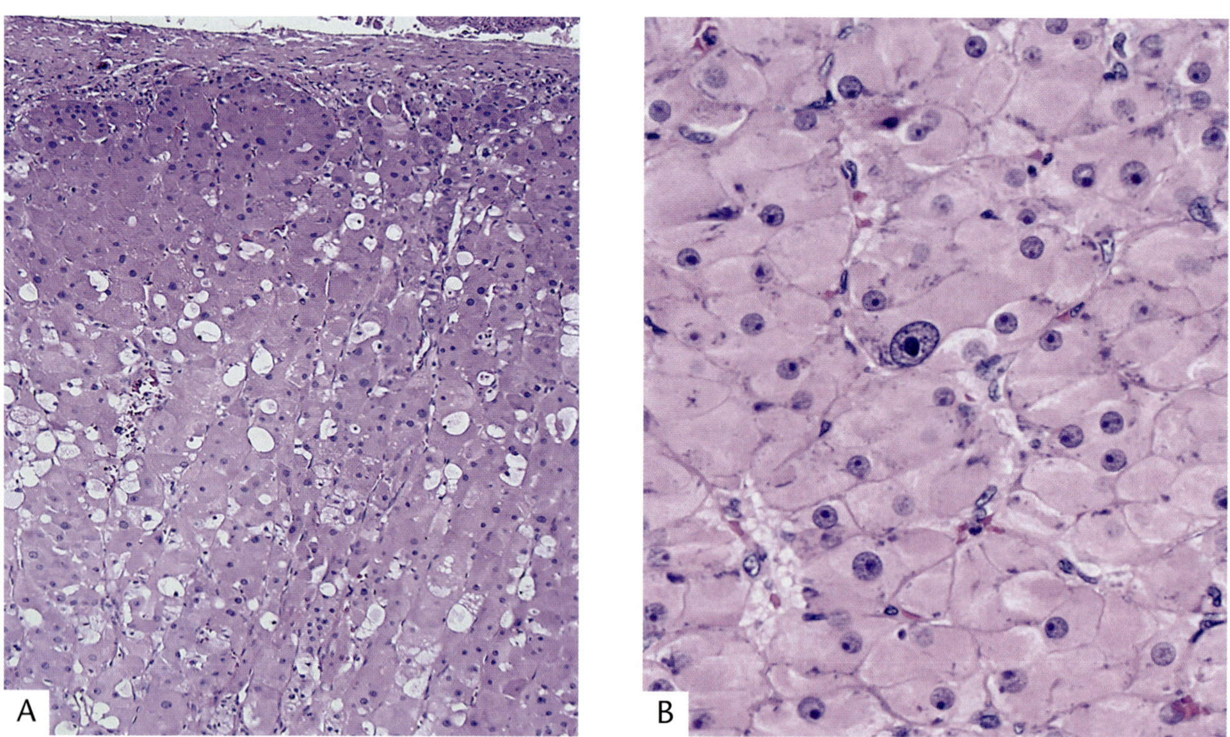

Figure 3-10

ECTOPIC ACTH SYNDROME

Large eosinophilic cells replace lipid-rich cells in the zona fasciculata of the adrenal cortex.

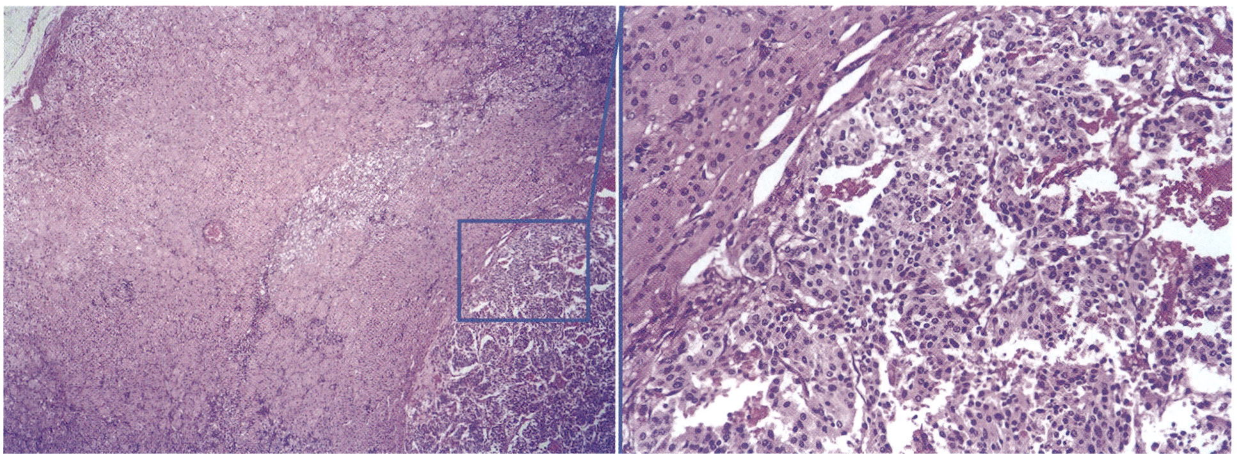

Figure 3-11

ECTOPIC ACTH SYNDROME

Diffuse ACTH-dependent adrenocortical hyperplasia from an adrenal metastasis of an ACTH-secreting lung neuroendocrine tumor.

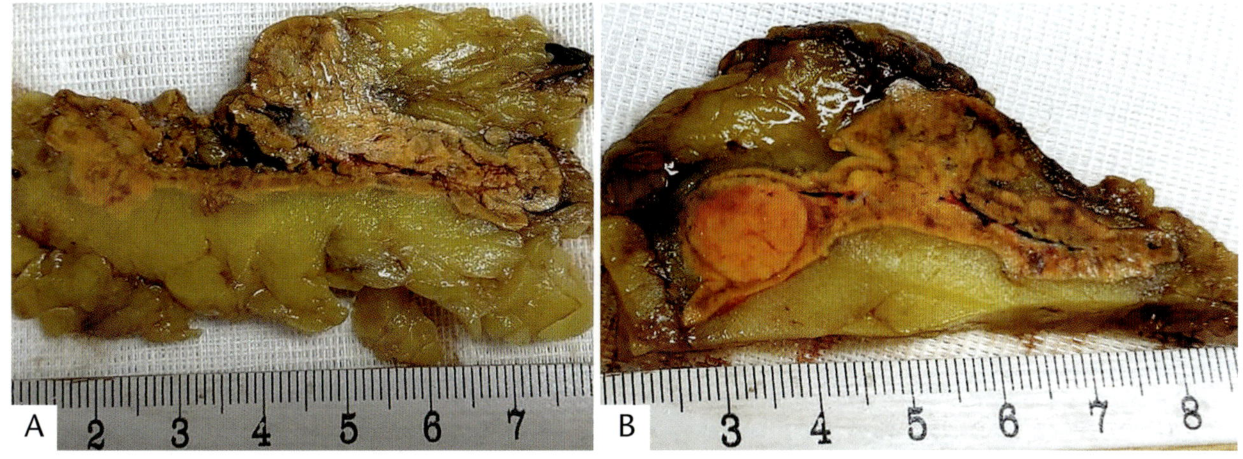

Figure 3-12

CUSHING DISEASE

Bilateral adrenal hyperplasia in a patient with ACTH-secreting pituitary tumor. There is mild and micronodular increase in size in the left (A) and right (B) adrenal glands. The right gland also has a macronodule (over 1 cm).

Clinical signs and symptoms are nonspecific and the onset may be subtle, with normokalemic hypertension, headache, fatigue, and weakness. In contrast, when present, hypokalemia can be insidious without obvious cause. More florid forms may present with cardiac dysrhythmias, polydipsia, polyuria, cramping, headaches, and weakness (17). Laboratory findings include increased plasma aldosterone levels and suppressed plasma renin activity resulting in an elevated plasma aldosterone-to-renin ratio. Aldosterone suppression tests are mandatory for a biochemical confirmation.

The most common causes of primary aldosteronism are bilateral aldosterone-producing diffuse hyperplasia and aldosterone-producing adenoma (18,19). The pathology of a solitary adrenal cortical adenoma causing this disorder is discussed in chapter 5. Bilateral idiopathic adrenal cortical hyperplasia accounts for 60 to

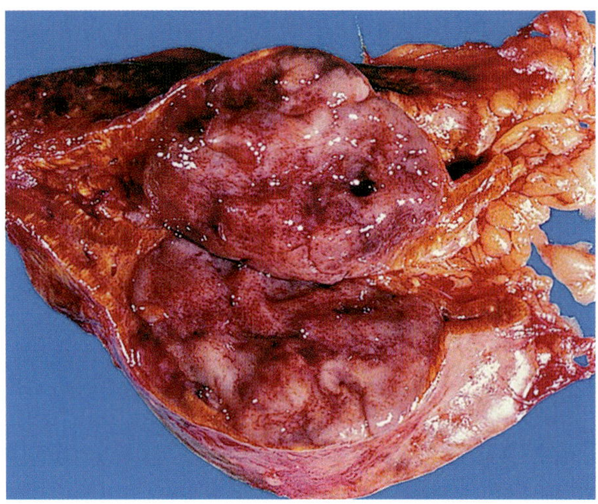

Figure 3-13

ECTOPIC ACTH SYNDROME DUE TO PHEOCHROMOCYTOMA

A 51-year-old white female presented with hypertension, facial swelling, and hirsutism. There were elevated steroid metabolites in the urine which could not be suppressed with high-dose dexamethasone. Skull X rays showed no abnormalities of the sella turcica. Following surgical removal of this 3-cm pheochromocytoma, hirsutism regressed and steroid levels normalized. Surgery was complicated by hypotensive episodes. The hyperplastic adrenal cortex was largely lipid depleted. Biopsy of the opposite hyperplastic adrenal gland was identical.

70 percent of cases (20), and is often reflected in subtle lesions, which are discussed in chapter 4. While most cases of primary hyperaldosteronism are sporadic, 1 to 5 percent are inherited familial forms, *familial aldosteronism* (FA) types I, II, III, and IV, transmitted as autosomal dominant traits (chapter 7) (21). Aldosterone-producing adrenal cortical carcinomas are exceptionally rare.

Imaging Features. The distinction of unilateral from bilateral disease is required for optimal therapeutic management (20). Since incidental nonfunctional adrenal cortical nodules are frequent in the general population, and not all radiologically identified adrenal lesions cause aldosterone excess in patients with primary aldosteronism (22), lateralization of the aldosterone excess is a critical step, and adrenal venous sampling has been pivotal, but it can be bypassed in patients with specific characteristics.

Molecular Pathologic Findings. Aldosterone is produced in the zona glomerulosa after activation of the angiotensin II type 2 receptor (ATR1) through a pathway that activates steroidogenic acute regulatory protein (StAR). Inositol-trisphosphate (IP3) is part of this pathway and its activation also stimulates intracellular Ca^{2+} levels, which subsequently activate calcium/calmodulin-dependent protein kinases (CaMK) and cAMP response element-binding (CREB) protein. Therefore, it is not surprising that mutations in genes encoding different ion channels are frequent in the pathogenesis of primary aldosteronism (23). Mutations in ion channels are the major cause of aldosterone-producing adenomas, and several mutations within the ion channel encoding genes have been identified.

Somatic mutations in four genes (*KCNJ5, ATP1A1, ATP2B3,* and *CACNA1D*) have been identified in nearly 60 percent of sporadic aldosterone-producing adenomas, while germline mutations in *KCNJ5* and *CACNA1H* have been reported in different subtypes of familial hyperaldosteronism. The cascade of steroidogenic enzymes that is involved in aldosterone synthesis include CYP11A1, 3βHSD, CYP21A, and CYP11B2; the latter is specific for aldosterone biosynthesis, and antibodies that can be applied for immunohistochemistry have paved the way for a much clearer understanding of the pathology responsible for the spectrum of disorders underlying primary hyperaldosteronism (24).

Treatment. The goals of treatment are to ameliorate or normalize blood pressure and eliminate excessive aldosterone production, hopefully reducing associated comorbidities, improving quality of life, and reducing mortality. The correct classification of adrenal lesions in primary aldosteronism is crucial since unilateral disease is cured by surgery whereas mineralocorticoid antagonists, such as spironolactone or eplerenone, are the treatment of choice for bilateral disease (20).

SEX STEROID EXCESS

The adrenal cortex produces some sex steroids, particularly androgens, such as dehydroepiandrosterone (DHEA) (which is sulfated in the liver to yield the more stable product DHEA sulfate [DHEAS]) as well as androstenedione (A4). Estrogens and progestogens can also be produced due to alterations in the enzymes that regulate steroid biosynthesis. The balance of steroidogenesis is normally in favor of glucocorticoids, with a small amount of sex steroid production, but this balance can be altered in pathologic conditions. It is most relevant in patients with congenital adrenal hyperplasia (chapter 2) but can also be seen in adrenal cortical neoplasms.

Although the production of androgens may be responsible for the hirsutism associated with Cushing syndrome, oversecretion of sex steroid hormones as the main feature of an adrenal cortical neoplasm is more frequent in adrenal cortical carcinoma; thus malignancy should be suspected in the presence of pronounced virilization in a woman and feminization of a man. Virilization is the most frequent hormonal manifestation of functioning adrenal cortical adenomas in the pediatric age group (25); it is relatively rare in postmenopausal women (26).

ADDISON DISEASE

Adrenal failure or insufficiency is commonly referred to as *Addison disease* or *Addison syndrome*, from the name of Thomas Addison who first described this syndrome in 1855 in his classic monograph entitled, "On the Constitutional and Local Effects of Disease of the Supra-Renal Capsules" (fig. 3-14) (27). The term has traditionally been applied to disorders resulting in widespread destruction with hypofunction of both adrenal cortices. The loss of function is typically slow in onset, with gradual decline ultimately resulting in nonspecific symptoms. These can range from fatigue to gastrointestinal nausea and diarrhea to hypotension and shock.

In the past, Addison disease was frequently the result of systemic infections such as tuberculosis, however, autoimmune diseases currently represent the most frequent cause in North America. This can be a solitary process affecting the adrenal glands or be accompanied by autoimmune conditions involving other endocrine and nonendocrine systems. In the autoimmune polyglandular forms, the disease is initially characterized by a chronic lymphoplasmacytic infiltrate of the adrenal glands; this progresses to fibrosis and atrophy and may result in near total loss of the adrenal cortex, leaving only a rim of fibrous tissue around the medulla. In other instances, the glands are destroyed following bilateral hemorrhage in the setting of systemic anticoagulation (heparin-induced thrombocytopenia [HIT]) antiphospholipid syndrome (APS), or in conjunction with the now familiar COVID (28).

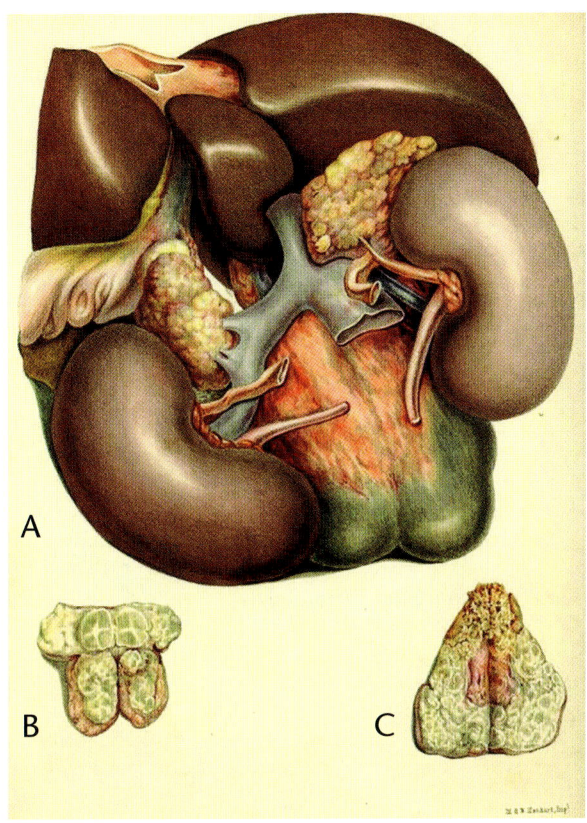

Figure 3-14

ORIGINAL DESCRIPTION OF ADDISON DISEASE BY THOMAS ADDISON

A: The liver of H. P., with the diseased supra-renal capsules in situ.

B,C: Sections of the diseased supra-renal capsules. (From his monograph, On the Constitutional and Local Effects of Disease of the Supra-renal Capsules.) (Hardin MD, University of Iowa; http://hardinmd.lib.uiowa.edu/ui/addisons/plate4.html)

REFERENCES

1. Ezzat S, de Herder WW, Volante M, Grossman A. The driver role of pathologists in endocrine oncology: what clinicians seek in pathology reports. Endocr Pathol 2023;34:437-54.
2. Cushing H. The basophil adenomas of the pituitary body and their clinical manifestations (pituitary basophilism). Obes Res 1994;2:486-508.
3. Hakami OA, Ahmed S, Karavitaki N. Epidemiology and mortality of Cushing's syndrome. Best Pract Res Clin Endocrinol Metab 2021;35:101521.
4. Chabre O. The difficulties of pseudo-Cushing's syndrome (or "non-neoplastic hypercortisolism"). Ann Endocrinol (Paris) 2018;79:138-45.
5. Fleseriu M, Auchus R, Bancos I, et al. Consensus on diagnosis and management of Cushing's disease: a guideline update. Lancet Diabetes Endocrinol 2021;9:847-75.
6. Nicolaides NC, Chrousos GP. Bilateral adrenal hyperplasia and NR3C1 mutations causing glucocorticoid resistance: is there an association? Eur J Endocrinol 2018;179:C1-4.
7. Nicolaides NC, Charmandari E. Primary generalized glucocorticoid resistance and hypersensitivity syndromes: a 2021 update. Int J Mol Sci 2021;22:10839.
8. Duan K, Hernandez KG, Mete O. Clinicopathological correlates of adrenal Cushing's syndrome. J Clin Pathol 2015;68:175-86.
9. Morani AC, Jensen CT, Habra MA, et al. Adrenocortical hyperplasia: a review of clinical presentation and imaging. Abdom Radiol (NY) 2020;45:917-27.
10. Michelle MA, Jensen CT, Habra MA, et al. Adrenal cortical hyperplasia: diagnostic workup, subtypes, imaging features and mimics. Br J Radiol 2017;90:20170330.
11. Elliott PF, Berhane T, Ragnarsson O, Falhammar H. Ectopic ACTH- and/or CRH-producing pheochromocytomas. J Clin Endocrinol Metab 2021;106:598-608.
12. Leader N, Ushinsky A, Malone CD. Adrenal vein sampling for ACTH-producing pheochromocytomas. Radiol Case Rep 2021;16:2672-5.
13. Daniel E, Aylwin S, Mustafa O, et al. Effectiveness of metyrapone in treating Cushing's syndrome: a retrospective multicenter study in 195 patients. J Clin Endocrinol Metab 2015;100:4146-54.
14. Assie G, Louiset E, Sturm N, et al. Systematic analysis of G protein-coupled receptor gene expression in adrenocorticotropin-independent macronodular adrenocortical hyperplasia identifies novel targets for pharmacological control of adrenal Cushing's syndrome. J Clin Endocrinol Metab 2010;95:E253-62.
15. Gao X, Yamazaki Y, Tezuka Y, et al. Pathology of aldosterone biosynthesis and its action. Tohoku J Exp Med 2021;254:1-15.
16. Kotliar C, Obregón S, Koretzky M, et al. Improved identification of secondary hypertension: use of a systematic protocol. Ann Transl Med 2018;6:293.
17. Hundemer GL, Vaidya A. Primary aldosteronism diagnosis and management: a clinical approach. Endocrinol Metab Clin North Am 2019;48:681-700.
18. Funder JW, Carey RM, Fardella C, et al. Case detection, diagnosis, and treatment of patients with primary aldosteronism: an endocrine society clinical practice guideline. J Clin Endocrinol Metab 2008;93:3266-81.
19. Duan K, Mete O. Clinicopathologic correlates of primary aldosteronism. Arch Pathol Lab Med 2015;139:948-54.
20. Reincke M, Bancos I, Mulatero P, Scholl UI, Stowasser M, Williams TA. Diagnosis and treatment of primary aldosteronism. Lancet Diabetes Endocrinol 2021;9:876-92.
21. Vaduva P, Bonnet F, Bertherat J. Molecular basis of primary aldosteronism and adrenal Cushing syndrome. J Endocr Soc 2020;4:bvaa075.
22. Nanba AT, Nanba K, Byrd JB, et al. Discordance between imaging and immunohistochemistry in unilateral primary aldosteronism. Clin Endocrinol (Oxf) 2017;87:665-72.
23. Omata K, Anand SK, Hovelson DH, et al. Aldosterone-producing cell clusters frequently harbor somatic mutations and accumulate with age in normal adrenals. J Endocr Soc 2017;1:787-99.
24. Mete O, Asa SL, Giordano TJ, Papotti M, Sasano H, Volante M. Immunohistochemical biomarkers of adrenal cortical neoplasms. Endocr Pathol 2018;29:137-49.
25. Lopes RI, Suartz CV, Neto RP, et al. Management of functioning pediatric adrenal tumors. J Pediatr Surg 2021;56:768-71.
26. Zhou WB, Chen N, Li CJ. A rare case of pure testosterone-secreting adrenal adenoma in a postmenopausal elderly woman. BMC Endocr Disord 2019;19:14.
27. Addison T. On the constitutional and local effects of disease of the supra-renal capsules. London: Highley; 1855.
28. Elhassan YS, Iqbal F, Arlt W, et al. COVID-19-related adrenal haemorrhage: multicentre UK experience and systematic review of the literature. Clin Endocrinol (Oxf) 2023;98:766-78.

4 ADRENAL CORTICAL NODULAR DISEASES

Adrenal cortical hyperplasia is a general term for a condition characterized by an increased number of cortical cells leading to an increase in size and weight of the gland. However, this term is currently considered a misnomer for almost all diseases that are defined as primary of the adrenal gland, except for congenital adrenal hyperplasia (CAH) (chapter 2) and is accepted for ACTH-dependent adrenal cortical hyperplasia (discussed in chapter 3). Within the group of adrenal lesions associated with primary aldosteronism in adults, the term "diffuse hyperplasia" indicates a lesion that may not be recognized histologically but is highlighted by a continuous linear pattern of positivity with CYP11B2 immunohistochemistry. Therefore, the term hyperplasia will not be used in this chapter unless old terminologies or a few specific conditions are mentioned, in line with the terminology proposed in the most recent World Health Organization (WHO) classification of endocrine and neuroendocrine tumors (1).

Adrenal cortical nodular diseases are classified according to defined macroscopic and microscopic criteria, and the presence or absence of a particular endocrine syndrome (2). They may be associated with cortisol hypersecretion and primary aldosteronism, or are biochemically and clinically nonfunctioning lesions. Adrenal cortical nodular diseases may be congenital or acquired. The identification of germline and sporadic alterations causing adrenal cortical nodular diseases associated with primary aldosteronism and Cushing syndrome enabled the development of a molecular classification that goes beyond pathology and allows a personalized approach to affected patients, including genetic screening for familial forms. A detailed description of hereditary conditions predisposing to adrenal cortical nodular diseases is covered in chapter 7.

Adrenal cortical proliferations, usually without clinical or biochemical evidence of hormone hypersecretion, have been described in the context of Carney triad (3). This is the combination of gastric gastrointestinal stromal tumor (GIST), pulmonary chondroma, and extra-adrenal paraganglioma (4). A genetic cause is assumed since the tumors are multicentric and multifocal, and usually affect young patients, although the causative gene(s) remains elusive and the syndrome does not appear to be hereditary. Epigenetic hypermethylation of the *SDHC* promoter is seen in many but not all cases. In Carney triad, adrenal lesions are usually seen with imaging or due to pathologic "adenoma-like" features, although sometimes the adrenal nodules are multiple, either synchronous or metachronous, or associated with histologically detected micronodules (5).

As stated in chapter 1, careful gross dissection of the adrenal glands, with removal of all periadrenal connective tissue and fat, is necessary for the accurate determination of the size and weight of the gland. Small nodules may not be apparent on external examination of the intact gland, and, therefore, transverse sectioning at about 3-mm intervals is recommended. There is usually an increase in size or weight of the adrenal glands, which may be symmetric or asymmetric. The "lumpy bumpy" contour of some glands may impede complete removal of adherent fat and connective tissue, and may make some recorded weights spuriously high. Nodular diseases are divided into micronodular and macronodular types based upon the size of the nodules, using 1 cm as the cut-off.

A similar pattern is observed in bilateral macronodular adrenal disease associated with primary aldosteronism. However, this condition is rarely seen in surgical material since it is more frequently treated by medical therapy.

Figure 4-1

NODULAR ADRENAL GLAND

Diagram of a transverse section shows cortical extrusions with various configurations. Some lie free in periadrenal adipose tissue area. A larger, dominant intracortical nodule is present (left). There is intermingling of nests of cortical cells with chromaffin cells, and both the cortex and medulla have a close anatomic relationship with the vascular space due to discontinuity of smooth muscle in tributaries of the central adrenal vein vessel walls.

SPORADIC NODULAR ADRENAL CORTICAL DISEASE

Epidemiology, Pathogenesis, and Clinical Findings. *Sporadic nodular adrenal cortical disease* is characterized by clinically nonfunctioning nodules made of adrenocortical cells that usually measure less than 1 cm (fig. 4-1). The nodules are most frequently discovered incidentally, may occur in all age groups, and are usually unilateral when clinically detected. However, morphologic evidence of nodularity is usually multifocal and bilateral. The nodules may be found on imaging, at autopsy, or during the histopathologic examination of adrenalectomy specimens undertaken for other reasons (6). They may be more common in elderly and hypertensive patients, but they rare in the pediatric population.

Prior to the availability of computerized tomography (CT) scanning and other sensitive imaging techniques, the "nonfunctioning" nodular adrenal gland was usually an incidental finding at autopsy. The incidence of cortical nodules at autopsy is difficult to estimate since there are no universally accepted morphologic criteria to define a cortical "nodule" in terms of size, number, and other distinguishing characteristics. Incidence data are largely based on studies of nonfunctioning lesions larger than 1 cm, thus simulating an adrenal cortical adenoma.

Currently, this disorder is usually detected as an incidental finding on high-resolution CT scan, magnetic resonance imaging (MRI), or during ultrasound examination for an unrelated problem or during the staging workup of a patient with a known malignancy elsewhere (fig. 4-2) (7). Adrenal "incidentalomas" are reported in around 5 percent of people subjected to imaging studies (7–9), although not all adrenal incidentalomas are of cortical origin. Moreover, the exact prevalence of sporadic nodular adrenal glands in surgical and radiologic series with complete endocrine workup remains to be determined since no homogeneous studies have been designed for this specific situation.

The etiology of sporadic nodular adrenocortical disease is largely unknown. Cigarette smoking may be a risk factor since the prevalence of both benign and malignant adrenal cortical neoplasms is higher in individuals with a history of tobacco use (10). An association between cortical nodularity and degenerative change in capsular vessels (capsular "arteriopathy") has been reported, suggesting that focal hyperplasia may be a response to localized ischemia (11). Also, sporadic nodular adrenocortical disease may be associated with hypertension.

Few studies have addressed the molecular characteristics of sporadic nodular adrenocortical disease. Sporadic small adrenocortical nodules have been suggested to be clonal (12,13), thus supporting the nomenclature of "nodular disease" as opposed to "nodular hyperplasia" (14). However, in a study of micro-dissected samples from both sporadic nodular adrenocortical disease and adrenal cortical adenomas using the methylation pattern of the androgen receptor alleles as a tool to investigate clonality, polyclonal patterns

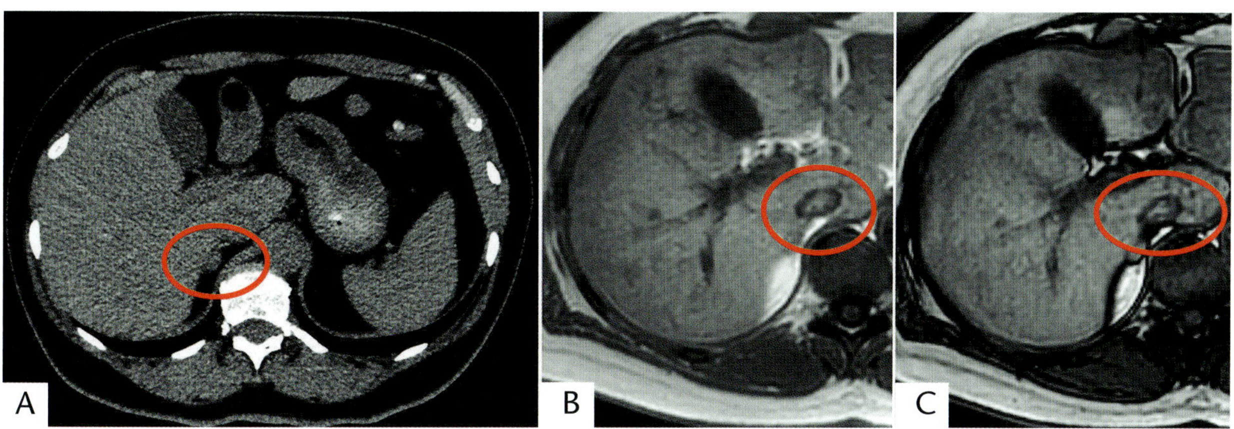

Figure 4-2

SPORADIC ADRENAL CORTICAL NODULE

An incidentally discovered, nonfunctioning adrenal cortical nodule, 0.6 cm in size on computerized tomography (CT) scan (A, 40 HU), with atypical imaging features at magnetic resonance imaging (MRI) ("in" phase, B, with no loss of signal in the "out" phase, C). (Courtesy of Prof. A. Veltri and Dr. F. Solitro, San Luigi Hospital, Turin, Italy)

were revealed in 14 of 18 sporadic nodules but only in 3 of 22 adenomas (15).

Pathologic Findings. Sporadic nodular adrenocortical disease appears as randomly distributed, small (usually less than 1 cm) yellow (fig. 4-3) to brown (fig. 4-4) cortical nodules. The histopathologic features of some adrenal cortical nodules in the setting of the sporadic nodular adrenal gland are indistinguishable from those of an adrenal cortical adenoma. However, the nodules tend to be less demarcated from the adjacent cortex, with no clear-cut capsule formation (figs. 4-5–4-7). The peri- or internodular cortex is not atrophic or hyperplastic, at variance with macronodular diseases associated with cortisol or aldosterone oversecretion. Cellular features may show a predominance of clear cells or eosinophilic cells but a mixed pattern is the most frequent (figs. 4-8, 4-9). Predominant oncocytic changes (fig. 4-10) or pigmentation (fig. 4-11) may be observed.

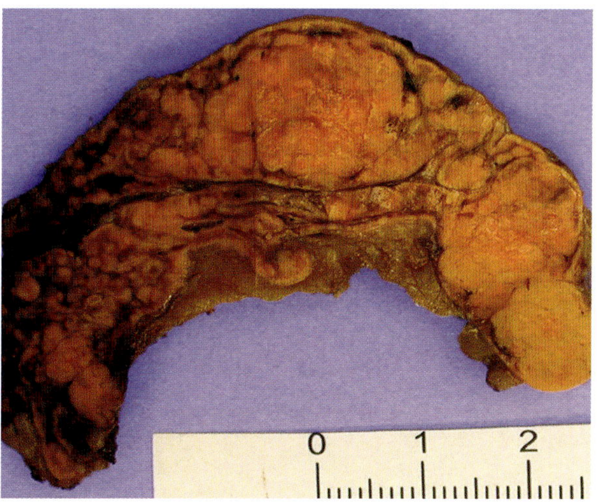

Figure 4-3

SPORADIC ADRENAL CORTICAL NODULAR DISEASE

Multiple homogeneous yellow nodules are visible, with a size ranging from a few millimeters to about 1 cm.

ADRENAL CORTICAL NODULAR DISEASES ASSOCIATED WITH INCREASED CORTISOL SECRETION

In the presence of endogenous causes of Cushing syndrome, *adrenal cortical lesions* are detectable in more than 80 percent of patients; most are associated with ACTH-dependent hypercortisolism (discussed in chapter 3). About 20 percent of cases are associated with ACTH-independent secretory mechanisms. In ACTH-independent lesions, up to 10 percent of primary adrenal Cushing syndrome cases are associated with bilateral adrenal cortical lesions; the rest are associated with unilateral adrenal cortical adenomas (2).

In bilateral cases, the clinical distinction of primary from secondary forms is based on the detection of low plasma ACTH levels (if

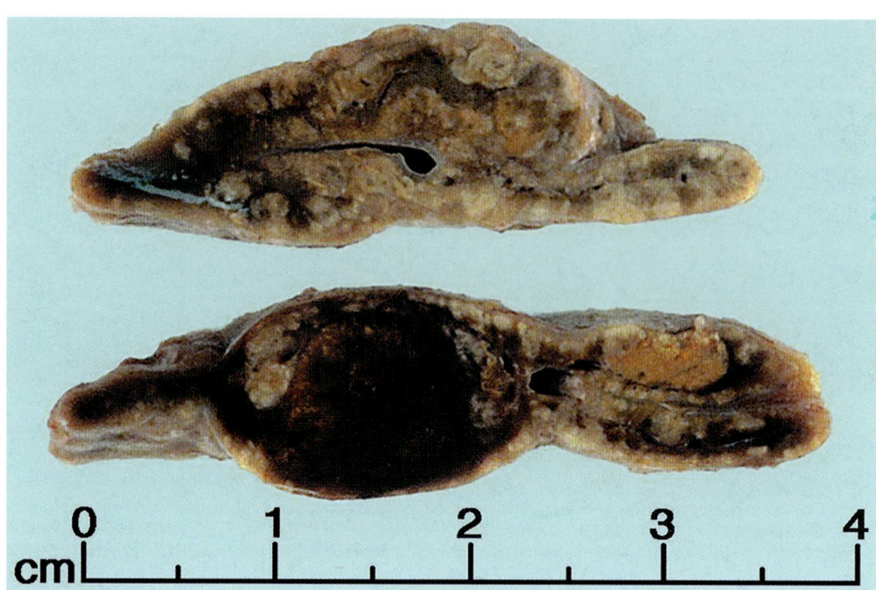

Figure 4-4

SPORADIC ADRENAL CORTICAL NODULAR DISEASE

An incidental pigmented macronodule is associated with a few subcentimetric nodules in the adjacent cortex visible on transverse section.

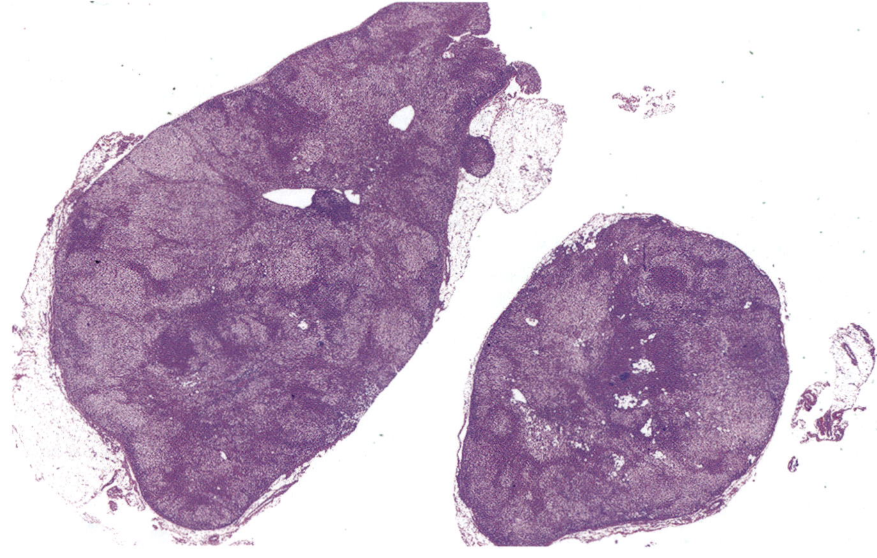

Figure 4-5

SPORADIC ADRENAL CORTICAL DISEASE

Diffuse and pseudonodular enlargement of the adrenal gland.

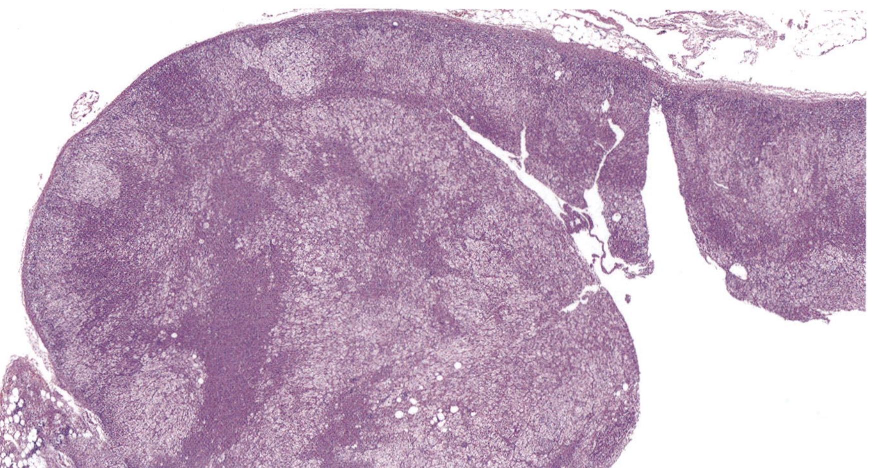

Figure 4-6

SPORADIC ADRENAL CORTICAL DISEASE

Nodular enlargement of the adrenal gland without encapsulation.

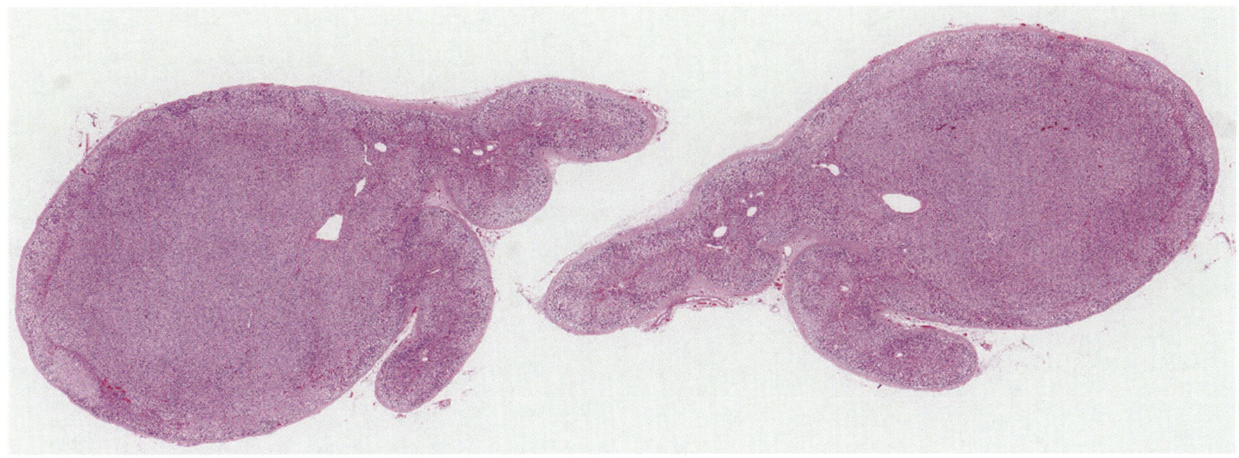

Figure 4-7

SPORADIC ADRENAL CORTICAL DISEASE

Nodular enlargement of the adrenal gland (less than 1 cm in size). At variance with adrenal cortical adenoma, there is absence of clear demarcation/encapsulation.

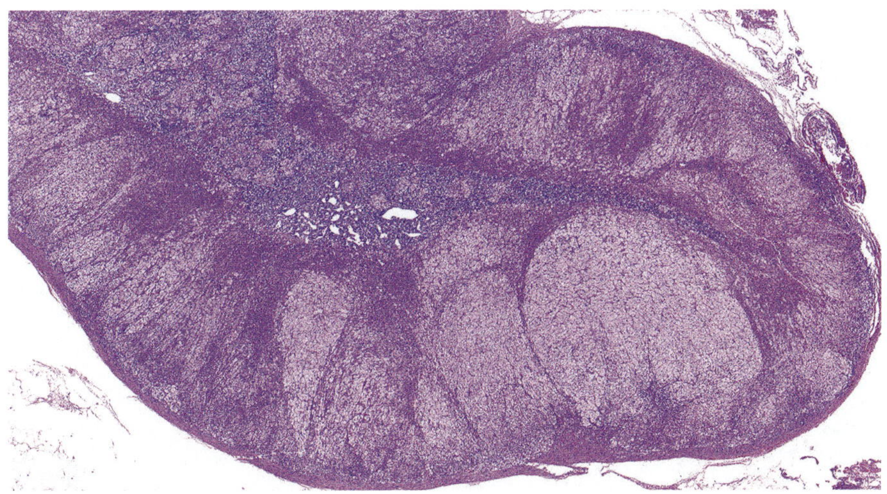

Figure 4-8

SPORADIC ADRENAL CORTICAL DISEASE

Diffuse and pseudonodular enlargement of the adrenal gland showing a predominant lipid-rich cell component. The adrenal medulla is not altered.

oversecretion of cortisol reaches sufficient levels). Bilateral primary causes include primary bilateral macronodular adrenal disease (PBMAD, previously known as primary bilateral macronodular adrenal hyperplasia [PBMAH]) and micronodular bilateral adrenal disease (MiBAD), further subclassified into primary pigmented nodular adrenocortical disease (PPNAD) and isolated micronodular adrenocortical disease (i-MAD). Distinguishing the different forms of primary adrenocortical nodular disease associated with cortisol oversecretion depends on the size of the nodules and the presence of specific histologic findings (Table 4-1).

Primary Bilateral Macronodular Adrenocortical Disease

Epidemiology, Pathogenesis, and Clinical Features. *Primary bilateral macronodular adrenocortical disease* (PBMAD) is characterized by multiple bilateral nodules that are larger than 1 cm. It is more frequent in adults, particularly in the 5th decade of life (mean age of 53 years), with a slight female predominance that is observed in sporadic but not in hereditary forms (16,17). It may also occur in the pediatric population, particularly in the first years of life, in strong association with the McCune-Albright syndrome.

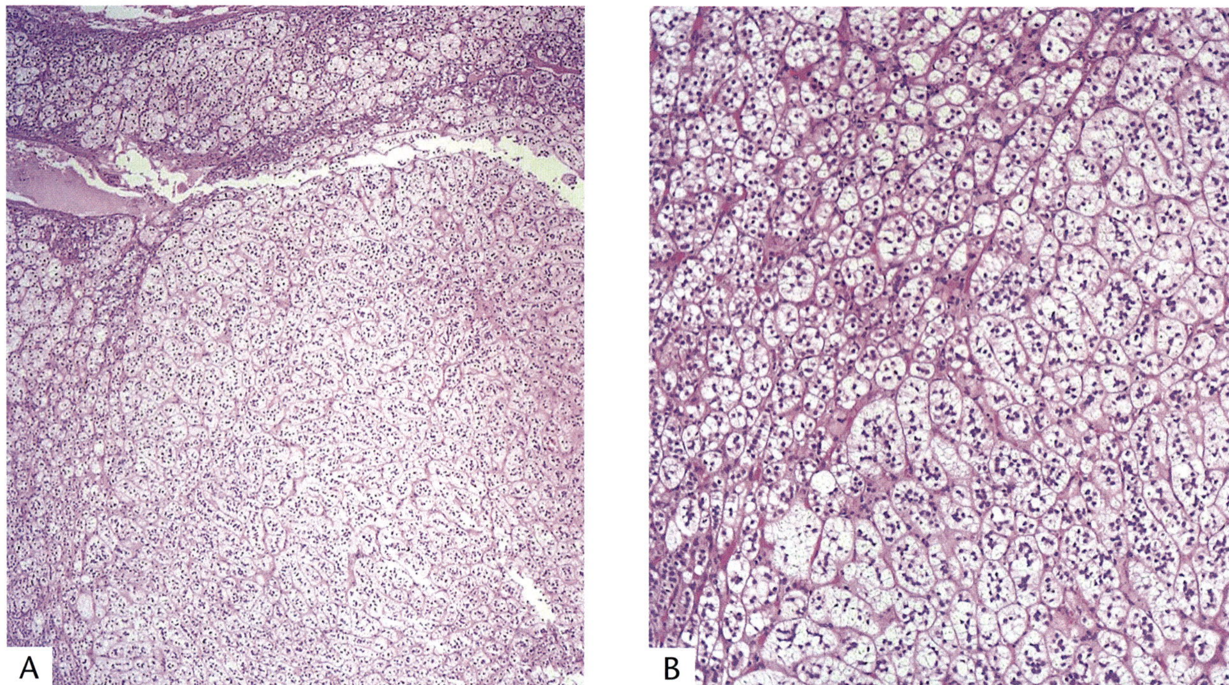

Figure 4-9

SPORADIC ADRENAL CORTICAL DISEASE

Nodular enlargement of the adrenal gland (less than 1 cm) incidentally found in a patient with a large renal cyst. Clear cell features are predominant. A and B show low and high magnifications of the same area, illustrating the architectural difference but cytological similarity between the nodule and adjacent normal cortex.

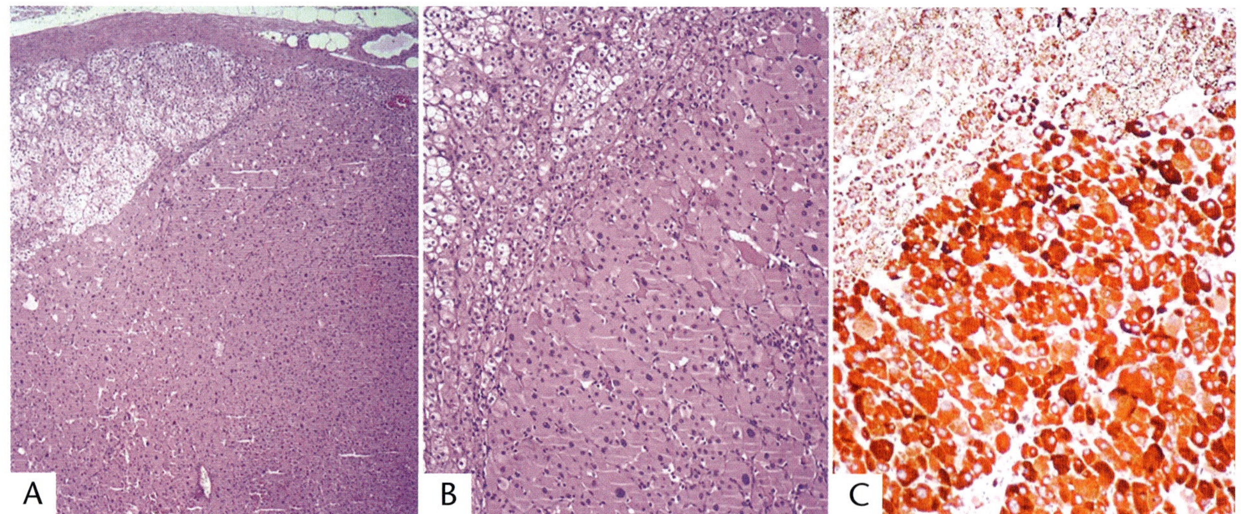

Figure 4-10

SPORADIC ADRENAL CORTICAL DISEASE

Nodular enlargement of the adrenal gland (less than 1 cm) with predominant oncocytic features and diffuse and intense Melan-A positivity (C) (same cases as in fig. 4-2). Oncocytic features are responsible for the atypical MRI images. A and B show low and high magnifications of the same area, illustrating both the architectural and cytological difference between the nodule and adjacent normal cortex.

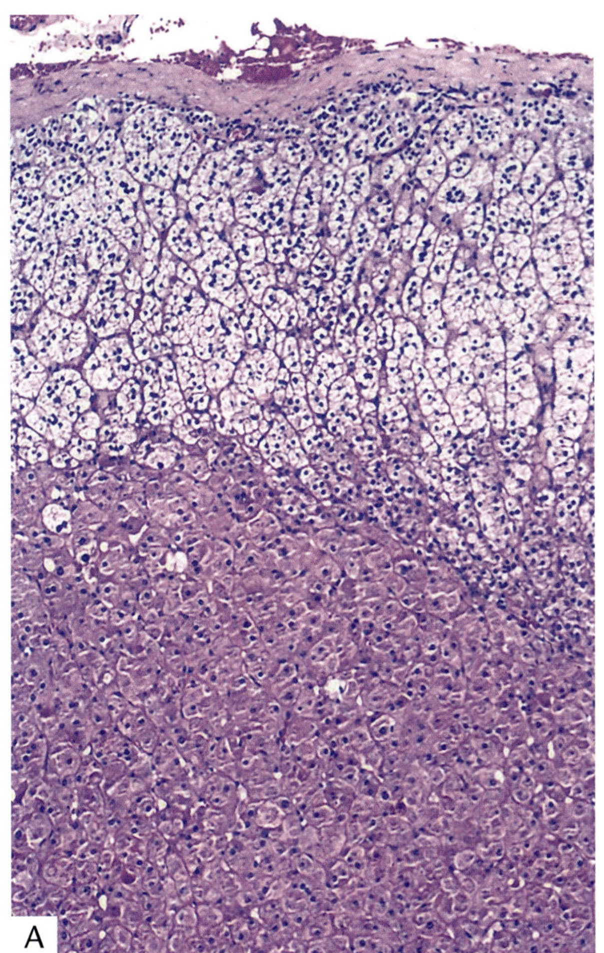

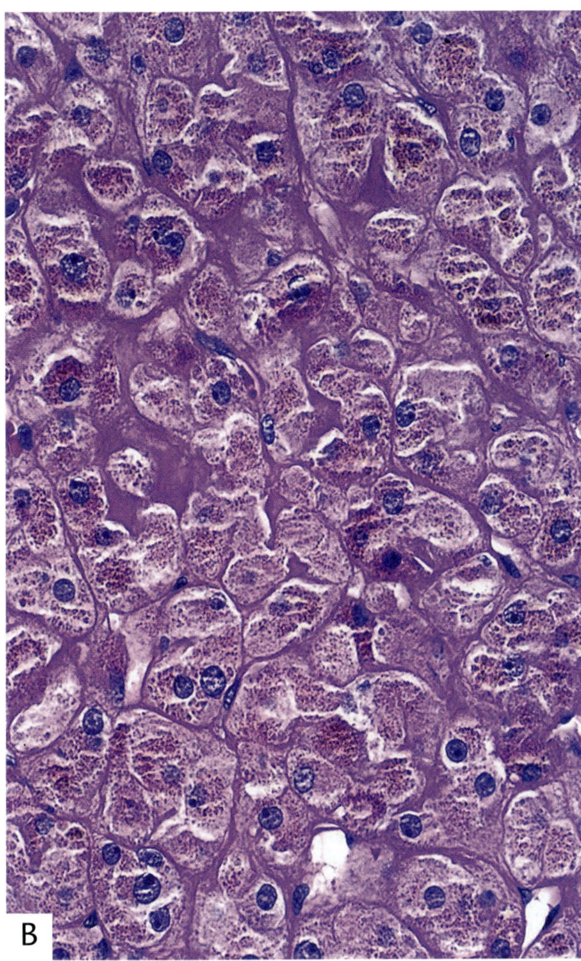

Figure 4-11

SPORADIC ADRENAL CORTICAL DISEASE WITH PIGMENTATION

A: Incidental pigmented nodule has its epicenter in the zona reticularis and expansile borders impinging slightly on the outer cortex and medulla.

B: Cortical cells contain a variable amount of granular, golden brown/dark brown pigment. Nuclei tend to be round to oval and vary only slightly in size.

Table 4-1

BILATERAL ADRENAL CORTICAL NODULAR DISEASES ASSOCIATED WITH CORTISOL OVERSECRETION

Disease	Main Clinical Characteristics	Main Pathologic Features
Bilateral macronodular adrenocortical disease	Middle aged; mild hypercortisolism (with or without mineralocorticoid excess); sporadic or familial (syndromic or isolated)	Nodules >1 cm, with or without internodular cortical atrophy
Primary pigmented nodular adrenocortical disease (PPNAD)	Children and young adults; overt hypercortisolism; isolated or hereditary (autosomal dominant)	Nodules <1 cm, with pigmentation and frequent internodular cortical atrophy
Isolated micronodular adrenocortical disease (i-MAD)	Children and young adults (and middle-aged adults in Carney complex), overt hypercortisolism; isolated or hereditary (autosomal dominant)	Nodules <1 cm, without pigmentation and with internodular cortical hyperplasia

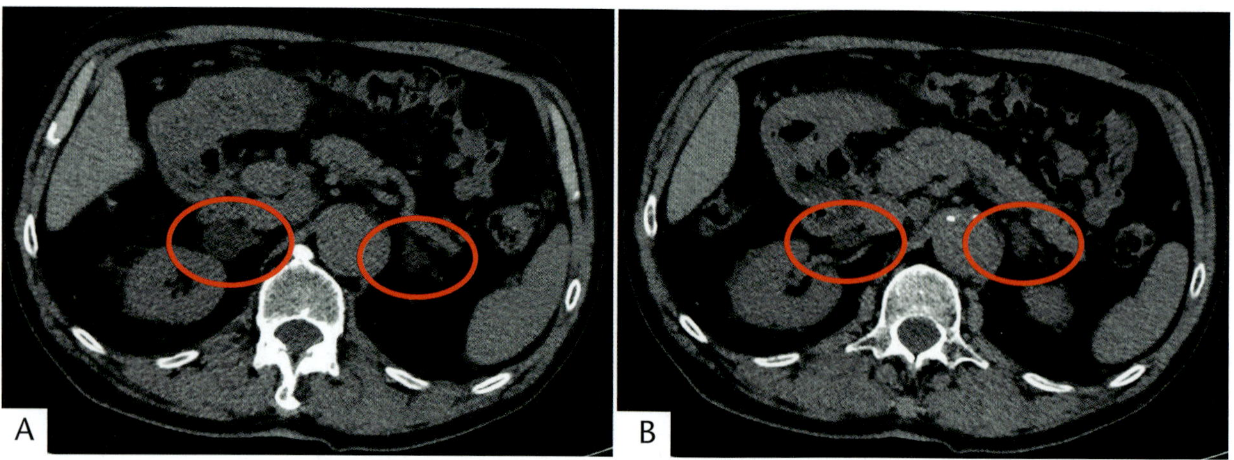

Figure 4-12

PRIMARY BILATERAL MACRONODULAR ADRENAL CORTICAL DISEASE

A 43-year-old male patient with marked hypercortisolism. CT scan (different levels shown in A and B) shows bilateral enlargement of the adrenal glands with both a diffuse and a pseudonodular pattern. The nodules had an heterogeneous density with lesions either <10 or >10 HU. (Courtesy of Prof. A. Veltri and Dr. F. Solitro, San Luigi Hospital, Turin, Italy)

This syndrome (polyostotic fibrous dysplasia, café-au-lait skin spots, and precocious puberty) is a genetically mosaic disorder with populations of mutant and normal cells in affected organs. In these children, spontaneous resolution of the Cushing syndrome may occur, but bilateral adrenalectomy is often required (18).

Cortisol oversecretion may be associated with overt Cushing syndrome but it may be also detected at biochemical workup for incidentally discovered lesions (2). Mixed biochemical profiles of cortisol and mineralocorticoid excess have been described (19). Pure virilization is almost exclusively present in congenital adrenal hyperplasia (CAH), whereas mixed cortisol and sex steroid hormone secretory profiles with feminization are observed in PBMAD (20).

CT scan usually shows marked enlargement of both adrenal glands, with multiple macronodules, although diffuse enlargement with no clear dominant nodules is also observed. Heterogeneous enhancement in different nodules of the same adrenal gland may occur, as well as an asymmetric appearance between the two glands (fig. 4-12), a feature that may complicate the distinction of this from a unilateral disease (21). Bilateral involvement may be asynchronous, with disease developing in one gland even years after the first clinical presentation. The internodular adrenal cortex may be normal or atrophic. MRI usually shows lesions that are hypointense on T1-weighted imaging and hyperintense on T2-weighted images (22). Uptake of FDG-PET is high, with SUV-max values above 3.1, approximating those of malignant tumors despite the benign nature of PBMAD (22).

PBMAD may arise in a hereditary context, associated with multiple endocrine neoplasia 1 (MEN1), familial adenomatous polyposis, hereditary leiomyomatosis, and renal cancer syndromes, among others (chapter 7). Germline mutations in the *ARMC5* gene have been detected in a large proportion (20 to 30 percent) of apparently sporadic cases of PBMAD (23). *ARMC5*-related macronodular disease has no gender predominance and often has a significantly higher level of cortisol secretion and absence of nonfunctional manifestations compared to *ARMC5* wild type disease (17). Somatic mutations of *GNAS1* gene have been detected in rare cases of clinically sporadic PBMAD (24). Other rare somatic mutations, described through whole exome sequencing techniques, include alterations in *DOT1L* and *HDAC9* genes (25).

These known molecular alterations are not enough to understand the pathophysiology of PBMAD, and the processes leading to cell growth and hypersecretion are complex. By means of transcriptome analysis, several genes encoding G-protein-coupled receptors (GPCR)

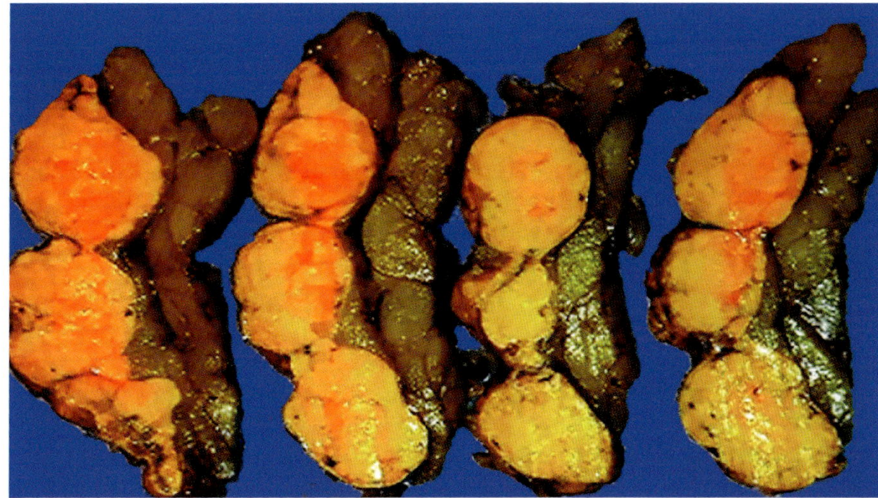

Figure 4-13

MACRONODULAR ADRENAL CORTICAL DISEASE ASSOCIATED WITH CORTISOL OVERSECRETION

Multiple macronodules in transverse section, with a thin rim of residual adrenal cortex.

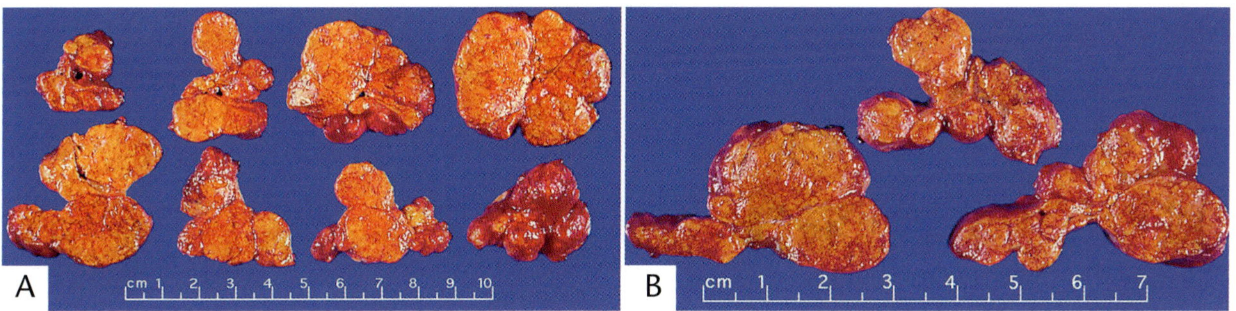

Figure 4-14

BILATERAL MACRONODULAR ADRENAL CORTICAL DISEASE

Transverse sections of both adrenal glands (shown separately in A and B) show marked thickening and nodularity of the golden yellow cortex. The nodules have ill-defined borders. The medulla is not evident.

were found to be overexpressed in PBMAH, including *MLNR, GABBR1,* and *ADRA2A*, which encode for motilin, gamma-aminobutyric acid, and alpha2 adrenergic receptors, respectively (26), whereas LH/hCG receptor overexpression, resulting from germline gene duplication, was observed in a case occurring in pregnancy (27). Whether aberrant GPCR expression is involved in initiation or secondary regulation of cortisol secretion in PBMAD, as well as the molecular mechanisms underlying the aberrant GCPR expression, are still unclear. PBMAD cells have been reported to secrete ectopic ACTH as a consequence of proopiomelanocortin (POMC) mRNA transcription and in association with GPCR aberrant overexpression (28).

Pathologic Findings. In macronodular disease associated with hypercortisolism, the lesions are bilateral, distorted due to coarse nodularity of the adrenal glands, and markedly enlarged, with individual weight reaching up to 200 g (17). On cut section, there are multiple unencapsulated nodules with a yellow to golden appearance and irregular light brown foci. Grossly, the nodules appear discrete and sharply demarcated, and some dominant macronodules give the impression of encapsulation (fig. 4-13). Prominent cortical nodules may compress and distort the adrenal medulla, and it may be difficult to recognize the medullary compartment in histologic sections (fig. 4-14).

Despite being clearly demarcated grossly, the nodules are usually less well-defined on microscopy, and blend almost imperceptibly with the adjacent cortex. Extrusions of adrenocortical tissue into the periadrenal fat is frequent and

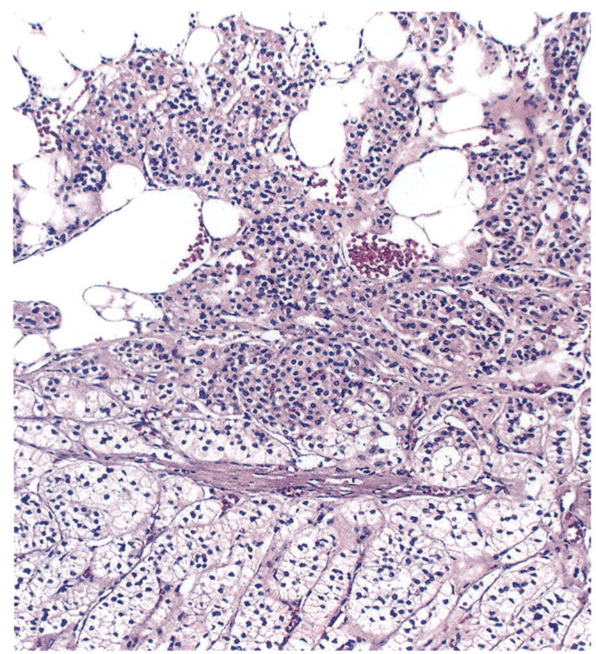

Figure 4-15

PRIMARY BILATERAL MACRONODULAR ADRENOCORTICAL DISEASE

Irregular extrusions of cortical cells into the periadrenal fat without any encapsulation.

should not be considered a sign of malignancy (fig. 4-15). There may be a structured capsule in dominant nodules. There is a variable mixture of pale-staining, lipid-rich cells and lipid-depleted cells with compact, eosinophilic cytoplasm (fig. 4-16). Some cortical cells have abundant lipid-rich cytoplasm, giving a "balloon" appearance. Occasionally, the nodules have an oncocytic cellular component.

Cortical nodules have many architectural and cytologic appearances (fig. 4-17). Some of the architectural patterns include trabecular, nesting, diffuse, gyriform, ribbon-like, and pseudoglandular. The latter pattern is rare but may have well-defined luminal borders and contents that consist of stringy amorphous eosinophilic or basophilic material or a few degenerated cortical cells. Nodules can be seen in every region of the cortex, and the degree of nodularity can vary widely from gland to gland. The nodules may be larger than 2 cm in diameter, and have areas of lipomatous, myelolipomatous, or even osseous metaplasia (fig. 4-18). The adrenal medulla may be deformed and difficult to identify due to compression by cortical nodules (fig. 4-19). The sclerosis may be patchy or involve much of

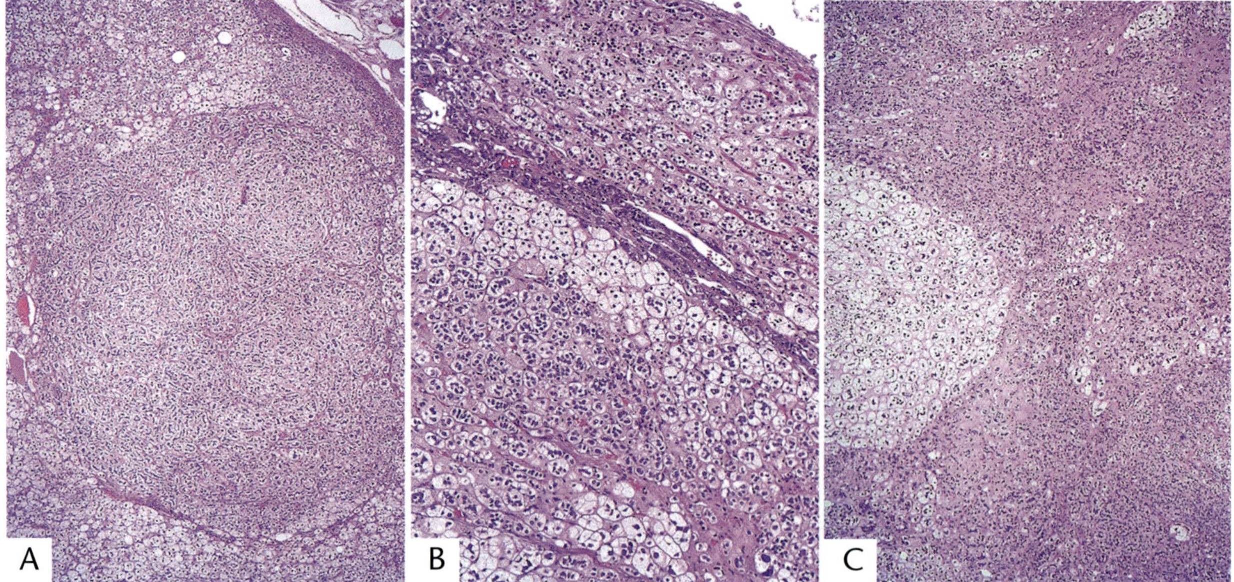

Figure 4-16

PRIMARY BILATERAL MACRONODULAR ADRENOCORTICAL DISEASE ASSOCIATED WITH CORTISOL OVERSECRETION

Heterogeneous architectural patterns with diffuse and pseudonodular lesions from the adrenal lesion seen in figure 4-12. (A: eosinophilic lipid-depleted cells; B: focal cytological atypia; C: clear lipid-rich cells)

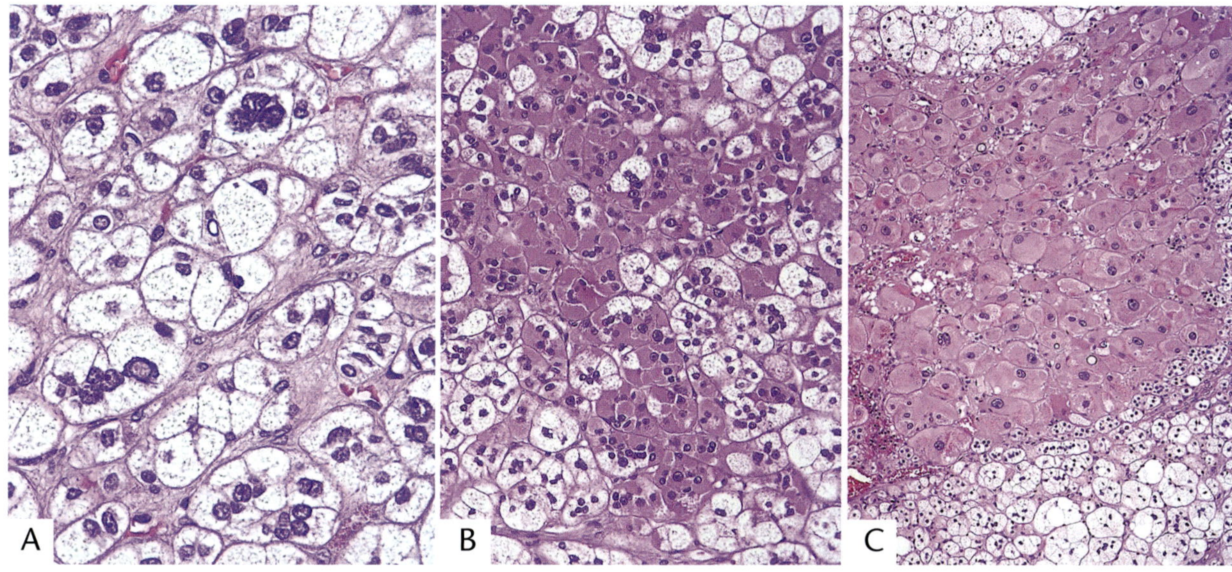

Figure 4-17

PRIMARY BILATERAL MACRONODULAR ADRENOCORTICAL DISEASE ASSOCIATED WITH CORTISOL OVERSECRETION

Heterogeneity of cellular features from the same adrenal lesion seen in figures 4-12 and 4-16. Clear cells with marked nuclear atypia (A), a mixed clear and eosinophilic cell pattern (B), and oncocytic component with pigmentation (C).

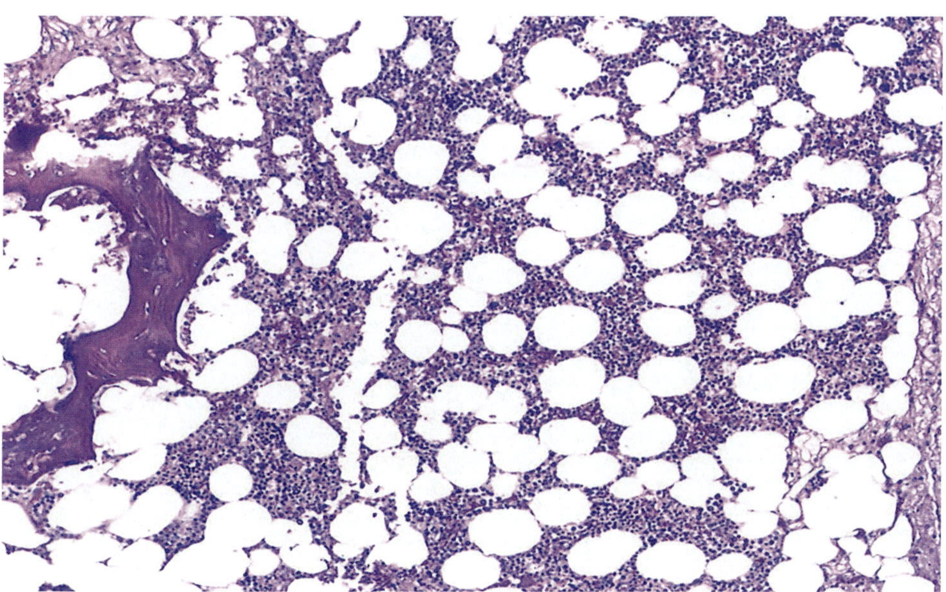

Figure 4-18

PRIMARY BILATERAL MACRONODULAR ADRENOCORTICAL DISEASE

Myelolipomatous metaplasia with bone formation.

the nodule (fig. 4-20), and may be accompanied by dystrophic calcification. Large nodules may show degenerative features such as cystic change, sclerosis, or old hemorrhage (fig. 4-21). Rarely, areas resembling amyloid are present.

In patients affected by McCune-Albright syndrome, a characteristic bimorphic pattern of diffuse and nodular proliferations and a distinctive form of cortical atrophy with apparent zona glomerulosa hyperplasia are typically observed, especially in young individuals. This peculiar pattern has been suggested to be the result of the mosaic distribution of mutant and normal cells in the adrenal glands and is absent in other forms of inherited or sporadic ACTH-independent Cushing syndromes (29).

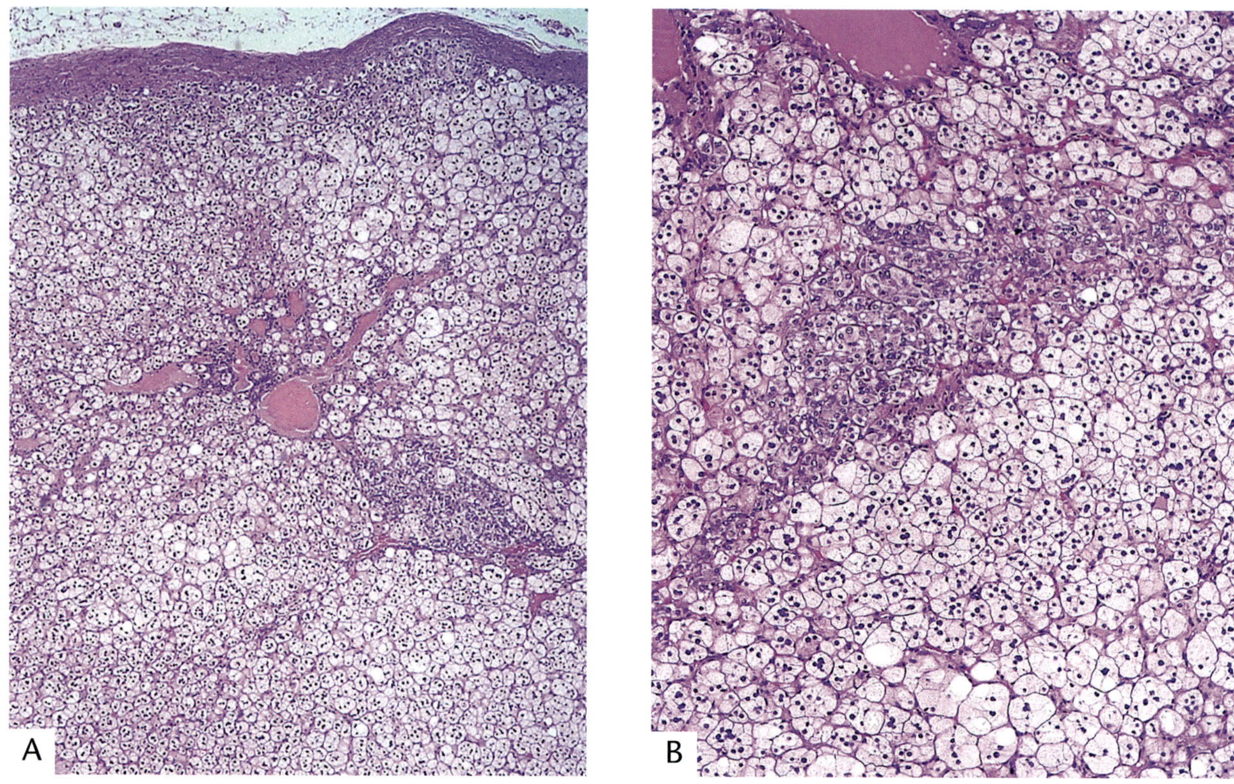

Figure 4-19

PRIMARY BILATERAL MACRONODULAR ADRENOCORTICAL DISEASE ASSOCIATED WITH CORTISOL OVERSECRETION

The adrenal medulla is compressed by the cortex. A and B show low and high magnifications of the same area, illustrating the increased amount of cortex and the somewhat compressed but cytologically unaltered medulla.

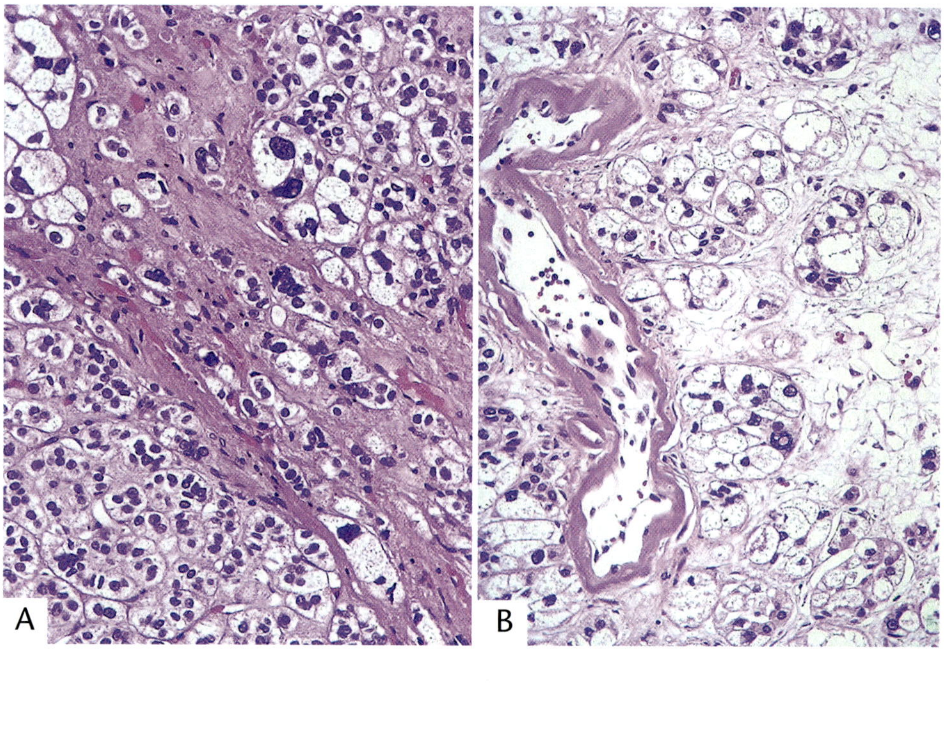

Figure 4-20

PRIMARY BILATERAL MACRONODULAR ADRENOCORTICAL DISEASE ASSOCIATED WITH CORTISOL OVERSECRETION

Mild (A) to moderate (B) intranodular fibrosis.

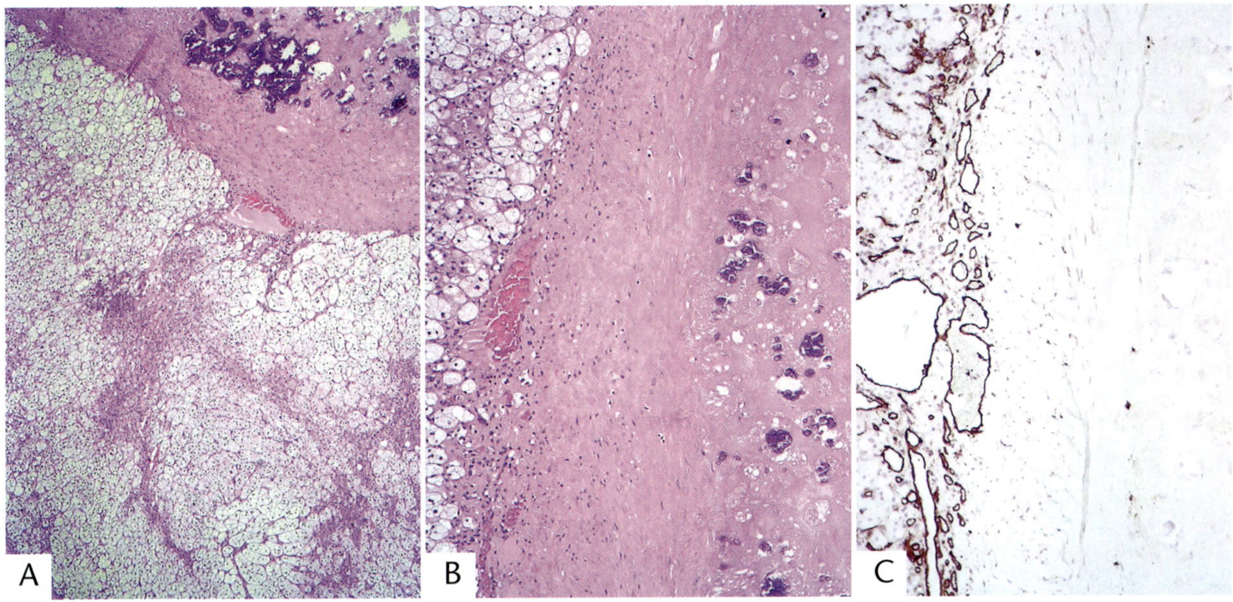

Figure 4-21

PRIMARY BILATERAL MACRONODULAR ADRENOCORTICAL DISEASE

Involutive changes including marked sclerosis and calcifications (A) and pseudocystic changes (B,C: same area marked with CD31).

Recent studies focused on genotype-phenotype correlations have subclassified PBMAD using histologic criteria. Violon et al. (30) proposed four histopathologic subtypes of PBMAD by combining cytomorphology (proportion of clear, compact, and oncocytic cells) and architecture (fibrous septa formation within nodules, residual adrenal cortex). Subtype 1 (enriched for clear cells, with lower amounts of compact cells and oncocytic cells) and subtype 2 (less enriched for clear cells, but still in the majority, with larger subsets of compact cells) correlate significantly with pathogenic *ARMC5* and *KDM1A* variants, respectively. Both subtypes 1 and 2 are distinguished from other subtypes by the presence of round fibrous septa formation within macronodules. In a morphologic and immunohistochemical analysis of a series of PBMADs characterized by pathogenic *ARMC5* mutations, de Arruda Bothelho et al. (31) found pseudoglandular and/or trabecular growth patterns and capsular extrusion of cellular proliferations were the best morphologic predictors of pathogenic *ARMC5* variants.

Primary Pigmented Nodular Adrenocortical Disease

Epidemiology, Pathogenesis, and Clinical Features. *Primary pigmented nodular adrenocortical disease* (PPNAD) is a descriptive term for a distinctive but rare form of ACTH-independent Cushing syndrome that represents the most frequent type of micronodular bilateral adrenal disease. It may be isolated (i-PPNAD) or, more frequently, represents an endocrine manifestation of *Carney complex*, where it occurs in around 57 percent of patients (32,33).

Carney complex consists of an array of diverse abnormalities, and has an autosomal dominant mode of inheritance in 70 percent of cases, the remainder being apparently sporadic (34). It was originally described by Carney in 1985 as a complex disease with co-occurrence of myxomas, spotty pigmentation, and endocrine overactivity (32). Revised diagnostic criteria of Carney complex are listed in Table 4-2 (34). If patients are identified with two or more elements of this complex (particularly PPNAD, bilateral large cell calcifying Sertoli cell tumors, or mucocutaneous pigmentation [fig. 4-22]), investigation for possible cardiac myxoma, frequently multiple, is recommended to permit early detection and treatment. Additional but more rare manifestations in Carney complex include cardiomyopathy, colonic polyps and adenocarcinoma, bronchogenic cysts, hepatocellular adenoma and carcinoma, gastric cancer, and pancreatic cancer. The genetic background of Carney complex and i-PPNAD is described in chapter 7.

Table 4-2

DIAGNOSTIC CRITERIA OF CARNEY COMPLEX[a]

Skin	Spotty skin pigmentation with typical periorificial distribution (lentigines) Myxomas
Heart	Myxomas
Adrenal gland	PPNAD with Cushing syndrome
Pituitary gland	Pituitary GH-secreting neuroendocrine tumor with acromegaly
Breast	Myxomatosis Ductal adenoma
Testis	Large cell calcifying Sertoli cell tumor
Thyroid gland	Follicular nodular disease Multifocal follicular adenomas with papillary architecture Follicular and papillary carcinoma
Soft tissue and bone	Psammomatous malignant melanotic nerve sheath tumor Osteochondromyxoma

[a]Two or more manifestations are needed for diagnosis.

In PPNAD, young individuals, more often females, are typically affected and the associated Cushing syndrome may be severe. PPNAD can be overt, subclinical, or latent, with no clinical or laboratory evidence of Cushing syndrome. The cyclic or intermittent hypersecretion of glucocorticoids that occurs in patients with PPNAD causes relapsing and remitting symptoms. Patients with PPNAD (as well as those with isolated-micronodular adrenocortical disease [i-MAD], see below) typically respond to dexamethasone with a paradoxical increase in 17-hydroxycorticosteroid and urinary free cortisol levels (so-called Liddle test), a feature that distinguishes PPNAD from Cushing syndrome caused by other primary adrenal disorders (35).

CT images with slice thickness of 5 or 10 mm obtained in patients with both PPNAD (and i-MAD) underestimate the presence of the disease and may be regarded as normal. If CT examinations are obtained with a slice thickness of 3 mm or less, before and after intravenous

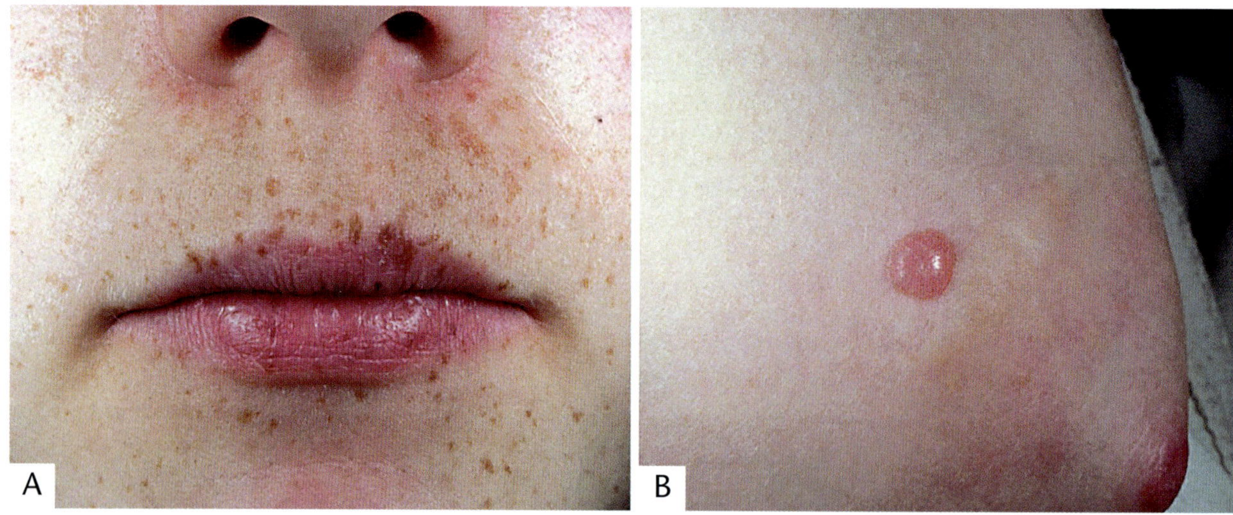

Figure 4-22

COMPLEX OF MYXOMA AND SPOTTY PIGMENTATION

A: A young girl with multiple pigmented lesions about the mouth. The patient also had a darkly pigmented lesion on the vulva, probably a blue nevus.

B: Cutaneous myxoma on the breast of the same patient, who also had a cardiac myxoma. There was no evidence of an adrenal abnormality or Cushing syndrome.

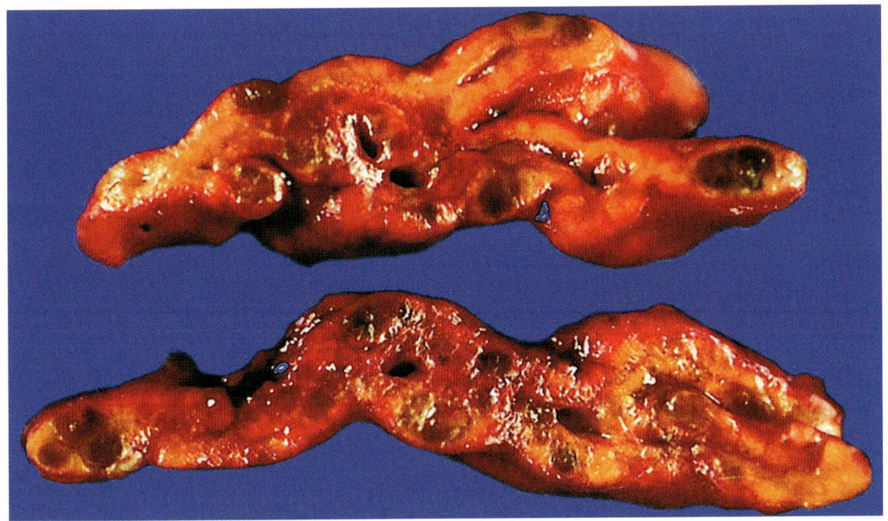

Figure 4-23

PRIMARY PIGMENTED NODULAR ADRENOCORTICAL DISEASE

There is no atrophy of the internodular cortex. The color of pigmented nodules can vary from brown to jet black.

injection of contrast material, numerous small (less than 3 mm), round, well-delineated and always hypodense lesions (as compared with the rest of the adrenal parenchyma in the precontrast and postcontrast studies) can be seen (36). However, CT and MRI findings in bilateral micronodular adrenocortical disease are variable, and may underestimate the presence of bilateral adrenal gland involvement. Half of affected patients may have abnormal imaging findings, including multiple, bilateral bead-like small nodules (33,37). Functional imaging studies using iodine-131-6-β-iodomethyl-19-norcholesterol (NP-59) scintigraphy/PET scans help to identify the bilateral nature of the disease in ACTH-independent Cushing disease with borderline or normal CT scans (38).

Familial forms of PPNAD are inherited in an autosomal-dominant manner (see chapter 7). Somatic mutations in the *CTNNB1* gene have been also identified in a subset of PPNAD patients, either co-occurring or not with germline *PRKAR1A* mutations, and associated with the development of "adenoma-like" macronodules (39).

Spontaneous regression of cortisol secretion has been reported. However, the treatment of choice for PPNAD is bilateral adrenalectomy (2).

Pathologic Findings. The adrenal surface may be apparently normal, with nodules recognizable only on sectioning the gland. Numerous nodules smaller than 1 cm are seen, usually measuring a few millimeters. There is a variable degree of pigmentation within the nodules due to the accumulation of lipofuscin. The adrenal glands may be small, normal in size, or mildly enlarged. The weights of the glands vary from 0.9 to 13.4 g, with an average combined weight of 9.6 g and a median weight of 4 g (40). Gross examination of the external surface of the intact gland may reveal scattered pigmented nodules, usually 1 to 3 mm in size, but also larger than 1 cm (as the result of the confluence of smaller micronodules), located beneath the capsule or bulging into periadrenal connective tissue. In transverse sections, there may be a wide array of small pigmented nodules studding the cortex, ranging in color from light gray, gray-brown, dark brown, to jet black. However, the degree of pigmentation of the nodules may be greatly variable (fig. 4-23). A few interspersed dominant nodules may be more yellow (fig. 4-24).

On histologic examination, the pigmented nodules are usually round to oval or have an irregular contour. Some have a silhouette that is reminiscent of an "hourglass," "string of beads," or "links of sausage." The nodules are unencapsulated, and are often deep within the cortex, centered on the zona reticularis, or straddle the corticomedullary junction with expansive borders (fig. 4-25). In contrast to i-MAD, internodular cortical atrophy is seen. Some nodules appear to occupy nearly the entire thickness of the cortex, and, occasionally, they extend into periadrenal fat without any encapsulation.

Most cells within the nodules have lipid-depleted, compact, eosinophilic cytoplasm (fig. 4-26),

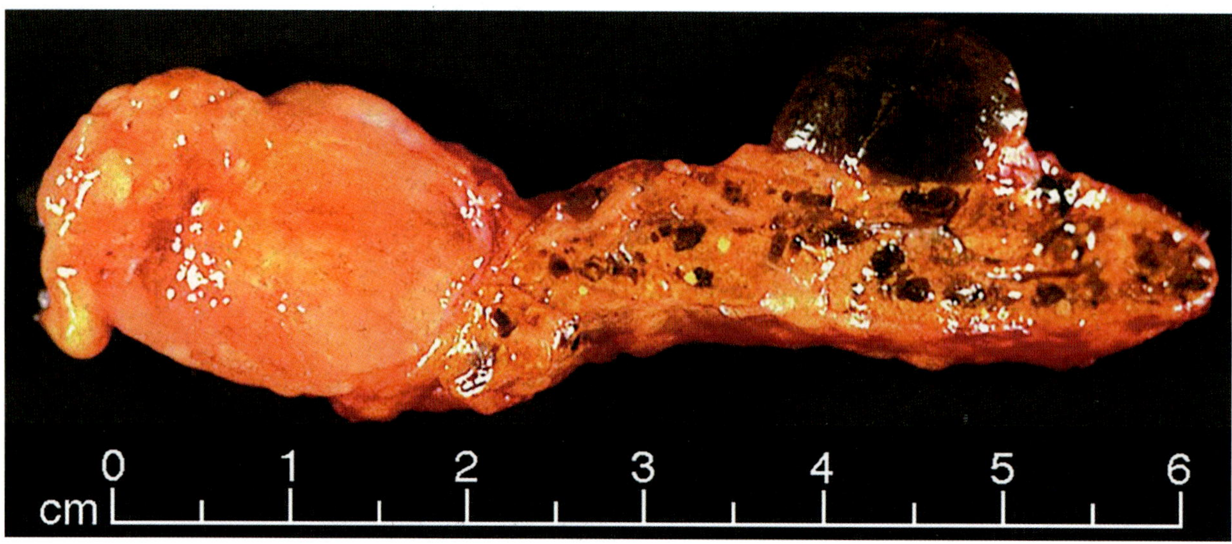

Figure 4-24

PRIMARY PIGMENTED NODULAR ADRENOCORTICAL DISEASE

A large pigmented macronodule and a larger pale yellow nodule with small areas of necrosis are seen. The patient was part of a family pedigree with the complex of myxomas, spotty pigmentation, and endocrine overactivity (Carney complex). There was a complex history of a retrogastric malignant melanotic nerve sheath tumor (formerly "melanotic schwannoma"), fibrolamellar hepatoma, hyperpigmentation of the inner canthus of both eyes, and most recently, an ovarian serous cystadenoma. The patient developed multiple pulmonary nodules, which histologically were consistent with an adrenal cortical origin (possible metastases), but paradoxically was in good health for several years. The patient apparently died of metastatic malignant melanotic nerve sheath tumor but autopsy was not performed.

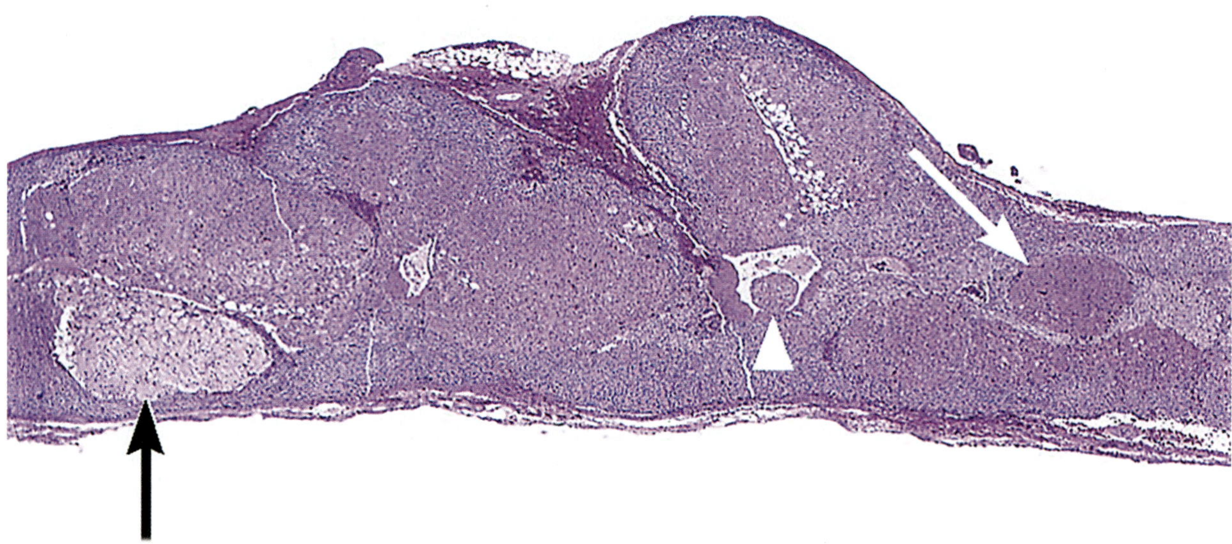

Figure 4-25

PRIMARY PIGMENTED NODULAR ADRENOCORTICAL DISEASE

Transverse section through the body of adrenal gland with numerous poorly defined micronodules. Some nodules straddle the corticomedullary junction and impinge on chromaffin cells (white arrow). A nodule is composed of lipid-rich cells (black arrow) and the small hyperplastic nodule protrudes into the vascular lumen between discontinuous bundles of smooth muscle (white arrowhead).

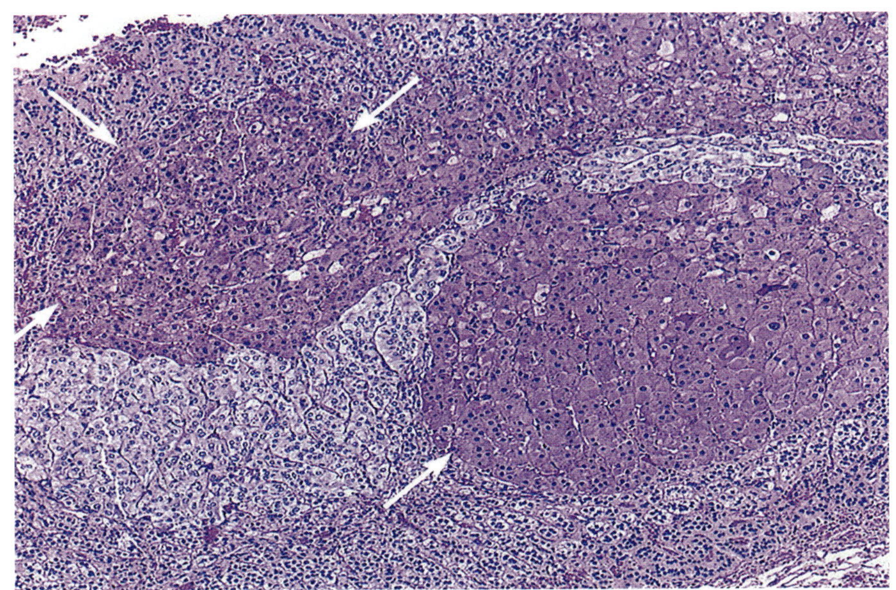

Figure 4-26

PRIMARY PIGMENTED NODULAR ADRENOCORTICAL DISEASE

Micronodules (arrows) are present at the corticomedullary junction on either side of the medulla. The nodules contain cells with compact, eosinophilic cytoplasm and focally prominent pigment.

although there may be an admixture of cells with more lipid-rich, vacuolated cytoplasm. Some nodules contain abundant lipid-rich cells that may show some "ballooning." Nuclear hyperchromasia and pleomorphism are uncommon, although occasionally cells are binucleated or multinucleated, or have enlarged hyperchromatic nuclei and prominent nucleoli.

Intraluminal projection of a nodule into tributaries of the central adrenal vein in association with discontinuities in bundles of smooth muscle may be observed but should not be interpreted as vascular invasion. Sparse lymphocytic infiltrates have been noted rarely, mainly around vessels, but also within the nodules. The intracytoplasmic pigment has the staining characteristics and ultrastructural morphology of lipofuscin. Myelolipomas are rare incidental findings in the context of PPNAD (39). Adrenal cortical adenomas (larger than 1 cm) (41,42) and adrenal cortical carcinomas (43,44) have rarely been reported in Carney complex and likely represent progression from PPNAD.

Isolated Micronodular Adrenocortical Disease

Isolated micronodular adrenocortical disease (i-MAD) is a rare subtype of bilateral micronodular adrenocortical disease that is distinguished from i-PPNAD based on the lack of pigmentation and internodular cortical atrophy. As in PPNAD, i-MAD shows variable degrees of hypercortisolism but Cushing syndrome is the most frequent clinical presentation, and exhibits the paradoxical response to dexamethasone suppression described above. I-MAD is characterized by multiple small nodules (less than 1 cm) that often range in size from 1 to 4 mm. Cortical micronodules are composed of compact (lipid-poor) adrenocortical cells, and are typically located in the zona fasciculata or in the interface of zona reticularis and zona fasciculata (45). Pigmentation and internodular atrophy are absent.

ADRENAL CORTICAL NODULAR LESIONS ASSOCIATED WITH PRIMARY ALDOSTERONISM

Idiopathic bilateral hyperplasia/nodular disease of adrenal cortical zona glomerulosa is the most frequent cause of primary aldosteronism, which is estimated to have a prevalence of about 10 percent in hypertensive individuals (46). Bilateral adrenal cortical nodular disease accounts for 60 to 70 percent of cases, with adrenal cortical adenoma responsible for almost all remaining cases (47). Recently, a consensus for the nomenclature and histopathology of adrenal cortical lesions in unilateral primary aldosteronism (HISTALDO classification) has been proposed (48). The scheme includes aldosterone-producing adrenal cortical adenoma, aldosterone-producing adrenal cortical carcinoma, aldosterone-producing nodule, aldosterone-producing micronodule (formerly termed aldosterone-producing cell cluster),

Table 4-3

TYPES OF ADRENAL CORTICAL LESIONS IN UNILATERAL PRIMARY ALDOSTERONISM[a]

Type	Macroscopy	Histology	CYP11B2 IHC[b]
Aldosterone-producing diffuse hyperplasia	Not recognizable	Rarely recognizable	Broad and uninterrupted zona glomerulosa
Aldosterone-producing micronodule	Not recognizable	Undetectable (recognizable on IHC, only); outer margin of the subcapsular zona glomerulosa	Positive with variable intensity[c]
Aldosterone-producing nodule	Visible	Recognizable on H&E, <10 mm, compact or clear or mixed cell histology; no specific location within the adrenal cortex	Positive with variable intensity[c]
Multiple aldosterone-producing nodules and micronodules	Variable, multiple nodules or macronodules (sometimes mixed)	Variable, multiple nodules or macronodules (sometimes mixed)	Positive with variable intensity[c]

[a]Adrenal cortical adenoma and carcinoma excluded.
[b]IHC = immunohistochemistry; H&E = hematoxylin and eosin.
[c]A polarized pattern with a strongest intensity at the periphery.

multiple aldosterone-producing nodules and micronodules, and aldosterone-producing diffuse hyperplasia (Table 4-3). This classification has some practical relevance because it may reflect different outcomes after unilateral adrenalectomy (49).

Non-classic primary aldosteronism (diffuse hyperplasia alone or concurrent with aldosterone-producing adenoma, or multiple aldosterone-producing nodules/micronodules) is associated with longer duration of hypertension, smaller mean nodule size, lower lateralization index (a ratio indicating the level of asymmetry of aldosterone production from the adrenal glands), higher aldosterone-to-renin ratio after surgery, and higher serum potassium concentrations compared to classic histology (classic aldosterone-producing adrenal cortical adenoma or single aldosterone-producing nodule). Moreover, rates of complete biochemical success are 81.9 percent in classic and 33.3 percent in non-classic histology groups (48).

The HISTALDO classification integrates macroscopy, histology, and immunohistochemical findings using stains for CYP11B2. This classification scheme also highlights the need for appropriate sampling. The whole adrenalectomy sample should be submitted for histopathologic examination, with close transverse sectioning to avoid misinterpretation of the size and location of the lesions.

Adrenal Cortical Lesions in Unilateral Primary Aldosteronism (HISTALDO Classification)

Apart from adrenal cortical adenoma (fig. 4-27) (chapter 5) and carcinoma (chapter 6), unilateral primary aldosteronism is divided into the following entities.

Aldosterone-Producing Nodule (APN). APN is an aldosterone-producing lesion of less than 1 cm composed of clear, compact, or mixed cell histology that is similar to an aldosterone-producing adenoma except for the size of the lesion. CYP11B2 polarized expression with decreasing intensity from the outer to the inner portion of the lesions is typical of APN and contrasts with the homogeneous or heterogeneous but not polarized pattern observed in an aldosterone-producing adenoma (50).

Aldosterone-Producing Micronodule (APM). APM is a lesion of less than 1 cm, unrecognizable on hematoxylin and eosin (H&E) staining but visible on CYP11B2 immunohistochemistry only, and composed by cells in the zona glomerulosa packed beneath the adrenal capsule. At variance with aldosterone-producing adrenal cortical adenoma and similar to APN, CYP11B2 immunohistochemistry shows the typical polarized expression. Enzymes involved in cortisol (CYP11B1) and sex steroid hormones (CYP17A1) production are negative. The number of APMs with features of autonomous and nonphysiologic

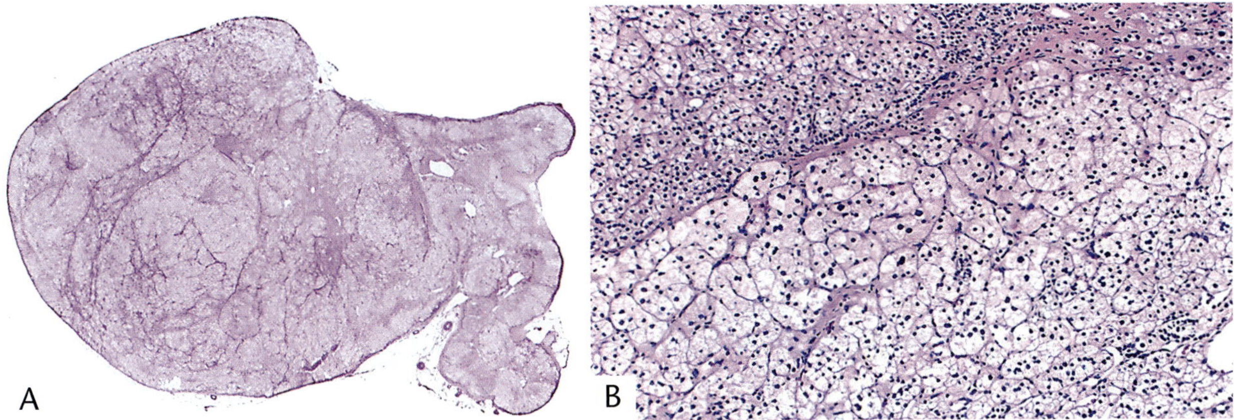

Figure 4-27

ALDOSTERONE-SECRETING ADRENAL CORTICAL ADENOMA

A: The tumor has expansile smooth borders that compress adjacent adrenal parenchyma and expand the adrenal capsule.
B: The borders of the adenoma are sharply demarcated from the adjacent cortex (left upper portion of field), although there is no well-developed fibrous capsule.

aldosterone secretion is increased in the elderly, thus possibly representing senescent changes of the adrenal cortex (51). APMs were first described in patients affected by primary aldosteronism but with negative imaging (52). In the clinical diagnostic setting, knowledge of their functionality is of relevance since they may influence the results of adrenal vein sampling (53). Approximately 35 percent of APMs harbor somatic mutations observed in aldosterone-secreting adenomas, mainly in *CACNA1D* and *ATP1A1*, and more rarely in *KCNJ5* (54,55).

Multiple Aldosterone-Producing Nodules or Micronodules. These two lesions are characterized by the identification of aldosterone-producing nodules or micronodules within an otherwise histologically normal zona glomerulosa (fig. 4-28). Nodules and micronodules may be detected by CYP11B2 immunohistochemical staining (fig. 4-29).

Aldosterone-Producing Diffuse Hyperplasia (APDH). At variance with APM, APDH is recognized by the presence of a broad band of hyperplastic zona glomerulosa cells (fig. 4-30) that are highlighted in a nonpolarized, uninterrupted "strip-like pattern" with CYP11B2 immunohistochemistry. APDH may be isolated or associated with the presence of aldosterone-producing nodules or micronodules in the same adrenal gland.

Bilateral Adrenal Cortical Nodular Disease in Primary Aldosteronism

Bilateral aldosterone excess is most commonly the result of aldosterone-producing adrenal cortical nodular disease (56), which requires lifelong antimineralocorticoid therapy, rather than surgery. Familial forms of bilateral adrenal disease associated with primary aldosteronism are detailed in chapter 7. At variance with aldosterone-secreting adenomas, the genetic causes of *sporadic bilateral adrenal nodular disease in primary aldosteronism* remain largely unknown. Analysis of adrenal glands from patients with bilateral adrenal nodular disease has shown that it may result from the accumulation or enlargement of aldosterone-producing micronodules harboring somatic mutations, particularly in the *CACNA1D* gene (57). In a recent study, somatic *CACNA1D* mutations were the most frequent genetic abnormality in aldosterone-producing adrenal cortical adenomas from patients with partial or absent success after adrenalectomy, suggesting common pathogenic mechanisms as the origin of aldosterone-producing adenomas with poor biochemical outcome and bilateral adrenal hyperplasia (58).

The histologic findings are similar to those occurring in association with cortisol oversecretion, but the glands are seldom seen as a surgical specimen. However, the functional status may sometimes be determined by examination of the

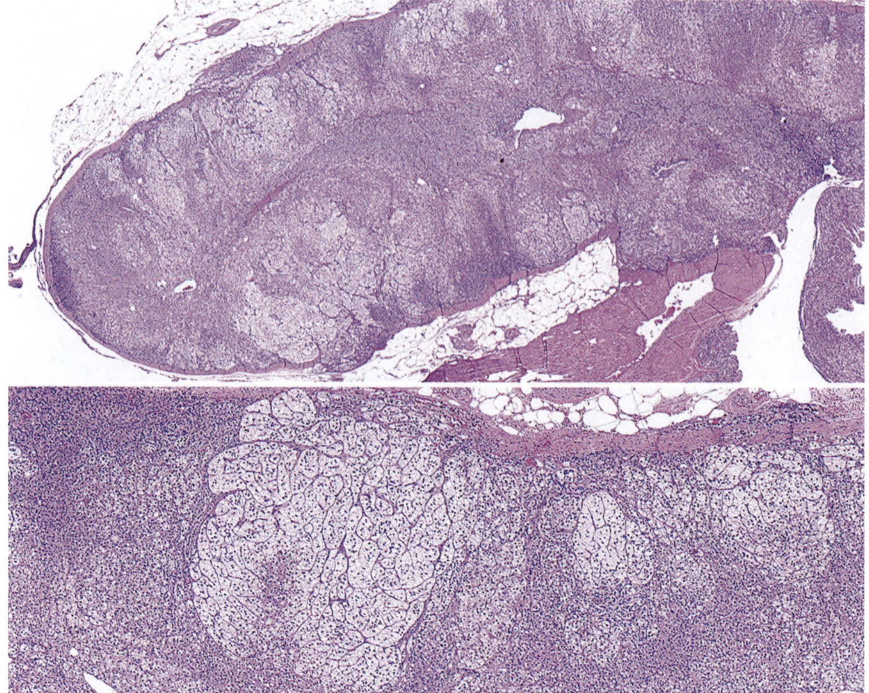

Figure 4-28

MULTIPLE ALDOSTERONE-PRODUCING NODULES

Right adrenal gland of 54-year-old man with Conn syndrome. Maximum size of nodules is 4 mm.

Figure 4-29

MULTIPLE ALDOSTERONE-PRODUCING MICRONODULES

CYP11B12 immunohistochemistry shows the presence of histologically unrecognizable micronodules of aldosterone-producing cortical cells.

internodular ipsilateral or contralateral cortex. The presence of cortical atrophy may indicate hypercortisolism while hyperplasia of the zona glomerulosa may indicate aldosterone secretion. Moreover, in primary aldosteronism the nodules are usually composed of predominantly of lipid-rich cells. In patients treated with the aldosterone antagonist spironolactone, small intracytoplasmic phospholipid-containing whorls, called "spironolactone bodies," are often found in aldosterone-producing cells and have historically been used as an aid to diagnosis (fig. 4-31). They are not seen in patients treated with the newer drug eplerenone (59).

DIFFERENTIAL DIAGNOSIS OF ADRENAL CORTICAL NODULAR DISEASES

In adrenal glands from autopsy material and in tissue obtained at surgery the presence of a cortical nodule can be a diagnostic challenge for the pathologist. The algorithm for the differential diagnosis among the different forms of adrenal cortical nodular diseases or with other

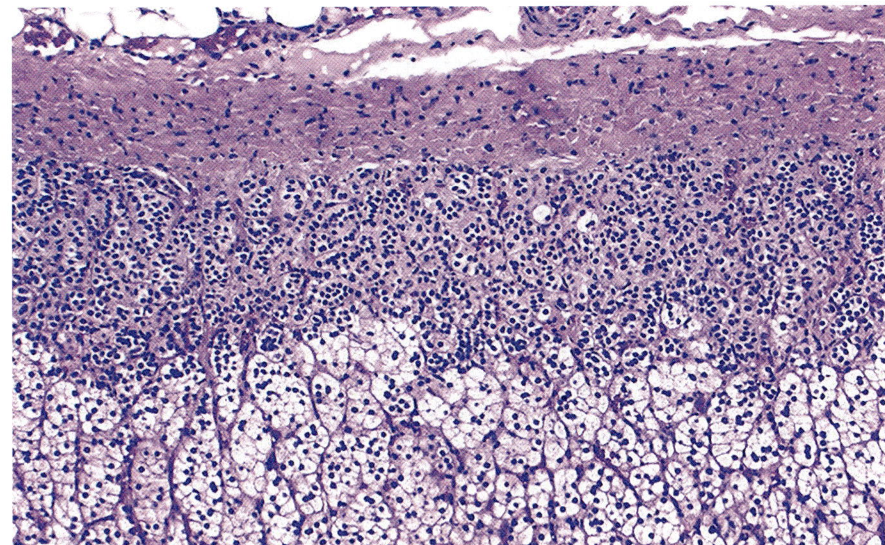

Figure 4-30

DIFFUSE ALDOSTERONE-PRODUCING HYPERPLASIA

A continuous zone of hyperplastic zona glomerulosa is present in non-neoplastic adrenal gland.

proliferative adrenal cortical lesions is based on the combination of macroscopic and histologic features (Table 4-4). This needs to be integrated in a complete clinical and functional characterization, to avoid the risk of misinterpretation of the pathologic findings with relevant consequences for the therapeutic management of the patients and their allocation into risk groups of hereditary diseases.

Only a few diseases have pathologic features specific enough to permit an unequivocal diagnosis. However, if considered in an appropriate clinical context, the majority of situations do not need ancillary tools apart from a careful histologic evaluation. Cortical origin of the lesions is easily assessed by morphology, and is proven if needed by the expression of cortical markers (SF1 and Melan-A among others; see fig. 4-10). These lesions do not generally suggest malignancy, although extracapsular extrusions may be concerning. However, silver staining clearly shows a preserved reticulin pattern (fig. 4-32) and the Ki-67 proliferation index is always low (fig. 4-33). An exception is aldosterone-producing unilateral lesions whose classification may require immunohistochemical analysis for steroidogenic enzymes, specifically CYP11B2. Other enzymes that also play pivotal roles in the process of aldosterone biosynthesis, such as HSD3B1 and HSD3B2, have been shown to be of potential usefulness (60–62), but may have lower success in diagnostic practice.

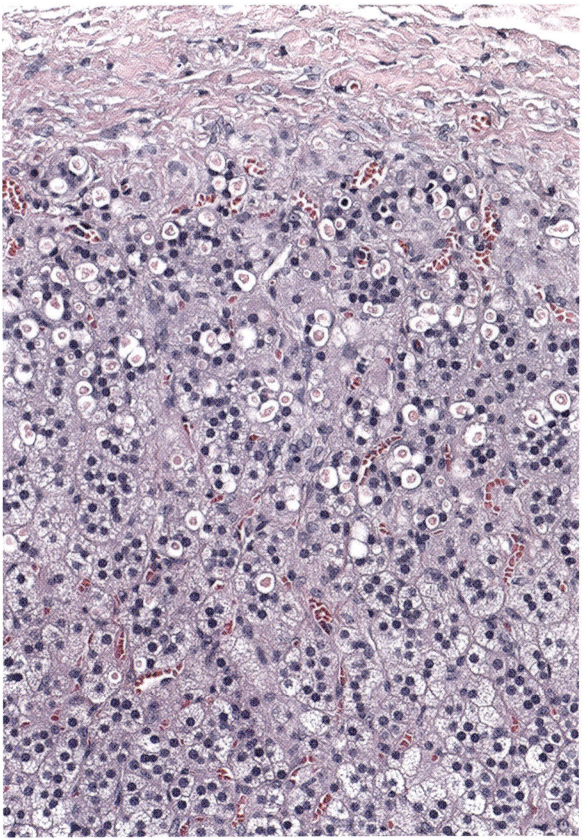

Figure 4-31

SPIRONOLACTONE BODIES

Numerous spironolactone bodies are seen in the cytoplasm of zona glomerulosa cells in this adrenal gland from a patient treated with spironolactone for primary hyperaldosteronism.

Table 4-4

PATHOLOGIC CLUES FOR THE DIFFERENTIAL DIAGNOSIS OF ADRENOCORTICAL NODULAR DISEASES

Parameter	Diagnostic Setting
Size <1 cm	In nonfunctioning lesions, size distinguishes sporadic nodular adrenocortical disease from adenoma In cortisol-secreting lesions, size distinguishes primary bilateral micronodular from macronodular disease In unilateral aldosterone-secreting lesions, size distinguishes aldosterone-producing nodule from aldosterone-producing adenoma
Multifocality	In general, it suggests adrenocortical nodular disease rather than adenoma In unilateral aldosterone-secreting lesions, it distinguishes multiple from single aldosterone-producing nodules or micronodules
Macroscopically evident nodular pattern	It is recognizable in sporadic nodular adrenocortical disease, primary bilateral macronodular adrenal disease associated with cortisol secretion, and single or multiple aldosterone-producing nodules
Encapsulation of a dominant nodule	In every functional setting, it is suggestive of adenoma
Pericapsular or extra-capsular nodules	It is suggestive of cortisol-producing lesions
Internodular cortical atrophy	In macronodular disease, it is suggestive of cortisol-producing lesions In primary micronodular disease, it distinguishes PPNAD[a] from i-MAD
Pigmentation	In primary micronodular disease, it distinguishes PPNAD from i-MAD
CYP11B2 polarized expression	In morphologically recognizable aldosterone-secreting lesions <1 cm, it is typical of aldosterone-producing nodule In morphologically unrecognizable aldosterone-secreting lesions, it distinguishes aldosterone-producing micronodule from aldosterone-producing diffuse hyperplasia

[a]PPNAD = primary pigmented nodular adrenocortical disease; i-MAD = isolated micronodular adrenocortical disease.

Irrespective of the functional situation, a dominant adrenal cortical nodule is impossible to distinguish from an adrenal cortical adenoma histologically. Cellular types and architectures may be heterogeneous in both situations.

The concept of macronodular and micronodular disease is different in cortisol-secreting and aldosterone- secreting groups of lesions. In the former, distinction between the two forms is based on a size cut-off of 1 cm, whereas in the latter, the concept is related to the morphologic recognition or the lack of morphologic evidence of the lesions. Multifocality and bilaterality per se may be equivocal. In the presence of a dominant nodule, separate associated lesions may be unnoticed (e.g., in aldosterone-producing lesions) and bilaterality may not be recognized clinically or histologically due to the presence of wide asymmetry or metachronous onset.

Some features may be suggestive of particular functional situations. The presence of overgrowth of cortical cells in isolated clusters or nodules at the periphery or outside the adrenal capsule is more common in cortisol-secreting lesions. The presence of pigmentation (isolated or in the context of PPNAD) and internodular cortical atrophy are hallmarks of cortisol oversecretion in ACTH-independent lesions.

Much of the confusion about the hyperplastic/nodular adrenal gland results from a combination of factors: vague and heterogeneous approaches to terminology, continuing difficulty in distinguishing an adrenal cortical neoplasm from a dominant macronodule, overlap in morphology of adrenal cortical lesions in patients with different endocrine syndromes, absence of correlation between pathology and clinical/endocrinologic data, and still incomplete understanding of the etiology and pathogenesis of the different forms of adrenal nodular lesions. However, pathologists should be aware of the different entities under the umbrella of the overall term, adrenal cortical

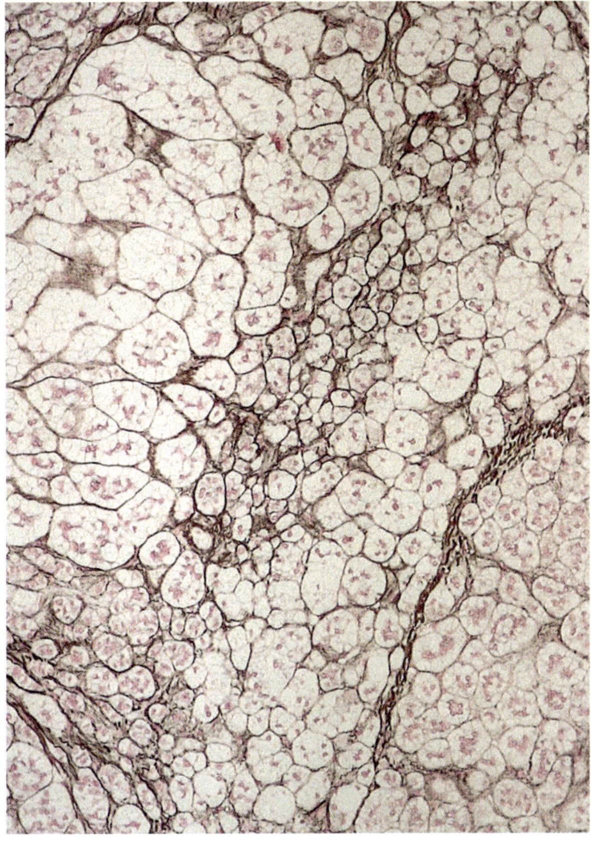

Figure 4-32

PRIMARY BILATERAL MACRONODULAR ADRENOCORTICAL DISEASE ASSOCIATED WITH CORTISOL OVERSECRETION

Preserved reticulin pattern, as highlighted by silver staining.

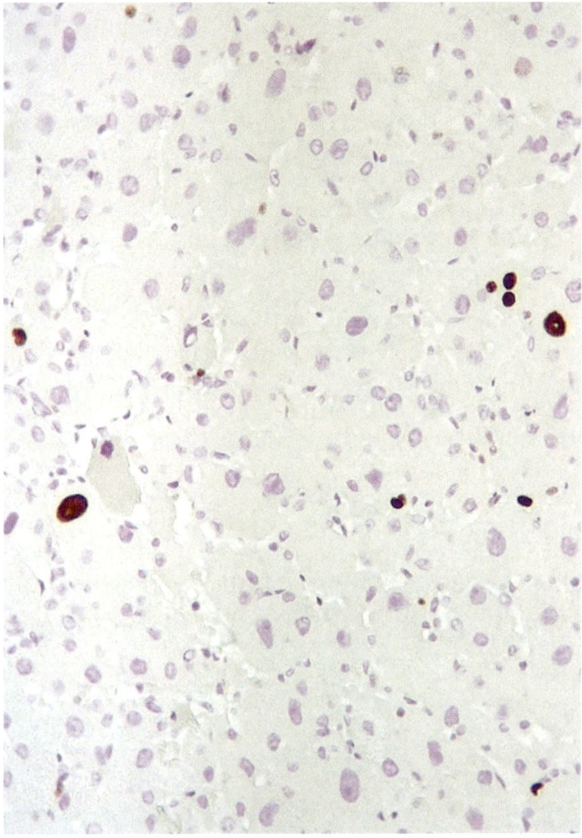

Figure 4-33

PRIMARY BILATERAL MACRONODULAR ADRENOCORTICAL DISEASE ASSOCIATED WITH CORTISOL OVERSECRETION

Low Ki-67 index (hot spot area).

nodular disease, to address the patient and the clinician to the most appropriate approach for treatment and follow-up. Despite the great advance in the understanding of the molecular background of these conditions, little is still known in a great proportion of sporadic cases, and the real proportion of familial forms is probably still underestimated.

REFERENCES

1. Mete O, Erickson LA, Juhlin CC, et al. Overview of the 2022 WHO Classification of Adrenal Cortical Tumors. Endocr Pathol 2022;33:155-96.
2. Bourdeau I, Parisien-La Salle S, Lacroix A. Adrenocortical hyperplasia: a multifaceted disease. Best Pract Res Clin Endocrinol Metab 2020;34: 101386.
3. LeBlanc M, Tabrizi M, Kapsner P, Hanson JA. Synchronous adrenocortical neoplasms, paragangliomas, and pheochromocytomas: syndromic considerations regarding an unusual constellation of endocrine tumors. Hum Pathol 2014;45:2502-6.
4. Carney JA. Carney triad. Front Horm Res 2013; 41:92-110.
5. Carney JA, Stratakis CA, Young WF Jr. Adrenal cortical adenoma: the fourth component of the Carney triad and an association with subclinical Cushing syndrome. Am J Surg Pathol 2013;37: 1140-9.
6. Hodgson A, Pakbaz S, Mete O. A diagnostic approach to adrenocortical tumors. Surg Pathol Clin 2019;12:967-95.
7. Muth A, Hammarstedt L, Hellström M, et al. Cohort study of patients with adrenal lesions discovered incidentally. Br J Surg 2011;98:1383-91.
8. Bovio S, Cataldi A, Reimondo G, et al. Prevalence of adrenal incidentaloma in a contemporary computerized tomography series. J Endocrinol Invest 2006;29:298-302.
9. Hammarstedt L, Muth A, Wängberg B, et al. Adrenal lesion frequency: a prospective, cross-sectional CT study in a defined region, including systematic re-evaluation. Acta Radiol 2010;51: 1149-56.
10. Yousaf A, Patterson J, Hobbs G, et al. Smoking is associated with adrenal adenomas and adrenocortical carcinomas: a nationwide multicenter analysis. Cancer Treat Res Commun 2020;25: 100206.
11. Dobbie JW. Adrenocortical nodular hyperplasia: the ageing adrenal. J Pathol 1969;99:1-18.
12. van Nederveen FH, de Krijger RR. Precursor lesions of the adrenal gland. Pathobiology 2007; 74:285-90.
13. Mete O, Asa SL. Precursor lesions of endocrine system neoplasms. Pathology 2013;45:316-30.
14. Juhlin CC, Bertherat J, Giordano TJ, Hammer GD, Sasano H, Mete O. What did we learn from the molecular biology of adrenal cortical neoplasia? From histopathology to translational genomics. Endocr Pathol 2021;32:102-33.
15. Díaz-Cano SJ, de Miguel M, Blanes A, Tashjian R, Galera H, Wolfe HJ. Clonality as expression of distinctive cell kinetics patterns in nodular hyperplasias and adenomas of the adrenal cortex. Am J Pathol 2000;156:311-9.
16. Duan K, Gomez Hernandez K, Mete O. Clinicopathological correlates of adrenal Cushing's syndrome. J Clin Pathol 2015;68:175-86.
17. Espiard S, Drougat L, Libé R, et al. ARMC5 mutations in a large cohort of primary macronodular adrenal hyperplasia: clinical and functional consequences. J Clin Endocrinol Metab 2015; 100:E926-35.
18. Kirk JM, Brain CE, Carson DJ, Hyde JC, Grant DB. Cushing's syndrome caused by nodular adrenal hyperplasia in children with McCune-Albright syndrome. J Pediatr 1999;134:789-92.
19. Suzuki S, Tatsuno I, Oohara E, et al. Germline deletion of ARMC5 in familial primary macronodular adrenal hyperplasia. Endocr Pract 2015; 21:1152-60.
20. Berthon A, Hannah-Shmouni F, Maria AG, Faucz FR, Stratakis CA. High expression of adrenal P450 aromatase (CYP19A1) in association with ARMC5-primary bilateral macronodular adrenocortical hyperplasia. J Steroid Biochem Mol Biol 2019;191:105316.
21. Morani AC, Jensen CT, Habra MA, et al. Adrenocortical hyperplasia: a review of clinical presentation and imaging. Abdom Radiol (NY) 2020;45:917-27.
22. Doppman JL, Chrousos GP, Papanicolaou DA, Stratakis CA, Alexander HR, Nieman LK. Adrenocorticotropin-independent macronodular adrenal hyperplasia: an uncommon cause of primary adrenal hypercortisolism. Radiology 2000;216:797-802.
23. Albiger NM, Regazzo D, Rubin B, et al. A multicenter experience on the prevalence of ARMC5 mutations in patients with primary bilateral macronodular adrenal hyperplasia: from genetic characterization to clinical phenotype. Endocrine 2017;55:959-68.
24. Fragoso MC, Domenice S, Latronico AC, et al. Cushing's syndrome secondary to adrenocorticotropin-independent macronodular adrenocortical hyperplasia due to activating mutations of GNAS1 gene. J Clin Endocrinol Metab 2003;88: 2147-51.
25. Cao Y, He M, Gao Z, et al. Activating hotspot L205R mutation in PRKACA and adrenal Cushing's syndrome. Science 2014;344:913-7.

26. Assie G, Louiset E, Sturm N, et al. Systematic analysis of G protein-coupled receptor gene expression in adrenocorticotropin-independent macronodular adrenocortical hyperplasia identifies novel targets for pharmacological control of adrenal Cushing's syndrome. J Clin Endocrinol Metab 2010;95:E253-62.
27. Chui MH, Ozbey NC, Ezzat S, Kapran Y, Erbil Y, Asa SL. Case report: adrenal LH/hCG receptor overexpression and gene amplification causing pregnancy-induced Cushing's syndrome. Endocr Pathol 2009;20:256-61.
28. Louiset E, Duparc C, Young J, et al. Intraadrenal corticotropin in bilateral macronodular adrenal hyperplasia. N Engl J Med 2013;369:2115-25.
29. Carney JA, Young WF, Stratakis CA. Primary bimorphic adrenocortical disease: cause of hypercortisolism in McCune-Albright syndrome. Am J Surg Pathol 2011;35:1311-26.
30. Violon F, Bouys L, Berthon A, et al. Impact of morphology in the genotype and phenotype correlation of bilateral macronodular adrenocortical disease (BMAD): a series of clinicopathologically well-characterized 35 cases. Endocr Pathol 2023;34:179-99.
31. de Arruda Botelho ML, Nishi MY, Ribeiro KB, Zerbini MC. Morphological harbingers of ARMC5-pathogenic variant-related bilateral macronodular adrenocortical disease. Endocr Pathol 2023;34:200-12.
32. Carney JA, Gordon H, Carpenter PC, Shenoy BV, Go VL. The complex of myxomas, spotty pigmentation, and endocrine overactivity. Medicine (Baltimore) 1985;64:270-83.
33. Espiard S, Vantyghem MC, Assié G, et al. Frequency and incidence of Carney complex manifestations: a prospective multicenter study with a three-year follow-up. J Clin Endocrinol Metab 2020;105:e436-46.
34. Stratakis CA, Kirschner LS, Carney JA. Clinical and molecular features of the Carney complex: diagnostic criteria and recommendations for patient evaluation. J Clin Endocrinol Metab 2001; 86:4041-6.
35. Louiset E, Stratakis CA, Perraudin V, et al. The paradoxical increase in cortisol secretion induced by dexamethasone in primary pigmented nodular adrenocortical disease involves a glucocorticoid receptor-mediated effect of dexamethasone on protein kinase A catalytic subunits. J Clin Endocrinol Metab 2009;94:2406-13.
36. Courcoutsakis NA, Tatsi C, Patronas NJ, Lee CC, Prassopoulos PK, Stratakis CA. The complex of myxomas, spotty skin pigmentation and endocrine overactivity (Carney complex): imaging findings with clinical and pathological correlation. Insights Imaging 2013;4:119-33.
37. Rockall AG, Babar SA, Sohaib SA, et al. CT and MR imaging of the adrenal glands in ACTH-independent cushing syndrome. Radiographics 2004;24:435-52.
38. Vezzosi D, Tenenbaum F, Cazabat L, et al. Hormonal, radiological, NP-59 scintigraphy, and pathological correlations in patients with cushing's syndrome due to primary pigmented nodular adrenocortical disease (PPNAD). J Clin Endocrinol Metab 2015;100:4332-8.
39. Tadjine M, Lampron A, Ouadi L, Horvath A, Stratakis CA, Bourdeau I. Detection of somatic beta-catenin mutations in primary pigmented nodular adrenocortical disease (PPNAD). Clin Endocrinol (Oxf) 2008;69:367-73.
40. Shenoy BV, Carpenter PC, Carney JA. Bilateral primary pigmented nodular adrenocortical disease. Rare cause of the Cushing syndrome. Am J Surg Pathol 1984;8:335-44.
41. Lowe KM, Young WF Jr, Lyssikatos C, Stratakis CA, Carney JA. Cushing syndrome in Carney complex: clinical, pathologic, and molecular genetic findings in the 17 affected Mayo Clinic patients. Am J Surg Pathol 2017;41:171-81.
42. Yoshida M, Sasano H, Kikumori T, et al. A case of subclinical Cushing syndrome due to primary pigmented nodular adrenocortical disease associated with adrenocortical adenoma. Endocrine 2011;40:144-5.
43. Morin E, Mete O, Wasserman JD, Joshua AM, Asa SL, Ezzat S. Carney complex with adrenal cortical carcinoma. J Clin Endocrinol Metab 2012;97:E202-6.
44. Anselmo J, Medeiros S, Carneiro V, et al. A large family with Carney complex caused by the S147G PRKAR1A mutation shows a unique spectrum of disease including adrenocortical cancer. J Clin Endocrinol Metab 2012;97:351-9.
45. Mete O, Duan K. The many faces of primary aldosteronism and cushing syndrome: a reflection of adrenocortical tumor heterogeneity. Front Med (Lausanne) 2018;5:54.
46. Duan K, Mete O. Clinicopathologic correlates of primary aldosteronism. Arch Pathol Lab Med 2015;139:948-4.
47. Reincke M, Bancos I, Mulatero P, Scholl UI, Stowasser M, Williams TA. Diagnosis and treatment of primary aldosteronism. Lancet Diabetes Endocrinol 2021;9:876-92.
48. Williams TA, Gomez-Sanchez CE, Rainey WE, et al. International histopathology consensus for unilateral primary aldosteronism. J Clin Endocrinol Metab 2021;106:42-54.
49. Nishimoto K, Umakoshi H, Seki T, et al. Diverse pathological lesions of primary aldosteronism and their clinical significance. Hypertens Res 2021;44:498-507.

50. Nishimoto K, Nakagawa K, Li D, et al. Adrenocortical zonation in humans under normal and pathological conditions. J Clin Endocrinol Metab 2010;95:2296-305.
51. Nanba K, Vaidya A, Williams GH, Zheng I, Else T, Rainey WE. Age-related autonomous aldosteronism. Circulation 2017;136:347-55.
52. Yamazaki Y, Nakamura Y, Omata K, et al. Histopathological classification of cross-sectional image-negative hyperaldosteronism. J Clin Endocrinol Metab 2017;102:1182-92.
53. Kometani M, Yoneda T, Aono D, et al. Impact of aldosterone-producing cell clusters on diagnostic discrepancies in primary aldosteronism. Oncotarget 2018;9:26007-18.
54. Omata K, Tomlins SA, Rainey WE. Aldosterone-producing cell clusters in normal and pathological states. Horm Metab Res 2017;49:951-6.
55. Omata K, Anand SK, Hovelson DH, et al. Aldosterone-producing cell clusters frequently harbor somatic mutations and accumulate with age in normal adrenals. J Endocr Soc 2017;1:787-99.
56. Zennaro MC, Fernandes-Rosa F, Boulkroun S, Jeunemaitre X. Bilateral idiopathic adrenal hyperplasia: genetics and beyond. Horm Metab Res 2015;47:947-52.
57. Omata K, Satoh F, Morimoto R, et al. Cellular and genetic causes of idiopathic hyperaldosteronism. Hypertension 2018;72:874-80.
58. Hacini I, De Sousa K, Boulkroun S, et al. Somatic mutations in adrenals from patients with primary aldosteronism not cured after adrenalectomy suggest common pathogenic mechanisms between unilateral and bilateral disease. Eur J Endocrinol 2021;185:405-12.
59. Patel KA, Calomeni EP, Nadasdy T, Zynger DL. Adrenal gland inclusions in patients treated with aldosterone antagonists (Spironolactone/Eplerenone): incidence, morphology, and ultrastructural findings. Diagn Pathol 2014;9:147.
60. Doi M, Satoh F, Maekawa T, et al. Isoform-specific monoclonal antibodies against 3β-hydroxysteroid dehydrogenase/isomerase family provide markers for subclassification of human primary aldosteronism. J Clin Endocrinol Metab 2014;99:E257-62.
61. Konosu-Fukaya S, Nakamura Y, Satoh F, et al. 3b-hydroxysteroid dehydrogenase isoforms in human aldosterone-producing adenoma. Mol Cell Endocrinol 2015;408:205-12.
62. Mete O, Asa SL, Giordano TJ, Papotti M, Sasano H, Volante M. Immunohistochemical biomarkers of adrenal cortical neoplasms. Endocr Pathol 2018;29:137-49.

5 ADRENAL CORTICAL ADENOMA

Adrenal cortical adenoma is a benign neoplasm composed of adrenal cortical steroid hormone-producing cells.

EPIDEMIOLOGY

Adrenal cortical adenomas are not rare, however, the calculation of their real incidence is biased by the use of the term "nodular hyperplasia" for the entity now called cortical nodular disease, which is essentially multiple adenomas. In addition, not all cases undergo surgery, thus pathologic confirmation is not always available.

Solitary adenomas occur at any age and in both sexes. Reports of large autopsy series have found adrenal adenomas greater than 2 to 5 mm in 1.5 to 5.7 percent of the population, and the incidence appears to increase with age (1). In the last decades, an increase in their incidence has been reported; this is attributed to the increasing use of computerized tomography (CT) or ultrasound (US) abdominal imaging (2). Adrenal cortical adenomas, with special reference to nonfunctioning adenomas, account for most adrenal "incidentalomas." A large retrospective surgical series, however, showed that when an appropriate endocrinologic workup is performed, up to 50 percent of these so-called incidental findings are recognized as functioning tumors.

The risk of malignancy associated with an adrenal cortical adenoma is approximately 10 percent, with a significant association with size (3). Patients with incidentally discovered lesions that were thought to be nonfunctioning develop subtle or overt endocrine hyperfunction during follow-up in up to 7 percent of cases, most frequently hypercortisolism, and significant mass enlargement occurs in 17.7 percent (4,5). Therefore, it is suggested that the implementation of a thorough preoperative workup within a multidisciplinary adrenal tumor board is desirable to reduce unnecessary surgical treatment of benign nonfunctioning adenomas while maintaining adequate follow-up. The most recent guidelines recommend that surgery is indicated in cases with a tumor size of 4 cm or larger with unenhanced HU (Hounsfield unit) 11-20, tumor size less than 4 cm with unenhanced HU over 20, or tumor size less than 4 cm with heterogeneous appearance. In cases not fulfilling these criteria, an individualized approach with discussion in a multidisciplinary team meeting is recommended, with additional imaging approaches according to the center expertise and availability (6).

ETIOLOGY

The hereditary context of adrenal cortical adenoma is discussed separately in chapter 7. No specific etiologic factors, including environmental and other predisposing conditions, are recognized in sporadic cases, although recently a population-based study identified an increased risk to develop adrenal cortical adenoma (and carcinoma) in smokers (7).

CLINICAL FEATURES

Adrenal cortical adenoma is usually solitary and unilateral. There is no side predilection. However, in a cross-sectional study of more than 1,300 cases, adrenal cortical adenomas were more likely detected in the left adrenal gland, possibly as the result of the more difficult imaging detection of lesions in the right adrenal if not of a significant size (8).

Adrenal cortical adenomas are classified based on hormonal status into clinical subtypes: aldosterone-producing adenomas, cortisol-producing adenomas, sex hormone-producing adenomas, and nonfunctioning adenomas (Table 5-1). Hypersecretion of adrenal steroids manifests as: a fully developed endocrine syndrome, i.e., primary hyperaldosteronism or Cushing syndrome;

Table 5-1

MAIN DIFFERENTIAL DIAGNOSTIC CHARACTERISTICS AMONG CLINICAL SUBTYPES OF ADRENAL CORTICAL ADENOMAS

	Aldosterone-Producing Adenoma	Cortisol-Producing Adenoma	Sex Hormone-Producing Adenoma	Nonfunctioning Adenoma
Clinical presentation	Hypertension (+/- hypokalemia), palpitation, polydipsia, polyuria, cramping, headaches, weakness	Weight gain, central obesity, facial roundness, hirsutism, easy bruising, poor wound healing, red to purple skin striae, weakness, hyperglycemia, hypertension, osteoporosis, immunodeficiency	Virilization or feminization	Mainly incidental
Age	4th-5th decade	4th-5th decade	Pediatric	5th-6th decade
M/F	Slight male predominance	Female predominance	Female predominance	Equal
Potential precursor lesions	Aldosterone-producing cell clusters	Unknown	Unknown	Unknown
Size	Usually <2 cm	3-4 cm	Usually <3 cm (mostly case reports, only)	3-4 cm
Peculiar histopathologic findings	More frequent pigmentation (and black adenoma variant) and cytologic atypia	Predominant clear cell type, spironolactone bodies (in treated patients), rare oncocytic variant	Predominant compact cell type, Reinke crystalloids in androgen-secreting tumors	None
Adjacent adrenal tissue	Normal or presence of small cortical nodules	Cortical atrophy	Normal	Normal
Somatic molecular genetics	*KCNJ5* (most frequent), *CACNA1D, ATP1A1, ATP2B3;* rare: *PRKACA, CTNNB1, CLCN2;* about 50% with unknown molecular alterations	*PRKACA* (most frequent), *PRKACB, PRKAR1A, PRKAR1B, GNAS1, CTNNB1;* about 50% with unknown molecular alterations	Unknown molecular alterations	*CTNNB1* (most frequent); about 50% with unknown molecular alterations

abnormal laboratory data; or cortical atrophy of the ipsilateral or contralateral adrenal cortex.

Aldosterone-Producing Adenoma

Aldosterone secretion by an adrenal cortical adenoma is the main cause of *primary hyperaldosteronism*, also called *Conn syndrome*. The key sign of hyperaldosteronism is hypertension, with or without hypokalemia, which is associated with a significant increase in the risk of cardiovascular disease. Nonspecific signs and symptoms include palpitations, polydipsia, polyuria, cramping, headaches, and weakness (9). Laboratory findings include suppressed plasma renin activity, increased plasma aldosterone levels, and a nonsuppressible aldosterone level in the blood or urine. The frequency of primary hyperaldosteronism as a cause of hypertension is influenced by the diagnostic algorithmic approach, but in large population-based studies it represents up to 12 percent of cases, and is the second most frequent cause of secondary hypertension (10).

Aldosterone-producing adenoma is not the only cause of primary hyperaldosteronism, and bilateral adrenal cortical nodular disease accounts for 60 to 70 percent of patients with primary hyperaldosteronism. A consensus for the nomenclature and histopathology of adrenal cortical lesions in unilateral primary aldosteronism (HISTALDO classification) has been recently proposed (11), and includes, together with aldosterone-producing adrenal cortical adenoma, aldosterone-producing adrenal cortical carcinoma, aldosterone-producing nodule, aldosterone-producing micronodule, multiple aldosterone-producing nodules and micronodules, and aldosterone-producing diffuse hyperplasia. This classification is discussed in more detail in chapter 3. While most cases of primary hyperaldosteronism are sporadic, 1 to 5 percent are inherited familial forms, transmitted as autosomal dominant traits (12). These familial forms are described in detail in chapter 7.

Cortisol-Producing Adenoma

Cushing syndrome is the manifestation of a cortisol-secreting adenoma. Elevated cortisol levels are detected in the serum, saliva, or urine. Cortisol secretion is adrenocorticotropic hormone (ACTH) independent and typical signs and symptoms include weight gain, central obesity, facial rounding (so called "moon facies"), hirsutism, easy bruising, poor wound healing, red to purple skin striae, weakness, hyperglycemia, hypertension, osteoporosis, and immunodeficiency leading to susceptibility to opportunistic infections. Based on the wide heterogeneity of clinical presentations, Cushing syndrome is often misdiagnosed or recognized after prolonged delays. Moreover, overt signs and symptoms of Cushing syndrome may be mild or absent despite increased cortisol levels, thus defining subclinical Cushing syndrome. Hereditary diseases leading to Cushing syndrome are usually associated with the development of multifocal disease (see chapters 4 and 7).

Cortisol and Aldosterone Co-Producing Adenoma

On rare occasion, adenomas are associated with concurrent primary hyperaldosteronism and hypercortisolism (13). These co-secreting adenomas present with mixed clinical features, including a higher proportion of cardiovascular complications, glucose intolerance/diabetes, and osteopenia/osteoporosis (14–16). The average diameter of a co-secreting adenoma is larger (26.2 mm) than an aldosterone-producing adenoma (17). While size may be the determining factor, subclinical hypercortisolism may also be more frequently found in association with smaller aldosterone-producing adenomas if it was investigated more often; this is consistent with in vitro studies revealing the capacity of aldosterone-secreting cells to co-secrete cortisol (18).

Lesions co-producing cortisol and aldosterone should be distinguished from synchronous separate adrenal cortical adenomas in patients with hyperaldosteronism and subclinical Cushing syndrome. In this context, CYP11B1 and CYP11B2 immunohistochemistry may help to recognize the functional status of each individual lesion, thus enabling the most appropriate clinical surveillance (see also below) (19).

Sex Hormone-Producing Adenoma

The hypersecretion of steroid hormones by an adrenal cortical neoplasm is more frequent in adrenal cortical carcinoma, thus malignancy should be suspected in the presence of virilization or feminization, as either part of a pure or mixed endocrine syndrome (e.g., Cushing syndrome and virilization), particularly in large neoplasms weighing over 100 g. In an old series of 190 adrenal tumors collected over a 30-year period, only 3 of 10 cases associated with virilization in female patients were benign (20). Feminization in a male patient is almost certainly the sign of malignancy and benign adrenal cortical tumors are exceptional in that setting (21). Virilization is the most frequent hormonal manifestation of functioning adrenal adenomas in the pediatric age (22), but is exceptional in postmenopausal women (23).

Nonfunctioning Adenoma

The clinical diagnosis of nonfunctioning adenoma is mostly related to the incidental discovery of an adrenal mass on radiologic imaging. As discussed above, a complete physical, radiologic, and endocrinologic evaluation is mandatory to exclude subclinical hormone hypersecretion. Many nonfunctioning tumors are capable of taking up radiolabeled steroid precursor, but are unable to secrete sufficient steroids to cause the signs or symptoms of hypersecretion. Surgery is not initially recommended for incidentally discovered adrenal lesions with homogeneous CT images, a size under 4 cm, and negative results at hormonal workup (24).

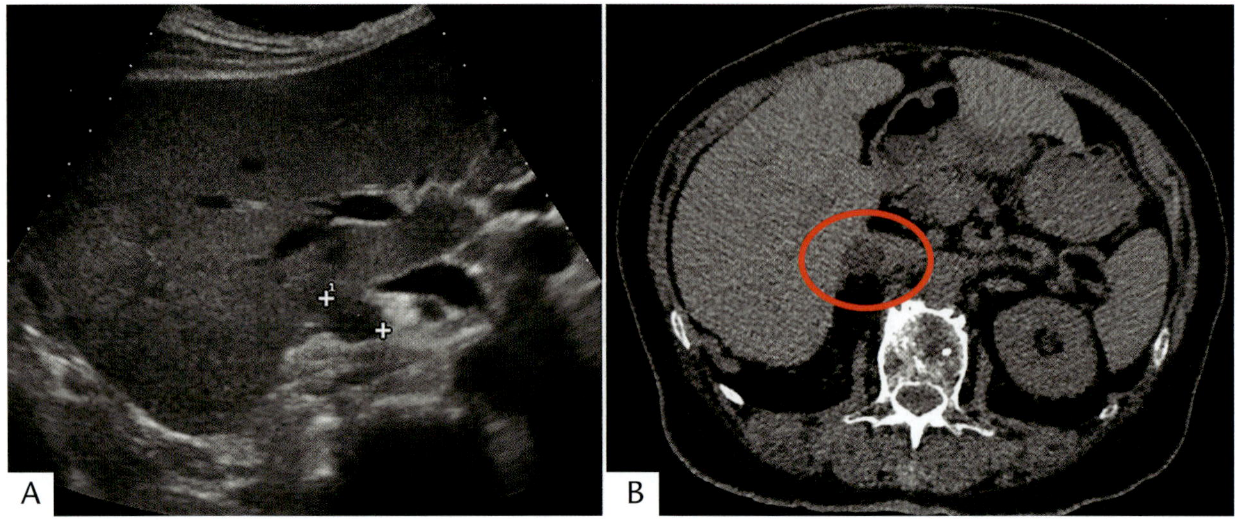

Figure 5-1

NONFUNCTIONAL ADRENAL CORTICAL ADENOMA

Ultrasound (US) scan (A) and computerized tomography (CT) (B, <10 HU) of a 1.7-cm lesion in the right adrenal gland. (Courtesy of Prof. A. Veltri and Dr. F. Solitro, San Luigi Hospital, Turin, Italy)

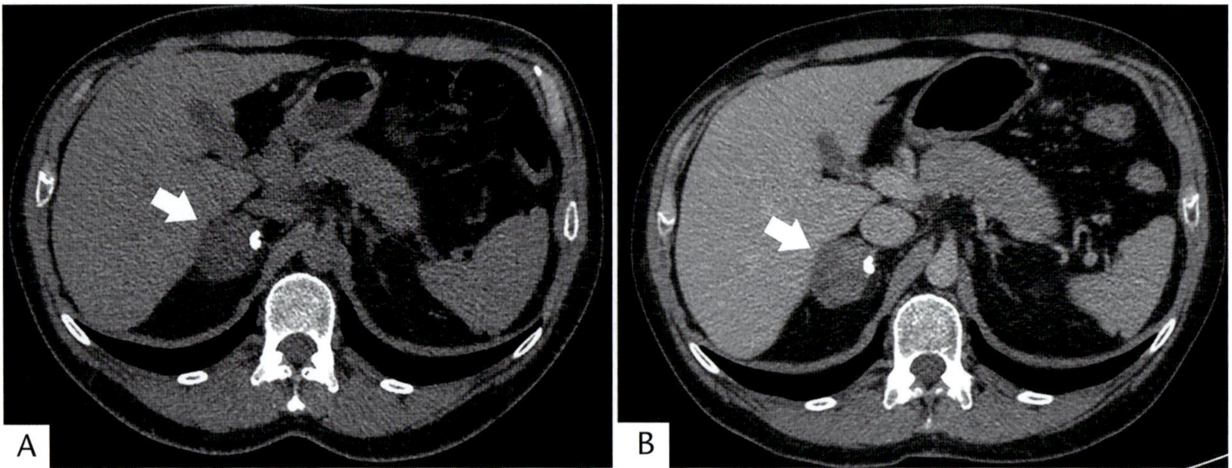

Figure 5-2

ADRENAL CORTICAL ADENOMA IN CUSHING SYNDROME

CT scan in basal phase (A) and poor uptake in the venous phase (B). (Courtesy of Prof. A. Veltri and Dr. F. Solitro, San Luigi Hospital, Turin, Italy)

PRECURSOR LESIONS

There are no definite precursor lesions for cortisol-secreting adenomas, except in hereditary situations. Aldosterone-producing nodules (also termed aldosterone-producing cell clusters) are a potential precursor of a subset of adenomas, although the precise mechanism of this transition has not yet been clarified (25).

IMAGING FEATURES

Adenomas are usually homogeneous on unenhanced and contrast-enhanced CT images, and the density is equal to or slightly lower than that of normal adrenal gland tissue. When the tumor is necrotic or cystic, the density is uneven (figs. 5-1–5-3). Adenomas can be divided on CT into two types: lipid-rich adenomas (70

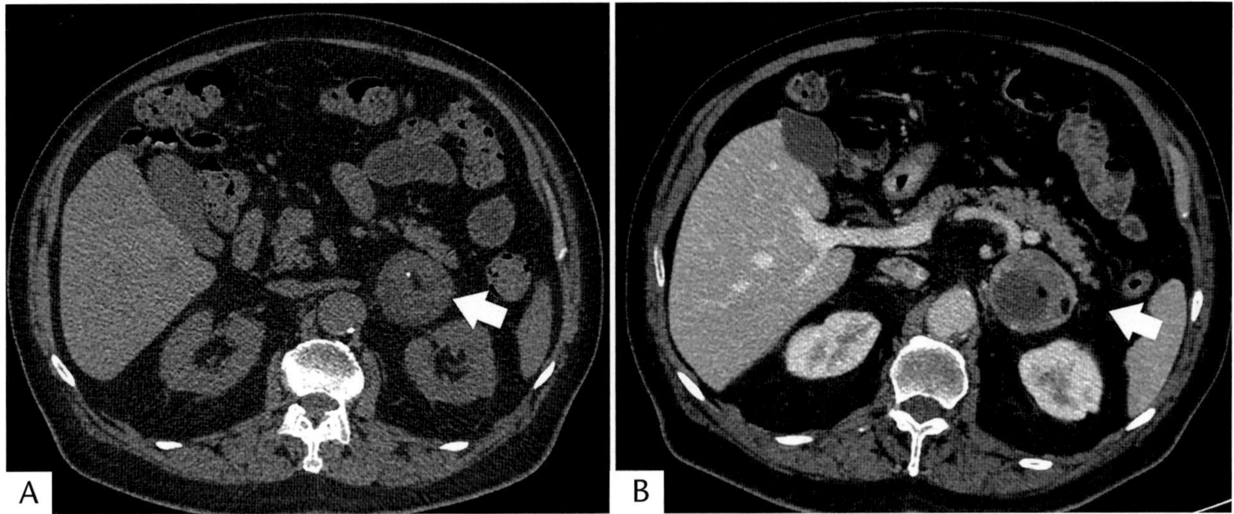

Figure 5-3

NONFUNCTIONAL ADRENAL CORTICAL ADENOMA

CT scan in basal phase (A) and nonhomogeneous uptake in the venous phase (B). (Courtesy of Prof. A. Veltri and Dr. F. Solitro, San Luigi Hospital, Turin, Italy)

percent, density less than 10 HU on precontrast-enhanced CT scan images) and lipid-poor adenomas (30 percent, with a density between 10 and 30 HUs). On unenhanced CT scans, the decrease in the density of the lesion is due to an increase in the amount of fat, and higher density is seen in lipid-poor than in lipid-rich adenomas (26). Although CT cannot distinguish functional from nonfunctional adenomas, it can suggest a functioning adenoma in the presence of atrophy of the contralateral adrenal gland.

With magnetic resonance imaging (MRI), as a result of the different precession frequencies of protons in water molecules and fat, many adenomas show high signal on in-phase imaging and the signal decreases in the out-of-phase imaging (figs. 5-4, 5-5) (27). Lipid-rich adenomas, in particular, show high intensity signal on T1- and T2-weighted images. After contrast injection, the majority of adenomas show relatively regular enhancement on immediate gadolinium-enhanced images (28). The high lipid content on MRI in nonfunctioning lesions and the lesion size are major criteria for identifying an adrenal lesion that warrants surveillance rather that surgery.

Recently, ^{18}F-FDG-PET (positron emission tomography)/CT imaging has been suggested to discriminate between benign and malignant adrenal lesions, SUVmax of 4.6 or greater being suggestive of malignancy. However, lower values did not reliably predict benign tumors, and thus are not specific for adenoma (29).

GROSS FINDINGS

Adrenal cortical adenoma is a solitary, well-circumscribed lesion, confined to the adrenal gland and surrounded by a thin capsule or well-demarcated from the surrounding adrenal tissue. Tumor size and weight do not usually exceed 5 cm and 100 g, respectively, although the size alone is not a criterion for distinguishing adenoma from carcinoma, since large adenomas occur, especially if clinically nonfunctioning. Some tumors have vague nodularity on cross section, but typically without coarse lobulation or broad fibrous bands, which are more typical of adrenal cortical carcinoma.

The color of the cut surface is partly dependent on the hormonal status, but it is usually yellow to orange due to steroid content, although cortisol-secreting adenomas may have a more brown color while aldosterone-producing tumors are characteristically bright yellow to orange (figs. 5-6–5-11). In the oncocytic variant, the color is more tan-brown to ochre (figs. 5-12, 5-13), whereas a dark brown to black color macroscopically defines the "black" adenoma

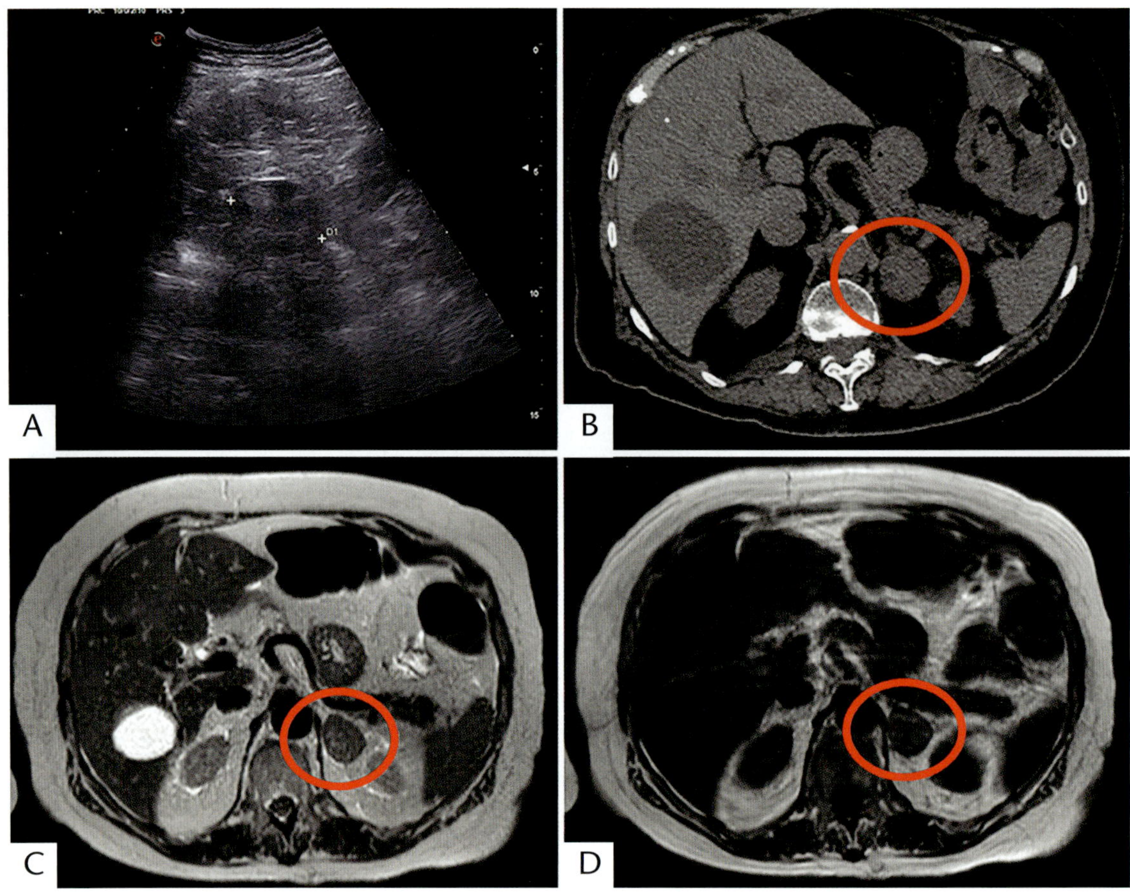

Figure 5-4

ADRENAL CORTICAL ADENOMA IN CUSHING SYNDROME

US scan (A), CT scan (B, 20 HU), and magnetic resonance imaging (MRI) ("in" phase, C, and "out" phase, D) show a lipid-poor pattern in a 4.5-cm adenoma. (Courtesy of Prof. A. Veltri and Dr. F. Solitro, San Luigi Hospital, Turin, Italy)

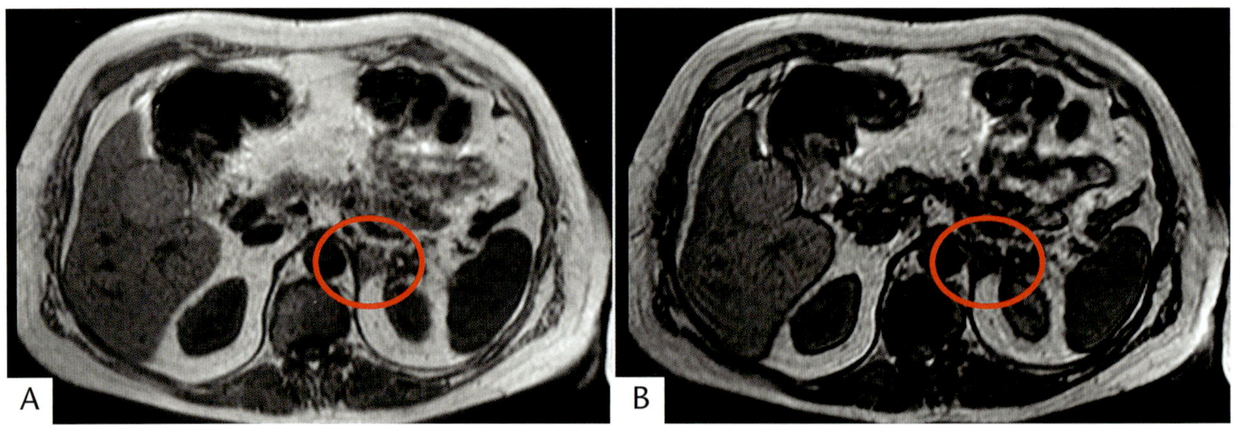

Figure 5-5

ADRENAL CORTICAL BLACK ADENOMA

"In" phase (A) and "out" phase (B). (Courtesy of Prof. A. Veltri and Dr. F. Solitro, San Luigi Hospital, Turin, Italy)

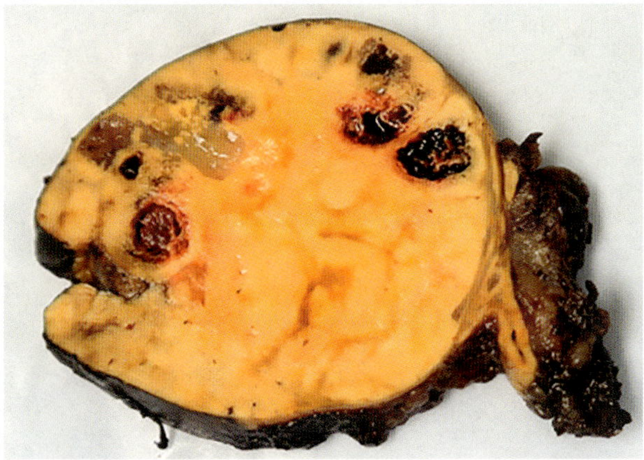

Figure 5-6

CORTISOL-SECRETING ADRENAL CORTICAL ADENOMA

Yellow cut surface with focal areas of hemorrhage. The residual adrenal gland (right) is atrophic.

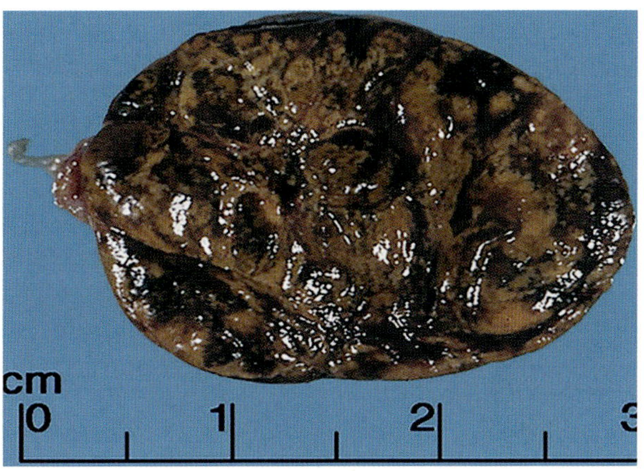

Figure 5-7

ADRENAL CORTICAL ADENOMA ASSOCIATED WITH CUSHING SYNDROME

This adrenal cortical adenoma from a 32-year-old woman with Cushing syndrome measured 2.1 cm. The geographic to mottled zones of dark pigmentation are due to lipid depletion of neoplastic cortical cells as well as an accumulation of lipofuscin.

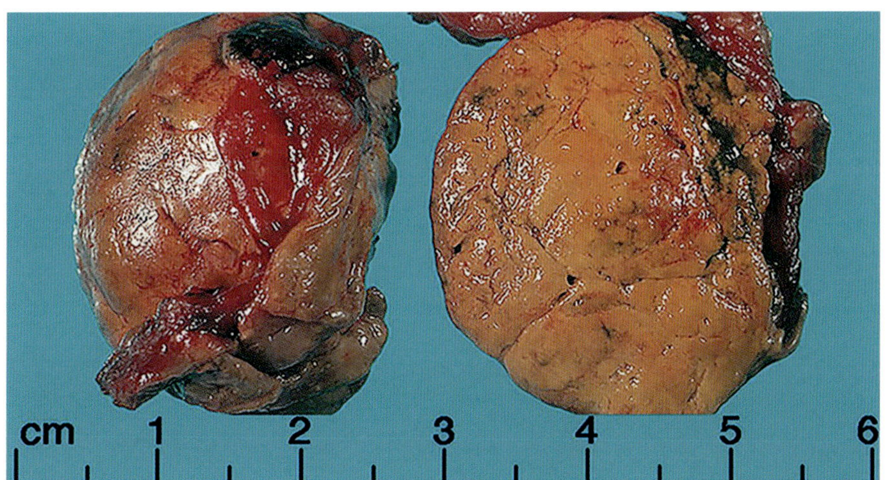

Figure 5-8

ADRENAL CORTICAL ADENOMA ASSOCIATED WITH CUSHING SYNDROME

This adrenal cortical adenoma is from a 33-year-old patient with Cushing syndrome. The tumor is yellow-orange on cross section and has vague lobulations. It measured 3.5 x 3.0 x 2.5 cm.

Figure 5-9

ADRENAL CORTICAL ADENOMA ASSOCIATED WITH CUSHING SYNDROME

This adrenal cortical adenoma is from a 35-year-old female with Cushing syndrome. The tumor has a confluent brown color due to cells with compact, lipid-depleted cytoplasm and intracellular lipochrome pigment. The attached adrenal cortex is markedly atrophic. The tumor weighed 16 g including the attached adrenal remnant.

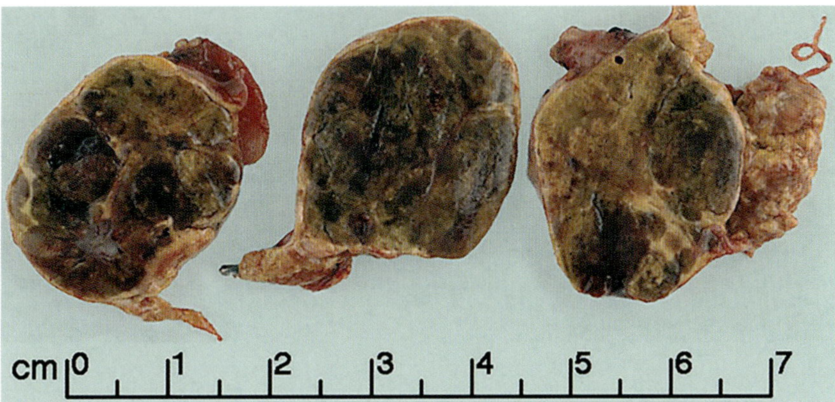

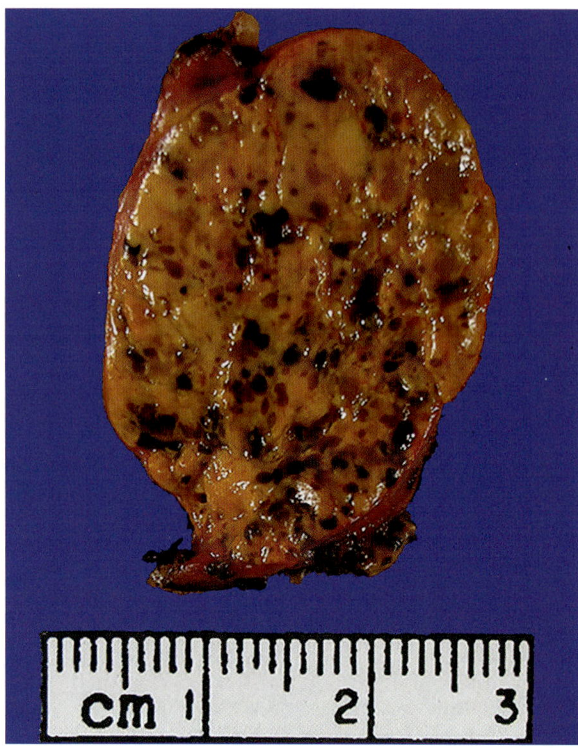

Figure 5-10

CORTISOL-SECRETING ADRENAL CORTICAL ADENOMA

The heterogeneous color ranges from yellow to dark brown, due to coexistence of lipid-rich cells and cells with lipid-depleted cytoplasm with lipochrome pigment.

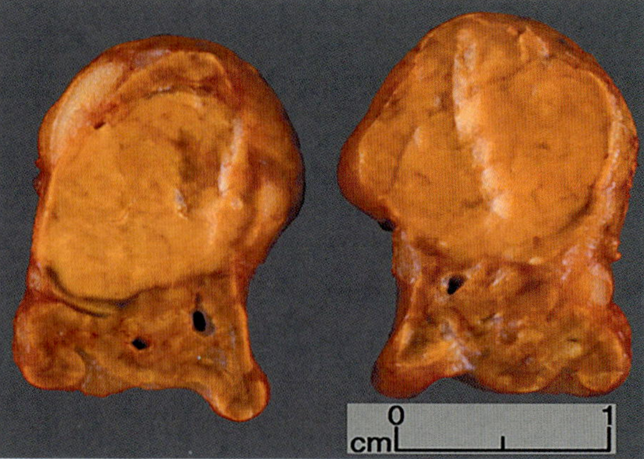

Figure 5-11

ALDOSTERONE-SECRETING ADRENAL CORTICAL ADENOMA

On cross section, the tumor is sharply demarcated and appears encapsulated, particularly where it expands the contours of the adrenal capsule. The tumor measured 1 cm in diameter and is a uniform yellow-orange.

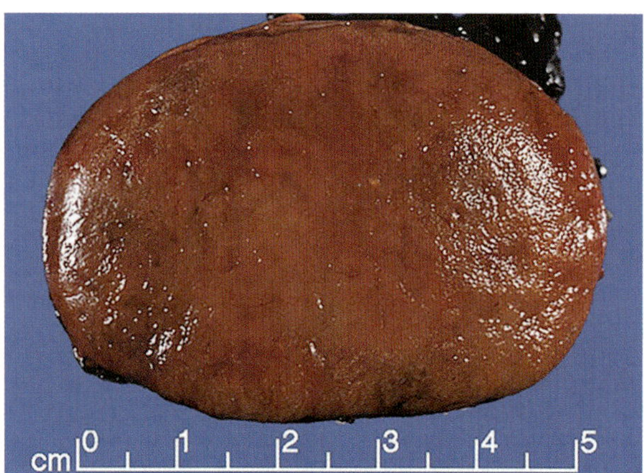

Figure 5-12

ONCOCYTIC ADRENAL CORTICAL ADENOMA

This nonfunctional oncocytic adrenal cortical adenoma weighed 70 g and was 4 cm in diameter. On cross section, the tumor is uniformly mahogany brown.

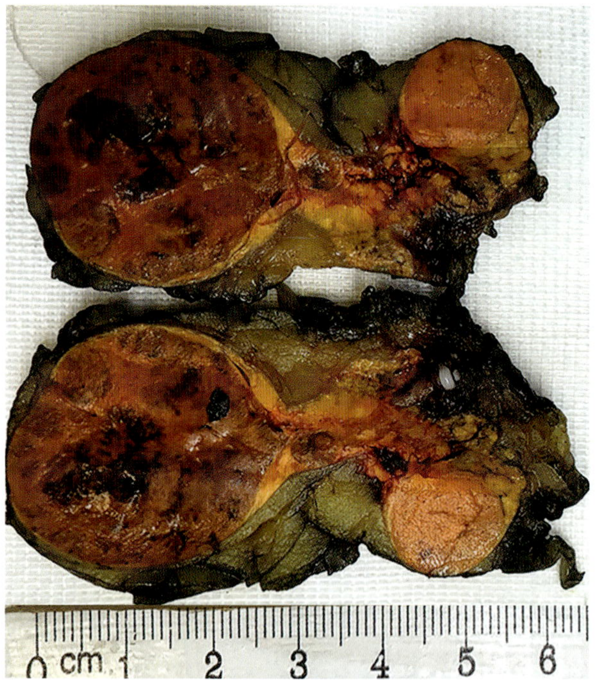

Figure 5-13

ONCOCYTIC ADRENAL CORTICAL ADENOMA COEXISTING WITH A CLEAR CELL CORTICAL NODULE

The dishomogeneous brown nodule, 2.7 cm in size (bottom), corresponded histologically to an oncocytic adenoma, well-circumscribed and with a thin capsule. The small yellow nodule 1 cm in size was poorly demarcated from the adjacent cortex and composed of lipid-rich cells.

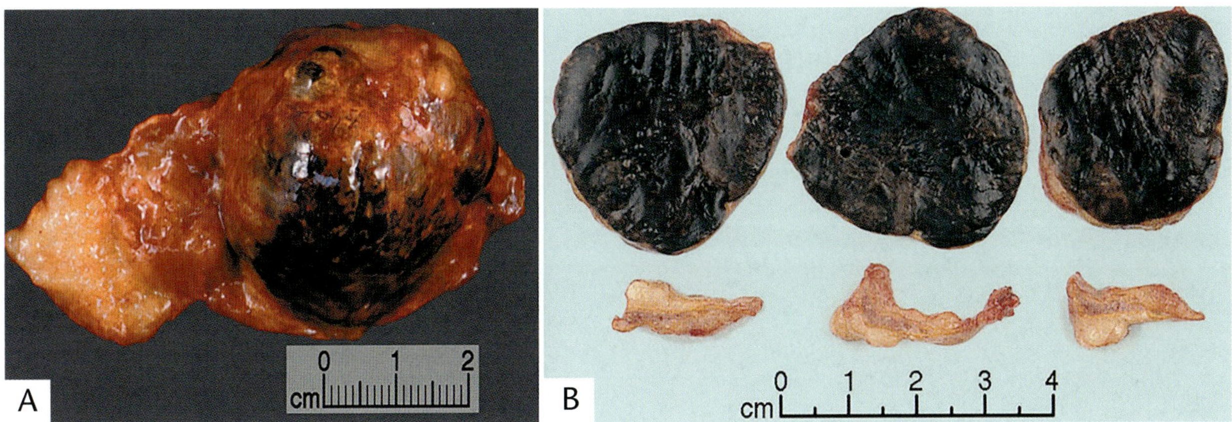

Figure 5-14

BLACK ADRENAL CORTICAL ADENOMA

A: The adenoma was resected from a 58-year-old man with Cushing syndrome. The dark exterior aspect of the tumor is evident through the intact adrenal capsule and thin investing fat.

B: The tumor measured 2.9 x 2.8 x 2.5 cm and weighed 30 g. On cross section, the tumor became even darker after exposure to air. The adrenal cortex at the bottom of the field is markedly atrophic. Fat could not be removed entirely because the atrophic gland was so friable.

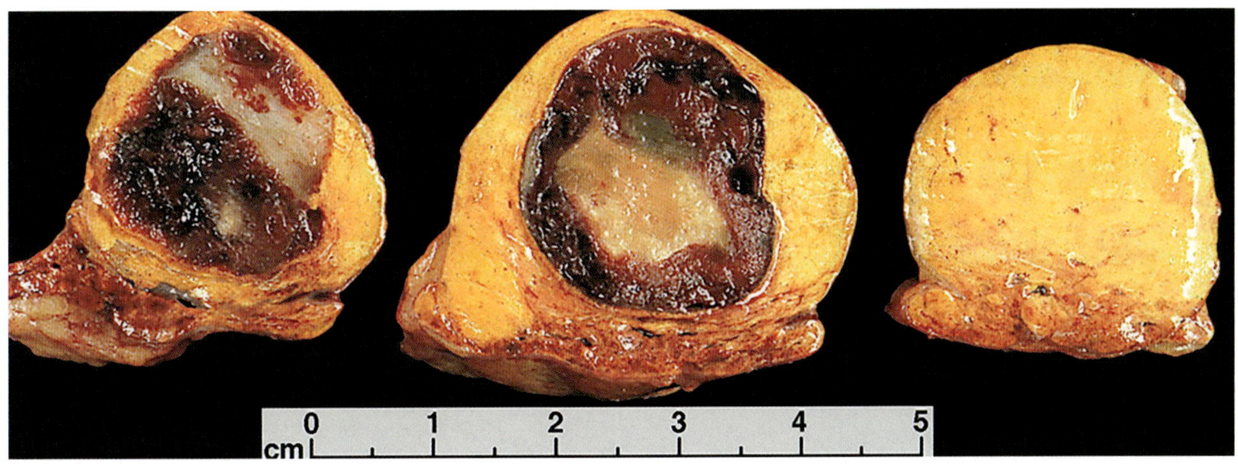

Figure 5-15

POST-FINE NEEDLE ASPIRATION BIOPSY INVOLUTIVE CHANGES IN AN INCIDENTAL ADRENAL CORTICAL ADENOMA

Incidental nonfunctional adrenal cortical adenoma in a 73-year-old man who had gallstones. Fine needle aspirate did not yield sufficient material for diagnosis. Right adrenalectomy was performed during elective cholecystectomy. The adrenal adenoma weighed 28 g and measured nearly 3.5 cm in diameter. A central organizing hematoma occurred following the attempt at aspiration.

variant (fig. 5-14). Similar black to brown areas are also observed focally in conventional adenomas. Irregular foci of dark discoloration also may be seen frequently in adrenal cortical adenomas, and correlate with the presence of recent or old hemorrhage, areas of lipid-depletion within the tumor, or the presence of increased lipofuscin.

Involutive changes may occur, especially in large lesions, including fibrotic changes, hemorrhage, and necrosis (fig. 5-15). Lipomatous and myelolipomatous changes also regularly occur, and they may be recognizable at macroscopy as gelatinous areas, more frequently at the periphery of the lesion (fig. 5-16). The adrenal cortex adjacent to the adenoma is usually normal in cases of Conn syndrome or nonfunctioning adenomas, but there is thinning of the cortex in cortisol-secreting adenomas due to the cortical atrophy caused by suppression of the hypothalamic-pituitary-adrenal axis.

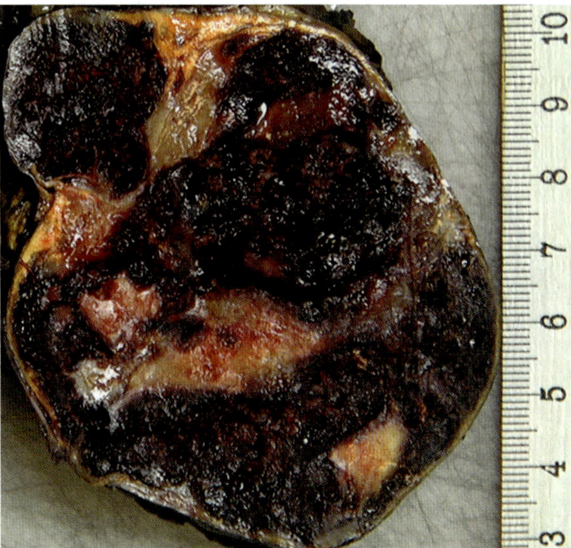

Figure 5-16

ADRENAL CORTICAL ADENOMA WITH MYELOLIPOMA

A 7.5-cm lesion with a peripheral cortical cell component is associated with a prominent myelolipomatous component with hemorrhage and infarct-type necrosis.

MICROSCOPIC FINDINGS

Adrenal cortical adenomas are mainly composed of two types of cells: clear and compact tumor cells. Clear cells are lipid-rich cells or zona fasciculata-like cells, and harbor relatively abundant lipid droplets that can be stained on fresh frozen tissue with oil red O (fig. 5-17). The compact cells, also called lipid-poor cells, are spherical with eosinophilic cytoplasm. These two cell types are proportionally different in different hormonally active adenomas (fig. 5-18). Nevertheless, the details of hormonal and biological activities of these cells are still largely unknown. Although there are some microscopic features that are more likely to be associated with tumors producing a particular endocrine syndrome, it is in general difficult to predict the associated endocrine syndrome based upon histologic features alone in individual cases, without correlation with clinical or biochemical data.

The architecture of adrenal cortical adenoma largely resembles that of the normal adrenal cortex, with cords, nests, or islands of cells in an alveolar arrangement, separated by a thin, delicate and well-organized vascular network, with sinusoidal structures (fig. 5-19) that are clearly highlighted by silver (reticulin) staining (fig. 5-20). Still, a wide heterogeneity of architectures may be observed even within the same

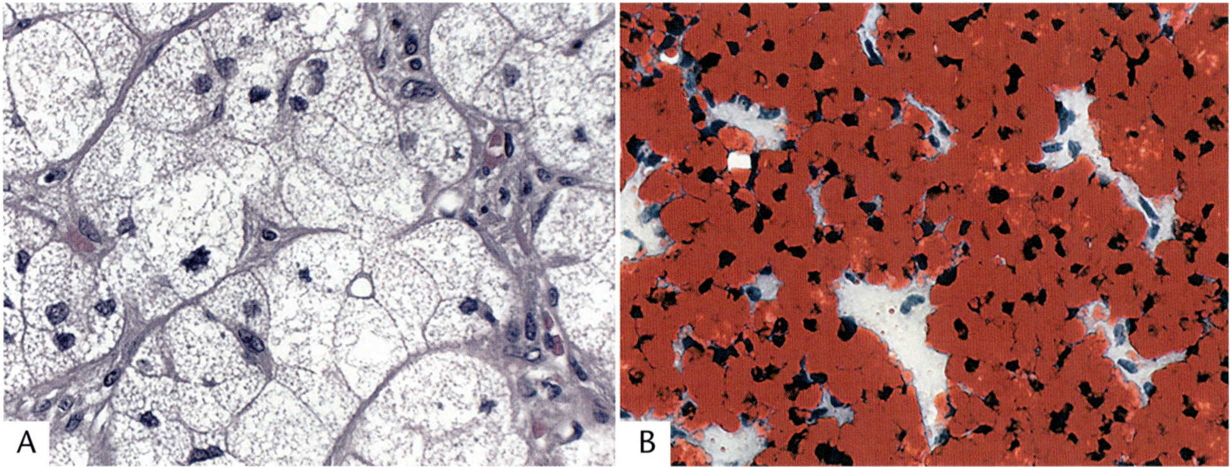

Figure 5-17

ADRENAL CORTICAL ADENOMA WITH "BALLOON CELLS"

A: The enlarged "balloon cells" in an adrenal cortical adenoma from a patient with Cushing syndrome have greatly expanded, lipid-rich cytoplasm. Some cells were nearly 10-fold larger than other cells in the adenoma.

B: Abundant neutral lipid content is highlighted by oil red-O stain.

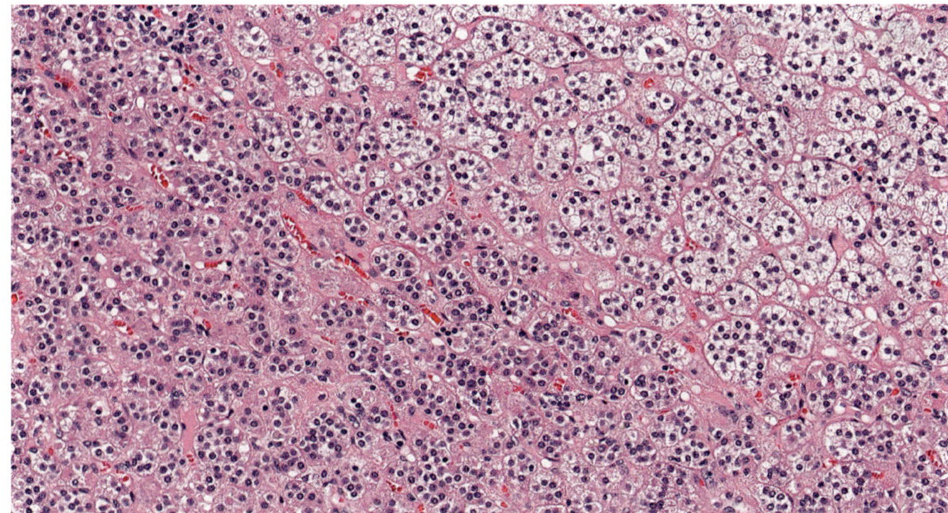

Figure 5-18

ADRENAL CORTICAL ADENOMA IN CUSHING SYNDROME

The tumor is composed of clear (upper right) and compact cells (left).

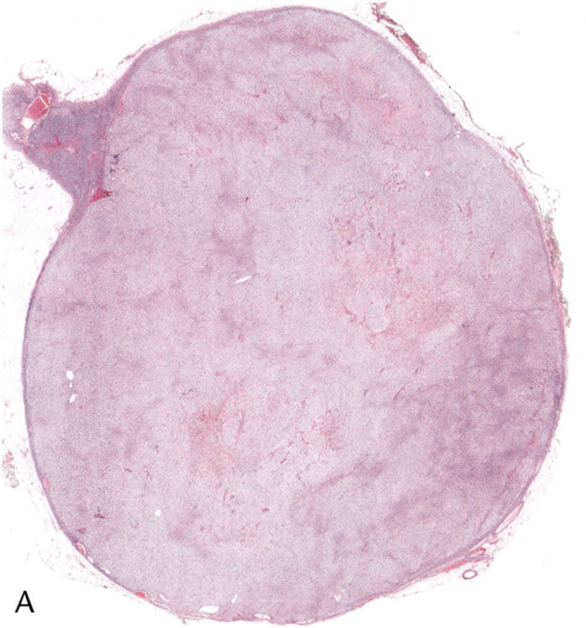

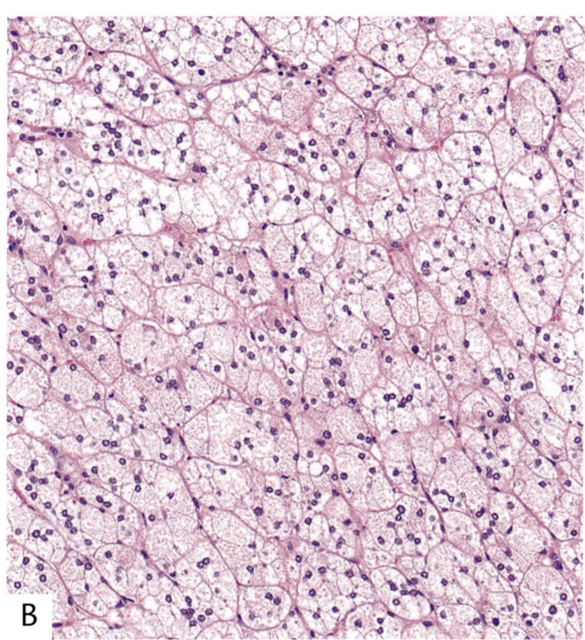

Figure 5-19

CORTISOL-SECRETING ADRENAL CORTICAL ADENOMA

A: This adenoma consists of a homogeneous population of predominantly clear cells with marked atrophy of the adjacent cortex. B: The clear cells have uniform small round nuclei and abundant cytoplasm.

lesion (fig. 5-21). Occasionally, the tumor cells are focally aligned in slender cords reminiscent of normal zona fasciculata, or have a spindle cell appearance (fig. 5-22), although rarely and focally, thus not posing differential diagnostic problems. The diffuse growth typical of adrenal carcinoma is usually not seen, although it may be present in the oncocytic variant.

Cell size differs based on the content of lipids, but is generally larger than normal adrenal cortical cells. Nuclei are small and uniform in most cells. Occasionally, nucleoli or nuclear polymorphism, including nuclear pseudoinclusions, occur, but in isolated cells or small clusters (figs. 5-23, 5-24). Hyaline globules are rarely encountered (fig. 5-25). The mitotic index

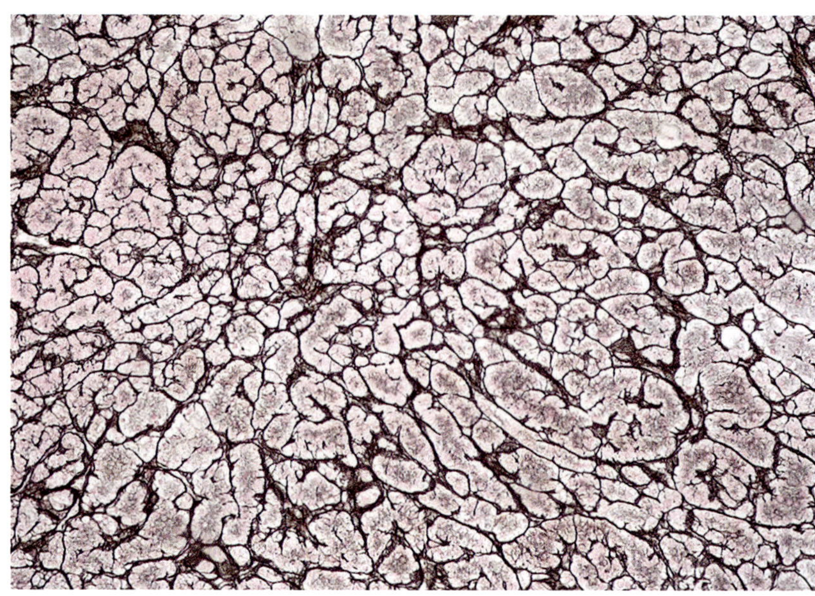

Figure 5-20

ADRENAL CORTICAL ADENOMA IN CUSHING SYNDROME

A reticulin stain highlights the regular architecture of the lesion.

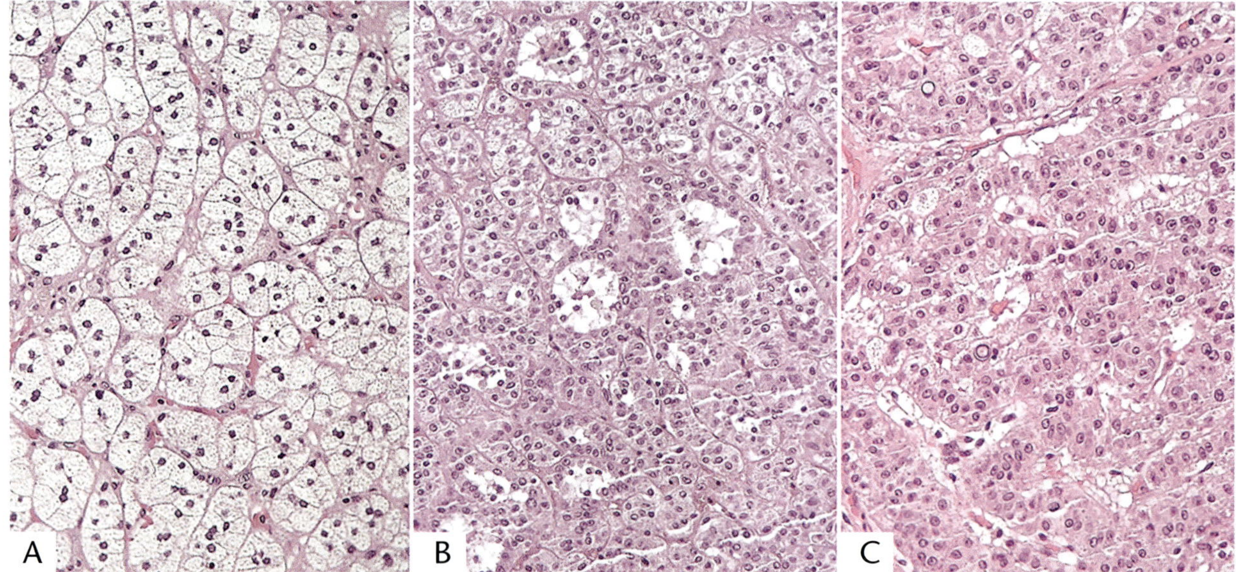

Figure 5-21

INCIDENTALLY DISCOVERED, NONFUNCTIONAL ADRENAL CORTICAL ADENOMA

The lesion measured 5 cm and presented heterogeneous morphologic features, including alveolar with lipid-rich clear cells (A), alveolar/pseudoglandular with lipid-poor compact cells (B), and trabecular with eosinophilic cells (C).

is much less than 1 in 10 mm^2 and atypical mitoses are absent.

Complete penetration of the tumor capsule and venous invasion are absent. Clusters of cortical cells may be present outside the adrenal capsule at the periphery of the lesion, as also described in the normal adrenal gland (see chapter 1) and in adrenal cortical hyperplasia, but these cells should not be considered as capsular invasion. These clusters do not show cytologic atypia, the adrenal capsule in contact with them has a bland appearance, and there is no desmoplastic reaction in the surrounding stroma.

Lipomatous or myelolipomatous changes are frequent (fig. 5-26), usually as focal peripheral areas. They may, however, predominate, and

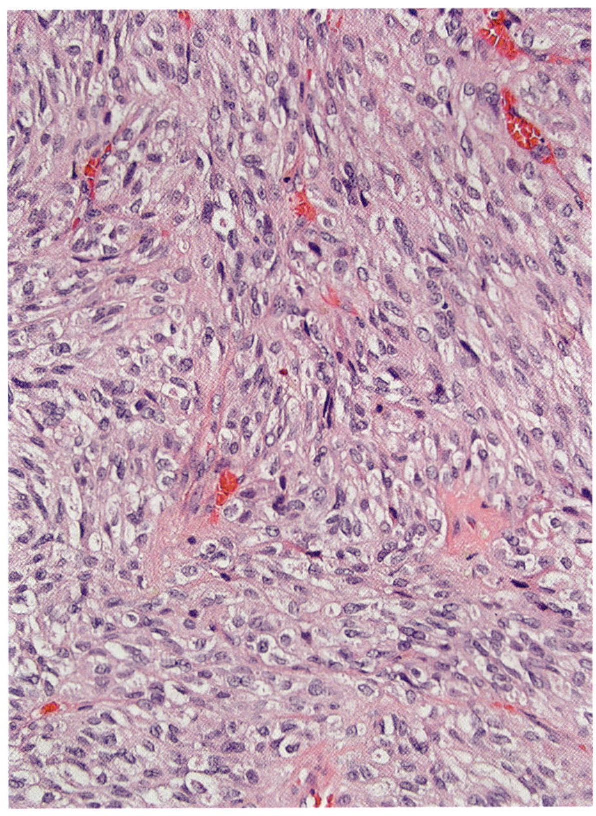

Figure 5-22

ADRENAL CORTICAL ADENOMA WITH SPINDLE CELL FEATURES

Areas of this adrenal adenoma show an ill-defined spindle cell pattern. More conventional histology was present in other fields.

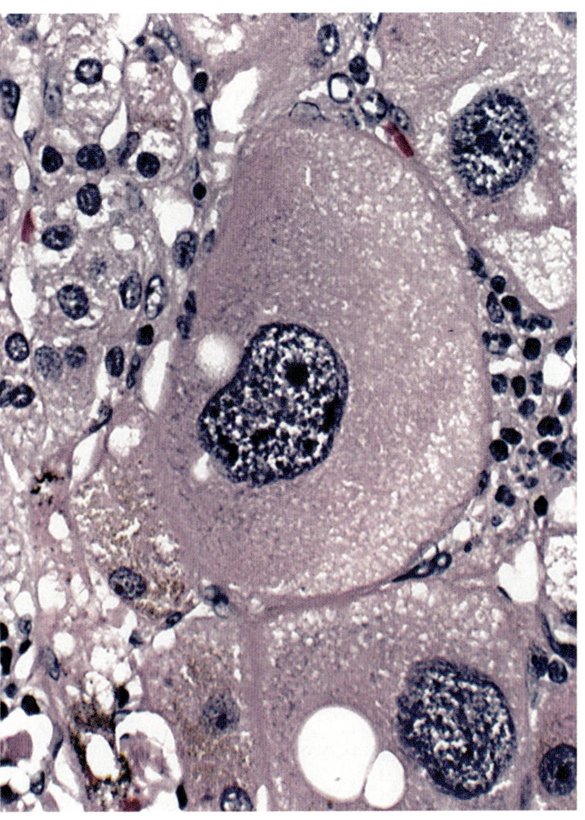

Figure 5-23

ADRENAL CORTICAL ADENOMA ASSOCIATED WITH CUSHING SYNDROME

Greatly enlarged pleomorphic nuclei are seen in some cells, which are markedly different than the nuclei of adjacent tumor cells. A few cells also contain finely granular brown lipofuscin pigment.

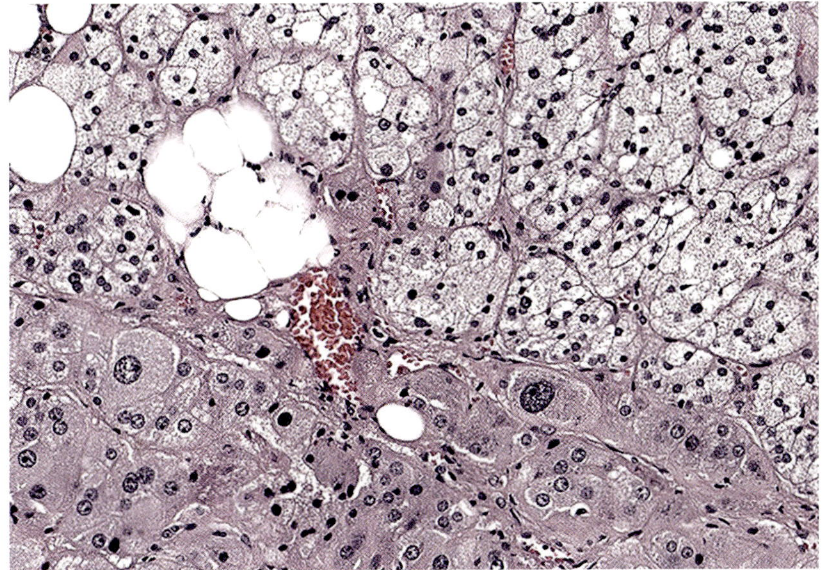

Figure 5-24

ATYPICAL CELLS IN ADRENAL CORTICAL ADENOMA

This cortical adenoma exhibits focal "endocrine atypia" and lipomatous change.

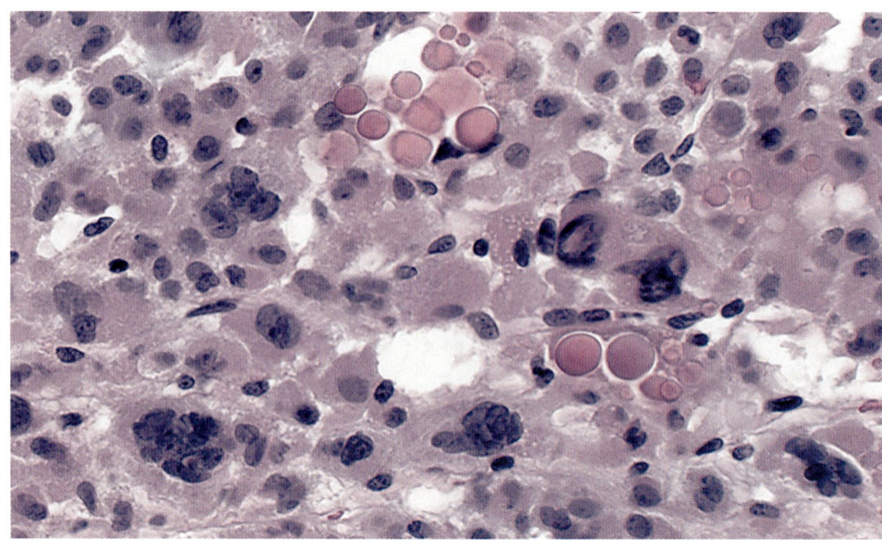

Figure 5-25

ADRENAL CORTICAL ADENOMA ASSOCIATED WITH CUSHING SYNDROME

The tumor cells have compact eosinophilic cytoplasm and contain eosinophilic hyaline globules.

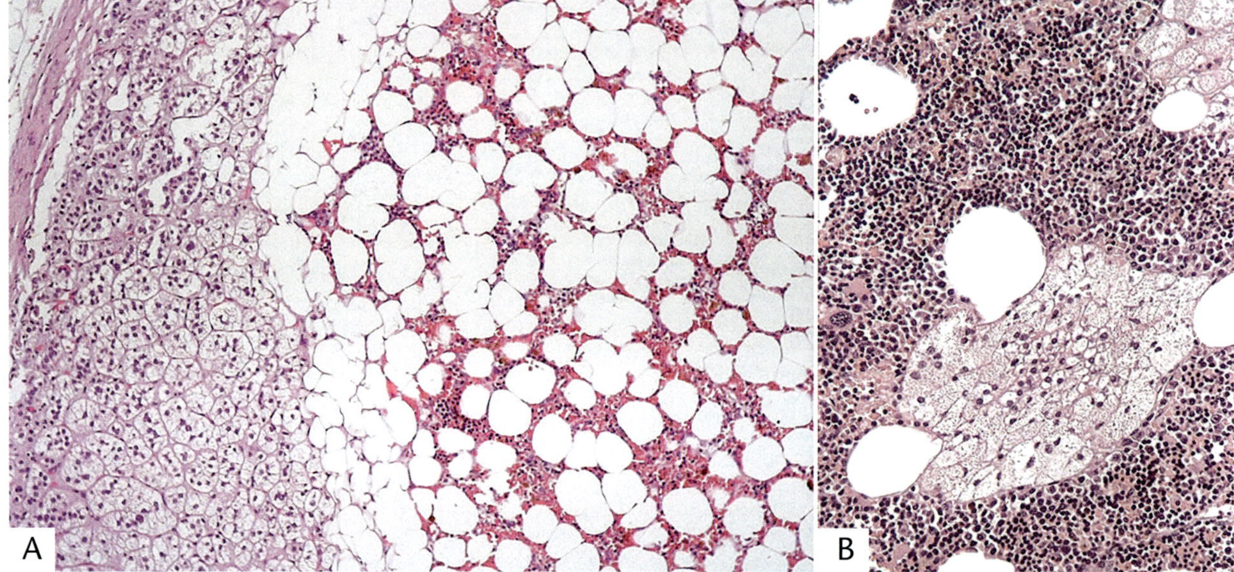

Figure 5-26

ADRENAL CORTICAL ADENOMA ASSOCIATED WITH CUSHING SYNDROME

Predominant myelolipomatous changes are present in association with an adenomatous component made of lipid-rich clear cells (same case as fig. 5-16).

in such cases the adenomatous tissue component may be difficult to distinguish from entrapped normal adrenal cortical tissue. Degenerative changes are more frequent in large tumors, and include fibrosis and edema, hyalinization, macro- or microcystic changes, hemorrhage with hemosiderin deposition, bone metaplasia, and rarely, necrosis of the infarction type (figs. 5-27–5-30). The presence of all of the above may create problems in the differential diagnosis with other entities (such as vascular tumors or other soft tissue tumors) or in the assessment of the benign versus malignant nature of the adrenal cortical lesion, since some of the definitional parameters of malignancy (i.e., included in the Weiss score [see chapter 6]) may be hard to assess, but preservation of the reticulin framework and low proliferation index are helpful in the differential diagnosis (fig. 5-31).

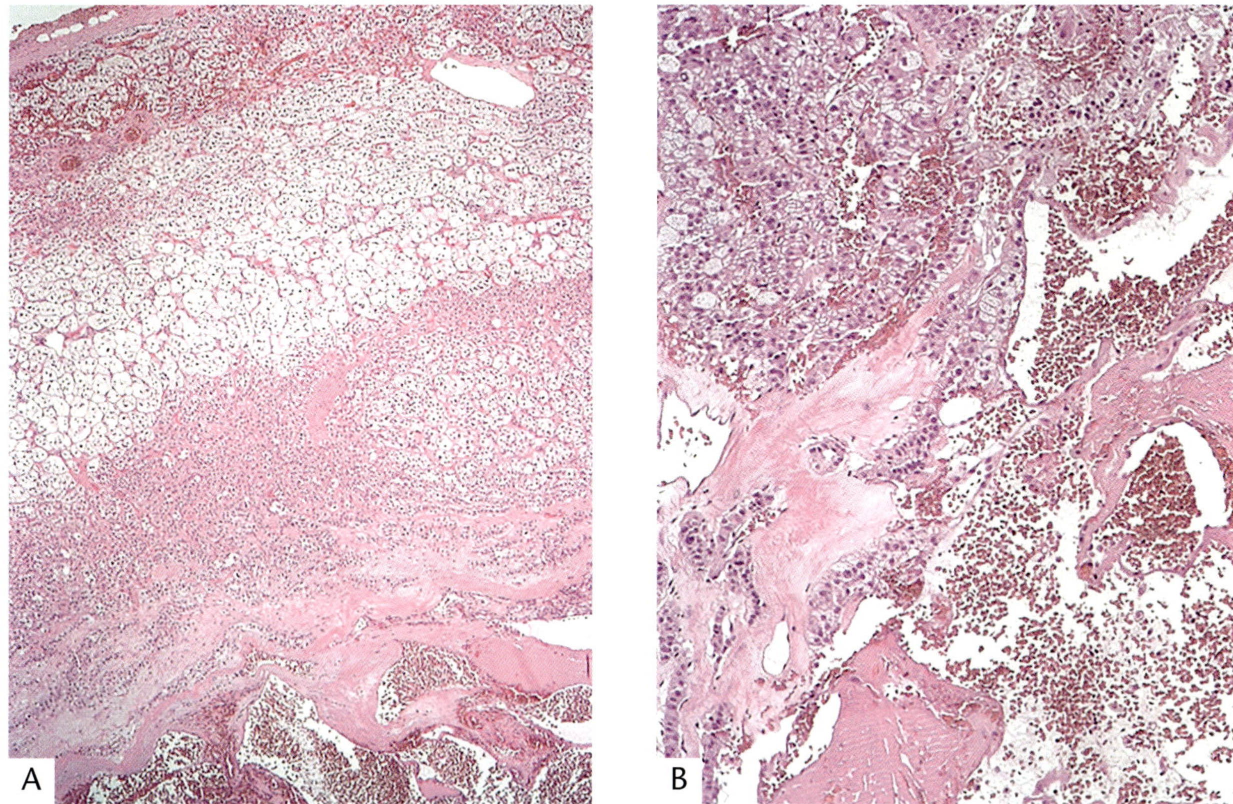

Figure 5-27

NONFUNCTIONAL ADRENAL CORTICAL ADENOMA

A: The central portion of this adenoma shows extensive degeneration and hemorrhage with neovascularization. B: Cavernous channels in the hemorrhagic area have features suggestive of a vascular neoplasm.

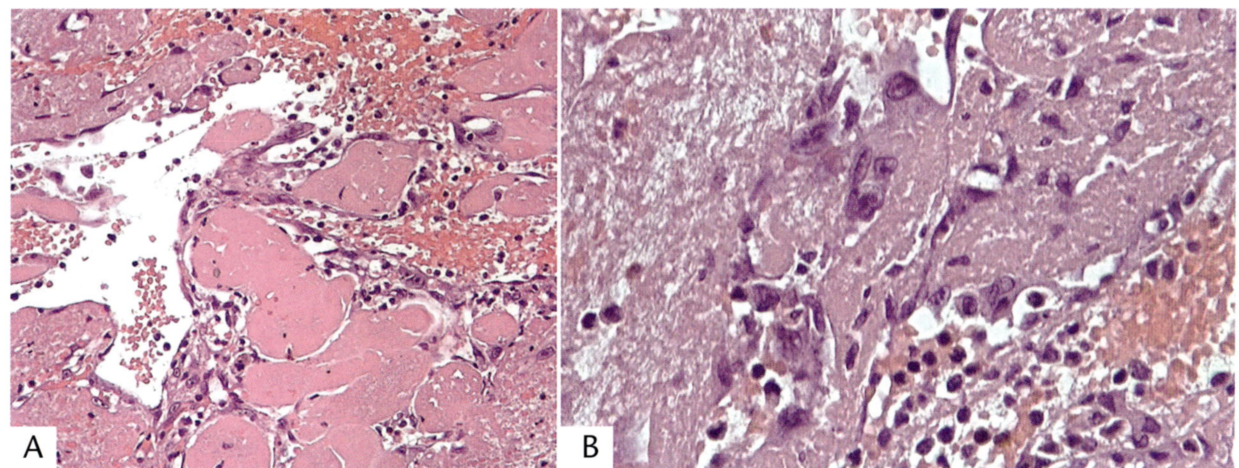

Figure 5-28

NONFUNCTIONAL ADRENAL CORTICAL ADENOMA

Endothelial hyperplasia associated with fibrin (A) and showing mild atypia (B) are present in hemorrhagic areas (same case as fig. 5-27).

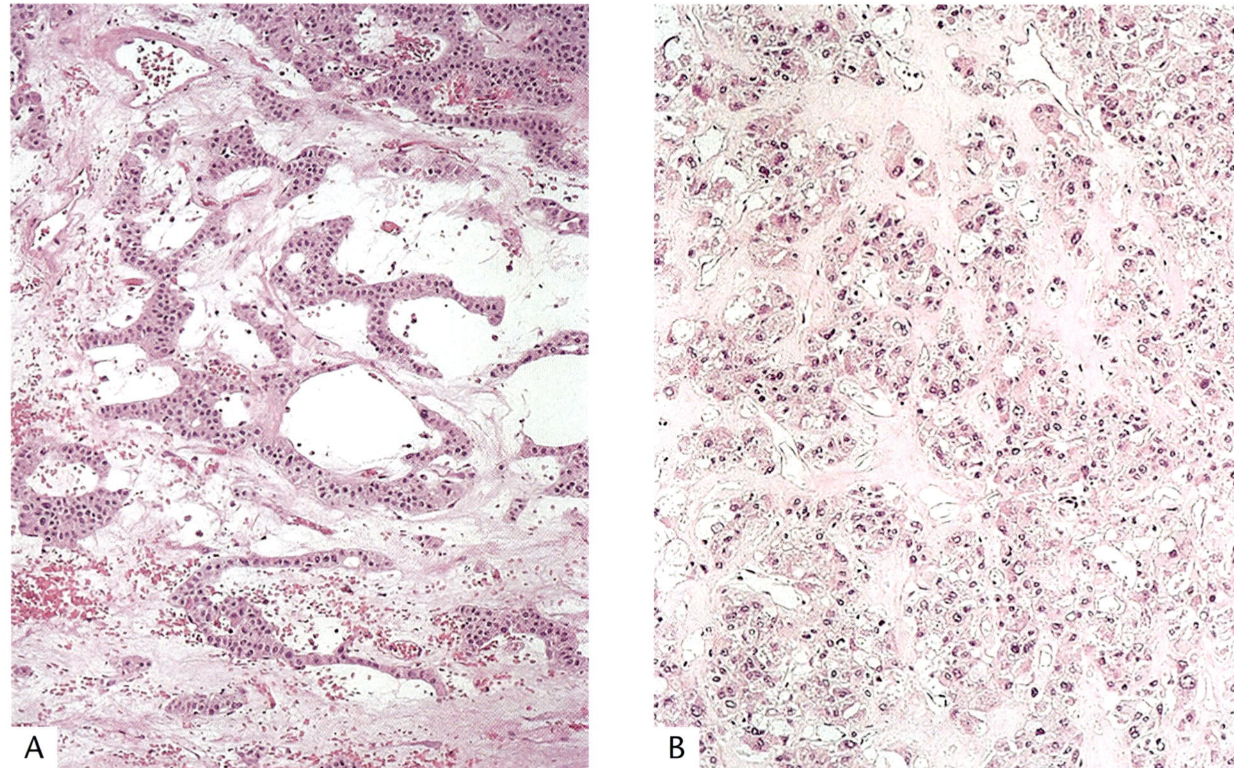

Figure 5-29

ADRENAL CORTICAL ADENOMA, STROMAL CHANGES

Some adenomas show focal or predominant areas with stromal changes, either as loose edematous stroma (A) or as more dense and collagenous depositions embedding small clusters of tumor cells (B).

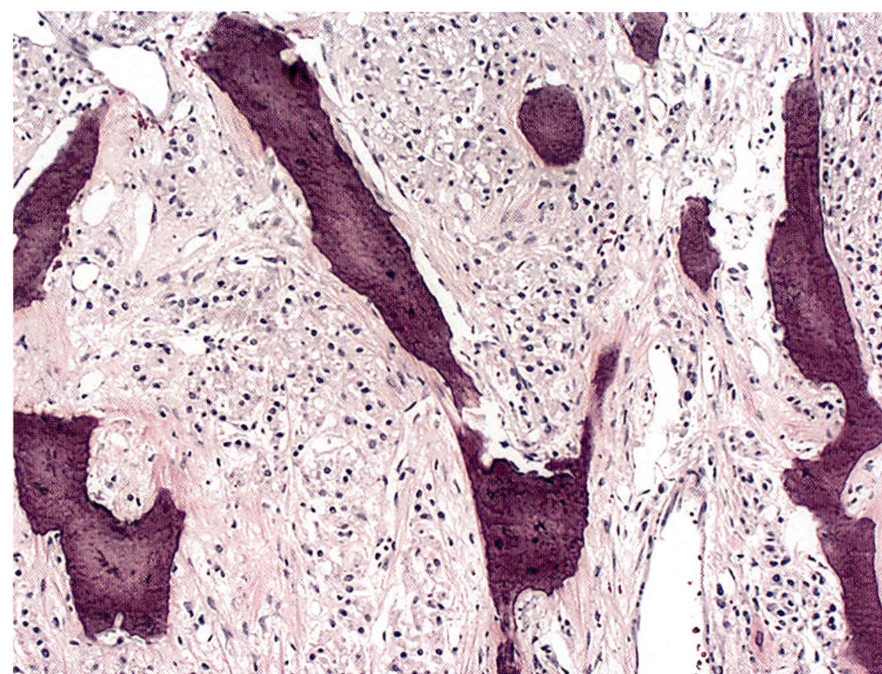

Figure 5-30

ADRENAL CORTICAL ADENOMA ASSOCIATED WITH CUSHING SYNDROME

An area shows metaplastic bone formation in association with a fibrotic stroma.

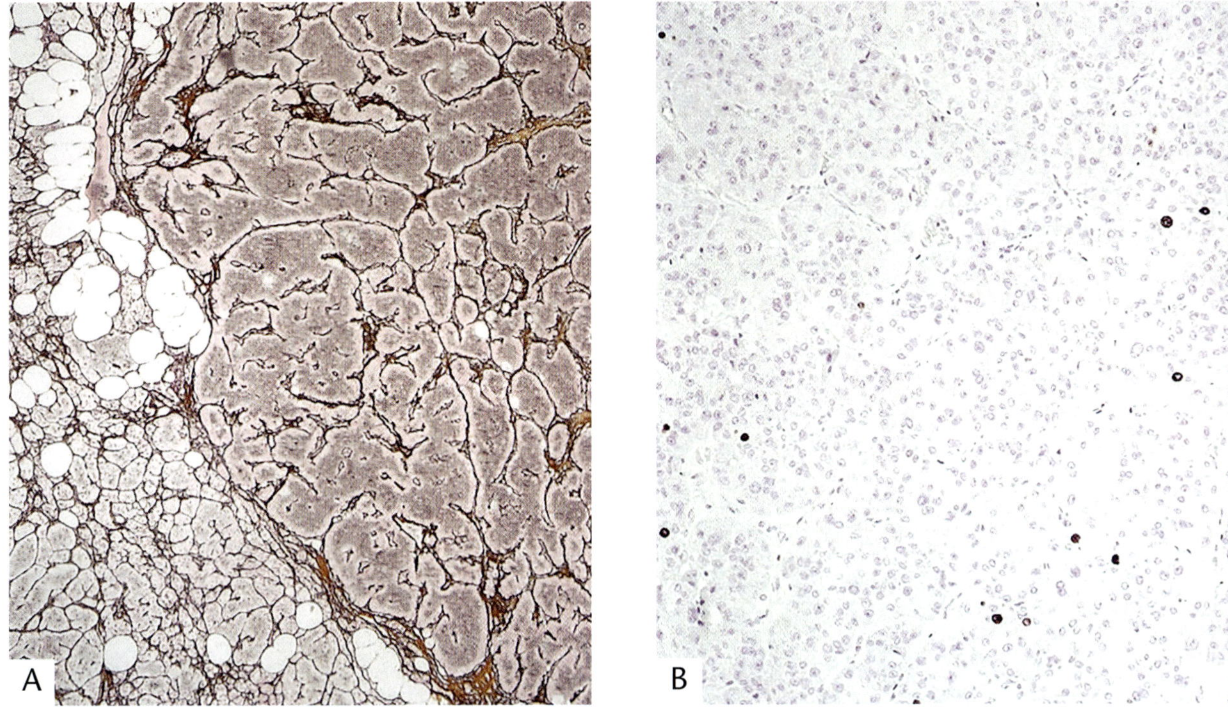

Figure 5-31

ANCILLARY STAINING IN THE DIFFERENTIAL DIAGNOSIS OF ADRENAL CORTICAL ADENOMA AND CARCINOMA

A diagnosis of adrenal cortical carcinoma was suspected because of the presence of predominant oncocytic cells with areas of marked atypia. The preserved reticulin pattern (A) and a low Ki-67 index (B) were supportive of a diagnosis of adenoma, together with the absence of other malignancy-related morphologic parameters.

Morphologic and Clinical Correlates

Although the morphology of adrenal cortical adenomas is variable, as discussed above, there are some specific clinical and pathologic correlates. In cortisol-secreting adrenal cortical adenomas, there is almost invariably some degree of adrenal cortical atrophy (see figs. 5-19 and 5-32). The presence of lipofuscin is more frequent (figs. 5-33, 5-34) as is the presence of intratumoral lymphocytic infiltration (fig. 5-35) and myelolipomatous changes.

In aldosterone-producing adenomas, the adjacent adrenal cortex is usually normal (fig. 5-36) but may contain small nodules in the subcapsular region (fig. 5-37), which are also observed in the contralateral adrenal gland, if available. Aldosterone-producing adenomas may be associated with aldosterone-producing diffuse hyperplasia, which may not be identified in hematoxylin and eosin (H&E) slides but appears as a nonpolarized, uninterrupted band when immunostained for CYP11B2. The occurrence of aldosterone-producing nodules and aldosterone-producing diffuse hyperplasia in association with an aldosterone-producing adenoma is associated with a lower rate of complete biochemical success after surgery (11).

In the aldosterone-producing adenomas of patients treated with spironolactone, the drug may cause the formation of eosinophilic, concentrically laminated, electron-dense inclusions, called spironolactone bodies, both in the tumor and in the adjacent adrenal cortex (particularly in the zona glomerulosa) (fig. 5-38). Due to their rich content of phospholipid, the inclusions may stain medium blue with the Luxol fast blue stain. Interestingly, these bodies are not observed in patients treated with newer aldosterone antagonists, such as eplerenone (30).

In sex hormone-secreting adenomas, there may be a predominance of cells with compact, eosinophilic cytoplasm, resembling cells of the zona reticularis (which elaborate sex steroids).

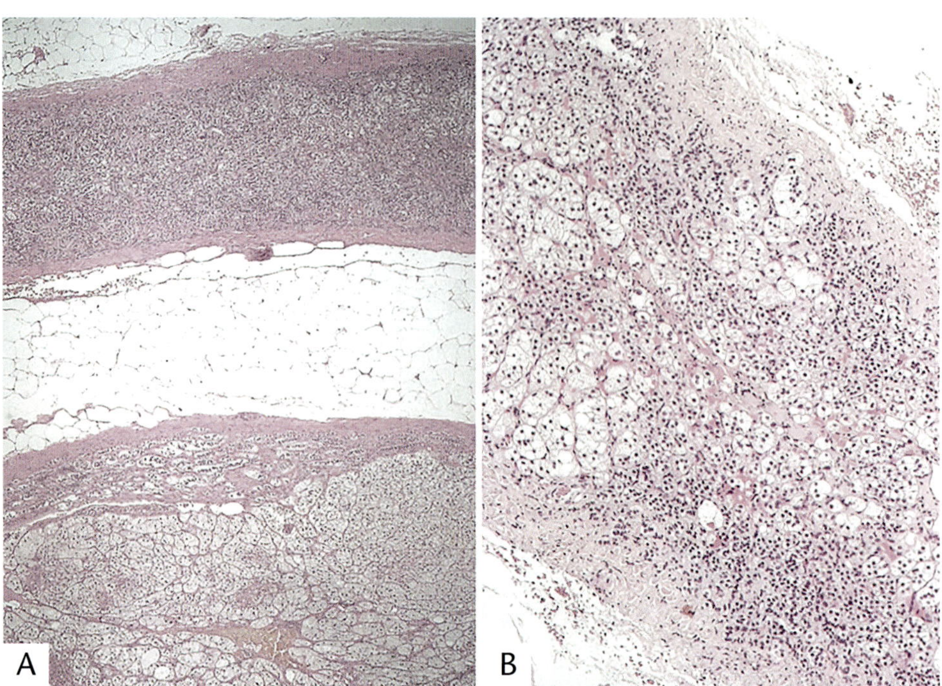

Figure 5-32

CORTICAL ATROPHY IN ADRENAL CORTICAL ADENOMA ASSOCIATED WITH CUSHING SYNDROME

Cortisol-secreting adrenal cortical adenoma (A, bottom) with atrophic perilesional adrenal cortex (A, top and B).

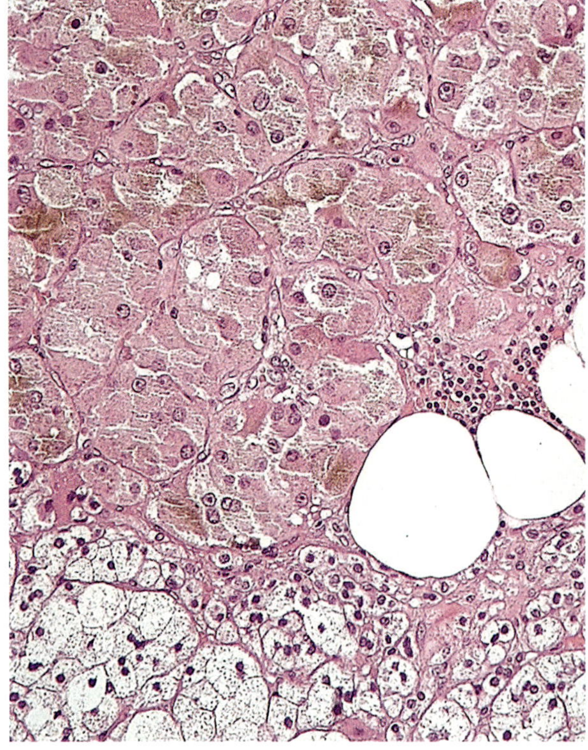

Figure 5-33

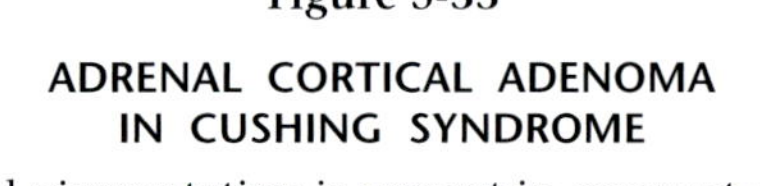

ADRENAL CORTICAL ADENOMA IN CUSHING SYNDROME

Focal pigmentation is present in compact cells.

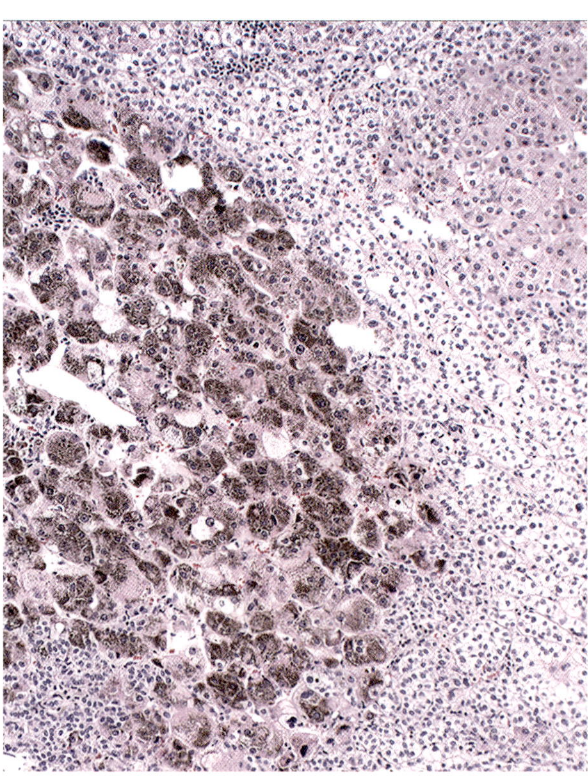

Figure 5-34

ADRENAL CORTICAL ADENOMA

There is dense pigment deposition in clusters of compact lipid-depleted cells.

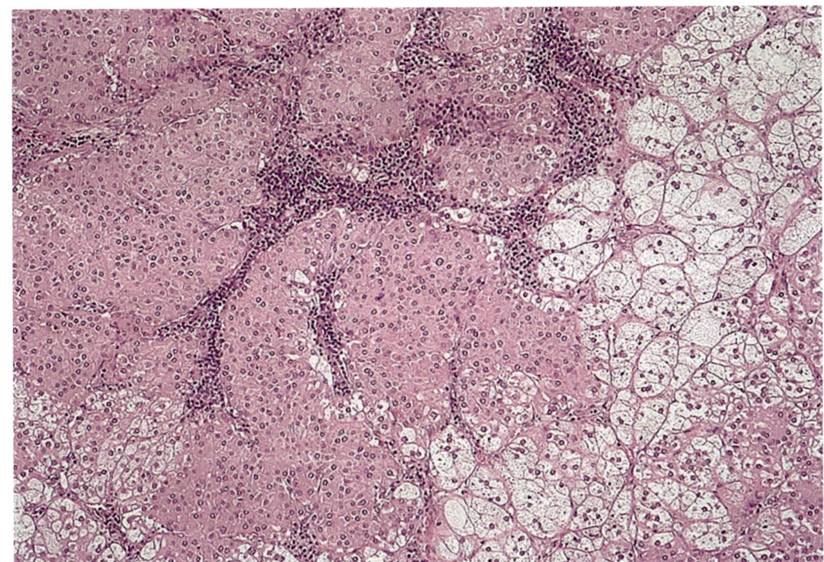

Figure 5-35

ADRENAL CORTICAL ADENOMA IN CUSHING SYNDROME

Intralesional lymphocytic infiltration.

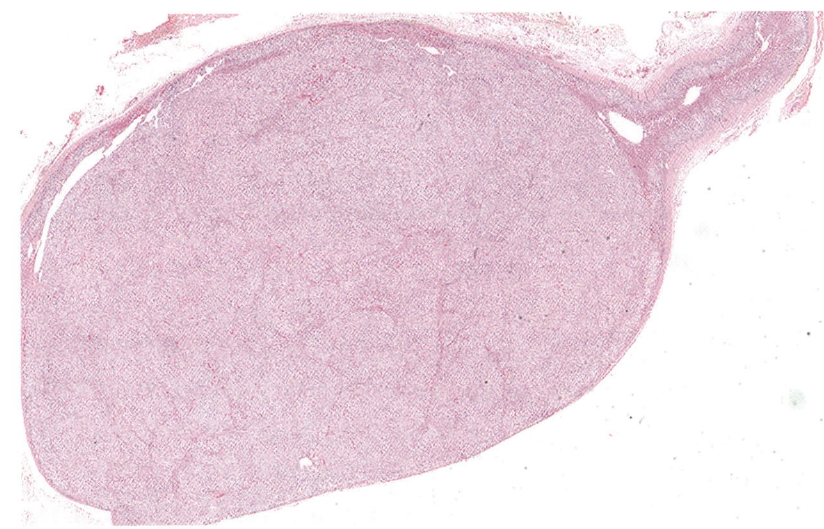

Figure 5-36

ALDOSTERONE-SECRETING ADRENAL CORTICAL ADENOMA

The lesion is composed of a predominant clear cell population; the peritumoral adrenal cortex is not atrophic.

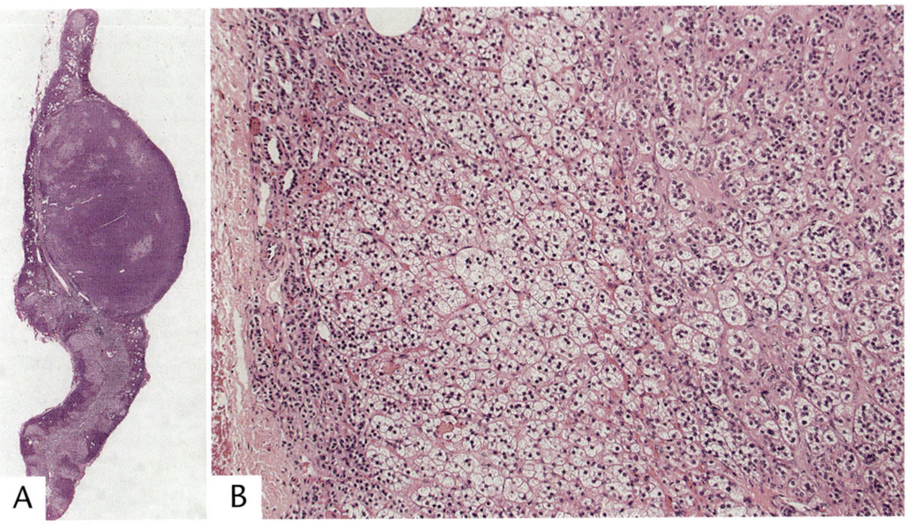

Figure 5-37

ALDOSTERONE-SECRETING ADRENAL CORTICAL ADENOMA

The lesion is composed of a population of both compact and clear cells; the peritumoral adrenal cortex is not atrophic.

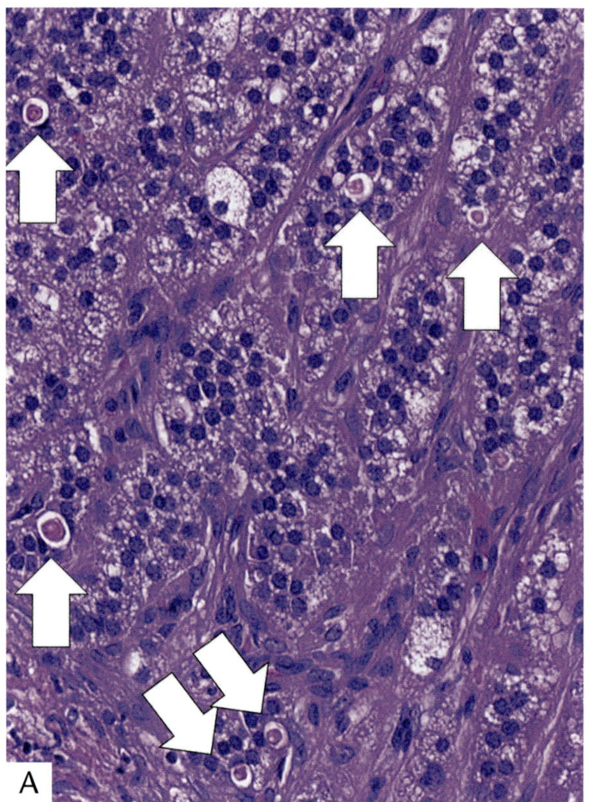

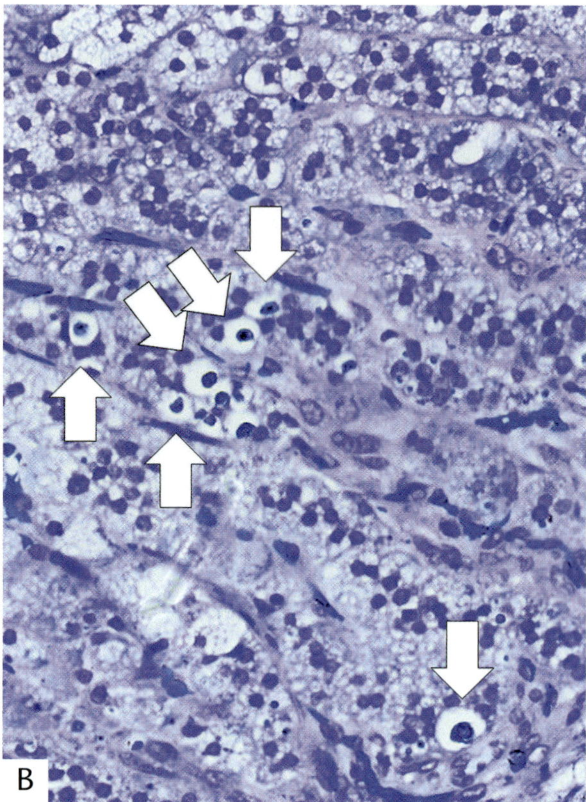

Figure 5-38

ALDOSTERONE-SECRETING ADRENAL CORTICAL ADENOMA

Spironolactone bodies appear as eosinophilic inclusions (arrows) within the cytoplasm of zona glomerulosa cells and are often surrounded by a clear halo (A). Due to their rich content of phospholipid, they stain blue with the Luxol fast blue stain (B) (arrows).

Reinke crystalloids have been identified in rare cases of virilizing adrenal adenomas (31). If the cases do not co-secrete glucocorticoids, the adjacent adrenal remnant or contralateral adrenal cortex is not atrophic.

Adrenal Cortical Adenoma Variants

Oncocytic Adenoma. Oncocytic adenomas represent about one third of oncocytic adrenal cortical neoplasms. They show a female predominance and are nonfunctioning in more than 50 percent of cases (32). In one study of nonfunctioning cases, there was no expression of the enzymes involved in steroidogenesis that were examined (33). Hormonally active tumors more frequently secrete cortisol or sex steroid hormones (34). Radiologically, oncocytic adenomas have the typical characteristics of lipid-poor adrenal tumors (35). They are composed of cells with abundant granular eosinophilic cytoplasm. The lipid content within these cells tends to be sparse or almost nonexistent. Cytologic, ultrastructural, and molecular features of the oncocytic cells are similar to oncocytes in other organs and tumors, including the presence of abundant mitochondria and the mitochondrial DNA common deletion (36).

While oncocytic changes may be frequent in adrenal cortical adenomas, they should be predominant (more than 75 percent) to define this variant. The histologic features are similar to those of other adrenal cortical adenomas, but these tumors more frequently show prominent and diffuse nuclear pleomorphism and a loose dyscohesive pattern of growth, which can pose problems in the differential diagnosis with oncocytic adrenal cortical carcinoma (figs. 5-39–5-41). The diagnostic procedures to distinguish oncocytic adenomas from carcinomas are detailed in chapter 6 and include specific

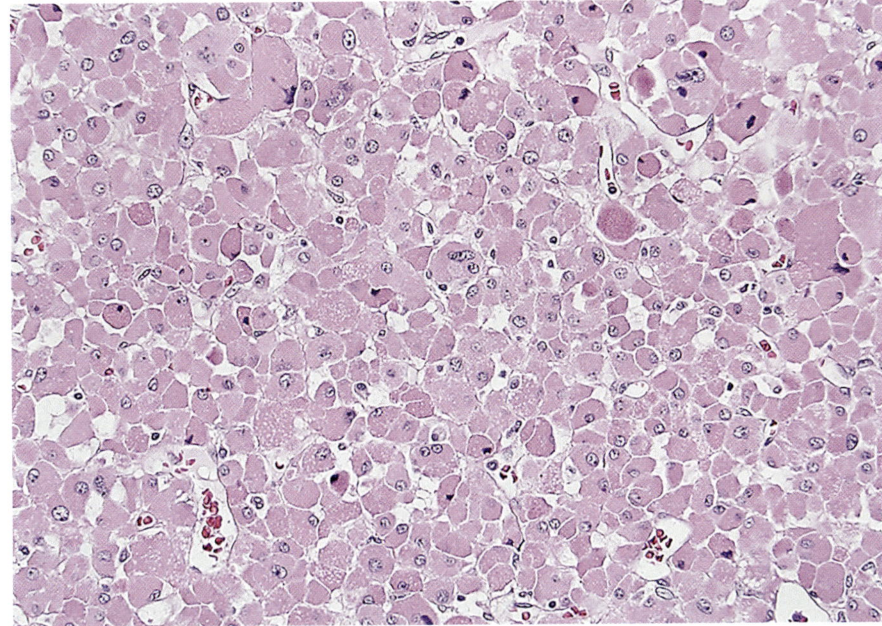

Figure 5-39

ADRENAL CORTICAL ADENOMA, ONCOCYTIC TYPE

Deeply eosinophilic large cells in a diffuse pattern of growth with prominent nucleoli.

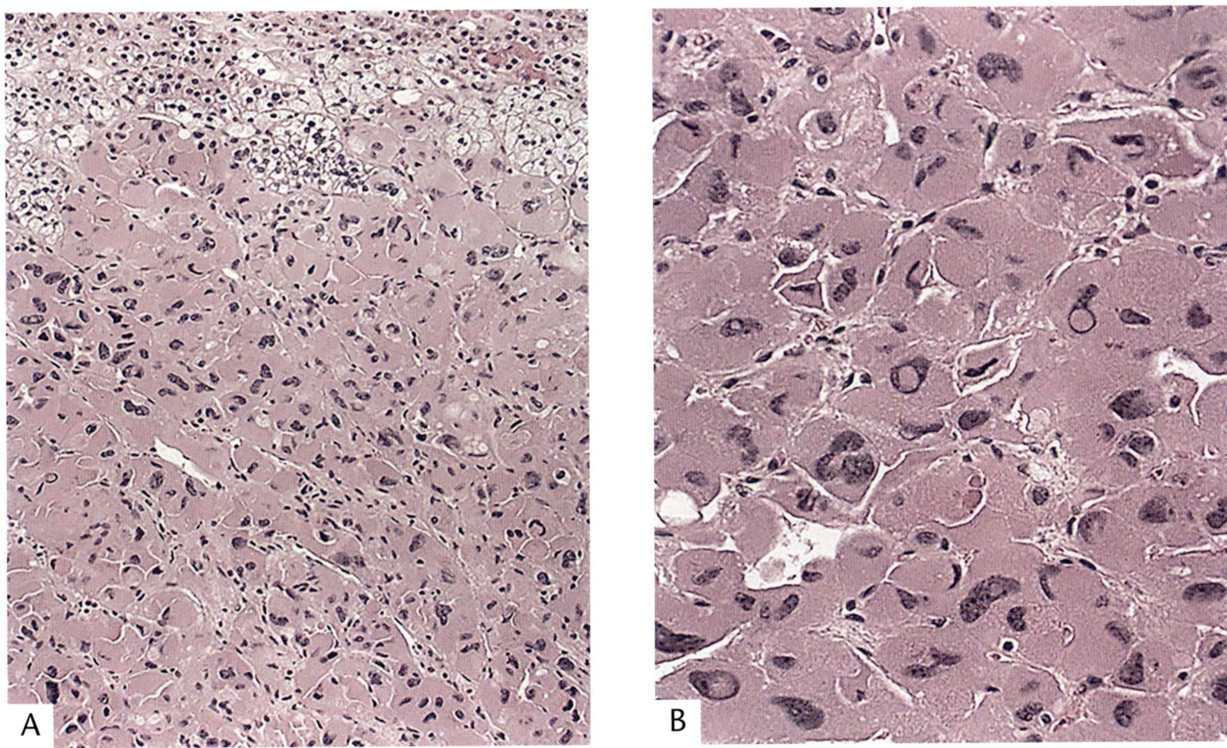

Figure 5-40

ADRENAL CORTICAL ADENOMA, ONCOCYTIC TYPE

A: The oncocytic adrenal cortical adenoma weighed 12 g and measured 3 cm. It was discovered incidentally on abdominal CT scan in a 67-year-old man who underwent surgical resection of a colonic adenocarcinoma. The tumor compresses the adjacent adrenal cortex.

B: The tumor cells have an oncocytic appearance, with abundant eosinophilic, granular cytoplasm. There is a marked degree of nuclear pleomorphism and numerous nuclear pseudoinclusions.

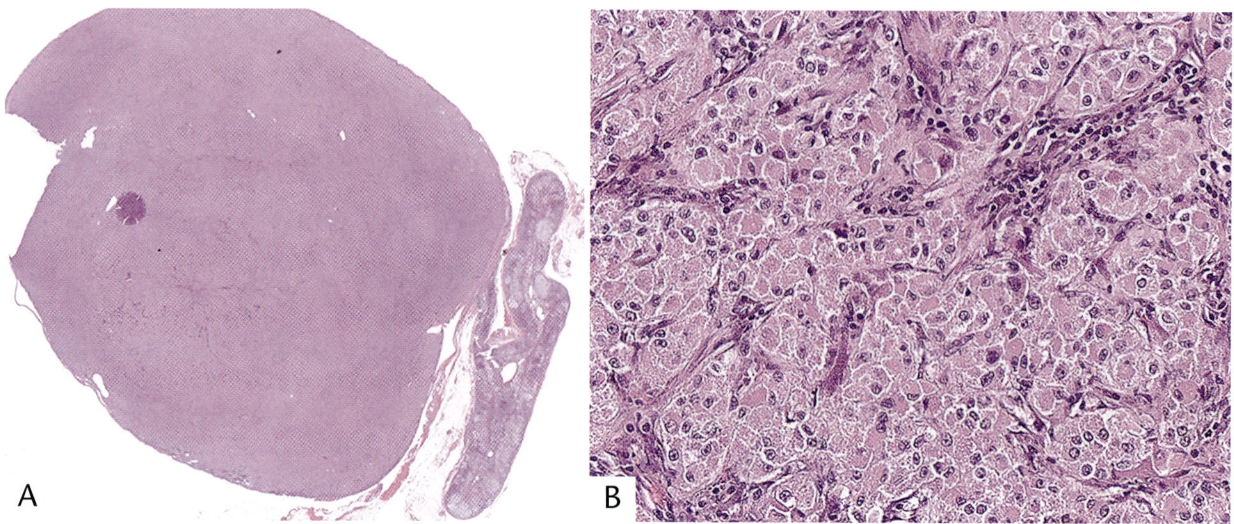

Figure 5-41

ADRENAL CORTICAL ADENOMA, ONCOCYTIC TYPE

A nonfunctioning lesion, well circumscribed, with peritumoral adrenal cortex that is not atrophic (A), is composed of large eosinophilic cells in a lobular architecture with rare intratumoral lymphocytes (B).

diagnostic algorithms, such as the Lin-Weiss-Bisceglia system (37). Another important differential diagnosis of oncocytic adenoma is pheochromocytoma, due to the deep eosinophilic cytoplasm, the large cell size, and the presence of marked nuclear pleomorphism (38). This differential diagnosis may be especially worrisome in case of ectopic retroperitoneal lesions (39).

Pigmented "Black" Adenoma. Most "black" adrenal cortical adenomas are associated with overt or subclinical Cushing syndrome. However, a few cases are associated with mixed cortisol and androgen secretion (40) or primary hyperaldosteronism (41). There is a distinct predilection for female patients, and most tumors are diagnosed in the 3rd to 5th decades of life. The tumors usually weigh less than 35 g, and measure 2 to 3 cm in diameter (42). Black adenoma may show false positive results for malignancy or inconclusive results for benignity with modern imaging modalities, including CT, MRI, adrenal scintigraphy with radiolabeled cholesterol and radiolabeled MIBG, and FDG-PET (43).

Histologically, the architectural patterns of black adenomas are comparable to other adrenal cortical adenomas without excessive lipochrome pigment, and most tumors are composed predominantly or entirely of cells with compact, eosinophilic cytoplasm, similar to, but larger than, cells of the zona reticularis. There is a variable amount of brown or golden brown pigment within the cytoplasm (fig. 5-42), which may spare the immediate perinuclear zone. Clusters of pale-staining, lipid-rich cells are found, and occasionally, there are foci of lipomatous or myelolipomatous metaplasia within the tumor (44) as well as lymphocytic infiltration (fig. 5-43). The deposits correspond to intracytoplasmic material with histochemical and electron microscopic characteristics of lipofuscin and not melanin or neuromelanin (45). Lipofuscin content may be demonstrated by histochemical staining, such as a positive argentaffin reaction with the Fontana-Masson stain that is susceptible to bleaching procedures, a red-orange tint with the periodic acid–Schiff (PAS) stain, and a negative reaction with the iron stain.

Black adenoma should be distinguished from other primary or secondary pigmented lesions in the adrenal gland. A potential source of diagnostic pitfall is the difficulty to evaluate immunohistochemical reactions in diffusely pigmented lesions, even when using red or blue chromogens (fig. 5-44). Among other primary adrenal lesions, pheochromocytomas with diffuse melanin pigmentation are on record (46)

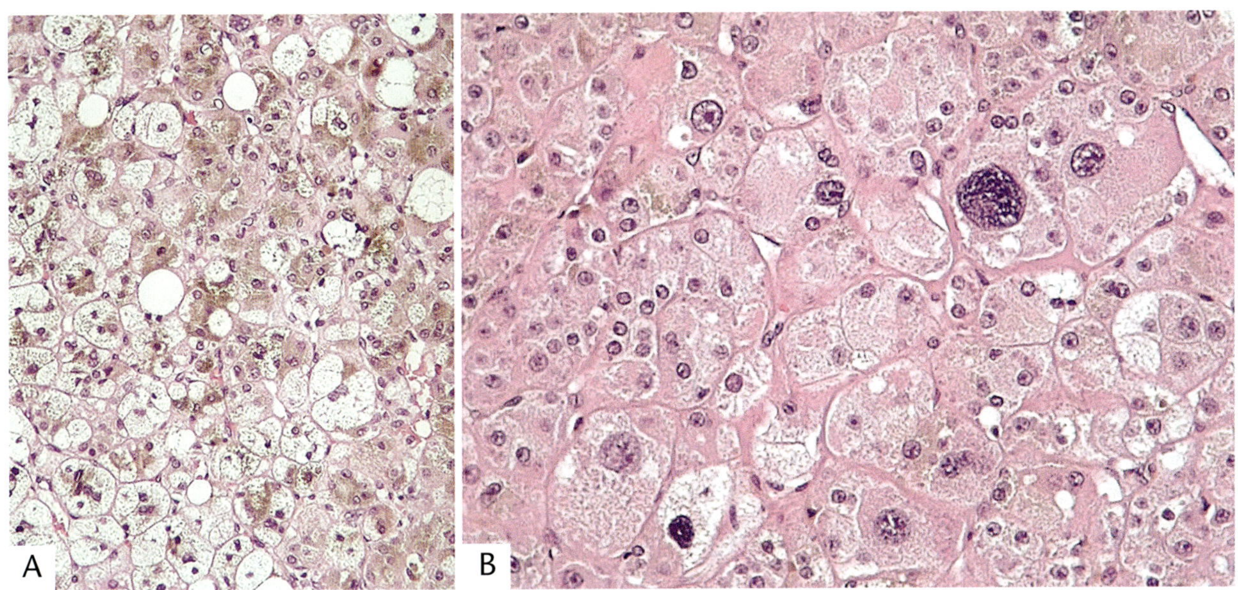

Figure 5-42

"BLACK" ADRENAL CORTICAL ADENOMA IN CUSHING SYNDROME

Pigment deposition is present in both clear and compact cells (A). Eosinophilic cells show focally marked nuclear atypia (B).

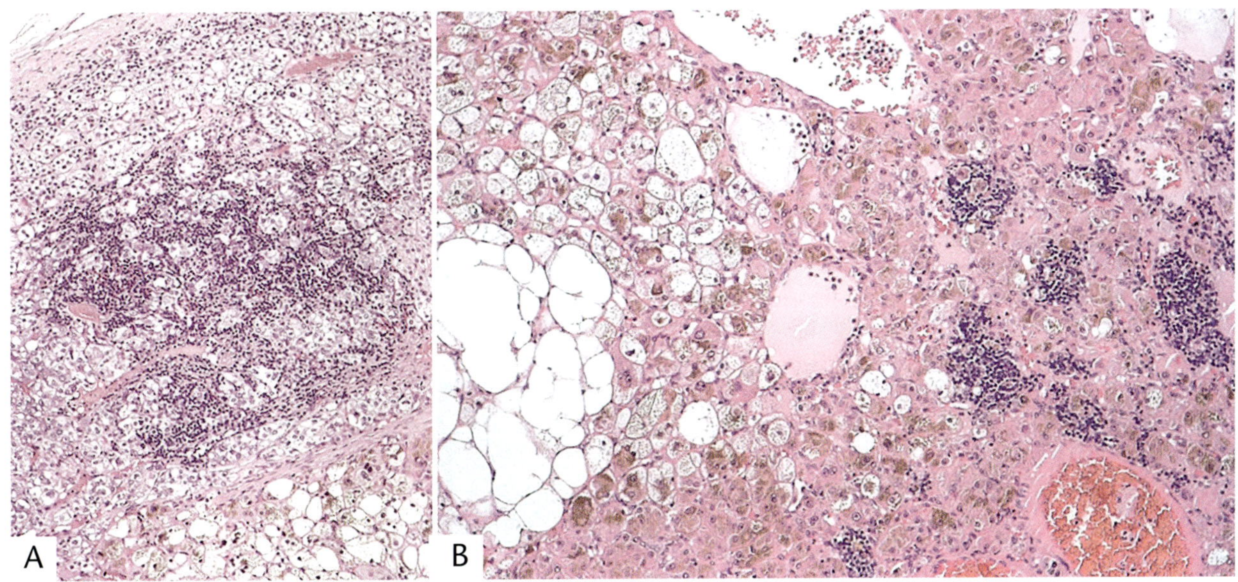

Figure 5-43

"BLACK" ADRENAL CORTICAL ADENOMA IN CUSHING SYNDROME

This black adenoma is characterized by a variably pigmented cell population (A), focally dense lymphocytic infiltration and adipose metaplasia (B).

and represent the most relevant differential diagnosis (see also chapter 9). In the group of secondary lesions, melanoma metastases represent the first differential diagnosis that should be taken into account, even in the absence of a previous clinical history of melanoma, since an adrenal metastasis may be the first clinical evidence of disease (47).

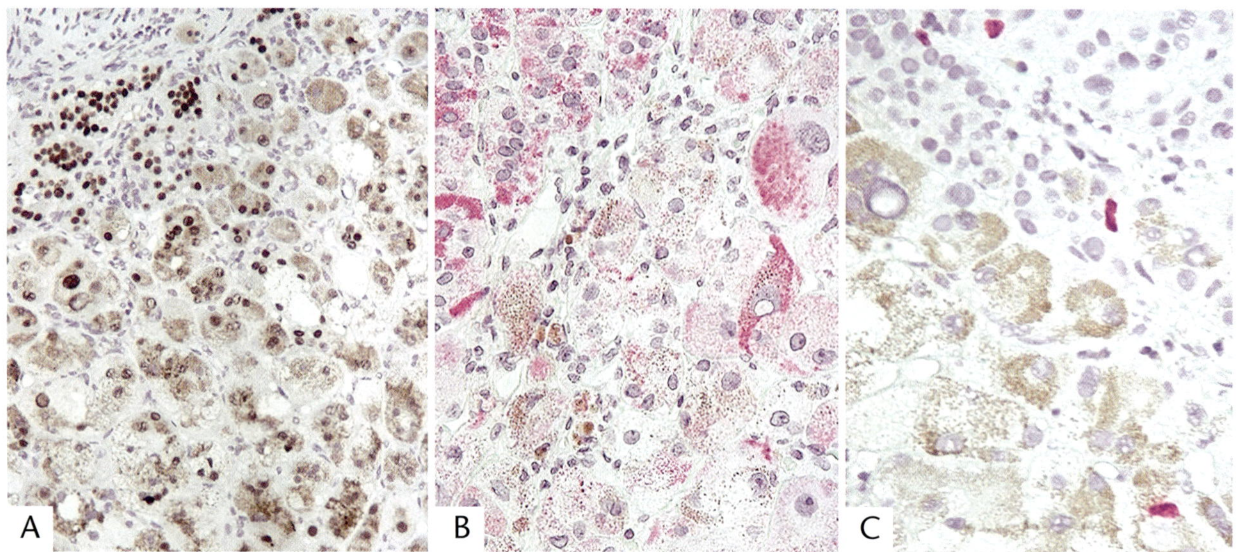

Figure 5-44

"BLACK" ADRENAL CORTICAL ADENOMA IN CUSHING SYNDROME

Tumor cells express steroidogenic factor 1 (SF1, A), Melan-A (red, B), and a low Ki-67 labeling index (red, C).

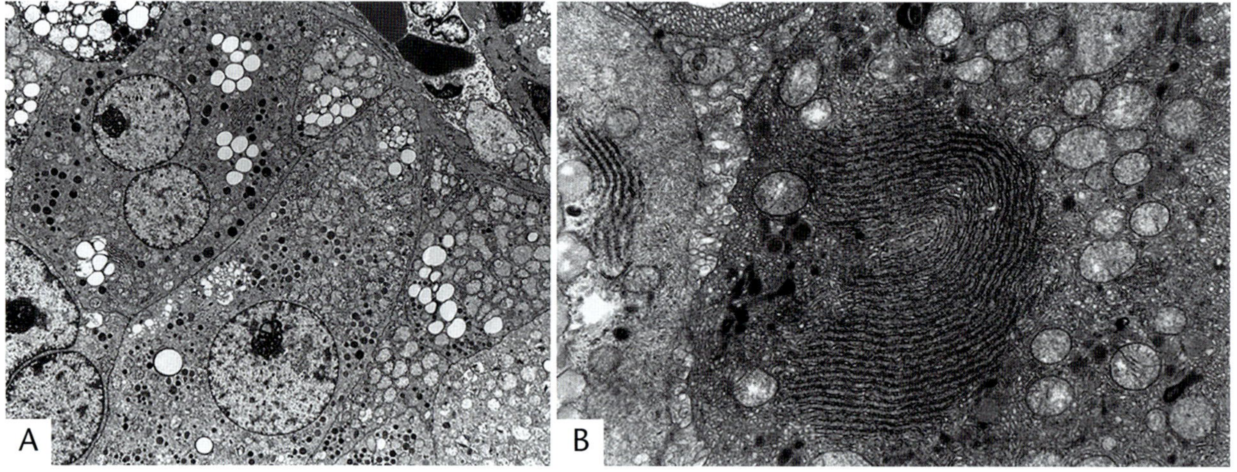

Figure 5-45

ADRENAL CORTICAL ADENOMA IN CUSHING SYNDROME

A: The tumor cells contain small to medium-sized electron-dense granules, which represent lipofuscin. Some cells contain sparse lipid droplets.

B: Stacks of rough endoplasmic reticulum and prominent smooth endoplasmic reticulum are seen. Clusters of dense granules probably represent primary lysosomes.

ULTRASTRUCTURAL FINDINGS

Electron microscopy no longer has a major role in the differential diagnosis of adrenal cortical adenomas, but ultrastructural studies have been fundamental to comprehend some functional characteristics of these tumors (figs. 5-45–5-49). Most tumor cells are remarkable for the abundant amount of intracytoplasmic lipid droplets, which vary in size and density from cell to cell, including cells that are totally depleted of lipids. As in the normal adrenal gland, there may

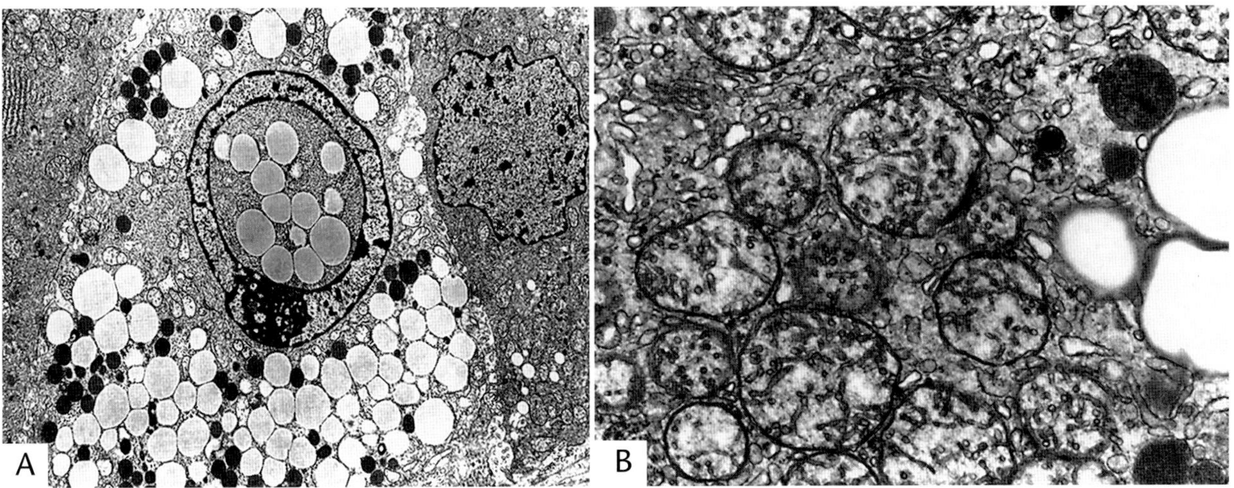

Figure 5-46

ADRENAL CORTICAL ADENOMA IN CUSHING SYNDROME

A: A nuclear pseudoinclusion viewed en face in an adrenal cortical adenoma from a patient with Cushing syndrome. The structure is nonspecific and results from irregular deep infolding of the nuclear membrane.

B: Cristae of mitochondria have a tubular or vesicular profile typical of steroid-producing cells. There is prominent smooth endoplasmic reticulum as well as a few free polyribosomes.

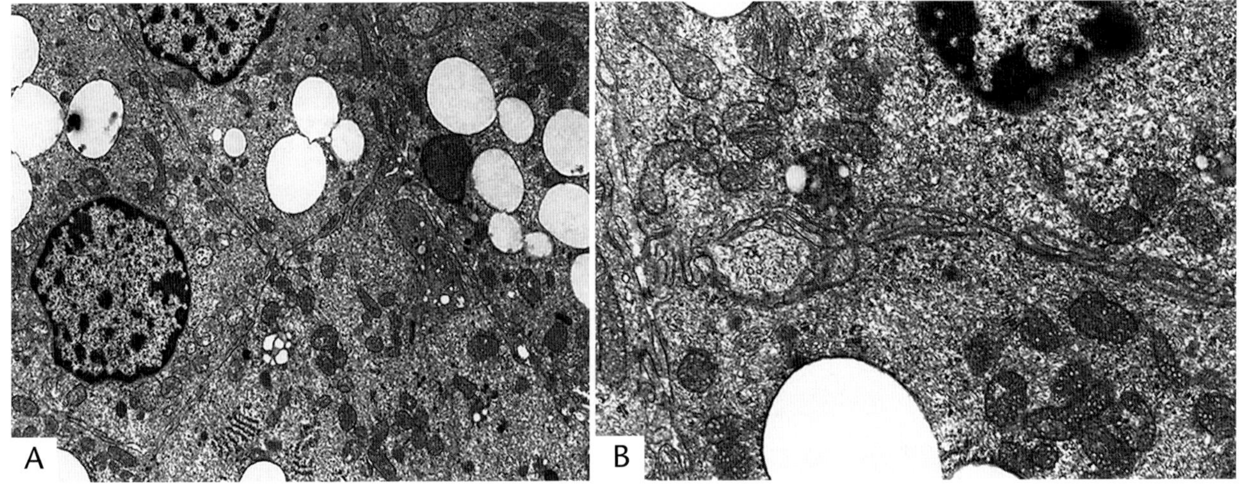

Figure 5-47

ALDOSTERONE-SECRETING ADRENAL CORTICAL ADENOMA

A: The tumor cells have much smooth endoplasmic reticulum and a few lipid droplets, which vary in size. Microvillous cytoplasmic extensions are evident along the cell borders. Mitochondria have flat plate-like cristae.

B: Most mitochondria have flat lamellar cristae, some of which are slightly dilated.

be prominent microvillous projections along the cell borders and abundant smooth endoplasmic reticulum. In the presence of nuclear pseudoinclusions, these are due to an irregularity in the nuclear membrane, with indentation and infolding of cell cytoplasm, often with the same complement and density of cellular organelles. Mitochondria may be prominent, and are usually round to oval, although some are elongated or distorted in shape. The mitochondrial cristae usually have a tubular or vesicular profile, similar to the cells of the zonae reticularis and

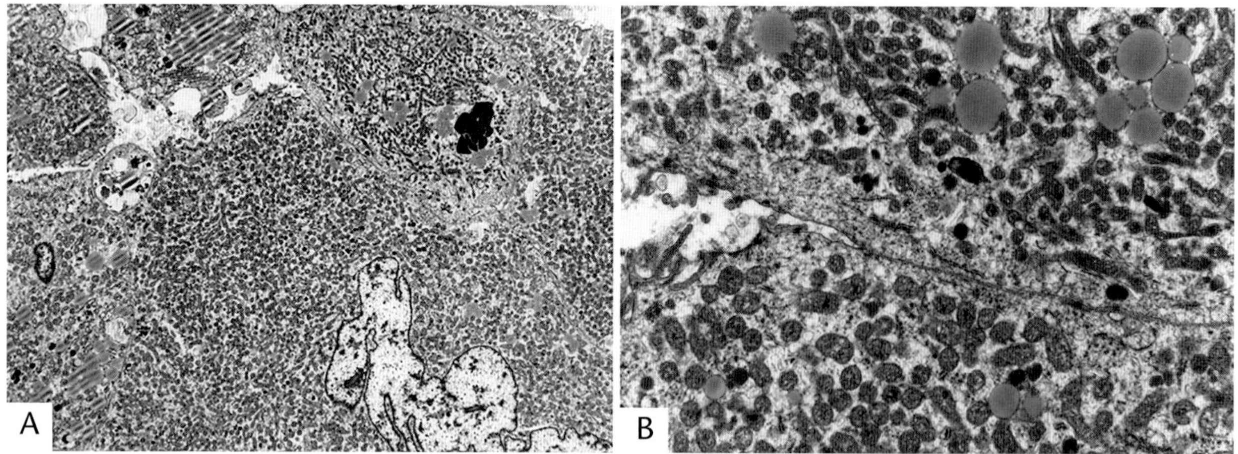

Figure 5-48

ONCOCYTIC ADRENAL CORTICAL ADENOMA

A: Cells of this oncocytic adrenal cortical adenoma contain numerous round to elongated mitochondria. There are a few small lipid droplets. The nuclear folds and clefts may form nuclear pseudoinclusions, detectable on routine light microscopy.

B: There are numerous mitochondria with predominantly tubular cristae. Few lipid droplets are present, along with a few primary lysosomes. Microvillous projections are seen along the cell borders on the left side.

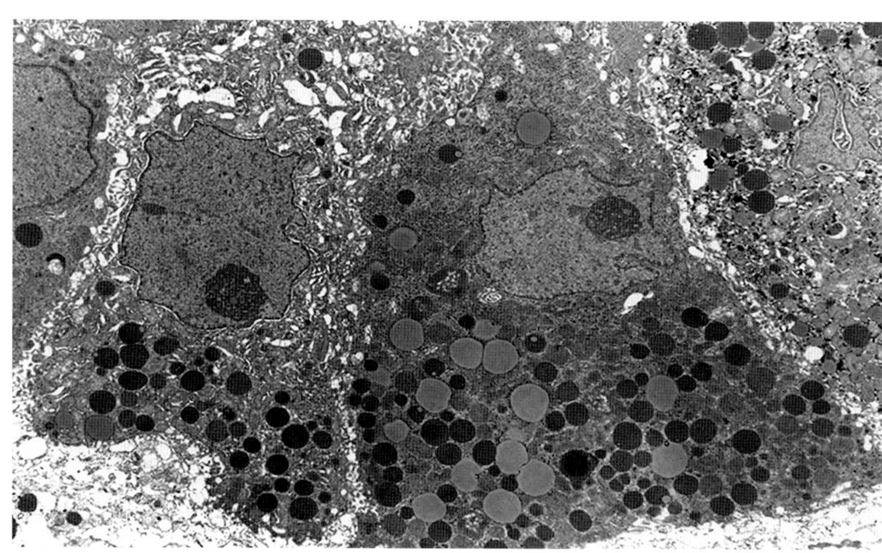

Figure 5-49

"BLACK" ADENOMA ASSOCIATED WITH CUSHING SYNDROME

The tumor cells contain abundant lipofuscin and have only sparse lipid droplets. Microvillous projections are seen along the cytoplasmic borders.

fasciculata. Aldosteronomas are characterized by mitochondria with lamellar cristae resembling the morphology of zona reticularis cells; prior to the availability of reliable antibodies for steroidogenic enzymes, ultrastructural examination was useful for this distinction (48).

IMMUNOHISTOCHEMICAL FINDINGS

Immunohistochemistry is indicated for adrenal cortical adenoma in three situations: the definition of the primary adrenal cortical origin, the description of functional characteristics, and the distinction of adenoma from carcinoma (49). This last aspect is discussed in chapter 6.

Primary Adrenal Cortical Origin

Several biomarkers have been proposed to confirm adrenal cortical origin. However, the specificity and sensitivity of various biomarkers should be recognized since the adrenal gland

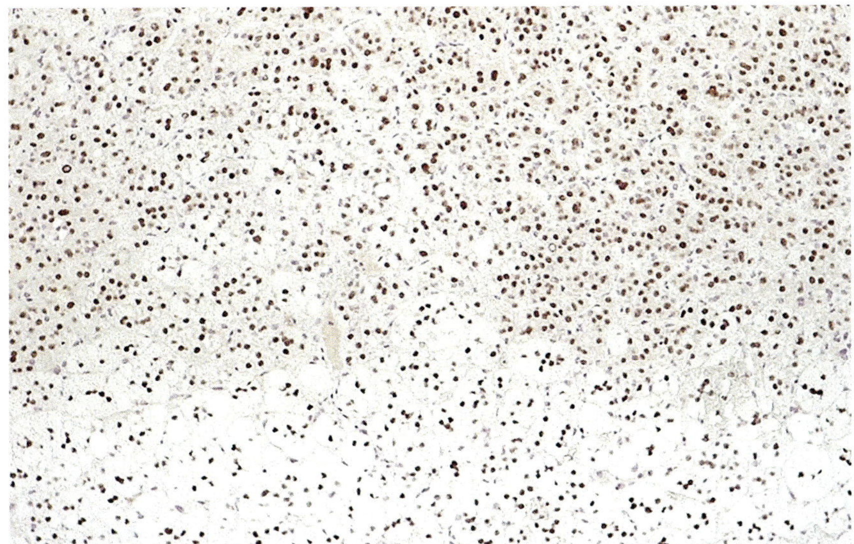

Figure 5-50

ADRENAL CORTICAL ADENOMA IN CUSHING SYNDROME

Steroidogenic factor 1 (SF1) is expressed in both compact (top) and clear (bottom) cells.

and retroperitoneum host various neoplasms that can pose diagnostic challenges during the preoperative workup. Steroidogenic factor 1 (SF1) is considered the most reliable and specific, since this nuclear transcription factor is expressed in normal adrenal cortex as well as in all types of adrenal cortical neoplasms (fig. 5-50). SF1 is a master regulator of steroidogenesis and is expressed, other than in the adrenal cortex, in steroidogenic cells of the gonads as well as in the gonadotrophs of the pituitary gland. Excluding these two other cell types, SF1 specificity for adrenal cortical origin is up to 100 percent, and its sensitivity reaches also 100 percent in benign adrenal cortical lesions (50). Melan-A is a marker for normal steroidogenic cells and tissues (51), and is also 100 percent sensitive in adrenal adenomas, at variance to carcinoma. However, it is less specific than SF1 since it also stains melanomas and PEComas (52). A similar profile is observed for other markers expressed in the adrenal cortex and steroidogenic tissues such as alpha-inhibin, calretinin, and podoplanin (stained with antibody D2–40). Notably, these latter two markers are also expressed in mesothelial cells and related neoplasms that can also manifest in the adrenal gland alone or in association with adrenal cortical adenoma (53).

Functional Characterization

As already discussed, the demonstration of atrophy of the adjacent adrenal cortex supports the presence of a glucocorticoid-secreting adrenal cortical proliferation in the absence of exogenous cortisol administration, whereas the identification of spironolactone bodies in an adrenal cortical neoplasm and adjacent paradoxical zona glomerulosa hyperplasia often confirms aldosterone secretion. However, patients may present with multiple cortical proliferations, and some of these are related to clinical, morphologic, and steroidogenic heterogeneity features that are also reflected in the molecular characteristics of adrenal cortical neoplasms. Therefore, immunophenotype may be a useful tool to accurately distinguish functional differentiation in these cases.

Several biomarkers have been suggested in the assessment of functional differentiation in adrenal cortical proliferations leading to primary aldosteronism (54). Among these, monoclonal antibodies against steroidogenic enzymes have garnered significant interest (55,56). Monoclonal antibodies not only distinguish these isoforms from each other, but more importantly, enable diagnosticians to visualize the cellular proliferation responsible for the synthesis of the corresponding steroid hormones (e.g., aldosterone and cortisol) in surgical specimens, depicting a wide functional heterogeneity. Among others, cytochrome P450 11B2 (CYP11B2) plays a pivotal role in aldosterone synthesis (fig. 5-51), while cytochrome P450 11B1 (CYP11B1) and cytochrome P450 17A1 (CYP17) are involved in cortisol synthesis in normal human adrenal glands. Interestingly,

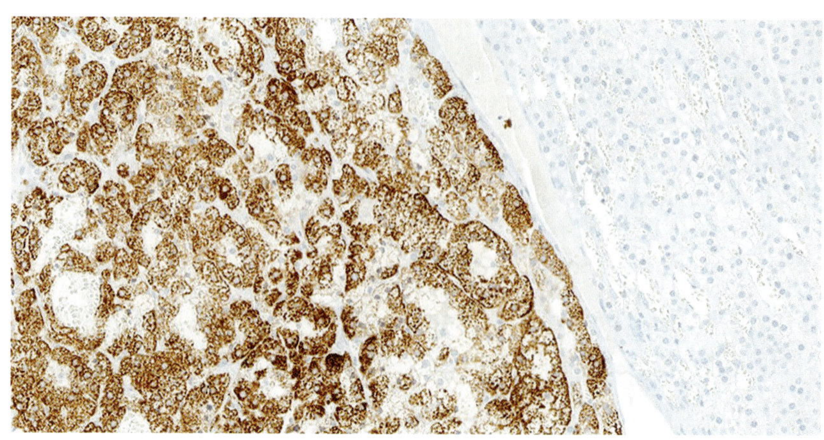

Figure 5-51

ALDOSTERONE-SECRETING ADRENAL CORTICAL ADENOMA

Diffuse nonpolarized CYP11B2-positive immunohistochemistry is compared to adjacent cortex devoid of aldosterone-producing cells.

detailed data on their expression in aldosterone-secreting adenomas clearly show a marked intratumoral heterogeneity, with double- or triple-positive hybrid cells, especially occurring in cases with cortisol co-secretion (57). 3β-hydroxysteroid dehydrogenase type 2 (HSD3B2) immunoreactivity is more abundant than type 1 (HSD3B1) in aldosterone-producing adenomas, but the status of HSD3B1 immunoreactivity in the tumor significantly correlates with that of CYP11B2 (58). Also, in cortisol-secreting adenomas, the distribution of expression of steroidogenic enzymes is heterogeneous: CYP21A, CYP11B1, and CYP17A1 immunoreactivity are all significantly higher in compact cells than in clear cells and also correlate with the genomic status of cases (59).

MOLECULAR GENETICS OF SPORADIC ADRENAL CORTICAL ADENOMAS

The molecular landscape of adrenal cortical adenomas largely differs according to the functional status (60). This was demonstrated in a large study including different benign adrenal cortical lesions, either sporadic or developing in a hereditary setting (61). These lesions were subclassified molecularly through a multimodal approach, which integrated transcriptomic, methylomic, miRNome, and genomic profiling, into four main groups. The largest group included most cortisol-producing lesions (including adrenal cortical adenomas and primary bilateral macronodular and micronodular adrenocortical diseases) that shared the activation of the cAMP/PKA pathway as a common hallmark. The second group was composed mainly of adrenal cortical adenomas, either nonfunctioning or with mild autonomous cortisol excess, and was characterized by *CTNNB1* mutations, as well as by 9q gains and a lower methylation of CpG islands. A third group was homogeneously composed of patients affected by *ARMC5*-related primary bilateral macronodular adrenocortical disease, with a high expression of MIR100 cluster and a global hypermethylation of CpG islands. The last group included aldosterone-producing adrenal cortical adenomas and was characterized by a high *CYP11B2* expression. These data are extremely informative and of interest in a pathophysiologic perspective, but still have a limited applicability mainly due to the complexity of the molecular approach. The paragraphs below restrict their focus on main somatic genomic characteristics of sporadic adrenal cortical adenomas, subdivided based on their functional characteristics.

Adenoma Producing Aldosterone

Whole exome sequencing performed on DNA from aldosterone-producing adenomas led to the identification of recurrent somatic mutations in genes coding for ion channels (*KCNJ5* and *CACNA1D*) and ATPases (*ATP1A1* and *ATP2B3*). These genes are essential for regulating intracellular ion homeostasis and cell membrane potential. All these mutations promote increased intracellular calcium signaling through cell membrane depolarization and opening of voltage-dependent calcium channels, or impaired intracellular calcium recycling, thereby leading to high aldosterone levels by constitutive expression of CYP11B2

(62). Somatic mutations in these genes are detectable in about 54 percent of cases, and the frequency varies among the different genes from 38 percent for *KCNJ5*, 9.3 percent for *CACNA1D*, 5.3 percent for *ATP1A1*, and 1.7 percent for *ATP2B3* (63). The relative prevalence of mutations is also influenced by race, with *KCNJ5* somatic mutations much more common in East Asians compared to Europeans (64,65) and *CACNA1D* most prevalent in Blacks (66).

KCNJ5 codes for an inwardly rectifying K+ channel, the G-protein-activated inward rectifier potassium channel GIRK4, which is mainly expressed in the zona glomerulosa of the adrenal cortex. The two most common somatic mutations map to the selectivity filter of GIRK4 (p.Gly151Arg and p.Leu168Arg). In addition to these two mutations, the majority of *KCNJ5* mutations are located within or near the selectivity filter, rendering the channel permeable to sodium, which leads to chronic cell membrane depolarization (67).

KCNJ5 mutations are associated with female sex, younger age, higher levels of plasma aldosterone, larger tumors, and better outcome (in terms of biochemical correction of blood pressure) in young patients with aldosterone-producing adenomas (63,68,69). Moreover, genotype-phenotype correlations show that *KCNJ5*-mutated aldosterone-producing adenomas have a higher proportion of clear lipid-rich cells than compact cells and a higher expression of CYP17A1, whereas both *ATP1A1*- and *CACNA1D*-mutated adenomas harbor more marked histologic intratumoral heterogeneity (70). *KCNJ5*-mutated aldosterone-producing adenomas have also specific transcriptome and methylome signatures with general DNA hypomethylation, and gene expression changes in Wnt signaling and inflammatory response pathways (71).

More than 20 mutations have been identified in the *CACNA1D* gene, all gain-of-function mutations. These lead to a decrease in the threshold of the voltage-dependent activation or impaired channel inactivation, which is followed by increased intracellular calcium concentrations and thereby an induction of aldosterone biosynthesis (72,73).

ATP1A1 and *ATP2B3* are members of the P-type family of ATPases and are composed of 10 transmembrane domains with intracellular N and C termini. *ATP1A1* codes for the Na^+ ATPase alpha-1 subunit. Mutations in this pump lead to a loss of its activity and affinity to K+ and to an inward proton or sodium leak, which has been proposed to induce aldosterone production through cell membrane depolarization and increased calcium influx (72,74). *ATP2B3* codes for the plasma membrane calcium-transporting ATPase 3 (PMCA3). Mutations are found in the transmembrane domain M4 and lead to reduced calcium export, which is due to the loss of the physiologic pump functions, and increased intracellular calcium signaling due to the depolarization-activated Ca^{2+} channels or to an increase in calcium influx by the opening of depolarization activated calcium channels (75).

Alterations of zinc transporters may also cause aldosterone-producing adenoma. In a recent multi-institutional study, recurrent in-frame deletions in *SLC30A1* were detected in about 1 percent of adenomas producing aldosterone (or aldosterone-producing nodules) (76). The identified *SLC30A1* variants are situated close to the zinc-binding site in transmembrane domain II and probably cause abnormal ion transport, leading to an aberrant Na^+ intracellular influx and an increase in cytosolic Ca^{2+} activity, which stimulates CYP11B2 mRNA expression and aldosterone production. More recently, somatic mutations in *CLCN2* and *CACNA1H* (whose germline mutations are associated with familial hyperaldosteronism types II and IV) have been found in about 1 percent and 4 percent, respectively, of sporadic aldosterone-producing adenomas lacking mutations in other known genetic drivers (77,78).

Aldosterone-producing adenomas may also harbor mutations that are more common in other types of adenoma. Mutations in *PRKACA* (encoding protein kinase cAMP-activated catalytic subunit a) have been identified in less than 2 percent of cases, and are associated with biochemical cortisol hypersecretion (79). Beta-catenin accumulation in both nuclear and cytoplasmic compartments is common in aldosterone-producing adenomas, with a prevalence of about 70 percent, and is functionally associated with increased transcription of *CYP11B2* (80). However, only 2 to 5 percent of aldosterone-producing adenomas carry somatic mutations in the beta-catenin coding gene, *CTNNB1* (81), suggesting that the Wnt/beta-catenin pathway is activated through other mechanisms.

Adenoma Producing Cortisol

Protein kinase A (PKA) is a tetrameric enzyme composed of a regulatory dimer and two catalytic subunits (82). There are four regulatory subunits (RIα, RIβ, RIIα, and RIIβ) and four catalytic subunits (Cα, Cβ, and Cγ, and PRKX) that act as serine threonine kinases. Activating somatic *PRKACA* mutations are the most common molecular alterations found in cortisol-producing adenomas, occurring in approximately 40 percent of cases (83–86), with the hotspot mutation L206R. In patients with cortisol-producing adenomas carrying a somatic *PRKACA* mutation, the clinical phenotype is more severe than wild-type ones, and the mutations are only found in cases responsible for overt Cushing syndrome.

An activating somatic mutation of the catalytic subunit beta of the PKA gene (*PRKACB*) has also been reported in cortisol-producing adenomas as responsible for overt Cushing syndrome, but this alteration is apparently rare (87). Somatic alterations of *PRKAR1A* were also described in cortisol-producing adenomas, with loss of heterozygosity (LOH) in the *PRKAR1A* locus 17q found in 7 of the 29 studied cases and somatic-inactivating mutations in 3 cases responsible for overt Cushing syndrome (88). Recently, *PRKAR1B* copy-number gains were found in 3 of 21 cortisol-secreting adenomas (89). Somatic-activating mutations of *GNAS1*, the gene coding for Gs protein alpha subunit, were also reported in a few rare cases of cortisol-producing adenoma (90).

Activation of cAMP/PKA signaling leads to different pathway alterations in cortisol-secreting adenomas. In *GNAS1*-mutated tumors, an overexpression of extracellular matrix receptor interaction and focal adhesion pathways is observed, while in *PRKAR1A*-mutated tumors, genes related to the Wnt signaling pathway are overexpressed (91). Genotypes in cortisol-secreting adenomas are also associated with steroidogenic activity, since CYP17A1 and 3βHSD enzymes are expressed at higher levels in *PRKACA*-mutated cases, whereas the serum DHEAS level is significantly higher in *GNAS1*-mutated cases (60).

Activation of the Wnt/beta-catenin pathway is reported in around 40 percent of adrenal cortical adenomas without any somatic mutation of *PRKAR1A*, most explained by the occurrence of a somatic-activating mutation of the beta-catenin gene (*CTNNB1*) (92). *CTNNB1* mutations are, however, more frequent in adenomas with subclinical Cushing syndrome and in nonfunctioning adrenal adenomas.

Adenoma Co-Producing Cortisol and Aldosterone

A specific molecular signature of this adenoma variant is not definitely depicted, although it is associated with the presence of *KCNJ5* mutations, as part of the spectrum of aldosterone-producing adenoma-specific genomic alterations (93). Epigenetic regulation of CYP11B1 by DNA methylation is associated with cortisol secretion in aldosterone-producing adenomas (94).

Sex Hormone-Producing Adenoma

No molecular data are available for this clinical subtype.

Nonfunctioning Adenoma

CTNNB1 mutations are the most frequent molecular alteration in nonfunctioning adenomas, accounting for more than 50 percent of cases (92). No additional recurrent molecular alterations have been described in this clinical subset.

DIFFERENTIAL DIAGNOSIS

The diagnostic algorithm of adrenal cortical adenoma takes into account macroscopic and histologic findings (fig. 5-52). Distinguishing between adrenal cortical adenoma and carcinoma is discussed in chapter 6. It is straightforward in most cases, but some tumors may be worrisome due to the presence of equivocal morphologic features including disruption of the reticulin framework and architectural and cytologic features in the absence of unequivocal signs of malignancy such as increased mitotic activity, atypical mitotic features, necrosis, or vascular and capsular invasion (fig. 5-53). Extensive sampling is mandatory in such cases.

In the presence of a large proportion of compact cells, or more importantly, in the presence of diffuse oncocytic features, adrenal

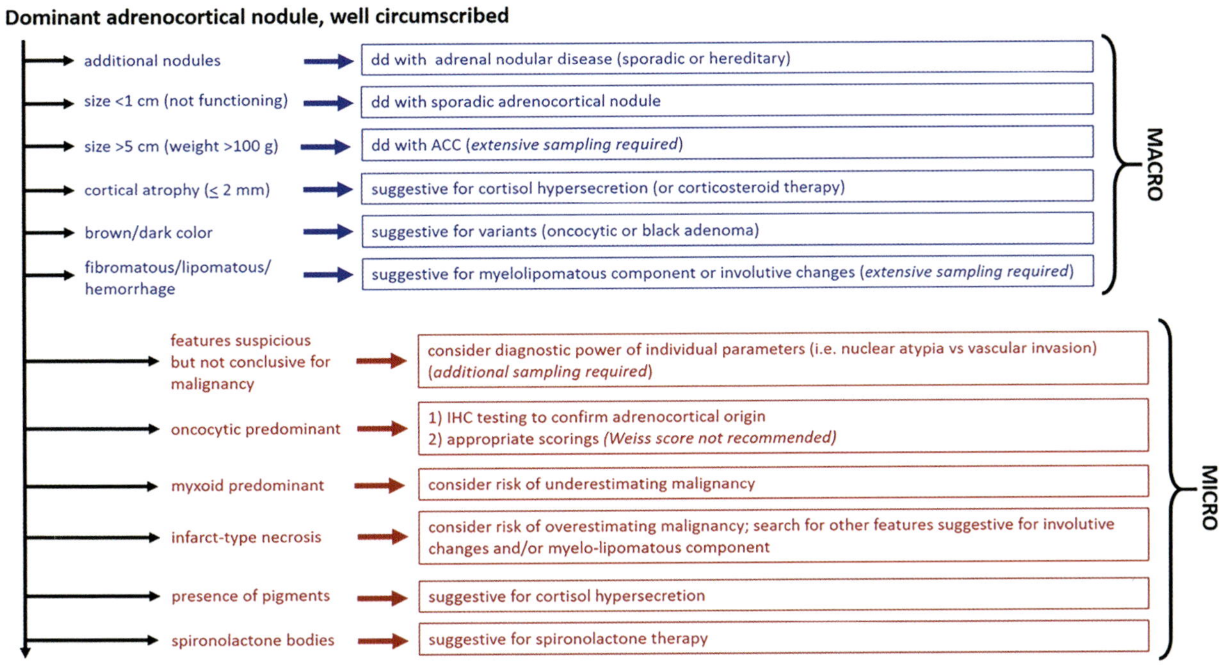

Dominant adrenocortical nodule, well circumscribed

MACRO

- additional nodules → dd with adrenal nodular disease (sporadic or hereditary)
- size <1 cm (not functioning) → dd with sporadic adrenocortical nodule
- size >5 cm (weight >100 g) → dd with ACC (*extensive sampling required*)
- cortical atrophy (≤ 2 mm) → suggestive for cortisol hypersecretion (or corticosteroid therapy)
- brown/dark color → suggestive for variants (oncocytic or black adenoma)
- fibromatous/lipomatous/ hemorrhage → suggestive for myelolipomatous component or involutive changes (*extensive sampling required*)

MICRO

- features suspicious but not conclusive for malignancy → consider diagnostic power of individual parameters (i.e. nuclear atypia vs vascular invasion) (*additional sampling required*)
- oncocytic predominant → 1) IHC testing to confirm adrenocortical origin 2) appropriate scorings *(Weiss score not recommended)*
- myxoid predominant → consider risk of underestimating malignancy
- infarct-type necrosis → consider risk of overestimating malignancy; search for other features suggestive for involutive changes and/or myelo-lipomatous component
- presence of pigments → suggestive for cortisol hypersecretion
- spironolactone bodies → suggestive for spironolactone therapy

Figure 5-52

MAIN MACROSCOPIC AND MICROSCOPIC FEATURES APPLIED IN THE DIAGNOSTIC ALGORITHM OF ADRENAL CORTICAL ADENOMA

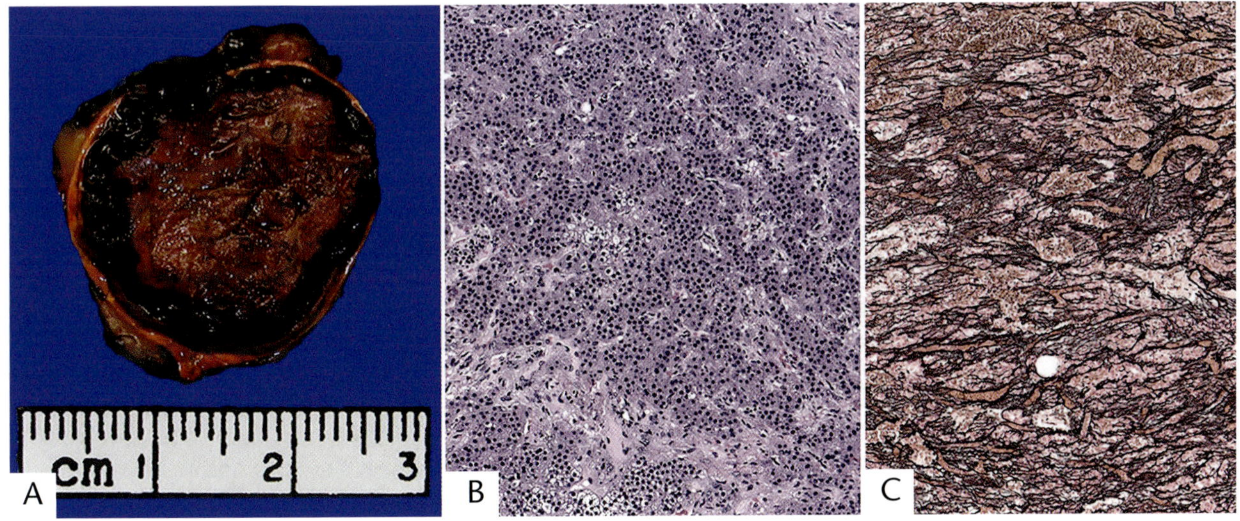

Figure 5-53

ADRENAL CORTICAL ADENOMA WITH WORRISOME MORPHOLOGIC FEATURES

This cortical adenoma was resected because of interval enlargement. The macroscopic specimen (A) shows the adrenal distended by a hemorrhagic infarcted mass with no definite tumor identifiable grossly. Focal viable tumor identified microscopically at the periphery of the mass is composed predominantly of lipid-depleted cells (B) and shows an altered reticulin pattern (C) but does not fully qualify to be diagnosed as carcinoma by either the Weiss criteria or reticulin algorithm. There was no evidence of invasion, and mitoses were not identified.

cortical adenoma should be distinguished from pheochromocytoma. The differential diagnosis is easy using the appropriate immunohistochemical panel but pathologists should be aware of this potential pitfall (38). Synaptophysin is expressed in both lesions, and the distinction should rely instead on other neuroendocrine (i.e., INSM1, chromogranin A) and adrenal cortical (SF1, Melan-A) markers. PEComa may arise in the adrenal gland, more frequently in the context of tuberous sclerosis (95). Melan-A immunohistochemistry is not helpful in this differential diagnosis, which should rely on other PEComa-specific markers, such as HMB-45 and smooth muscle actin.

Metastases are frequent in the adrenal gland and may come to the pathologist's attention even without knowledge of a previous or synchronous malignancy. In this context, metastases should be distinguished morphologically in the majority of cases from adrenal cortical neoplasia, but clear cell neoplasms may pose problems in the differential diagnosis with adenoma. Among those, renal cell carcinoma may grow in alveolar and nesting patterns and may have mild atypia, with no or poor evidence of nucleoli. Immunohistochemistry is helpful and a panel including additional renal cell carcinoma-specific markers (i.e., PAX 8 and cytokeratin 7) is advisable.

Some markers that are generally used for renal cell carcinoma diagnosis may be expressed in adrenal cortical lesions. CD10 is not specific for renal cell carcinoma since it is expressed in more than 50 percent of adrenal cortical tumors, both adenomas and carcinomas (96), and CAIX is expressed in about 60 percent of carcinomas and about 10 percent of adenomas (97). A subset of renal cell carcinomas, namely, Xp11 translocation renal cell carcinomas, which are characterized by *TFE3* gene rearrangements and protein overexpression, can also express Melan-A (98). Many adrenal cortical neoplasms also express TFE3 protein (although without the presence of *TFE3* fusion) thus representing a potential diagnostic pitfall (99) that can be solved by genomic or fluorescence in situ hybridization (FISH) analysis of the *TFE3* gene.

PROGNOSIS

Adrenal cortical adenomas are biologically benign, thus with no potential to metastasize. However, their clinical behavior is strongly influenced by hormonal status and severity of symptoms related to adrenal cortical hormone hypersecretion. In the presence of ACTH suppression due to cortisol hypersecretion, the negative feedback to the hypothalamic-pituitary-adrenal axis causes adrenal cortical atrophy and when surgery is performed to remove the lesion, patients require glucocorticoid supplementation that can be gradually tapered as the suppression is reversed.

PATHOLOGY REPORTING

No current consensus, guidelines, or recommendations exist for pathology reporting of adrenal cortical adenoma. It is recommended, however, that the pathology report provide some microscopic descriptive features. These include any parameter that is considered in the Weiss score, if present, and the morphologic aspects of the nontumorous adrenal cortex, including the presence/absence of atrophy or of cortical nodules (48). Pathologists may also consider including a microscopic description of the relative proportion of lipid-rich clear versus compact cells if mixed cell types are present.

Immunohistochemistry for adrenal cortical markers is not mandatory in all cases, but if required in the differential diagnosis with noncortical lesions, the results of the panel of markers employed should be mentioned in the report. The role of immunohistochemical detection of steroidogenic enzymes in routine clinical diagnosis of adrenal cortical adenomas is currently limited to CYP11B2 staining for patients with aldosterone excess; genomic profiling is not routinely required.

REFERENCES

1. Mansmann G, Lau J, Balk E, Rothberg M, Miyachi Y, Bornstein SR. The clinically inapparent adrenal mass: update in diagnosis and management. Endocr Rev 2004;25:309-40.
2. Bovio S, Cataldi A, Reimondo G, et al. Prevalence of adrenal incidentaloma in a contemporary computerized tomography series. J Endocrinol Invest 2006;29:298-302.
3. Gaujoux S, Aimé A, Assié G, et al. Adrenalectomy for incidentaloma: lessons learned from a single-centre series of 274 patients. ANZ J Surg 2018;88:468-73.
4. Morelli V, Reimondo G, Giordano R, et al. Long-term follow-up in adrenal incidentalomas: an Italian multicenter study. J Clin Endocrinol Metab 2014;99:827-34.
5. Falcetta P, Orsolini F, Benelli E, et al. Clinical features, risk of mass enlargement, and development of endocrine hyperfunction in patients with adrenal incidentalomas: a long-term follow-up study. Endocrine 2021;71:178-88.
6. Fassnacht M, Tsagarakis S, Terzolo M, et al. European Society of Endocrinology clinical practice guidelines on the management of adrenal incidentalomas, in collaboration with the European Network for the Study of Adrenal Tumors. Eur J Endocrinol 2023;189:G1-2.
7. Yousaf A, Patterson J, Hobbs G, et al. Smoking is associated with adrenal adenomas and adrenocortical carcinomas: a nationwide multicenter analysis. Cancer Treat Res Commun 2020;25: 100206.
8. Hao M, Lopez D, Luque-Fernandez MA, et al. The lateralizing asymmetry of adrenal adenomas. J Endocr Soc 2018;2:374-85.
9. Hundemer GL, Vaidya A. Primary aldosteronism diagnosis and management: a clinical approach. Endocrinol Metab Clin North Am 2019;48:681-700.
10. Kotliar C, Obregón S, Koretzky M, et al. Improved identification of secondary hypertension: use of a systematic protocol. Ann Transl Med 2018;6:293.
11. Williams TA, Gomez-Sanchez CE, Rainey WE, et al. International Histopathology Consensus for Unilateral Primary Aldosteronism. J Clin Endocrinol Metab 2021;106:42-54.
12. Vaduva P, Bonnet F, Bertherat J. Molecular basis of primary aldosteronism and adrenal cushing syndrome. J Endocr Soc 2020;4:bvaa075.
13. Inoue K, Yamazaki Y, Tsurutani Y, et al. Evaluation of cortisol production in aldosterone-producing adenoma. Horm Metab Res 2017;49:847-53.
14. Tang L, Li X, Wang B, et al. Clinical characteristics of aldosterone- and cortisol-coproducing adrenal adenoma in primary aldosteronism. Int J Endocrinol 2018;2018:4920841.
15. Shigematsu K, Nishida N, Sakai H, et al. Primary aldosteronism with aldosterone-producing adenoma consisting of pure zona glomerulosa-type cells in a pregnant woman. Endocr Pathol 2009; 20:66-72.
16. Willenberg HS, Späth M, Maser-Gluth C, et al. Sporadic solitary aldosterone- and cortisol-co-secreting adenomas: endocrine, histological and genetic findings in a subtype of primary aldosteronism. Hypertens Res 2010;33:467-72.
17. Späth M, Korovkin S, Antke C, Anlauf M, Willenberg HS. Aldosterone- and cortisol-co-secreting adrenal tumors: the lost subtype of primary aldosteronism. Eur J Endocrinol 2011;164:447-55.
18. Stowasser M, Bachmann AW, Tunny TJ, Gordon RD. Production of 18-oxo-cortisol in subtypes of primary aldosteronism. Clin Exp Pharmacol Physiol 1996;23:591-3.
19. Stenman A, Shabo I, Ramström A, Zedenius J, Juhlin CC. Synchronous aldosterone- and cortisol-producing adrenocortical adenomas diagnosed using CYP11B immunohistochemistry. SAGE Open Med Case Rep 2019; 7:2050313X19883770.
20. Del Gaudio AD, Del Gaudio GA. Virilizing adrenocortical tumors in adult women. Report of 10 patients, 2 of whom each had a tumor secreting only testosterone. Cancer 1993;72:1997-2003.
21. Kidd MT, Karlin NJ, Cook CB. Feminizing adrenal neoplasms: case presentations and review of the literature. J Clin Oncol 2011;29:e127-30.
22. Lopes RI, Suartz CV, Neto RP, et al. Management of functioning pediatric adrenal tumors. J Pediatr Surg 2021;56:768-71.
23. Zhou WB, Chen N, Li CJ. A rare case of pure testosterone-secreting adrenal adenoma in a postmenopausal elderly woman. BMC Endocr Disord 2019;19:14.
24. Fassnacht M, Tsagarakis S, Terzolo M, et al. European Society of Endocrinology clinical practice guidelines on the management of adrenal incidentalomas, in collaboration with the european network for the study of adrenal tumors. Eur J Endocrinol 2023;189:G1-42.
25. Omata K, Tomlins SA, Rainey WE. Aldosterone-producing cell clusters in normal and pathological states. Horm Metab Res 2017;49: 951-6.
26. Wang F, Liu J, Zhang R, et al. CT and MRI of adrenal gland pathologies. Quant Imaging Med Surg 2018;8:853-75.

27. Platzek I, Sieron D, Plodeck V, Borkowetz A, Laniado M, Hoffmann RT. Chemical shift imaging for evaluation of adrenal masses: a systematic review and meta-analysis. Eur Radiol 2019;29:806-17.
28. d'Amuri FV, Maestroni U, Pagnini F, et al. Magnetic resonance imaging of adrenal gland: state of the art. Gland Surg 2019;8(Suppl 3):S223-32.
29. Vos EL, Grewal RK, Russo AE, et al. Predicting malignancy in patients with adrenal tumors using ^{18} F-FDG-PET/CT SUVmax. J Surg Oncol 2020;122:1821-6.
30. Patel KA, Calomeni EP, Nadasdy T, Zynger DL. Adrenal gland inclusions in patients treated with aldosterone antagonists (Spironolactone/Eplerenone): incidence, morphology, and ultrastructural findings. Diagn Pathol 2014;9:147.
31. Ryan JJ, Rezkalla MA, Rizk SN, Peterson KG, Wiebe RH. Testosterone-secreting adrenal adenoma that contained crystalloids of Reinke in an adult female patient. Mayo Clin Proc 1995;70:380-3.
32. Kanitra JJ, Hardaway JC, Soleimani T, Koehler TJ, McLeod MK, Kavuturu S. Adrenocortical oncocytic neoplasm: a systematic review. Surgery 2018;164:1351-9.
33. Sasano H, Suzuki T, Sano T, Kameya T, Sasano N, Nagura H. Adrenocortical oncocytoma. A true nonfunctioning adrenocortical tumor. Am J Surg Pathol 1991;15:949-56.
34. Hong Y, Hao Y, Hu J, Xu B, Shan H, Wang X. Adrenocortical oncocytoma: 11 case reports and review of the literature. Medicine (Baltimore) 2017;96:e8750.
35. De Leo A, Mosconi C, Zavatta G, et al. Radiologically defined lipid-poor adrenal adenomas: histopathological characteristics. J Endocrinol Invest 2020;43:1197-204.
36. Duregon E, Volante M, Cappia S, et al. Oncocytic adrenocortical tumors: diagnostic algorithm and mitochondrial DNA profile in 27 cases. Am J Surg Pathol 2011;35:1882-93.
37. Bisceglia M, Ludovico O, Di Mattia A, et al. Adrenocortical oncocytic tumors: report of 10 cases and review of the literature. Int J Surg Pathol 2004;12:231-43.
38. Duregon E, Volante M, Bollito E, et al. Pitfalls in the diagnosis of adrenocortical tumors: a lesson from 300 consultation cases. Hum Pathol 2015;46:1799-807.
39. Saygin I, Cakir E, Ercin ME, Eyüboglu I. Incidental retroperitoneal oncocytoma (ectopic oncocytic adrenocortical adenoma): case report and review of the literature. Indian J Pathol Microbiol 2019;62:132-5.
40. Tanaka S, Tanabe A, Aiba M, et al. Glucocorticoid- and androgen-secreting black adrenocortical adenomas: unique cause of corticotropin-independent Cushing syndrome. Endocr Pract 2011;17:e73-8.
41. Cohen RJ, Brits R, Phillips JI, Botha JR. Primary hyperaldosteronism due to a functional black (pigmented) adenoma of the adrenal cortex. Arch Pathol Lab Med 1991;115:813-5.
42. Lack EE, Travis WD, Oertel JE. Adrenal cortical neoplasms. In: Lack EE, ed. Pathology of the adrenal glands. New York: Churchill Livingston; 1990:115-71.
43. Nakajo M, Nakajo M, Kajiya Y, et al. A black adrenal adenoma difficult to be differentiated from a malignant adrenal tumor by CT, MRI, scintigraphy and FDG PET/CT examinations. Ann Nucl Med 2011;25:812-7.
44. Armand R, Cappola AR, Horenstein RB, Drachenberg CB, Sasano H, Papadimitriou JC. Adrenal cortical adenoma with excess black pigment deposition, combined with myelolipoma and clinical Cushing's syndrome. Int J Surg Pathol 2004;12:57-61.
45. Kameyama K, Takami H. Pigmented granules in functional black adenoma of the adrenal gland: a histochemical and ultrastructural study. Endocr Pathol 1999;10:353-7.
46. Kakkar A, Kaur K, Kumar T, et al. Pigmented pheochromocytoma: an unusual variant of a common tumor. Endocr Pathol 2016;27:42-5.
47. Blanco R, Rodríguez Villar D, Fernández-Pello S, et al. Massive bilateral adrenal metastatic melanoma of occult origin: a case report. Anal Quant Cytopathol Histpathol 2014;36:51-4.
48. Mete O, Asa SL. Morphological distinction of cortisol-producing and aldosterone-producing adrenal cortical adenomas: not only possible but a critical clinical responsibility. Histopathology 2012;60:1015-7.
49. Mete O, Asa SL, Giordano TJ, Papotti M, Sasano H, Volante M. Immunohistochemical biomarkers of adrenal cortical neoplasms. Endocr Pathol 2018;29:137-49.
50. Sbiera S, Schmull S, Assie G, et al. High diagnostic and prognostic value of steroidogenic factor-1 expression in adrenal tumors. J Clin Endocrinol Metab 2010;95:E161-71.
51. Busam KJ, Iversen K, Coplan KA, et al. Immunoreactivity for A103, an antibody to melan-A (Mart-1), in adrenocortical and other steroid tumors. Am J Surg Pathol 1998;22:57-63.
52. Mete O, van der Kwast TH. Epithelioid angiomyolipoma: a morphologically distinct variant that mimics a variety of intra-abdominal neoplasms. Arch Pathol Lab Med 2011;135:665-70.
53. Taskin OC, Gucer H, Mete O. An unusual adrenal cortical nodule: composite adrenal cortical adenoma and adenomatoid tumor. Endocr Pathol 2015;26:370-3.

54. Seccia TM, Caroccia B, Gomez-Sanchez EP, Vanderriele PE, Gomez-Sanchez CE, Rossi GP. Review of markers of zona glomerulosa and aldosterone-producing adenoma cells. Hypertension 2017;70:867-74.
55. Doi M, Satoh F, Maekawa T, et al. Isoform-specific monoclonal antibodies against 3β-hydroxysteroid dehydrogenase/isomerase family provide markers for subclassification of human primary aldosteronism. J Clin Endocrinol Metab 2014;99:E257-62.
56. Gomez-Sanchez CE, Qi X, Velarde-Miranda C, et al. Development of monoclonal antibodies against human CYP11B1 and CYP11B2. Mol Cell Endocrinol 2014;383:111-7.
57. Nakamura Y, Kitada M, Satoh F, et al. Intratumoral heterogeneity of steroidogenesis in aldosterone-producing adenoma revealed by intensive double- and triple-immunostaining for CYP11B2/B1 and CYP17. Mol Cell Endocrinol 2016;422:57-63.
58. Konosu-Fukaya S, Nakamura Y, Satoh F, et al. 3β-hydroxysteroid dehydrogenase isoforms in human aldosterone-producing adenoma. Mol Cell Endocrinol 2015;408:205-12.
59. Gao X, Yamazaki Y, Tezuka Y, et al. Intratumoral heterogeneity of the tumor cells based on in situ cortisol excess in cortisol-producing adenomas; ~An association among morphometry, genotype and cellular senescence~. J Steroid Biochem Mol Biol 2020;204:105764.
60. Vaduva P, Bonnet F, Bertherat J. Molecular basis of primary aldosteronism and adrenal cushing syndrome. J Endocr Soc 2020;4:bvaa075.
61. Faillot S, Foulonneau T, Néou M, et al. Genomic classification of benign adrenocortical lesions. Endocr Relat Cancer 2021;28:79-95.
62. El Zein RM, Boulkroun S, Fernandes-Rosa FL, Zennaro MC. Molecular genetics of Conn adenomas in the era of exome analysis. Presse Med 2018;47:e151-8.
63. Fernandes-Rosa FL, Williams TA, Riester A, et al. Genetic spectrum and clinical correlates of somatic mutations in aldosterone-producing adenoma. Hypertension 2014;64:354-61.
64. Nanba K, Rainey WE. Genetics in endocrinology: impact of race and sex on genetic causes of aldosterone-producing adenomas. Eur J Endocrinol 2021;185:R1-11.
65. Williams TA, Monticone S, Mulatero P. KCNJ5 mutations are the most frequent genetic alteration in primary aldosteronism. Hypertension 2015;65:507-9.
66. Nanba K, Omata K, Gomez-Sanchez CE, et al. Genetic characteristics of aldosterone-producing adenomas in Blacks. Hypertension 2019;73:885-92.
67. Choi M, Scholl UI, Yue P, et al. K+ channel mutations in adrenal aldosterone-producing adenomas and hereditary hypertension. Science 2011;331:768-72.
68. Boulkroun S, Beuschlein F, Rossi GP, et al. Prevalence, clinical, and molecular correlates of KCNJ5 mutations in primary aldosteronism. Hypertension 2012;59:592-8.
69. Lenzini L, Rossitto G, Maiolino G, Letizia C, Funder JW, Rossi GP. A meta-analysis of somatic KCNJ5 K(+) channel mutations in 1636 patients with an aldosterone-producing adenoma. J Clin Endocrinol Metab 2015;100:E1089-95.
70. Ono Y, Yamazaki Y, Omata K, et al. Histological characterization of aldosterone-producing adrenocortical adenomas with different somatic mutations. J Clin Endocrinol Metab 2020;105:e282-9.
71. Murakami M, Yoshimoto T, Nakabayashi K, et al. Molecular characteristics of the KCNJ5 mutated aldosterone-producing adenomas. Endocr Relat Cancer 2017;24:531-41.
72. Azizan EA, Poulsen H, Tuluc P, et al. Somatic mutations in ATP1A1 and CACNA1D underlie a common subtype of adrenal hypertension. Nat Genet 2013;45:1055-60.
73. Scholl UI, Goh G, Stölting G, et al. Somatic and germline CACNA1D calcium channel mutations in aldosterone-producing adenomas and primary aldosteronism. Nat Genet 2013;45:1050-4.
74. Beuschlein F, Boulkroun S, Osswald A, et al. Somatic mutations in ATP1A1 and ATP2B3 lead to aldosterone-producing adenomas and secondary hypertension. Nat Genet 2013;45:440-4.
75. Tauber P, Aichinger B, Christ C, et al. Cellular pathophysiology of an adrenal adenoma-associated mutant of the plasma membrane Ca(2+)-ATPase ATP2B3. Endocrinology 2016;157:2489-99.
76. Rege J, Bandulik S, Nanba K, et al. Somatic SLC30A1 mutations altering zinc transporter ZnT1 cause aldosterone-producing adenomas and primary aldosteronism. Nat Genet 2023;55:1623-31.
77. Dutta RK, Arnesen T, Heie A, et al. A somatic mutation in CLCN2 identified in a sporadic aldosterone-producing adenoma. Eur J Endocrinol 2019;181:K37-41.
78. Nanba K, Blinder AR, Rege J, et al. CACNA1H mutation as a cause of aldosterone-producing adenoma. Hypertension 2020;75:645-9.
79. Rhayem Y, Perez-Rivas LG, Dietz A, et al. PRKACA somatic mutations are rare findings in aldosterone-producing adenomas. J Clin Endocrinol Metab 2016;101:3010-7.
80. Berthon A, Drelon C, Ragazzon B, et al. WNTβ-catenin signalling is activated in aldosterone-producing adenomas and controls aldosterone production. Hum Mol Genet 2014;23:889-905.

81. Åkerström T, Maharjan R, Sven Willenberg H, et al. Activating mutations in CTNNB1 in aldosterone producing adenomas. Sci Rep 2016;6:19546.
82. Calebiro D, Bathon K, Weigand I. Mechanisms of aberrant PKA activation by Cα subunit mutations. Horm Metab Res 2017;49:307-14.
83. Beuschlein F, Fassnacht M, Assié G, et al. Constitutive activation of PKA catalytic subunit in adrenal Cushing's syndrome. N Engl J Med 2014; 370:1019-28.
84. Cao Y, He M, Gao Z, et al. Activating hotspot L205R mutation in PRKACA and adrenal Cushing's syndrome. Science 2014;344:913-7.
85. Goh G, Scholl UI, Healy JM, et al. Recurrent activating mutation in PRKACA in cortisol-producing adrenal tumors. Nat Genet 2014;46:613-7.
86. Sato Y, Maekawa S, Ishii R, et al. Recurrent somatic mutations underlie corticotropin-independent Cushing's syndrome. Science 2014;344:917-20.
87. Espiard S, Knape MJ, Bathon K, et al. Activating PRKACB somatic mutation in cortisol-producing adenomas. JCI Insight 2018;3:e98296.
88. Bertherat J, Groussin L, Sandrini F, et al. Molecular and functional analysis of PRKAR1A and its locus (17q22-24) in sporadic adrenocortical tumors: 17q losses, somatic mutations, and protein kinase A expression and activity. Cancer Res 2003;63:5308-19.
89. Drougat L, Settas N, Ronchi CL, et al. Genomic and sequence variants of protein kinase A regulatory subunit type β? (PRKAR1B) in patients with adrenocortical disease and Cushing syndrome. Genet Med 2021;23:174-82.
90. Kobayashi H, Usui T, Fukata J, Yoshimasa T, Oki Y, Nakao K. Mutation analysis of Gsalpha, adrenocorticotropin receptor and p53 genes in Japanese patients with adrenocortical neoplasms: including a case of Gsalpha mutation. Endocr J 2000;47:461-6.
91. Almeida MQ, Azevedo MF, Xekouki P, et al. Activation of cyclic AMP signaling leads to different pathway alterations in lesions of the adrenal cortex caused by germline PRKAR1A defects versus those due to somatic GNAS mutations. J Clin Endocrinol Metab 2012;97:E687-93.
92. Ronchi CL, Di Dalmazi G, Faillot S, et al. Genetic landscape of sporadic unilateral adrenocortical adenomas without PRKACA p.Leu206Arg mutation. J Clin Endocrinol Metab 2016;101:3526-38.
93. Tezuka Y, Yamazaki Y, Kitada M, et al. 18-Oxocortisol synthesis in aldosterone-producing adrenocortical adenoma and significance of KCNJ5 mutation status. Hypertension 2019;73:1283-90.
94. Kometani M, Yoneda T, Demura M, et al. Genetic and epigenetic analyses of aldosterone-producing adenoma with hypercortisolemia. Steroids 2019;151:108470.
95. Torres Luna N, Mosquera JE, Comba IY, Kinaan M, Otoya J. A primary adrenal epithelioid angiomyolipoma (PEComa) in a patient with tuberous sclerosis complex: report of a case and review of the literature. Case Rep Med 2020;2020:5131736.
96. Mete O, Kapran Y, Güllüoglu MG, et al. Anti-CD10 (56C6) is expressed variably in adrenocortical tumors and cannot be used to discriminate clear cell renal cell carcinomas. Virchows Arch 2010;456:515-21.
97. Donato DP, Johnson MT, Yang XJ, Zynger DL. Expression of carbonic anhydrase IX in genitourinary and adrenal tumours. Histopathology 2011;59:1229-39.
98. Xia QY, Wang XT, Zhan XM, et al. Xp11 Translocation renal cell carcinomas (RCCs) with RBM10-TFE3 gene fusion demonstrating melanotic features and overlapping morphology with t(6;11) RCC: interest and diagnostic pitfall in detecting a paracentric inversion of TFE3. Am J Surg Pathol 2017;41:663-76.
99. Wang X, Ng CS, Yin W, Liang L. Application of TFE3 immunophenotypic and TFE3 mRNA expressions in diagnosis and prognostication of adrenal cortical neoplasms and distinction from kidney tumors. Appl Immunohistochem Mol Morphol 2023;31:9-16.

6 ADRENAL CORTICAL CARCINOMA

Adrenal cortical carcinoma (ACC) is a malignant neoplasm originating from the steroidogenic epithelial cells of the adrenal cortex.

EPIDEMIOLOGY

ACC accounts for the majority of deaths attributable to primary adrenal neoplasia, but it is a rare disease, meeting the criteria for an "orphan" disease designation in the European Union and in the United States (less than 50 cases per 100,000 population and less than 64 cases per 10,000 population, respectively) (1). The annual incidence ranges between 0.7 to 2.0 cases per million in the United States (2), with no significant changes over the last four decades. This is in accordance with a population-based study from the Netherlands, which reported an incidence of 1 case per million (3).

Like other adrenal diseases, ACC is more frequent in females than males (4). The female to male ratio ranges from 1.5–2.5 to 1.0; women have a higher prevalence of hormone-secreting tumors and a younger age at diagnosis (5). The reasons for the increased prevalence of ACC in females is not clearly understood but it has been postulated that the predominance may reflect the sexual dimorphism in the adrenal cortex and the influence of gonadal hormones in many aspects of adrenal physiology (6).

Some children with ACC have peculiar clinical and pathologic characteristics, which are described later in this chapter. In adult patients, a broad age-span peak in the 5th and 6th decades is observed, although ACC can occur at any age. There does not appear to be a significant racial predilection, although some epidemiologic studies show an increased rate among Blacks (7) whereas others indicate an increased incidence in non-Hispanic Whites (2,8).

ETIOLOGY

The few cases attributable to genetic susceptibility syndromes are discussed in detail in chapter 7. Among sporadic cases, cigarette smoking is the most significant etiologic factor. Tobacco use is associated with adrenal cortical neoplasia (9,10) and specifically with ACC (8), with a 50 percent greater risk in smokers than in nonsmokers, more significantly in males. Oral contraceptive use before age 25 years has also been suggested as a risk factor for ACC in females (9), supported by the in vitro evidence of antiproliferative effects of estrogen and progesterone inhibition on ACC cells (11,12); the known occurrence and rapid growth of ACC in pregnancy, however, contradicts these in vitro studies. Alcohol consumption is associated with reduced risk in males (8). Environmental causes, while not considered as relevant as hereditary predisposition (13), also play a role. For example, in Brazil, *TP53* R337H carriers living in an agricultural region had a lower risk of developing pediatric adrenal cortical tumors than those living in industrial and large urban regions (14).

ACC is associated with a significant history of previous or subsequent associated cancers, thus suggesting heterogeneous underlying cancer predisposition mechanisms. In a recent epidemiologic study, 9 percent of patients had a history of other malignancies and 3 percent developed other malignancies after ACC diagnosis (8). Associated malignancies were extremely variable, including different types of carcinomas (prostate, bladder, lung, thyroid gland, breast, endometrial, cervical, renal, cutaneous nonmelanomatous) as well as testicular germ cell tumors, melanomas, lymphomas, and sarcomas.

PRECURSOR LESIONS

ACC is generally considered to arise de novo. However, data from patients with Carney complex (15,16) and multiple endocrine neoplasia 1 (MEN1)-associated ACC suggest that there may be progression from benign to malignant adrenal cortical nodules (17,18). Moreover, even in the sporadic setting, some pathologic

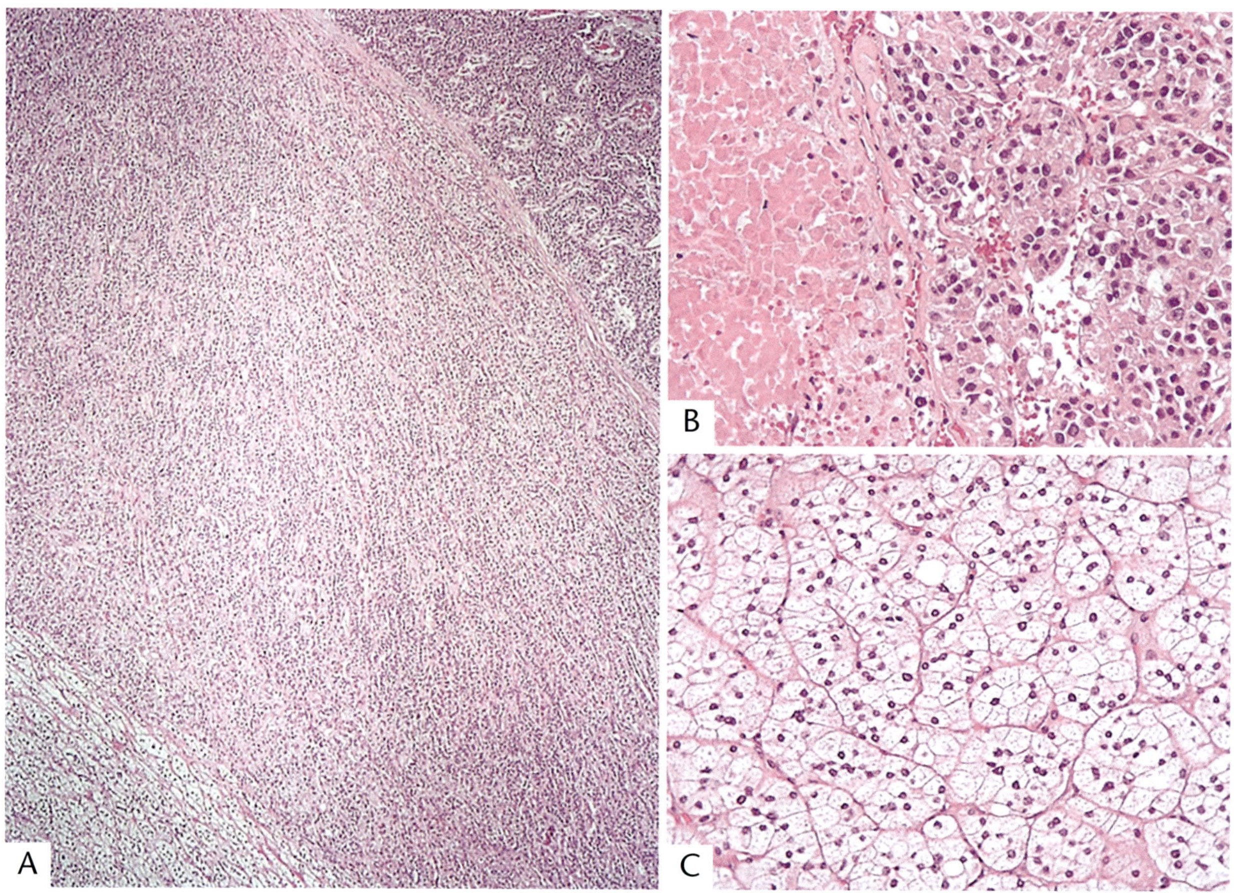

Figure 6-1

ADRENAL CORTICAL CARCINOMA ARISING IN ADRENAL CORTICAL ADENOMA

A: Transition from adenomatous areas to carcinoma within the same lesion (from bottom to top).

B,C: Carcinomatous areas have atypical eosinophilic cells with necrosis (B), whereas adenomatous areas have the typical nesting pattern without atypia and with clear cell cytoplasm (C).

and molecular evidence supports that at least a subset of ACCs derive from preexisting adrenal cortical adenomas. Description of cases with a bimodal appearance due to the coexistence within a single adrenal cortical tumor of morphologically benign and malignant components, bearing consistent molecular alterations in the two distinct tumor populations, are on record in the literature (figs. 6-1, 6-2) (19).

Molecular data favor a multistep progression from normal adrenal cortex to adenoma and subsequently to carcinoma. Some of the gene mutations that are common in ACCs are also prevalent in adenomas, including somatic *CTNNB1* and *MEN1* mutations (20). A genome-wide gene expression study documented an overlap of about 36 percent among genes differentially expressed in adenoma and carcinoma compared to normal adrenal cortex, mainly involving downregulated genes (21). In another study using high-resolution single nucleotide polymorphism arrays, more than 70 percent of copy number gains detected in adenomas were also present in carcinomas.

Among the signaling pathways involved in both benign and malignant tumors, the Notch signaling pathway is the most frequently altered (22). Therefore, it may be postulated that some ACCs characterized by specific molecular features originate from benign tumors, but the clinical meaning of this pathogenetic hypothesis is not well established, nor does it influence

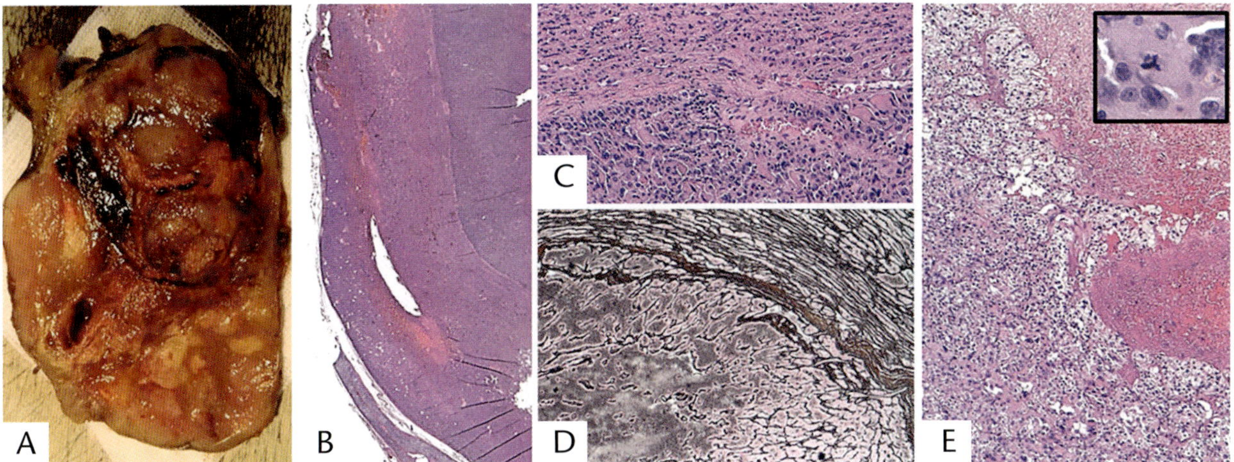

Figure 6-2

ADRENAL CORTICAL CARCINOMA

This adrenal cortical carcinoma shows a gross nodule-in-nodule appearance (A). Peripheral areas are well circumscribed with oncocytic features (B). A thin capsule (C) separates the peripheral areas from a central component with an altered reticulin pattern (D), and high degree of atypia with necrosis (E) and atypical mitotic figures (inset).

the current clinical management of adrenal cortical tumors.

CLINICAL FEATURES

Primary Localization

Most ACCs manifest with unilateral adrenal gland involvement. For unknown reasons, the left adrenal gland is slightly more frequently affected than the right. One hypothesis is that the left adrenal has a larger volume than the right adrenal, and therefore ACC has a higher chance to occur (23). Synchronous or metachronous bilateral involvement may occur, comprising roughly 2 to 10 percent of cases (24), including functioning cases (25), but the possibility of a contralateral adrenal metastasis should be considered.

Exceptional cases occur in heterotopic/ectopic locations, more frequently in the retroperitoneum and pelvis (26–32). Adrenohepatic fusion/adhesion or heterotopia/ectopia may explain the occurrence of rare examples of hepatic ACCs (33).

Clinical and Biochemical Characteristics

ACC is associated with highly variable clinical presentations. It may be an incidental finding, or may cause signs and symptoms related to the presence of an abdominal mass. Some present with clinically overt metastatic disease or features of hormone hypersecretion (34). These different clinical scenarios require a multidisciplinary approach for the formulation of ACC diagnostic algorithms and characterization (35).

Approximately 10 to 15 percent of ACCs are incidentally detected during imaging studies for unrelated causes. In a recent series, ACC incidence in adrenal incidentalomas was 1.7 percent, with an increased risk in males and in tumors with a size greater than 4.6 cm, increased Hounsfield density on computerized tomography (CT) scan, and progressive tomography increase in size of more than 0.6 cm/year (36).

Nonspecific local signs and symptoms are present in about one third of patients, and most frequently include abdominal pain and other local compressive symptoms related to tumor growth and invasion of surrounding structures (37). ACC may infiltrate the periadrenal fat and adjacent organs, including kidney, liver, spleen, and large veins (38). Approximately 25 percent of these cases present with metastases at the time of diagnosis. Common sites of metastasis are lung, liver, and lymph nodes, and less frequently, bone (39).

Systemic symptoms also may be present (40), including prominent weight loss, intermittent low-grade fever (possibly related to extensive tumor necrosis), and anorexia. Hypoglycemia and hypercalcemia have been reported in association with ACC, with the latter attributed to elaboration of a hypercalcemic hormone (41,42). Other unusual presentations include overwhelming disseminated intravascular coagulation (43) and hemorrhagic shock due to spontaneous rupture of the ACC (44).

Biochemically functional tumors account for approximately 60 percent of all ACC cases, although symptoms caused by hormone hypersecretion are present in only about 40 percent of cases. The frequency of endocrinologically functioning ACC depends upon whether "function" is defined as a clinically recognizable "pure" syndrome, such as Cushing syndrome, a "mixed" endocrine syndrome, or any biochemical evidence of excess steroid production. In a review of clinically "silent" adrenal cortical neoplasms, the tumors were shown to be capable of forming precursor steroids without hormonal activity and, therefore, the tumors were not actually "nonfunctional" in biochemical terms (45).

Data suggest that ACCs are less efficient in steroidogenesis due to decreased activity of 3-beta-hydroxysteroid dehydrogenase, 17-hydroxylase, and 17,20-desmolase (46,47). At variance with adrenal cortical adenomas, the expression of enzymes involved in steroidogenesis is highly heterogeneous within different tumor regions, and even within individual tumor cells. Culture of tumor cells in vitro has demonstrated the coexistence of multiple pathways of steroidogenesis, including formation of corticosteroids, mineralocorticoids, androgens, and estrogens, with expression of all major enzyme systems required for their synthesis (48).

Immunohistochemistry of steroidogenic enzyme expression (see below) is a valuable tool to characterize the hormonal properties of ACC at the tissue level (49). However, immunohistochemical expression of these enzymes presents a different pattern from that of adrenal cortical adenoma, with marked intratumoral heterogeneity; this may account for the overproduction and oversecretion of precursor steroids (50). In addition, at variance with adrenal cortical adenomas, a discrepancy between expression levels of mRNA and corresponding proteins has been reported for several steroidogenic enzymes, such as P450c17 (50).

Cortisol hypersecretion is recognizable in 30 to 40 percent of cases, androgen hypersecretion in 20 to 30 percent, estrogen hypersecretion in 6 to 10 percent, and aldosterone hypersecretion in up to 2.5 percent (49). Apart from an initial clinical characterization, the hormone levels may serve as useful biomarkers for recurrence during follow-up after surgery. However, consistent with the heterogeneous expression patterns of steroidogenic enzymes discussed above, detailed biochemical characterization shows that ACCs are usually associated with the overproduction and hypersecretion of multiple hormones (more frequently cortisol and sex steroid hormones) and precursors, rather than a single hormone (49). Therefore, excessive production or secretion of a single adrenocortical steroid is probably more indicative of the benign nature of an incidentally discovered adrenal cortical neoplasm.

Patients with hypercortisolism are younger than those with nonfunctioning tumors or tumors associated with hypersecretion of other hormones (51). Overt signs and symptoms of Cushing syndrome may be present (52,53), but mild autonomous glucocorticoid secretion (subclinical Cushing syndrome) can also occur (54). The diagnosis may be delayed in the presence of less specific clinical findings, such as hypertension, osteoporosis, metabolic syndrome, and diabetes mellitus. Markedly rapid-onset symptoms, with accentuated hypertension and poorly regulated blood glucose with increased precursor steroids such as DHEAS or others, may be clues to an underlying cortisol-producing ACC (55). Multiple opportunistic infections (56) or Kaposi sarcoma (57), possibly caused by cortisol-related immunosuppression, are also described in ACC.

In adults, hormonal syndromes related to oversecretion of sex steroid hormones due to an adrenal mass, with special reference to feminization, are generally considered a sign of malignancy (58), although rare adrenal cortical adenomas may secrete sex steroid hormones. Androgen-secreting ACC in females are associated with clinical signs of virilization, while the hormonal excess may be masked in adult males due to the relatively higher testosterone levels produced by the testes (59). Physiologically, the adrenal cortex has little capacity to produce estrogens because of the lack of aromatase. However, some studies show high aromatase cytochrome P450 (CYP19A1) mRNA and aromatase activity in ACCs producing estrogens (60). Estrogen production in male patients is associated with gynecomastia, whereas adult female patients are relatively asymptomatic (60–62).

Aldosterone-producing ACC is a rare cause of hypertension, typically associated with hypokalemia (63–65). These tumors present with pure or mixed secretory profiles, the latter usually with cortisol co-secretion (66). Metastases are present in 10 percent of cases at initial diagnosis, and an additional 48 percent at follow-up (67). Aldosterone-producing ACCs are associated with an increased perioperative mortality rate, thus emphasizing the need for accurate biochemical presurgical characterization (66). Hypersecretion of 11-deoxycorticosterone is also a rare cause of mineralocorticoid-induced hypertension in ACC, with no cortisol or aldosterone secretion (68).

Urine Steroid Metabolomics. The rationale for this approach for diagnosis is based on the findings discussed above that ACCs are relatively inefficient steroid producers, due to the dysregulated expression of steroidogenic enzymes. This results in the excessive secretion of a range of steroid hormone precursors, instead of the end products of steroid hormone biosynthesis; the precursors can be quantified using mass spectrometry-based urinary steroid metabolite profiling of 24-hour urine samples (69–75). A seminal paper retrospectively compared 45 patients with ACC to 102 patients with benign adrenal adenomas, and showed that the former had significantly greater urinary excretion of androgen precursor metabolites, metabolites of active androgens, deoxycorticosterone, and glucocorticoid precursors (76). These data have been validated in the EURINE-ACT prospective study enrolling an international cohort of 2,017 patients, including 98 ACC patients (77). In this study, the combination of tumor diameter, imaging characteristics, and urine steroid metabolomics demonstrated very high positive (76.5 percent) and negative (99.7 percent) predictive values for diagnosing ACC.

Urine steroid metabolomics has been proposed as a tool for the detection of ACC recurrence after surgery (72). Despite the clear benefits of this noninvasive assay, the widespread implementation of this technology into routine clinical practice is currently limited by both the cost and availability of mass spectrometry equipment, as well as the need for interlaboratory cross-validation of assay results.

IMAGING FEATURES

Imaging methods are the cornerstone of evaluation in patients with suspected ACC. They provide information about the malignant potential of adrenal masses in the presence of localized lesions, and accurately identify the presence of distant metastases. However, despite their widespread use in routine assessment, there is insufficient evidence for the diagnostic value of individual imaging tests in distinguishing benign from malignant adrenal masses (Table 6-1) (35,78). In the context of incidental adrenal mass lesions, the highest specificity has been demonstrated for CT noncontrast tumor density (over 10 Hounsfield units [HU]) and positron emission tomography (PET) ALR SUVmax (corresponding to the ratio of SUVmax of the adrenal gland compared with the liver) (79).

In the follow-up of a patient with known ACC, each individual imaging modality offers different sensitivity and specificity that are distinct from the primary diagnostic setting. Therefore, each imaging approach should be calibrated and integrated according to the patient-specific clinical scenario.

Ultrasonography

Ultrasonography (US) is useful for the detection of large adrenal masses, but further characterization may be limited by body habitus and operator skills. The most common US features of ACC are a rounded to oval shape, sometimes surrounded by a thick partial or complete echogenic rim, corresponding to the presence of a tumor capsule (80). Smaller tumors have a homogeneous US pattern, similar to renal cortical echogenicity, whereas larger lesions vary in texture, having a heterogeneous appearance, with focal or scattered echopenic or echogenic zones representing areas of tumor necrosis, hemorrhage, and/or calcifications (81). On color doppler, ACC is usually a hypervascular mass, which may demonstrate an afferent blood vessel (82). Contrast-enhanced US reveals a hypervascularized lesion with early

Table 6-1

IMAGING CHARACTERISTICS OF BENIGN VERSUS MALIGNANT ADRENAL MASSES[a]

Characteristic	Likely Benign	Suspicious for Malignancy
Size	Small (<4 cm)	Large (>4 cm)
Shape	Round or oval	Irregular
Margins	Smooth	Irregular
Texture	Homogeneous	Heterogeneous
Calcifications/necrosis/hemorrhage	Rare	Common
Growth	Slow	Rapid
Density of noncontrast CT[b]	<10 HU	>10 HU
Contrast washout on CT protocol at 10–15 minutes	Absolute >60% Relative >40%	Absolute <60% Relative >40%
MRI chemical shift	Yes	No
FDG-PET	Non-avid	Avid

[a]Modified Table 1 from reference 35.
[b]CT = computerized tomography; FDG-PET = fluorodeoxyglucose positron emission tomography; HU = Hounsfield units; MRI = magnetic resonance imaging.

enhancement similar to that of pheochromocytoma (83).

Computerized Tomography

The most commonly used first-line modality for a suspected ACC is an abdominal CT scan (84). CT features suggestive of an ACC rather than a benign adrenal mass include the presence of irregular borders, hemorrhage, or calcifications (35,58). In a recent study, necrosis was the most important imaging feature significantly associated with ACC (odds ratio 35, 95 percent confidence interval 5.1-241.6), present in all ACC cases (85). Invasion into periadrenal fat, adjacent organs, or vasculature, such as the inferior vena cava, may also be seen (86). Size is a significant parameter associated with the risk of malignancy in an adrenal mass, with a progressive increase from 2 to 25 percent for adrenal lesions under 4 cm and over 6 cm, respectively (87). Calculation of tissue density as an approximation of intracellular lipid content on unenhanced CT, by measuring HU, is also used to distinguish between ACC and benign adrenal cortical adenomas (figs. 6-3–6-5). Lipid-rich adenomas typically exhibit low attenuation on unenhanced CT, and a HU threshold of less than 10 is suggested as indicative of a benign lesion in current ENSAT/ESE and European Society for Medical Oncology (ESMO) guidelines (88,89). In contrast, a HU over 10 is suggested as being indicative of malignancy.

Magnetic Resonance Imaging

Heterogeneous signal intensity is characteristic of ACC both in T1- and T2-weighted magnetic resonance images (MRI), as a consequence of the presence of hemorrhage and necrotic areas. Hemorrhage is reflected by intrinsic T1-shortening and hemorrhage by T2 prolongation (24). Chemical shift with signal loss on out-phase imaging and corresponding to intracellular lipid content is rare in ACC and, if present, usually focal (fig. 6-6) (90). Diffusion-weighted imaging is not accurate enough to distinguish benign from malignant incidental adrenal lesions because of largely overlapping diffusion coefficients in these two groups (91).

Metabolic profiles as assessed using magnetic resonance spectroscopy (MRS) have been demonstrated to distinguish ACC from adrenal cortical adenoma, pheochromocytoma, and metastases with high specificity and sensitivity, particularly because of a high choline compound content (92,93). An added value of MRI in ACC characterization is related to its better soft tissue resolution as compared to CT, thus helping to better identify vascular involvement (94).

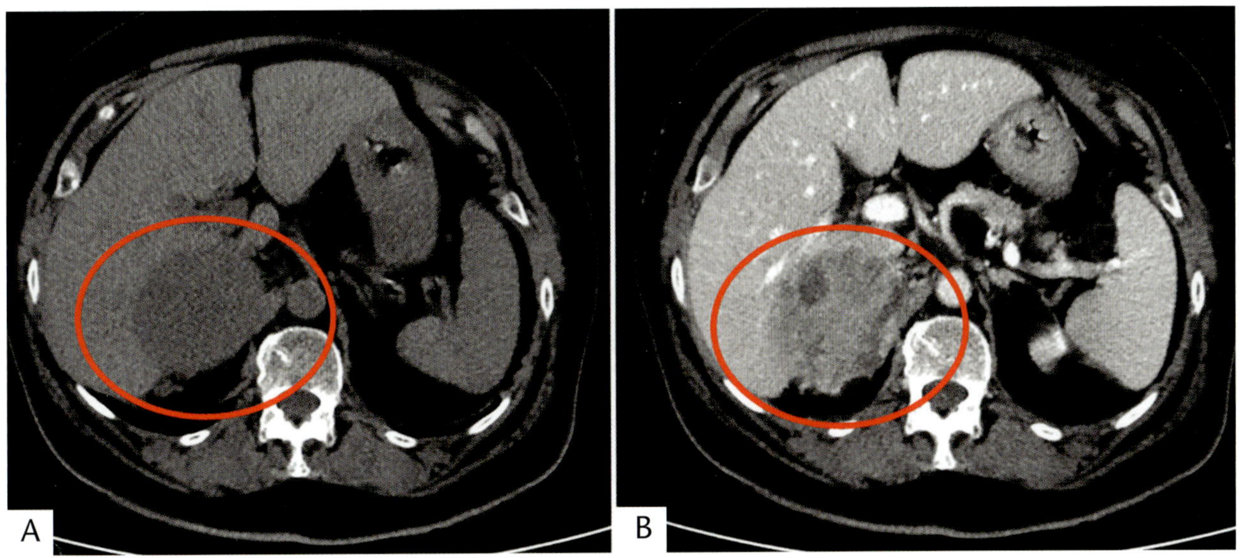

Figure 6-3

ADRENAL CORTICAL CARCINOMA

Computerized tomography (CT) scan of a 10-cm lesion in the right adrenal gland adjacent to liver.
A: Basal uptake (>20 HU) with poorly defined separation from the liver.
B: Heterogeneous uptake in the venous phase. (Courtesy of Prof. A. Veltri and Dr. F. Solitro, San Luigi Hospital, Turin, Italy)

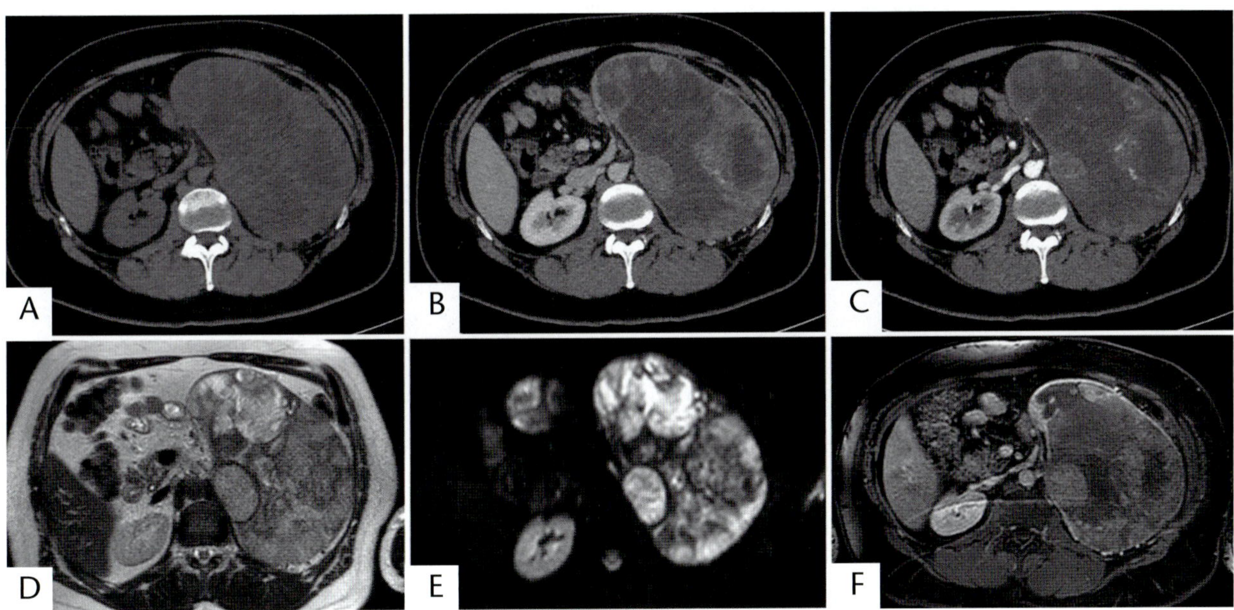

Figure 6-4

ADRENAL CORTICAL CARCINOMA

This 19-cm adrenal lesion in the left adrenal gland has marked intratumoral heterogeneity. Upper panels: CT scan. Basal (A, >20 HU); heterogeneous uptake in the venous (B) and arterial (C) phases. Lower panels: MRI scan. T2 phase (D); in diffusion (E) and with contrast (F). (Courtesy of Prof. A. Veltri and Dr. F. Solitro, San Luigi Hospital, Turin, Italy)

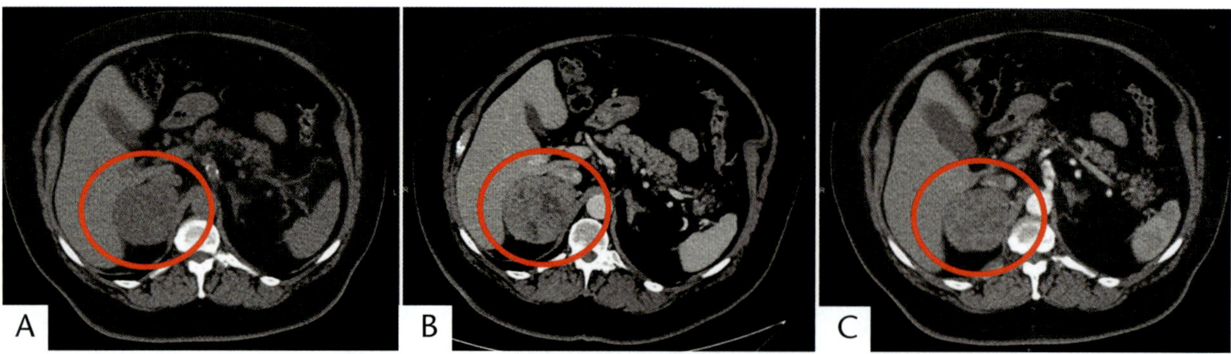

Figure 6-5

ONCOCYTIC ADRENAL CORTICAL CARCINOMA

CT scan of 10-cm adrenal lesion. Basal (20 HU), clearly defined borders (A); poor and heterogeneous uptake in the venous (B) and arterial (C) phases. (Courtesy of Prof. A. Veltri and Dr. F. Solitro, San Luigi Hospital, Turin, Italy)

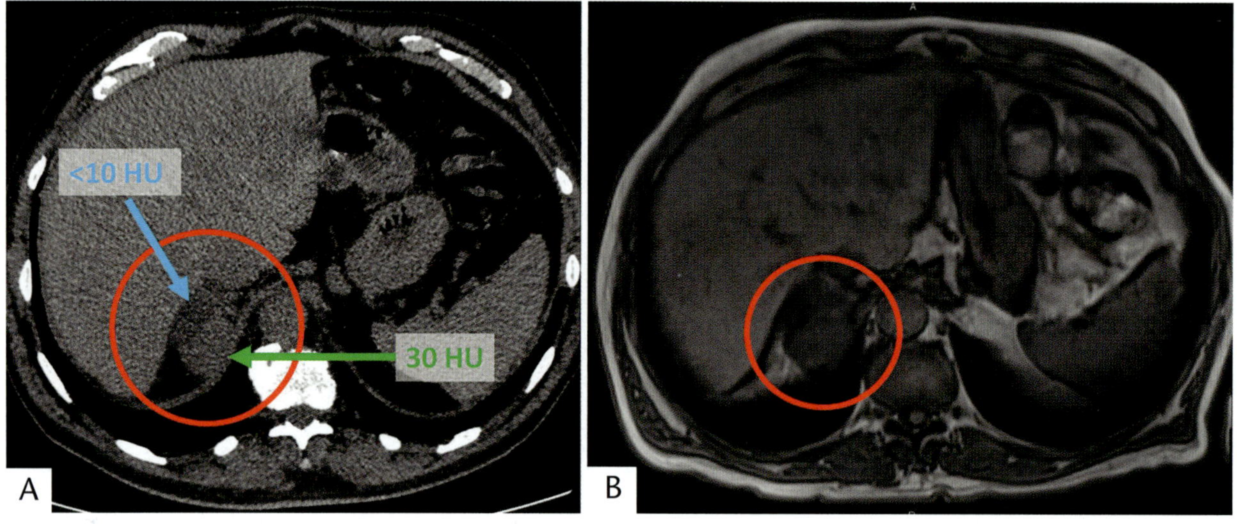

Figure 6-6

ADRENAL CORTICAL CARCINOMA ARISING IN ADRENAL CORTICAL ADENOMA

This adrenal lesion, 11 cm in size (same case as shown in fig. 6-1), has two distinct areas: a benign one with <10 HU and the malignant one at 30 HU. (A: CT scan, basal; B: MRI, "out" phase image). (Courtesy of Prof. A. Veltri and Dr. F. Solitro, San Luigi Hospital, Turin, Italy)

Positron Emission Tomography-Computerized Tomography

Functional imaging studies with FDG-PET scan can provide additional value for the prediction of malignancy risk in adrenal cortical neoplasms. The adrenal uptake is generally compared to the background activity in the liver by measuring the standardized uptake value (SUV) within the region of interest. Normal adrenal glands have an FDG-PET uptake equal to or less than the liver. Using this approach, FDG-PET has been proposed to discriminate benign and malignant adrenal masses, with sensitivity and specificity values ranging from 92 to 100 percent and from 80 to 100 percent, respectively (95–97). An integrated model for malignancy risk assessment based on tumor size and tumor-to-liver uptake SUV_{max} ratio has been proposed recently to calculate an individual malignancy risk for a patient-based decision-making process (98).

Currently, the American College of Radiology committee on incidental findings suggests utilizing FDG-PET in patients with a prior cancer history and indeterminate adrenal masses that are less than 4 cm (99). The negative predictive value and sensitivity of FDG-PET in characterization of adrenal masses are much higher than the positive predictive value and specificity (34). False positive findings may be detected in functional adenomas (mostly cortisol-secreting) and pheochromocytomas (100). Oncocytic adenomas also show marked uptake on FDG-PET imaging (101,102), possibly as the result of GLUT-1 overexpression (103). False negative results are much rarer, and may be the result of intratumoral heterogeneity leading to discordant results in primary versus metastatic lesions (104). This occurrence decreases the sensitivity of FDG-PET compared to CT in the detection of metastases (105).

The degree of FDG-PET uptake does not correlate with survival of patients (106), although FDG-PET better defines progression than enhanced CT in patients treated with chemotherapy (107). Additional functional characterization may use metomidate, a potent inhibitor of the adrenal enzymes CYP11B1 (11β-hydroxylase) and CYP11B2 (aldosterone synthase). Given its selectivity for these adrenal-specific targets, it is not useful to distinguish benign from malignant adrenal tumors. Metomidate labelled with ^{11}C for PET imaging or ^{123}I for single-photon emission computerized tomography (SPECT)/CT is an alternative tracer for functional adrenal imaging to distinguish ACC from nonadrenocortical malignancies, such as metastatic carcinomas (108,109). Moreover, ^{123}I labeled metomidate has been tested for selective radionuclide-based systematic therapy in patients with metastatic ACC (110).

Radiomics and Texture Analysis

The radiomic approach in adrenal pathology has multiple potential applications, from diagnosing benign and adrenal lesions, to the prediction of treatment outcomes, and even the prediction of histopathologic subtypes. Machine learning and radiomic feature extraction from contrast-enhanced CT scans (111,112) or MRI (113) distinguish with high sensitivity and specificity benign from malignant lesions in a group of otherwise indeterminate adrenal tumors. Importantly, such an approach has been shown to possess a higher mean accuracy compared to a morphology-based evaluation by experienced radiologists (114). Radiomic features derived from preoperative contrast-enhanced CT images have been reported to predict the Ki-67 index in the corresponding surgical samples of patients with ACC (115).

GROSS FINDINGS

The macroscopy of ACC reflects its high heterogeneity, in terms of both variable degrees of propensity to invade and the different cellular/histologic features that usually coexist within a single lesion. Consequently, ACC may present macroscopically either as an encapsulated lesion or a mass infiltrating surrounding structures (figs. 6-7–6-9), and most often shows combined aspects on the cut surface, all representative of peculiar histologic characteristics such as hemorrhagic or necrotic areas (fig. 6-10). The overall appearance usually gives the unmistakable impression of malignancy, but it may not be specific enough to distinguish ACC from other malignant neoplasms in the adrenal gland, including pheochromocytoma, sarcoma, or metastasis. Heterogeneity of macroscopic features also depends on tumor size, with small lesions usually showing a more homogeneous and less suspect appearance, with clear demarcation and no apparent invasion of capsular and extracapsular structures. A detailed macroscopic analysis of an adrenal lesion suspected to be ACC is the key to guide an accurate and exhaustive sampling procedure.

ACC is often a bulky neoplasm. Tumor weight is determined after adipose tissue at the periphery or other organs are removed. The average weight in large retrospective series varies from 365 to 428 g, with values ranging from 5 to 3,100 g (116,117). Although tumor weight has been reported in the past to predict biologic behavior, there is no evidence that weight accurately predicts malignancy, except in the setting of oncocytic adrenal cortical tumors. Nevertheless, although weight alone is not a criterion for malignancy, larger lesions are more likely to be malignant and require careful attention to sampling.

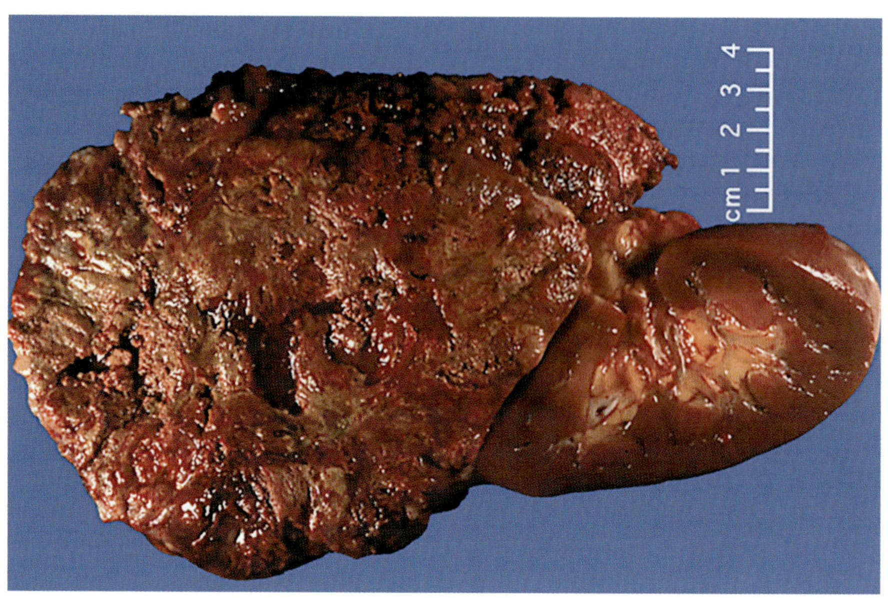

Figure 6-7

ADRENAL CORTICAL CARCINOMA

A large adrenal cortical carcinoma (ACC) from a 57-year-old woman simulated a renal cell carcinoma. The patient had no recognizable endocrine syndrome. Over half of the neoplasm was necrotic.

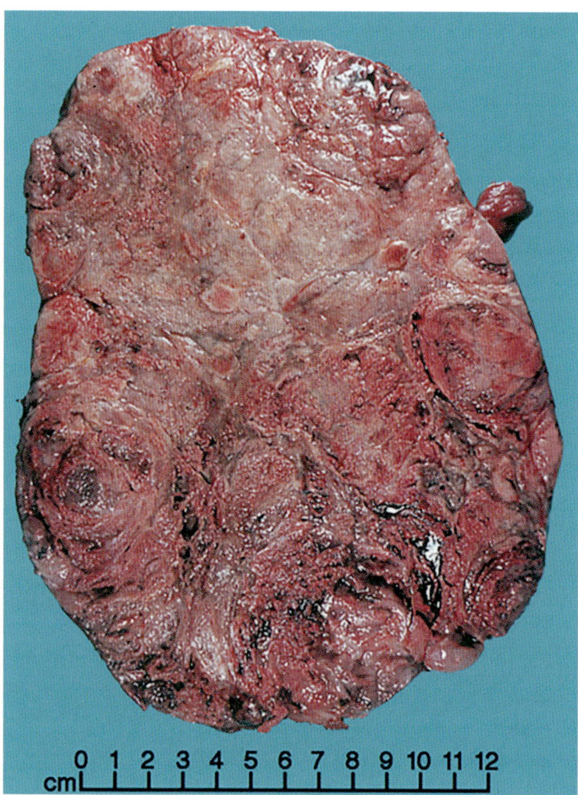

Figure 6-8

ADRENAL CORTICAL CARCINOMA

An ACC from a 32-year-old female weighed 2,200 g and measured 21 cm in diameter. There were extensive areas of necrosis with foci of vascular invasion. The patient also had pulmonary metastases at the time of diagnosis.

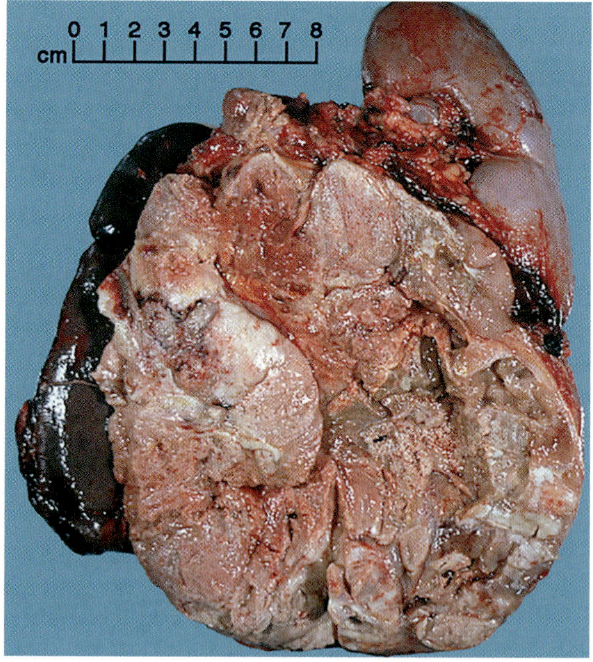

Figure 6-9

ADRENAL CORTICAL CARCINOMA

This large ACC was resected from a 27-year-old woman with virilization. The tumor measured 17 cm in diameter and invaded both kidney and spleen, which necessitated en bloc removal of these organs along with the tumor.

ACCs are usually larger than their benign counterparts, with mean sizes of about 10.5 to 11.4 cm (compared to 4 cm in adrenal cortical adenomas), and ranging between 1.6 to 30 cm (116,117). Size optimally is assessed in three dimensions, but at least the greatest dimension should be carefully determined, as it also represents a relevant parameter for tumor staging. Hormonally inactive carcinomas may occasionally reach impressive proportions: tumors as large as 38 cm have been reported (118). Tumor size also influences the surgical approach: cases undergoing laparoscopic resection are significantly smaller than those undergoing open surgery (119). The laparoscopic approach is associated with artifacts that preclude accurate assessment of the tumor capsule and with a higher risk of positive resection margins (120).

On external examination and in cross section, ACC often has a coarsely lobulated appearance, with soft, bulging nodules ranging from yellow-orange to tan. The color is determined by the relative proportions of clear or eosinophilic cells, including oncocytes, and by the stromal characteristics. A more pronounced nodular pattern is related to the presence of intersecting broad fibrous bands. A well-defined, usually thick tumor capsule is almost invariably present, and may be interrupted in fronts of invasion or penetration into adjacent tissues by the tumor. Necrosis and hemorrhage are characteristic but may not be evident in some cases and need to be carefully sought during sampling. Necrotic areas appear pale tan or yellow-white, with a putty-like consistency. Calcifications and cystic degeneration are common.

ACC can invade adjacent organs or tissues, and en bloc surgical removal of involved structures may be necessary for gross total resection. Gross extension into large veins, extra-adrenal adipose tissue, and adjacent organs are features of stages III and IV disease, and sampling should accurately identify all relationships between the tumor and surrounding tissues for appropriate tumor staging and for the correct definition of the status of the resection margins.

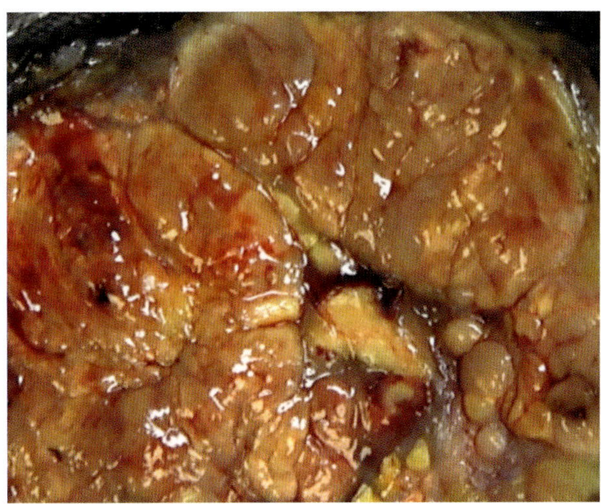

Figure 6-10

ADRENAL CORTICAL CARCINOMA

Heterogeneous cut surface with irregular lobulation, and areas of hemorrhage and necrosis.

MICROSCOPIC FINDINGS

The diagnosis of ACC is based on a combination of architectural and cytologic features, necrosis, mitotic activity, and parameters related to invasion; however, they are almost all fraught with complexity of interpretation (Table 6-2). Although the coexistence of these characteristics is indicative of ACC in most cases at first glance, they have been variably included into scoring systems or diagnostic algorithms for the formulation of a conclusive diagnosis.

Architecture

A variety of architectural patterns are found in ACC, frequently admixed in the same neoplasm (fig. 6-11). Irrespective of the tumor architecture, a consistent finding is the loss of the well-organized alveolar pattern seen in nontumorous adrenal cortical tissue and in adrenal cortical adenoma. This architectural disruption is reflected in a loss of the reticulin framework on silver stain-based histochemistry, which is the basis for the reticulin algorithm described below (117).

One of the more characteristic architectural patterns is a broad trabecular growth pattern with anastomosing columns and cords of cells, often 10 to 20 or more cells wide, separated by delicate, gaping sinusoids lined by an attenuated endothelial layer. Occasionally, the trabeculae are oriented in a more parallel, elongated array, or have an irregular contour with a daisy-like or clover-leaf configuration. Where anastomosing trabeculae are cut in various planes of section,

Table 6-2

CHARACTERISTICS AND DIAGNOSTIC PITFALLS FOR MALIGNANCY-RELATED MORPHOLOGIC PARAMETERS ASSESSED FOR ADRENAL CORTICAL CARCINOMA DIAGNOSIS

Parameter	Key Features	Pitfalls
Architecture	Any architecture different from well-organized alveolar growth is suggestive of ACC[a]; it may be highlighted by silver staining	Unusual patterns (storiform, uniform trabecular, or "carcinoid-like") may mimic noncortical tumors
Necrosis	Only coagulative-type necrosis is to be considered	It may be focal and missed if sampling is not adequate; infarct-type cell necrosis in the presence of hemorrhage or involutive changes (possibly as the consequence of fine needle biopsy) may be erroneously interpreted
Cell cytoplasm	Eosinophilic homogeneous cytoplasm devoid of vacuoles is a morphologic sign of loss of differentiation	Eosinophilic cytoplasm should be distinguished from oncocytic changes characterized by deeper eosinophilia and a finely granular appearance
Nuclear grade	To be assessed based on the presence and size of nucleoli	Nuclear pleomorphism is nonspecific and usually found in ACC mimics such as pheochromocytoma
Mitotic index and atypical mitotic figures	Mitotic index should be determined in 10 mm^2; even a single but unequivocal atypical mitotic figure indicates malignancy	Reproducibility of mitotic index is poor in cases with low mitotic activity; no consensus on the method for mitotic count estimation; morphologic recognition of typical and atypical mitotic figures may be difficult in tumor tissue with inadequate fixation
Invasion	Vascular, capsular, and lymphatic (sinusoidal) invasion should be sought at the periphery of the lesion	The probability of finding invasive foci largely depends on the extent of sampling

[a]ACC = adrenal cortical carcinoma.

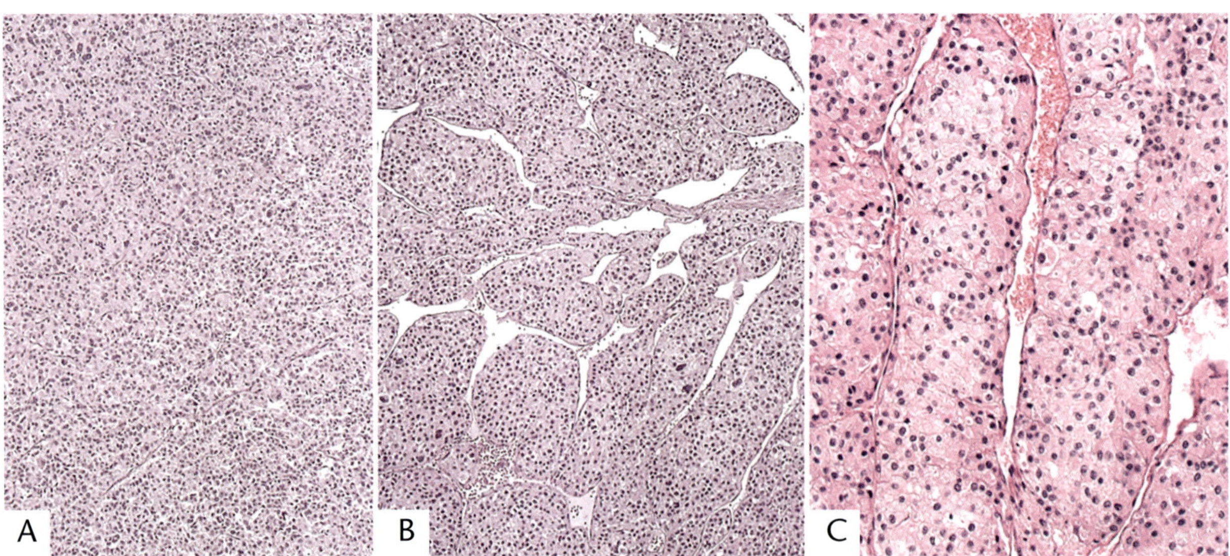

Figure 6-11

ADRENAL CORTICAL CARCINOMA: ARCHITECTURAL PATTERNS

Tumor architecture may be extremely variable, with patterns including the solid (A), "insular" (B), or large trabecular (C), frequently present within the same lesion.

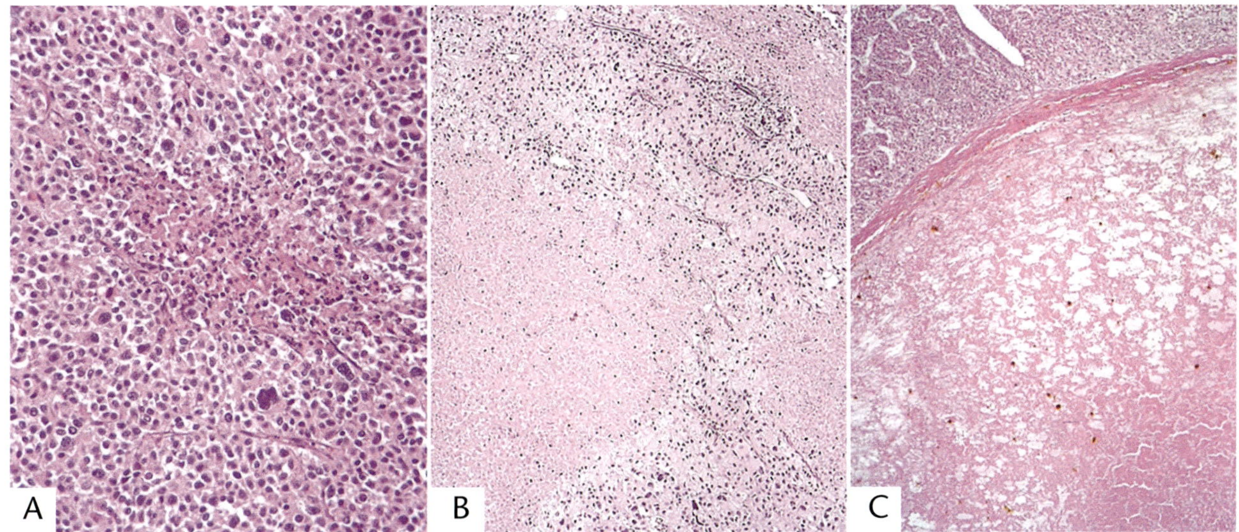

Figure 6-12

NECROSIS IN ADRENAL CORTICAL CARCINOMA

Necrotic areas may be focal punctate (A) or extensive (B), in some cases involving most of the tumor (C).

they may appear "free-floating" and unattached. This gives the impression of vascular or sinusoidal invasion. The trabecular pattern can also be seen as an alignment of cells in slender elongated cords, sometimes with myxoid stromal change or fibrosis. In rare cases, there is a serpiginous arrangement of cells as might be seen in neuroendocrine tumors.

Another architectural pattern is a nesting or alveolar arrangement of cells, which is usually not predominant and blends with the trabecular arrangement noted above. Another pattern is diffuse or solid, which may be focal or predominant, and is typically associated with loss of the reticulin framework. A pseudopapillary arrangement of cells is an uncommon architectural pattern and may be artifactual due to inappropriate fixation. A storiform pattern is typical of the sarcomatoid variant but may be focally seen also in ACCs with conventional histology. The residual nontumorous adrenal cortex is recognizable, if present, as a thin and compressed rim of tissue with fibrosis and cellular atrophy.

The presence of cortical atrophy in adjacent adrenal tissue not directly involved by the tumor usually indicates hypercortisolism with Cushing syndrome or a mixed endocrine syndrome such as Cushing syndrome and virilization. In the rare cases of ACC associated with primary hyperaldosteronism, there may be hyperplasia of the zona glomerulosa in the attached adrenal remnant.

Necrosis

Necrosis is occasionally focal and punctate, however, in most cases, it is extensive, with confluent foci and with a comedo-like pattern when occurring in areas with solid growth (fig. 6-12). There may be broad confluent areas of necrosis, and occasionally, most of the tumor is necrotic. Necrotic areas also undergo calcification (fig. 6-13). A perithelial arrangement of viable tumor cells may be present in association with tissue necrosis. Coagulative necrosis associated with malignancy should be separated from ischemic-type necrosis that may occur as a consequence of fine needle aspiration biopsy procedures. Although not specifically described for adrenal lesions, this phenomenon is well known in other endocrine organs, especially oncocytic tumors in the thyroid gland (121).

Cytologic Findings

Most ACCs contain a predominance of cells with lipid-depleted or acidophilic, compact cytoplasm (fig. 6-14). Occasionally, coarse granular pigment, representing lipofuscin, is seen within the cytoplasm. In rare tumors,

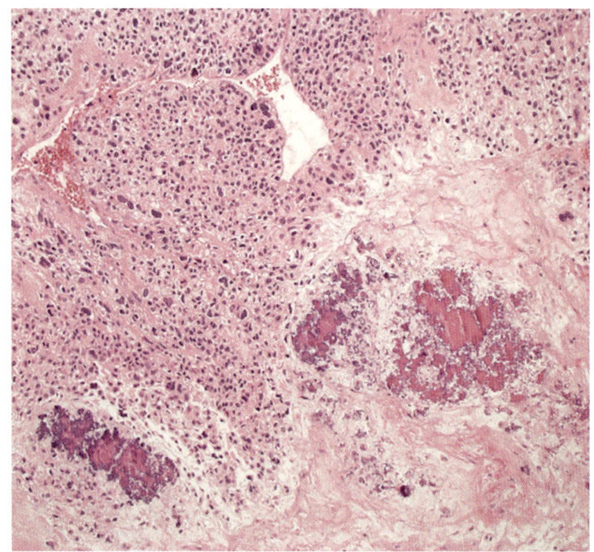

Figure 6-13

NECROSIS IN ADRENAL CORTICAL CARCINOMA

Calcifications may occur in extensive necrotic areas.

lipid-rich, pale-staining cells predominate. The lipid content may be finely dispersed into small vacuoles or more sharply defined with peripheral nuclear displacement (fig. 6-15). Intracytoplasmic hyaline globules may be present as round to oval, slightly refractile, and deeply eosinophilic (periodic acid–Schiff [PAS]-positive) vacuoles (fig. 6-16). However, they are also present in adrenal cortical adenomas and are of poor diagnostic value.

The cytoplasmic features of ACC cells do not significantly correlate with any specific endocrine syndrome. Cell borders are often fairly well defined. Nuclear pleomorphism and hyperchromasia are almost always present and their degree may vary widely within a single lesion, including bizarre and multinucleated cells (figs. 6-17, 6-18). Rare cases have rhabdoid features, a finding that is apparently not associated with worse biologic or clinical behavior (122). Solitary or multiple nuclear pseudoinclusions may be seen and appear as sharply outlined structures that stain as the remaining cell

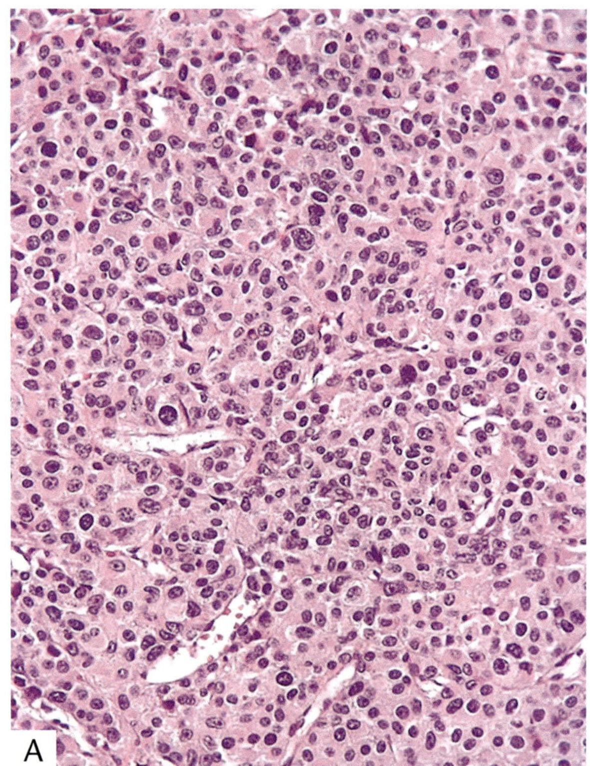

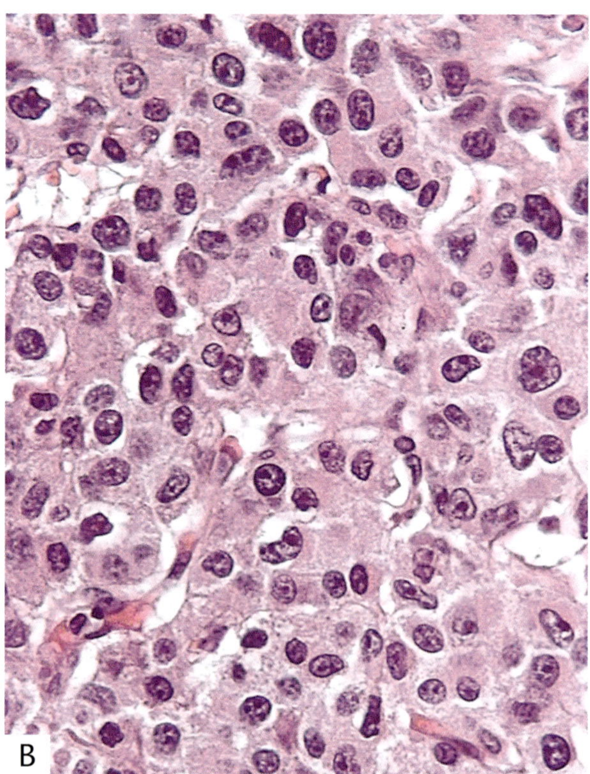

Figure 6-14

ADRENAL CORTICAL CARCINOMA

In most cases, a population of eosinophilic lipid-poor cells predominates (A), with irregular nuclei and prominent nucleoli (B).

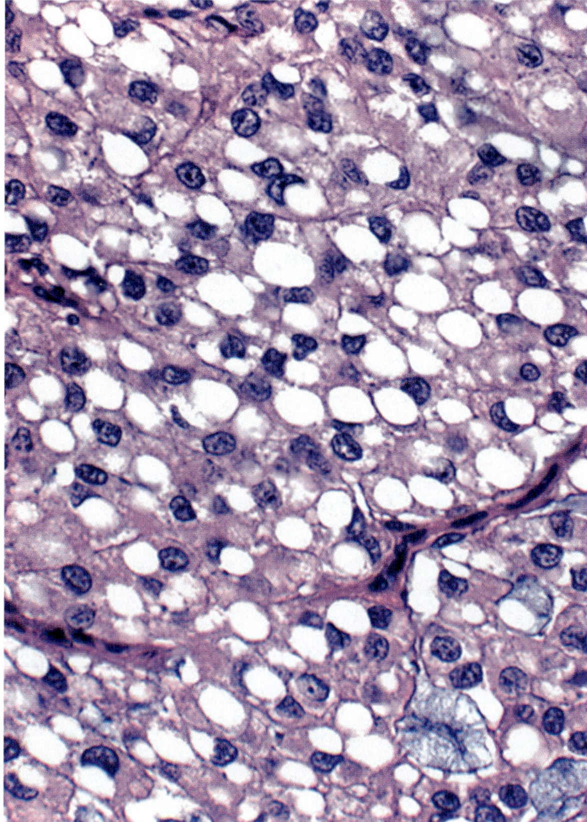

Figure 6-15

ADRENAL CORTICAL CARCINOMA

Most tumor cells in this case have large, coarse cytoplasmic vacuoles, which distend many cells and displace the nucleus to the periphery. Some cells have a "signet ring" appearance.

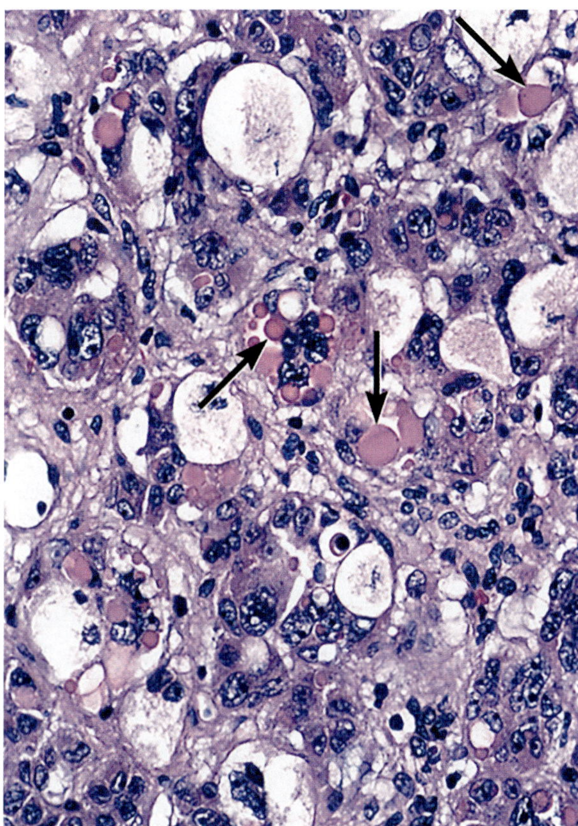

Figure 6-16

ADRENAL CORTICAL CARCINOMA WITH HYALINE GLOBULES

This ACC is composed of cells with predominantly eosinophilic granular cytoplasm although some cells are vacuolated with pale-staining cytoplasm. Numerous intracytoplasmic hyaline globules of various sizes are seen (arrows), with some as large or larger than the nucleus.

cytoplasm or show pallor or vacuolar change. These pseudoinclusions result from irregularities or infoldings of the nuclear membrane and intranuclear extension of the cell cytoplasm.

Nuclear atypia may occur in benign adrenal cortical lesions and is therefore a nonspecific feature. A more ACC-specific nuclear feature is the presence of one or more prominent nucleoli, usually centrally located, which represent a key diagnostic feature according to the Weiss scoring system (see below).

Mitotic Activity

Mitotic figures are common in ACC, and all diagnostic systems incorporate mitotic index as an essential parameter, with a cut-off for malignancy of more than 5 mitoses per 10 mm^2 (50 high-power fields) (123–126). The distribution of mitotic figures is heterogeneous, and they usually cluster in hot-spots. Most ACCs have mitotic counts beyond the cut-off for malignancy, but cases with low mitotic activity exist, and the assessment of the mitotic index in such cases may be more problematic and less reproducible. Phospho-histone-H3 immunohistochemistry may help highlight mitoses in carcinomas with low mitotic counts (127), or in those cases with tissue artifacts. However, the use of this marker needs a reformulation of the mitotic index cut-off for malignancy, since the correlation between immunohistochemistry- and morphology-based counting methods is fairly good but not perfect.

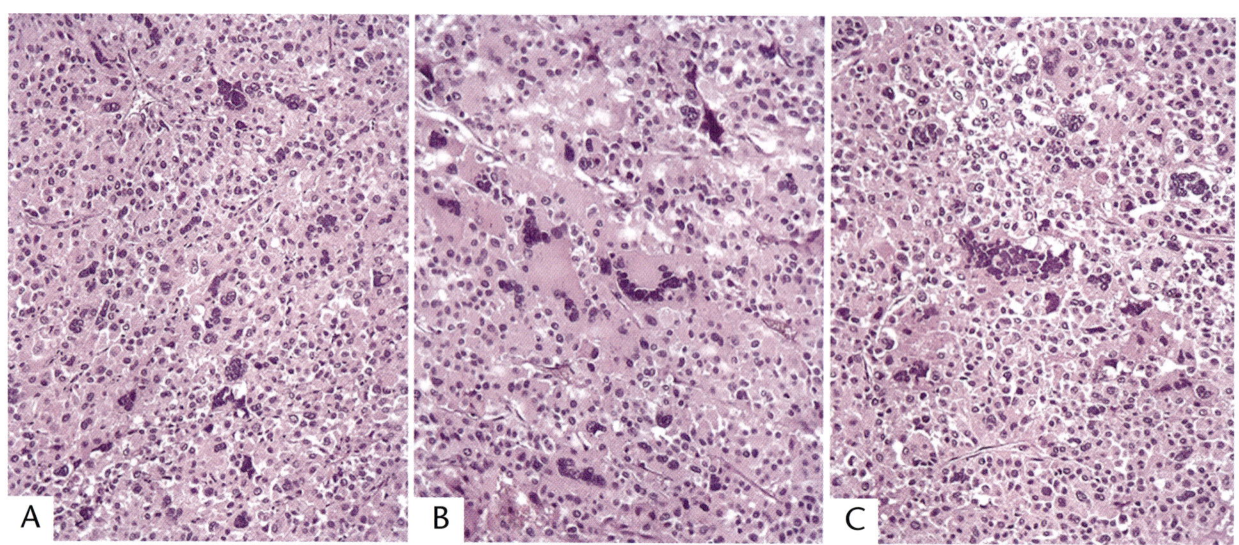

Figure 6-17

ADRENAL CORTICAL CARCINOMA: NUCLEAR ATYPIA

A–C: Pleomorphic cells are variably present, including multinucleated large cells.

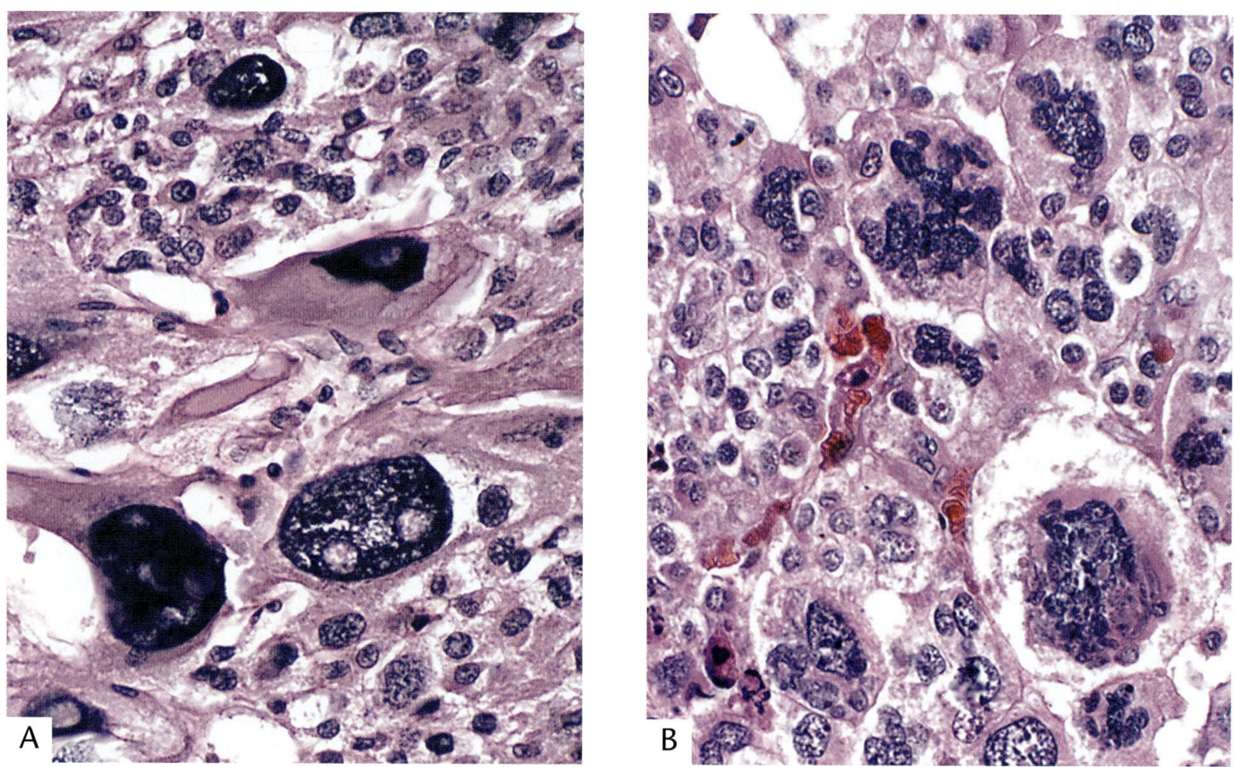

Figure 6-18

ADRENAL CORTICAL CARCINOMA: NUCLEAR ATYPIA

There is marked nuclear hyperchromasia, and great disparity in nuclear size and shape. Some cells show nuclear pseudoinclusions (A). Some tumor cells have enlarged, hyperlobated nuclei and multinucleation (B).

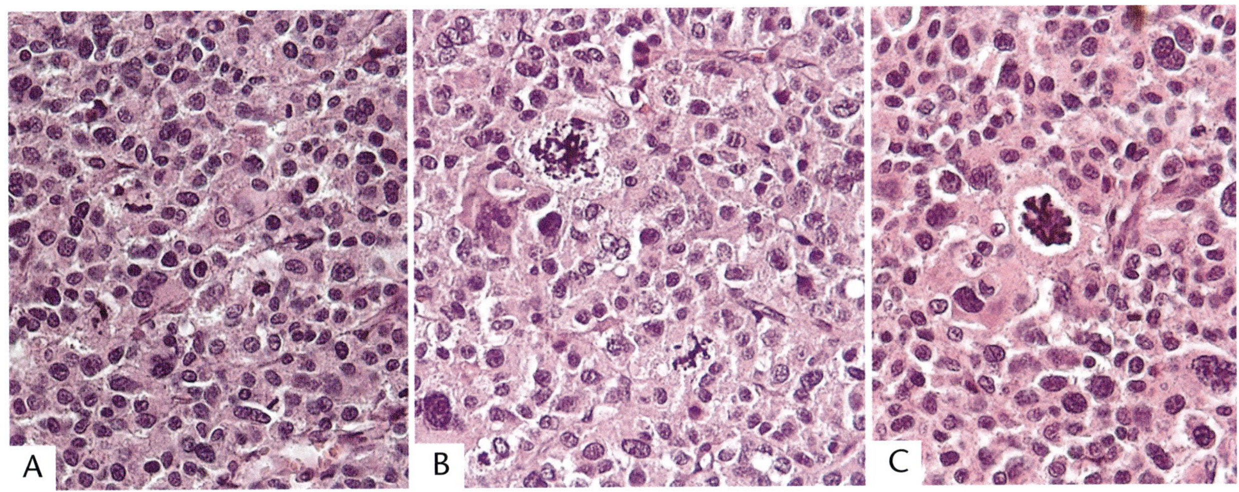

Figure 6-19

ADRENAL CORTICAL CARCINOMA: ATYPICAL MITOSES

A–C: Atypical mitotic figures in varied shapes are a pathologic hallmark of ACC.

Atypical mitotic figures are important findings suggestive of abnormal chromosome content (aneuploidy), and when present, represent a hallmark of malignancy, even when a single unequivocal mitotic figure is identified (fig. 6-19) (128). However, they are recorded in only one third of ACCs (116), thus they represent a poorly sensitive parameter for malignancy.

Invasion

Invasion in ACC is divided into capsular, vascular, and sinusoidal. Capsular invasion is defined as complete penetration of the tumor capsule. The tumor capsule itself may be formed by the tissue response to tumor growth or by the preexisting adrenal capsule. Capsular invasion can be difficult to recognize since the tumor capsule may be irregular and distorted by fibrous septa that extend into the tumor. Invasion of adjacent soft tissues or organs such as kidney or liver is a priori evidence of malignancy as it represents extension beyond the adrenal capsule (fig. 6-20). Capsular invasion is identifiable in more than half of ACC cases but rarely is it essential for establishing the malignant nature of the tumor.

Vascular invasion is an ominous finding in an adrenal cortical neoplasm (fig. 6-21). It is unusual to recognize angioinvasion on gross examination of the resected tumor or in preoperative imaging studies, although some tumors have been detected as invading large veins such as the inferior vena cava (fig. 6-22) and even extending into the right atrium and ventricle (129). The tumor may extend in a sausage-like fashion into the vena cava without firm attachment to the wall. Invasion of the inferior vena cava can be associated with malignant ascites and lower extremity edema and, if unrecognized, may pose a danger in surgical resection if intraoperative mechanical manipulation results in a major tumor embolus. However, vascular invasion is more frequently seen in histologic sections as loose plugs of tumor within the lumen of large veins or sometimes in smaller vascular spaces.

Data to inform criteria for assessing vascular invasion are limited in ACC, as for other endocrine tumors. However, the identification of tumor cells invading through a vessel wall and forming a thrombus/fibrin-tumor complex or intravascular tumor cells admixed with platelet thrombus/fibrin is thought to be the most reliable feature to assess vascular invasion (130). In one study, angioinvasion using these strict criteria was the best predictor of metastatic disease and worse prognosis (131).

Sinusoidal invasion is equivocally considered either as the presence of tumor cells in thin-walled vascular spaces within the tumor or, more consistent with current guidelines for

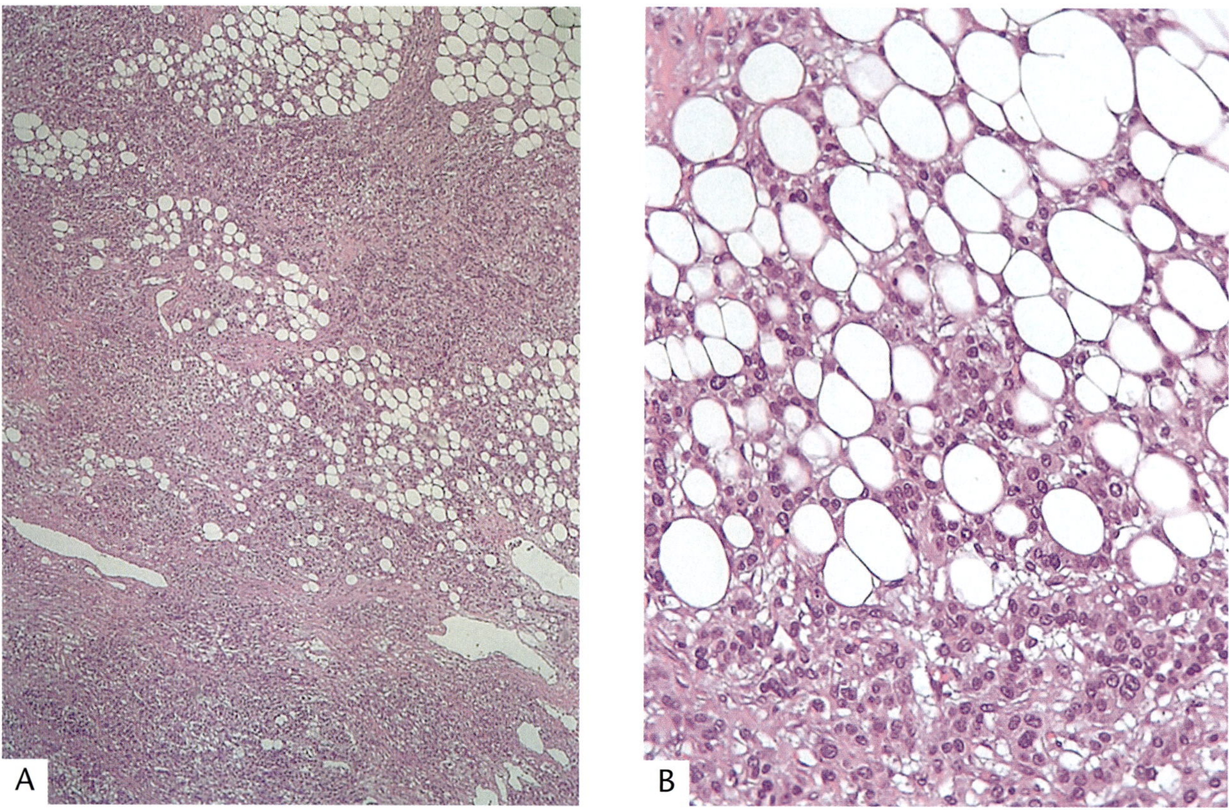

Figure 6-20

INVASIVE ADRENAL CORTICAL CARCINOMA

Extra-adrenal extension with invasion of the periadrenal fat, both as large nodules (A) and as an infiltrative "Indian file" pattern (B).

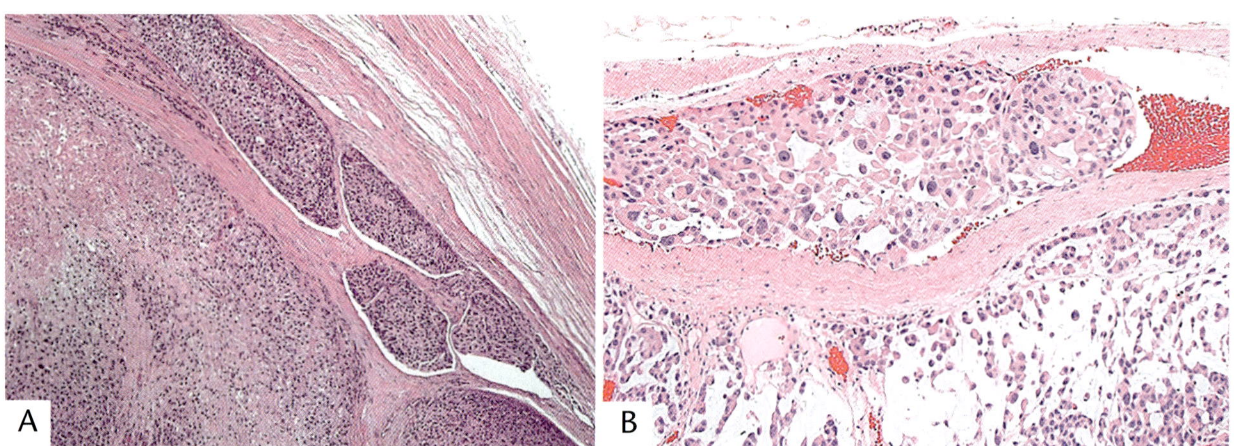

Figure 6-21

SINUSOIDAL AND VASCULAR INVASION IN ADRENAL CORTICAL CARCINOMA

Patterns of vascular invasion in ACC include sinusoidal invasion (A) and angioinvasion with tumor cells in vascular channels associated with fibrin thrombi (B).

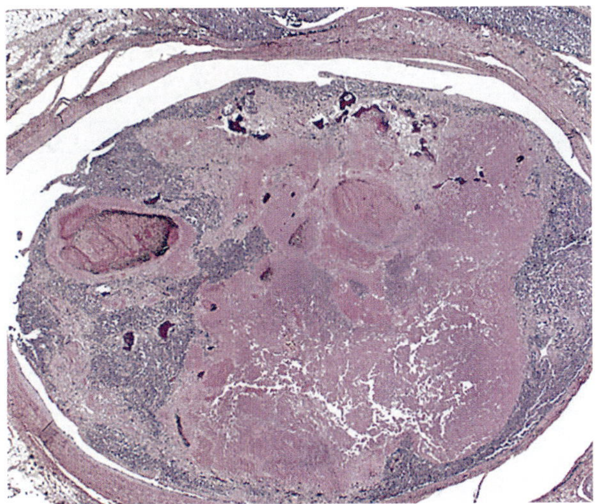

Figure 6-22

LARGE VESSEL INVASION IN ADRENAL CORTICAL CARCINOMA

This ACC caused almost complete occlusion of the lumen of the inferior vena cava. The tumor shows extensive necrosis. One portion is attached to the wall of the vein. A few areas of dystrophic calcification are present.

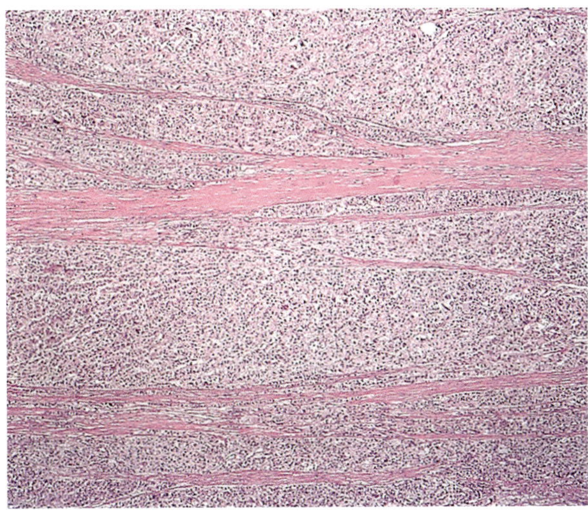

Figure 6-23

ADRENAL CORTICAL CARCINOMA: FIBROUS BANDS

Fibrous bands are frequent, seen as irregularly interconnecting septa.

pathologic reporting (132), as the invasion of lymphatic vessels at the periphery of the tumor. Equivocal interpretation of definitional criteria for sinusoidal invasion render this parameter poorly conclusive for diagnosing ACC and the worst morphologic sign of malignancy in terms of reproducibility (133). The lack of consensus criteria for the definition of sinusoidal invasion makes it difficult to recognize "true" invasion from artifactual images caused by sample manipulation during the surgical procedure.

Stroma

Broad fibrous bands may intersect the tumor, subdividing it into irregular macroscopic nodules (fig. 6-23). Irregular foci of dense dystrophic calcifications may be present, usually following tumor necrosis. Calcifications are detected by US or CT scan of the abdomen in up to 20 percent of cases (80). Calcifications appear as dust-like aggregates associated with individual necrotic cells. Metaplastic bone formation is rarely seen. Lipomatous or myelolipomatous metaplasia can occur, but is usually more rare and focal than in adrenal cortical adenoma. Extensive myxoid changes are characteristic of the myxoid variant, but they may be focal in conventional ACC.

Lymphoid cells at the periphery or within the tumor are seen in ACC but are usually not prominent, except for the oncocytic variant. A recent study aimed at characterizing the immune infiltrate in ACC showed that 86.3 percent of cases have tumor-infiltrating T cells (although with a low number of cells/high-power field) including T helper cells, cytotoxic T cells, and Tregs. Interestingly, the number of tumor-infiltrating lymphocytes is positively associated with better overall survival and negatively with cortisol hypersecretion by the tumor (134).

Histologic Variants

The presence of histologic variants underlines the extreme heterogeneity of the histologic features of ACC, since variants may be almost pure but frequently are combined with even minimal components of conventional ACC (135). The most characterized variants, in decreasing order of frequency, are oncocytic, myxoid, and sarcomatoid carcinomas (Table 6-3) (136).

Oncocytic ACC. Among oncocytic adrenal cortical neoplasms, oncocytic ACC represents 24 percent of cases, with an additional 41 percent of uncertain malignant potential (137). The age of onset of this variant is similar to conventional ACC but there is no female predominance. Functioning oncocytic ACC represents about

Table 6-3

MAIN CLINICAL FEATURES OF ADRENAL CORTICAL CARCINOMA HISTOLOGIC SUBTYPES COMPARED TO THE CONVENTIONAL SUBTYPE IN ADULTS[a]

Characteristic	Conventional	Oncocytic	Myxoid	Sarcomatoid
Number of reported cases	>8000	56	47	28
Age	5th-6th decades	4th decade	5th decade	6th-7th decades
Rate of functioning tumors	50%	50%	57%	11%
Most common hormone produced	Cortisol	Sex steroid hormones	Cortisol	-
Rate of metastases	26-35%	13%	68%	75%
Median survival	17-35 months	60 months	29 months	7 months

[a]Modified Table 1 from Lam AK. Adrenocortical carcinoma: updates of clinical and pathological features after renewed World Health Organisation classification and pathology staging. Biomedicines 2021;9:175.

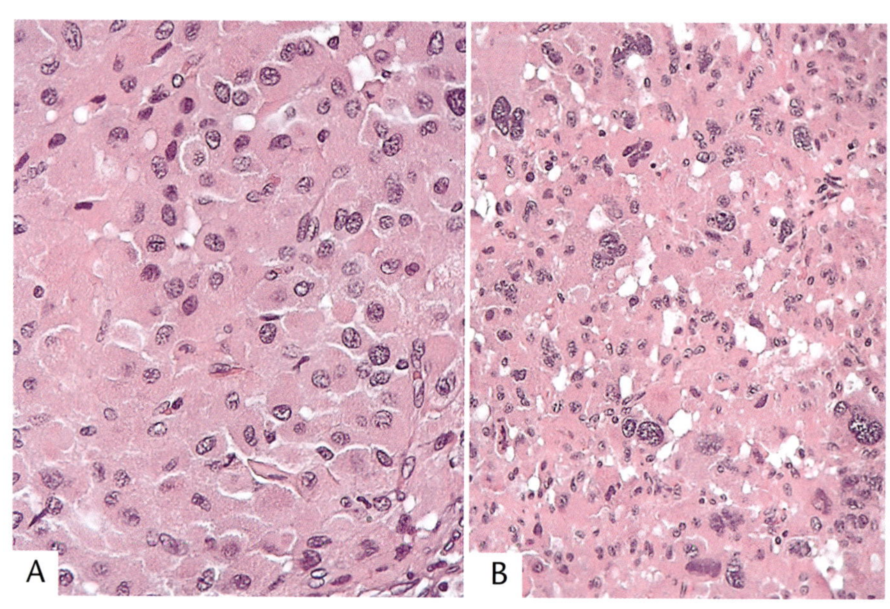

Figure 6-24

ADRENAL CORTICAL CARCINOMA: ONCOCYTIC VARIANT

Large eosinophilic cells with abundant granular cytoplasm (A) and nuclear atypia with prominent nucleoli (B) are characteristic of this variant.

half of the cases, and sex steroid hormone secretion is predominant followed by cortisol secretion (137). In contrast, aldosterone-secreting oncocytic ACC is rare.

The macroscopic features of oncocytic ACC are peculiar, with a tan to brown color and frequent fibrotic scars and necrosis. Oncocytic ACC is histologically defined as the presence of a pure (over 90 percent) population of oncocytes (fig. 6-24) (132,138) together with malignancy-related parameters. However, the prevalence of oncocytes may be variable and mixed conventional-oncocytic ACC exists. Oncocytic cells are distinguished from eosinophilic cells in conventional ACC by their large deeply eosinophilic and granular cytoplasm.

Ultrastructural assessment of mitochondrial accumulation is not necessary for the diagnosis of oncocytic ACC, but staining with antimitochondrial antibodies may be helpful (139). Intratumoral lymphocytes are more frequently observed than in conventional ACC, usually forming small aggregates (fig. 6-25), or exceptionally, germinal centers.

In tumors of pure form, the assessment of malignancy cannot be performed using the Weiss scoring system because three parameters (eosinophilic cytoplasm, diffuse growth pattern, and presence of nucleoli) are invariably present in oncocytic adrenal cortical tumors irrespective of their biological behavior. Therefore, a specific diagnostic algorithm has been proposed for

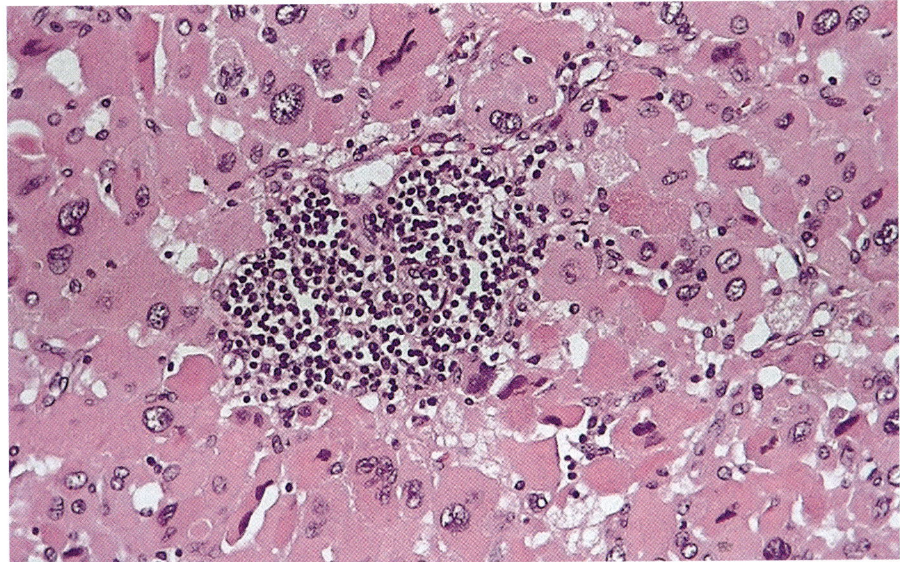

Figure 6-25

ADRENAL CORTICAL CARCINOMA: ONCOCYTIC VARIANT

Intratumoral lymphocytes are frequently seen.

these tumors, the Lin-Weiss-Bisceglia system (138). Other diagnostic systems/scorings that do not consider the three parameters listed above, such as the Helsinki score or the reticulin algorithm, may be indicated in the differential diagnosis (see below) (117,118,140).

Myxoid ACC. This variant has been only recently recognized as a specific ACC variant. It is rare, with less than 50 cases described in the literature since its original description in 1979 (141). Functioning tumors represent 57 percent of cases, with a predominance of cortisol-secreting tumors; sex steroid hormone secretion has never been described.

Grossly, the tumor is characterized by the presence of gelatinous material, to a variable extent, that corresponds to the abundant extracellular connective tissue mucin that characterizes its histology (fig. 6-26) (142,143). Myxoid features may also be observed in metastatic deposits. A myxoid ACC with extensive lipomatous metaplasia was reported (144). There is no definitive evidence on how much myxoid change is necessary to classify an ACC as myxoid, although the presence of prominent myxoid areas is suggested (130). Focal stromal myxoid changes may also occur in an otherwise conventional ACC and should be noted in the pathology report. Moreover, neoplastic cells in myxoid ACC are usually smaller than in conventional ACC, growing in cords or thin trabeculae, and with less pronounced cytologic atypia. In myxoid ACC, some Weiss parameters (such as diffuse growth pattern, lymphatic invasion, and nuclear atypia) may be absent or difficult to assess, thus increasing the risk of an underdiagnosis of malignancy.

Myxoid ACC may mimic various primary retroperitoneal myxoid tumors, including chordoma, myxoma, extraskeletal myxoid chondrosarcoma, lipoma, liposarcoma, benign or malignant nerve sheath tumors, myxoid leiomyoma, myxoid leiomyosarcoma, or gastrointestinal stromal tumor. The comprehensive histopathologic examination should be complemented by an appropriate immunohistochemical panel (i.e., including SF1, see below) to correctly identify these other lesions. Moreover, especially in cytologic samples, myxoid adrenal cortical carcinomas may be misdiagnosed as mucinous adenocarcinomas.

Sarcomatoid ACC. Formerly named carcinosarcoma, sarcomatoid ACC is very rare, with less than 30 cases reported in the literature (135). Most cases are not hormonally functioning and this variant is not described in the pediatric population. The few hormone-secreting tumors are characterized by sex steroid hormone or aldosterone excess.

Macroscopic features of this variant are nonspecific, but it is usually associated with extensive involvement of extra-adrenal tissues. Most cases histologically show a biphasic morphology, with an epithelioid component consistent with a

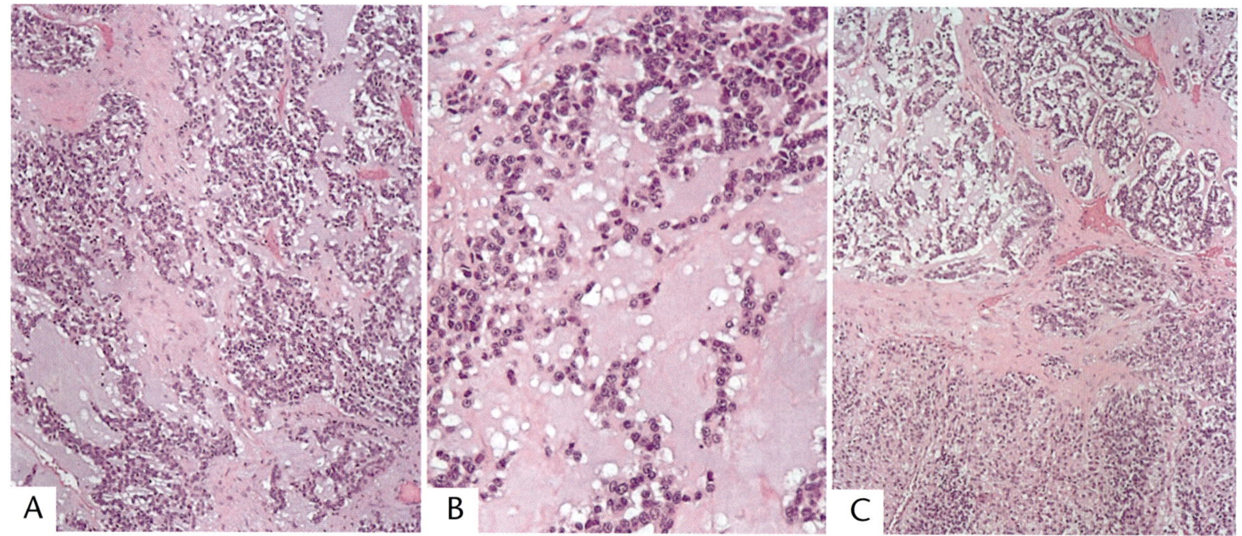

Figure 6-26

ADRENAL CORTICAL CARCINOMA: MYXOID VARIANT

A–C: Myxoid stromal changes are variably in extent and distribution. Neoplastic cells are usually small, grow in cords or thin trabeculae, and have less pronounced cytologic atypia.

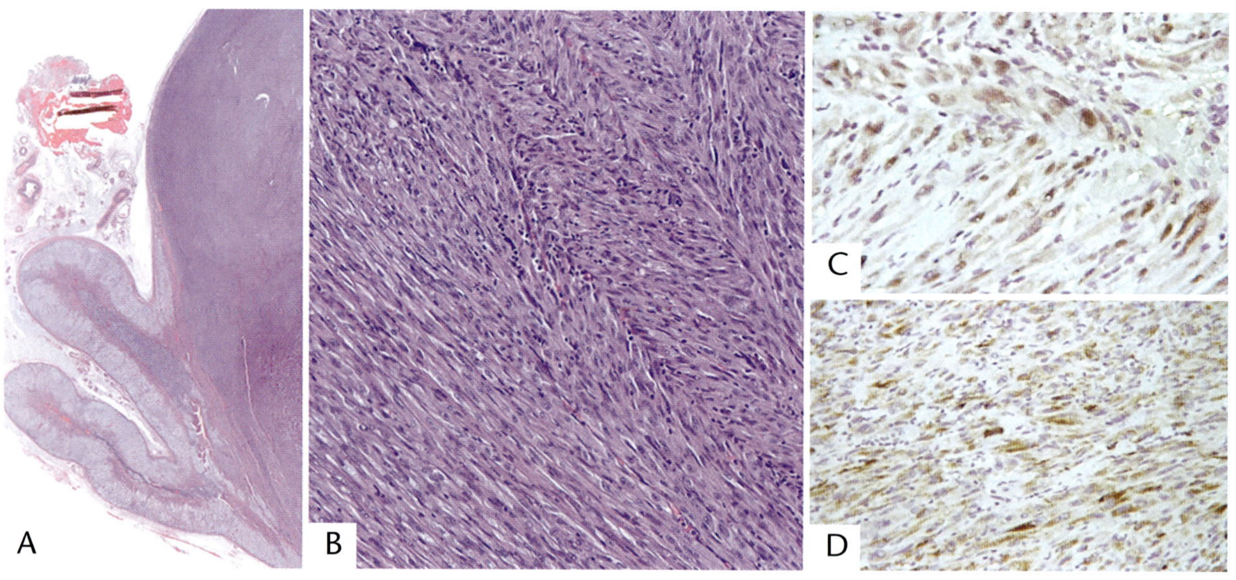

Figure 6-27

ADRENAL CORTICAL CARCINOMA: SARCOMATOID VARIANT

This tumor is well-circumscribed within the adrenal (A). It has a pure storiform architectural pattern (B) and consists of spindle cells (B) expressing SF1 (C) and Melan-A (D).

conventional ACC admixed to a variable extent with an often predominant sarcomatoid pattern, characterized by spindle cell architecture and highly pleomorphic cells (145). When monophasic, sarcomatoid ACC may be indistinguishable from true sarcomas of the adrenal gland but adrenocortical markers are usually positive although often only focally (fig. 6-27) (146).

Table 6-4

PRINCIPLES OF MOST WIDELY APPLIED SCORING/DIAGNOSTIC SYSTEMS IN ADULT ADRENAL CORTICAL CARCINOMA DIAGNOSIS

Parameter Considered	Weiss Score	Helsinki Score	Reticulin Algorithm	Lin-Weiss-Bisceglia Classification
Cytologic features	Nuclear grade 3-4[a]; mitotic index >5x50 HPF[b]; atypical mitotic figures	Mitotic index >5x50 HPF[b]	Mitotic index >5x50 HPF[b]	Mitotic index >5x50 HPF[b]; atypical mitotic figures
Tumor architecture	Clear cell component <25%; presence of necrosis; diffuse architecture in >30% of the tumor	Presence of necrosis	Presence of necrosis	Presence of necrosis
Invasiveness	Sinusoidal invasion; vascular invasion; capsular invasion	None	Vascular invasion	Sinusoidal invasion; vascular invasion; capsular invasion
Additional features	None	Exact value of Ki-67 in hot spots	Silver stain-based reticulin evaluation	Size >10 cm; weight >200 g
Principles of scoring	Each parameter, if present, is scored as 1	3 for mitotic index + 5 for presence of necrosis + exact value of Ki-67	No numerical scoring; algorithmic approach	Mitotic index, atypical mitoses, and vascular invasion are major criteria; all others are minor criteria
Score/pattern for malignancy	Score value =/or >3	Score value >8.5	Altered reticulin staining + at least one of mitotic index, presence of necrosis, and/or vascular invasion	Malignant if one or more major criteria; of uncertain malignant potential (borderline) if one or more minor criteria only
Main advantages	The most widely adopted and validated; no need for additional staining other than conventional H&E	Easy to assess and reproducible	Easy to assess and reproducible	The most widely adopted and validated in oncocytic tumors; no need for additional staining other than conventional H&E
Main limitations	Poor reproducibility for some parameters; not applicable in the oncocytic variant or in the pediatric age range; risk of underestimating malignancy in the myxoid variant	Need to assess Ki-67; to be validated in large independent cohorts of cases	Need to assess reticulin pattern; to be validated in large independent cohorts of cases	To be adopted in oncocytic tumors, only; need to assess weight

[a]According to Fuhrman grading for renal cell carcinoma.
[b]Preferably expressed as 10 mm^2.

SCORING SYSTEMS/ DIAGNOSTIC ALGORITHMS

The histologic features of ACC are assessed within multifactorial algorithms/scoring systems, which are useful and practical approaches for the diagnostic assessment of malignancy in an adrenal cortical lesion (Table 6-4). No single system has been shown to be completely sensitive or specific in all settings, nor have the systems proven to have a complete observational concordance in individual lesions. These systems are not always needed for diagnosis, and there is an ongoing debate about their validation and reproducibility. It is advisable that pathologists use their judgment to select the appropriate system or multiple systems for use in their practice depending on the morphologic features of the lesion they are observing. Nevertheless, standardized use of these systems would have the great benefit of recording as

Table 6-5

WEISS SCORING PARAMETERS

Nuclear grade 3 or 4, according to Fuhrman grading for renal cell carcinoma
Mitotic index >5/10mm^2 (50 HPF)
Presence of atypical mitotic figures
Lipid-rich (clear) cell content <25 percent of the tumor
Presence of a diffuse architecture in >30 percent of the tumor
Presence of tumor necrosis
Presence of venous invasion
Presence of sinusoidal invasion
Presence of capsular invasion

consistently as possible the individual data that contribute to the scoring systems, in the hope that they may be further refined and validated (132). Moreover, the integration of multiple approaches may be advisable to characterize adrenal cortical lesions that lack clear-cut morphologic signs of malignancy and fall into the gray zone between benign and malignant tumors.

Weiss and Modified Weiss Scores

The Weiss score was the first scoring system for assessing malignancy in ACC and is by far the most widely adopted and validated (147). It was built to predict the occurrence of metastasis in adrenal cortical tumors. The Weiss score consists of nine parameters, each if present counting for 1 point (range, 0 to 9). Malignancy is defined by a score of 3 points or more (Table 6-5).

The Weiss score is also used as a prognostic parameter, usually with a malignancy cut-off of 6 (118,148). Its independent prognostic value, however, is called into question and may fail to maintain statistical significance in multivariable analysis (149).

The modified Weiss score restricts the number of parameters to 5, excluding diffuse architecture, nuclear grade, and vascular and sinusoidal invasion, and was aimed at simplifying the scoring algorithm while maintaining a high concordance with the original Weiss system (150). Two parameters (mitotic index and content of lipid-rich cells) have an individual value of 2. Therefore, the score ranges from 1 to 7, with a cut-off of malignancy of 3 or more. The major disadvantages of both Weiss and modified Wiess scores is the inadequate applicability to the oncocytic and myxoid variants, overestimating or underestimating malignancy, respectively. Also, despite its wide use, the Weiss score is poorly reproducible (133). It is not used in pediatric adrenal cortical tumors (see below).

Helsinki Score

The Helsinki score is the most recent scoring system proposed, and it has the prime aim to identify adrenal cortical lesions with significant metastatic potential (126). The principle of this scoring proposal is to simplify the Weiss scoring system, limiting the number of variables to apply (thus improving reproducibility), and to integrate Ki-67 as an additional tool based on the strong evidence-based data on its prognostic role. The proliferation index is measured by computer-assisted image analysis.

The identification and relative impact of each parameter are assessed by means of stepwise regression analysis. The score is based on the following: 3 points for a mitotic count over 5 per 50 high-power fields/10 mm^2, plus 5 points for tumor necrosis if present, plus the numeric value of the Ki-67 proliferative index in the most proliferative area of the tumor. A Helsinki score over 8.5 points is associated with metastatic potential with 100 percent sensitivity and 99.4 percent specificity, and warrants the diagnosis of ACC.

The Helsinki score has been largely evaluated and validated in independent series for both conventional ACC and ACC variants (140). In a recent mono-institutional large series of over 500 ACCs, the Helsinki score outperformed the Weiss system to detect malignancy (151). In addition to its value in assessing malignancy, the Helsinki score has been also shown to be prognostic in terms of survival, with a threshold of 17 points in the original study (126) and of 28.5 points in a validation study on a large series of cases (118).

Reticulin Algorithm

The reticulin algorithm proposal stemmed from the evidence generated by molecular data showing that the vascular network, as well as stromal deposition, are almost invariably regular in adrenal cortical adenoma, closely mimicking the normal adrenal cortex, whereas they

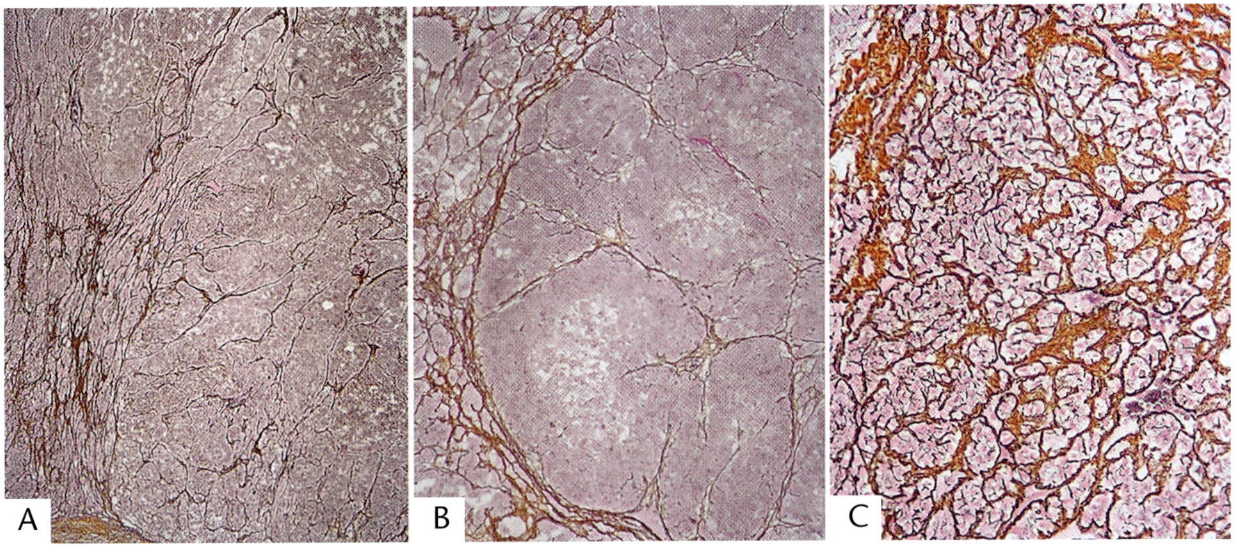

Figure 6-28

ADRENAL CORTICAL CARCINOMA: ALTERED RETICULIN PATTERNS

Irregular reticulin pattern as highlighted by silver staining techniques, either in the form of "quantitative" changes (large nests disrupting the regular alveolar pattern typical of benign conditions (A,B) or "qualitative" changes, in the form of thickened fibers embedding single cells or small clusters (C).

are disrupted at a variable degree of distribution and quality in ACC (152). These differences are reflected in the reticulin framework, which is easily highlighted by silver-based staining procedures that were therefore proposed as the first step of the algorithm. In the first study, an altered reticulin pattern was defined by the loss of continuity of reticular fibers or basal membrane network evaluated in one high-power field (HPF) (400X, assessing a single HPF = 0.2 mm^2), extended to at least one third of the lesion (116). This definition was then modified in a large validation study (117), adding the concept of qualitative disruptive changes characterized by an apparently intact reticulin network made of fibers having variable and irregular thickness, with a frayed appearance, surrounding single cells or, more rarely, small groups of cells. In more general terms, all patterns that differ from the reticulin network of a normal adrenal gland have to be recorded as "altered," and this easy picture explains the very high reproducibility of this evaluation (fig. 6-28) (117).

It is important to recognize sites of degeneration or hemorrhage that may show a disrupted reticulin framework (58). If the recognition of an altered reticulin pattern is the inclusion criterion for the algorithm, a diagnosis of ACC is rendered in the presence of at least one of the following criteria: mitotic index over 5 in 50 HPFs (10 mm^2), presence of necrosis, or presence of vascular invasion. In the original study this algorithm showed a 100 percent sensitivity and specificity for recognizing adrenal tumors coded as malignant according to the Weiss system, but with easier and more practical applicability (116). The diagnostic performance was also validated by other groups and proved to be applicable in both conventional ACC and in ACC variants (131,140,153–155).

Silver-stained sections may also be the master for automated digital imaging analysis. A proof-of-concept study demonstrated that the reticulin pattern can be quantitatively analyzed by computerized morphometry (156), opening its potential incorporation into machine learning-based diagnostic algorithms. The reticulin algorithm in the original description was also associated with a specific risk stratification, using stage and mitotic index as parameters to identify low-, intermediate-, and high-risk subgroups that were significantly associated with a diverse overall survival (116), but this finding has not been clinically validated in independent studies.

Lin-Weiss-Bisceglia System

In 2004, a group of pathologists highlighted the need for a specific scoring system to be applied in oncocytic adrenal cortical tumors to overcome the limitations of the Weiss and modified Weiss systems (138). By definition, this diagnostic approach should be restricted to pure (greater than 90 percent) oncocytic adrenal cortical tumors. By correlating pathologic features with outcome in a series of pure oncocytic cases, three major criteria definitional for malignancy and four other minor criteria were identified that, in the absence of major criteria, would lead to a diagnosis of tumor with uncertain malignant potential (borderline).

The major criteria include: mitotic index as for the Weiss system, presence of atypical mitotic figures, and venous invasion. The presence of one single major criterion is enough to assess malignant potential. The minor criteria include: presence of necrosis, presence of capsular invasion, size over 10 cm, and weight over 200 g. Also in this case, the presence of one single minor criterion alone is enough to reach a diagnosis of borderline tumor. The diagnostic performance of this system has been validated in independent series (140,155).

The Lin-Weiss-Bisceglia system is easy to apply but tumor size and weight need to be available. This may be a limitation in cases that are fragmented or lack sample integrity.

LESIONS AT THE THRESHOLD OF MALIGNANCY

The use of scoring systems or diagnostic algorithms is of particular value in adrenal cortical tumors lacking clear-cut clinical or morphologic signs of malignancy. Some adrenal cortical tumors may not be easily placed into a benign or malignant category because of the presence of worrisome but insufficient features to accurately predict the biologic behavior. A borderline diagnostic category is coded in the Lin-Weiss-Bisceglia system (138), but even in the other diagnostic schemes, some cases fit into categories uncertain for malignancy (i.e., cases with a Weiss score of 2 or Helsinki scores close to the threshold of malignancy). A practical principle is that whenever a case presents morphologic features suspicious of, but not sufficient for, a definitive diagnosis of ACC or equivocal signs of malignancy (i.e., equivocal vascular invasion), additional sampling and, whenever indicated, additional sectioning are required.

Usually, in cases with scores falling short of the cut-off for malignancy, the points for scoring are not clear-cut signs of malignancy (such as increased mitotic activity or invasion), but rather less specific features are present, such as cytoplasmic eosinophilia, diffuse growth, or nuclear atypia. Necrosis itself, in the absence of increased mitotic activity or invasion, may not be completely specific for malignancy. Although these cases are encountered in clinical diagnostic practice, their clinical outcome and prognosis are almost completely unknown. In published case series, such cases are not represented or are present in very small numbers that do not allow any conclusion of clinical value. It is almost certain that some of these cases may present an aggressive clinical course. In a series of eight oncocytic tumors of uncertain malignant potential, two had disease progression, one leading to the patient's death (155). In a recent molecular study on ACC prognostic stratification, 8.5 percent of cases with a Weiss score 2 (and 36.5 percent of cases with a Weiss score 3) had disease-related adverse events during follow-up (157). Therefore, lesions at the threshold of malignancy should be sampled adequately and although they are generally considered of low risk of malignancy, an adverse clinical course cannot be excluded. Future studies on this specific population of adrenal cortical tumors are needed to clarify their pathologic and clinical characteristics.

HISTOPATHOLOGIC GRADING

Despite the wide range of morphologic parameters assessed for diagnostic scoring systems, mitotic index is the only parameter considered for grading to date. Alternative approaches to define tumor grade have been proposed, analyzing the prognostic influence of the Weiss score, the Ki-67 index, or a combination of both (158), but are not applied in clinical practice. The independent prognostic impact of mitotic index has been consistent since the original report by Weiss et al. (147), up to and including several subsequent studies (116,118,126,127,159), although overall these papers use different cut-off values.

At variance with the "diagnostic value" of mitotic index based on a threshold for malignancy

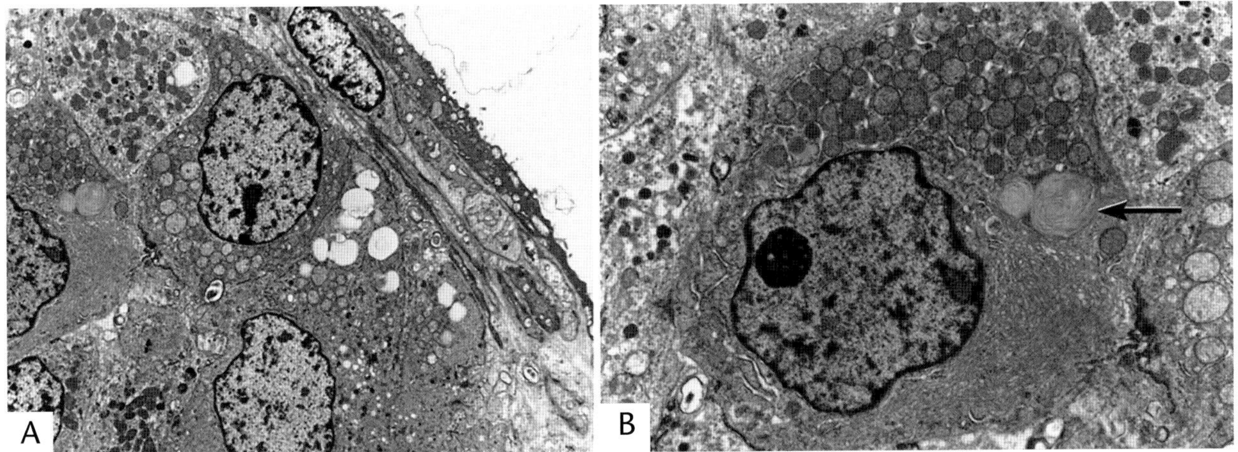

Figure 6-29

ADRENAL CORTICAL CARCINOMA: ULTRASTRUCTURE

A: Tumor cells contain sparse lipid droplets; some have small, irregular microvillous extensions of cytoplasm along the cell border. The cells contain ample mitochondria and smooth endoplasmic reticulum. The delicate vasculature is illustrated by a vascular channel in the right upper corner that is lined by endothelial cells and partially surrounded by pericytes.

B: The tumor cell is essentially devoid of lipid droplets but contains abundant smooth endoplasmic reticulum and some profiles of rough endoplasmic reticulum. A few concentric whorls of smooth endoplasmic reticulum are also present (arrow).

at more than 5 mitotic figures in 10 mm^2, irrespective of the scoring system applied, ACCs are classically subdivided on the basis of mitotic activity into low grade (≤20 mitoses/10 mm^2) and high grade (>20 mitoses/10 mm^2) (160,161). However, a mitotic counting-based grading approach is notably time consuming and subject to interobserver variability, although this variability may decrease with specific training (133). The independent clinical value of mitotic index-based grading in ACC is still debated and needs prospective validation on a large scale.

ULTRASTRUCTURAL FINDINGS

Electron microscopy has currently a limited role in the differential diagnosis and investigation of ACC, although specific ultrastructural features characterize lesions with different morphologies and functional properties. Ultrastructural findings in ACC reflect the morphologic and functional heterogeneity of ACC tumor cells (162,163). The distinctive ultrastructural features are usually present irregularly in tumor cells of a given lesion, with extensive variability among different cases.

The abundance of cells with compact, eosinophilic cytoplasm indicates a cell population with little to no intracytoplasmic lipid droplets (fig. 6-29), as opposed to the lipid-rich cortical cells in adenomas and normal cortical cells of the zona fasciculata. Mitochondria are abundant and are responsible for the finely granular appearance on hematoxylin and eosin (H&E) staining. They may vary in size and shape, but are often small and round to oval. They may have tubular, vesicular, or lamellar cristae; the identification of tubulovesicular cristae is helpful only to characterize a lesion as derived from steroidogenic lineage. A granular matrix may be prominent within individual mitochondria.

The cytoplasmic eosinophilia is attributable to a prominent smooth endoplasmic reticulum forming concentric whorls or a tangled skein, or arranged in sparse lamellated structures resembling myelin figures. Most tumors contain cells with short, flattened profiles of rough endoplasmic reticulum, occasionally arranged in small stacks. As observed with light microscopy, the nucleus may be predominant in some cells, and may have prominent nucleoli and marginated chromatin dispersed on the nuclear membrane or densely clumped. Stubby, microvillous projections may be observed, but to a lesser extent than in adrenal cortical adenomas. Simple intercellular junctions are occasionally seen, but they tend to be sparse in number. Small intracytoplasmic lakes of glycogen are rarely seen. Cells of virilizing ACC can resemble the compact cells of the zona reticularis with tubular cristae, lipofuscin pigment, and scant lipid, but it is virtually impossible to accurately predict the presence or absence of any particular endocrine syndrome solely from the ultrastructural appearance.

In the oncocytic variant, electron microscopy is characterized by dense packaging of mitochondria (164), a feature that is the hallmark of oncocytic neoplasms irrespective of their primary localization. Mitochondria exhibit either typical lamellar cristae, as seen in zona glomerulosa cells of normal adrenal cortex, or tubulovesicular cristae, as seen in zona fasciculata and reticularis cells. Some amorphous intramitochondrial inclusions may be seen. As in conventional ACC, additional features are the presence of smooth endoplasmic reticulum and scarce lipid droplets (138).

In the myxoid variant, tumor cells display the typical features of steroidogenic cells, with abundant smooth endoplasmic reticulum, mitochondria with tubulovesicular cristae, and lipid droplets. Moreover, the extracellular matrix is characterized by proteoglycan particles, fibrillary and flocculent material, and membrane debris (165).

Recently, electron microscopy helped to identify specific ultrastructural characteristics associated with responsiveness to mitotane. ACC cells resistant to mitotane treatment showed the lack of morphologic ultrastructural changes, whereas nonresistant cells showed a wide range of mitotane-induced alterations, including mitochondrial swelling with loss of cristae, irregular nuclear shapes, large lipid droplets surrounded by concentric layers of rough endoplasmic reticulum, necrotic cells, and intracellular protein deposits (166). The absence of mitochondrial damage in resistant cells supports the hypothesis that the mechanism of resistance in this model acts upstream of mitochondrial damage.

IMMUNOHISTOCHEMICAL FINDINGS

There are three reasons to use immunohistochemical markers for ACC characterization: the identification of primary adrenocortical origin, the distinction of ACC from benign lesions, and the definition of prognostic subgroups (Table 6-6) (167). The first two are discussed here; the third is discussed in more detail below in the section on prognosis.

ACC is strongly and diffusely positive for vimentin, a feature that is of little help in the diagnostic workup. Conversely, ACC is usually negative for epithelial membrane antigen (EMA) and only focally positive for cytokeratins, a clue that may help to distinguish ACC from metastatic carcinomas. To prove adrenocortical differentiation, a wide panel of markers may be used but with different specificities and sensitivities that should be taken into account. The best marker in this setting is steroidogenic factor 1 (SF1), which is considered the most reliable and specific, since it is expressed in normal adrenal cortex as well as in all types of adrenal cortical neoplasms (fig. 6-30) (168,169).

SF1 is a master regulator of steroidogenesis and is expressed also in steroidogenic cells of the gonads as well as in the gonadotrophs of the pituitary gland (170). Excluding these two types of cells, SF1 specificity for adrenal cortical origin is up to 100 percent and its sensitivity is about 95 percent, thus it is a very reliable marker of adrenocortical origin in extra-adrenal locations to diagnose metastatic ACC (fig. 6-31). While various commercially available antisera provide comparable results, monoclonal antibodies against SF1 are proven to be more accurate (171). A major disadvantage of SF1 is the loss of antigenicity in samples with inadequate fixation, which may limit its evaluability, and the loss of expression in a minority of ACCs, as compared to adrenal cortical adenomas, which are almost invariably positive.

Several less specific biomarkers may also be used in a panel approach, but their variable positive rates in ACC as well as their immunoreactivity in some ACC mimics should be considered. Among these markers, Melan-A (fig. 6-32), synaptophysin, alpha-inhibin (fig. 6-33), and calretinin are the most widely adopted, however, all are expressed in other tumors that must be distinguished from ACC (167,172). Other markers that may be positive in ACC and represent potential diagnostic pitfalls are CD10 (173), GATA3 (in rare cases and focally) (174), and CD56 (175).

The distinction of a benign adenoma from low-grade ACC confined to the adrenal gland

Table 6-6

MOST RELEVANT IMMUNOHISTOCHEMICAL MARKERS FOR ADRENAL CORTICAL CARCINOMA DIAGNOSIS AND CHARACTERIZATION

Marker	Use	Main Advantages	Potential Pitfalls
SF1	Definition of primary adreno-cortical origin; additional prognostic value	Very high sensitivity and specificity	Sensitive to fixation
Melan-A	Definition of primary adreno-cortical origin	High sensitivity but not complete specificity	Positive in melanoma, PEComas, and a subset of renal cell carcinomas
Synaptophysin	Definition of primary adreno-cortical origin	High sensitivity but not complete specificity	Positive in pheochromocytoma and other neuroendocrine tumors
Alpha-inhibin	Definition of primary adrenocortical origin	Low sensitivity and moderate specificity	It may be positive in different types of carcinomas, pheochromocytoma, and steroidogenic cells; sensitive to fixation
Calretinin	Definition of primary adrenocortical origin	Low sensitivity and low specificity	Positive in mesothelioma and different types of carcinomas
D2-40	Definition of adrenocortical origin; useful to assess lymphatic (sinusoidal) invasion	Low sensitivity and low specificity	Positive in mesothelioma and different types of carcinomas
Ki-67	Diagnostic in borderline lesions; strongly prognostic	Robust validation	Universal cut-off values still debated; reproducibility low
IGF2	Diagnostic in borderline lesions	High specificity but low sensitivity	Limited use in surgical pathology practice; sensitive to fixation
p53	Diagnostic in borderline lesions; prognostic	Easy to assess (complete loss to be determined in the presence of positive internal control) or diffuse positivity	Incomplete validation; sensitive to fixation
Beta-catenin	Prognostic	Easy to assess	Incomplete validation; sensitive to fixation

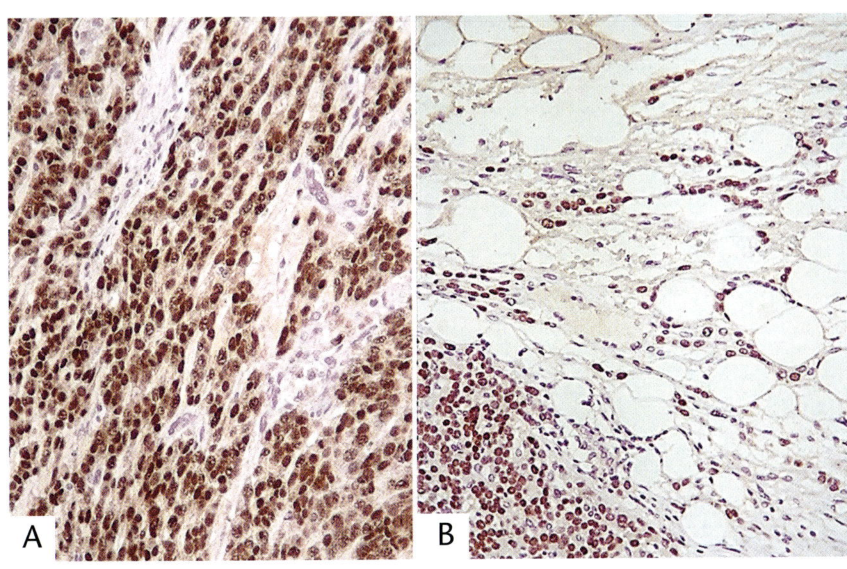

Figure 6-30

SF1 IN ADRENAL CORTICAL CARCINOMA

Diffuse SF1-positive nuclear staining (A) highlights single neoplastic cells invading the periadrenal adipose tissue (B).

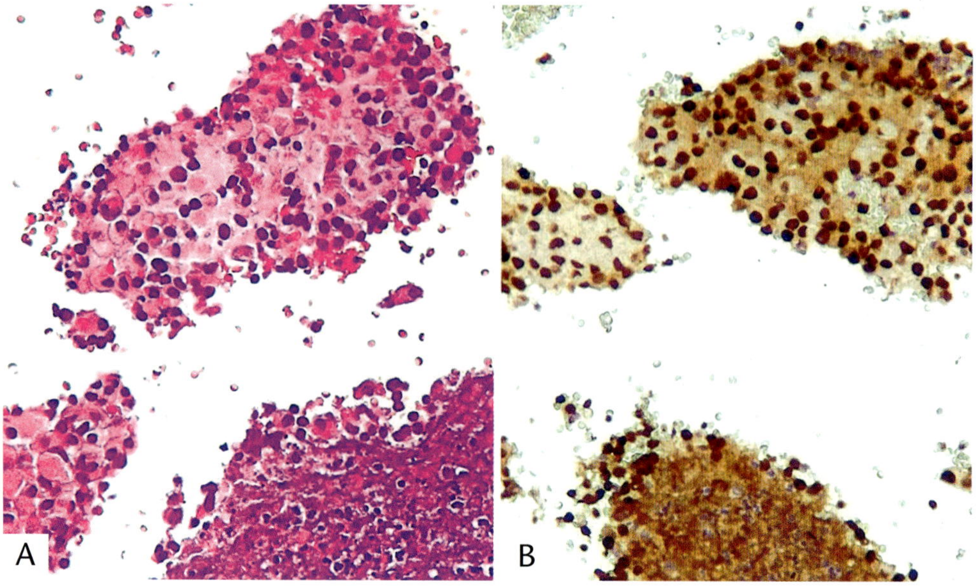

Figure 6-31

SF1 IN ADRENAL CORTICAL CARCINOMA

The diagnosis of metastatic adrenal cortical carcinoma in this transbronchial fine needle aspiration biopsy of a perihilar lymph node (A) is supported by positive nuclear staining for SF1 (B).

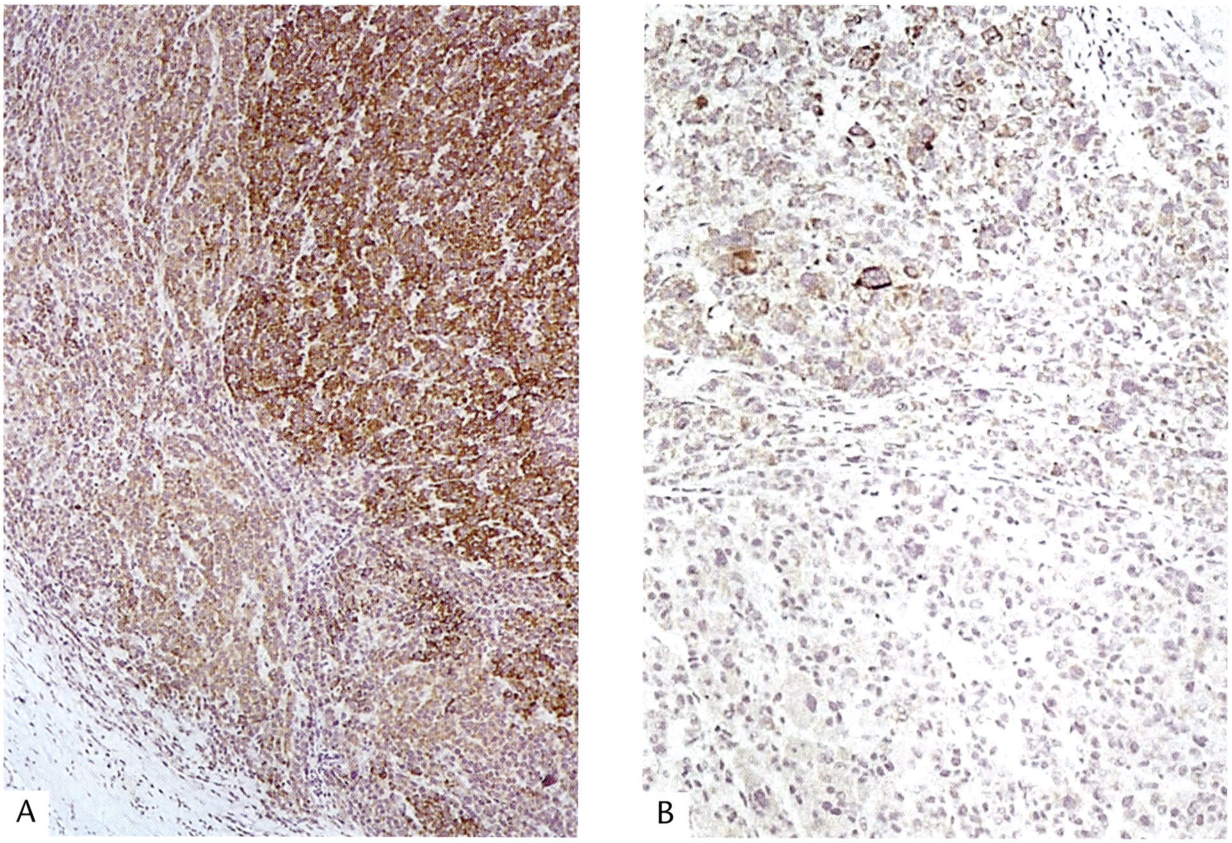

Figure 6-32

MELAN-A IN ADRENAL CORTICAL CARCINOMA

Melan-A expression in ACC is usually heterogeneous, either diffuse (A) or focal (B) even within the same lesion.

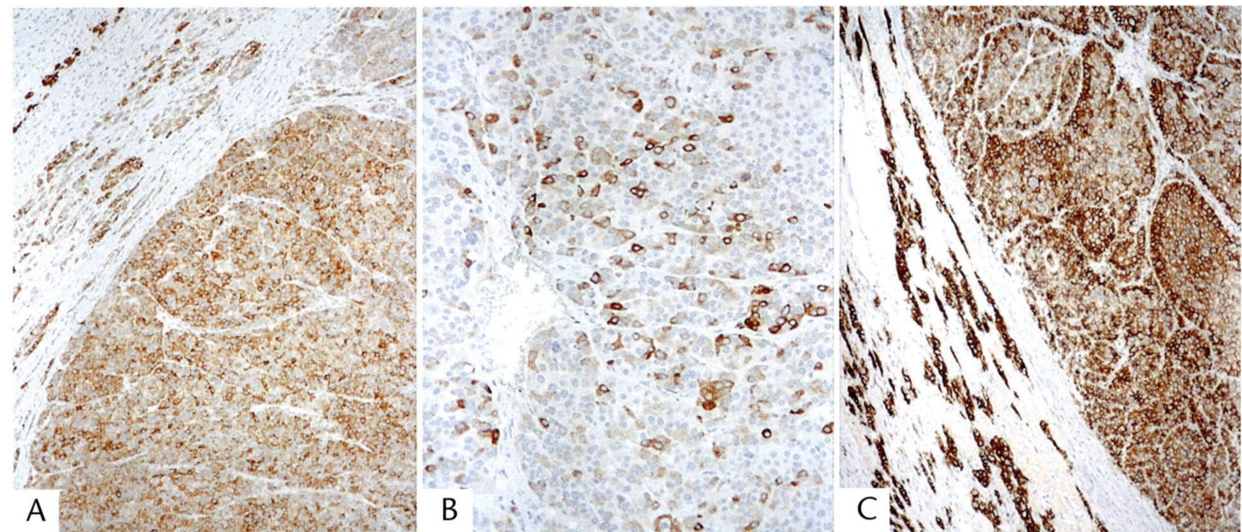

Figure 6-33

SYNAPTOPHYSIN, INHIBIN, AND CD56 IN ADRENAL CORTICAL CARCINOMA

Less specific adrenal cortical markers are synaptophysin (A), alpha-inhibin (B), and CD56 (C).

or with adenoma-like regions may require the use of ancillary biomarkers in addition to the accepted multifactorial systems. IGF2 paranuclear immunoreactivity is the most reliable translational biomarker in the distinction of ACC from adenoma, correlating with the high prevalence of *IGF2* gene overexpression in ACC (176,177). Although reported in a few adrenal cortical adenomas, IGF2, when evaluated in a rigid methodologic context, is a specific marker. Altered expression of beta-catenin (in the form of nuclear/cytoplasmic staining as opposed to the "wild-type" membrane pattern) is an expression of Wnt pathway alterations (fig. 6-34). In a biomarker study, altered expression of beta-catenin was observed in none of 50 adrenal cortical adenomas and in 6 of 42 ACCs, thus showing a high specificity but low sensitivity to predict malignancy (131). *CTNBB1* mutations have been detected in adrenal cortical adenomas, thus questioning the role of this marker to discriminate malignant from benign lesions (178).

Another translation biomarker is p53. The pattern of p53 protein expression in the absence of a *TP53* mutation is variable nuclear staining. In the presence of *TP53* mutations, p53 expression is altered, either as overexpression, which is the most frequent pattern, or as global loss. Altered p53 expression has been widely used

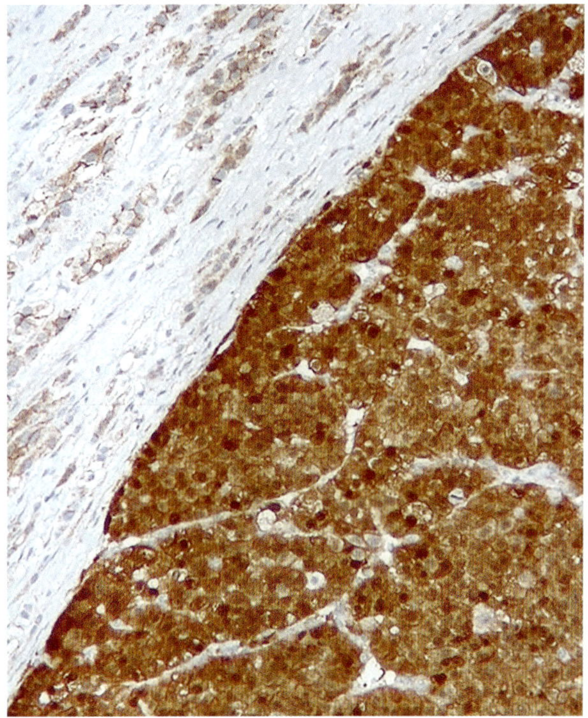

Figure 6-34

BETA-CATENIN IN ADRENAL CORTICAL CARCINOMA

Altered nuclear and cytoplasmic beta-catenin expression in ACC may be focal with a "clonal" pattern or more frequently, diffuse as in the figure. Membrane pattern is visible in peritumoral non-neoplastic adrenal cortical tissue (top left).

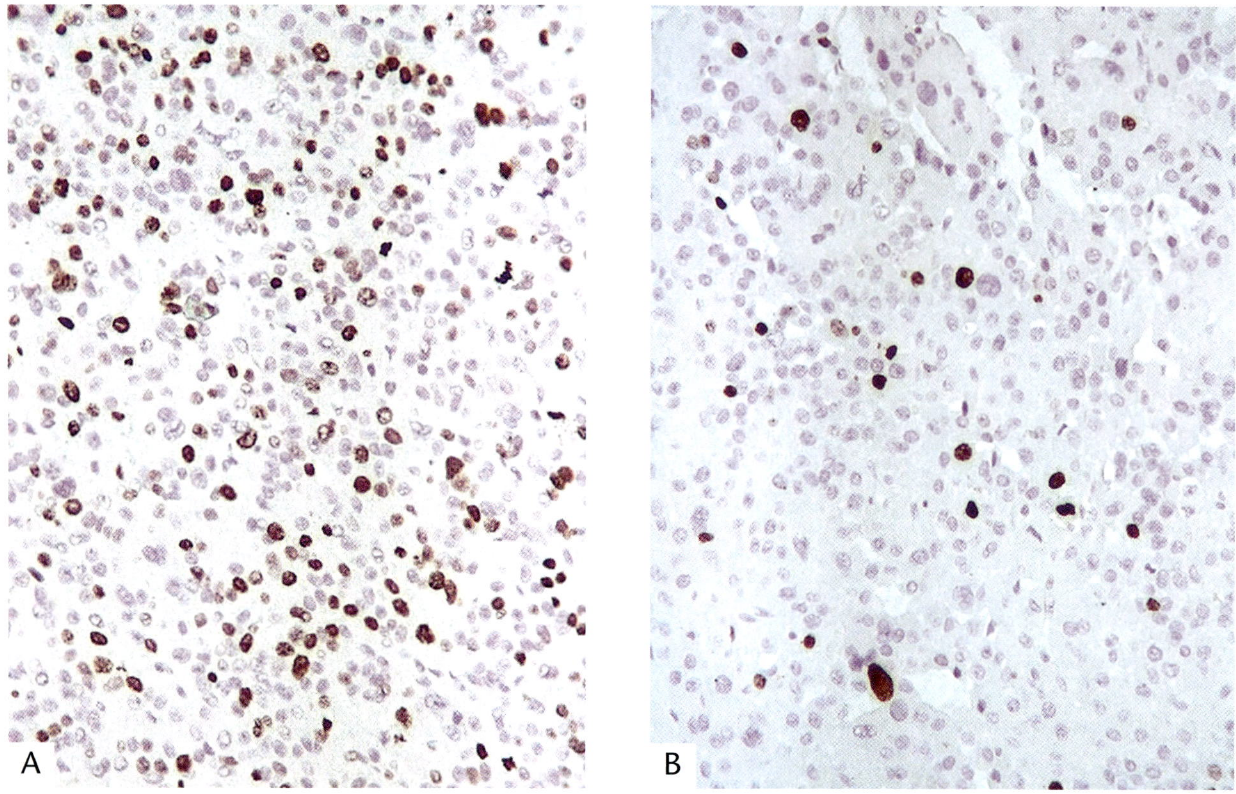

Figure 6-35

KI-67 IN ADRENAL CORTICAL CARCINOMA

The Ki-67 proliferative index is highly variable in ACC, as illustrated in the two tumors shown in (A) and (B).

to support the diagnosis of ACC (130,179), although not all ACCs have *TP53* mutations, reducing the sensitivity of this marker. p53 is frequently assessed in ACC because of its prognostic value (131,180).

Metalloproteinase type 2 expression has been proposed to distinguish malignant from benign adrenal cortical tumors both in adults (181) and in children (182). Also, Ki-67 has been proposed as a diagnostic marker in this setting (fig. 6-35). Several studies indicate that most ACCs have a Ki-67 labelling index greater than 5 percent (177,183,184) although the Helsinki score only integrates proliferation index in a diagnostic algorithm (126).

MOLECULAR GENETICS OF SPORADIC ACC

The hereditary context of ACC is discussed in chapter 7. However, several genes that are involved in germline predisposition may also harbor somatic mutations and contribute to the molecular basis of sporadic ACC. The overall molecular features of sporadic ACC combine the integration of genomic, transcriptomic, and epigenetic profiling that represent the summary of a large number of innovative studies conducted in the last decades (185).

Genomic Alterations

ACC-related common driver alterations affect several signaling pathways (186) including: 1) Wnt signaling (*CTNNB1* mutations and *ZNRF3* deletions); 2) cell cycle regulation (*TP53, RB1, MDM2, CDK4, CDKN2A* mutations); and 3) chromosome maintenance and chromatin remodeling (*DAXX, ATRX, MEN1, TERT* mutations and *TERF2* amplifications). Mutations in pathways commonly altered in adrenal cortical adenomas, such as ion channel genes or cyclic AMP signaling-associated pathway are absent or rare (187), and generally the genomic alterations in ACC are not as strongly associated with specific functional status as in adrenal cortical adenomas and nodular disease.

The genes with a prevalence of somatic mutations above 5 percent are *TP53* (18 percent), *CTNNB1* (15 percent), *NF1* (6 percent), and *ATRX* (5 percent) (185). Impairment of the *TP53* pathway is also caused by mutations in *RB1* and *CDK2NA* in some cases (188). Deletions in Wnt pathway repressors *ZNRF3* and *KREMEN1* are usually mutually exclusive with *CTNNB1* mutations (189,190). Interestingly, *TP53* and Wnt pathway alterations are often mutually exclusive, but enriched in tumors with a poor prognosis (186,188,189). *TERT* is altered in a relevant proportion of cases; promoter mutations are rare but gene amplifications are detectable in more than 20 percent of cases and strongly associated with an adverse outcome (191).

Chromosomal Alterations

Nuclear pleomorphism and atypical mitotic figures are consistent with the complex karyotype of ACC (192). Recurrent regions of gain include 9q34, encompassing *NR5A1* encoding SF1, and 5p15 where the *TERT* gene is located. Loss of genetic material, on the other hand, is regularly noted at loci encoding important tumor suppressors, such as *MEN1*, *TP53*, and *RB1*.

Loss of the maternal 11p15 allele (with synchronous duplication of the paternal allele) is a frequent finding in ACCs (193,194). This locus contains the imprinted *H19* and *IGF2* genes, which are exclusively expressed from maternal and paternal alleles, respectively (195). The selective loss of the maternal allele and duplication of the paternal allele is therefore reflected in ACC as evident downregulation of H19 (a long noncoding RNA with tumor suppressive properties) and overexpression of IGF2.

Global Transcriptome Profiling

Transcriptome profiling was one of the first molecular screening approaches to identify molecular biomarkers in ACC. A seminal paper by Giordano et al. (196) identified a panel of genes differentially expressed in ACCs as compared to adenomas, with *IGF2, TOP2A,* and *SPP1* among the most significant. In a subsequent study, two different clusters of gene expression were identified, one enriched in IGF2 characterizing ACC cases, and the second enriched for steroidogenic genes typical of adrenal cortical adenomas (197). The evolution of transcriptome profiling was dedicated to the discovery of sets of genes linked to ACC prognosis (198,199). The transcriptomic profiling approach has also been validated by means of RNA sequencing on formalin-fixed, paraffin-embedded samples, thus allowing its technical feasibility on a larger scale (200).

Epigenetic Modifications

Hyper- or hypomethylation of CpG sites in promoter regions of genes is a relevant mechanism regulating the promotion and progression of cancer. Global methylome profiles in ACC show a general hypomethylation, compared to normal adrenal tissue and adenomas (201). A further global methylation analysis subdivided ACC into three groups based on the status of CpG island methylator phenotypes (CIMP): a non-CIMP class with methylation levels similar to adrenal cortical adenomas and with a better prognosis; a CIMP-low class with increased methylation and intermediate prognosis; and a CIMP-high class with the highest level of methylation and the worst prognosis (202). Targeted methylation analysis in paraffin-embedded material proved feasible in a large series of cases, and identified a panel of genes (*PAX5, PYCARD, PAX6,* and *G0S2*) whose hypermethylation status is associated with an aggressive clinical course (203).

Pangenomic Characterization

Extensive genomic studies of ACC, including transcriptome and next-generation sequencing analyses, have identified several molecular clusters with prognostic relevance (186,188, 189), which are described more detail in the section on prognosis. The most validated algorithm divides tumors into two principal classes, based upon the mutational signatures, mRNA expression profiles, DNA methylation patterns, and copy number landscape (186,188). One class is characterized by tumors with a low disease progression rate, and enriched for cases that lack *TP53* or Wnt pathway mutations, are adherent to an mRNA expression clusters traditionally associated with more indolent disease, exhibit low levels of CIMP, and do not display whole-genome doubling. The second

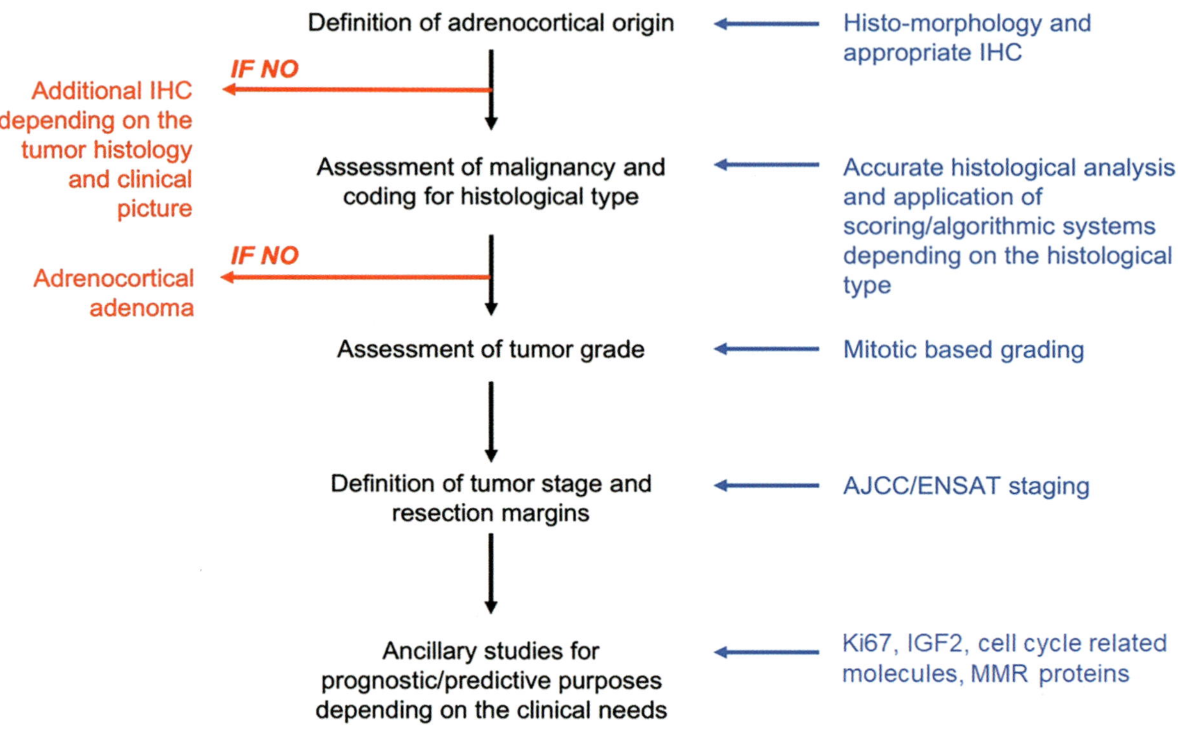

Figure 6-36

SCHEMATIC DIAGNOSTIC ALGORITHM FOR ADRENAL CORTICAL CARCINOMA DIAGNOSIS

class includes high-risk ACC cases with frequent *TP53* and Wnt pathway gene mutations, an expression pattern often associated with worse outcome, CIMP-high phenotype, and numerous nonrecurrent alterations with potential whole-genome doubling. This integrated approach has been recently validated in a more focused target approach with a significant impact on prognosis (see also below) (157).

DIFFERENTIAL DIAGNOSIS

The diagnostic flow in a suspected ACC case integrates morphology and immunohistochemistry to deal with all possible aspects of a differential diagnosis (fig. 6-36). The adrenal cortical nature of a lesion suspected to be ACC should be assessed in any case in which morphology is not completely conclusive. In fact, the erroneous misinterpretation of the tissue of origin in this context represents the most frequent source of diagnostic failure (204). Pheochromocytoma is the most relevant entity in the differential diagnosis in this context, with special reference to lesions with pure oncocytic features, since the nuclear features (nuclear pleomorphism and nuclear pseudoinclusions) may be very similar. Other primary adrenal lesions to be excluded are soft tissue tumors with epithelioid appearance and possible malignancy related features, such as angiosarcoma, solitary fibrous tumor, and PEComas.

This large differential diagnosis underscores the importance of appropriate immunohistochemical assessment. Potential immunohistochemical pitfalls are focal positive cytokeratin staining in angiosarcoma, synaptophysin and inhibin expression in pheochromocytoma, and Melan-A expression in PEComa. Spindle cell sarcomas may be indistinguishable from sarcomatoid ACC, unless an associated component of conventional ACC is present. The detection of sarcoma-associated molecular alterations, which are negative in sarcomatoid ACC, may assist in this setting (205).

Extra-adrenal malignant neoplasms are also in the differential diagnosis. Eosinophilic neoplasms with an epithelioid appearance may give rise to adrenal metastases, from malignant melanoma to renal cell carcinomas of different

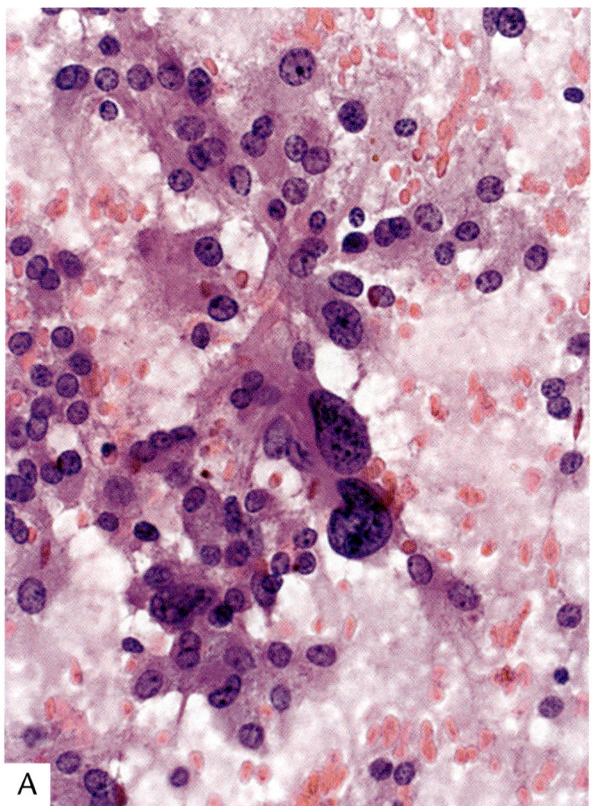

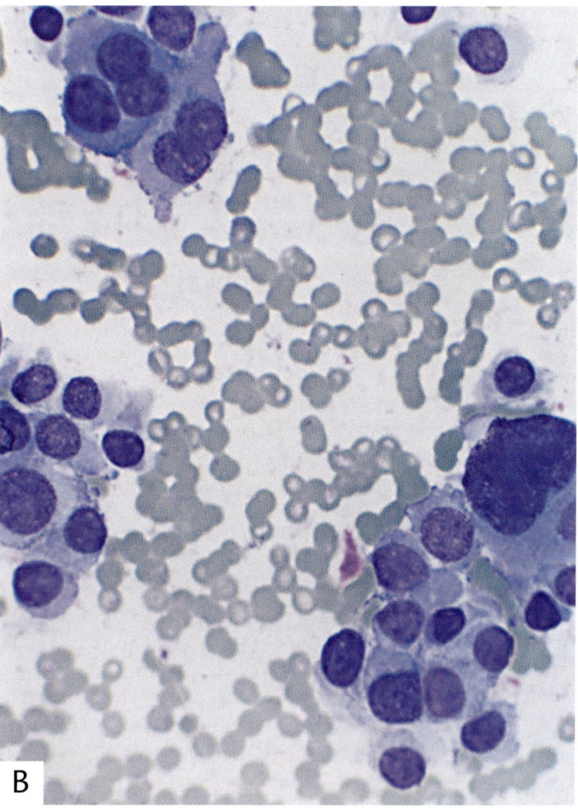

Figure 6-37

ADRENAL CORTICAL CARCINOMA

Aspirate smear preparation of an ACC shows cells with compact, eosinophilic cytoplasm (A: hematoxylin and eosin [H&E] stain). Marked variation in nuclear size is visible in tumor cells dispersed or loosely aggregated in small clusters (B: Diff Quik stain).

histotypes, to hepatocellular carcinoma, to different kinds of poorly differentiated carcinomas with solid growth lacking morphologic signs of differentiation (i.e., glandular or squamous features). Melan-A is positive in melanoma and may be positive in some renal cell carcinomas (Xp11/TFE3-rearranged renal cell carcinoma) (206). Another renal cell carcinoma and hepatocellular carcinoma marker, CD10, is also expressed in ACC (173).

Diffuse and intense pancytokeratin expression is not typical of ACC and should suggest metastatic carcinoma. The myxoid variant should be also distinguished from myxoid neoplasms, such as chordoma, extraskeletal myxoid chondrosarcoma, or myxoid liposarcoma, but the maintenance of adrenal cortical markers in myxoid ACC usually easily resolves the issue. In ectopic locations, ACC may be very difficult to distinguish from other steroidogenic tumors, including rare intra-abdominal or pelvic gonadal neoplasms that express SF1, such as sex cord stromal tumors, Leydig cell tumors, and steroid cell tumors that are extremely similar to ACC (32,207).

The preoperative diagnosis of ACC using fine needle aspiration biopsy or core biopsy is not routinely recommended and should be restricted to selected cases (fig. 6-37). Although for core biopsy a specificity of 88 percent and a sensitivity of 86 percent have been reported in diagnosing adrenal masses (208), the major role of preoperative diagnosis in adrenal lesions is to diagnose adrenal metastases of a known primary, to confirm a clinical suspicion of a hematologic disorder, or to exclude non-neoplastic conditions such as infections. Classification schemes for cytology reporting of adrenal lesions have been also proposed but currently lack clinical validation (209).

Table 6-7

ADRENAL CORTICAL CARCINOMA TNM STAGING[a]

Clinical Stage	T Stage	N Stage	M Stage	Overall Survival Probability at 5 Years (%)
I	T1 (≤5 cm, no infiltration of surrounding tissues)	N0	M0	66.7
II	T2 (>5 cm, no infiltration of surrounding tissues)	N0	M0	54.4
III	T3 (invasion of the surrounding tissues) or T4 (invasion of adjacent organs or vena cava or renal vein invasion)	N0 or N1 if T3-T4, N1 if T1-T2 (N1: any positive lymph node, irrespective of number and site)	M0	44.1
IV	Any T	Any N	M1 (any metastasis, irrespective of number and site)	9.2

[a]Tumor staging by the 8th edition of the American Joint Committee on Cancer TNM Staging System and European Network for the Study of Adrenal Tumors.

STAGING

Current TNM staging of ACC is based on integration of the American Joint Committee on Cancer (AJCC) 8th edition of the TNM system and the staging proposal by the European Network for the Study of Adrenal Tumors (ENSAT) (Table 6-7). Compared to the AJCC 7th edition, the ENSAT proposal restricts stage IV disease to patients with metastatic disease only. Such an approach demonstrated a better capability to stratify stages III and IV patients into different prognostic groups (210). According to the AJCC 8th edition, the 5-year overall survival rates in stages I, II, III, and IV were 66.7, 54.4, 44.1, and 9.2 percent, respectively. More favorable survival rates have been described in other series (211).

A proposal for a modification of pT stage was formulated in a study aimed at incorporating lymphovascular invasion (212), but it was not included in the current staging scheme due to lack of validation. Another recent study (213) divided patients with stage IV ACC into single (IVA) versus multiple (IVB) sites of metastases. This stage IV subdivision proved to be significantly associated with differing overall survival. The most common sites of metastases are liver, lung, retroperitoneum, and lymph nodes. Bone metastases are usually lytic. Rare sites of metastases include the brain and skin (figs. 6-38, 6-39).

PROGNOSIS AND PREDICTION

Several parameters or tumor characteristics have been associated with ACC prognosis, either alone or combined into multiparametric systems or integrated classes (Table 6-8). Considering all these variables as well as the heterogeneity of clinical strategies that potentially depend on them, it is currently recommended that all patients with ACC be followed by tertiary center experts to guarantee accessibility to the most complete clinical management and to the best therapeutic strategies (35).

The most relevant clinical parameters associated with adverse clinical outcome are advanced tumor stage and tumor functionality. In terms of hormone secretion, cortisol hypersecretion has been shown to significantly impact prognosis (214). The negative impact of cortisol in ACC prognosis probably depends on a combination of adverse metabolic effects of cortisol secretion (including immunosuppression) and biologic characteristics of cortisol-secreting ACCs that have to be underpinned by future research (215).

The main prognostic pathologic findings include incompleteness of resection (i.e., positive margins), histologic type, mitotic index, and presence of vascular invasion. The latter is a well-recognized adverse feature, and also is the best predictor of disease-free survival in patients with low-grade ACC (131). Among Weiss

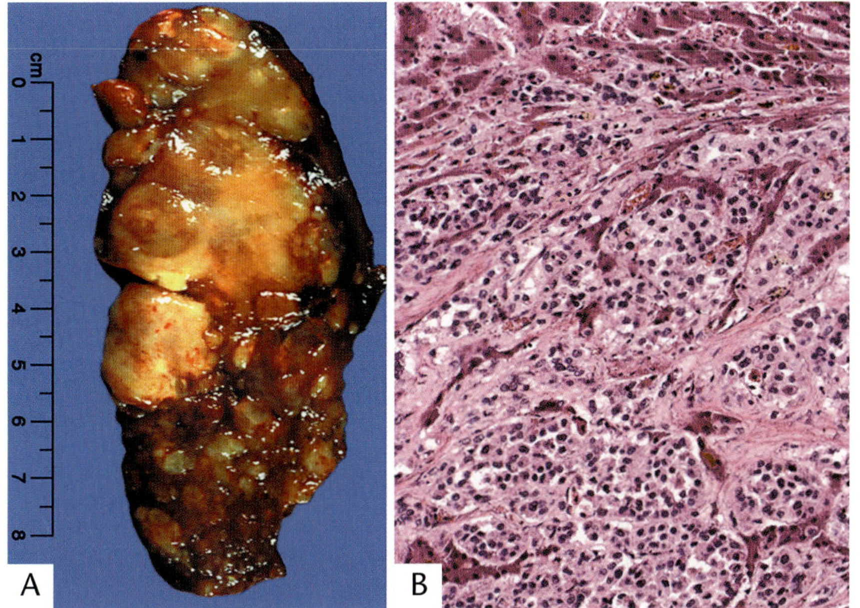

Figure 6-38

LIVER METASTASIS FROM ADRENAL CORTICAL CARCINOMA

A: Hepatic lobectomy specimen is largely replaced by metastatic ACC. The tumor is mottled tan-yellow.

B: Metastatic ACC disrupts the liver parenchyma and has a large nesting pattern.

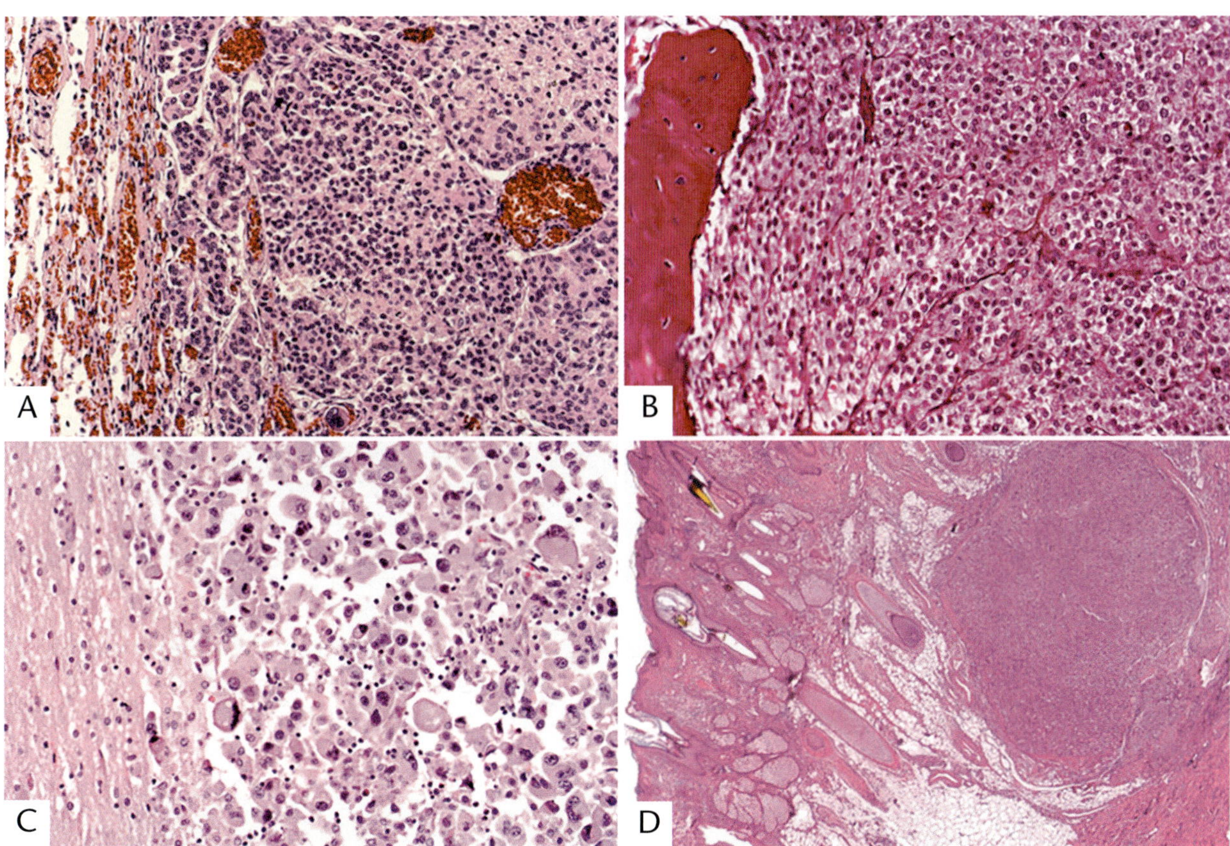

Figure 6-39

SITES OF METASTASIS FROM ADRENAL CORTICAL CARCINOMA

Lung (A), bone (B), brain (C), and skin (a surgically excised nodule on the scalp, D) metastases from ACC.

Table 6-8

MOST RELEVANT ADVERSE PROGNOSTIC PARAMETERS IN ADULT ADRENAL CORTICAL CARCINOMA

Clinical	Histologic	Immunophenotypic	Molecular
Advanced tumor stage	Sarcomatoid, myxoid, conventional, and oncocytic types, in order of aggressiveness	Ki67 (two- or three-tier prognostic groups)	*TERT* alterations
Cortisol hypersecretion	Mitotic-based grading	Beta-catenin (nuclear staining)	Hypermethylation of *GOS2*
	Vascular invasion	p53 (overexpression or complete loss)	Dysregulation of miR195 and miR-483-5p
	Status of the resection margins	SF1 overexpression	Integrated molecular subclasses
		ATRX loss	
		ZNRF3 loss	

parameters, necrosis is the most powerful adverse prognostic parameter (216). Oncocytic histologic type is associated with the best clinical outcome (140). Recently, a combined clinical and pathologic score based on tumor grade, resection status, age, and symptoms of hormone hypersecretion (GRAS score) has been shown to significantly stratify patients in terms of overall survival and disease-free survival after resection (217).

The most relevant immunohistochemical prognostic marker in ACC is the Ki-67 index. Consensus on prognostic cut-offs, however, is not reached and both three-tier and two-tier categorization approaches are described. Moreover, even similar approaches may use different cut-offs as the result of the proliferation index distribution within the different series analyzed. In a three-tier prognostic stratification, categorization is based on cut-offs of <10 percent, 10 to 19 percent, ≥20 percent or <15 percent, 15 to 30 percent, >30 percent or <20 percent, 20 to 50 percent, and >50 percent (127,218,219). This complexity is compounded by the fact that reproducibility and standardization of Ki-67 index evaluation is poor (219). Manual counting or automated image analysis algorithms should be applied in the assessment of Ki-67 proliferation index (219, 220). In ACC, Ki-67 can also help rationalize the need for adjuvant mitotane therapy in select patients (34).

Other immunohistochemical prognostic markers are surrogates of molecular evidence that some driver gene alterations confer more aggressive biologic behavior. The use of these immunohistochemical markers still needs adequate clinical validation, however. p53 and beta-catenin are considered as prognostic biomarkers since most carcinomas with adverse molecular clusters tend to show nuclear beta-catenin expression and/or aberrant p53 staining (overexpression or global loss) (167). Other markers that correlate with poorer clinical outcome are high expression of SF1 (169,171) and loss of *ATRX* and *ZNRF3* expression (221).

As already mentioned above, transcriptome (222) and pangenomic (157) studies clearly underscore the role of molecular risk stratification in ACC. A specific methylation signature has been established in ACC with a particularly poor prognosis, suggesting that methylation patterns may predict clinical outcome (202). The main drawback of these studies is the lack of translation validation and of well-established biomarkers to be assessed in routine clinical work. Among the few, the reduced expression of *BUB1B* and *PINK1* predicted prolonged survival (198). Hypermethylation of the *G0/G1 Switch 2* (*G0S2*) gene is strongly associated with the CIMP-high group of poor prognostic and fatal ACCs that are clustered as CoC-III tumors as defined by the Cancer Genome Atlas (TCGA) data (223).

Dysregulation of micro-RNA (miRNA) may also be associated with prognosis. For example, downregulation of miR-195 and overexpression of miR-483-5p are evident in cases with increased mortality (224). Among altered genes that act as molecular drivers in ACC pathogenesis, *TERT* gene alterations (including in order

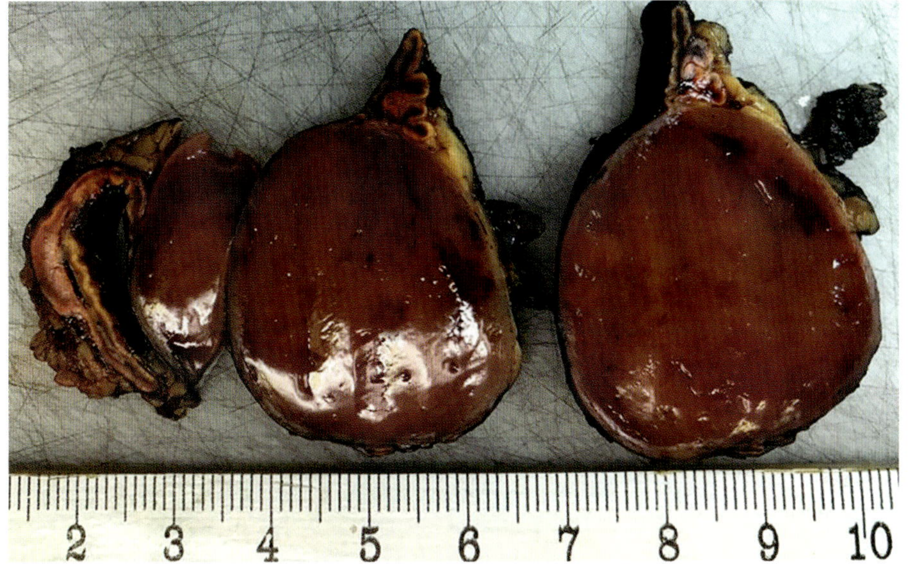

Figure 6-40

PEDIATRIC ADRENAL CORTICAL TUMOR: BENIGN

Adrenal tumor in a 6-year-old boy associated with virilization, classified as benign behavior by means of the Wieneke score. The cut surface was homogeneously brown, with no evidence of necrosis or hemorrhage.

of frequency gene amplifications, promoter mutations, and rearrangements) are strongly associated with the presence of metastases (191).

In terms of prediction of response to therapy, a couple of markers have been associated with response to mitotane but are not validated in independent studies nor are currently used. RRM1 overexpression has been shown to reduce mitotane responsiveness (225), possibly interfering with its metabolization (226). Cytochrome P450 is thought to interfere with mitotane metabolization and activity. CYP2W1 overexpression in mitotane-treated patients is associated with increased survival (227), whereas CYP11B1 has no role in mitotane action and metabolism in ACC cells (228). Germline variants of CYP2W1 and CYP2B6 (namely CYP2W1*6 and CYP2B6*6) may alter mitotane blood levels and profiles of responsiveness (229). Finally, MMR proteins and PD-L1 immunohistochemistry may help determine the eligibility for immunotherapy in select patients (230), although these markers have no clear indication to date.

PEDIATRIC ADRENAL CORTICAL TUMORS

Pediatric adrenal cortical tumors are discussed in this chapter because of their frequent metastatic behavior and their pronounced histopathologic abnormalities. Both malignant and benign pediatric cortical tumors are included, since most epidemiologic and histopathologic studies have focused on the entire group.

ACC in patients under 20 years of age is very rare; the incidence is lower than in the adult population, with an estimated rate of 0.2 to 0.3 cases per million (231). In Brazil, however, the incidence is about 15 times higher compared to the United States and Europe, probably as the consequence of the prevalence of the *TP53* p.R337H mutation in that population (see chapter 7) (232). Therefore, most data on the clinical and pathologic features of childhood adrenal cortical tumors derive from Brazilian cohorts of patients, even if there are no apparent differences from those diagnosed elsewhere in the world (233).

Childhood adrenal cortical tumors frequently occur before the age of 5 years, with a second peak in adolescence, and females are more often affected (234). Functional tumors are more frequent than in adults, accounting for up to 85 percent of cases (136). Virilization due to androgen secretion is the most common functional manifestation (80 percent of functioning cases) (fig. 6-40), either pure or associated with cortisol secretion (235). It manifests with pubic hair, accelerated growth, skeletal maturation, enlarged penis and clitoris, hirsutism, and acne. Pure cortisol secretion leading to Cushing syndrome is rarer and associated with larger tumor size and older age (236), and aldosterone secretion is exceptional. Estrogen-secreting adrenal cortical tumors are quite rare, accounting for 0.37 to 2.0 percent. Feminization is more common in children 8 years of age or younger, with a median age at diagnosis of 6 years and a better prognosis than in adult males (237).

Clinical diagnostic workups include detailed hormonal characterization and imaging procedures, but sensitivity and specificity of radiology methods to predict malignant behavior are lower than in adults (80). Malignancy can be determined before surgery only in the presence of metastases.

The pathologic classification of pediatric tumors is a challenge since the correlation between tumor behavior and histopathologic findings is more inconsistent than in adult cases (238). The Weiss score has been shown to have good specificity to detect cases with an aggressive clinical course (239), but generally overestimates malignant potential in pediatric tumors. As a specific diagnostic tool for pediatric adrenal cortical tumors, the Wieneke/Armed Forces Institute of Pathology (AFIP) criteria were established in 2003 (240) and validated in several subsequent studies (Table 6-9). The presence of 0 to 2 criteria defines a benign lesion (fig. 6-41), a score of greater than 3 is indicative of malignancy (fig. 6-42), and a score of 3 correspond to a category indeterminate for malignancy. Increased scores are correlated with worse overall and disease-free survival (241). These criteria are not 100 percent sensitive or specific.

Table 6-9

AFIP CRITERIA FOR DIAGNOSING PEDIATRIC ADRENAL CORTICAL TUMORS

Tumor weight >400 g
Tumor size >10.5 cm
Extra-adrenal extension
Invasion into the vena cava
Venous invasion
Capsular invasion
Tumor necrosis
Mitotic index >15/20 HPF
Atypical mitotic figures

Other scoring systems and diagnostic algorithms have been validated in pediatric adrenal cortical tumors in individual studies. In a study of 28 cases, both the Helsinki score (at the cut-off score of 24) and the reticulin algorithm showed high accuracy rates (80.77 percent, and 80.0 percent, respectively), with the highest specificity (81.25 percent) for the Helsinki score (242). Moreover, similar to tumors in adults, a score incorporating clinical and histologic parameters has been designed, the pediatric S-GRAS (pS-GRAS) score (243), which is derived from the sum

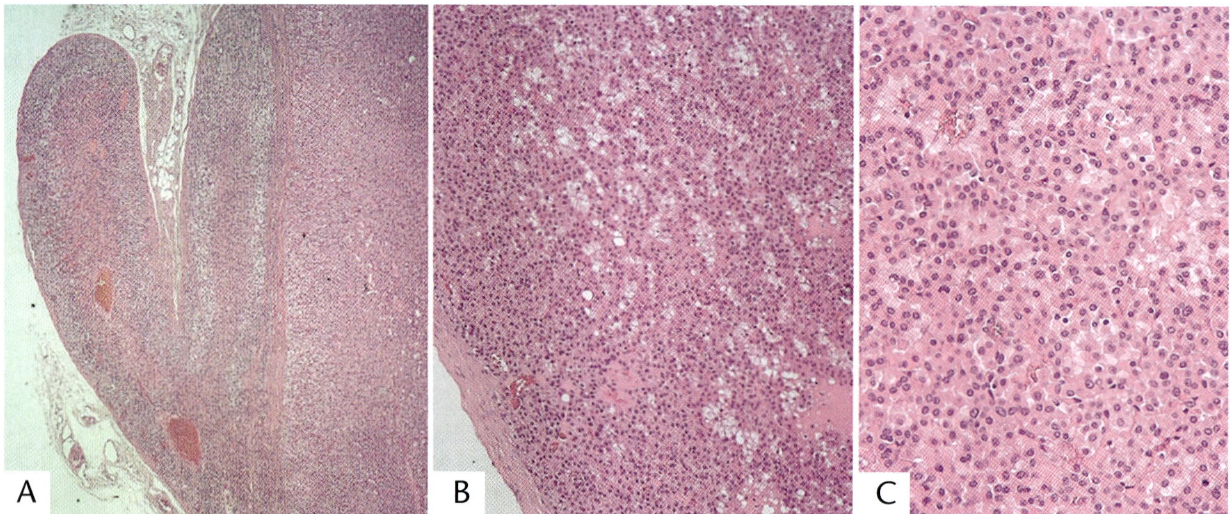

Figure 6-41

PEDIATRIC ADRENAL CORTICAL TUMOR: BENIGN

A–C: The tumor is shown at progressively higher magnification. It lacks necrosis and signs of invasion and showed low mitotic and proliferation indexes. Final Wieneke score was 0. (Same case as fig. 6-40.)

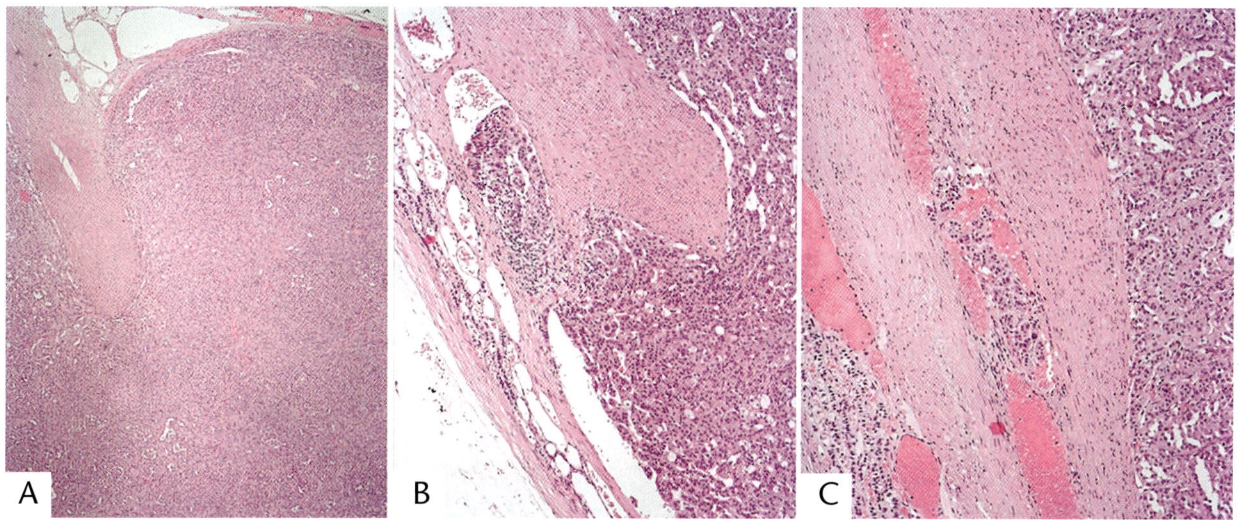

Figure 6-42

PEDIATRIC ADRENAL CORTICAL CARCINOMA

This tumor showed capsular (A) and vascular invasion (B,C). Increased mitotic activity and rare atypical mitotic figures were also found. Cytologic atypia was mild as it is usually in pediatric adrenal cortical tumors irrespective of the biologic behavior. Final Wieneke score was 4. (Courtesy of Prof. M. Papotti, University of Turin, Turin, Italy)

of tumor stage, grade (based on Ki-67 index and mitotic rate), resection status, age (with a cut-off of 4 years), and hormone-related symptoms.

The use of immunohistochemical markers for diagnosis of tumors of primary adrenal cortical origin is similar to that in adult cases. Recently, a Ki-67 labeling index over 15 percent has been proposed to identify malignant cases (241,244). Additional immunohistochemical markers that are associated with risk of malignancy include p53, SF1, IGF2, matrix metalloproteinase 2, and human leukocyte antigen class II (fig. 6-43) (182,245).

Staging of pediatric ACC has been developed by the International Pediatric Adrenocortical Tumor Registry (IPACTR) and the Children's Oncology Group (COG) (136). The different stages are: stage I: a completely resected ACC with R0 margins, weight ≤200 g, and no metastases; stage II: a completely resected ACC with R0 margins, weight >200 g, and no metastases; stage III: residual or inoperable ACC; and stage IV: presence of metastases at presentation. Lungs, liver, bone, and brain (fig. 6-44) are the most common sites of metastases (246).

Clinical adverse prognostic parameters include metastatic disease at presentation, a delay in diagnosis, and cortisol secretion (247). Age is a prognostic parameter, since metastatic disease is more common in children aged over 12 years than those under 4 years (246). Pathologic parameters include stage, tumor weight up to 200 g, extra-adrenal extension, and initial noncomplete surgical resection (244). All these parameters have been confirmed in a recent meta-analysis collecting data from more than 1,000 patients (234).

PATHOLOGY REPORTING

With the refinement of ACC classification and staging and the implementation of available prognostic parameters, standardization of pathology reporting is essential to incorporate all tumor features that may have diagnostic and/or clinical relevance. In June 2021, the College of American Pathologists (CAP) posted the most recent version of the Protocols for the Examination of Specimens from patients with Carcinoma of the Adrenal Gland [https://documents.cap.org/protocols/Adrenal_4.2.0.0.REL_CAPCP.pdf]. The International Collaboration of Cancer Reporting (ICCR) developed a universal dataset for standardization of reporting for ACC (132). The ICCR dataset includes 23 core items that should be included in the pathology report, together with non-core additional information

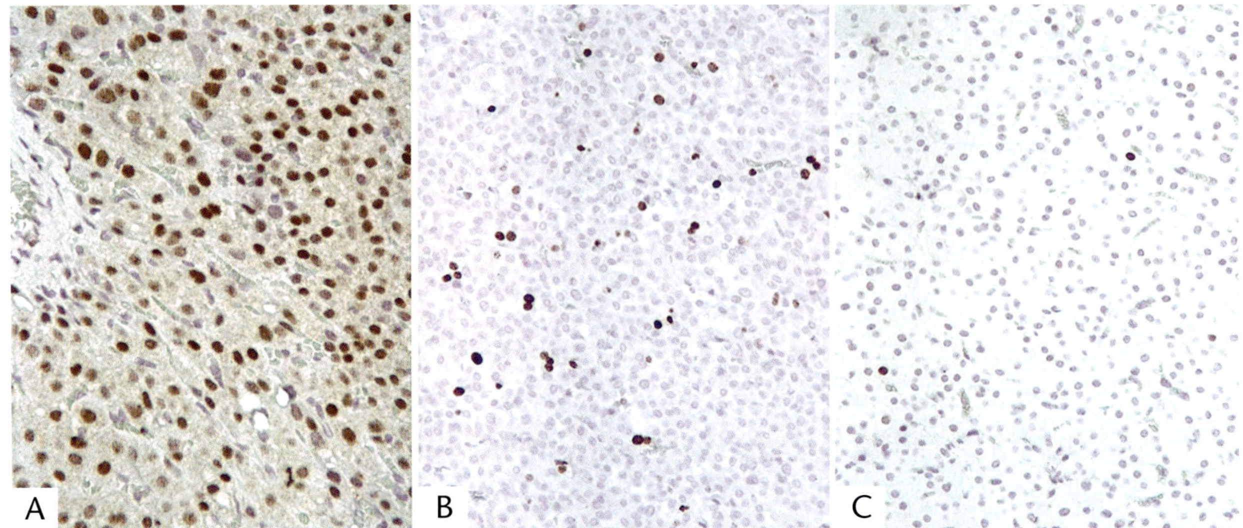

Figure 6-43

PEDIATRIC ADRENAL CORTICAL TUMOR: BENIGN

The tumor diffusely expressed SF1 (A). Ki-67 index (B) was 5 percent, whereas p53 was expressed in rare nuclei (C). (Same case as figs. 6-40 and 6-41.)

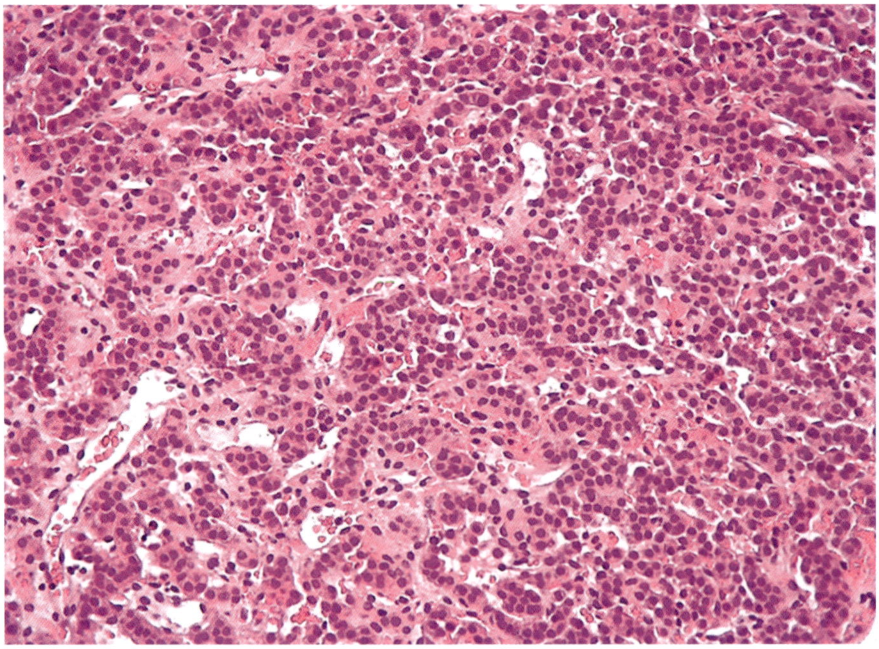

Figure 6-44

PEDIATRIC ADRENAL CORTICAL CARCINOMA

Synchronous brain metastasis from a malignant pediatric adrenal cortical tumor. (Same case as fig. 6-42.)

that may be added to further identify peculiar features that may have clinical importance. All these elements are summarized in Table 6-10. It is desirable that all pathology societies support the use of such schemes (adapted to local needs and realities where necessary) to promote standardization of pathology reporting, increase diagnostic reproducibility, ensure adequate clinical management of the patients, and increase the overall knowledge of this disease.

Table 6-10

INTERNATIONAL COLLABORATION ON CANCER REPORTING (ICCR) DATASET FOR PATHOLOGY REPORTING OF ADRENAL CORTICAL CARCINOMA[a]

Core Elements	Non-Core Elements
Clinical notes	-
Operative procedure	-
Type of specimen(s) submitted	-
Tumor site	-
Specimen integrity	-
Tumor size (largest single dimension)	Additional two dimensions
Tumor weight	-
Histologic tumor type	-
Extent of invasion	-
Tumor architecture	-
Lipid-rich cell content	-
Capsular invasion	-
Lymphatic invasion	-
Vascular invasion	-
Atypical mitotic figures	-
Necrosis	Extent of necrosis
Nuclear grade	-
Mitotic index and histologic tumor grade	-
Ki-67 proliferation index	-
Margin status	Distance of the tumor to the closest margin
Lymph node status	Extranodal extension
Histologically confirmed distant metastases	-
Pathologic staging	-
-	Multifactorial scoring systems
-	Ancillary studies (including reticulin staining)
-	Coexistent adrenal pathology

[a]Modified Table 1 from Giordano TJ, Berney D, de Krijger RR, et al. Data set for reporting of carcinoma of the adrenal cortex: explanations and recommendations of the guidelines from the International Collaboration on Cancer Reporting. Hum Pathol 2021;110:50-61.

REFERENCES

1. Shariq OA, McKenzie TJ. Adrenocortical carcinoma: current state of the art, ongoing controversies, and future directions in diagnosis and treatment. Ther Adv Chronic Dis 2021;12: 20406223211033103.
2. Sharma E, Dahal S, Sharma P, et al. The characteristics and trends in adrenocortical carcinoma: a United States population based study. J Clin Med Res 2018;10:636-40.
3. Kerkhofs TM, Verhoeven RH, Van der Zwan JM, et al. Adrenocortical carcinoma: a population-based study on incidence and survival in the Netherlands since 1993. Eur J Cancer 2013;49:2579-86.
4. Audenet F, Méjean A, Chartier-Kastler E, Rouprêt M. Adrenal tumours are more predominant in females regardless of their histological subtype: a review. World J Urol 2013;31:1037-43.
5. Scollo C, Russo M, Trovato MA, et al. Prognostic factors for adrenocortical carcinoma outcomes. Front Endocrinol (Lausanne) 2016;7:99.
6. Lyraki R, Schedl A. The sexually dimorphic adrenal cortex: implications for adrenal disease. Int J Mol Sci 2021;22:4889.
7. Correa P, Chen VW. Endocrine gland cancer. Cancer 1995;75:338-52.
8. Habra MA, Sukkari MA, Hasan A, et al. Epidemiological risk factors for adrenocortical carcinoma: a hospital-based case-control study. Int J Cancer 2020;146:1836-40.
9. Hsing AW, Nam JM, Co Chien HT, McLaughlin JK, Fraumeni JF Jr. Risk factors for adrenal cancer: an exploratory study. Int J Cancer 1996;65:432-6.
10. Yousaf A, Patterson J, Hobbs G, et al. Smoking is associated with adrenal adenomas and adrenocortical carcinomas: a nationwide multicenter analysis. Cancer Treat Res Commun 2020;25: 100206.
11. Sirianni R, Zolea F, Chimento A, et al. Targeting estrogen receptor-α reduces adrenocortical cancer (ACC) cell growth in vitro and in vivo: potential therapeutic role of selective estrogen receptor modulators (SERMs) for ACC treatment. J Clin Endocrinol Metab 2012;97:E2238-50.
12. Tamburello M, Abate A, Rossini E, et al. Preclinical evidence of progesterone as a new pharmacological strategy in human adrenocortical carcinoma cell lines. Int J Mol Sci 2023;24:6829.
13. Portnov BA, Barchana M, Dubnov J. Exploratory analysis of potential risk factors of a rare disease: spatial distribution of adrenocortical carcinoma in Israel as a case study. Sci Total Environ 2009; 407:1738-43.
14. Costa TE, Gerber VK, Ibañez HC, et al. Penetrance of the TP53 R337H mutation and pediatric adrenocortical carcinoma incidence associated with environmental influences in a 12-year observational cohort in southern Brazil. Cancers (Basel) 2019;11:1804.
15. Anselmo J, Medeiros S, Carneiro V, et al. A large family with Carney complex caused by the S147G PRKAR1A mutation shows a unique spectrum of disease including adrenocortical cancer. J Clin Endocrinol Metab 2012;97:351-9.
16. Morin E, Mete O, Wasserman JD, Joshua AM, Asa SL, Ezzat S. Carney complex with adrenal cortical carcinoma. J Clin Endocrinol Metab 2012; 97:E202-6.
17. Langer P, Cupisti K, Bartsch DK, et al. Adrenal involvement in multiple endocrine neoplasia type 1. World J Surg 2002;26:891-6.
18. Gatta-Cherifi B, Chabre O, Murat A, et al. Adrenal involvement in MEN1. Analysis of 715 cases from the Groupe d'etude des Tumeurs Endocrines database. Eur J Endocrinol 2012;166:269-79.
19. Bernard MH, Sidhu S, Berger N, et al. A case report in favor of a multistep adrenocortical tumorigenesis. J Clin Endocrinol Metab 2003;88:998-1001.
20. Zheng GY, Zhang XB, Li HZ, Zhang YS, Deng JH, Wu XC. Sum of high-risk gene mutation (SHGM): a novel attempt to assist differential diagnosis for adrenocortical carcinoma with benign adenoma, based on detection of mutations of nine target genes. Biochem Genet 2021;59:902-18.
21. Gara SK, Wang Y, Patel D, et al. Integrated genome-wide analysis of genomic changes and gene regulation in human adrenocortical tissue samples. Nucleic Acids Res 2015;43:9327-39.
22. Ronchi CL, Sbiera S, Leich E, et al. Single nucleotide polymorphism array profiling of adrenocortical tumors—evidence for an adenoma carcinoma sequence? PLoS One 2013;8:e73959.
23. Lam KY, Chan AC, Lo CY. Morphological analysis of adrenal glands: a prospective analysis. Endocr Pathol 2001;12:33-8.
24. Bharwani N, Rockall AG, Sahdev A, et al. Adrenocortical carcinoma: the range of appearances on CT and MRI. AJR Am J Roentgenol 2011;196: W706-14.
25. Claimon A, Tantranont N, Claimon T. ^{18}F-fluoro-2-deoxy-D-glucose positron emission tomography/computed tomography for preoperative planning in a rare case of hyperfunctional bilateral adrenocortical carcinoma and review of literatures. World J Nucl Med 2020;19:301-5.

26. Akishima-Fukasawa Y, Yoshihara A, Ishikawa Y, et al. Malignant adrenal rest tumor of the retroperitoneum producing adrenocortical steroids. Endocr Pathol 2011;22:112-7.
27. Yokoyama H, Adachi T, Tsubouchi K, Tanaka M, Sasano H. Non-functioning adrenocortical carcinoma arising in an adrenal rest: immunohistochemical study of an adult patient. Tohoku J Exp Med 2013;229:267-70.
28. Cornejo KM, Afari HA, Sadow PM. Adrenocortical carcinoma arising in an adrenal rest: a case report and review of the literature. Endocr Pathol 2017; 28:165-70.
29. Wright JP, Montgomery KW, Tierney J, Gilbert J, Solórzano CC, Idrees K. Ectopic, retroperitoneal adrenocortical carcinoma in the setting of Lynch syndrome. Fam Cancer 2018;17:381-5.
30. Salle L, Mas R, Teissier-Clément MP. Ectopic adrenocortical carcinoma of the ovary: an unexpected outcome. Ann Endocrinol (Paris) 2020;81:516-8.
31. Tsai WH, Chen TC, Dai SH, Zeng YH. Case report: ectopic adrenocortical carcinoma in the ovary. Front Endocrinol (Lausanne) 2021;12:662377.
32. Falco EC, Daniele L, Metovic J, et al. Adrenal rests in the uro-genital tract of an adult population. Endocr Pathol 2021;32:375-84.
33. Permana H, Darmawan G, Ritonga E, Kusumawati M, Miftahurachman M, Soetedjo NN. An interesting case of hepatic adrenocortical carcinoma. Acta Med Indones 2018;50:257-9.
34. Fassnacht M, Dekkers OM, Else T, et al. European Society of Endocrinology Clinical Practice Guidelines on the management of adrenocortical carcinoma in adults, in collaboration with the European Network for the Study of Adrenal Tumors. Eur J Endocrinol 2018;179:G1-46.
35. Kiseljak-Vassiliades K, Bancos I, Hamrahian A, et al. American Association of Clinical Endocrinology Disease State Clinical Review on the evaluation and management of adrenocortical carcinoma in an adult: a practical approach. Endocr Pract 2020;26:1366-83.
36. Kahramangil B, Kose E, Remer EM, et al. A modern assessment of cancer risk in adrenal incidentalomas: analysis of 2219 patients. Ann Surg 2022;275:e238-44.
37. Else T, Kim AC, Sabolch A, et al. Adrenocortical carcinoma. Endocr Rev 2014;35:282-326.
38. Marincola Smith P, Kiernan CM, Tran TB, et al. Role of additional organ resection in adrenocortical carcinoma: analysis of 167 patients from the U.S. Adrenocortical Carcinoma Database. Ann Surg Oncol 2018;25:2308-15.
39. Kumar T, Nigam JS, Sharma S, Kumari M, Pandey J. Uncommon metastasizing site of adrenocortical carcinoma. Cureus 2021;13:e15267.
40. Allolio B, Fassnacht M. Clinical review: adrenocortical carcinoma: clinical update. J Clin Endocrinol Metab 2006;91:2027-37.
41. Orland SM, Stewart AF, Livolsi VA, Wein AJ. Detection of the hypercalcemic hormone of malignancy in an adrenal cortical carcinoma. J Urol 1986;136:1000-2.
42. Haissaguerre M, Louiset E, et al. Immunohistochemical characterization of a steroid secreting oncocytic adrenal carcinoma responsible for paraneoplastic hyperparathyroidism. Eur J Endocrinol 2023;188:K11-6.
43. Lee KW, Chon SB, Kim DY, et al. Adrenal cortical carcinoma initially presented with overwhelming disseminated intravascular coagulation. Ann Hematol 2003;82:596-8.
44. Stamoulis JS, Antonopoulou Z, Safioleas M. Haemorrhagic shock from the spontaneous rupture of an adrenal cortical carcinoma. A case report. Acta Chir Belg 2004;104:226-8.
45. Lewinsky BS, Grigor KM, Symington T, Neville AM. The clinical and pathologic features of "non-hormonal" adrenocortical tumors. Report of twenty new cases and review of the literature. Cancer 1974;33:778-90.
46. D'Agata R, Malozowski S, Barkan A, Cassorla F, Loriaux D. Steroid biosynthesis in human adrenal tumors. Horm Metab Res 1987;19:386-8.
47. Uchida T, Nishimoto K, Fukumura Y, et al. Disorganized steroidogenesis in adrenocortical carcinoma, a case study. Endocr Pathol 2017;28:27-35.
48. Gazdar AF, Oie HK, Shackleton CH, et al. Establishment and characterization of a human adrenocortical carcinoma cell line that expresses multiple pathways of steroid biosynthesis. Cancer Res 1990;50:5488-96.
49. Nakamura Y, Yamazaki Y, Felizola SJ, et al. Adrenocortical carcinoma: review of the pathologic features, production of adrenal steroids, and molecular pathogenesis. Endocrinol Metab Clin North Am 2015;44:399-410.
50. Sasano H. Localization of steroidogenic enzymes in adrenal cortex and its disorders. Endocr J 1994; 41:471-82.
51. Berruti A, Fassnacht M, Haak H, et al. Prognostic role of overt hypercortisolism in completely operated patients with adrenocortical cancer. Eur Urol 2014;65:832-8.
52. Duan K, Gomez Hernandez K, Mete O. Clinicopathological correlates of adrenal Cushing's syndrome. J Clin Pathol 2015;68:175-86.
53. Ahn CH, Kim JH, Park MY, Kim SW. Epidemiology and comorbidity of adrenal cushing syndrome: a nationwide cohort study. J Clin Endocrinol Metab 2021;106:e1362-72.

54. Rosset A, Greenman Y, Osher E, Stern N, Tordjman K. Revisiting Cushing syndrome, milder forms are now a common occurrence: a single-center cohort of 76 subjects. Endocr Pract 2021;27:859-65.
55. Davenport E, Lennard T. Acute hypercortisolism: what can the surgeon offer? Clin Endocrinol (Oxf) 2014;81:498-502.
56. Song BG, Lim MG, Bae JH, et al. Multiple opportunistic infections related to hypercortisolemia due to adrenocortical carcinoma: a case report. Infect Chemother 2021;53:797-801.
57. Bala M, Ronchi CL, Pichl J, et al. Suspected metastatic adrenocortical carcinoma revealing as pulmonary Kaposi sarcoma in adrenal Cushing's syndrome. BMC Endocr Disord 2014;14:63.
58. Hodgson A, Pakbaz S, Mete O. A diagnostic approach to adrenocortical tumors. Surg Pathol Clin 2019;12:967-95.
59. Elhassan YS, Idkowiak J, Smith K, et al. Causes, patterns, and severity of androgen excess in 1205 consecutively recruited women. J Clin Endocrinol Metab 2018;103:1214-23.
60. Wu L, Xie J, Jiang L, et al. Feminizing adrenocortical carcinoma: the source of estrogen production and the role of adrenal-gonadal dedifferentiation. J Clin Endocrinol Metab 2018;103:3706-13.
61. Moreno S, Guillermo M, Decoulx M, Dewailly D, Bresson R, Proye Ch. Feminizing adreno-cortical carcinomas in male adults. A dire prognosis. Three cases in a series of 801 adrenalectomies and review of the literature. Ann Endocrinol (Paris) 2006;67:32-8.
62. Hatano M, Takenaka Y, Inoue I, et al. Feminizing adrenocortical carcinoma with distinct histopathological findings. Intern Med 2016;55:3301-7.
63. Morioka M, Furukawa Y, Kobayashi T, Tanaka H, Ohashi Y, Jin TX. Aldosterone-producing adrenal cortical cancer: a case report and analysis of steroidogenic enzymes in the tumor. Endocr J 1997;44:547-52.
64. Griffin AC, Kelz R, LiVolsi VA. Aldosterone-secreting adrenal cortical carcinoma. A case report and review of the literature. Endocr Pathol 2014;25:344-9.
65. Mouat IC, Omata K, McDaniel AS, et al. Somatic mutations in adrenocortical carcinoma with primary aldosteronism or hyperreninemic hyperaldosteronism. Endocr Relat Cancer 2019;26:217-25.
66. Kendrick ML, Curlee K, Lloyd R, et al. Aldosterone-secreting adrenocortical carcinomas are associated with unique operative risks and outcomes. Surgery 2002;132:1008-12.
67. Seccia TM, Fassina A, Nussdorfer GG, Pessina AC, Rossi GP. Aldosterone-producing adrenocortical carcinoma: an unusual cause of Conn's syndrome with an ominous clinical course. Endocr Relat Cancer 2005;12:149-59.
68. Müssig K, Wehrmann M, Horger M, Maser-Gluth C, Häring HU, Overkamp D. Adrenocortical carcinoma producing 11-deoxycorticosterone: a rare cause of mineralocorticoid hypertension. J Endocrinol Invest 2005;28:61-5.
69. Velikanova LI, Shafigullina ZR, Lisitsin AA, et al. Different types of urinary steroid profiling obtained by high-performance liquid chromatography and gas chromatography-mass spectrometry in patients with adrenocortical carcinoma. Horm Cancer 2016;7:327-35.
70. Schweitzer S, Kunz M, Kurlbaum M, et al. Plasma steroid metabolome profiling for the diagnosis of adrenocortical carcinoma. Eur J Endocrinol 2019;180:117-25.
71. Fanelli F, Di Dalmazi G. Serum steroid profiling by mass spectrometry in adrenocortical tumors: diagnostic implications. Curr Opin Endocrinol Diabetes Obes 2019;26:160-5.
72. Chortis V, Bancos I, Nijman T, et al. Urine steroid metabolomics as a novel tool for detection of recurrent adrenocortical carcinoma. J Clin Endocrinol Metab 2020;105:e307-18.
73. Rossi C, Cicalini I, Verrocchio S, Di Dalmazi G, Federici L, Bucci I. The potential of steroid profiling by mass spectrometry in the management of adrenocortical carcinoma. Biomedicines 2020;8:314.
74. Suzuki S, Minamidate T, Shiga A, et al. Steroid metabolites for diagnosing and predicting clinicopathological features in cortisol-producing adrenocortical carcinoma. BMC Endocr Disord 2020;20:173.
75. Mizdrak M, Ticinovic Kurir T, Božic J. The role of biomarkers in adrenocortical carcinoma: a review of current evidence and future perspectives. Biomedicines 2021;9:174.
76. Arlt W, Biehl M, Taylor AE, et al. Urine steroid metabolomics as a biomarker tool for detecting malignancy in adrenal tumors. J Clin Endocrinol Metab 2011;96:3775-84.
77. Bancos I, Taylor AE, Chortis V, et al. Urine steroid metabolomics for the differential diagnosis of adrenal incidentalomas in the EURINE-ACT study: a prospective test validation study. Lancet Diabetes Endocrinol 2020;8:773-81.
78. Ahmed AA, Thomas AJ, Ganeshan DM, et al. Adrenal cortical carcinoma: pathology, genomics, prognosis, imaging features, and mimics with impact on management. Abdom Radiol (NY) 2020;45:945-63.
79. Dinnes J, Bancos I, Ferrante di Ruffano L, et al. Management of endocrine disease: imaging for the diagnosis of malignancy in incidentally discovered adrenal masses: a systematic review and meta-analysis. Eur J Endocrinol 2016;175:R51-64.

80. Hamper UM, Fishman EK, Hartman DS, Roberts JL, Sanders RC. Primary adrenocortical carcinoma: sonographic evaluation with clinical and pathologic correlation in 26 patients. AJR Am J Roentgenol 1987;148:915-9.
81. Lattin GE Jr, Sturgill ED, Tujo CA, et al. From the radiologic pathology archives: adrenal tumors and tumor-like conditions in the adult: radiologic-pathologic correlation. Radiographics 2014;34:805-29.
82. Fan J, Tang J, Fang J, et al. Ultrasound imaging in the diagnosis of benign and suspicious adrenal lesions. Med Sci Monit 2014;20:2132-41.
83. Friedrich-Rust M, Glasemann T, Polta A, et al. Differentiation between benign and malignant adrenal mass using contrast-enhanced ultrasound. Ultraschall Med 2011;32:460-71.
84. Young WF Jr. Conventional imaging in adrenocortical carcinoma: update and perspectives. Horm Cancer 2011;2:341-7.
85. Garay-Lechuga D, Pérez-Soto RH, Hernández-Acevedo JD, et al. Computed tomography (CT) scan identified necrosis, but is it a reliable single parameter for discerning between malignant and benign adrenocortical tumors? Surgery 2022;171:104-10.
86. Rowe SP, Lugo-Fagundo C, Ahn H, Fishman EK, Prescott JD. What the radiologist needs to know: the role of preoperative computed tomography in selection of operative approach for adrenalectomy and review of operative techniques. Abdom Radiol (NY) 2019;44:140-53.
87. Iñiguez-Ariza NM, Kohlenberg JD, Delivanis DA, et al. Clinical, biochemical, and radiological characteristics of a single-center retrospective cohort of 705 large adrenal tumors. Mayo Clin Proc Innov Qual Outcomes 2017;2:30-9.
88. Fassnacht M, Arlt W, Bancos I, et al. Management of adrenal incidentalomas: European Society of Endocrinology Clinical Practice Guideline in collaboration with the European Network for the Study of Adrenal Tumors. Eur J Endocrinol 2016; 175:G1-4.
89. Fassnacht M, Assie G, Baudin E, et al. Adrenocortical carcinomas and malignant phaeochromocytomas: ESMO-EURACAN Clinical Practice Guidelines for diagnosis, treatment and follow-up. Ann Oncol 2020;31:1476-90.
90. Shin YR, Kim KA. Imaging features of various adrenal neoplastic lesions on radiologic and nuclear medicine imaging. AJR Am J Roentgenol 2015;205:554-63.
91. Sandrasegaran K, Patel AA, Ramaswamy R, et al. Characterization of adrenal masses with diffusion-weighted imaging. AJR Am J Roentgenol 2011;197:132-8.
92. Imperiale A, Elbayed K, Moussallieh FM, et al. Metabolomic profile of the adrenal gland: from physiology to pathological conditions. Endocr Relat Cancer 2013;20:705-16.
93. Melo HJ, Goldman SM, Szejnfeld J, et al. Application of a protocol for magnetic resonance spectroscopy of adrenal glands: an experiment with over 100 cases. Radiol Bras 2014;47:333-41.
94. Elbanna KY, Khalili K, O'Malley M, Chawla T. Imaging and implications of tumor thrombus in abdominal malignancies: reviewing the basics. Abdom Radiol (NY) 2020;45:1057-68.
95. Boland GW, Dwamena BA, Jagtiani Sangwaiya M, et al. Characterization of adrenal masses by using FDG PET: a systematic review and meta-analysis of diagnostic test performance. Radiology 2011;259:117-26.
96. Kim SJ, Lee SW, Pak K, Kim IJ, Kim K. Diagnostic accuracy of 18F-FDG PET or PET/CT for the characterization of adrenal masses: a systematic review and meta-analysis. Br J Radiol 2018;91:20170520.
97. He X, Caoili EM, Avram AM, Miller BS, Else T. 18F-FDG-PET/CT evaluation of indeterminate adrenal masses in noncancer patients. J Clin Endocrinol Metab 2021;106:1448-59.
98. Salgues B, Guerin C, Amodru V, et al. Risk stratification of adrenal masses by [18 F]FDG PET/CT: changing tactics. Clin Endocrinol (Oxf) 2021;94:133-40.
99. Mayo-Smith WW, Song JH, Boland GL, et al. Management of incidental adrenal masses: a white paper of the ACR Incidental Findings Committee. J Am Coll Radiol 2017;14:1038-44.
100. Akkus G, Güney IB, Ok F, et al. Diagnostic efficacy of 18F-FDG PET/CT in patients with adrenal incidentaloma. Endocr Connect 2019;8:838-45.
101. Acar C, Akkas BE, Sen I, Sozen S, Kitapci MT. False positive 18F-FDG PET scan in adrenal oncocytoma. Urol Int 2008;80:444-7.
102. Kim DJ, Chung JJ, Ryu YH, Hong SW, Yu JS, Kim JH. Adrenocortical oncocytoma displaying intense activity on 18F-FDG-PET: a case report and a literature review. Ann Nucl Med 2008;22:821-4.
103. Sato N, Nakamura Y, Takanami K, et al. Case report: adrenal oncocytoma associated with markedly increased FDG uptake and immunohistochemically positive for GLUT1. Endocr Pathol 2014;25:410-5.
104. Ghander C, Tissier F, Tenenbaum F, et al. A concomitant false-negative ^{18}F-FDG PET imaging in an adrenocortical carcinoma and a high uptake in a corresponding liver metastasis. J Clin Endocrinol Metab 2012;97:1096-7.

105. Mackie GC, Shulkin BL, Ribeiro RC, et al. Use of [18F]fluorodeoxyglucose positron emission tomography in evaluating locally recurrent and metastatic adrenocortical carcinoma. J Clin Endocrinol Metab 2006;91:2665-71.
106. Tessonnier L, Ansquer C, Bournaud C, et al. (18)F-FDG uptake at initial staging of the adrenocortical cancers: a diagnostic tool but not of prognostic value. World J Surg 2013;37:107-12.
107. Takeuchi S, Balachandran A, Habra MA, et al. Impact of ^{18}F-FDG PET/CT on the management of adrenocortical carcinoma: analysis of 106 patients. Eur J Nucl Med Mol Imaging 2014;41:2066-73.
108. Wong KK, Miller BS, Viglianti BL, et al. Molecular imaging in the management of adrenocortical cancer: a systematic review. Clin Nucl Med 2016;41:e368-82.
109. Chen Cardenas SM, Santhanam P. ^{11}C-metomidate PET in the diagnosis of adrenal masses and primary aldosteronism: a review of the literature. Endocrine 2020;70:479-87.
110. Hahner S, Kreissl MC, Fassnacht M, et al. [131I] iodometomidate for targeted radionuclide therapy of advanced adrenocortical carcinoma. J Clin Endocrinol Metab 2012;97:914-22.
111. Yu H, Parakh A, Blake M, McDermott S. Texture analysis as a radiomic marker for differentiating benign from malignant adrenal tumors. J Comput Assist Tomogr 2020;44:766-71.
112. Moawad AW, Ahmed A, Fuentes DT, Hazle JD, Habra MA, Elsayes KM. Machine learning-based texture analysis for differentiation of radiologically indeterminate small adrenal tumors on adrenal protocol CT scans. Abdom Radiol (NY) 2021;46:4853-63.
113. Stanzione A, Cuocolo R, Verde F, et al. Handcrafted MRI radiomics and machine learning: classification of indeterminate solid adrenal lesions. Magn Reson Imaging 2021;79:52-8.
114. Elmohr MM, Fuentes D, Habra MA, et al. Machine learning-based texture analysis for differentiation of large adrenal cortical tumours on CT. Clin Radiol 2019;74:818.e1-818.e7.
115. Ahmed AA, Elmohr MM, Fuentes D, et al. Radiomic mapping model for prediction of Ki-67 expression in adrenocortical carcinoma. Clin Radiol 2020;75:479.e17-479.e22.
116. Volante M, Bollito E, Sperone P, et al. Clinicopathological study of a series of 92 adrenocortical carcinomas: from a proposal of simplified diagnostic algorithm to prognostic stratification. Histopathology 2009;55:535-43.
117. Duregon E, Fassina A, Volante M, et al. The reticulin algorithm for adrenocortical tumor diagnosis: a multicentric validation study on 245 unpublished cases. Am J Surg Pathol 2013;37:1433-40.
118. Duregon E, Cappellesso R, Maffeis V, et al. Validation of the prognostic role of the "Helsinki Score" in 225 cases of adrenocortical carcinoma. Hum Pathol 2017;62:1-7.
119. Autorino R, Bove P, De Sio M, et al. Open versus laparoscopic adrenalectomy for adrenocortical carcinoma: a meta-analysis of surgical and oncological outcomes. Ann Surg Oncol 2016;23:1195-202.
120. Nakanishi H, Miangul S, Wang R, et al. Open versus laparoscopic surgery in the management of adrenocortical carcinoma: a systematic review and meta-analysis. Ann Surg Oncol 2023;30:994-1005.
121. Conti L, Vatrano S, Bertero L, et al. Mitochondrial DNA "common deletion" in post-fine needle aspiration infarcted oncocytic thyroid tumors. Hum Pathol 2017;69:23-30.
122. Weissferdt A, Phan A, Suster S, Moran CA. Primary rhabdoid adrenocortical carcinoma: a clinicopathological and immunohistochemical study of three cases. Histopathology 2013;62:771-7.
123. Hough AJ, Hollifield JW, Page DL, Hartmann WH. Prognostic factors in adrenal cortical tumors. A mathematical analysis of clinical and morphologic data. Am J Clin Pathol 1979;72:390-9.
124. Weiss LM. Comparative histologic study of 43 metastasizing and nonmetastasizing adrenocortical tumors. Am J Surg Pathol 1984;8:163-9.
125. van Slooten H, Schaberg A, Smeenk D, Moolenaar AJ. Morphologic characteristics of benign and malignant adrenocortical tumors. Cancer 1985;55:766-73.
126. Pennanen M, Heiskanen I, Sane T, et al. Helsinki score-a novel model for prediction of metastases in adrenocortical carcinomas. Hum Pathol 2015;46:404-10.
127. Duregon E, Molinaro L, Volante M, et al. Comparative diagnostic and prognostic performances of the hematoxylin-eosin and phospho-histone H3 mitotic count and Ki-67 index in adrenocortical carcinoma. Mod Pathol 2014;27:1246-54.
128. Sarkar S, Sahoo PK, Mahata S, et al. Mitotic checkpoint defects: en route to cancer and drug resistance. Chromosome Res 2021;29:131-44.
129. Abdelnabi M, Almaghraby A, Saleh Y, Abd El-Samad S. Right atrial and ventricular invasion by adrenal carcinoma: a case report. J Cardiol Cases 2019;20:42-4.
130. Mete O, Erickson LA, Juhlin CC, et al. Overview of the 2022 WHO classification of adrenal cortical tumors. Endocr Pathol 2022;33:155-96.

131. Mete O, Gucer H, Kefeli M, Asa SL. Diagnostic and prognostic biomarkers of adrenal cortical carcinoma. Am J Surg Pathol 2018;42:201-13.
132. Giordano TJ, Berney D, de Krijger RR, et al. Data set for reporting of carcinoma of the adrenal cortex: explanations and recommendations of the guidelines from the International Collaboration on Cancer Reporting. Hum Pathol 2021; 110:50-61.
133. Tissier F, Aubert S, Leteurtre E, et al. Adrenocortical tumors: improving the practice of the Weiss system through virtual microscopy: a National Program of the French Network INCa-COMETE. Am J Surg Pathol 2012;36:1194-201.
134. Landwehr LS, Altieri B, Schreiner J, et al. Interplay between glucocorticoids and tumor-infiltrating lymphocytes on the prognosis of adrenocortical carcinoma. J Immunother Cancer 2020;8:e000469.
135. de Krijger RR, Papathomas TG. Adrenocortical neoplasia: evolving concepts in tumorigenesis with an emphasis on adrenal cortical carcinoma variants. Virchows Arch 2012;460:9-18.
136. Lam AK. Adrenocortical carcinoma: updates of clinical and pathological features after renewed World Health Organisation classification and pathology staging. Biomedicines 2021;9:175.
137. Kanitra JJ, Hardaway JC, Soleimani T, Koehler TJ, McLeod MK, Kavuturu S. Adrenocortical oncocytic neoplasm: a systematic review. Surgery 2018;164:1351-9.
138. Bisceglia M, Ludovico O, Di Mattia A, et al. Adrenocortical oncocytic tumors: report of 10 cases and review of the literature. Int J Surg Pathol 2004;12:231-43.
139. Wong DD, Spagnolo DV, Bisceglia M, Havlat M, McCallum D, Platten MA. Oncocytic adrenocortical neoplasms—a clinicopathologic study of 13 new cases emphasizing the importance of their recognition. Hum Pathol 2011;42:489-99.
140. Renaudin K, Smati S, Wargny M, et al. Clinicopathological description of 43 oncocytic adrenocortical tumors: importance of Ki-67 in histoprognostic evaluation. Mod Pathol 2018; 31:1708-16.
141. Tang CK, Harriman BB, Toker C. Myxoid adrenal cortical carcinoma: a light and electron microscopic study. Arch Pathol Lab Med 1979; 103:635-8.
142. Hsieh MS, Chen JH, Lin LW. Myxoid adrenal cortical carcinoma presenting as primary hyperaldosteronism: case report and review of the literature. Int J Surg Pathol 2011;19:803-7.
143. Papotti M, Volante M, Duregon E, et al. Adrenocortical tumors with myxoid features: a distinct morphologic and phenotypical variant exhibiting malignant behavior. Am J Surg Pathol 2010;34:973-83.
144. Izumi M, Serizawa H, Iwaya K, Takeda K, Sasano H, Mukai K. A case of myxoid adrenocortical carcinoma with extensive lipomatous metaplasia. Arch Pathol Lab Med 2003;127:227-30.
145. Papathomas TG, Duregon E, Korpershoek E, et al. Sarcomatoid adrenocortical carcinoma: a comprehensive pathological, immunohistochemical, and targeted next-generation sequencing analysis. Hum Pathol 2016;58:113-22.
146. Hayashi T, Gucer H, Mete O. A mimic of sarcomatoid adrenal cortical carcinoma: epithelioid angiosarcoma occurring in adrenal cortical adenoma. Endocr Pathol 2014;25:404-9.
147. Weiss LM, Medeiros LJ, Vickery AL Jr. Pathologic features of prognostic significance in adrenocortical carcinoma. Am J Surg Pathol 1989; 13:202-6.
148. Liang J, Liu Z, Zhou L, et al. The clinical utility of 'GRAS' parameters in stage I-III adrenocortical carcinomas: long-term data from a high-volume institution. Endocrine 2020;67:449-56.
149. Lacombe AM, Soares IC, Mariani BM, et al. Sterol O-Acyl transferase 1 as a prognostic marker of adrenocortical carcinoma. Cancers (Basel) 2020;12:247.
150. Aubert S, Wacrenier A, Leroy X, et al. Weiss system revisited: a clinicopathologic and immunohistochemical study of 49 adrenocortical tumors. Am J Surg Pathol 2002;26:1612-9.
151. Minner S, Schreiner J, Saeger W. Adrenal cancer: relevance of different grading systems and subtypes. Clin Transl Oncol 2021;23:1350-7.
152. Volante M, Rapa I, Metovic J, et al. Differential expression profiles of cell-to-matrix-related molecules in adrenal cortical tumors: diagnostic and prognostic implications. J Pers Med 2021;11:378.
153. Oliveira RC, Martins MJ, Moreno C, et al. Histological scores and tumor size on stage II in adrenocortical carcinomas. Rare Tumors 2021; 13:20363613211026494.
154. Angelousi A, Kyriakopoulos G, Athanasouli F, et al. The role of immunohistochemical markers for the diagnosis and prognosis of adrenocortical neoplasms. J Pers Med 2021;11:208.
155. Duregon E, Volante M, Cappia S, et al. Oncocytic adrenocortical tumors: diagnostic algorithm and mitochondrial DNA profile in 27 cases. Am J Surg Pathol 2011;35:1882-93.
156. Dalino Ciaramella P, Vertemati M, Petrella D, et al. Analysis of histological and immunohistochemical patterns of benign and malignant adrenocortical tumors by computerized morphometry. Pathol Res Pract 2017;213:815-23.

157. Assié G, Jouinot A, Fassnacht M, et al. Value of molecular classification for prognostic assessment of adrenocortical carcinoma. JAMA Oncol 2019;5:1440-7.
158. Libé R, Borget I, Ronchi CL, et al. Prognostic factors in stage III-IV adrenocortical carcinomas (ACC): an European Network for the Study of Adrenal Tumor (ENSAT) study. Ann Oncol 2015;26:2119-25.
159. Sohail S, Azmat U, Khawaja S, et al. Clinical and histopathological variables and prognostic factors of adrenocortical carcinoma. Cureus 2021;13:e15721.
160. Miller BS, Gauger PG, Hammer GD, Giordano TJ, Doherty GM. Proposal for modification of the ENSAT staging system for adrenocortical carcinoma using tumor grade. Langenbecks Arch Surg 2010;395:955-61.
161. Giordano TJ. The argument for mitotic rate-based grading for the prognostication of adrenocortical carcinoma. Am J Surg Pathol 2011; 35:471-3.
162. Mackay B, el-Naggar A, Ordonez NG. Ultrastructure of adrenal cortical carcinoma. Ultrastruct Pathol 1994;18:181-90.
163. Silva EG, Mackay B, Samaan NA, Hickey RC. Adrenocortical carcinomas: an ultrastructural study of 22 cases. Ultrastruct Pathol 1982;3:1-7.
164. Ali AE, Raphael SJ. Functional oncocytic adrenocortical carcinoma. Endocr Pathol 2007;18: 187-9.
165. Karim RZ, Wills EJ, McCarthy SW, Scolyer RA. Myxoid variant of adrenocortical carcinoma: report of a unique case. Pathol Int 2006;56:89-94.
166. Seidel E, Walenda G, Messerschmidt C, et al. Generation and characterization of a mitotane-resistant adrenocortical cell line. Endocr Connect 2020;9:122-34.
167. Mete O, Asa SL, Giordano TJ, Papotti M, Sasano H, Volante M. Immunohistochemical biomarkers of adrenal cortical neoplasms. Endocr Pathol 2018;29:137-49.
168. Enriquez ML, Lal P, Ziober A, Wang L, Tomaszewski JE, Bing Z. The use of immunohistochemical expression of SF-1 and EMA in distinguishing adrenocortical tumors from renal neoplasms. Appl Immunohistochem Mol Morphol 2012;20:141-5.
169. Sbiera S, Schmull S, Assie G, et al. High diagnostic and prognostic value of steroidogenic factor-1 expression in adrenal tumors. J Clin Endocrinol Metab 2010;95:E161-71.
170. Mete O, Cintosun A, Pressman I, Asa SL. Epidemiology and biomarker profile of pituitary adenohypophysial tumors. Mod Pathol 2018; 31:900-9.
171. Duregon E, Volante M, Giorcelli J, Terzolo M, Lalli E, Papotti M. Diagnostic and prognostic role of steroidogenic factor 1 in adrenocortical carcinoma: a validation study focusing on clinical and pathologic correlates. Hum Pathol 2013;44:822-8.
172. Mete O, Pakbaz S, Lerario AM, Giordano TJ, Asa SL. Significance of alpha-inhibin expression in pheochromocytomas and paragangliomas. Am J Surg Pathol 2021;45:1264-73.
173. Mete O, Kapran Y, Güllüoglu MG, et al. Anti-CD10 (56C6) is expressed variably in adrenocortical tumors and cannot be used to discriminate clear cell renal cell carcinomas. Virchows Arch 2010;456:515-21.
174. Perrino CM, Ho A, Dall CP, Zynger DL. Utility of GATA3 in the differential diagnosis of pheochromocytoma. Histopathology 2017;71:475-9.
175. Peppelman M, Timmers HJ, Lenders JW, Hermus AR, Küsters B. CD56 immunohistochemistry does not discriminate between cortisol-producing and aldosterone-producing adrenal cortical adenomas. Histopathology 2011;58:994-6.
176. Schmitt A, Saremaslani P, Schmid S, et al. IGFII and MIB1 immunohistochemistry is helpful for the differentiation of benign from malignant adrenocortical tumours. Histopathology 2006;49:298-307.
177. Soon PS, Gill AJ, Benn DE, et al. Microarray gene expression and immunohistochemistry analyses of adrenocortical tumors identify IGF2 and Ki-67 as useful in differentiating carcinomas from adenomas. Endocr Relat Cancer 2009;16:573-83.
178. Tadjine M, Lampron A, Ouadi L, Bourdeau I. Frequent mutations of beta-catenin gene in sporadic secreting adrenocortical adenomas. Clin Endocrinol (Oxf) 2008;68:264-70.
179. Martins AC, Cologna AJ, Tucci S Jr, Suaid HJ, Falconi RA. Clinical features and immunoexpression of p53, MIB-1 and proliferating cell nuclear antigen in adrenal neoplasms. J Urol 2005;173:2138-42.
180. Hescot S, Faron M, Kordahi M, et al. Screening for prognostic biomarkers in metastatic adrenocortical carcinoma by tissue micro arrays analysis identifies p53 as an independent prognostic marker of overall survival. Cancers (Basel) 2022;14:2225.
181. Volante M, Sperone P, Bollito E, et al. Matrix metalloproteinase type 2 expression in malignant adrenocortical tumors: diagnostic and prognostic significance in a series of 50 adrenocortical carcinomas. Mod Pathol 2006;19:1563-9.

182. Magro G, Esposito G, Cecchetto G, et al. Pediatric adrenocortical tumors: morphological diagnostic criteria and immunohistochemical expression of matrix metalloproteinase type 2 and human leucocyte-associated antigen (HLA) class II antigens. Results from the Italian Pediatric Rare Tumor (TREP) Study project. Hum Pathol 2012;43:31-9.
183. Stojadinovic A, Brennan MF, Hoos A, et al. Adrenocortical adenoma and carcinoma: histopathological and molecular comparative analysis. Mod Pathol 2003;16:742-51.
184. Martins-Filho SN, Almeida MQ, Soares I, et al. Clinical impact of pathological features including the ki-67 labeling index on diagnosis and prognosis of adult and pediatric adrenocortical tumors. Endocr Pathol 2021;32:288-300.
185. Juhlin CC, Bertherat J, Giordano TJ, Hammer GD, Sasano H, Mete O. What did we learn from the molecular biology of adrenal cortical neoplasia? From histopathology to translational genomics. Endocr Pathol 2021;32:102-33.
186. Zheng S, Cherniack AD, Dewal N, et al. Comprehensive pan-genomic characterization of adrenocortical carcinoma. Cancer Cell 2016; 29:723-36.
187. Mouat IC, Omata K, McDaniel AS, et al. Somatic mutations in adrenocortical carcinoma with primary aldosteronism or hyperreninemic hyperaldosteronism. Endocr Relat Cancer 2019;26:217-25.
188. Assié G, Letouzé E, Fassnacht M, et al. Integrated genomic characterization of adrenocortical carcinoma. Nat Genet 2014;46:607-12.
189. Juhlin CC, Goh G, Healy JM, et al. Whole-exome sequencing characterizes the landscape of somatic mutations and copy number alterations in adrenocortical carcinoma. J Clin Endocrinol Metab 2015;100:E493-502.
190. Vatrano S, Volante M, Duregon E, et al. Detailed genomic characterization identifies high heterogeneity and histotype-specific genomic profiles in adrenocortical carcinomas. Mod Pathol 2018;31:1257-69.
191. Gupta S, Won H, Chadalavada K, et al. TERT copy number alterations, promoter mutations and rearrangements in adrenocortical carcinomas. Endocr Pathol 2022;33:304-14.
192. Dohna M, Reincke M, Mincheva A, Allolio B, Solinas-Toldo S, Lichter P. Adrenocortical carcinoma is characterized by a high frequency of chromosomal gains and high-level amplifications. Genes Chromosomes Cancer 2000;28:145-2.
193. Henry I, Jeanpierre M, Couillin P, et al. Molecular definition of the 11p15.5 region involved in Beckwith-Wiedemann syndrome and probably in predisposition to adrenocortical carcinoma. Hum Genet 1989;81:273-7.
194. Gicquel C, Bertagna X, Schneid H, et al. Rearrangements at the 11p15 locus and overexpression of insulin-like growth factor-II gene in sporadic adrenocortical tumors. J Clin Endocrinol Metab 1994;78:1444-53.
195. Sasaki H, Ishihara K, Kato R. Mechanisms of Igf2/H19 imprinting: DNA methylation, chromatin and long-distance gene regulation. J Biochem 2000;127:711-5.
196. Giordano TJ, Thomas DG, Kuick R, et al. Distinct transcriptional profiles of adrenocortical tumors uncovered by DNA microarray analysis. Am J Pathol 2003;162:521-31.
197. de Fraipont F, El Atifi M, Cherradi N, et al. Gene expression profiling of human adrenocortical tumors using complementary deoxyribonucleic acid microarrays identifies several candidate genes as markers of malignancy. J Clin Endocrinol Metab 2005;90:1819-29.
198. de Reyniès A, Assié G, Rickman DS, et al. Gene expression profiling reveals a new classification of adrenocortical tumors and identifies molecular predictors of malignancy and survival. J Clin Oncol 2009;27:1108-15.
199. Giordano TJ, Kuick R, Else T, et al. Molecular classification and prognostication of adrenocortical tumors by transcriptome profiling. Clin Cancer Res 2009;15:668-76.
200. Plaska SW, Liu CJ, Lim JS, et al. Targeted RNAseq of formalin-fixed paraffin-embedded tissue to differentiate among benign and malignant adrenal cortical tumors. Horm Metab Res 2020;52: 607-13.
201. Rechache NS, Wang Y, Stevenson HS, et al. DNA methylation profiling identifies global methylation differences and markers of adrenocortical tumors. J Clin Endocrinol Metab 2012; 97:E1004-13.
202. Barreau O, Assié G, Wilmot-Roussel H, et al. Identification of a CpG island methylator phenotype in adrenocortical carcinomas. J Clin Endocrinol Metab 2013;98:E174-84.
203. Lippert J, Altieri B, Morrison B, et al. Prognostic role of targeted methylation analysis in paraffin-embedded samples of adrenocortical carcinoma. J Clin Endocrinol Metab 2022;107:2892-9.
204. Duregon E, Volante M, Bollito E, et al. Pitfalls in the diagnosis of adrenocortical tumors: a lesson from 300 consultation cases. Hum Pathol 2015;46:1799-807.
205. Saeger W, Mohren W, Behrend M, Iglauer P, Wilczak W. Sarcomatoid adrenal carcinoma: case report with contribution to pathogenesis. Endocr Pathol 2017;28:139-45.

206. Akgul M, Williamson SR, Ertoy D, et al. Diagnostic approach in TFE3-rearranged renal cell carcinoma: a multi-institutional international survey. J Clin Pathol 2021;74:291-9.
207. Alshaikh OM, Laframboise S, Asa SL, Clarke B, Mete O, Ezzat S. Malignant ovarian steroid cell tumor causing severe hyperandrogenism: case report and review of the literature. AACE Clinical Case Rep. 2017;3:e269-74.
208. Villelli NW, Jayanti MK, Zynger DL. Use and usefulness of adrenal core biopsies without FNA or on-site evaluation of adequacy: a study of 204 cases for a 12-year period. Am J Clin Pathol 2012;137:124-31.
209. Cantley RL. Approach to fine needle aspiration of adrenal gland lesions. Adv Anat Pathol 2022;29:373-9.
210. Fisher SB, Habra MA, Chiang YJ, et al. Comparative performance of the 7th and 8th editions of the American Joint Committee on Cancer Staging manual for adrenocortical carcinoma. World J Surg 2020;44:544-51.
211. Kostiainen I, Hakaste L, Kejo P, et al. Adrenocortical carcinoma: presentation and outcome of a contemporary patient series. Endocrine 2019;65:166-74.
212. Poorman CE, Ethun CG, Postlewait LM, et al. A novel T-stage classification system for adrenocortical carcinoma: proposal from the US Adrenocortical Carcinoma Study Group. Ann Surg Oncol 2018;25:520-7.
213. Abdel-Rahman O. Revisiting the AJCC staging system of adrenocortical carcinoma. J Endocrinol Invest 2022;45:89-94.
214. Berruti A, Fassnacht M, Haak H, et al. Prognostic role of overt hypercortisolism in completely operated patients with adrenocortical cancer. Eur Urol 2014;65:832-8.
215. Vanbrabant T, Fassnacht M, Assie G, Dekkers OM. Influence of hormonal functional status on survival in adrenocortical carcinoma: systematic review and meta-analysis. Eur J Endocrinol 2018;179:429-36.
216. Luconi M, Cantini G, van Leeuwaarde RS, et al. Prognostic value of microscopic tumor necrosis in adrenal cortical carcinoma. Endocr Pathol 2023;34:224-33.
217. Baechle JJ, Marincola Smith P, Solórzano CC, et al. Cumulative GRAS score as a predictor of survival after resection for adrenocortical carcinoma: analysis from the U.S. Adrenocortical Carcinoma Database. Ann Surg Oncol 2021;28:6551-61.
218. Beuschlein F, Weigel J, Saeger W, et al. Major prognostic role of Ki67 in localized adrenocortical carcinoma after complete resection. J Clin Endocrinol Metab 2015;100:841-9.
219. Papathomas TG, Pucci E, Giordano TJ, et al. An international Ki67 reproducibility study in adrenal cortical carcinoma. Am J Surg Pathol 2016;40:569-76.
220. Yamazaki Y, Nakamura Y, Shibahara Y, et al. Comparison of the methods for measuring the Ki-67 labeling index in adrenocortical carcinoma: manual versus digital image analysis. Hum Pathol 2016;53:41-50.
221. Brondani VB, Lacombe AM, Mariani BM, et al. Low protein expression of both ATRX and ZNRF3 as novel negative prognostic markers of adult adrenocortical carcinoma. Int J Mol Sci 2021; 22:1238.
222. Yan X, Guo ZX, Yu DH, et al. Identification and validation of a novel prognosis prediction model in adrenocortical carcinoma by integrative bioinformatics analysis, statistics, and machine learning. Front Cell Dev Biol 2021;9:671359.
223. Mohan DR, Lerario AM, Else T, et al. Targeted assessment of g0s2 methylation identifies a rapidly recurrent, routinely fatal molecular subtype of adrenocortical carcinoma. Clin Cancer Res 2019;25:3276-88.
224. Soon PS, Tacon LJ, Gill AJ, et al. miR-195 and miR-483-5p identified as predictors of poor prognosis in adrenocortical cancer. Clin Cancer Res 2009;15:7684-92.
225. Volante M, Terzolo M, Fassnacht M, et al. Ribonucleotide reductase large subunit (RRM1) gene expression may predict efficacy of adjuvant mitotane in adrenocortical cancer. Clin Cancer Res 2012;18:3452-61.
226. Germano A, Rapa I, Volante M, et al. RRM1 modulates mitotane activity in adrenal cancer cells interfering with its metabolization. Mol Cell Endocrinol 2015;401:105-10.
227. Ronchi CL, Sbiera S, Volante M, et al. CYP2W1 is highly expressed in adrenal glands and is positively associated with the response to mitotane in adrenocortical carcinoma. PLoS One 2014;9:e105855.
228. Germano A, Saba L, De Francia S, et al. CYP11B1 has no role in mitotane action and metabolism in adrenocortical carcinoma cells. PLoS One 2018;13:e0196931.
229. Altieri B, Sbiera S, Herterich S, et al. Effects of germline CYP2W1*6 and CYP2B6*6 single nucleotide polymorphisms on mitotane treatment in adrenocortical carcinoma: a multicenter ENSAT study. Cancers (Basel) 2020;12:359.
230. Araujo-Castro M, Pascual-Corrales E, Molina-Cerrillo J, Alonso-Gordoa T. Immunotherapy in adrenocortical carcinoma: predictors of response, efficacy, safety, and mechanisms of resistance. Biomedicines 2021;9:304.

231. McAteer JP, Huaco JA, Gow KW. Predictors of survival in pediatric adrenocortical carcinoma: a Surveillance, Epidemiology, and End Results (SEER) program study. J Pediatr Surg 2013; 48:1025-31.
232. Ribeiro RC, Sandrini F, Figueiredo B, et al. An inherited p53 mutation that contributes in a tissue-specific manner to pediatric adrenal cortical carcinoma. Proc Natl Acad Sci U S A 2001;98:9330-5.
233. Pinto EM, Chen X, Easton J, et al. Genomic landscape of paediatric adrenocortical tumours. Nat Commun 2015;6:6302.
234. Zambaiti E, Duci M, De Corti F, et al. Clinical prognostic factors in pediatric adrenocortical tumors: a meta-analysis. Pediatr Blood Cancer 2021;68:e28836.
235. Pinto EM, Zambetti GP, Rodriguez-Galindo C. Pediatric adrenocortical tumours. Best Pract Res Clin Endocrinol Metab 2020;34:101448.
236. Stratakis CA. Cushing syndrome caused by adrenocortical tumors and hyperplasias (corticotropin- independent Cushing syndrome). Endocr Dev 2008;13:117-32.
237. Vuralli D, Gönç N, Özön A, et al. Feminizing adrenocortical tumors as a rare etiology of isosexual/contrasexual pseudopuberty. J Clin Res Pediatr Endocrinol 2022;14:17-28.
238. Dehner LP, Hill DA. Adrenal cortical neoplasms in children: why so many carcinomas and yet so many survivors? Pediatr Dev Pathol 2009; 12:284-91.
239. Riedmeier M, Thompson LD, Molina CA, et al. Prognostic value of the Weiss and Wieneke (AFIP) scoring systems in pediatric ACC—a mini review. Endocr Relat Cancer 2023;30:e220259.
240. Wieneke JA, Thompson LD, Heffess CS. Adrenal cortical neoplasms in the pediatric population: a clinicopathologic and immunophenotypic analysis of 83 patients. Am J Surg Pathol 2003; 27:867-81.
241. Martins-Filho SN, Almeida MQ, Soares I, et al. Clinical impact of pathological features including the Ki-67 labeling index on diagnosis and prognosis of adult and pediatric adrenocortical tumors. Endocr Pathol 2021;32:288-300.
242. Jangir H, Ahuja I, Agarwal S, et al. Pediatric adrenocortical neoplasms: a study comparing three histopathological scoring systems. Endocr Pathol 2023;34:213-23.
243. Riedmeier M, Decarolis B, Haubitz I, et al. Assessment of prognostic factors in pediatric adrenocortical tumors: a systematic review and evaluation of a modified S-GRAS score. Eur J Endocrinol 2022;187:751-63.
244. Picard C, Orbach D, Carton M, et al. Revisiting the role of the pathological grading in pediatric adrenal cortical tumors: results from a national cohort study with pathological review. Mod Pathol 2019;32:546-59.
245. Guntiboina VA, Sengupta M, Islam N, et al. Diagnostic and prognostic utility of SF1, IGF2 and p57 immunoexpression in pediatric adrenal cortical tumors. J Pediatr Surg 2019;54:1906-12.
246. Gupta N, Rivera M, Novotny P, Rodriguez V, Bancos I, Lteif A. Adrenocortical carcinoma in children: a clinicopathological analysis of 41 patients at the Mayo clinic from 1950 to 2017. Horm Res Paediatr 2018;90:8-18.
247. Michalkiewicz E, Sandrini R, Figueiredo B, et al. Clinical and outcome characteristics of children with adrenocortical tumors: a report from the International Pediatric Adrenocortical Tumor Registry. J Clin Oncol 2004;22:838-45.

7 HEREDITARY PREDISPOSITION TO ADRENAL CORTICAL LESIONS

Adrenal proliferations (including nodular diseases and tumors) can occur in the context of familial or syndromic diseases, with a wide variety of clinical presentations and molecular backgrounds (Table 7-1). The main features and morphofunctional characteristics of pathologic lesions related to each familial syndrome are depicted in the specific chapters describing adrenal cortical nodular diseases (chapter 4), adenoma (chapter 5), and carcinoma (chapter 6). The Beckwith-Wiedemann syndrome is specifically described in chapter 2 dealing with adrenal congenital anomalies.

Beyond the pathologic and clinical features of these diseases, the identification of a germline mutation is of paramount importance for patient management and follow-up. The implementation of next-generation sequencing technologies in molecular genetics laboratories has built a new era of genetic testing for adrenal tumors, favoring the rapid sequencing of a large number of genes in a single assay. The knowledge gained by this approach has led to changes in the recommendations for genetic testing at initial diagnosis.

As detailed in this chapter, genetic testing should be considered for patients diagnosed with rare adrenal lesions that have a strong association with genetic predisposition, such as primary macronodular adrenal disease (i.e., screening of *AMRC5)* and primary pigmented micronodular adrenocortical disease (screening of *PRKAR1A*). Patients with early-onset or severe primary hyperaldosteronism should be screened for genes responsible for familial forms of the disease. All patients with adrenal cortical carcinoma should be tested for the presence of germline *TP53* mutations or microsatellite instability, irrespective of age of onset and the clinical presentation.

In all cases, the identification of a germline mutation strongly modifies clinical surveillance and prompts genetic testing in relatives. Therapeutic strategies in adrenal cortical carcinoma are influenced by the genetic status, with a potential benefit of immunotherapy in mismatch repair-deficient tumors and the indication for limiting radio- and chemotherapy in Li-Fraumeni-associated tumors.

FAMILIAL PRIMARY HYPERCORTISOLISM ASSOCIATED WITH ADRENAL CORTICAL PROLIFERATIONS

Bilateral adrenal cortical nodules associated with adrenocorticotropic hormone (ACTH)-independent cortisol secretion represent a heterogeneous group of diseases. They are divided into two subgroups: primary bilateral micronodular adrenal cortical disease and primary bilateral macronodular adrenal cortical disease. They do not constitute two different stages of a single disease but rather are distinct entities based on their genetic characteristics, biochemical and clinical profiles, and pathologic features (see chapter 4).

Primary Bilateral Macronodular Adrenal Cortical Disease

This disease is characterized by bilateral adrenal enlargement due to multiple macroscopic (i.e., exceeding 1 cm) nodules enriched in clear cells. Despite the homogeneous pathologic features, *primary bilateral macronodular adrenal cortical disease* has a complex pathogenesis, with several molecular mechanisms involved, including some that are responsible for genetic susceptibility. This disease may be encountered in the spectrum of multiple hereditary tumor syndromes, such as multiple endocrine neoplasia 1 (MEN1), hereditary leiomyomatosis and renal cell carcinoma syndrome (HLRCC),

Table 7-1

HEREDITARY SYNDROMES ASSOCIATED WITH ADRENAL CORTICAL PROLIFERATIONS

Syndrome	Gene(s)	Adrenal Cortical Phenotype	Relative Risk of Adrenal Cortical Cancer	Main Other Associated Features
Beckwith-Wiedemann syndrome[a]	*IGF2* locus	Cysts, adenoma (carcinoma in case reports)	<1%	Carcinomas in childhood, Wilms tumor, hepatoblastoma, rhabdomyosarcoma, neuroblastoma
Primary macronodular adrenal disease (excluding MEN1[b], HLRCC, and FAP)	*ARMC5* (*PDE11A* rare)	Primary macronodular adrenal disease	<1%	—
Primary micronodular adrenal disease	*PDE11A, PDE8B, PRKACA*	Isolated micronodular adrenal disease	<1%	—
Carney complex and isolated familial PPNAD	*PRKAR1A* (*PDE11A, PDE8B* rare)	PPNAD (carcinoma in case reports)	<1%	Pituitary and thyroid tumors, cardiac myxomas, peripheral nerve sheath tumors
Familial primary hyperaldosteronism (types I to IV and other variants)	*CYP11B1-CYP11B2; CLCN2; KCNJ5; CACN1AH; CACN1AD*[a]	Bilateral micro- and macronodular disease and adenoma	<1%	Seizure and neurologic abnormalities[a]
Multiple endocrine neoplasia type 1 (MEN1)	*MEN1*	Primary macronodular adrenal disease, adenoma, carcinoma	<2%	Parathyroid hyperplasia/adenoma, pituitary tumors, foregut neuroendocrine tumors
Familial adenomatous polyposis (FAP)	*APC*	Adenoma, primary macronodular adrenal disease and carcinoma	<1%	Colon carcinoma, duodenal adenoma, thyroid carcinoma
Hereditary leiomyomatosis and renal cell carcinoma (HLRCC)	*FH*	Primary macronodular adrenal disease (adenoma and carcinoma in case reports)	<1%	Multiple cutaneous leiomyomas, uterine leiomyoma, papillary renal cell carcinoma, pheochromocytoma
Birt-Hogg-Dube syndrome	*FLCN*	Oncocytic adenoma (in case reports)	<1%	Skin hamartomas, pulmonary cysts, oncocytoid renal neoplasms (oncocytoma and renal cell carcinoma)
Li-Fraumeni syndrome	*TP53*	Carcinoma	3-5% (adults) 50-80% (children)	Gliomas, breast and lung carcinomas, sarcomas, leukemias, choroid plexus tumor
Lynch syndrome	*MSH2, MLH1, PMS2, MSH6, EPCAM*	Carcinoma	3% (adults)	Colorectal, endometrial, small bowel, urothelial, pancreatic, prostatic, sebaceous carcinomas
Neurofibromatosis type 1	*NF1*	Carcinoma (in children)	<1%	Gliomas, malignant peripheral nerve sheet tumors, benign neural tumors
BRCA2	*BRCA2*	Carcinoma	<1%	Breast, ovarian, prostate, pancreatic, gastric cancers

[a]See chapter 2.
[b]MEN1 = multiple endocrine neoplasia 1; HLCC = hereditary leiomyomatosis and renal cell carcinoma; FAP = familial adenomatous polyposis; PPNAD = primary pigmented micronodular adrenal disease.

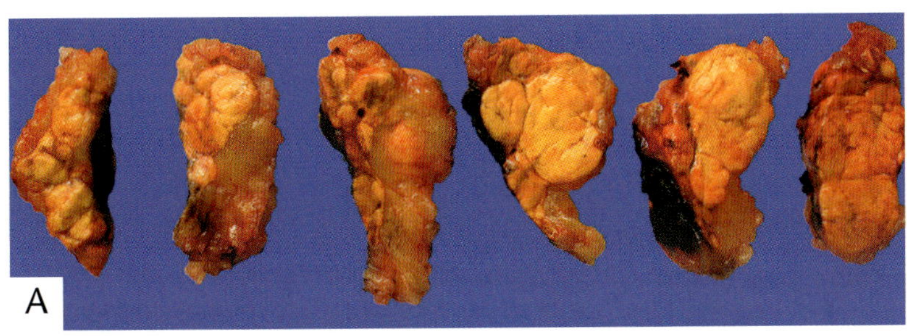

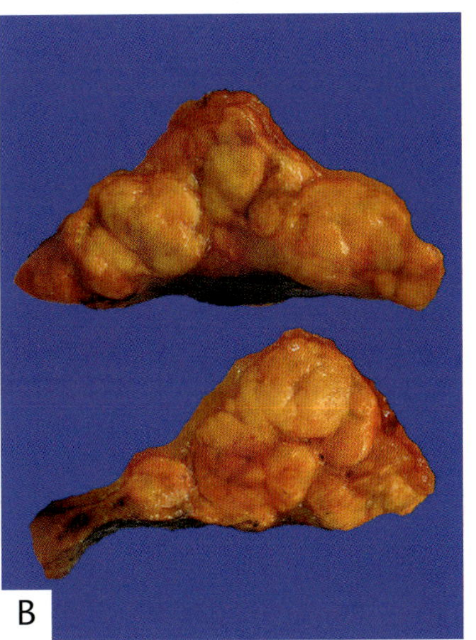

Figure 7-1

PRIMARY BILATERAL MACRONODULAR ADRENAL CORTICAL DISEASE IN A PATIENT WITH A KNOWN *ARMC5* MUTATION AND SUBCLINICAL CUSHING SYNDROME

A: Multiple adrenal cortical nodules of varying size are present throughout the adrenal gland (representative sections).

B: Higher magnification of two additional sections showing that the residual adrenal cortex away from nodules is atrophic, consistent with autonomous function of the nodular areas.

and familial adenomatous polyposis (FAP). It is also detected in the context of the nonhereditary McCune-Albright syndrome, a germline disorder caused by postzygotic somatic *GNAS* mutation mosaicism (1).

Heterozygous germline mutations of the armadillo repeat containing 5 (*ARMC5*) gene are the most frequent molecular alterations associated with hypercortisolism caused by primary bilateral macronodular adrenal cortical disease (fig. 7-1). Their prevalence in different series varies from 25 to 55 percent (2–4). Around 50 percent of patients' relatives and 30 percent of patients with apparently sporadic primary bilateral macronodular adrenal cortical disease have germline *ARMC5* mutations. Patients with *ARMC5*-mutated primary bilateral macronodular adrenal cortical disease display overt clinical and biochemical Cushing syndrome more frequently than patients with wild-type disease; they have larger adrenal lesions and more nodules (3). Rare cases of adrenal cortical carcinoma occur in patients with germline *ARMC5* mutations (fig. 7-2). Interestingly, *ARMC5* germline variants have also been detected in patients with MEN1 syndrome, but in that setting there is no significant association with the presence of adrenal lesions (5).

Despite the high prevalence of germline *ARMC5* alterations in primary bilateral macronodular adrenal cortical disease, the functional impact of these mutations remains largely unclear (6). Transcriptomic data have shown that primary bilateral macronodular adrenal cortical disease cases can be classified into two groups depending on the *ARMC5* genotype (2). *ARMC5* silencing in *ARMC5* wild-type cells isolated from primary bilateral macronodular adrenal cortical disease decreased steroidogenesis-related genes and increased proliferative capacity, without affecting cell viability; in contrast, *ARMC5* overexpression induced cell death in *ARMC5*-mutated cell cultures (7). Reduced steroidogenesis in *ARMC5*-mutant in vitro models is, however, in contrast to the association between *ARMC5*-inactivating mutations and higher cortisol levels in patients. The relative inefficiency to produce cortisol in *ARMC5*-mutated cells is probably compensated by the larger adrenal mass observed in *ARMC5*-mutated patients. Somatic *ARMC5* mutations have been detected in nodules of primary bilateral macronodular adrenal cortical disease in addition to the germline mutations, supporting a "two-hit" model for *ARMC5* as a tumor suppressor gene (2).

In addition to germline *ARMC5* mutations and syndrome-related primary bilateral macronodular adrenal cortical diseases (MEN1, FAP, or HLRCC), germline variants of *PDE11A* are also associated with increased genetic susceptibility to this disease. In 46 patients with primary bilateral macronodular adrenal cortical disease, germline *PDE11A* variants were detected at a frequency significantly higher (28 percent) than in 192 controls (7.2 percent) (8). In this

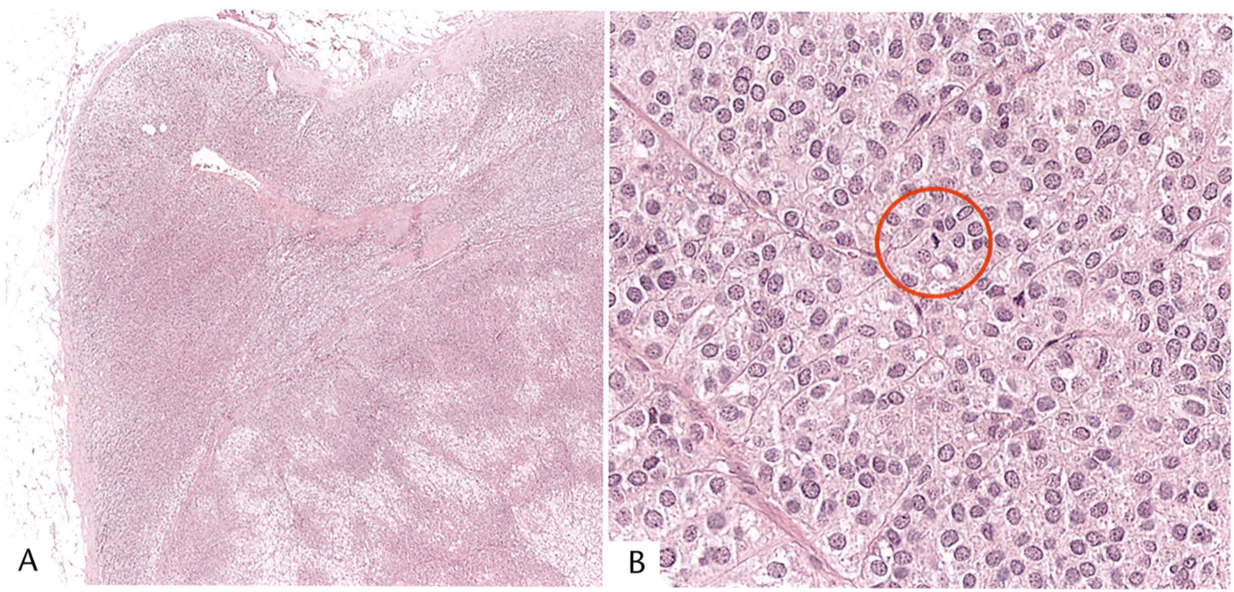

Figure 7-2

ADRENAL CORTICAL TUMOR IN A PATIENT HARBORING *ARMC5* GERMLINE VARIANT

The adrenal mass (A) is composed of a peripheral benign macronodular component (A, upper left), and is associated with a central component made of eosinophilic cells (B, lower right) showing malignancy-related features, including increased mitotic activity (B, red circle) for a total Weiss score of 4.

same study, functional data provided evidence that *PDE11A* variants detected in patients were significantly effective in increasing c-AMP intracellular levels and activity.

Primary Bilateral Micronodular Adrenal Cortical Disease

This disease is further divided into main entities, *primary pigmented micronodular adrenal cortical disease (PPNAD)* and *isolated micronodular adrenocortical disease*. PPNAD can occur as an isolated adrenal lesion or in the context of Carney complex (9). These conditions can present as sporadic or familial forms, but familial PPNAD as part of the Carney complex is the most common presentation.

Carney complex was first described in 1985 as a sporadic or familial disease with autosomal dominant inheritance (10). It is a rare multiple neoplasia syndrome characterized by cardiac, endocrine, cutaneous, and neural tumors, as well as a variety of pigmented lesions of the skin and mucosae (11). PPNAD is the most common endocrine manifestation of Carney complex, occurring in 60 percent of cases. Up to 90 percent of PPNAD cases develop in association with other manifestations of Carney complex, whereas only 10 percent of PPNAD patients have isolated adrenal disease.

In the familial setting, PPNAD is frequently associated with germline pathogenetic variants in genes encoding proteins acting in the PKA signaling pathway. PPNAD is significantly associated with pathogenic variants of the *PRKAR1A* gene: more than 80 percent of patients affected by PPNAD or Carney complex have inactivating *PRKAR1A* gene mutations, detectable both in leukocytes and adrenal lesions (12,13). This gene encodes the regulatory subunit 1A of protein kinase A, a central component of the cAMP signaling pathway. Most mutations are base substitutions, small deletions, and insertions, and most are unique, with a few identified in more than three unrelated pedigrees (14). Since loss of the unmutated allele is not detected in adrenal tissues of affected individuals, it has been postulated that germline *PRKAR1A* gene mutations are sufficient to augment PKA activity, leading to tumorigenesis. Approximately 70 percent of *PRKAR1A*-mutated PPNAD cases are familial; the remainder carry de novo mutations, and the penetrance approximates 100 percent (9).

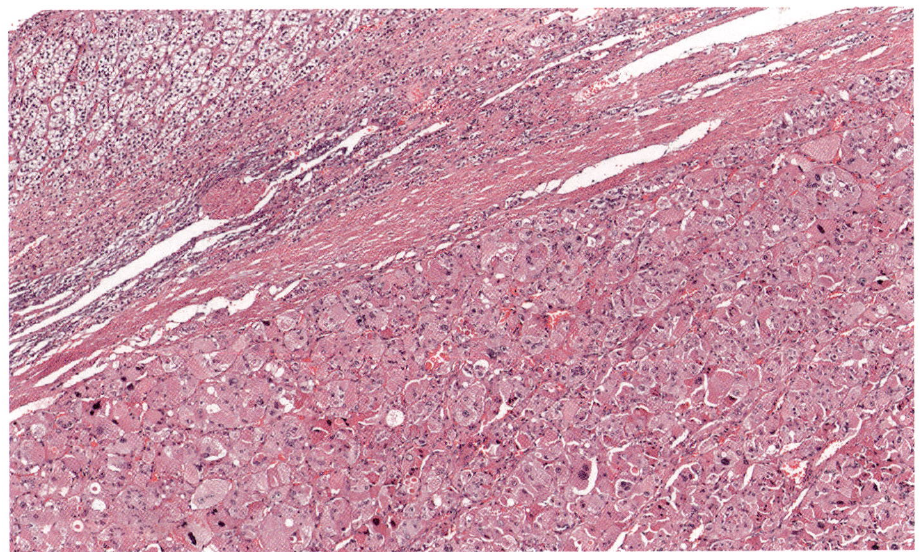

Figure 7-3

ADRENAL CORTICAL CARCINOMA IN A PATIENT WITH PRIMARY PIGMENTED MICRONODULAR ADRENOCORTICAL DISEASE (PPNAD)

Highly atypical oncocytic cells with a fibrous capsule separate the tumor from the peritumoral residual adrenal cortex and present as a tiny eosinophilic micronodule in the pericapsular location.

Macronodules may occur in PPNAD and classification as typical adenoma may result in misdiagnosis. Adrenal cortical carcinomas arising in PPNAD in the context of Carney complex are extremely rare but have been reported, with rapid local recurrence or metastasis (fig. 7-3) (15,16). This suggests that adrenal cortical carcinomas in Carney complex develop from adrenal adenomas in a multistep model of tumor progression through benign nodule to cancer, but the exact mechanisms remain to be elucidated (17). One patient had both *PRKAR1A* mutation and a heterozygous *TP53* polymorphic substitution (16), suggesting multifactorial pathogenesis. In *PRKAR1A* wild-type cases, germline *PDE11A* and *PDE8B* variants or germline *PRKACA* duplication have been identified (18).

The co-occurrence of *PDE11A* and *PRKAR1A* germline variants in Carney complex patients is associated with a higher frequency of PPNAD, male sex, and the occurrence of large cell calcifying Sertoli cell tumors, suggesting that *PDE11A* is a genetic modifying factor for the development of testicular and adrenal tumors in patients with germline *PRKAR1A* mutations (19). Moreover, *ARMC5* variants are also described in PPNAD (20). *ARMC5* polymorphic variants may interfere with the onset and clinical severity of hypercortisolism in *PRKAR1A*-mutated PPNAD cases (21). Finally, germline variants of the *PRKACB* gene, another player in PKA signaling, are associated with unusual phenotypes in subjects with PPNAD who did not have other PKA gene-related defects (22).

At variance with PPNAD, the genetic characteristics of isolated micronodular adrenal cortical disease are as yet poorly defined. A genetic predisposition is postulated by the very early onset of the disease and bilateral involvement. Germline *PDE11A* and *PDE8B* mutations have been identified (23,24), and germline *PRKACA* amplification has been described in two cases (25).

FAMILIAL ALDOSTERONISM SYNDROMES

There are four forms of familial primary aldosterone excess based on the underlying genetic defects. *Familial aldosteronism type I* (also known as *glucocorticoid remediable aldosteronism*) is an autosomal dominant disease that accounts for 0.5 to 1.0 percent of primary hyperaldosteronism in the adult hypertensive population. Its phenotype is characterized by aldosterone suppression and hypertension that resolve with dexamethasone treatment. Familial hyperaldosteronism type I is usually due to bilateral nodular disease, but in rare cases, to adrenal adenomatous nodules. The molecular defect is a chimeric *CYP11B1/CYP11B2* hybrid gene (26) that includes the promoter region of *CYP11B1* and a large part of the *CYP11B2* coding sequence, making aldosterone production dependent on ACTH regulation.

Familial aldosteronism types II, III, and IV are not responsive to glucocorticoids. *Familial hyperaldosteronism type II* accounts for 1.2 to 6.0 percent of primary hyperaldosteronism in the adult hypertensive population and is clinically and biochemically indistinguishable from sporadic forms of primary hyperaldosteronism. There are no notable differences in age of disease onset, or plasma renin, serum aldosterone, or potassium levels compared to sporadic disease. It is only diagnosed based on the presence of two or more affected family members. It may be associated with both adrenal cortical adenoma and bilateral adrenal nodular disease, with a high phenotypic variability, even within the same family. Familial hyperaldosteronism type II is caused by different gain-of-function mutations in the *CLCN2* gene, encoding for chloride channel protein 2, which is expressed in adrenal glomerulosa cells (27,28). The penetrance of this genetic disorder is suggested to be incomplete.

Familial hyperaldosteronism type III is a rare and usually severe form of primary aldosteronism associated with massive bilateral adrenal nodular disease, caused by mutations in the *KCNJ5* gene (29), coding for the G protein-activated inward rectifier potassium channel GIRK4. The most common clinical presentation includes childhood onset, severe hypertension associated with hypokalemia, aldosterone levels greater than 100 ng/dL, and renin activity less than 1.0 ng/mL/h (30–32). Adequate clinical management requires either lifelong mineralocorticoid antagonist therapy or bilateral adrenalectomy.

A wide phenotypic variability is present in type III disease, from spironolactone-responsive hyperaldosteronism to massive adrenal hypertrophy with drug-resistant hypertension. Different genotypes are associated with clinical variability. For example, patients with the G151E variant show a milder phenotype, with less pronounced nodular disease, explained by enhanced sodium conductance of the channel encoded by the mutated gene and high cell lethality (33). Germline *KCNJ5* mutations have been described in 2 out of 26 patients with apparent essential hypertension without primary hyperaldosteronism exhibiting ACTH-dependent aldosterone hypersecretion and responsiveness to treatment with mineralocorticoid receptor antagonists (34). Recently, mosaicism of a *KCNJ5* defect has been described in a young patient with bilateral adrenocortical nodular disease and primary aldosteronism, milder than that usually observed in familial hyperaldosteronism type III (35).

Familial hyperaldosteronism type IV is caused by germline mutations of *CACNA1H* gene, encoding calcium channels, which are expressed in adrenal glomerulosa cells. These mutations are responsible for the diverse clinical presentations of primary hyperaldosteronism and may be associated with either normal-appearing adrenal glands, a single adenoma, or bilateral nodular disease (36,37). Germline mutations of *CACNA1D*, a gene altered by somatic mutations in a subset of aldosterone-secreting adenomas, have been identified in young patients with primary hyperaldosteronism and neuromuscular abnormalities (38).

The spectrum of germline variants of primary aldosteronism also includes other genes. Germline *ARMC5* variants are associated with primary aldosteronism by some authors (39) but not by others (40). Additional genes include *PDE2A, PDE3B* (41), and *ATP2B4* (42). These findings indicate that the overall genetic susceptibility of primary aldosteronism is probably still underestimated.

MULTIPLE ENDOCRINE NEOPLASIA TYPE 1 SYNDROME

Multiple endocrine neoplasia type 1 (MEN1) is a rare autosomal dominant hereditary tumor syndrome with a predisposition to a multitude of endocrine and nonendocrine tumors. It is caused by inactivating mutations of the tumor suppressor gene *MEN1* (43). The syndrome is classically characterized by neoplasms of parathyroid gland, pituitary gland, and duodenal and pancreatic neuroendocrine cells. There is a high degree of penetrance and of clinical presentations, of which parathyroid gland lesions are the most frequent (up to 90 percent of patients by age 40) (44). Other endocrine tumors noted with increased frequency in MEN1 patients are gastric, thymic, and bronchial neuroendocrine tumors.

Adrenal tumors also occur in MEN1 patients. Pheochromocytomas are rare (less than 1 percent) whereas adrenal cortical lesions may be detected in more than 50 percent of patients, depending on the imaging modality used for screening (45). A large cohort study of over 700 MEN1 patients followed by the Groupe d'etude des Tumeurs Endocrines (GTE) reported adrenal enlargement in more than 20 percent of cases (46), including tumors larger than 1 cm in about 10 percent of the cohort, and lesions larger than 4 cm in 2 percent. Forty-five patients underwent adrenal surgery, with benign cortical adenomas or nodular disease in 35 cases and 10 adrenal cortical carcinomas (1.4 percent of the cohort) (figs. 7-4–7-6). In two additional cases, adrenal metastases from pancreaticoduodenal neuroendocrine tumors were identified. Adrenal cortical carcinomas were characterized by a size less than 5cm, stage I or II, in 90 percent of cases.

In another study (47), adrenal lesions were identified in 18 of 67 (26.8 percent) MEN1 patients. Among those, bilateral adrenal involvement was recognized in 8 patients. Interestingly, patients with mutations in exons 2 and 10 developed adrenal tumors significantly more often than patients with other mutations. An incidence of adrenal cortical carcinoma similar to the adult population has been described in MEN1 patients younger than 21 years, with a case of carcinoma detected at the age of 3 years (48).

The histology of adrenal cortical carcinoma in MEN1 is not specific and includes histologic variants, such as the sarcomatoid (46) and myxoid variant (49). Data from different series report that adrenal cortical carcinomas may develop after a long follow-up of radiologically stable small adrenal nodules, thus suggesting progression from benign to malignant nodules (46,47).

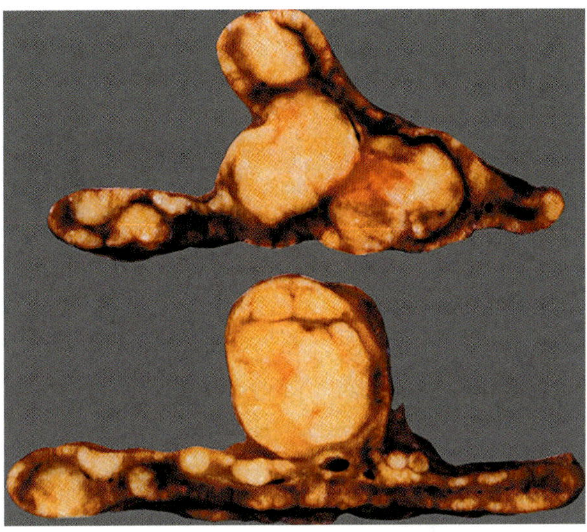

Figure 7-4

ADRENAL CORTICAL NODULAR DISEASE IN A PATIENT WITH MULTIPLE ENDOCRINE NEOPLASIA 1 (MEN1)

Transverse sections of adrenal gland with micro- and macronodular nodular disease. The pale yellow cortical nodules range in size from 1.5 to 0.5 cm to over 1.0 cm. There are a few areas of capsular extrusion.

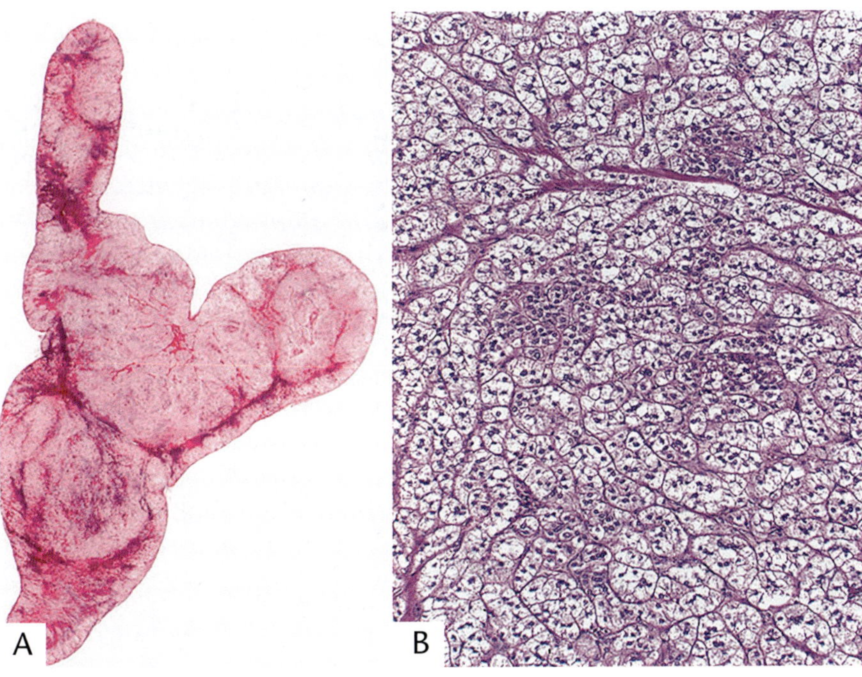

Figure 7-5

ADRENAL CORTICAL NODULAR DISEASE IN A PATIENT WITH MEN1 ASSOCIATED WITH CUSHING SYNDROME

A: Marked adrenal enlargement with irregular expansile cortical nodules merging with adjacent hyperplastic cortex.

B: The nodules consist predominantly of clear cells as in cortical adenomas.

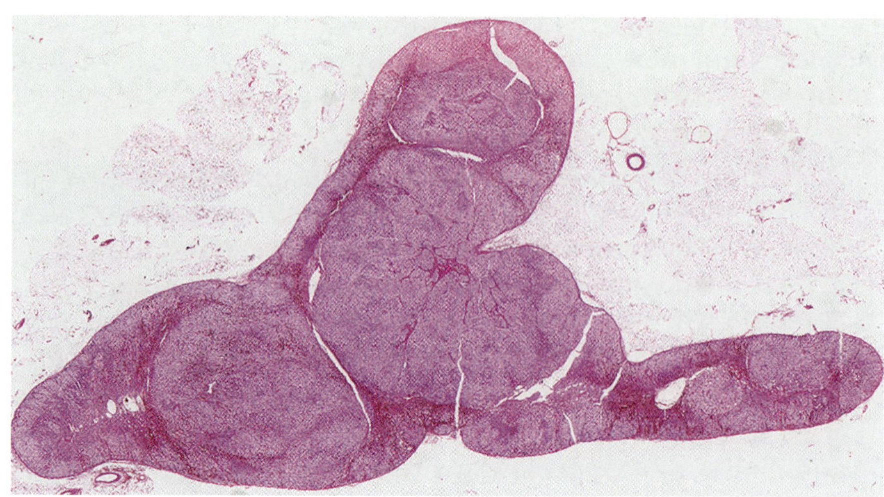

Figure 7-6

ADRENAL CORTICAL NODULAR DISEASE IN A PATIENT WITH MEN1 ASSOCIATED WITH CUSHING SYNDROME

The adrenal gland shows both micronodular and macronodular arrangements.

Most adrenal lesions in MEN1 are hormonally silent; the overall prevalence of endocrine hypersecretion among MEN1 patients with adrenal tumors is about 15 percent (46). Compared to patients with sporadic incidentalomas, MEN1 patients more frequently present with primary hyperaldosteronism (including both benign and malignant adrenal lesions) (46). MEN1 patients may also present with overt ACTH-independent Cushing syndrome, which can be associated with macronodular disease, adenoma, or carcinoma (46,50,51). In a study by Langer et al. (47), however, adrenal cortical carcinomas were more frequently hormonally active than benign lesions, with 3 out of 4 cases associated with hormone hypersecretion (although with no specific information of the type of hormone produced or associated clinical syndrome). Sex hormone secretion can be also detected with special reference to adrenal cortical carcinomas (48,52).

FAMILIAL ADENOMATOUS POLYPOSIS SYNDROME

Familial adenomatous polyposis (FAP) is an autosomal dominant family cancer syndrome caused by germline mutations in the adenomatous polyposis coli (*APC*) gene on chromosome 5q21. It is responsible for about 1 percent of all colorectal adenomas and carcinomas and other neoplasms, including upper digestive tract adenomas and carcinomas, thyroid gland carcinomas (i.e., the cribriform-morular histotype), desmoid tumors, and hepatoblastomas. *APC* encodes a protein that is involved in cell cycle control and downregulation of beta-catenin through the Wnt signaling pathway; mutation results in activation of beta-catenin, a phenomenon that is important in adrenal cortical carcinoma, thus it is no surprising that adrenal cortical lesions are associated with FAP, including nodular disease, adenoma, and carcinoma.

In a study of 107 patients affected by FAP, 13 percent had an adrenal mass 1 cm or larger (bilateral in one case) with lack of clinical evidence of endocrine disturbance or hypertension in all cases. All cases except one (a pheochromocytoma) were classified as benign at histology or based upon clinical and radiologic findings (53). In a more recent clinical chart review of more than 300 patients affected by FAP, adrenal masses were reported in about 16 percent of cases, with adenomas representing up to 80 percent (54). The functional status is heterogeneous, and includes ACTH-independent Cushing syndrome (55), aldosterone secretion, or virilization. Most cases in the Shiroky et al. series (54), however, were nonfunctioning.

Biallelic inactivation of *APC* gene has been demonstrated in tumor DNA of FAP-associated adrenal lesions (56). Somatic alterations in genes other than *APC* may occur, and are associated with the development and functionality of adrenal lesions, such as somatic *KCNJ5* mutations in aldosterone-secreting nodules (57). Adrenal cortical carcinomas may arise from small, clinically indolent lesions after a long follow-up period (58) or may be incidentally discovered at initial clinical workup of FAP-affected patients (59), thus questioning the most appropriate follow-up strategies for adrenal lesions in FAP patients.

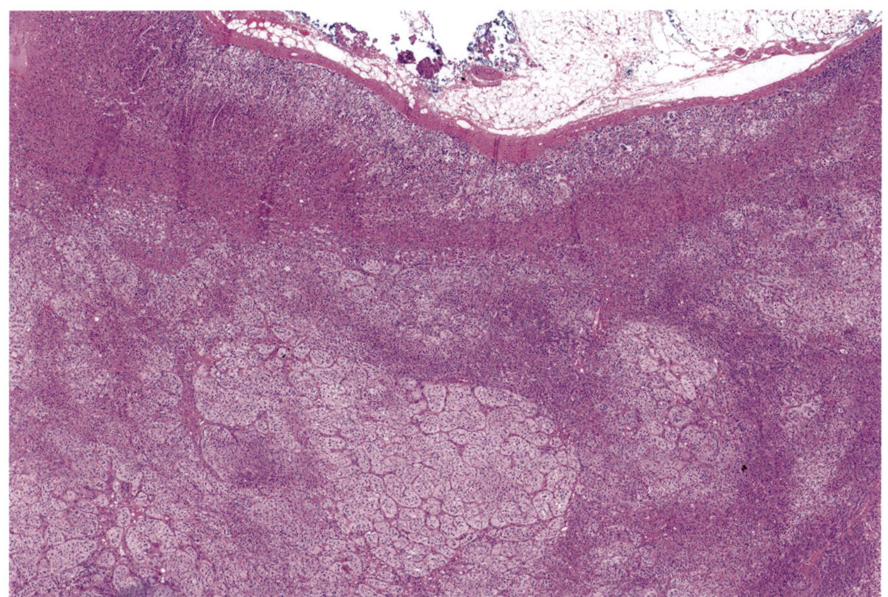

Figure 7-7

ADRENAL CORTICAL ADENOMA IN A PATIENT WITH *FH* GERMLINE MUTATION

This patient had a previous renal cell carcinoma and a Leydig cell tumor. The adenoma shows a nodular aspect without cytologic atypia or other worrisome cellular or architectural features. Preexistent cortex is present at the top of the image.

HEREDITARY LEIOMYOMATOSIS AND RENAL CELL CARCINOMA SYNDROME

Hereditary leiomyomatosis and renal cell carcinoma (HLRCC) syndrome is a rare disease caused by heterozygous germline mutations in the fumarate hydratase (*FH*) gene (60). The presence of multiple cutaneous leiomyomas, symptomatic uterine leiomyoma, and papillary renal cell carcinoma is the most frequent clinical phenotype. However, about 8 percent of patients present with an adrenal lesion, in a minority of cases also associated with ACTH-independent hypercortisolism (61).

Pathologic findings are mostly consistent with macronodular adrenal cortical disease (fig. 7-7). Interestingly, despite the benign nature of the adrenal lesions, 70 percent show abnormal uptake greater than the liver on FDG-PET imaging. This phenomenon is probably due to the peculiar metabolic profile of *FH*-deficient cells that have impairment of the Krebs cycle and a metabolic shift towards glycolysis (62).

An adrenal nodule diagnosed as adenoma was recently described in a HLRCC patient with multiple cutaneous leiomyomas, a cardiac myxoma, and subclinical Cushing syndrome. Expression patterns of steroidogenic enzymes in this lesion revealed distinctive findings that were similar to the findings of primary pigmented nodular adrenal cortical disease (63). Adrenal cortical carcinomas in HLRCC are exceptional and described in case reports only (64,65).

BIRT-HOGG-DUBE SYNDROME

Birt-Hogg-Dube (BHD) syndrome is an autosomal dominant condition resulting from germline mutations in the folliculin (*FLCN*) gene, a tumor suppressor gene located on chromosome 17p11.2. Most common lesions include benign skin hamartomas (fibrofolliculomas, trichodiscomas, and angiofibromas), pulmonary cysts, and recurrent pneumothoraces. Most seriously, there is an increased risk of renal tumors (66). Benign adrenal tumors are on record, expanding the spectrum of BHD-related tumors. The oncocytic variant of adrenal cortical adenoma predominates (67,68), which is consistent with the oncocytic/hybrid features encountered in renal cell tumors in the syndrome (69).

The adrenal manifestations of BHD have been expanded recently by the description of an adrenal cortical carcinoma occurring in a patient with a germline likely pathogenic variant in the *FLCN* gene (p.Gln232Ter) (70). The tumor was characterized by extensive invasion (vascular, capsular, and extra-adrenal) and high mitotic index.

LI-FRAUMENI SYNDROME

Li-Fraumeni syndrome is an autosomal dominant disease caused by germline mutations in the *TP53* gene (71). It is characterized by a strong familial aggregation of malignant neoplasms, early onset of tumors, and a wide

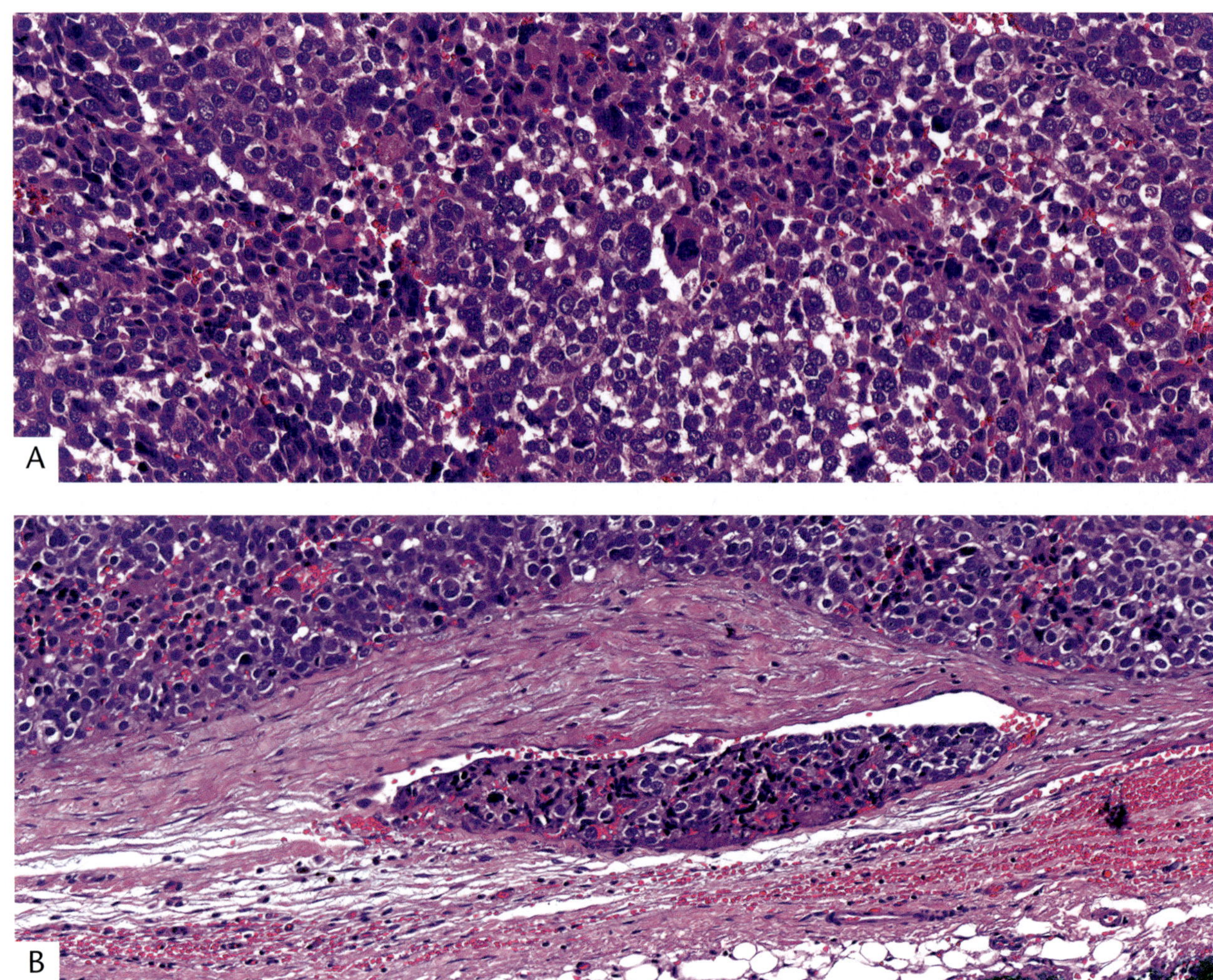

Figure 7-8

ADRENAL CORTICAL CARCINOMA IN A YOUNG PATIENT WITH LI-FRAUMENI SYNDROME

Nine-month-old boy with 8-cm right adrenal tumor. No signs of hormonal overproduction. Resection specimen shows nuclear pleomorphism and increased mitotic activity, in addition to atypical mitoses (A), vasoinvasive growth (B), and necrosis (not shown), leading to Wieneke score of 5, compatible with adrenal cortical carcinoma.

variety of histologies. Core Li-Fraumeni cancers include: soft tissue sarcomas, osteosarcomas, adrenal cortical carcinoma, central nervous system tumors, and breast cancer in patients under the age of 31 years (72). Family history of cancer is often mute, and germline *TP53* mutation carriers bear a wide diversity of clinical presentations, leading to a more extended concept of the *TP53*-related cancer syndrome (73). De novo mutations account for up to 20 percent of cases (74).

Adrenal cortical carcinomas in Li-Fraumeni syndrome more frequently occur in children. Patients diagnosed with adrenal cortical carcinoma before the age of 18 years have an 80 percent probability of harboring a germline *TP53* mutation, regardless of the family history (75), and 3 to 10 percent of children with Li-Fraumeni syndrome develop adrenal cortical carcinoma (fig. 7-8) (74). Rare pediatric adrenal tumors with morphologic parameters consistent with a benign lesion are also on record (76). In the presence of adrenal cortical carcinoma in children negative for *TP53* germline testing, the presence of mosaic *TP53* alterations should be considered (77).

In the adult population, the prevalence of germline *TP53* mutations is much lower. A 3 percent prevalence has been reported in patients older than 18 years, and in most cases, there was a strong family history of cancer (78). A slightly higher prevalence, up to 5.3 percent, has been described in another series (79).

There is no standardization of screening for adrenal cortical carcinoma in patients affected by Li-Fraumeni syndrome. Cancers at an earlier stage of disease and with improved long-term survival have been identified using a combined multimodal radiologic and biochemical screening program (80). No studies specifically compared the clinical and pathologic features of adrenal cortical carcinomas developing in the Li-Fraumeni syndrome or as sporadic forms, but *TP53*-related cases do not seem to possess peculiar morphologic or biologic characteristics, including hormone secretion properties.

Special attention has been dedicated to the *TP53* p.R337H variant. Since its first description in an adrenal cortical carcinoma of a female patient at the age of 3 years, it was questioned whether this variant was germline or somatic (81). This same mutation was then identified as germline in a very high percentage of pediatric cases of adrenal cortical tumors from the Paranà region in southern Brazil (82). In these cases, there was no evidence of an increased incidence of cancer in relatives and the few families with other cancers did not fulfill the criteria for Li-Fraumeni syndrome. These same results were also confirmed by other groups (83). In large population studies in south Brazil, the *TP53* p.R337H variant was identified in about 0.3 percent of the population (84,85). Despite the original hypothesis that this variant acts in an adrenal tissue specific manner, other tumors are associated with it in the Brazilian population, such as choroid plexus carcinoma in children (86). In adults, this variant has been linked to other cancer types, such as breast cancer and sarcomas, but with a prevalence rate much lower than for adrenal cortical tumors and choroid plexus tumors in children (87).

LYNCH SYNDROME

Lynch syndrome (also known as *hereditary nonpolyposis colorectal cancer [HNPCC] syndrome)* is an autosomal dominant cancer predisposition attributed to deleterious germline mutations in genes involved in the DNA mismatch repair (MMR) pathway. These include *MLH1, MSH2, MSH6,* and *PMS2;* loss of MLH1 protein expression may be caused also by deletion of *EPCAM* gene (88). Biallelic inactivation of these genes leads to a defect in DNA mismatch repair, resulting in genomic instability that in tumors is expressed as high levels of instability of microsatellite loci.

Adrenal tumors in the Lynch syndrome are almost exclusively adrenal cortical carcinomas. They may occur in more than one member in Lynch families (89). In a study of 94 Lynch syndrome-affected patients, the prevalence of adrenal cortical carcinomas reached 3.2 percent, a percentage similar to that of other Lynch syndrome-associated malignancies, such as colorectal and endometrial cancers. These adrenal cortical carcinoma cases were characterized by loss of MMR proteins but not by microsatellite instability (90). However, in a large pan-cancer study, 43 percent of adrenal cortical carcinomas had an unstable microsatellite pattern, and 10 percent were associated with Lynch syndrome (91). In contrast to these data, a prevalence of 0.47 percent of adrenal cortical carcinomas has been recently described among 634 patients from 220 Lynch families (92). All cases in this latter series harbored mutations in the *MSH2* gene, as for the vast majority of reported cases quoted above, suggesting a genotype-phenotype correlation that needs to be further investigated.

The pathologic and clinical characteristics of adrenal cortical carcinomas associated with Lynch syndrome are poorly defined and need to be evaluated in future studies. In a recent report, a patient with adrenal cortical carcinoma and a family and clinical history consistent with Lynch syndrome was found to harbor concurrent germline mutations in *MSH2* and *RET*, although with no phenotypic characteristics of MEN2A (93); the clinical significance of this observation requires validation.

NEUROFIBROMATOSIS TYPE 1 SYNDROME

Neurofibromatosis type 1 (NF1) is an autosomal dominant disease with an incidence of about 1 in 3,000. It is characterized by multiple café-au-lait spots, intertriginous freckling, and neurofibromas. Cerebral and spinal tumors, skeletal dysplasia, and ophthalmologic abnormalities

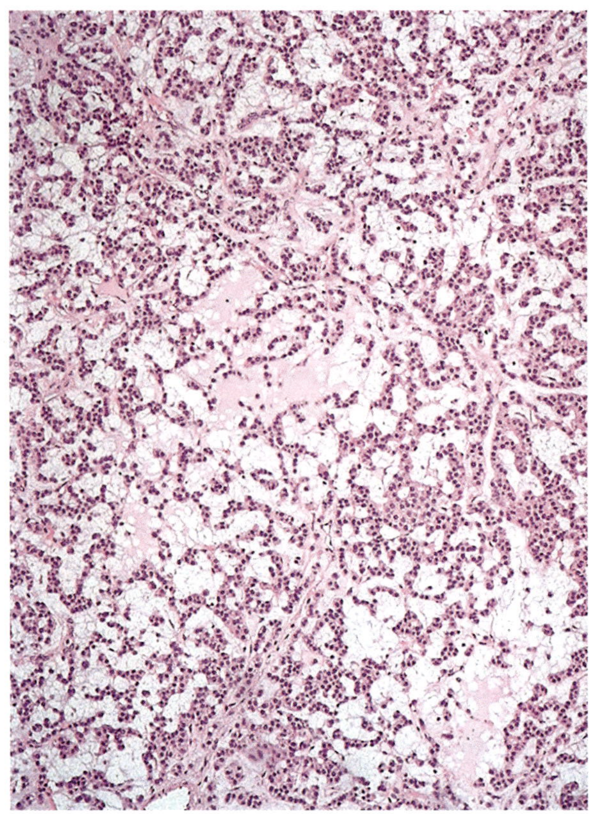

Figure 7-9

ADRENAL CORTICAL CARCINOMA IN A MALE PATIENT WITH *BRCA2* GERMLINE MUTATION

The patient had a previous diagnosis of prostate cancer. The adrenal lesion showed predominant features of the myxoid variant.

are also found (94). NF1 is well known to be associated with pheochromocytoma.

Adrenal cortical lesions in NF1 are rare and are almost invariably adrenal cortical carcinomas. They arise both in children (95) and adults (96). In a study of Danish NF1 patients with a follow-up of over 40 years, 2 adrenal cortical carcinomas were diagnosed among 212 malignant tumors (97). Loss of heterozygosity of the *NF1* locus has been demonstrated in adrenal cortical cancer tissue, proving the pathogenetic role of the *NF1* gene in tumorigenesis (98).

BRCA1 AND BRCA2 SYNDROMES

BRCA1 and *BRCA2 syndromes* are caused by germline mutations in the *BRCA1* and *BRCA2* tumor suppressor genes, respectively, which result in a significant increase in the likelihood of developing particular types of epithelial malignancies, namely breast, ovarian, pancreatic, and prostate cancer, among others. Germline *BRCA1* and *BRCA2* gene mutations are inherited in an autosomal dominant fashion and make up roughly half of the cancer cases related to inherited genetic risk (99). Adrenal cortical carcinomas have not been reported in BRCA1 syndrome, to the best of our knowledge, however they have been reported in *BRCA2*-germline mutation carriers. In one report, loss of heterozygosity of *BRCA2* in adrenal cortical cancer tumoral DNA was detected, suggesting a causal link between the *BRCA2* 8765delAG mutation harbored by the patient and the development of adrenal cancer (100). No description of the pathologic features of the adrenal cortical cancer is available for this patient. A case of adrenal cortical carcinoma in a male patient with germline *BRCA2* mutation is reported by the authors, showing the predominant features of the myxoid variant (fig. 7-9).

By targeted next-generation sequencing analysis, a prevalence of 2.8 percent each of *BRCA1* and *BRCA2* gene mutations was detected in tumor DNA in a series of over 100 adrenal cortical cancers (101). However, no testing was performed in this study to assess the somatic or germline nature of these mutations.

OTHER RARE GERMLINE VARIANTS ASSOCIATED WITH ADRENAL CORTICAL CARCINOMA

Individual case studies/series describe additional germline variants associated with the development of adrenal cortical carcinoma. Given how statistically improbable the concurrence of adrenal cortical carcinoma and pathogenic germline mutations is expected to be, these observations widen the spectrum of hereditary syndromes that may have adrenal cortical carcinoma as a phenotypic manifestation. Four unrelated adrenal cortical carcinomas, all cortisol secreting, were found to harbor *SDHx* mutations (two *SDHA* and two *SDHC*) (102).

BER-related genomic instability in adrenal cortical carcinoma due to germline inactivating mutations in *MUTYH* have been described in 3 percent of adrenal cortical carcinomas (4 out of 136 patients overall analyzed) (103), and this particular genotype is associated with a high tumor mutational burden (104). Pathogenic germline mutations in genes encoding cell cycle checkpoint molecules have also been described in adrenal cortical carcinoma, including *CHEK2* (105) and its activator *ATM* (106).

SCREENING APPROACH FOR FAMILIAL FORMS OF ADRENAL CORTICAL TUMORS

At variance with other endocrine tumors, such as medullary carcinoma, pheochromocytoma, or paraganglioma, specific screening recommendations for familial forms of adrenal cortical tumors are not clearly defined (107). Although several types of hereditary adrenal cortical lesions develop in a suggestive clinical context or in kindreds already characterized at the genetic level (i.e., in MEN1 patients) or in the pediatric population, some hereditary cases, with special reference to adrenal cortical carcinomas, present as clinically sporadic tumors with no suggestive family history, isolated and in adults.

Some rare conditions, such as PPNAD, have a strong familial predisposition associated with very specific and peculiar clinical and pathologic features, strongly indicating a need for genetic counseling in these patients. Other diseases, such as bilateral macronodular adrenal cortical disease, possess less specific characteristics for association to a given familial disease, but are frequently associated with genetic predisposition and are therefore evaluated in an appropriate genetic context. Other diseases occur in an apparently nonspecific clinical and pathologic background. The typical example is adrenal cortical carcinoma, which is associated with genetic predisposition in up to 10 percent of cases. However, due to the rarity of the disease, most patients are at risk for being not correctly identified as developing in a hereditary context.

The pathologic features of adrenal cortical tumors in the setting of familial forms are not specific enough to suggest the presence of a hereditary predisposition. Therefore, a multidisciplinary approach is needed for the complete clinical workup to identify hereditary adrenal cortical tumors. Pathologists are part of the initial screening since for several conditions the presence of specific germline defects may be identified by testing the corresponding proteins.

The clinical recommendation is emerging to screen patients with adrenal cortical carcinomas in the pediatric population for the presence of Li-Fraumeni syndrome and in the adult population for the presence of both Li-Fraumeni and Lynch syndromes. In 2001, modified Li-Fraumeni syndrome criteria were coded to identify patients likely to carry *TP53* mutations. According to these criteria, a subject with adrenal cortical carcinoma is likely to carry a germline *TP53* mutation regardless of age at diagnosis (108). Because of the high frequency of de novo mutations, *TP53* genetic testing should be performed even in the absence of a suggestive family history. In particular, *TP53* germline testing should be considered prior to radiation or chemotherapy to avoid the risk of subsequent primary cancers (72). The immunohistochemical detection of p53 protein accumulation or total loss is not reliable enough to screen for the presence of *TP53* mutation since the concordance between protein expression and molecular status is low (78).

Lynch syndrome-associated tumor screening by immunohistochemical detection of MMR proteins, with or without microsatellite instability assessment by polymerase chain reaction (PCR), is recommended in all patients affected by colorectal and endometrial cancer since many Lynch patients are missed if clinical criteria only are applied. Recently, assessment of MMR protein expression has been also indicated for less frequent tumor types of Lynch syndrome, such as urologic malignancies (109). Immunohistochemistry for MMR proteins is probably better indicated to screen for the presence of Lynch syndrome in adrenal cortical carcinoma, due to incompletely consistent results of microsatellite instability in Lynch-associated cases. The specificity of panels conventionally used for colorectal and endometrial cancer screening may not be indicated for other tumor types, since optimal microsatellite panels may differ among different cancer types (110). In general, MMR protein deficiency may be associated with different immunohistochemical profiles in adrenal cortical carcinomas, but an MSH2/MSH6 loss is the most frequent phenotype in Lynch associated cases, whereas MLH1/PMS2 loss is less frequent than in colon and endometrial cancers (fig. 7-10).

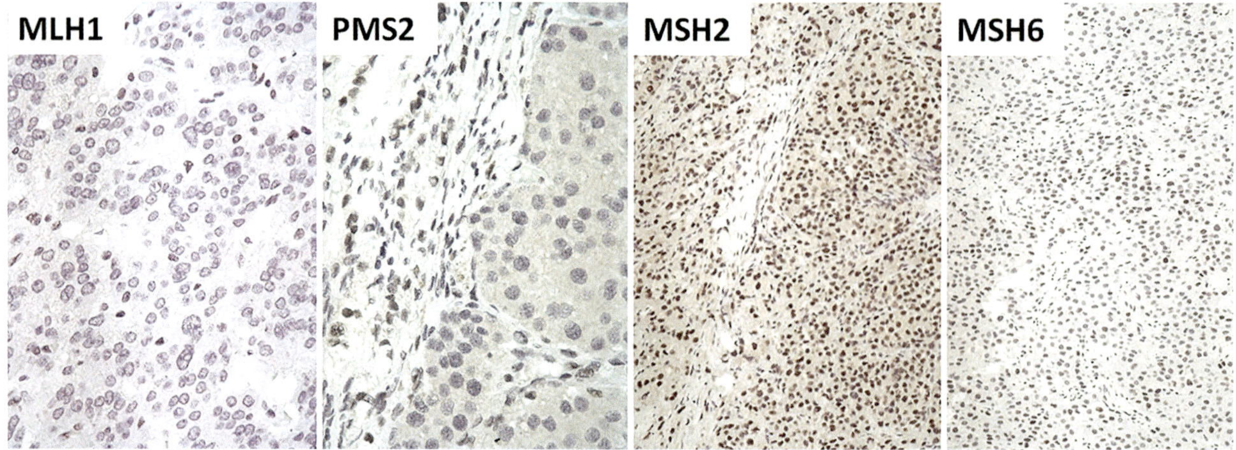

Figure 7-10

ADRENAL CORTICAL CARCINOMA IN A PATIENT WITH LYNCH SYNDROME

The pattern of expression of mismatch repair (MMR) proteins showed loss of MLH1 and PMS2 and retained expression of MSH2 and MSH6.

REFERENCES

1. Hamajima T, Maruwaka K, Homma K, Matsuo K, Fujieda K, Hasegawa T. Unilateral adrenalectomy can be an alternative therapy for infantile onset Cushing's syndrome caused by ACTH-independent macronodular adrenal hyperplasia with McCune-Albright syndrome. Endocr J 2010;57:819-24.
2. Assié G, Libé R, Espiard S, et al. ARMC5 mutations in macronodular adrenal hyperplasia with Cushing's syndrome. N Engl J Med 2013;369:2105-14.
3. Espiard S, Drougat L, Libé R, et al. ARMC5 mutations in a large cohort of primary macronodular adrenal hyperplasia: clinical and functional consequences. J Clin Endocrinol Metab 2015; 100:E926-35.
4. Faucz FR, Zilbermint M, Lodish MB, et al. Macronodular adrenal hyperplasia due to mutations in an armadillo repeat containing 5 (ARMC5) gene: a clinical and genetic investigation. J Clin Endocrinol Metab 2014;99:E1113-9.
5. Damjanovic SS, Antic JA, Elezovic-Kovacevic VI, et al. ARMC5 alterations in patients with sporadic neuroendocrine tumors and multiple endocrine neoplasia type 1 (MEN1). J Clin Endocrinol Metab 2020;105:e4531-42.
6. Stratakis CA, Berthon A. Molecular mechanisms of ARMC5 mutations in adrenal pathophysiology. Curr Opin Endocr Metab Res 2019;8:104-11.
7. Cavalcante IP, Nishi M, Zerbini MC, et al. The role of ARMC5 in human cell cultures from nodules of primary macronodular adrenocortical hyperplasia (PMAH). Mol Cell Endocrinol 2018;460:36-46.
8. Vezzosi D, Libé R, Baudry C, et al. Phosphodiesterase 11A (PDE11A) gene defects in patients with acth-independent macronodular adrenal hyperplasia (AIMAH): functional variants may contribute to genetic susceptibility of bilateral adrenal tumors. J Clin Endocrinol Metab 2012;97:E2063-9.
9. Tirosh A, Valdés N, Stratakis CA. Genetics of micronodular adrenal hyperplasia and Carney complex. Presse Med 2018;47:e127-37.
10. Carney JA, Gordon H, Carpenter PC, Shenoy BV, Go VL. The complex of myxomas, spotty pigmentation, and endocrine overactivity. Medicine (Baltimore) 1985;64:270-83.
11. Stratakis CA, Kirschner LS, Carney JA. Clinical and molecular features of the Carney complex: diagnostic criteria and recommendations for patient evaluation. J Clin Endocrinol Metab 2001; 86:4041-6.

12. Groussin L, Kirschner LS, Vincent-Dejean C, et al. Molecular analysis of the cyclic AMP-dependent protein kinase A (PKA) regulatory subunit 1A (PRKAR1A) gene in patients with Carney complex and primary pigmented nodular adrenocortical disease (PPNAD) reveals novel mutations and clues for pathophysiology: augmented PKA signaling is associated with adrenal tumorigenesis in PPNAD. Am J Hum Genet 2002;71:1433-42.
13. Bertherat J, Horvath A, Groussin L, et al. Mutations in regulatory subunit type 1A of cyclic adenosine 5'-monophosphate-dependent protein kinase (PRKAR1A): phenotype analysis in 353 patients and 80 different genotypes. J Clin Endocrinol Metab 2009;94:2085-91.
14. Cazabat L, Ragazzon B, Groussin L, Bertherat J. PRKAR1A mutations in primary pigmented nodular adrenocortical disease. Pituitary 2006;9:211-9.
15. Anselmo J, Medeiros S, Carneiro V, et al. A large family with Carney complex caused by the S147G PRKAR1A mutation shows a unique spectrum of disease including adrenocortical cancer. J Clin Endocrinol Metab 2012;97:351-9.
16. Morin E, Mete O, Wasserman JD, Joshua AM, Asa SL, Ezzat S. Carney complex with adrenal cortical carcinoma. J Clin Endocrinol Metab 2012;97:E202-6.
17. Bertherat J. Adrenocortical cancer in Carney complex: a paradigm of endocrine tumor progression or an association of genetic predisposing factors? J Clin Endocrinol Metab 2012;97:387-90.
18. Kamilaris CD, Hannah-Shmouni F, Stratakis CA. Adrenocortical tumorigenesis: lessons from genetics. Best Pract Res Clin Endocrinol Metab 2020;34:101428.
19. Libé R, Horvath A, Vezzosi D, et al. Frequent phosphodiesterase 11A gene (PDE11A) defects in patients with Carney complex (CNC) caused by PRKAR1A mutations: PDE11A may contribute to adrenal and testicular tumors in CNC as a modifier of the phenotype. J Clin Endocrinol Metab 2011;96:E208-14.
20. Faillot S, Foulonneau T, Néou M, et al. Genomic classification of benign adrenocortical lesions. Endocr Relat Cancer 2021;28:79-95.
21. Maria AG, Tatsi C, Berthon A, et al. ARMC5 variants in PRKAR1A-mutated patients modify cortisol levels and Cushing's syndrome. Endocr Relat Cancer 2020;27:509-17.
22. Espiard S, Drougat L, Settas N, et al. PRKACB variants in skeletal disease or adrenocortical hyperplasia: effects on protein kinase A. Endocr Relat Cancer 2020;27:647-56.
23. Horvath A, Boikos S, Giatzakis C, et al. A genome-wide scan identifies mutations in the gene encoding phosphodiesterase 11A4 (PDE11A) in individuals with adrenocortical hyperplasia. Nat Genet 2006;38:794-800.
24. Horvath A, Mericq V, Stratakis CA. Mutation in PDE8B, a cyclic AMP-specific phosphodiesterase in adrenal hyperplasia. N Engl J Med 2008;358:750-2.
25. Carney JA, Lyssikatos C, Lodish MB, Stratakis CA. Germline PRKACA amplification leads to Cushing syndrome caused by 3 adrenocortical pathologic phenotypes. Hum Pathol 2015;46:40-9.
26. Lifton RP, Dluhy RG, Powers M, et al. Hereditary hypertension caused by chimaeric gene duplications and ectopic expression of aldosterone synthase. Nat Genet 1992;2:66-74.
27. Fernandes-Rosa FL, Daniil G, Orozco IJ, et al. A gain-of-function mutation in the CLCN2 chloride channel gene causes primary aldosteronism. Nat Genet 2018;50:355-61.
28. Scholl UI, Stölting G, Schewe J, et al. CLCN2 chloride channel mutations in familial hyperaldosteronism type II. Nat Genet 2018;50:349-54.
29. Choi M, Scholl UI, Yue P, et al. K+ channel mutations in adrenal aldosterone-producing adenomas and hereditary hypertension. Science 2011;331:768-72.
30. Geller DS, Zhang J, Wisgerhof MV, Shackleton C, Kashgarian M, Lifton RP. A novel form of human mendelian hypertension featuring nonglucocorticoid-remediable aldosteronism. J Clin Endocrinol Metab 2008;93:3117-23.
31. Scholl UI, Nelson-Williams C, Yue P, et al. Hypertension with or without adrenal hyperplasia due to different inherited mutations in the potassium channel KCNJ5. Proc Natl Acad Sci U S A 2012;109:2533-8.
32. Charmandari E, Sertedaki A, Kino T, et al. A novel point mutation in the KCNJ5 gene causing primary hyperaldosteronism and early-onset autosomal dominant hypertension. J Clin Endocrinol Metab 2012;97:E1532-9.
33. Adachi M, Muroya K, Asakura Y, Sugiyama K, Homma K, Hasegawa T. Discordant genotype-phenotype correlation in familial hyperaldosteronism type III with KCNJ5 gene mutation: a patient report and review of the literature. Horm Res Paediatr 2014;82:138-42.
34. Sertedaki A, Markou A, Vlachakis D, et al. Functional characterization of two novel germline mutations of the KCNJ5 gene in hypertensive patients without primary aldosteronism but with ACTH-dependent aldosterone hypersecretion. Clin Endocrinol (Oxf) 2016;85:845-51.
35. Maria AG, Suzuki M, Berthon A, et al. Mosaicism for KCNJ5 causing early-onset primary aldosteronism due to bilateral adrenocortical hyperplasia. Am J Hypertens 2020;33:124-30.

36. Scholl UI, Stölting G, Nelson-Williams C, et al. Recurrent gain of function mutation in calcium channel CACNA1H causes early-onset hypertension with primary aldosteronism. Elife 2015;4:e06315.
37. Daniil G, Fernandes-Rosa FL, Chemin J, et al. CACNA1H mutations are associated with different forms of primary aldosteronism. EBioMedicine 2016;13:225-36.
38. Scholl UI, Goh G, Stölting G, et al. Somatic and germline CACNA1D calcium channel mutations in aldosterone-producing adenomas and primary aldosteronism. Nat Genet 2013;45:1050-4.
39. Zilbermint M, Xekouki P, Faucz FR, et al. Primary aldosteronism and ARMC5 variants. J Clin Endocrinol Metab 2015;100:E900-9.
40. Mulatero P, Schiavi F, Williams TA, et al. ARMC5 mutation analysis in patients with primary aldosteronism and bilateral adrenal lesions. J Hum Hypertens 2016;30:374-8.
41. Rassi-Cruz M, Maria AG, Faucz FR, et al. Phosphodiesterase 2A and 3B variants are associated with primary aldosteronism. Endocr Relat Cancer 2021;28:1-13.
42. Hattangady NG, Foster J, Lerario AM, et al. Molecular and electrophysiological analyses of ATP2B4 gene variants in bilateral adrenal hyperaldosteronism. Horm Cancer 2020;11:52-62.
43. Chandrasekharappa SC, Guru SC, Manickam P, et al. Positional cloning of the gene for multiple endocrine neoplasia-type 1. Science 1997;276:404-7.
44. Kamilaris CD, Stratakis CA. Multiple endocrine neoplasia type 1 (MEN1): an update and the significance of early genetic and clinical diagnosis. Front Endocrinol (Lausanne) 2019;10:339.
45. Waldmann J, Bartsch DK, Kann PH, Fendrich V, Rothmund M, Langer P. Adrenal involvement in multiple endocrine neoplasia type 1: results of 7 years prospective screening. Langenbecks Arch Surg 2007;392:437-43.
46. Gatta-Cherifi B, Chabre O, Murat A, et al. Adrenal involvement in MEN1. Analysis of 715 cases from the Groupe d'etude des Tumeurs Endocrines database. Eur J Endocrinol 2012;166:269-79.
47. Langer P, Cupisti K, Bartsch DK, et al. Adrenal involvement in multiple endocrine neoplasia type 1. World J Surg 2002;26:891-6.
48. Goudet P, Dalac A, Le Bras M, et al. MEN1 disease occurring before 21 years old: a 160-patient cohort study from the Groupe d'étude des Tumeurs Endocrines. J Clin Endocrinol Metab 2015;100:1568-77.
49. Harada K, Yasuda M, Hasegawa K, Yamazaki Y, Sasano H, Otsuka F. A novel case of myxoid variant of adrenocortical carcinoma in a patient with multiple endocrine neoplasia type 1. Endocr J 2019;66:739-44.
50. Yoshida M, Hiroi M, Imai T, et al. A case of ACTH-independent macronodular adrenal hyperplasia associated with multiple endocrine neoplasia type 1. Endocr J 2011;58:269-77.
51. Simonds WF, Varghese S, Marx SJ, Nieman LK. Cushing's syndrome in multiple endocrine neoplasia type 1. Clin Endocrinol (Oxf) 2012;76:379-86.
52. Skogseid B, Larsson C, Lindgren PG, et al. Clinical and genetic features of adrenocortical lesions in multiple endocrine neoplasia type 1. J Clin Endocrinol Metab 1992;75:76-81.
53. Smith TG, Clark SK, Katz DE, Reznek RH, Phillips RK. Adrenal masses are associated with familial adenomatous polyposis. Dis Colon Rectum 2000; 43:1739-42.
54. Shiroky JS, Lerner-Ellis JP, Govindarajan A, Urbach DR, Devon KM. Characteristics of adrenal masses in familial adenomatous polyposis. Dis Colon Rectum 2018;61:679-85.
55. Hosogi H, Nagayama S, Kanamoto N, et al. Biallelic APC inactivation was responsible for functional adrenocortical adenoma in familial adenomatous polyposis with novel germline mutation of the APC gene: report of a case. Jpn J Clin Oncol 2009;39:837-46.
56. Gaujoux S, Pinson S, Gimenez-Roqueplo AP, et al. Inactivation of the APC gene is constant in adrenocortical tumors from patients with familial adenomatous polyposis but not frequent in sporadic adrenocortical cancers. Clin Cancer Res 2010;16:5133-41.
57. Vouillarmet J, Fernandes-Rosa F, Graeppi-Dulac J, et al. Aldosterone-producing adenoma with a somatic KCNJ5 mutation revealing APC-dependent familial adenomatous polyposis. J Clin Endocrinol Metab 2016;101:3874-8.
58. Gagnon N, Boily P, Alguire C, et al. Small adrenal incidentaloma becoming an aggressive adrenocortical carcinoma in a patient carrying a germline APC variant. Endocrine 2020;68:203-9.
59. Agarwal S, Sharma A, Sharma D, Sankhwar S. Incidentally detected adrenocortical carcinoma in familial adenomatous polyposis: an unusual presentation of a hereditary cancer syndrome. BMJ Case Rep 2018;2018:bcr2018226799.
60. Tomlinson IP, Alam NA, Rowan AJ, et al. Germline mutations in FH predispose to dominantly inherited uterine fibroids, skin leiomyomata and papillary renal cell cancer. Nat Genet 2002; 30:406-10.
61. Shuch B, Ricketts CJ, Vocke CD, et al. Adrenal nodular hyperplasia in hereditary leiomyomatosis and renal cell cancer. J Urol 2013;189:430-5.
62. Tong WH, Sourbier C, Kovtunovych G, et al. The glycolytic shift in fumarate-hydratase-deficient kidney cancer lowers AMPK levels, increases anabolic propensities and lowers cellular iron levels. Cancer Cell 2011;20:315-27.

63. Suda K, Fukuoka H, Yamazaki Y, et al. Cardiac myxoma caused by fumarate hydratase gene deletion in patient with cortisol-secreting adrenocortical adenoma. J Clin Endocrinol Metab 2020;105:dgaa163.
64. Guo X, Chen H, Fu H, Wu H. Hereditary leiomyomatosis and renal cell carcinoma syndrome combined with adrenocortical carcinoma on 18F-FDG PET/CT. Clin Nucl Med 2017;42:692-4.
65. Silverman E, Addasi N, Azzawi M, et al. Recurrent Cushing syndrome from metastatic adrenocortical carcinoma with fumarate hydratase allelic variant. AACE Clin Case Rep 2022;8:259-63.
66. Steinlein OK, Ertl-Wagner B, Ruzicka T, Sattler EC. Birt-Hogg-Dubé syndrome: an underdiagnosed genetic tumor syndrome. J Dtsch Dermatol Ges 2018;16:278-83.
67. Raymond VM, Long JM, Everett JN, et al. An oncocytic adrenal tumour in a patient with Birt-Hogg-Dubé syndrome. Clin Endocrinol (Oxf) 2014;80:925-7.
68. Ramsingh J, Watson C. Oncocytoma of the adrenal gland in Birt-Hogg-Dube syndrome. BMJ Case Rep 2018;2018:bcr2018224283.
69. Furuya M, Hasumi H, Yao M, Nagashima Y. Birt-Hogg-Dubé syndrome-associated renal cell carcinoma: histopathological features and diagnostic conundrum. Cancer Sci 2020;111:15-22.
70. Hofstedter R, Sanabria-Salas MC, Di Jiang M, Ezzat S, Mete O, Kim RH. FLCN-driven functional adrenal cortical carcinoma with high mitotic tumor grade: extending the endocrine manifestations of Birt-Hogg-Dubé syndrome. Endocr Pathol 2023;34;257-64.
71. Malkin D, Li FP, Strong LC, et al. Germ line p53 mutations in a familial syndrome of breast cancer, sarcomas, and other neoplasms. Science 1990;250:1233-8.
72. Frebourg T, Bajalica Lagercrantz S, Oliveira C, et al. Guidelines for the Li-Fraumeni and heritable TP53-related cancer syndromes. Eur J Hum Genet 2020;28:1379-86.
73. Bougeard G, Renaux-Petel M, Flaman JM, et al. Revisiting Li-Fraumeni syndrome from TP53 mutation carriers. J Clin Oncol 2015;33:2345-52.
74. Wasserman JD, Novokmet A, Eichler-Jonsson C, et al. Prevalence and functional consequence of TP53 mutations in pediatric adrenocortical carcinoma: a children's oncology group study. J Clin Oncol 2015;33:602-9.
75. Gonzalez KD, Noltner KA, Buzin CH, et al. Beyond Li Fraumeni syndrome: clinical characteristics of families with p53 germline mutations. J Clin Oncol 2009;27:1250-6.
76. Dall'Igna P, Virgone C, De Salvo GL, et al. Adrenocortical tumors in Italian children: analysis of clinical characteristics and P53 status. Data from the national registries. J Pediatr Surg 2014; 49:1367-71.
77. Renaux-Petel M, Charbonnier F, Théry JC, et al. Contribution of de novo and mosaic TP53 mutations to Li-Fraumeni syndrome. J Med Genet 2018;55:173-80.
78. Herrmann LJ, Heinze B, Fassnacht M, et al. TP53 germline mutations in adult patients with adrenocortical carcinoma. J Clin Endocrinol Metab 2012;97:E476-85.
79. Raymond VM, Else T, Everett JN, Long JM, Gruber SB, Hammer GD. Prevalence of germline TP53 mutations in a prospective series of unselected patients with adrenocortical carcinoma. J Clin Endocrinol Metab 2013;98:E119-25.
80. Villani A, Shore A, Wasserman JD, et al. Biochemical and imaging surveillance in germline TP53 mutation carriers with Li-Fraumeni syndrome: 11 year follow-up of a prospective observational study. Lancet Oncol 2016;17:1295-305.
81. Varley JM, McGown G, Thorncroft M, et al. Are there low-penetrance TP53 alleles? Evidence from childhood adrenocortical tumors. Am J Hum Genet 1999;65:995-1006.
82. Ribeiro RC, Sandrini F, Figueiredo B, et al. An inherited p53 mutation that contributes in a tissue-specific manner to pediatric adrenal cortical carcinoma. Proc Natl Acad Sci U S A 2001;98: 9330-5.
83. Latronico AC, Pinto EM, Domenice S, et al. An inherited mutation outside the highly conserved DNA-binding domain of the p53 tumor suppressor protein in children and adults with sporadic adrenocortical tumors. J Clin Endocrinol Metab 2001;86:4970-3.
84. Custódio G, Parise GA, Kiesel Filho N, et al. Impact of neonatal screening and surveillance for the TP53 R337H mutation on early detection of childhood adrenocortical tumors. J Clin Oncol 2013;31:2619-26.
85. Seidinger AL, Caminha IP, Mastellaro MJ, et al. TP53 p.Arg337His geographic distribution correlates with adrenocortical tumor occurrence. Mol Genet Genomic Med 2020;8:e1168.
86. Custodio G, Taques GR, Figueiredo BC, et al. Increased incidence of choroid plexus carcinoma due to the germline TP53 R337H mutation in southern Brazil. PLoS One 2011;6:e18015.
87. Pinto EM, Zambetti GP. What 20 years of research has taught us about the TP53 p.R337H mutation. Cancer 2020;126:4678-86.
88. Moreira L, Balaguer F, Lindor N, et al. Identification of Lynch syndrome among patients with colorectal cancer. JAMA 2012;308:1555-65.

89. Challis BG, Kandasamy N, Powlson AS, et al. Familial adrenocortical carcinoma in association with Lynch syndrome. J Clin Endocrinol Metab 2016;101:2269-72.
90. Raymond VM, Everett JN, Furtado LV, et al. Adrenocortical carcinoma is a Lynch syndrome-associated cancer. J Clin Oncol 2013;31:3012-8.
91. Latham A, Srinivasan P, Kemel Y, et al. Microsatellite instability is associated with the presence of Lynch syndrome pan-cancer. J Clin Oncol 2019; 37:286-95.
92. Domènech M, Grau E, Solanes A, et al. Characteristics of adrenocortical carcinoma associated with Lynch syndrome. J Clin Endocrinol Metab 2021;106:318-25.
93. Raygada M, Raffeld M, Bernstein A, et al. Case report of adrenocortical carcinoma associated with double germline mutations in MSH2 and RET. Am J Med Genet A 2021;185:1282-7.
94. Miller DT, Freedenberg D, Schorry E, et al. Health supervision for children with neurofibromatosis type 1. Pediatrics 2019;143:e20190660.
95. Wagner AS, Fleitz JM, Kleinschmidt-Demasters BK. Pediatric adrenal cortical carcinoma: brain metastases and relationship to NF-1, case reports and review of the literature. J Neurooncol 2005; 75:127-33.
96. Minkiewicz I, Wilbrandt-Szczepanska E, Jendrzejewski J, Sworczak K, Korwat A, Sledzinski M. Co-occurrence of adrenocortical carcinoma and gastrointestinal stromal tumor in a patient with neurofibromatosis type 1 and a history of endometrial cancer. Acta Endocrinol (Buchar) 2020;16:353-8.
97. Sørensen SA, Mulvihill JJ, Nielsen A. Long-term follow-up of von Recklinghausen neurofibromatosis. Survival and malignant neoplasms. N Engl J Med 1986;314:1010-5.
98. Menon RK, Ferrau F, Kurzawinski TR, et al. Adrenal cancer in neurofibromatosis type 1: case report and DNA analysis. Endocrinol Diabetes Metab Case Rep 2014;2014:140074.
99. Palma M, Ristori E, Ricevuto E, Giannini G, Gulino A. BRCA1 and BRCA2: the genetic testing and the current management options for mutation carriers. Crit Rev Oncol Hematol 2006;57:1-23.
100. El Ghorayeb N, Grunenwald S, Nolet S, et al. First case report of an adrenocortical carcinoma caused by a BRCA2 mutation. Medicine (Baltimore) 2016;95:e4756.
101. Lippert J, Appenzeller S, Liang R, et al. Targeted molecular analysis in adrenocortical carcinomas: a strategy toward improved personalized prognostication. J Clin Endocrinol Metab 2018; 103:4511-23.
102. Else T, Lerario AM, Everett J, et al. Adrenocortical carcinoma and succinate dehydrogenase gene mutations: an observational case series. Eur J Endocrinol 2017;177:439-44.
103. Pilati C, Shinde J, Alexandrov LB, et al. Mutational signature analysis identifies MUTYH deficiency in colorectal cancers and adrenocortical carcinomas. J Pathol 2017;242:10-5.
104. Landwehr LS, Schreiner J, Appenzeller S, et al. A novel patient-derived cell line of adrenocortical carcinoma shows a pathogenic role of germline MUTYH mutation and high tumour mutational burden. Eur J Endocrinol 2021;184:823-35.
105. Xie C, Tanakchi S, Raygada M, Davis JL, Del Rivero J. Case report of an adrenocortical carcinoma associated with germline CHEK2 mutation. J Endocr Soc 2018;3:284-90.
106. Torres MB, Diggs LP, Wei JS, et al. Ataxia telangiectasia mutated germline pathogenic variant in adrenocortical carcinoma. Cancer Genet 2021; 256-257:21-5.
107. Petr EJ, Else T. Adrenocortical carcinoma (ACC): when and why should we consider germline testing? Presse Med 2018;47:e119-25.
108. Frebourg T, Abel A, Bonaiti-Pellie C, et al. [Li-Fraumeni syndrome: update, new data and guidelines for clinical management]. Bull Cancer 2001;88:581-7. [French]
109. Ju JY, Mills AM, Mahadevan MS, et al. Universal Lynch syndrome screening should be performed in all upper tract urothelial carcinomas. Am J Surg Pathol 2018;42:1549-55.
110. Long DR, Waalkes A, Panicker VP, Hause RJ, Salipante SJ. Identifying optimal loci for the molecular diagnosis of microsatellite instability. Clin Chem 2020;66:1310-8.

8 PHEOCHROMOCYTOMA AND ADRENAL MEDULLARY HYPERPLASIA

The largest paraganglion is the adrenal medulla, a sympathetic organ that secretes adrenaline (epinephrine) and noradrenaline (norepinephrine). The adrenal medulla has long been recognized to be structurally and functionally distinct from the adrenal cortex; in older literature, the development of brown coloration in the presence of chromate salts was reported by Bertholdus Werner in 1857 (1). Alfred Kohn coined the terms "chromaffin reaction" for the color change and "chromaffin cells" for the reactive cells. He went on to map the chromaffin tissues in other parts of the body, and conceived the unifying concept of the paraganglionic system based on analogy of these tissues to sympathetic ganglia (2).

While the chromaffin reaction is now obsolete as a diagnostic tool, the cells of the adrenal medulla still are called "chromaffin cells" and their corresponding tumors are called pheochromocytomas, alluding to the color change imparted by the chromaffin reaction. The name derives from the Greek *phaios*, dusky, and *chroma*, color. However, a pheochromocytoma is also an intra-adrenal paraganglioma, a diagnosis that was often rendered before "pheochromocytoma" took hold (3).

Current World Health Organization (WHO) nomenclature maintains the term *pheochromocytoma* for the intra-adrenal location while structurally and functionally related extra-adrenal tumors are classified as *paragangliomas* (4,5). This arbitrary convention dates to the 1950 Armed Forces Institute of Pathology (AFIP) adrenal gland Fascicle and was intended to prevent confusion caused by promiscuous diagnoses of paraganglioma (6). Until recently, however, it did little to prevent inconsistent diagnoses of "pheochromocytoma" (7). Acceptance of the arbitrary distinction between pheochromocytoma and paraganglioma has increased since the publication of the 2017 WHO classification of tumors of endocrine organs (8). However, reconsideration of the entire paraganglioma nomenclature may be beneficial in view of current understanding of both the commonalities and differences between paragangliomas in different locations.

A more fundamental conceptual change, first introduced in the 2017 WHO classification and maintained in the more recent 5th edition, is elimination of the previous separate classifications of benign and malignant pheochromocytoma (9). This concept has now been abandoned, since it is evident that this distinction is not currently possible or reliable. Instead, as with other neuroendocrine tumors, these neoplasms are all considered to have malignant potential, and those that do give rise to metastasis are classified as "metastatic." Moreover, with increasing awareness of germline mutations providing genetic susceptibility to multifocal disease, it has become apparent that some of the previous literature may have confused metastasis and multifocal primary lesions, which occasionally occur in unusual locations, such as lung and liver (10). Therefore, some of the features traditionally considered to be diagnostic of malignancy are instead indicative of germline disease.

PHEOCHROMOCYTOMA

Clinical Features

Pheochromocytoma is the prototypical sympathetic paraganglioma and the clinical features that it causes are usually attributable to catecholamine excess (11). The most diagnostic features are paroxysmal episodes of tachycardia, diaphoresis, pallor, headache, and anxiety as well as hypertension that can be more persistent. Sustained hypertension is reflective of noradrenaline (norepinephrine) secretion whereas tumors with predominant adrenaline (epinephrine) secretion tend to be associated with episodic hypertension (12). Paroxysms may be spontaneous or they may be provoked by triggers such as exercise, large meals (including specific foods, such as cheeses with tyramine), alcohol, medications, stress, and even abdominal pressure. These manifestations may result in hypertensive crisis, myocardial infarction, arrhythmias, Takotsubo cardiomyopathy, and acute heart failure. Tumors that secrete only epinephrine may be associated with hypotension (12). The classic features occur in only just over half of patients while some are asymptomatic (13).

The diagnosis of pheochromocytoma must be considered in the workup of any patient with an adrenal mass. Failure to consider this in the differential diagnosis may result in a catecholamine crisis precipitated by anesthesia or tumor manipulation during biopsy or surgical resection, with resulting stroke and even possibly death.

Epidemiology

The first report of a pheochromocytoma was likely by Charles Sugrue who in 1800 described the clinical findings and gross appearance of bilateral adrenal tumors (14). In 1886, Felix Fraenkel reported the case of a young woman named Minna Roll who died suddenly, and at autopsy, performed by Professor Rudolf Maier, was found to have bilateral adrenal tumors that had a specific brown appearance after exposure to chromate-containing Mueller fixative. These tumors were examined histologically by Professor Max Schottelius who provided an elegant microscopic description. Fraenkel postulated that the tumors were responsible for the patient's premature myocardial infarction due to secretion of a chemical that caused hypertension (15).

The first report of surgical resection of a pheochromocytoma came from the Mayo clinic in 1927 where Mother Mary Joachim, a Catholic nun, was the first to be cured of this disease by successful surgery (16). The disease was mainly the subject of case reports, but also played a role in history: it was the likely cause of erratic blood pressure in President Eisenhower (17), and may have been responsible for the famous American 30-year Hatfield-McCoy feud (18).

Pheochromocytomas are rare neoplasms. In the last century, the few epidemiologic reports indicated an average annual incidence of 8.0 to 9.5 cases/million persons per year in the United States (19), 2.1 cases/million in Sweden (20), and 1.9 cases/million in Denmark (21). However, like other neuroendocrine tumors, they appear to be increasing in incidence. In a population study carried out in the Netherlands (22), the age-standardized incidence rate for pheochromocytoma increased from 0.29 per 100,000 person years in 1995 to 1999 to 0.46 per 100,000 person years in 2011 to 2015; interestingly, this increase was associated with a concomitant decrease in tumor size and increase in age at diagnosis. The same study reported a systematic search that yielded only three papers reporting 530 cases of sympathetic paraganglioma (not only adrenal pheochromocytoma) that showed combined annual incidence rates that varied from 0.04 to 0.21 per 100,000 person-years.

With new technologies and the recognition of familial predisposition syndromes, the profile of this disease has changed: more are being discovered as incidental adrenal masses during imaging and during surveillance of kindreds with known hereditary predisposition (13). In a British study (23), the incidence of pheochromocytoma identified in a series of adrenalectomies was 5.30 cases/1 million population. These cases were divided into incidental adrenal masses, genetically screened patients, and symptomatic presentations. Among the incidental lesions, 33.34 percent had a mean delay in diagnosis of 22.95 months (4 to 120 months). Among those with symptoms, patients were younger and more often female, had larger tumors, and had higher preoperative hormone levels. Those with a genetic predisposition were the youngest group, as would be expected because of screening.

Earlier autopsy studies have shown that these lesions were commonly discovered incidentally and were associated with myocardial infarction or cardiovascular catastrophe, often after unrelated surgery (24). These older data may no longer be applicable, since diagnosis has improved over the last 40 years. Currently, paragangliomas are identified in 0.1 to 0.6 percent of the hypertensive patient population, and most of these are pheochromocytomas (25).

Pheochromocytomas can occur at any age but the peak age at diagnosis is in the fifth decade. Presentation in a young patient should prompt investigation of genetic predisposition (26,27), and genetic surveillance of members of known kindreds may detect tumors in children or adolescents (28). There is some evidence of more aggressive behavior in younger patients (27). Pheochromocytomas occur in both males and females, with only a minor predilection for either sex in most series. In some studies, the right adrenal gland is involved more frequently than the left, possibly due to the slightly larger size of the medulla in that gland (29,30). In adults, there is an average delay in the diagnosis of about 3 years (13).

The old teaching that pheochromocytomas have a "ten percent" rule (10 percent bilateral, 10 percent malignant, and 10 percent familial, as well as other criteria such as 10 percent extra-adrenal and 10 percent in childhood) is no longer considered to be correct. While these numbers change with each study, it is clear that a much greater proportion is familial, the familial ones are the ones that tend to be bilateral, and the term pheochromocytoma is no longer applied to extra-adrenal paragangliomas. Since malignancy is no longer distinguished, that percentage no longer exists.

Biochemistry and Signaling Pathways

Biochemical testing to assess catecholamine production, as discussed in chapter 1, is the first and most definitive way of confirming the diagnosis of pheochromocytoma (13). The current technologies have moved to measurement of plasma and urinary free metanephrine and normetanephrine; the older technique of measuring urinary vanillylmandelic acid (VMA) is no longer recommended. Measurement of plasma 3-methoxytyramine, the O-methylated metabolite of dopamine, is not usually required for diagnosis of pheochromocytomas but can be helpful in detecting metastases and extra-adrenal paragangliomas (31).

Some laboratories still measure deconjugated metanephrines in 24-hour urine; this assay detects a mix of both free and conjugated metanephrines that can cause false positive results, therefore, assays for free metanephrines are recommended. Liquid chromatography-mass spectrometry is the preferred assay method since it is more specific than immunoassays, which can underestimate values and may have interference from drugs. Dietary amines can increase plasma free and urinary 3-methoxytyramine but do not alter plasma or urinary free metanephrines.

Biochemical testing can identify profiles that predict the molecular type of pheochromocytoma (13). Alterations in the "pseudohypoxic pathway" involving mutations in *SDHx, VHL, HIF2a, FH, MDH2, PHD1/EGLN2,* and *PHD2/EGLN1* are classified as "cluster 1" disease. This is characterized by stabilization of hypoxia-inducible factors and epigenetic silencing of phenylethanolamine N-methyltransferase, the enzyme that converts noradrenaline into adrenaline, therefore, these tumors are characterized by predominant dopaminergic and/or noradrenergic secretory profiles. In contrast, tumors associated with mutations of *RET, NF1,* and related genes involved in kinase signaling (*TMEM127, MAX, KIF1Bβ*) comprise "cluster 2." This is associated with both adrenergic and noradrenergic secretion, resulting in elevations of metanephrine and normetanephrine. "Cluster 3" disease is attributed to somatic mutations and *MAML3* fusions that alter the WNT signaling pathway and is usually associated with both adrenergic and noradrenergic secretion.

Considerable diversity exists within these clusters according to the specific mutated gene and tumor location. A common denominator underlying the pathobiology of major cluster 1 tumors is impaired proteolytic degradation of hypoxia-inducible transcription factors, especially HIF2a. In normal cells, HIF2a protein levels increase in response to hypoxia and decrease by proteolytic degradation in response to oxygen (32). VHL protein mediates ubiquitination of HIF2a, leading to its proteolytic degradation under normoxic conditions. Signaling pathways in pseudohypoxic tumors caused by *VHL* mutations therefore predominantly reflect a direct effect of the loss of function of *VHL*. In contrast, tumors caused by mutations of *SDHx* genes have normal VHL protein but manifest multiple abnormalities caused by oncometabolites that result from TCA cycle disruption. In cluster 2, a function analogous to that of VHL protein has been identified for TMEM127 as a mediator of RET degradation (33).

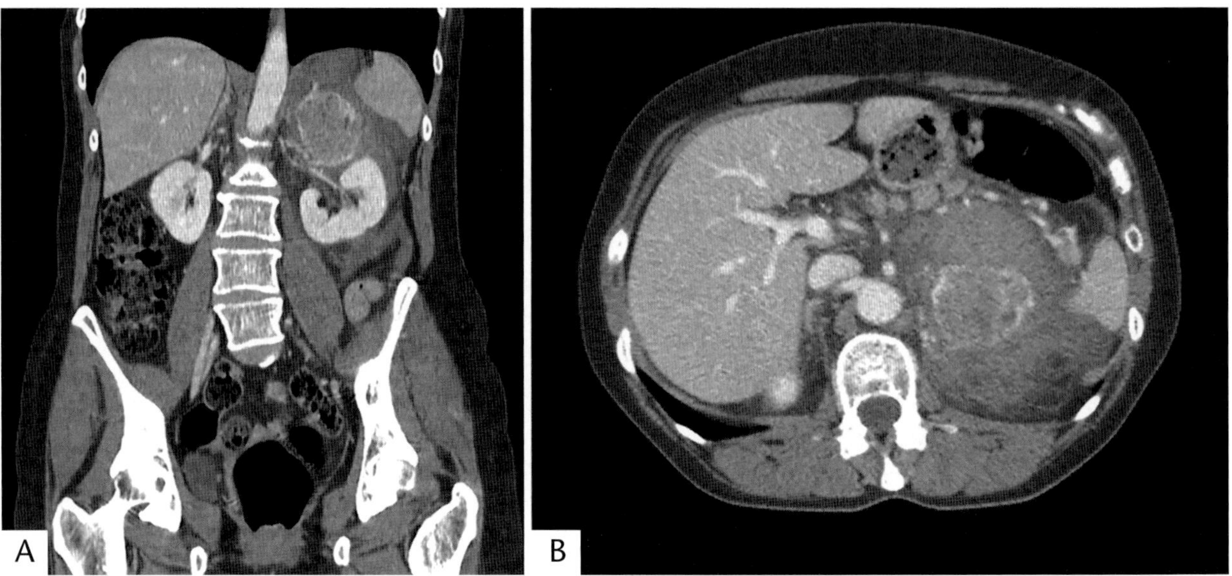

Figure 8-1

COMPUTERIZED TOMOGRAPHY (CT) SCAN

A pheochromocytoma in the left adrenal gland of an adult measures 5.3 cm and has focal hemorrhage (A: coronal view; B: transverse view).

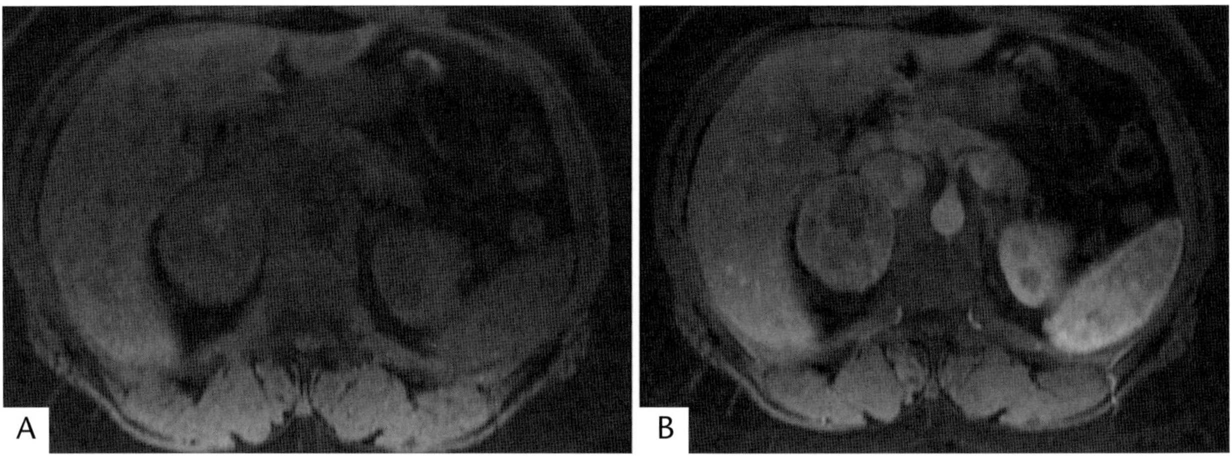

Figure 8-2

MAGNETIC RESONANCE IMAGING (MRI)

A: T1-weighted MRI of the abdomen shows a rounded pheochromocytoma on the right side with a low signal intensity. B: Tumor shows enhancement with gadolinium.

Preoperative Localization and Imaging

Anatomic imaging can be performed using computerized tomography (CT) or magnetic resonance imaging (MRI). On CT scan, these tumors are homogeneous with soft tissue density but they may show hemorrhage and necrosis with central low density (fig. 8-1). On MRI they are isointense with liver on T1-weighted images, have high signal intensity on T2-weighted images, and enhance with gadolinium (fig. 8-2). All patients with adrenal masses with density of greater than 10 Hounsfield units (HU) should be screened for pheochromocytoma even in the absence of hypertension or other symptoms.

Functional imaging can be used to verify the diagnosis (13). The adrenal medulla is able to take up ^{123}I-metaiodobenzylguanidine (^{123}I-MIBG), and imaging of pheochromocytomas with this ligand has been helpful for many years (fig. 8-3). ^{123}I-MIBG is less useful for tumors with *SDHx* mutations, which may express only low levels of norepinephrine transporter. However, they usually express somatostatin receptors (34) and are amenable to somatostatin receptor imaging by positron emission tomography (PET/CT) scans (fig. 8-4).

Currently, the major imaging agent is gallium 68-DOTATATE, in which a chelating agent (DOTA, [dodecanetetraacetic acid]) binds the radioisotope ^{68}Ga and is also linked to the somatostatin agonist analog octreotate. Variations of this "keychain" arrangement testing different agonist and antagonist analogs, and different radioisotopes and DOTA substitutes are under development (35). Somatostatin receptor imaging is not specific for pheochromocytomas or paragangliomas, since it also binds other neuroendocrine tumors that may be metastatic to the adrenal gland. An advantage of functional over anatomic imaging, in addition to greater specificity, is applicability to "theranostic" approaches in which the radioisotope used to diagnose and localize a tumor (e.g., ^{123}I or ^{68}Ga) is switched to a different isotope (e.g., ^{131}I or ^{177}Lu) to achieve cell killing in patients with metastatic or unresectable tumor (34).

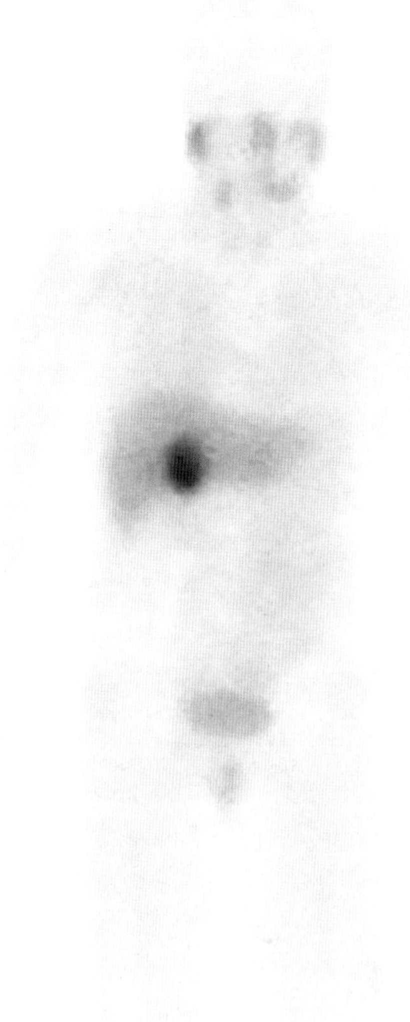

Figure 8-3

123I-METAIODOBENZYLGUANIDINE (123I-MIBG) SCAN

Pheochromocytoma of the right adrenal gland is visualized scintigraphically using this technique.

Gross Findings

The usual appearance of a sporadic pheochromocytoma is a solitary, round or oval mass that distorts the adrenal gland (fig. 8-5). The nontumorous gland is identified as an attached remnant (fig. 8-6) or attenuated around the tumor. Identification of the adrenal gland may be difficult with extremely large tumors. Tumor size varies widely, and tumors may measure more than 10 cm, but this is rare. Similarly, the weight can vary and tumors may weigh up to several hundred grams. There is no preferential localization or origin in either the head or the body of the gland, the tail, or the wings, despite the variable amounts of normal medulla in these areas.

On section, the tumor is usually well circumscribed and may even appear encapsulated, but these lesions do not usually have a capsule; there may be a fibrous pseudocapsule but it is usually incomplete. Pheochromocytomas rarely have a homogeneous appearance; when they do, they are firm and tan-gray, resembling nontumorous medulla, but they may also be tan brown when they are exposed to air and oxidize (fig. 8-6). More commonly, they have a variegated tan brown and red soft cut surface (fig. 8-7). Areas of

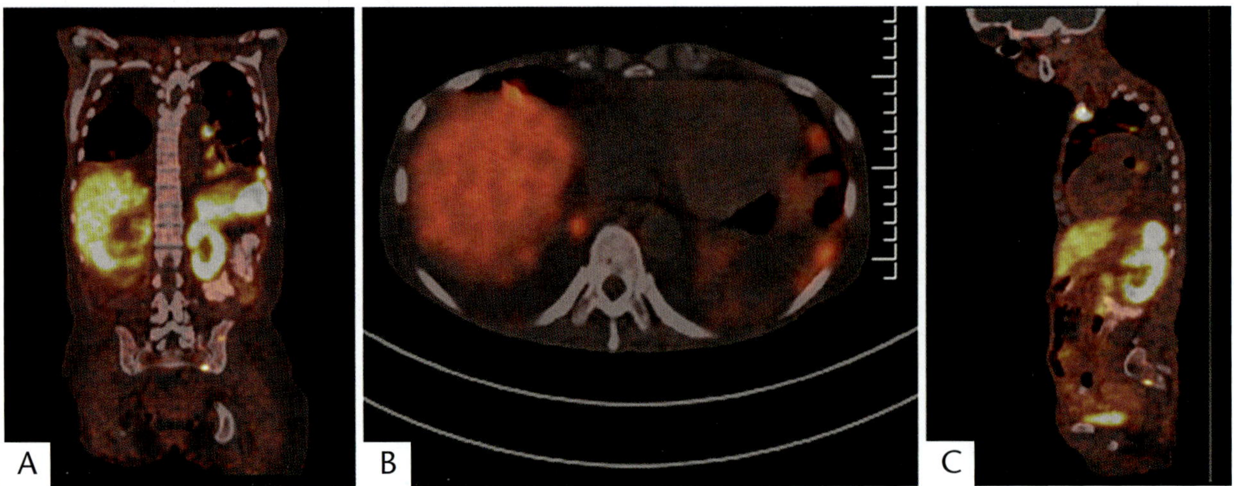

Figure 8-4

GALLIUM 68-DOTATATE POSITRON EMISSION TOMOGRAPHY (PET/CT) SCAN

Coronal (A), horizontal (B), and sagittal (C) views. A markedly ^{68}Ga-DOTATATE avid large right suprarenal mass likely invading the liver has metastasized to the pleura, celiac axis lymph node, and multiple skeletal sites. (Courtesy of Dr. S. Ezzat, Toronto, Canada)

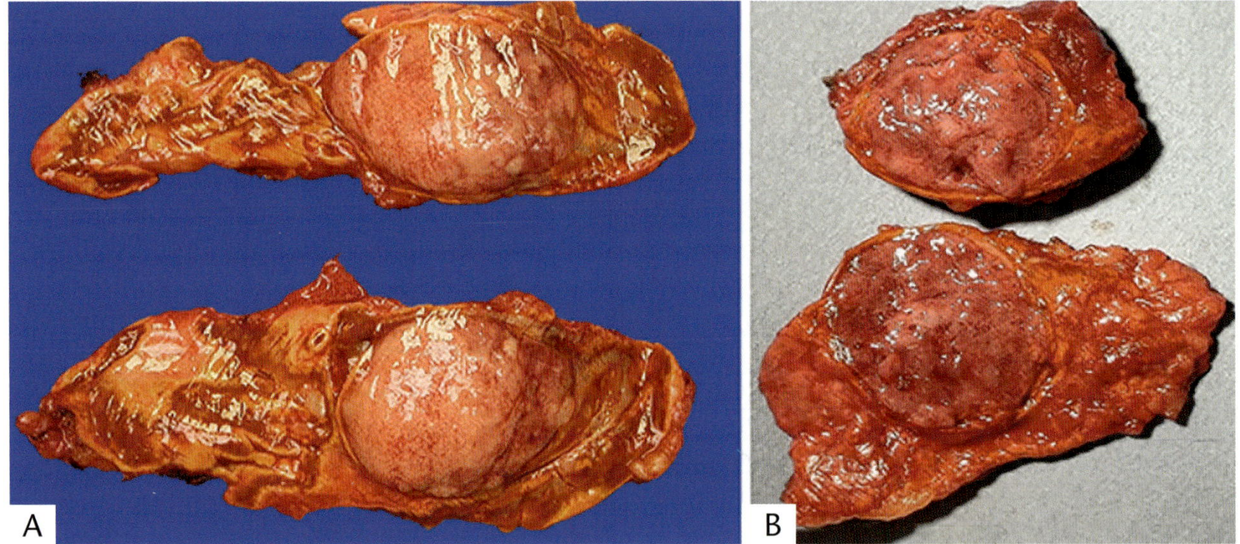

Figure 8-5

PHEOCHROMOCYTOMA

A: The cut surface of pheochromocytoma may be firm and gray-tan, resembling medulla.
B: The mass may distort the entire gland, which is attenuated around the lesion.

mottled to confluent congestion, and even frank hemorrhage, are frequent (fig. 8-8), sometimes extending into the surrounding adipose tissue.

Central degeneration is common in larger tumors that have necrosis, fibrosis, or cystic change. Cystic degeneration gives rise to nonhomogeneous areas on CT scan or ultrasound (figs. 8-9, 8-10) that may result in clinical confusion. The cyst contents may be thin blood-tinged fluid or thick red-brown material. Extreme examples are unusual but may even exhibit dystrophic calcification (fig. 8-11).

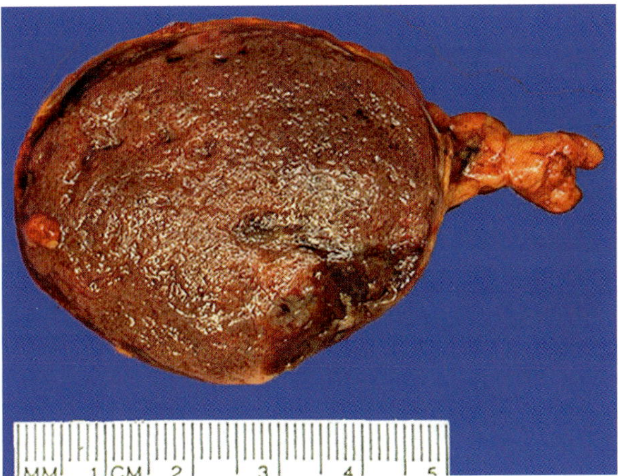

Figure 8-6

PHEOCHROMOCYTOMA

The tumor mass is sometimes a dusky brown color when it oxidizes in air. The tumor is located in one wing of the gland, with the adjacent nontumorous gland identified at one edge.

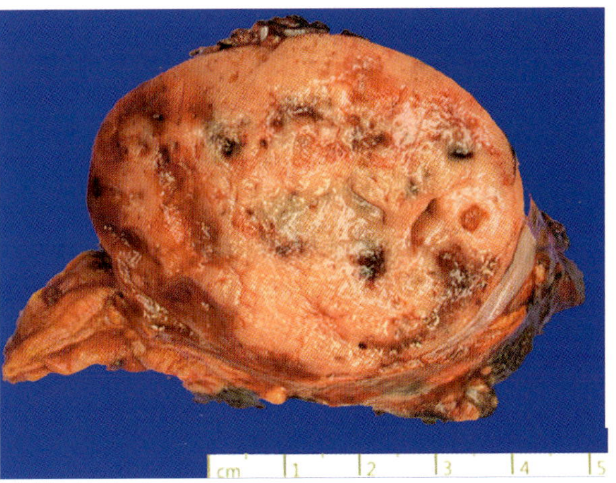

Figure 8-7

PHEOCHROMOCYTOMA

The cut surface of pheochromocytoma is often variegated, with areas of red hemorrhagic and dark brown old hemorrhages admixed with the dusky brown tissue that is characteristic of these tumors.

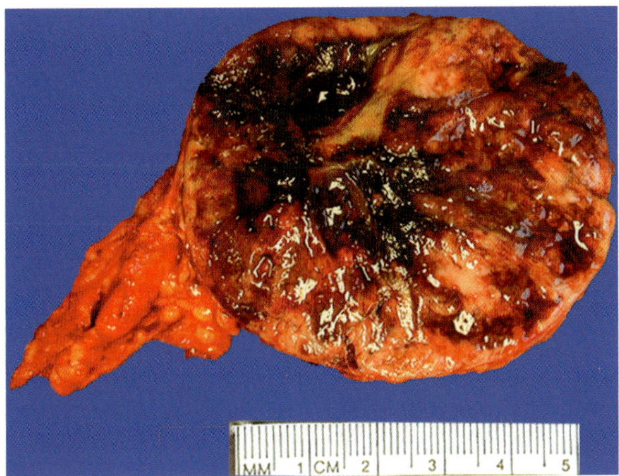

Figure 8-8

PHEOCHROMOCYTOMA

Large areas of hemorrhage and glistening edema are prominent.

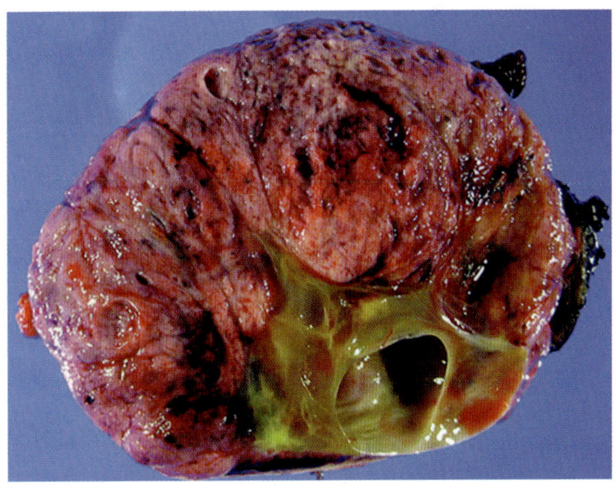

Figure 8-9

PHEOCHROMOCYTOMA: CYSTIC DEGENERATION

The large cyst at bottom of this 15-cm tumor most likely results from old hemorrhage or infarction and contains only thin blood-tinged fluid. The small cyst to its right contains degenerating blood.

An unusual finding is lipid degeneration that can mimic cortical adenoma (fig. 8-12) (36); these unusual lipid-rich tumors are due to the extreme stromal edema and vacuolization of cytoplasm seen in patients with von Hippel-Lindau syndrome, and may be bilateral due to the genetic predisposition (36). Pheochromocytomas in patients with germline predisposition, especially MEN2-related tumors, may present as multiple nodules on a background of diffusely expanded medulla or as single or multiple tumors in a single gland without an obviously expanded background (fig. 8-13).

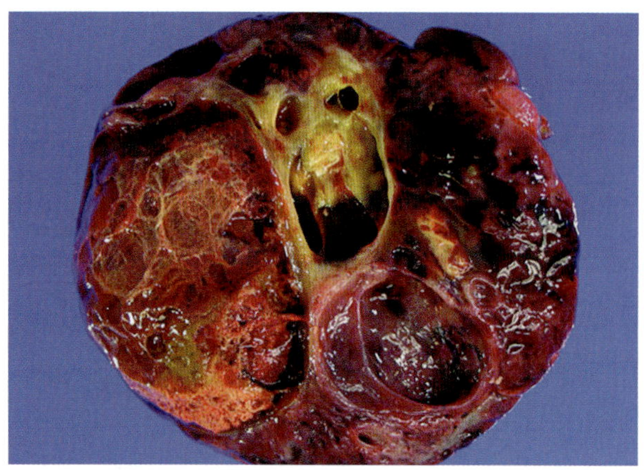

Figure 8-10

PHEOCHROMOCYTOMA: EVIDENCE OF MULTIPLE EPISODES OF INFARCTION AND HEMORRHAGE

The necrotic areas of varying age suggest the patient experienced episodic signs and symptoms.

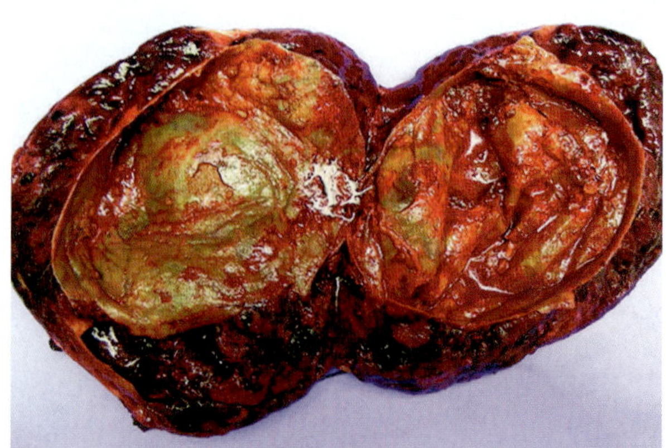

Figure 8-11

PHEOCHROMOCYTOMA: EXTENSIVE CYSTIC DEGENERATION

This lesion was identified as a large cyst on imaging; focal calcification is present in the wall.

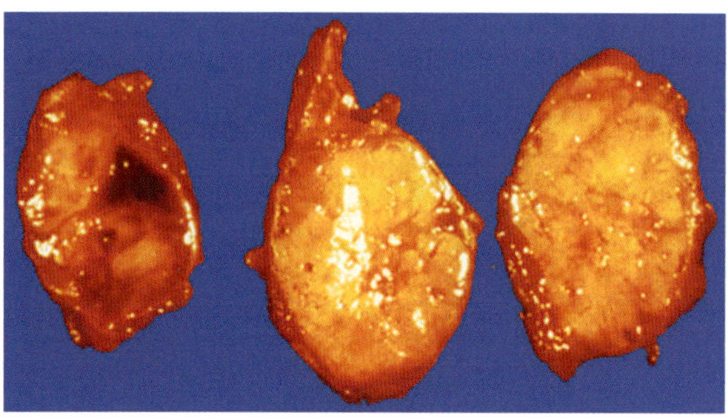

Figure 8-12

PHEOCHROMOCYTOMA ASSOCIATED WITH VON HIPPEL-LINDAU DISEASE (VHL): EXTENSIVE LIPID DEGENERATION

The gross appearance of this tumor is bright yellow, resembling an adrenal cortical neoplasm. This appearance is due to abundant lipid accumulation, which is seen in pheochromocytomas of patients with VHL disease.

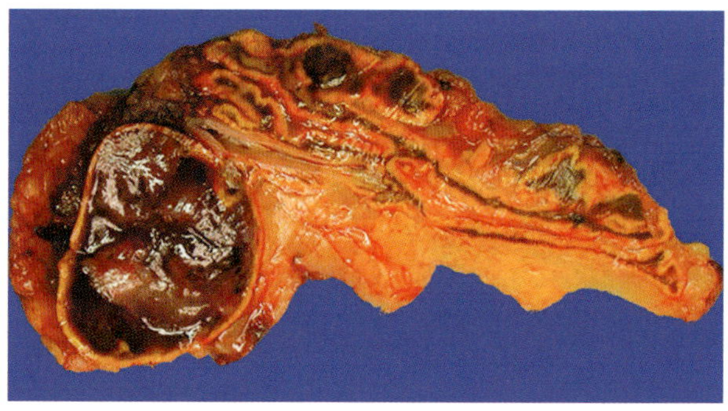

Figure 8-13

MULTIFOCAL PHEOCHROMOCYTOMA ASSOCIATED WITH MULTIPLE ENDOCRINE NEOPLASIA TYPE 2 (MEN2)

The adrenal gland contains a dominant pheochromocytoma as well as two smaller nodules (top) and an expanded pale tan adrenal medulla (right).

Microscopic Findings

Pheochromocytomas classically have a characteristic architecture featuring well-defined nests of tumor cells, known as "zellballen," surrounded by thin strands of fibrovascular stroma (fig. 8-14). The zellballen vary in size and shape, and in some tumors, other patterns, including anastomosing cords (fig. 8-15) or solid sheets (fig. 8-16), are prominent. The periphery of the lesion is usually delineated from the surrounding gland, but most tumors do not have a well-defined capsule or pseudocapsule, and commonly there is intermingling of tumor cells and cortical cells (fig. 8-17).

There may be areas of degeneration associated with hemorrhage, fibrosis, and hemosiderin deposition. This is particularly prominent in tumors that have been biopsied, but may also be spontaneous; it is evident in tumors with extensive hemorrhage and cystic change. A focal inflammatory infiltrate is common and may

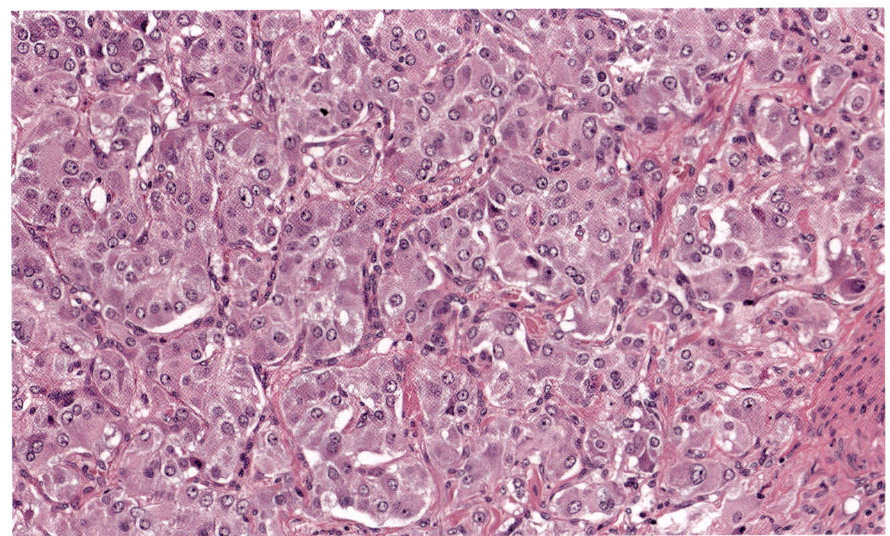

Figure 8-14

PHEOCHROMOCYTOMA: CLASSIC ZELLBALLEN

This pheochromocytoma has the characteristic "zellballen" architecture, with tumor cells in well-delineated nests surrounded by fibrovascular stroma.

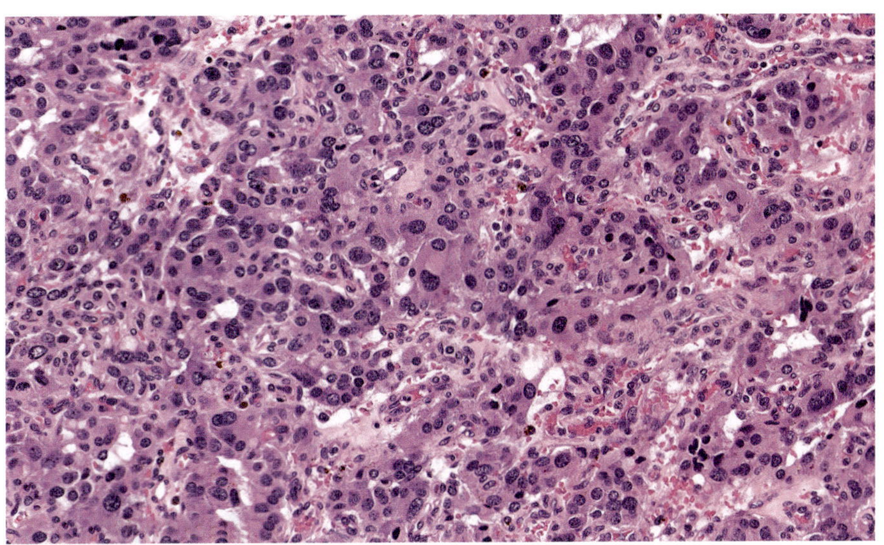

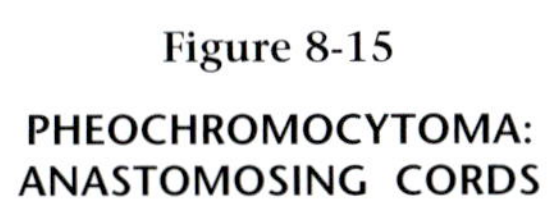

Figure 8-15

PHEOCHROMOCYTOMA: ANASTOMOSING CORDS

In some pheochromocytomas the tumor cells form anastomosing cords rather than well-defined nests.

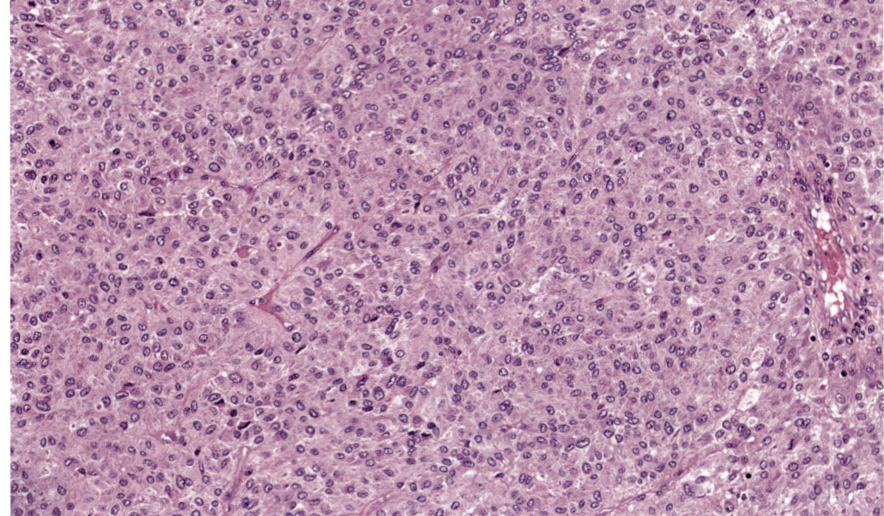

Figure 8-16

PHEOCHROMOCYTOMA: SOLID SHEETING ARCHITECTURE

Some pheochromocytomas have a more solid architecture, with solid sheets of cells and less prominent stroma.

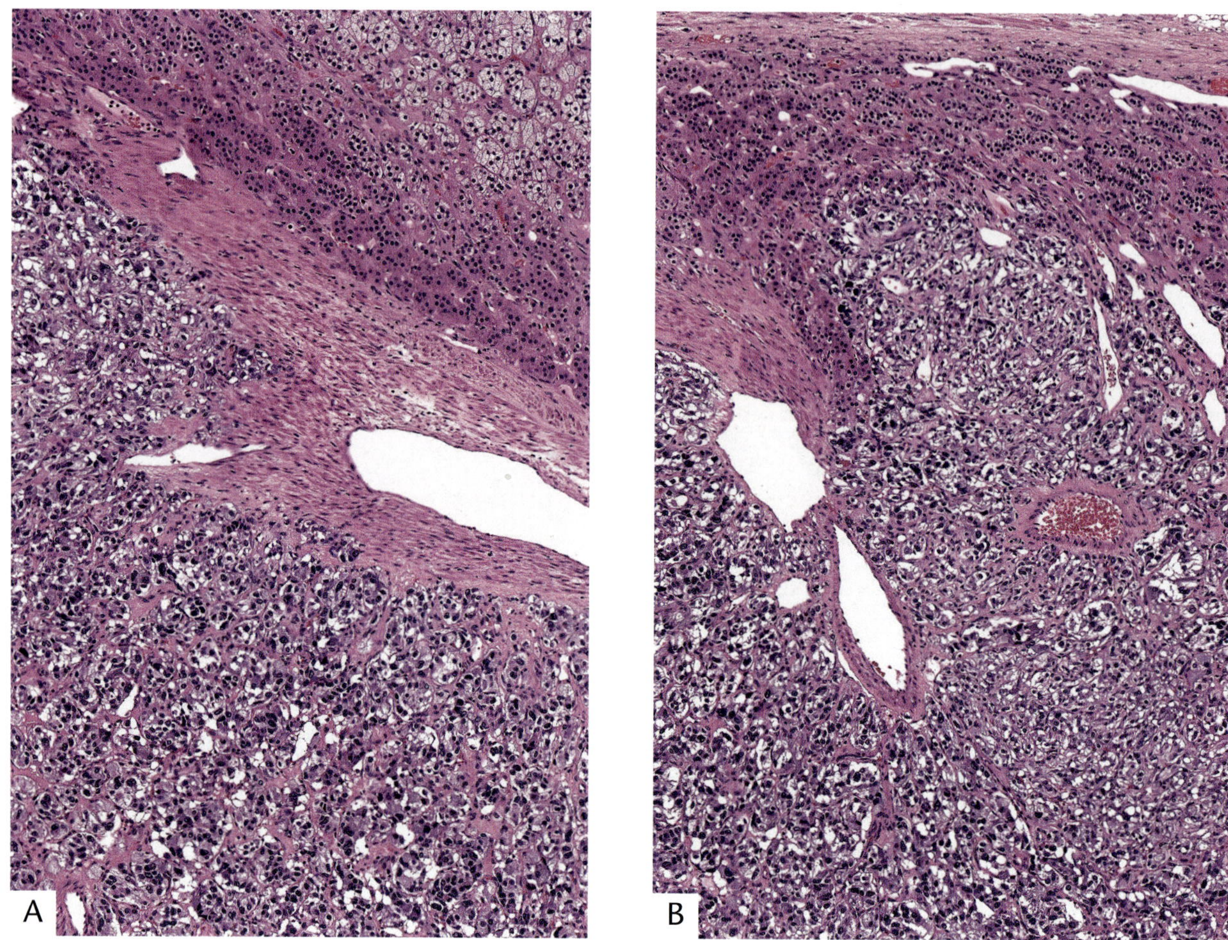

Figure 8-17

PHEOCHROMOCYTOMA: PERIPHERY

The periphery of these tumors is extremely variable, with some showing clear encapsulation (A) and others lacking a capsule, and instead showing interdigitation with the surrounding adrenal cortex (B). Both patterns are seen in the same tumor in different areas in these figures.

not be associated with degeneration, suggesting alternative mechanisms (fig. 8-18). The presence of a myxoid, hyalinized stroma with edema (fig. 8-19) should prompt consideration of von Hippel-Lindau as a pathogenetic mechanism (37); in this setting, there may be a thick fibrous capsule and numerous small blood vessels. Amyloid has been reported (38–40) but is rare (40).

The cytology of these lesions can be highly variable. The neoplastic cells may be large and polygonal with abundant granular basophilic cytoplasm and indistinct cell borders (fig. 8-20) or they may be lightly eosinophilic (fig. 8-21). Occasionally, they are round with chromophobic or amphophilic cytoplasm, and can be mistaken for an epithelial neuroendocrine tumor (fig. 8-22). They may have elongated processes that interdigitate with and surround adjacent cells. Occasional tumors have a spindle cell morphology (figs. 8-23, 8-24). The tumor cell nuclei are usually round to oval or elongated; they have prominent nucleoli that may be multiple (fig. 8-25). There may be nuclear inclusions that result from intranuclear cytoplasmic invaginations (fig. 8-25). Some tumors exhibit significant nuclear pleomorphism (fig. 8-26), forming bizarre giant hyperchromatic nuclei and sometimes multinucleated cells (figs. 8-27, 8-28). Occasional tumors have prominent mitotic figures (fig. 8-29).

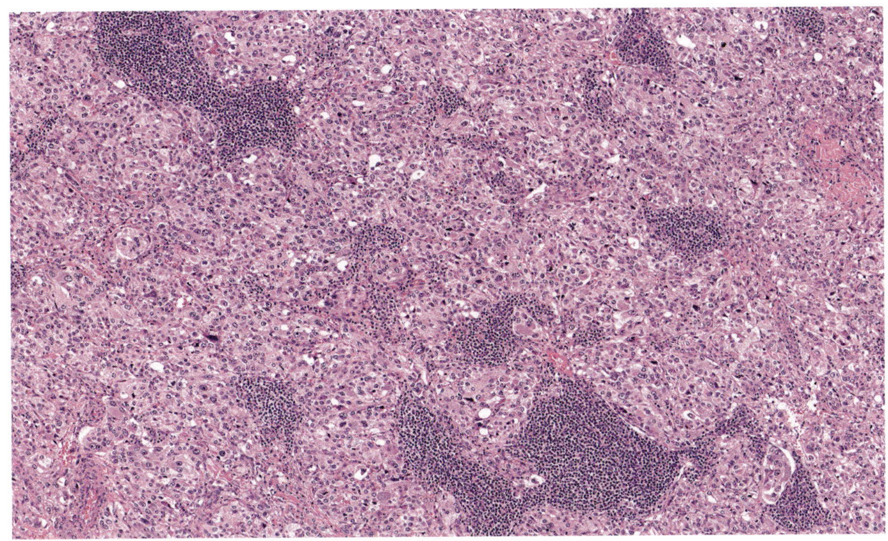

Figure 8-18

PHEOCHROMOCYTOMA: FOCAL CHRONIC INFLAMMATION

In occasional pheochromocytomas there is a focal lymphoplasmacytic infiltrate.

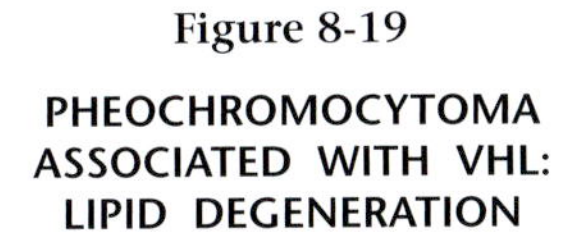

Figure 8-19

PHEOCHROMOCYTOMA ASSOCIATED WITH VHL: LIPID DEGENERATION

There is extensive lipid degeneration of the stroma (see gross appearance in fig. 8-12).

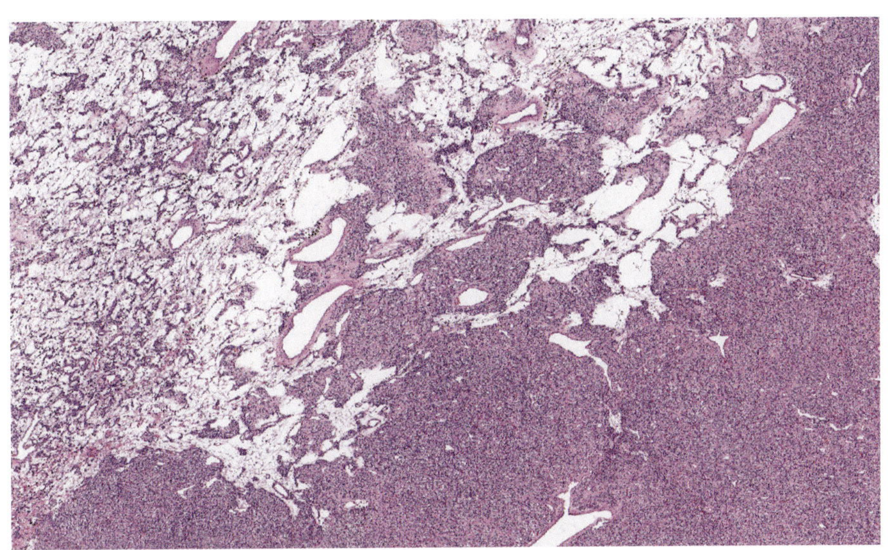

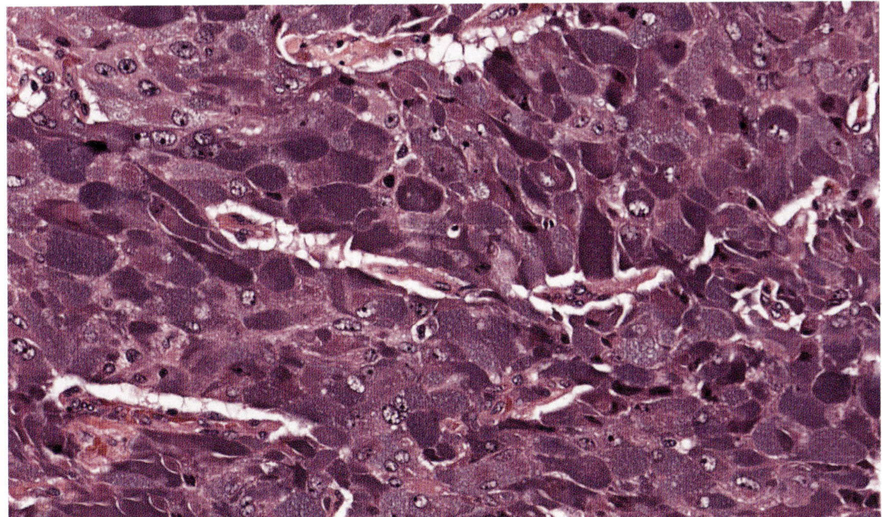

Figure 8-20

PHEOCHROMOCYTOMA

The tumor cells are large and polygonal, with abundant granular basophilic cytoplasm and indistinct cell borders.

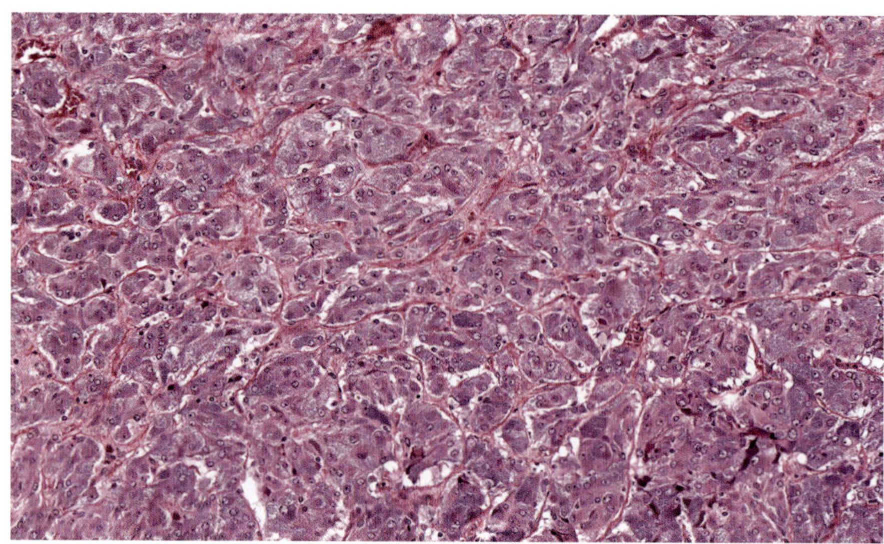

Figure 8-21

PHEOCHROMOCYTOMA

Large and polygonal tumor cells have pale eosinophilic cytoplasm and indistinct cell borders.

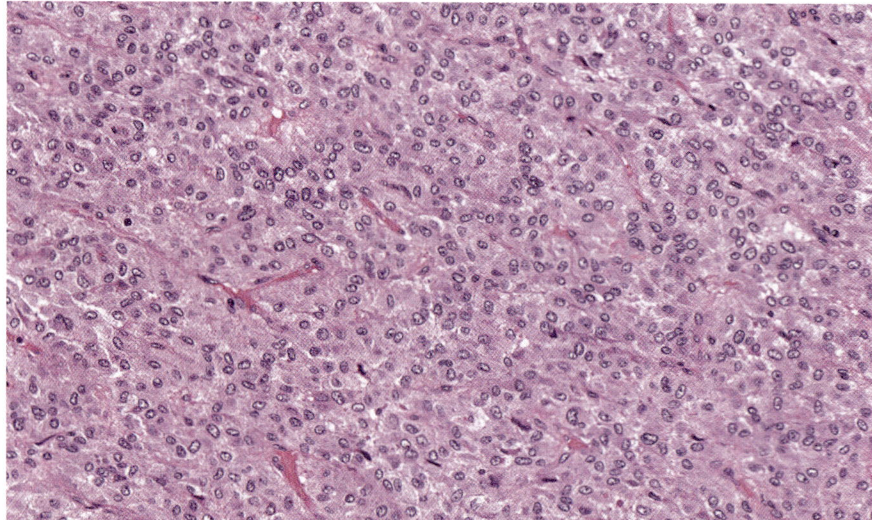

Figure 8-22

PHEOCHROMOCYTOMA

The tumor cells are round and regular, with abundant cytoplasm, resembling cells of an epithelial neuroendocrine tumor.

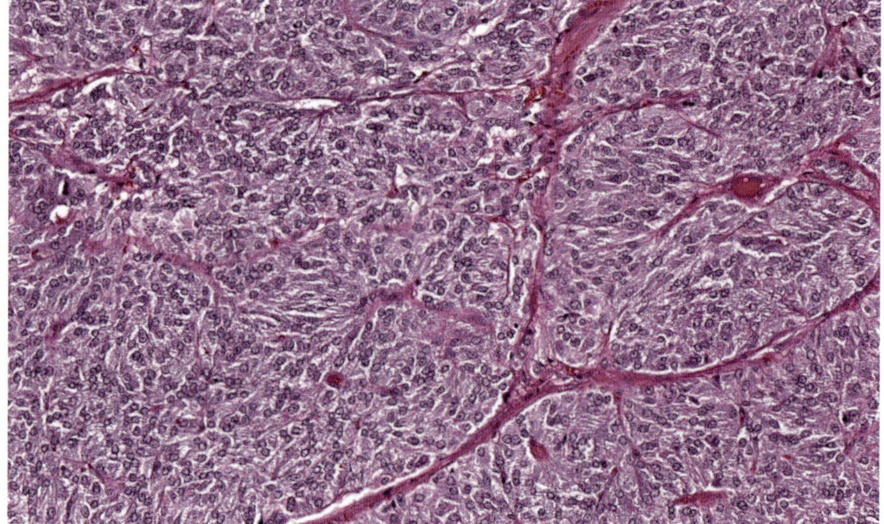

Figure 8-23

PHEOCHROMOCYTOMA

Discohesive, elongated, and spindle-shaped tumor cells with ill-defined borders.

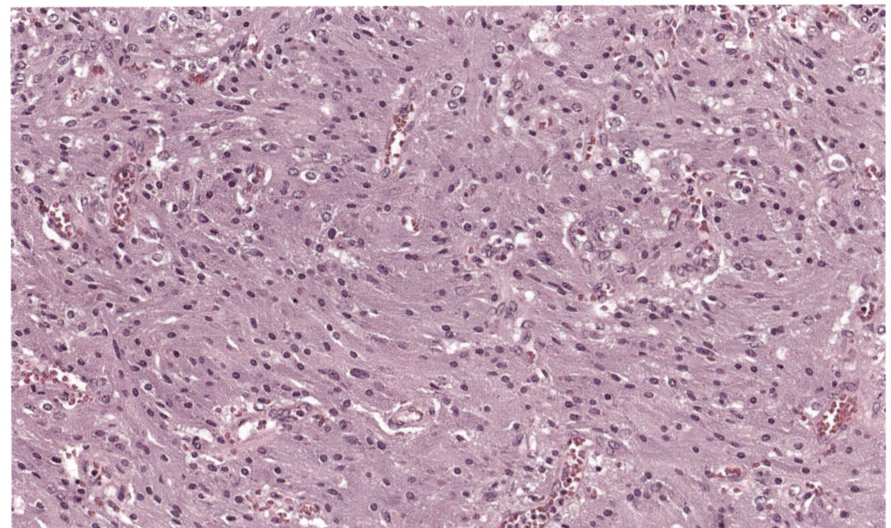

Figure 8-24

PHEOCHROMOCYTOMA

Spindle cell morphology is also seen in tumors with solid cohesive growth.

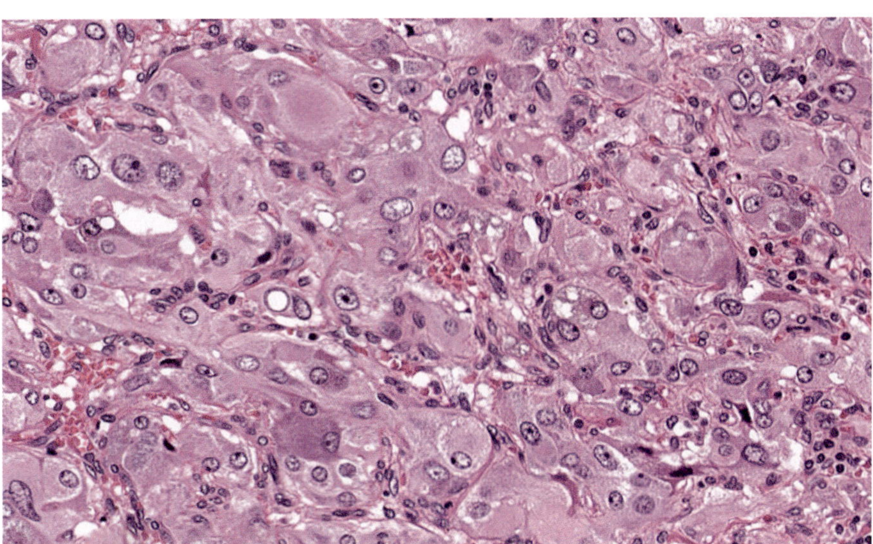

Figure 8-25

PHEOCHROMOCYTOMA

The tumor cell nuclei usually have prominent and occasionally multiple nucleoli, and some have clear pseudoinclusions.

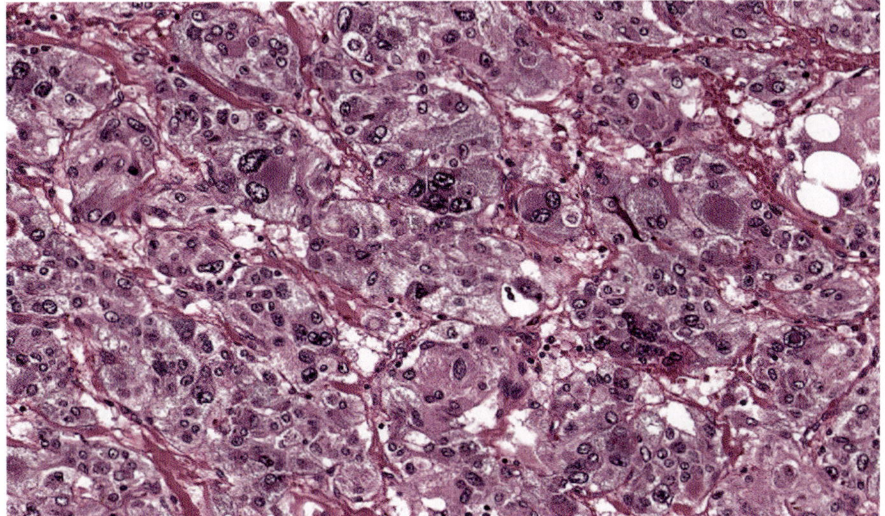

Figure 8-26

PHEOCHROMOCYTOMA

Some tumors have significant nuclear pleomorphism.

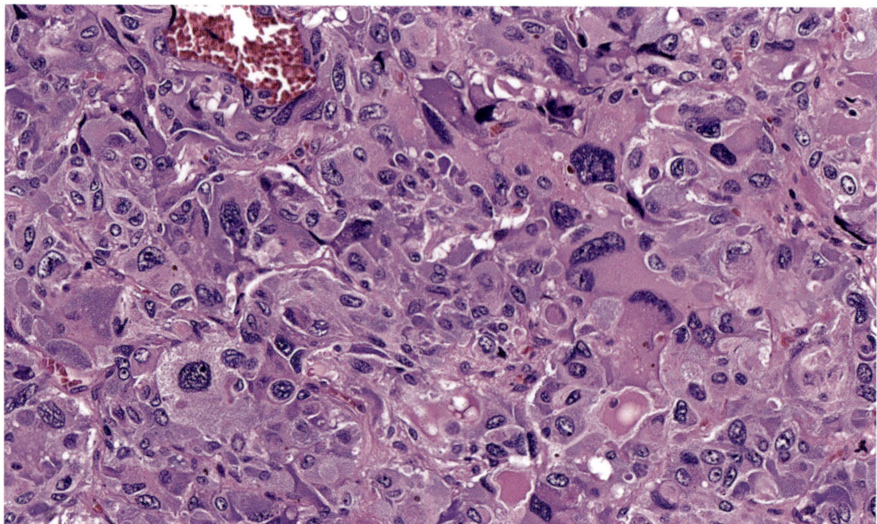

Figure 8-27

PHEOCHROMOCYTOMA

There may be large multinucleated giant cells.

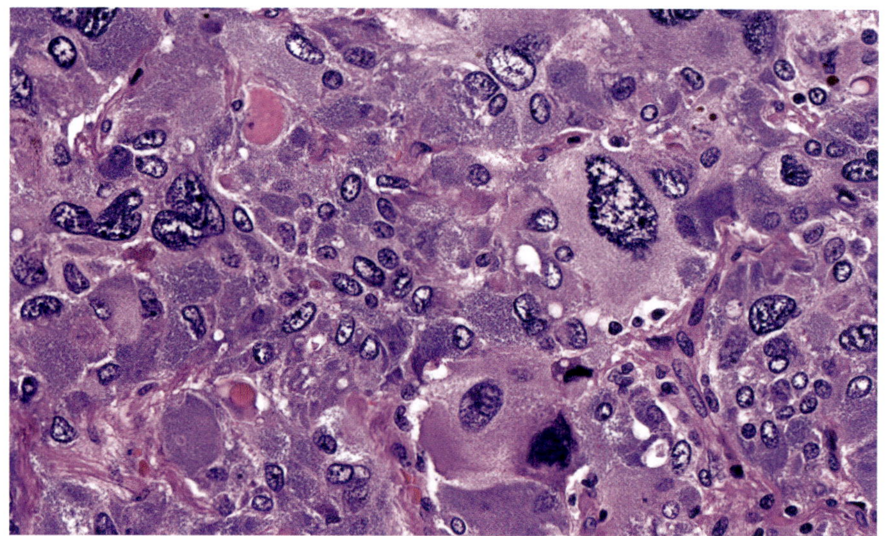

Figure 8-28

PHEOCHROMOCYTOMA

Some tumor cells are large, with multinucleated forms and striking nuclear atypia.

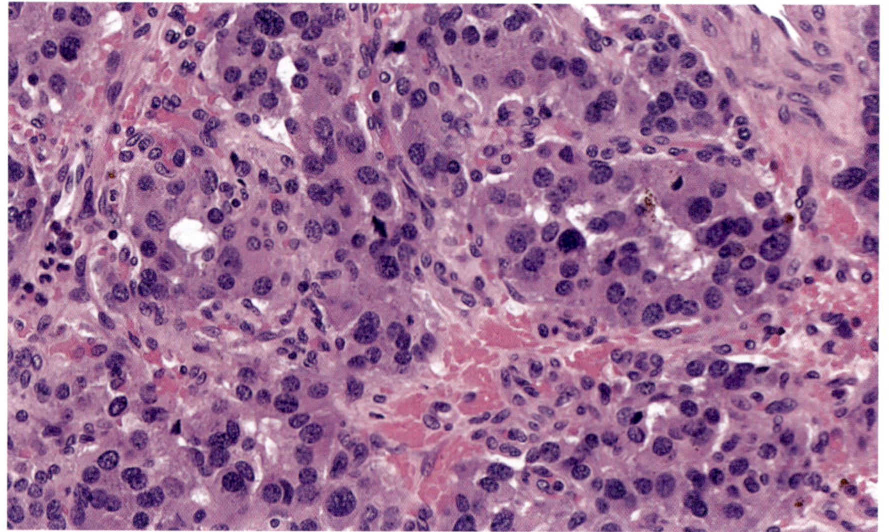

Figure 8-29

PHEOCHROMOCYTOMA: MITOSES

Mitoses are prominent in this tumor.

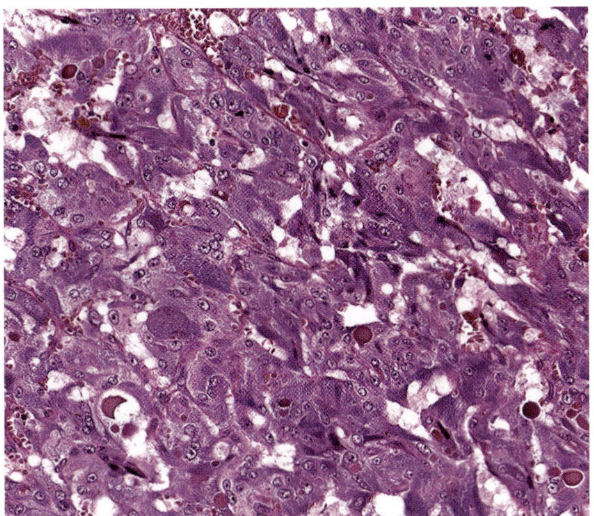

Figure 8-30

PHEOCHROMOCYTOMA: INTRACYTOPLASMIC HYALINE GLOBULES

Scattered tumor cells harbor large round cytoplasmic globules with a hyaline appearance.

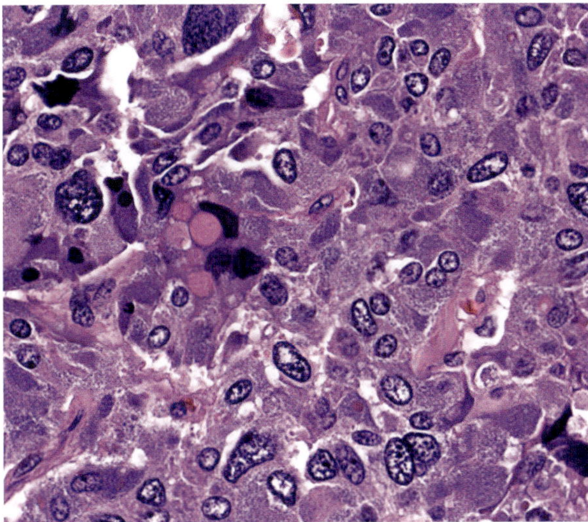

Figure 8-31

PHEOCHROMOCYTOMA: INTRACYTOPLASMIC HYALINE GLOBULES

These hyaline structures may be found in a few scattered cells or, rarely, as in this case, may be numerous.

Intracytoplasmic hyaline globules (figs. 8-30, 8-31) are another enigmatic finding since their etiology and significance are not known; they are common as an occasional feature that may be easily missed on hematoxylin and eosin (H&E)-stained slides but can be highlighted with periodic acid–Schiff (PAS) and are resistant to diastase predigestion. These globules have historically been associated with a better prognosis (41). They were absent in patients with von Hippel-Lindau disease (VHL) compared with those with multiple endocrine neoplasia type 2 (MEN2) in one study (37) but these associations were not reproduced in other studies.

Melanin pigment can also be found in these tumors, and when abundant, gives rise to a "black pheochromocytoma" (fig. 8-32) (42,43). In most cases, pigmented cells in pheochromocytomas contain neuromelanin or lipofuscin pigment. However, the presence of melanosomes and premelanosomes in these tumors and the formation of melanin pigment are attributed to the common neural crest origin of melanocytes and adrenal medulla. Immunohistochemical stains to demonstrate neuroendocrine markers and rule out melanoma should be employed in questionable cases to ensure that the more common diagnosis of metastatic melanoma is not missed (see chapter 12).

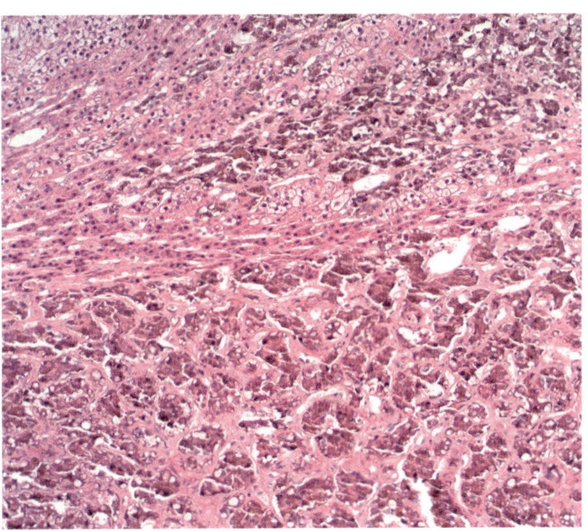

Figure 8-32

PIGMENTED PHEOCHROMOCYTOMA

The accumulation of melanin in the tumor cells can mimic melanoma and these tumors can appear black on gross examination.

Occasional cells resembling single neurons or ganglion cells with Nissl substance (fig. 8-33) are common and not of diagnostic significance.

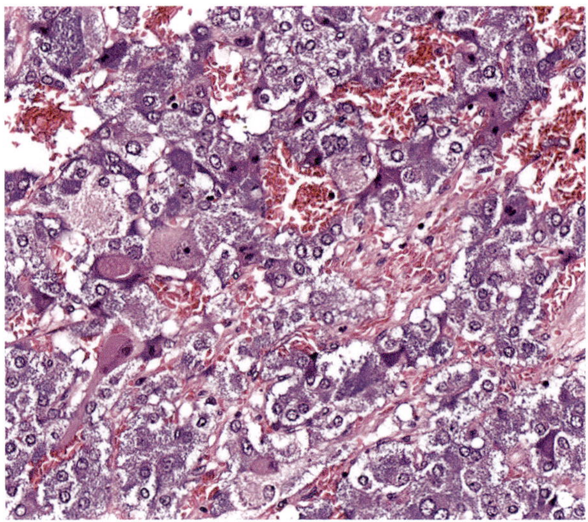

Figure 8-33

PHEOCHROMOCYTOMA: GANGLION CELLS

The presence of a few scattered single neurons is a feature of some pheochromocytomas and should not alter the diagnosis; when more conspicuous, they warrant classification as a composite tumor (see fig. 8-53).

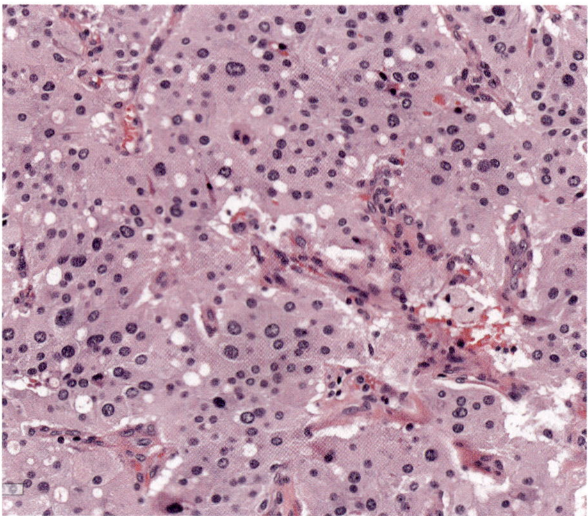

Figure 8-34

PHEOCHROMOCYTOMA: ONCOCYTIC CHANGE IN SDHX-RELATED DISEASE

Some pheochromocytomas in patients with SDHx-related disease have abundant granular eosinophilic cytoplasm.

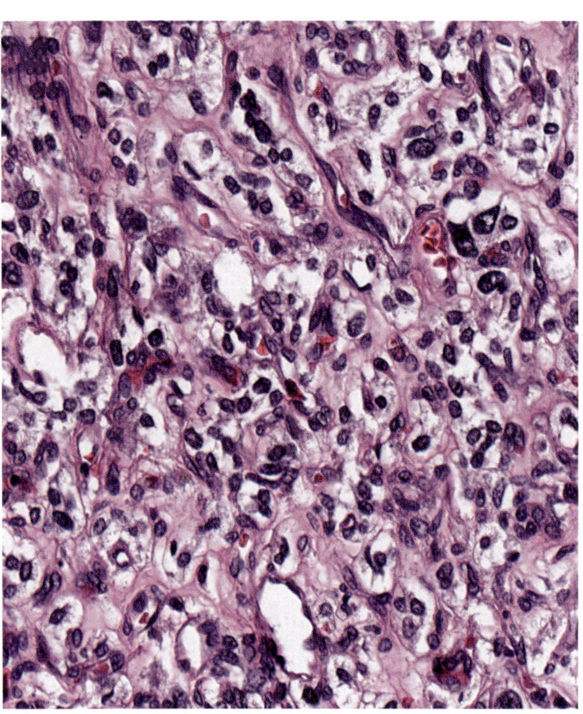

Figure 8-35

PHEOCHROMOCYTOMA: CLEAR CELL CHANGE IN VHL DISEASE

The presence of clear cell change in a pheochromocytoma is associated with VHL disease.

However, these features should prompt a search for more complete features of neuroblastoma, ganglioneuroblastoma, or ganglioneuroma that characterize composite tumors (see below). Areas with these features may be often be identified with the aid of immunohistochemistry, which shows relatively weak, often punctate staining for chromogranin A and dense staining for S-100 protein and neurofilament proteins (44).

Some cytologic features are suggestive of a pathogenetic mechanism. Tumors with *SDHx* mutations may be composed of large cells with abundant eosinophilic granular cytoplasm (fig. 8-34). The presence of abundant cytoplasm with lipid vacuoles filling the cytoplasm is a feature of tumors associated with VHL (fig. 8-35) (36,37). The presence of multifocal disease should prompt assessment of genetic predisposition, and the identification of adrenal medullary hyperplasia raises the possibility of MEN2 (45). An individual MEN2 tumor may also have a multinodular gross appearance (fig. 8-36).

Immunohistochemical Findings

Pheochromocytoma has a characteristic immunohistochemical profile. The tumor cells exhibit strong and diffuse positivity for the neuroendocrine markers INSM1, synaptophysin,

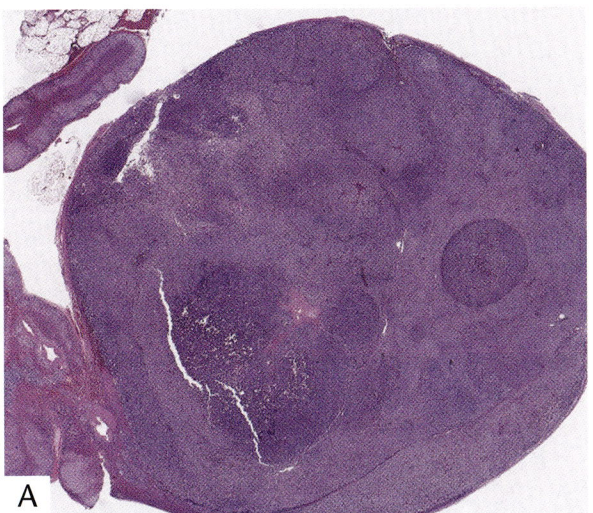

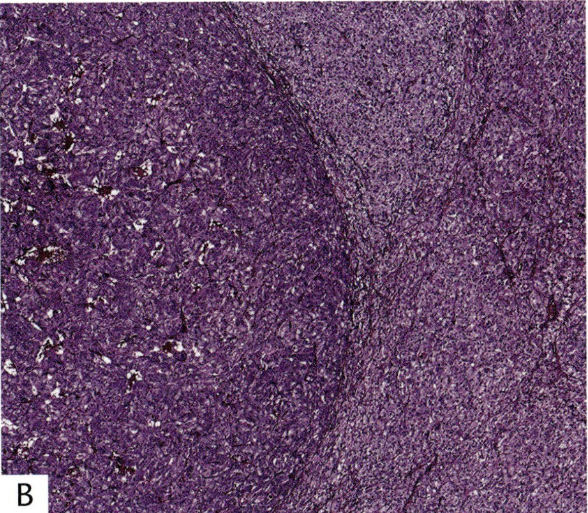

Figure 8-36

MULTIFOCAL PHEOCHROMOCYTOMA IN MEN2

A: There are multiple nodules with distinct characteristics.

B: The confluent morphologically distinct nodules may represent continued expansion of multiple independent clones already present at the nodular hyperplasia stage.

and chromogranin A (fig. 8-37) and are negative for keratins. Paraganglionic chief cells usually express the transcription factor GATA3; while this biomarker is expressed in many nonendocrine tissues, the only other neuroendocrine cells that express GATA3 are normal and neoplastic parathyroid and pituitary cells. The most specific markers are cytoplasmic tyrosine hydroxylase (fig. 8-38) and other catecholamine pathway enzymes such as dopamine β-hydroxylase (DBH) and phenylethanolamine N-methyltransferase (PNMT). They also may show variably intense nuclear and cytoplasmic positivity for S-100 protein, but the value of this stain is in highlighting sustentacular cells that surround the zellballen of tumor cells (fig. 8-39A).

Sustentacular cells are difficult to identify on H&E staining but are present in paraganglia, including the nontumorous medulla and in paragangliomas including pheochromocytomas; the loss of sustentacular cells has been suggested to be a feature of aggressive behavior (46,47) and metastatic foci may completely lack this component. Another biomarker for these cells is SOX10 (fig. 8-39B), a nuclear transcription factor that also distinguishes these cells that resemble glia and Schwann cells (48). Sustentacular cells, however, are not the hallmark of paragangliomas and pheochromocytomas since they may also be found in epithelial neuroendocrine tumors (48–50). Another cell type found in pheochromocytomas is the perivascular monocyte, which can mimic sustentacular cells in shape and distribution (51).

Immunohistochemistry also identifies other hormones produced by pheochromocytoma. The normal adrenal medulla is known to produce peptides, including enkephalin, somatostatin, substance P, and vasoactive intestinal peptide (VIP), which are normally identified in subsets of adrenal neurons (52). Serotonin can also be found in normal adrenal medulla, largely as a result of uptake (53). Ectopic expression of these and other hormones is relevant in patients who manifest the clinical features of hormone excess. Pheochromocytomas produce excess adrenocorticotropic hormone (ACTH) (fig. 8-40) and/or corticotropin-releasing hormone (CRH) resulting in Cushing syndrome (54,55), growth hormone-releasing hormone (GHRH) (fig. 8-41) causing acromegaly (56), and vasoactive intestinal peptide (VIP) giving rise to the watery diarrhea, hypokalemia, and alkalosis syndrome (57). In the case of local ACTH excess, the surrounding cortex is usually massively hyperplastic and lipid depleted (fig. 8-40).

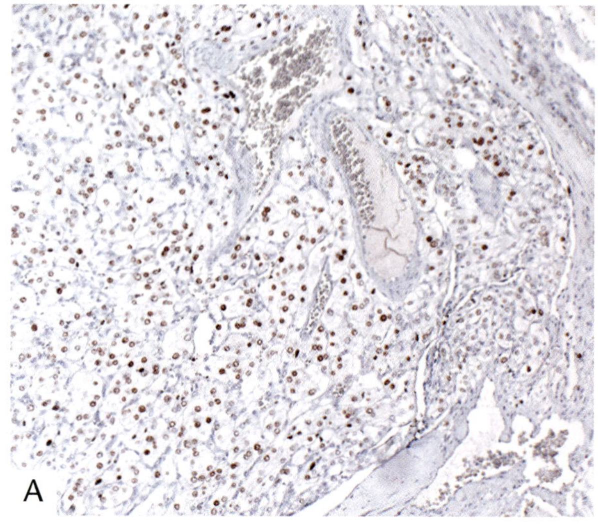

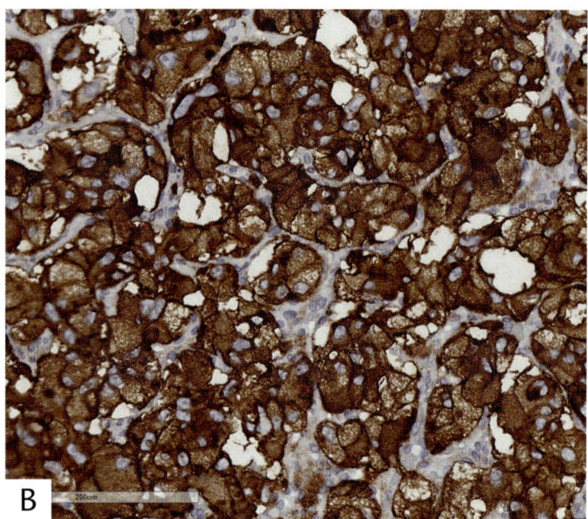

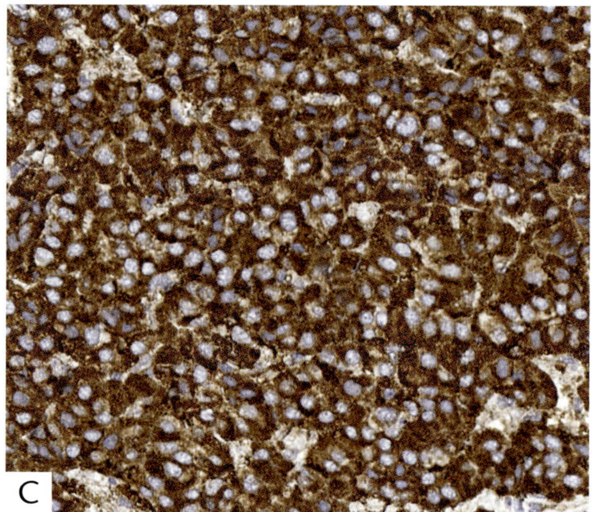

Figure 8-37

PHEOCHROMOCYTOMA: NEUROENDOCRINE MARKERS

The tumor cells express nuclear INSM1 (A), cytoplasmic synaptophysin (B), and cytoplasmic chromogranin A (C). These biomarkers indicate that this is a neuroendocrine neoplasm but are not specific for pheochromocytoma.

Ki-67 labeling can be used to assess proliferation as in other neuroendocrine neoplasms (fig. 8-42). There is evidence that a higher labeling index portends a worse prognosis, but the literature on this is limited (58–61).

Immunohistochemistry can and should be used to guide assessment of genetic predisposition. The most widely used tool is immunostaining for SDHB. This test is useful because it has been found that mutation in any SDH-related gene (*SDHA, SDHB, SDHC, SDHD*, or *SDH-AF2*) results in destabilization of the SDH complex and loss of SDHB expression in tumor cells, with retained positivity in the stromal internal controls (fig. 8-43). Because of the high incidence of *SDHx*-related disease among paragangliomas, including pheochromocytomas, this screen is very useful (62–65). The technical optimization of this stain is critical to ensure the highest sensitivity and specificity; some investigators have reported pitfalls such as blushy positivity in tumors with *SDHD* mutations, and it has also been shown that tumors with *VHL* mutations have reduced SDHB staining, misinterpreted as lost. Rarely, SDHB loss reflects a somatic mutation in the tumor (62). If SDHB is lost, some laboratories reflex to staining for SDHA which is lost in patients with *SDHA* mutations (66).

Other tests for germline predisposition depend on the biochemical profile of the patient

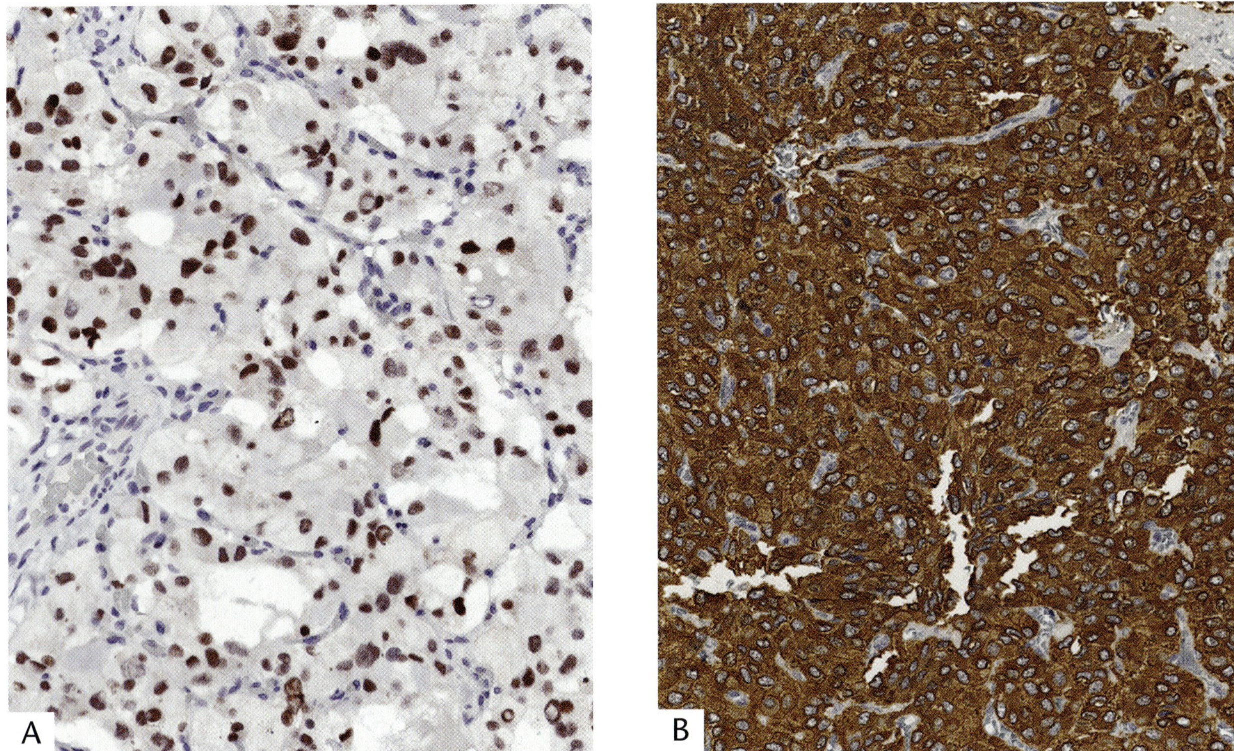

Figure 8-38

BIOMARKERS OF PHEOCHROMOCYTOMA

The most specific biomarkers of pheochromocytoma are nuclear GATA3 (A) and cytoplasmic tyrosine hydroxylase (B).

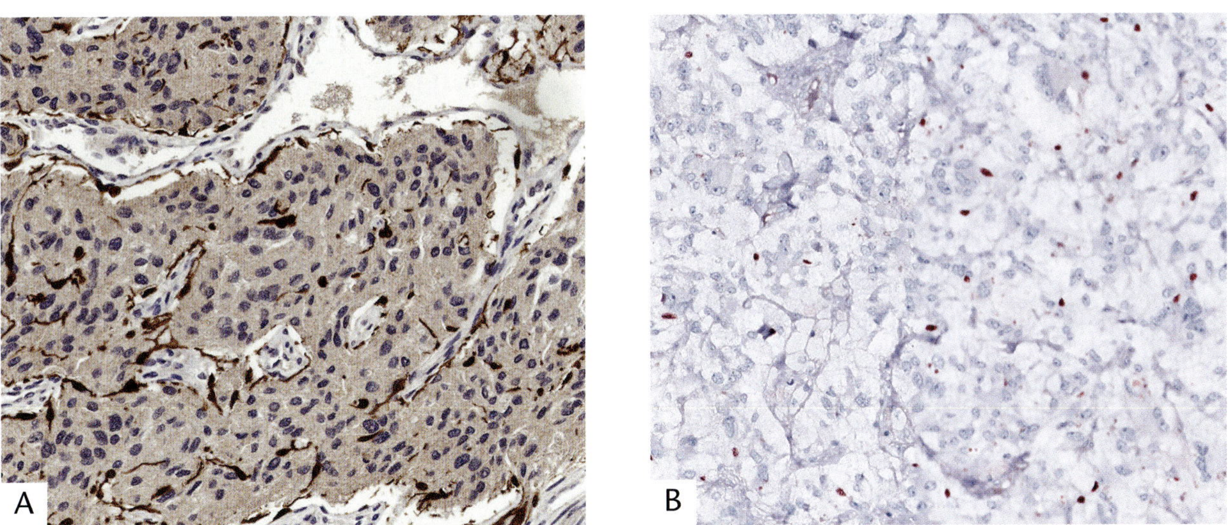

Figure 8-39

PHEOCHROMOCYTOMA: SUSTENTACULAR CELLS

A: Antibodies to S-100 protein show strong nuclear and cytoplasmic positivity in sustentacular cells, identified easily even if the tumor cells also show diffuse reactivity.

B: The nuclei of sustentacular cells are positive for SOX10.

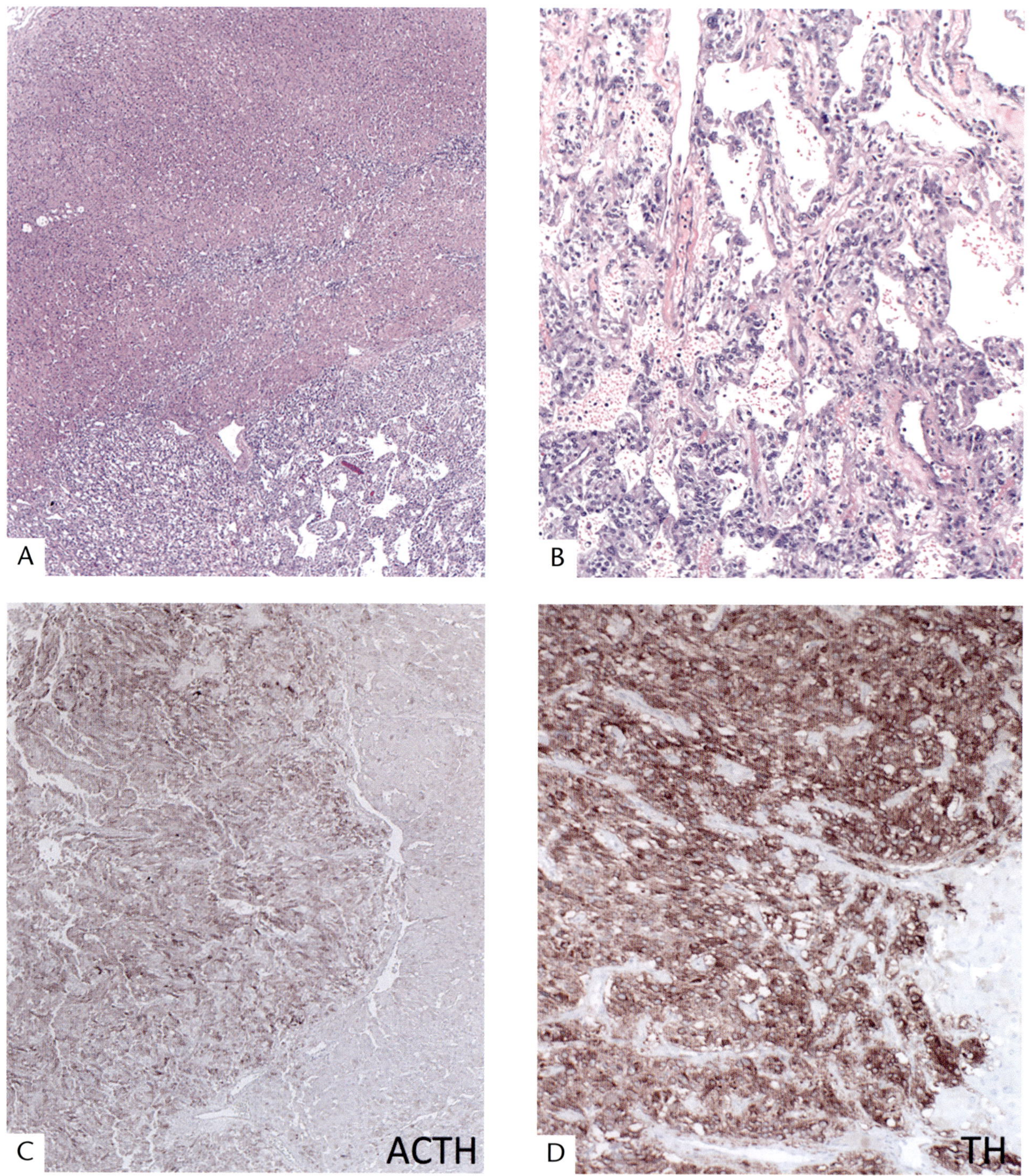

Figure 8-40

PHEOCHROMOCYTOMA PRODUCING ADRENOCORTICOTROPIC HORMONE (ACTH) AND CAUSING CUSHING SYNDROME

A: The adrenal cortex is markedly thickened and lipid depleted.
B: The tumor is a highly vascular pheochromocytoma, not a cortical tumor that can cause Cushing syndrome.
C: The expression of tyrosine hydroxylase confirms that this is pheochromocytoma.
D: The pheochromocytoma expresses ACTH, explaining the clinical presentation of ectopic ACTH syndrome.

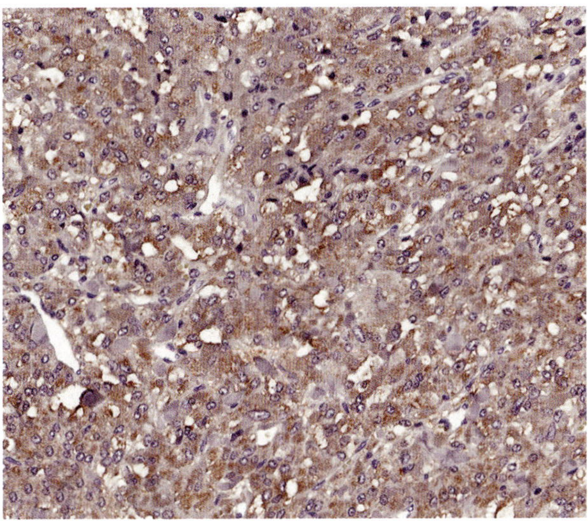

Figure 8-41

PHEOCHROMOCYTOMA PRODUCING GROWTH HORMONE-RELEASING HORMONE (GHRH) AND CAUSING ACROMEGALY

The presentation of this patient with acromegaly prompted staining of the tumor for GHRH; the tumor is diffusely positive and resection of the adrenal lesion resulted in cure of the acromegaly.

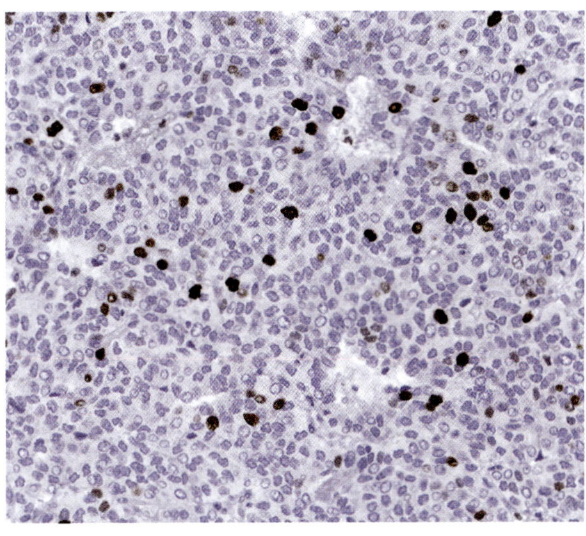

Figure 8-42

PHEOCHROMOCYTOMA: KI-67 LABELING

The Ki-67 labeling index is 12.7 percent as determined by an automated image analysis algorithm.

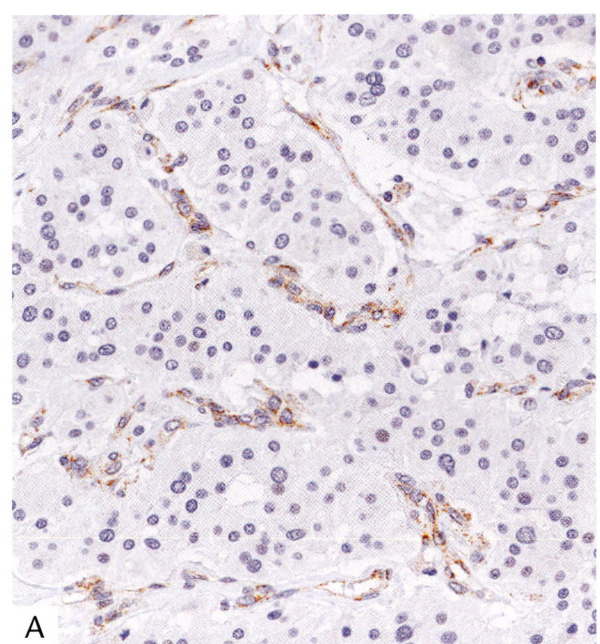

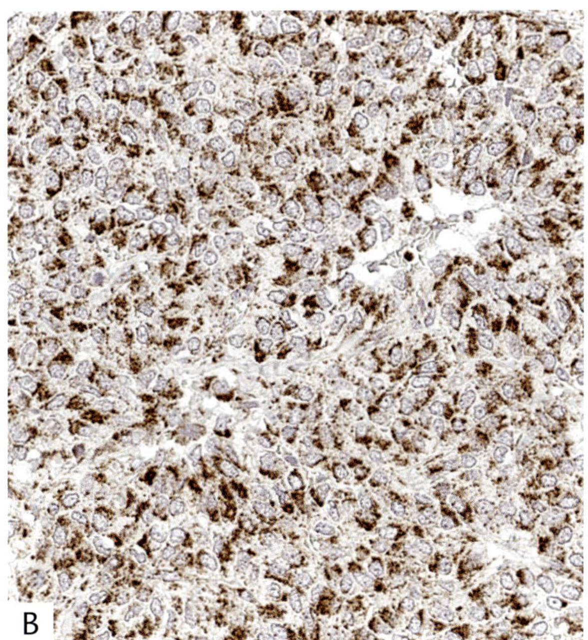

Figure 8-43

PHEOCHROMOCYTOMA: SDHB STAINING

A: Loss of SDHB immunoreactivity with retained positivity in the stroma serving as an internal control provides strong evidence of SDHX-related disease and should prompt genetic investigation.

B: The presence of intact reactivity for SDHB suggests that the tumor is not related to *SDHx* mutation.

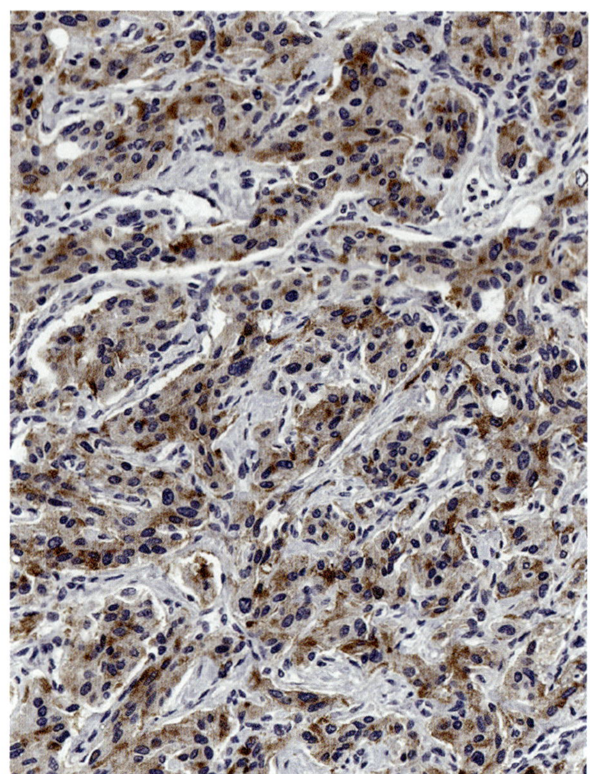

Figure 8-44

PHEOCHROMOCYTOMA: INHIBIN STAINING

The presence of inhibin positivity is associated with any hypoxic pathway disease, including SDHx- and VHL-related disease and can be confirmed by staining for SDHB and CAIX.

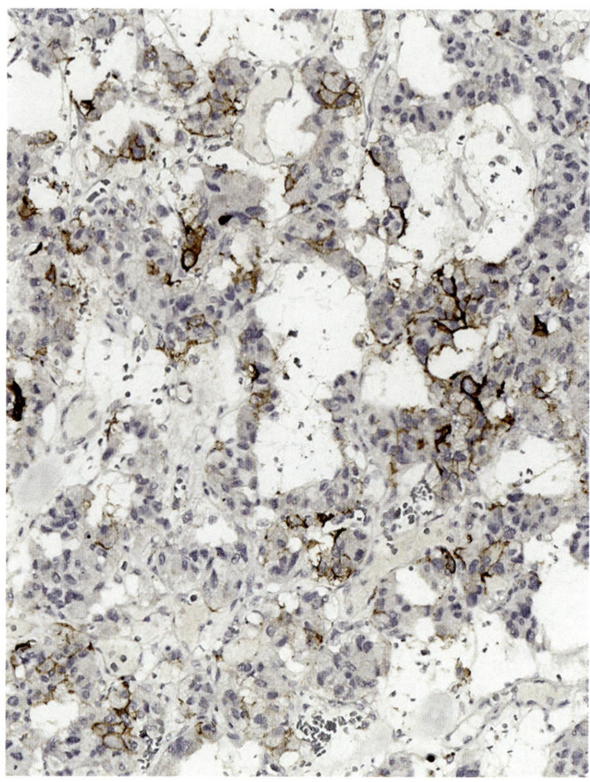

Figure 8-45

VHL-ASSOCIATED PHEOCHROMOCYTOMA: CAIX STAINING

CAIX positivity is associated with VHL-related disease; it may be diffuse or focal as in this case.

and associated morphologic findings. For example, expression of inhibin A (fig. 8-44) is a feature of tumors associated with cluster 1 disease (67). Tumors with stromal edema with vacuolated lipid-rich cytoplasm in the tumor can be stained for inhibin and CAIX (68); the former in association with intact SDHB indicates another form of cluster 1 disease and the latter is more specific for VHL-associated disease (fig. 8-45), although it may reflect somatic *VHL* mutation. There are no diagnostic morphologic features of fumarate hydratase (FH) deficiency on H&E, but a patient with a cluster 1 biochemical phenotype and no evidence of SDHx or VHL disease can be identified using immunohistochemical staining for FH; there is loss of cytoplasmic reactivity with retained staining in the stroma and positive staining for 2-succinocysteine (2SC) (fig. 8-46) (4).

Molecular Genetic Findings

The genetic alterations in pheochromocytoma have been well studied (60,69–71). These tumors have an exceptionally high incidence of genetic predisposition, with 30 to 40 percent attributed to germline mutations in genes that play a role in adrenal medullary cell metabolism (4); it appears that this high prevalence of genetic predisposition is a phenomenon of western society and is lower in India and China (72,73). The familial syndromes and the relevant genes are discussed in detail in chapter 9.

In sporadic tumors, the genes that are commonly mutated include *HRAS, RET, NF1, EPAS1, VHL, ATRX,* and *CSDE1*; fusions involving *MAML3, BRAF, NGFR,* and *NF1* have also been described (60). These tumors have a low mutational burden, with only rare tumors showing multiple

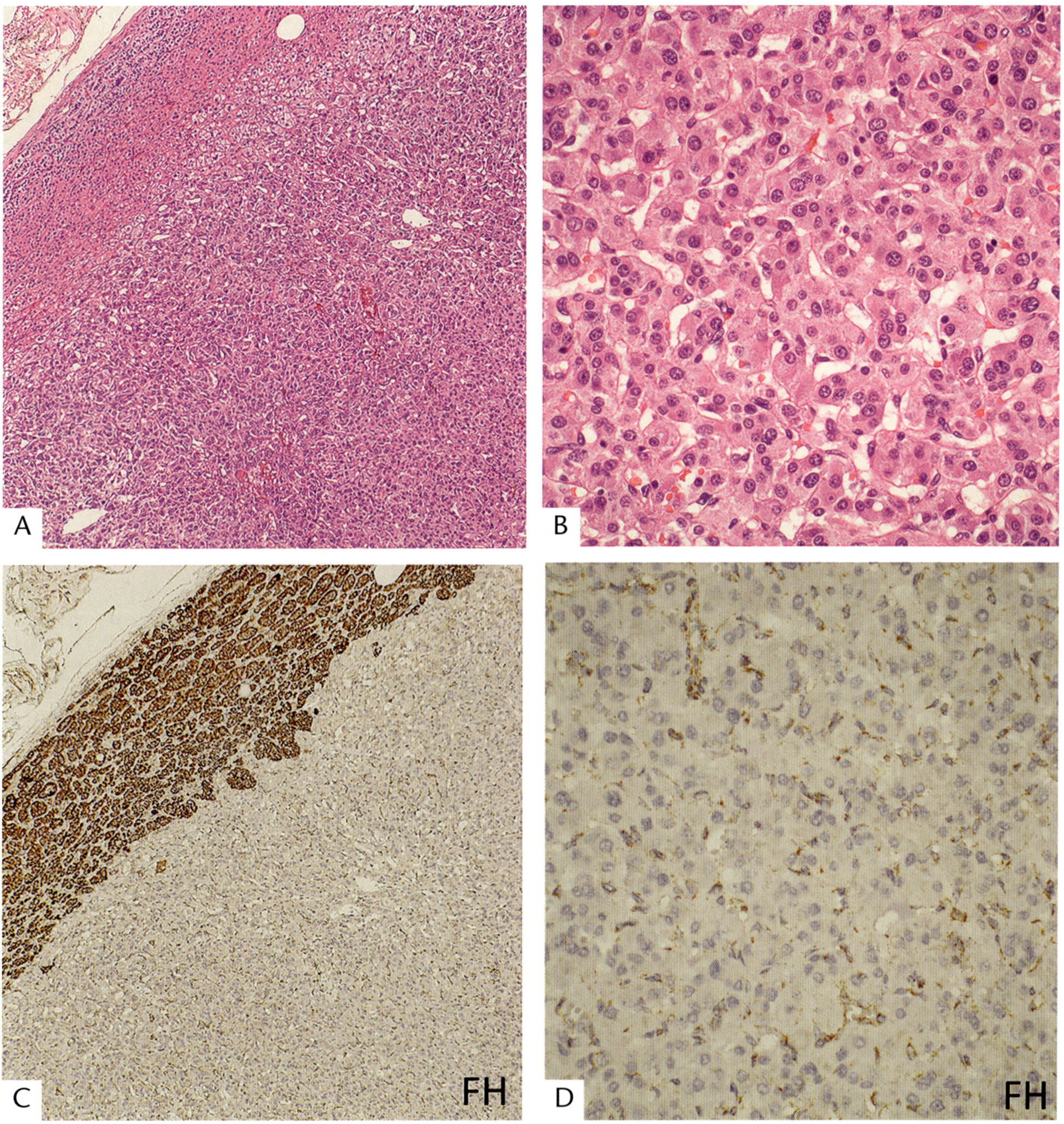

Figure 8-46

FUMARATE HYDRATASE (FH)-ASSOCIATED PHEOCHROMOCYTOMA

A,B: The histologic features of FH-related pheochromocytoma are not distinctive.
C,D: Immunohistochemical staining for FH shows loss of cytoplasmic reactivity with retained staining in the stroma.

mutations, most frequently *ATRX* somatic mutation in a tumor with a germline *SDHx* alteration. Loss of heterozygosity is common in hotspots related to the critical tumor suppressor genes.

Genetic alterations that predict aggressive disease include *MAML3* fusion, *SDHB* germline mutation, somatic mutation in *SETD2* or ATRX, high somatic mutation burden, a WNT-altered

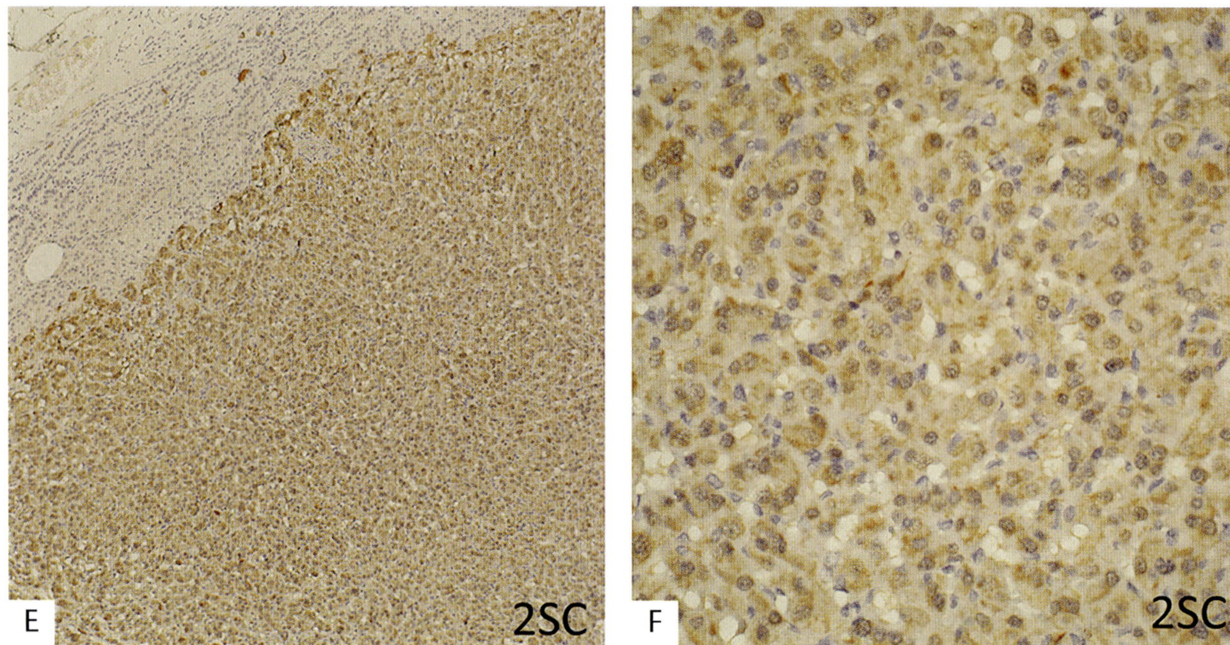

Figure 8-46, continued
E,F: There is accumulation of 2-succinocysteine (2SC). (Courtesy of Dr. A. Gill, Sydney, Australia)

mRNA subtype, and a hypermethylation subtype (60). Interestingly, ethnic differences show mutations in *RAS* and *FGFR1* more common in Asians and *NF1* mutations more common in Europeans (73).

A syndrome of polycythemia associated with paraganglioma (including pheochromocytoma) and somatostatinoma has been identified to be caused by *HIF2A* mutations that may be somatic, only in the tumors, or germline mosaic (74,75). The mutations cluster adjacent to an oxygen-sensing proline residue, resulting in proteins with increased stability compared to wild-type HIF2A, due to reduced HIF2a hydroxylation by prolyl hydroxylase and binding to the VHL protein, hence, resembling VHL syndrome.

Other Associated Features

Brown fat is thermogenic and catecholamines have been implicated in the transformation of adipose tissue into brown fat. Some studies have shown an increase in periadrenal brown fat in patients with pheochromocytoma, however, this is not a consistent finding and remains speculative (30). The association of pheochromocytoma with a tumor of brown fat, known as hibernoma, may be coincidental.

Cardiomyopathy may occur in patients with prolonged catecholamine excess, causing arrhythmia and cardiac failure. Weight loss, ileus, abdominal distention, pain, and other features of increased adrenergic tone should all be considered as manifestations of pheochromocytoma and may be found in the history of a patient with an unexpected pheochromocytoma at autopsy.

Differential Diagnosis

The identification of a chromogranin A-positive tumor in the adrenal gland is not sufficient to confirm the diagnosis of pheochromocytoma. The differential diagnosis includes a neuroendocrine tumor (NET) metastatic to the adrenal gland; the distinction relies on showing negative staining for keratin, and positivity for tyrosine hydroxylase and GATA3. In contrast, NETs commonly express keratins and also express transcription factors appropriate to their site of origin (e.g., TTF1 for lung and thyroid gland, CDX2 for small bowel, SATB2 for colon, and various others for pancreas).

As indicated above, some pheochromocytomas can mimic adrenal cortical neoplasms. Synaptophysin and inhibin can be expressed

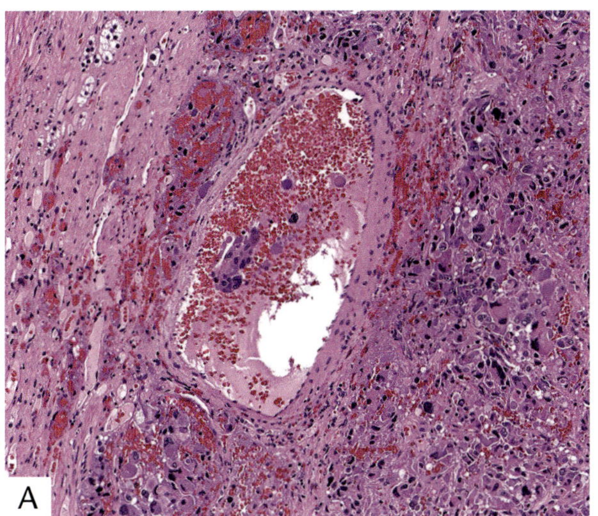

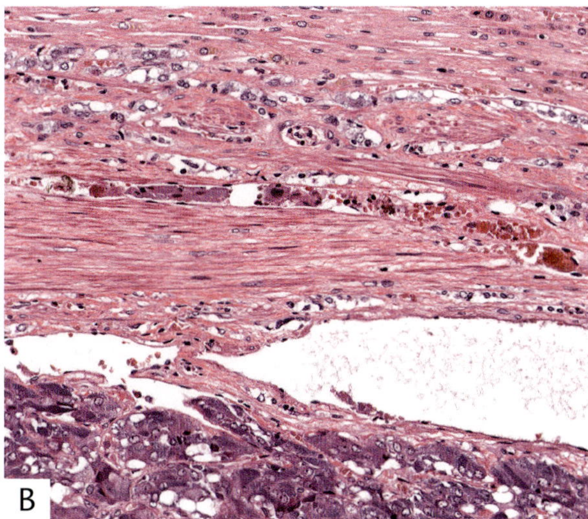

Figure 8-47

PHEOCHROMOCYTOMA: ANGIOINVASION

A,B: The presence of vascular invasion is a worrisome finding, but unlike in other neoplasms, it is not reliably associated with the development of metastatic disease.

by both these entities and are therefore not useful in the distinction. Adrenal cortical tumors express nuclear SF1 as the most specific biomarker (76).

Treatment and Prognosis

Most pheochromocytomas can be surgically removed, however, some large tumors invade critical structures, precluding complete resection. While all pheochromocytomas are considered potentially malignant, the prediction of aggressive behavior and metastasis has been the subject of a number of studies. Preoperative clinical features portending more aggressive disease include tumor size greater than 5 cm, radiologic evidence of gross vessel invasion, and germline SDHB pathogenic variant.

The Pheochromocytoma of the Adrenal Gland Scaled Score (PASS) was developed to address the histopathologic features that are associated with the development of metastatic disease (77). This score incorporates multiple parameters based on architecture and cellular morphology, invasion, necrosis, and proliferation. Invasion includes vessels (score = 1) (fig. 8-47), capsule (score = 1), and periadrenal adipose tissue (score = 2). Architectural and cytologic parameters are large nests or diffuse growth (score = 2), high cellularity (score = 2), cellular monotony (score = 2), tumor cell spindling (score = 2), profound nuclear pleomorphism (score = 1), and hyperchromasia (score = 1). Focal or confluent necrosis is important (score = 2) (fig. 8-48). Proliferative features include increased mitotic figures (more than 3 per 10 high-power fields; score = 2) and atypical mitotic figures (score = 2) (fig. 8-49). Calculating these weighted features identifies aggressive tumors with a score 4 or more whereas those that behave in a benign fashion have a score of less than 4. This scoring system offers promise but has not proven to be highly reproducible due to interobserver variability (78,79), and while it usually identifies tumors that will likely be cured by surgery, it overestimates risk for tumors with a score of 4 or more (80).

Another proposal, the Grading System for Adrenal Phaeochromocytoma and Paraganglioma (GAPP), was designed to also include extra-adrenal tumors and incorporate proliferation using Ki-67 (59). This system evaluates histologic pattern, cellularity (fig. 8-50), comedo-type necrosis, capsular/vascular invasion, Ki-67 labeling index, and catecholamine type, classifying tumors as score 0 to 10, with subsequent grading as well-differentiated (0-2 points), moderately

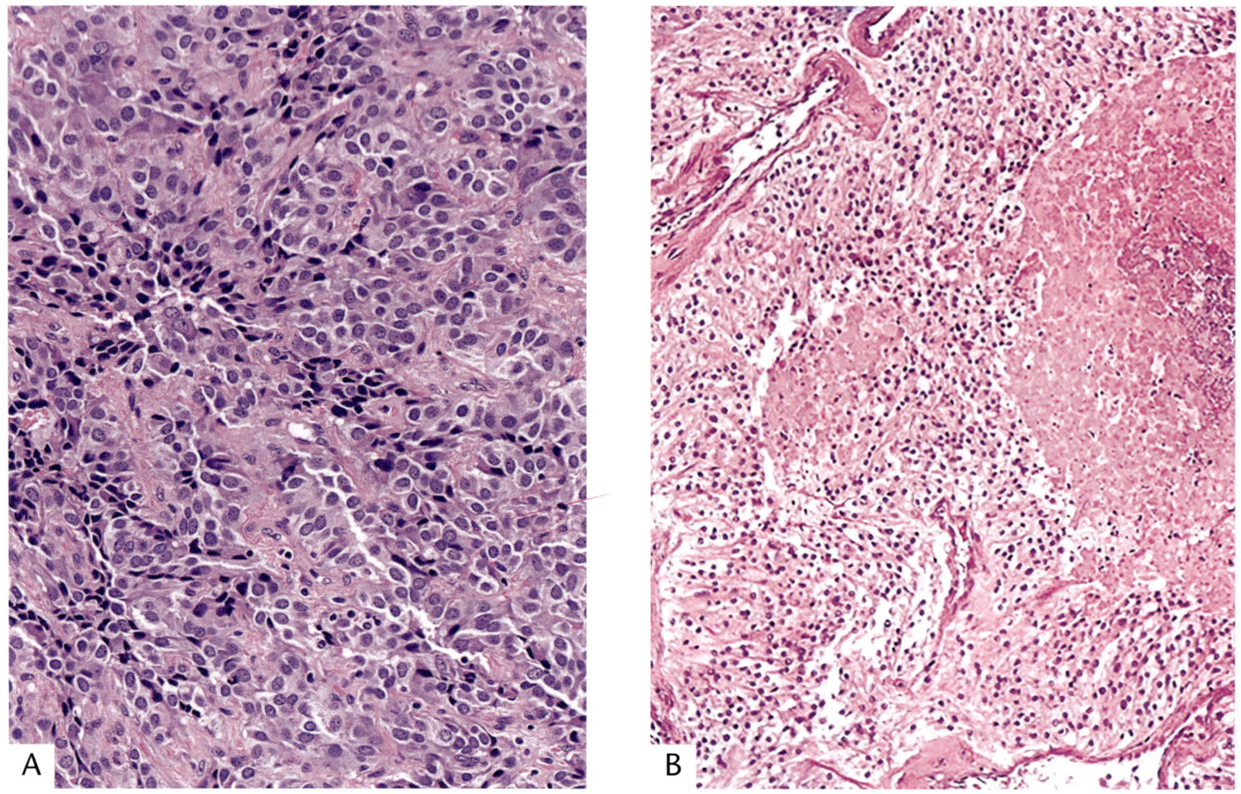

Figure 8-48

PHEOCHROMOCYTOMA: NECROSIS

The presence of tumor cell necrosis (A) is a worrisome feature but not significant alone; comedonecrosis (B) is considered to be indicative of more aggressive disease.

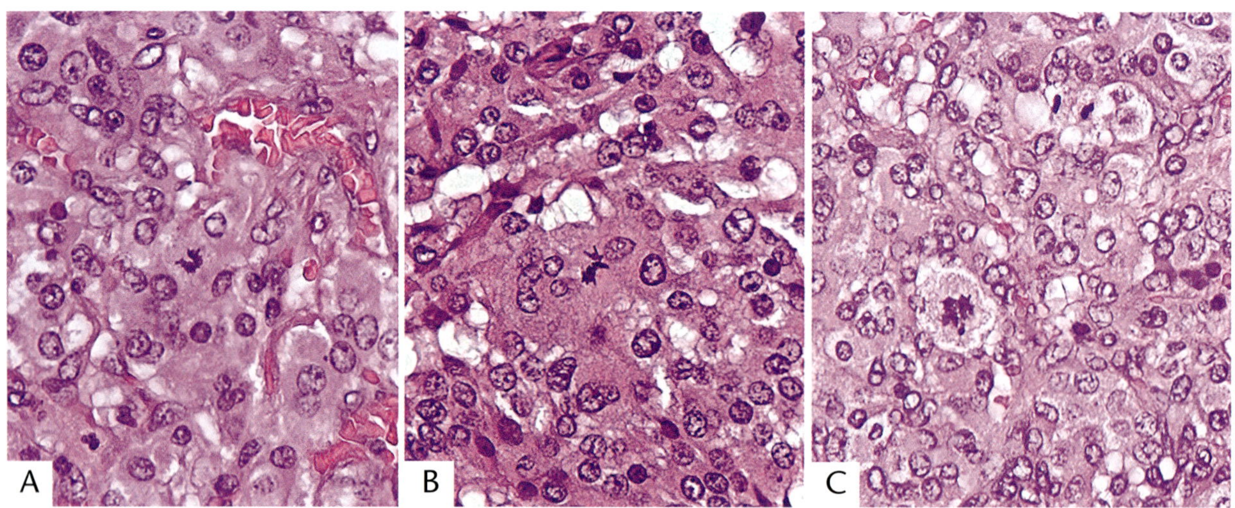

Figure 8-49

PHEOCHROMOCYTOMA: ATYPICAL MITOSES

A–C: The presence of atypical mitoses should prompt careful examination for other features suggestive of aggressive behavior.

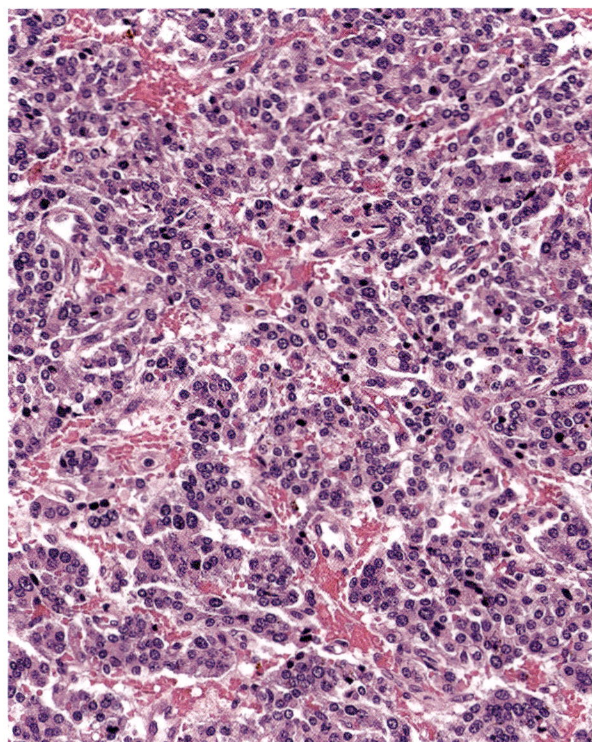

Figure 8-50

PHEOCHROMOCYTOMA: CELLULARITY

Increased cellularity due to crowded small tumor cells is a histologic feature of aggressive pheochromocytomas.

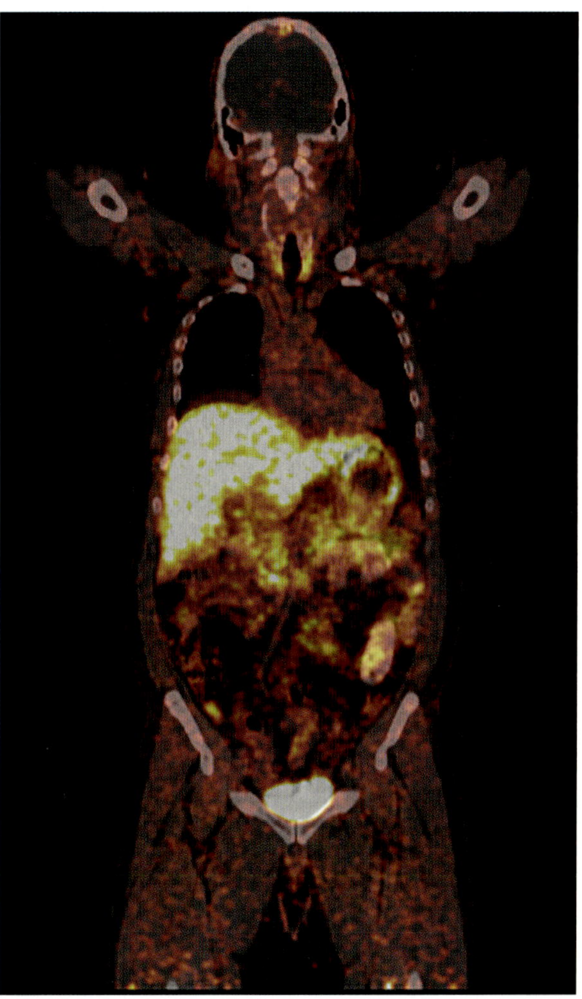

Figure 8-51

PHEOCHROMOCYTOMA: ^{68}GA DOTATATE SCAN

Multifocal osteolytic skeletal lesions demonstrate dotatate avidity consistent with a somatostatin receptor-positive tumor. There is also activity in the retroperitoneum showing recurrent disease around the resection margin of the primary tumor.

differentiated (3-6 points), and poorly differentiated (7-10 points). Again, the proposal suffers from a lack of clear criteria for morphologic features and does not incorporate molecular data that are helpful, since *SDHB* mutation is an important correlate of metastatic behavior (81). Even tumors that rank low on the scoring system have metastatic disease in almost 4 percent of cases (80).

A third proposal builds on the PASS and GAPP scores with the addition of SDHB immunohistochemistry (61). The composite pheochromocytoma/paraganglioma prognostic score (COPPS) scoring system assesses tumor size, necrosis, and vascular invasion, as well as loss of S-100 protein and SDHB immunoreactivities, adding yet another immunohistochemical biomarker to the calculation. Since the diagnosis of pheochromocytoma no longer separates malignant from benign disease, it remains to be seen which if any of these are best at predicting metastasis or more aggressive local growth in the case of unresectable primary disease.

Molecular data suggest that aggressiveness in pheochromocytoma is associated with *MAML3* fusions, and *ATRX* and *CSDE1* mutations (60,82). Somatic *ATRX* mutations and *TERT* activation are common in tumors with hereditary *SDHB* mutations, and when present, may be additional independent risk factors for metastasis (83).

When metastatic spread occurs, it is either via lymphatics or hematogenous routes to involve lymph nodes, bone, liver, and lung (figs. 8-51, 8-52). The 5-year survival rate of patients with metastatic disease is variable, with reports ranging from 50 to 70 percent (84,85).

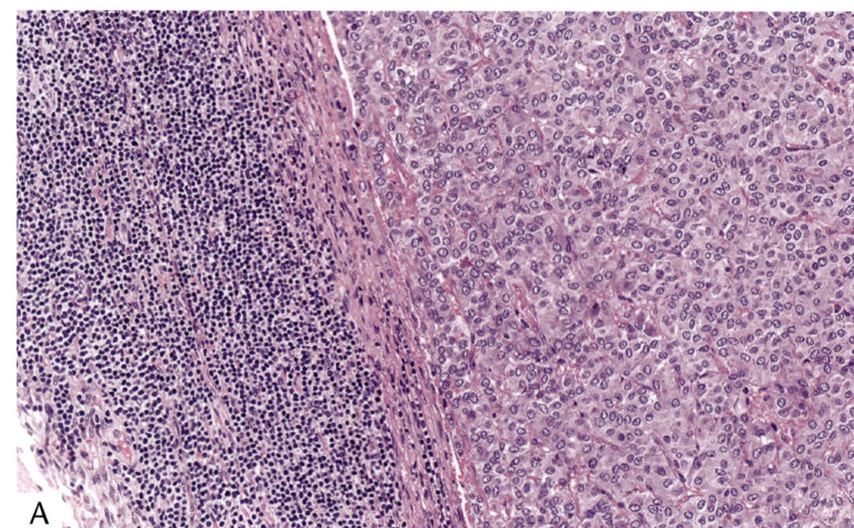

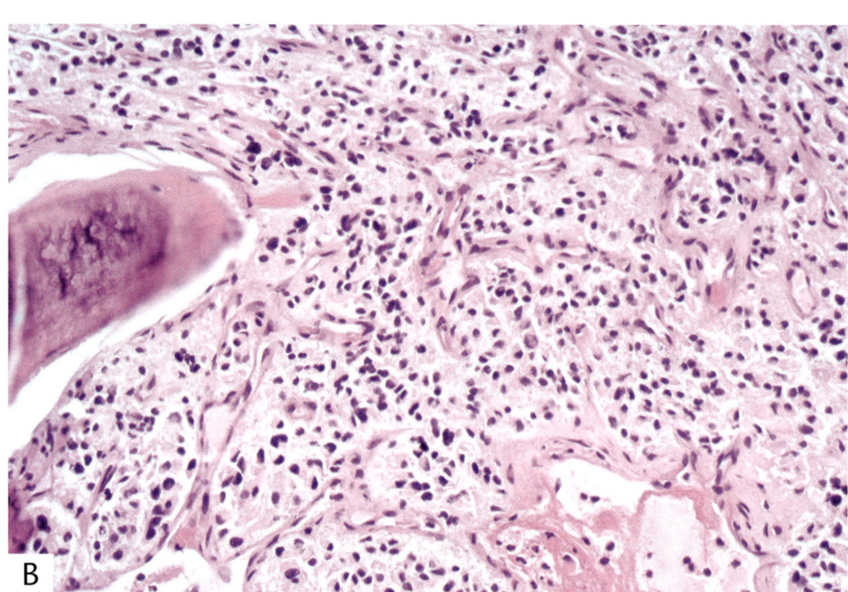

Figure 8-52

METASTATIC PHEOCHROMOCYTOMA

The presence of paraganglioma in lymph node (A) or bone (B) is unequivocal evidence of metastatic spread, since these tissues do not harbor paraganglia and therefore the disease cannot represent multifocal primary lesions in a genetically predisposed individual.

Distinction Between Primary and Metastatic Tumors

The diagnosis of metastatic disease must be made with great caution, especially in patients with known germline predisposition who are likely to develop multifocal disease in various locations, including liver and lung, where these tumors have long been erroneously thought to always be metastatic (10). Since paraganglia occur everywhere except in brain, bone, and lymph nodes, these are the only sites that can be accepted a priori. Further, lymph node metastasis must be confirmed histologically and not by imaging alone, since the distribution of lymph nodes in many locations is similar to the distribution of paraganglia.

Since sustentacular cells are usually sparse or absent in metastases, their presence in substantial numbers may help to support a diagnosis of primary tumor versus metastasis. However, their sparsity does not necessarily confirm the opposite because the numbers of these cells can vary markedly both within and between primary tumors (fig. 8-53). Sustentacular cells are currently believed to be non-neoplastic (86), and the factors that recruit them to tumors are currently unknown.

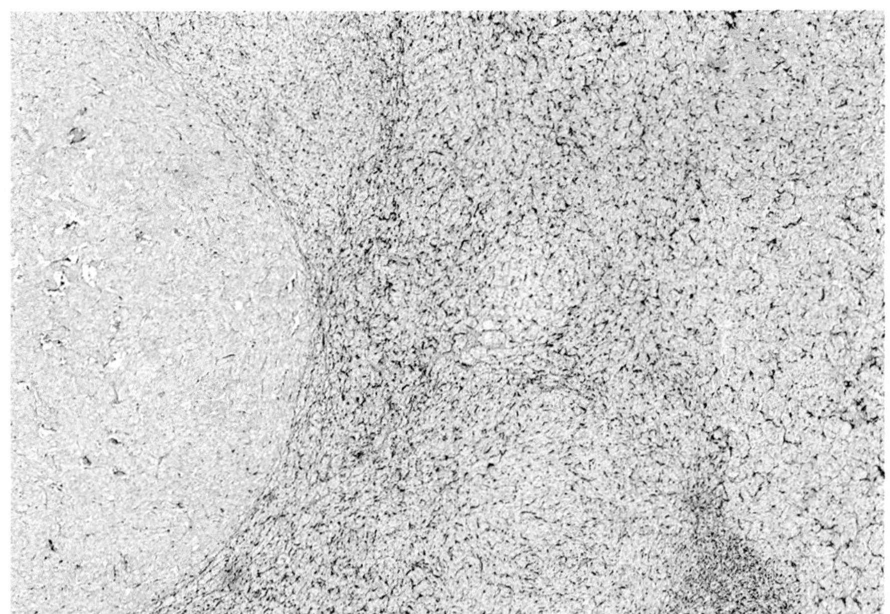

Figure 8-53

PHEOCHROMOCYTOMA: VARIATION IN SUSTENTACULAR CELLS

Immunohistochemical stain for S-100 protein shows intratumoral variation in density of sustentacular cells in a pheochromocytoma from a patient with MEN2A (same patient as figure 8-36). The sharply circumscribed nodule at the left is almost totally devoid of sustentacular cells, providing a cautionary note that the absence of these cells does not necessarily indicate that a tumor is metastatic or aggressive.

Synoptic Reporting and Staging

Synoptic reporting has been recommended for completeness of data. Pathologists should appreciate that it also offers the ability to capture data points for analysis by tumor registries, biobanks, and other investigational activities. Experts in the field created a synoptic report for these tumors (87) and this has been updated recently by the International Collaboration on Cancer Reporting (ICCR) (88). This dataset should be used for pheochromocytomas as well as paragangliomas (see chapter 9), including composite ones, however, neuroblastoma, ganglioneuroblastoma, and ganglioneuroma have a separate synoptic worksheet and should not be included in this protocol.

The American Joint Committee on Cancer (AJCC) 8th edition included the first staging system for pheochromocytomas (and paragangliomas) (Table 8-1). It incorporates the clinical and pathologic features that have been shown to predict metastasis and survival, including tumor size, location, and presence and location of metastatic disease. Because it was published before the 2017 WHO classification, it still distinguishes benign and malignant tumors, but despite this limitation and others discussed in chapter 9, it represents a starting point to collect information that may help determine prognosis. In a retrospective study, it was shown to correlate with the presence of metastases, disease-related death, and PASS scores as well as mutational and expression clusters (89).

COMPOSITE PHEOCHROMOCYTOMA

Composite pheochromocytomas are rare tumors composed of pheochromocytoma admixed with other elements including ganglioneuroma, ganglioneuroblastoma, neuroblastoma, or peripheral nerve sheath tumor (41,44,90–95). These neoplasms are all derived from related embryonic lineages in the neural crest. Their normal cellular counterparts are present in normal adrenal glands and can individually give rise to tumors alone, without pheochromocytoma. The morphology of these individual lesions is discussed in detail in chapter 11.

Composite pheochromocytomas with neuronal or neuroblastic components are described in adults and in children (96,97) but are most common in adults. They occur sporadically or in patients with hereditary pheochromocytoma syndromes, the latter most often neurofibromatosis type 1. They have been reported in a few patients with MEN2 (95,97–99) and one with a germline *MAX* mutation (100). There is no proven case with SDH-related disease (101). A sporadic case has been reported with mutation of *ATRX* (102). Composite features are often unexpected and found only upon histologic examination of

Table 8-1

AMERICAN JOINT COMMITTEE ON CANCER (AJCC)
8TH EDITION STAGING FOR PHEOCHROMOCYTOMA (AND PARAGANGLIOMA)

Tumor (T)

T Category	T Criteria
TX	Primary tumor cannot be assessed
T1	PCC size <5 cm in greatest dimension, no extra-adrenal invasion
T2	PCC size ≥5 cm, sympathetic PGL of any size, no extra-adrenal invasion
T3	Tumor of any size with invasion of surrounding tissues (e.g., liver, pancreas, spleen, kidneys)

Regional Lymph Node (N)

N Category	N Criteria
NX	Regional lymph nodes cannot be assessed
N0	No lymph node metastasis
N1	Lymph node positive

Distant Metastasis (M)

M Category	M Criteria
M0	No distant metastasis
M1	Distant metastasis
M1a	Distant metastasis to only bone
M1b	Distant metastasis to only lymph nodes/liver or lung
M1c	Distant metastasis to bone plus multiple other sites

Anatomic Stage and Prognostic Groups

When T is...	And N is...	And M is...	Then the stage group is...
T1	N0	M0	Stage I
T2	N0	M0	Stage II
T1	N1	M0	Stage III
T2	N1	M0	Stage III
T3	N0	M0	Stage III
T3	N1	M0	Stage III
Any T	Any N	M1	Stage IV

a clinically typical pheochromocytoma. Consequently, patients usually present with signs of catecholamine excess, but VIP hypersecretion may also be present (103,104).

The presence of scattered ganglion cells, neuroblasts, or Schwann cells does not constitute sufficient evidence of a composite tumor; there must be architectural evidence of a distinct tumor type defined by the same morphologic features applied to the corresponding pure tumor. Adequate sampling is necessary to ensure appropriate assessment of the proportion of each component, including the amount of schwannian stroma and the presence or absence of neuroblastic differentiation within the mass to ensure accurate subtyping of these tumors.

The most common type of composite pheochromocytoma is ganglioneuroma within a tumor that is predominantly pheochromocytoma (fig. 8-54). As in pure ganglioneuroma, large polygonal ganglion cells, often containing cytoplasmic granular basophilic Nissl substance, are present together with schwannian stroma. The neurons may exhibit nuclear NeuN and cytoplasmic tyrosine hydroxylase, often express VIP (fig. 8-55), and show variable but usually weak or absent staining for chromogranin A; Schwann cells express nuclear SOX10. Nuclear and cytoplasmic staining for S-100 protein but not SOX10 is seen in both cell types, making SOX10 the more specific marker, but it is generally more selective for the Schwann cell lineage. Unusual examples include a composite pheochromocytoma in which the neural component consists of mature neurons and bundles of naked axons with few or no Schwann cells (fig. 8-56).

In *composite ganglioneuroblastoma,* ganglion cells and immature neurons are dispersed in

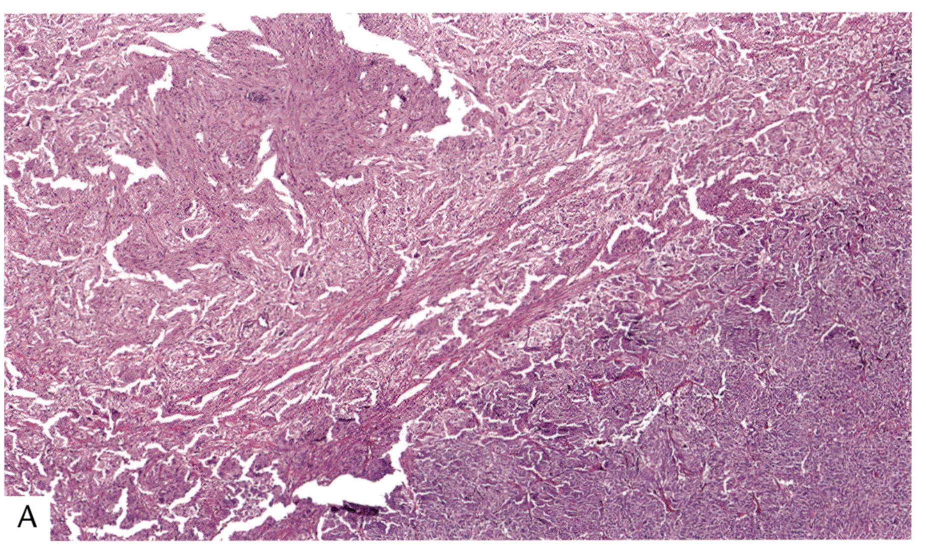

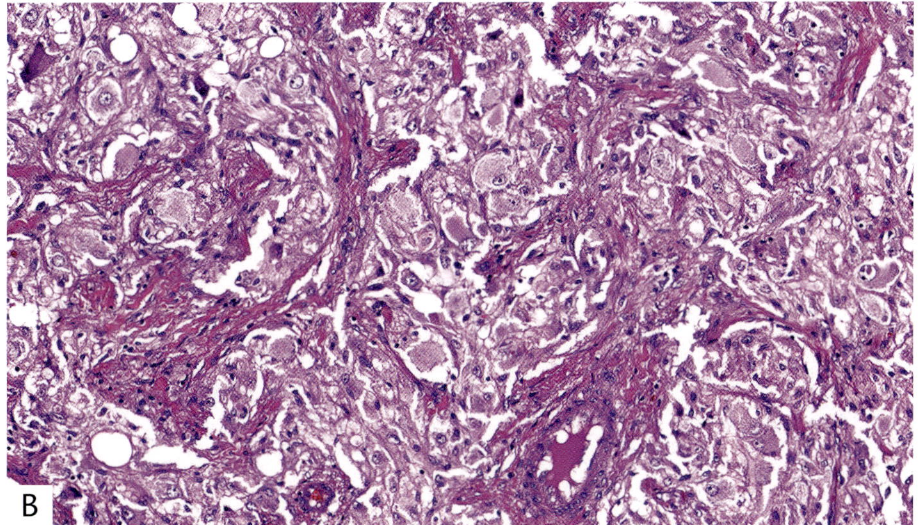

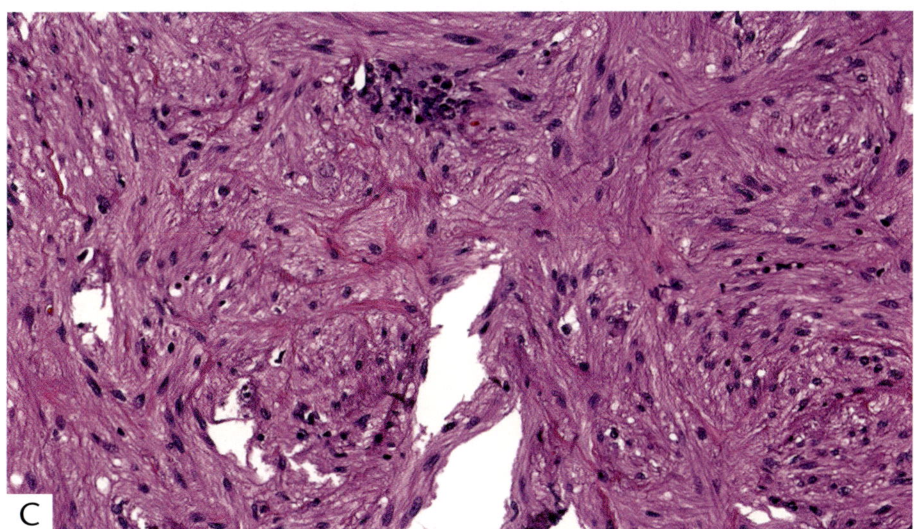

Figure 8-54

COMPOSITE PHEOCHROMOCYTOMA AND GANGLIONEUROMA

A central ganglioneuroma (A) is surrounded by pheochromocytoma (B). The ganglioneuroma (C) has clusters of large neurons within a schwannian stroma. There are areas of neuroma that are devoid of neurons.

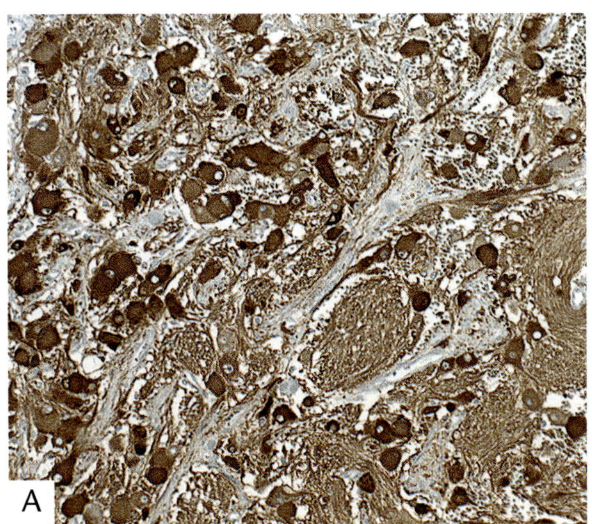

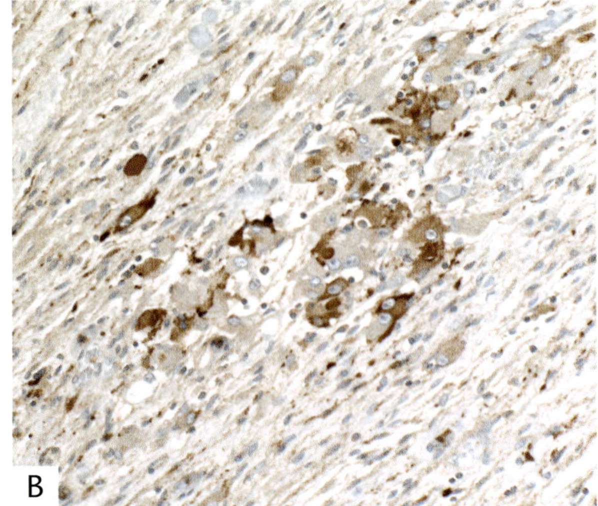

Figure 8-55

COMPOSITE PHEOCHROMOCYTOMA AND GANGLIONEUROMA

The ganglion cells stain for tyrosine hydroxylase (A) as well as VIP (B); the latter highlights axonal processes.

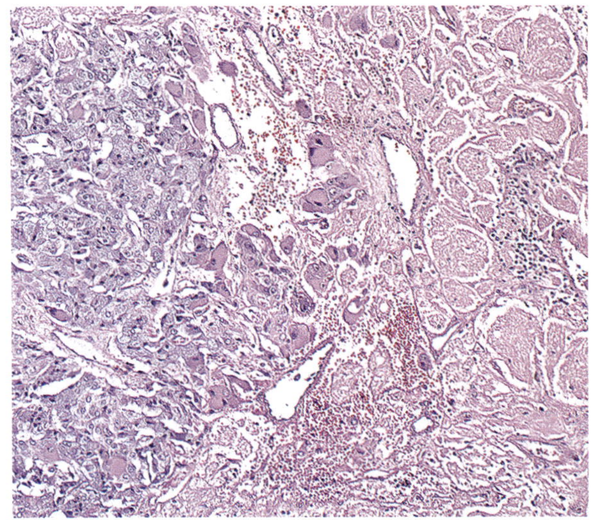

Figure 8-56

COMPOSITE PHEOCHROMOCYTOMA AND GANGLIONEUROMA

This unusual tumor contains a pheochromocytoma and clusters of large mature neurons that are "naked," that is, devoid of schwannian stroma.

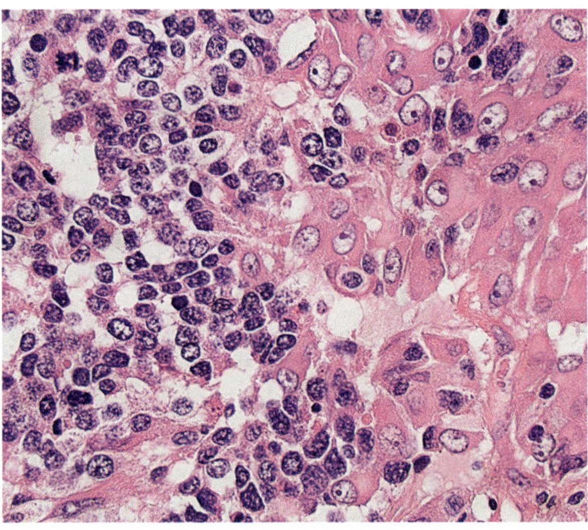

Figure 8-57

COMPOSITE PHEOCHROMOCYTOMA AND GANGLIONEUROBLASTOMA

schwannian stroma or in neuropil, while neuroblastoma is composed only of immature small round cells with or without neuropil (fig. 8-57). The use of stains for PHOX2B and the NB84 antibody assist in identifying neuroblastoma (105,106), and this component usually has an increased Ki-67 labeling index. Rare cases are composed of pheochromocytoma with peripheral nerve sheath tumor (fig. 8-58). These are composed of spindle cells that express SOX10 but lose S-100 protein staining and show increased mitoses. An unusual tumor also showed transdifferentiation to rhabdomyosarcoma (107).

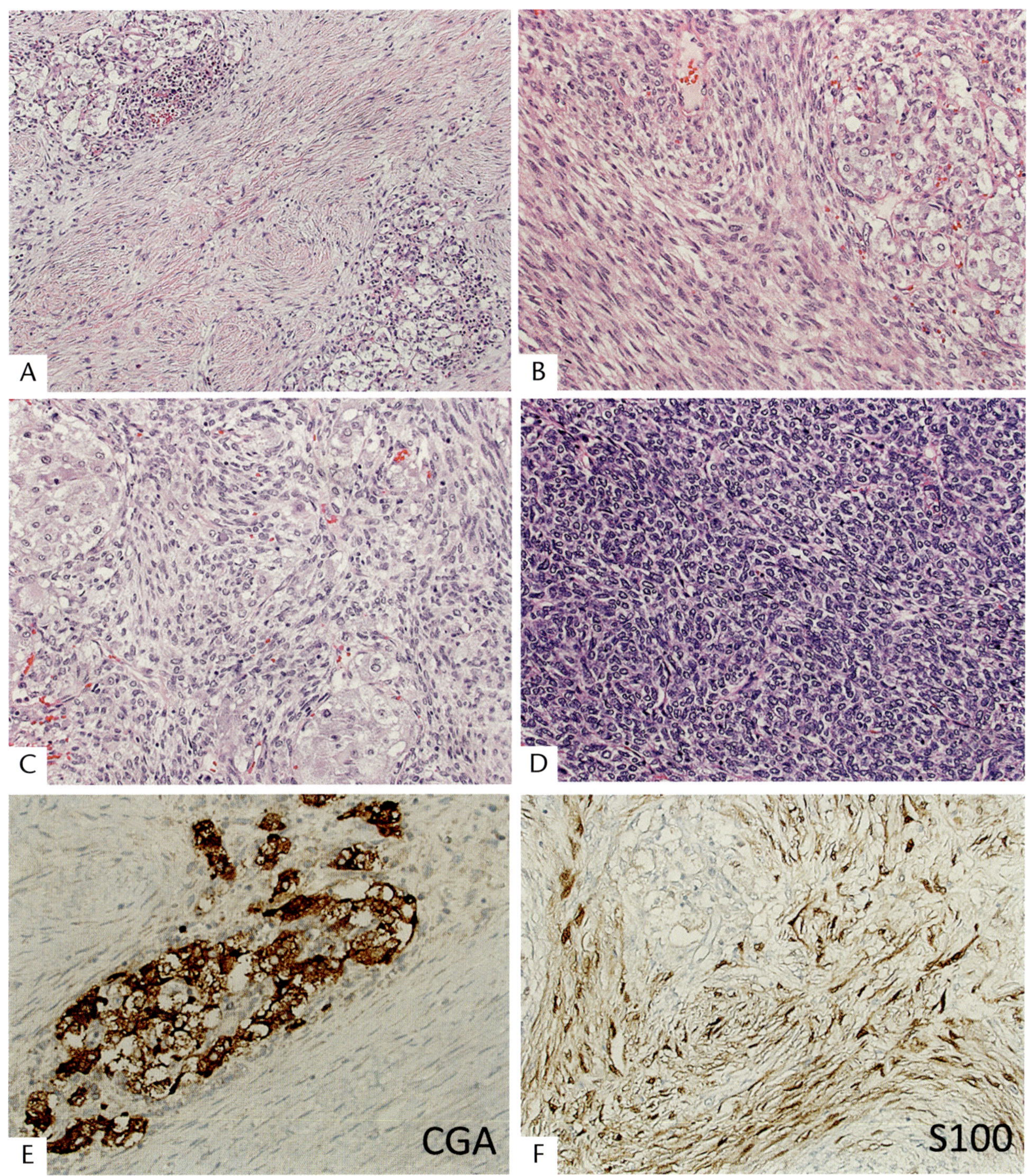

Figure 8-58

COMPOSITE PHEOCHROMOCYTOMA WITH MALIGNANT PERIPHERAL NERVE SHEATH TUMOR (MPNST)

This tumor is composed of an admixture of pheochromocytoma with a highly cellular nerve sheath tumor (A–D). The pheochromocytoma component expresses chromogranin A (CGA) (E) and the MPNST expresses S-100 protein (F) and CD34 (G). The Ki-67 labeling index of the MPNST is extremely high (H), confirming the malignant nature of this component.

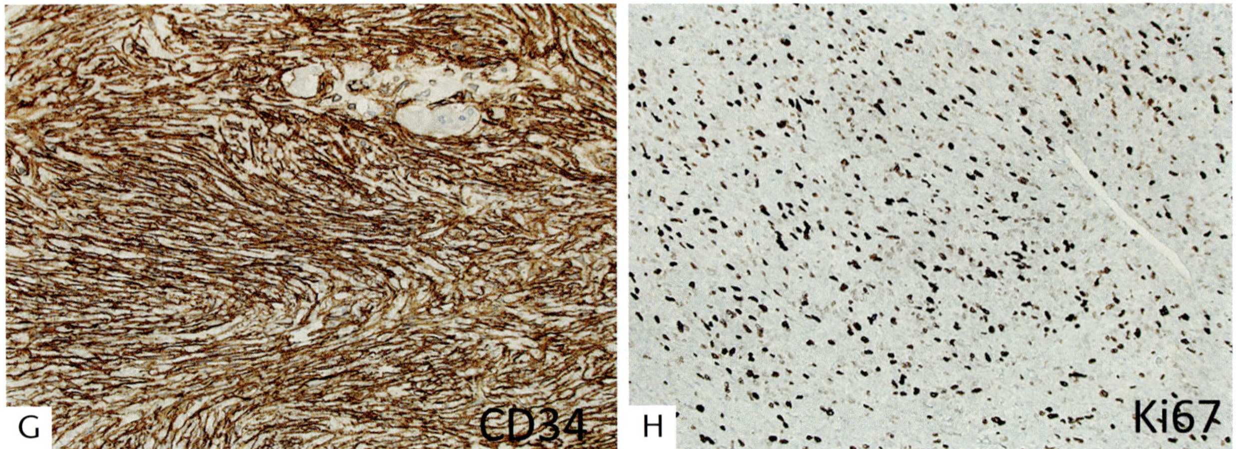

Figure 8-58, continued

The presence of neuroblastoma or ganglioneuroblastoma within a pheochromocytoma can portend a more aggressive tumor than pheochromocytoma alone. However, the likely effect on prognosis depends on the age of the patient and specific characteristics of the tumor cells. In children, these composite tumors tend to behave like conventional pediatric neuroblastoma with respect to metastatic potential and clinical course, while in adults they are more indolent, and if completely excised usually do not metastasize even after years of follow-up (108). In light of these differences it has been suggested that in order to better reflect tumor biology, the adult tumors should best be considered as "pheochromocytomas with divergent components of neuroblastoma differentiation," and the pediatric tumors as "neuroblastomas with divergent components of pheochromocytoma differentiation" (97). When metastases do occur, they are usually composed of ganglioneuroblastoma or neuroblastoma. One case with metastases containing both components has been reported (109).

The prognosis of patients with a composite tumor with ganglioneuroma is the same as for pheochromocytoma or paraganglioma. The presence of a malignant peripheral nerve sheath tumor imparts a worse prognosis (110).

Pheochromocytomas occasionally have been reported in association with adrenal cortical tumors, a phenomenon also known as *corticomedullary mixed tumor* (111,112). The cortical lesion is usually an adenoma. A rare event is a mixed pheochromocytoma and neuroendocrine tumor (113).

ADRENAL MEDULLARY HYPERPLASIA

Adrenal medullary hyperplasia is defined as an increase in the number of adrenal medullary cells that results in expansion of the medulla in proportion relative to the cortex and extension into parts of the adrenal gland where it is not normally present: the tail of the gland and the distal parts of the two wings (see chapter 1). Morphometric analysis is usually required for definitive diagnosis (114,115), since some examples of prominent medulla represent normal anatomic variation (116). Also, although chromaffin cells in some normal adult rodent adrenal glands proliferate in response to physiologic signals (117), the same has not been demonstrated for normal human chromaffin cells in vivo or in vitro.

True reactive hyperplasia is rare. Medullary expansion is most often due to germline disorders and reflects a spectrum from hyperplasia to neoplasia. To this point, molecular studies in patients with MEN2 have shown genetic abnormalities in both diffuse and nodular "hyperplasia" that were identical to those in pheochromocytoma (118).

Adrenal medullary hyperplasia may be diffuse, nodular, or both, and symmetric, involving both glands, or asymmetric. It may be associated with clinically detectable pheochromocytoma (see figs. 8-13, 8-36). As stated, this disorder is rare: in a series of 936 adrenalectomies, only 19

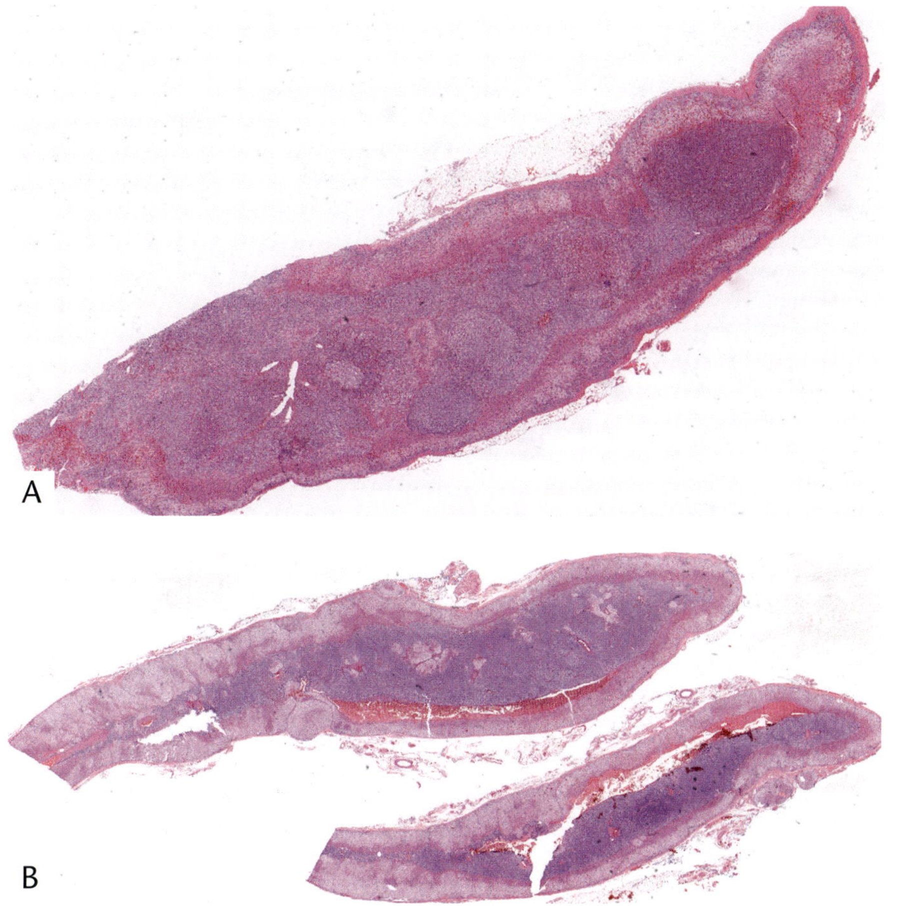

Figure 8-59

ADRENAL MEDULLARY HYPERPLASIA

Adrenal glands from patients with MEN2 (A) and SDHB-related disease (B) show diffuse expansion of the medulla that is much larger than the adrenal cortex on either side. While this is considered to be diffuse hyperplasia, there is a suggestion of slight nodularity in the adrenal gland in A, suggesting the early emergence of neoplastic nodules.

(2.1 percent) had adrenal medullary hyperplasia and only 2 were not associated with other adrenal pathology (119). Most cases are detected when screening individuals with known hereditary susceptibility to pheochromocytoma, for which it is a precursor lesion (45). It is most frequent in patients with MEN2 (115,120) but has also been reported with *SDHB, MAX*, and *TMEM127* germline mutations (121–124) and is seen in patients with Beckwith-Wiedemann syndrome (125). The signs and symptoms are typically less severe than with pheochromocytoma (119).

Rarely, adrenal medullary hyperplasia, particularly diffuse hyperplasia, is sporadic or reactive. Some cases are detected in patients with "pseudopheochromocytoma," i.e., unexplained hypertension, diaphoresis, tachycardia, and elevated urinary catecholamine and metanephrine levels, who may show increased uptake on MIBG scanning but without a focal lesion on CT or MRI. It has been reported in pediatric patients with cystic fibrosis, where it is thought to be a secondary phenomenon, on the basis of qualitative assessment of single adrenal sections associated with increased adrenal catecholamine content (126). Adrenal medullary hyperplasia was reported in sudden infant death syndrome (127) but is not a consistent finding (128). Although it is found in rats subjected to hypobaric hypoxia (129), there is no evidence of adrenal medullary hyperplasia in people who live at high altitudes.

Diffuse hyperplasia results in enlargement of the medulla, altering the cortex-to-medulla ratio (fig. 8-59). As a rough criterion, the maximum thickness of normal medulla is usually one third of the gland on cut section, with the surrounding cortex representing one third on each side; in hyperplasia, the medulla increases to more than one third of the gland. Another criterion is extension into the tail of the gland which is normally devoid of medullary tissue. It is important to ensure that the cortex itself

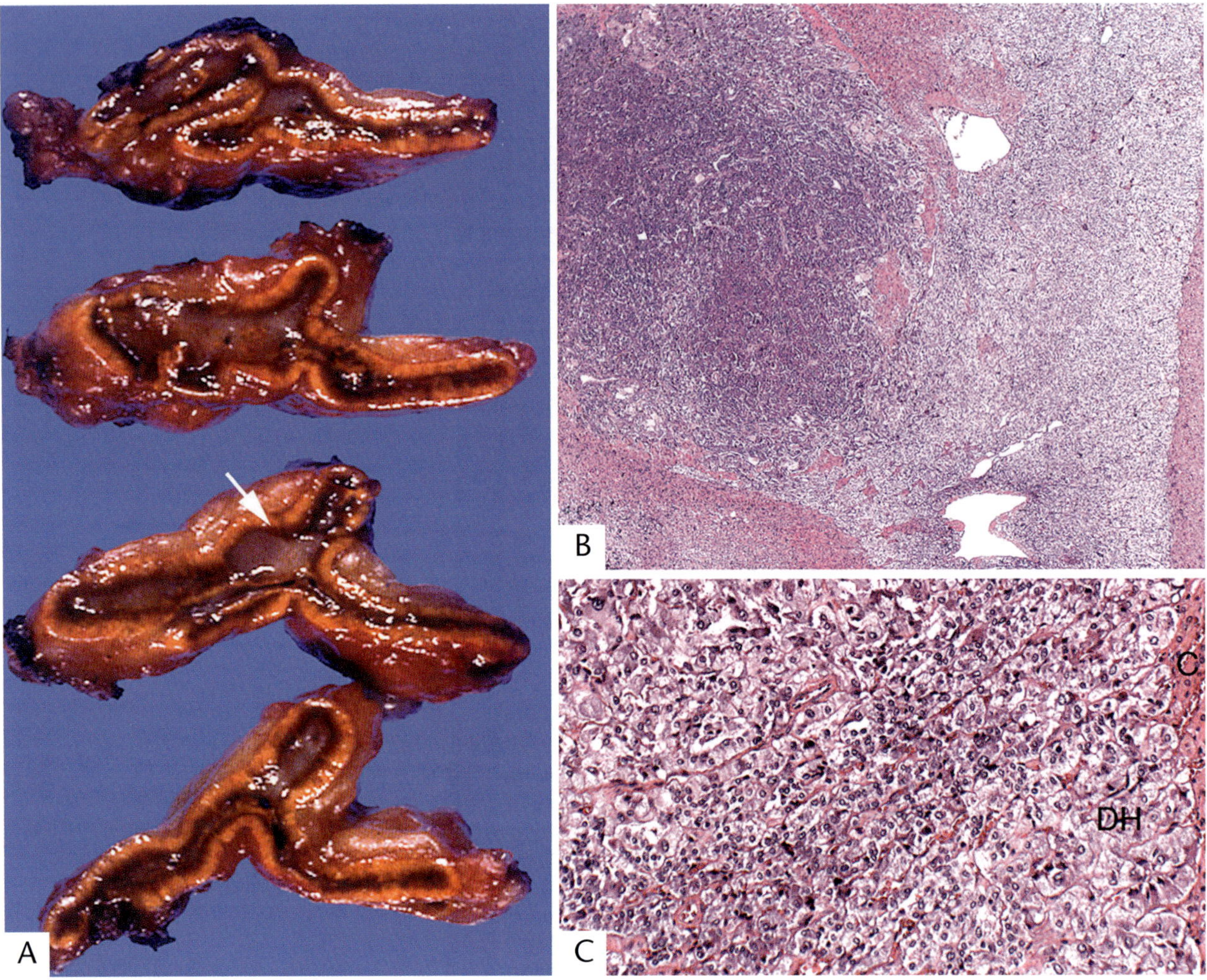

Figure 8-60

ADRENAL MEDULLARY HYPERPLASIA IN A PATIENT WITH MEN2A

A: The adrenal medulla is mildly, diffusely thickened and extends into the alae of the gland. A single grossly discernible nodule (arrow) is the clue to the hereditary disease.

B: Histologic section of the nodule marked by arrow in A shows two poorly formed nodules compressed against each other, the second with a smaller pinker area at top.

C: Higher magnification of an area showing the interface between adrenal cortex (C), diffuse hyperplasia (DH), and two poorly formed nodules (small blue cells with sparse cytoplasm at center, larger pink cells with prominent cytoplasm).

is of normal thickness before using this as the measure of adrenal size. Anatomic variations can be misleading using these criteria (116).

Diffuse hyperplasia is usually associated with the development of nodular hyperplasia (fig. 8-60). Lesions previously classified as nodular hyperplasia (fig. 8-61) are essentially precursor lesions to pheochromocytoma; the distinction of hyperplasia from neoplasia is historically based on an arbitrary size cut-off of 1 cm. However, molecular studies indicate that nodules that are smaller are indistinguishable from pheochromocytomas and are more appropriately considered micropheochromocytomas (118). The nodules may be multiple and may merge with each other.

The histologic appearance of adrenal medullary hyperplasia is variable. The enlarged medulla may have the microscopic appearance of normal medulla, but there may be changes

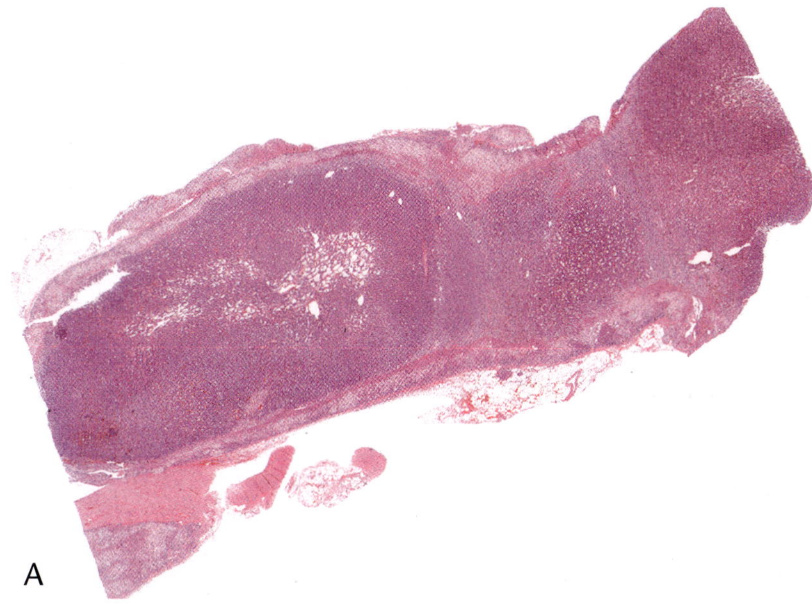

Figure 8-61

NODULAR ADRENAL MEDULLARY HYPERPLASIA IN A PATIENT WITH MEN2A

A: The adrenal medulla is diffusely thickened and contains multiple nodules.

B: At higher magnification there are three nodules that can be distinguished based on their architecture and cytology.

C: Higher magnification shows the distinct cytology and architecture of the three nodules: the left lesion has large pale eosinophilic cells, the middle small nodule is composed of basophilic large cells, and the nodule on the right has a slightly more epithelioid appearance.

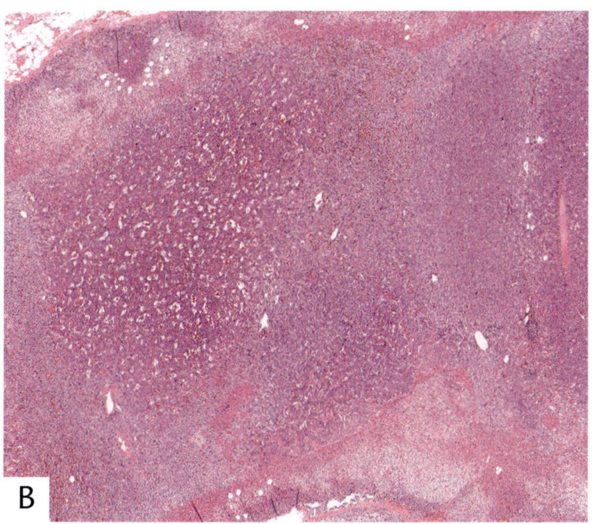

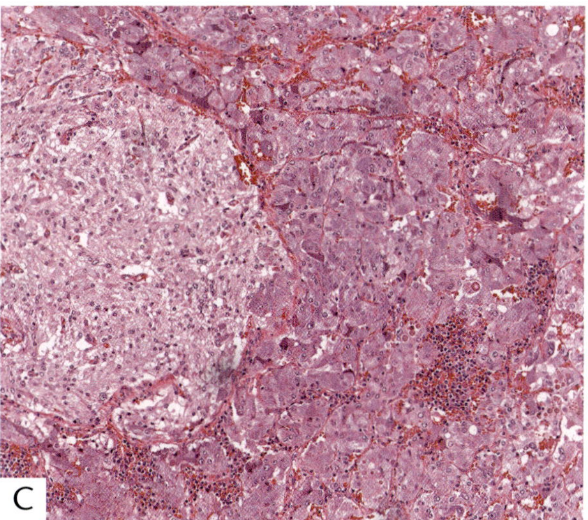

in architecture and cytology. The cells may proliferate in nests or trabeculae that resemble the architectural patterns of pheochromocytoma. The cells may be enlarged and may exhibit nuclear hyperchromasia and pleomorphism; spindle cell morphology may be found. Mitoses may be seen but are generally not numerous. Hyaline globules can be found.

REFERENCES

1. Carmichael SW. The history of the adrenal medulla. Rev Neurosci 1989;2:83-100.
2. Kohn A. Die paraganglien. Arch Mikr Anat 1903; 52:262-365.
3. Herde M. Zur lehre der paragangliome der nebenniere. Arch Klin Chir 1912;97:937-51.
4. Turchini J, Cheung VK, Tischler AS, de Krijger RR, Gill AJ. Pathology and genetics of phaeochromocytoma and paraganglioma. Histopathology 2018;72:97-105.
5. Lloyd RV, Osamura RY, Kloppel G, Rosai J. WHO classification of tumours of endocrine organs, 4th ed. Lyon: IARC; 2017.
6. Karsner HT. Tumors of the adrenal. Atlas of Tumor Pathology, 1st Series, Fascicle 29. Washington DC: Armed Forces Institute of Pathology; 1950.
7. Neumann HP, Young WF Jr, Eng C. Pheochromocytoma and paraganglioma. N Engl J Med 2019; 381:552-5.
8. DeLellis RA, Lloyd RV, Heitz PU, Eng C. Pathology and genetics of tumours of endocrine organs. Lyon: IARC Press; 2004.
9. Mete O, Asa SL, Gill AJ, Kimura N, de Krijger RR, Tischler A. Overview of the 2022 WHO classification of paragangliomas and pheochromocytomas. Endocr Pathol 2022;33:90-114.
10. Asa SL, Ezzat S, Mete O. The diagnosis and clinical significance of paragangliomas in unusual locations. J Clin Med 2018;7:280.
11. Tischler AS. Pheochromocytoma and extra-adrenal paraganglioma: updates. Arch Pathol Lab Med 2008;132:1272-84.
12. Y-Hassan S, Falhammar H. Cardiovascular manifestations and complications of pheochromocytomas and paragangliomas. J Clin Med 2020;9: 2435.
13. Lenders JW, Kerstens MN, Amar L, et al. Genetics, diagnosis, management and future directions of research of phaeochromocytoma and paraganglioma: a position statement and consensus of the Working Group on Endocrine Hypertension of the European Society of Hypertension. J Hypertens 2020;38:1443-56.
14. Cronin C. Charles Sugrue, M.D., of Cork (1775-1816) and the first description of a classical medical condition: phaeochromocytoma. Ir J Med Sci 2008;177:171-5.
15. Bausch B, Tischler AS, Schmid KW, Leijon H, Eng C, Neumann HP. Max Schottelius: pioneer in pheochromocytoma. J Endocr Soc 2017;1:957-64.
16. van Heerden JA. First encounters with pheochromocytoma. The story of Mother Joachim. Am J Surg 1982;144:277-9.
17. Messerli FH, Loughlin KR, Messerli AW, Welch WR. The president and the pheochromocytoma. Am J Cardiol 2007;99:1325-9.
18. Atuk NO, Stolle C, Owen JA Jr, Carpenter JT, Vance ML. Pheochromocytoma in von Hippel-Lindau disease: clinical presentation and mutation analysis in a large, multigenerational kindred. J Clin Endocrinol Metab 1998;83:117-20.
19. Beard CM, Sheps SG, Kurland LT, Carney JA, Lie JT. Occurrence of pheochromocytoma in Rochester, Minnesota, 1950 through 1979. Mayo Clin Proc 1983;58:802-4.
20. Stenstrom G, Svardsudd K. Pheochromocytoma in Sweden 1958-1981. An analysis of the National Cancer Registry Data. Acta Med Scand 1986; 220:225-32.
21. Andersen GS, Toftdahl DB, Lund JO, Strandgaard S, Nielsen PE. The incidence rate of phaeochromocytoma and Conn's syndrome in Denmark, 1977-1981. J Hum Hypertens 1988;2:187-9.
22. Berends AM, Buitenwerf E, de Krijger RR, et al. Incidence of pheochromocytoma and sympathetic paraganglioma in the Netherlands: a nationwide study and systematic review. Eur J Intern Med 2018;51:68-73.
23. Cvasciuc IT, Gull S, Oprean R, Lim KH, Eatock F. Changing pattern of pheochromocytoma and paraganglioma in a stable UK population. Acta Endocrinol (Buchar) 2020;16:78-85.
24. Sutton MG, Sheps SG, Lie JT. Prevalence of clinically unsuspected pheochromocytoma. Review of a 50-year autopsy series. Mayo Clin Proc 1981; 56:354-60.
25. Pappachan JM, Tun NN, Arunagirinathan G, Sodi R, Hanna FW. Pheochromocytomas and hypertension. Curr Hypertens Rep 2018;20:3.
26. Virgone C, Andreetta M, Avanzini S, et al. Pheochromocytomas and paragangliomas in children: data from the Italian Cooperative Study (TREP). Pediatr Blood Cancer 2020;67:e28332.
27. de Tersant M, Genere L, Freycon C, et al. Pheochromocytoma and paraganglioma in children and adolescents: experience of the French Society of Pediatric Oncology (SFCE). J Endocr Soc 2020; 4:bvaa039.
28. Rowland KJ, Chernock RD, Moley JF. Pheochromocytoma in an 8-year-old patient with multiple endocrine neoplasia type 2A: implications for screening. J Surg Oncol 2013;108:203-6.
29. Lack EE. Tumors of the adrenal gland and extra-adrenal paraganglia. AFIP Atlas of Tumor Pathology, 3rd Series, Fascicle 19. Washington, DC: American Registry of Pathology; 1997.

30. Lack EE. Tumors of the adrenal gland and extraadrenal paraganglia. Atlas of Tumor Pathology, 4th Series, Fascicle 8. Silver Spring: ARP Press; 2007.
31. Eisenhofer G, Lenders JW, Siegert G, et al. Plasma methoxytyramine: a novel biomarker of metastatic pheochromocytoma and paraganglioma in relation to established risk factors of tumour size, location and SDHB mutation status. Eur J Cancer 2012;48:1739-49.
32. Kaelin WG Jr. Von Hippel-Lindau disease: insights into oxygen sensing, protein degradation, and cancer. J Clin Invest 2022;132:e162480.
33. Guo Q, Cheng ZM, Gonzalez-Cantu H, et al. TMEM127 suppresses tumor development by promoting RET ubiquitination, positioning, and degradation. Cell Rep 2023;42:113070.
34. Fishbein L, Del RJ, Else T, et al. The North American Neuroendocrine Tumor Society consensus guidelines for surveillance and management of metastatic and/or unresectable pheochromocytoma and paraganglioma. Pancreas 2021;50:469-93.
35. Nock BA, Kanellopoulos P, Joosten L, Mansi R, Maina T. Peptide radioligands in cancer theranostics: agonists and antagonists. Pharmaceuticals (Basel) 2023;16:674.
36. Ramsay JA, Asa SL, van Nostrand AW, Hassaram ST, de Harven EP. Lipid degeneration in pheochromocytomas mimicking adrenal cortical tumors. Am J Surg Pathol 1987;11:480-6.
37. Koch CA, Mauro D, Walther MM, et al. Pheochromocytoma in von hippel-lindau disease: distinct histopathologic phenotype compared to pheochromocytoma in multiple endocrine neoplasia type 2. Endocr Pathol 2002;13:17-27.
38. Schimke RN, Hartmann WH. Familial amyloid-producing medullary thyroid carcinoma and pheochromocytoma: a distinct genetic entity. Ann Intern Med 1965;63:1027-37.
39. Steinhoff MM, Wells SA Jr, DeSchryver-Kecskemeti K. Stromal amyloid in pheochromocytomas. Hum Pathol 1992;23:33-6.
40. Miranda RN, Wu CD, Nayak RN, Kragel PJ, Medeiros LJ. Amyloid in adrenal gland pheochromocytomas. Arch Pathol Lab Med 1995;119:827-30.
41. Linnoila RI, Keiser HR, Steinberg SM, Lack EE. Histopathology of benign versus malignant sympathoadrenal paragangliomas: clinicopathologic study of 120 cases including unusual histologic features. Hum Pathol 1990;21:1168-80.
42. Bellezza G, Giansanti M, Cavaliere A, Sidoni A. Pigmented "black" pheochromocytoma of the adrenal gland: a case report and review of the literature. Arch Pathol Lab Med 2004;128:e125-8.
43. Kakkar A, Kaur K, Kumar T, et al. Pigmented pheochromocytoma: an unusual variant of a common tumor. Endocr Pathol 2016;27:42-5.
44. Franquemont DW, Mills SE, Lack EE. Immunohistochemical detection of neuroblastomatous foci in composite adrenal pheochromocytoma-neuroblastoma. Am J Clin Pathol 1994;102:163-70.
45. Mete O, Asa SL. Precursor lesions of endocrine system neoplasms. Pathol 2013;45:316-30.
46. Bialas M, Okon K, Dyduch G, et al. Neuroendocrine markers and sustentacular cell count in benign and malignant pheochromocytomas—a comparative study. Pol J Pathol 2013;64:129-35.
47. Zhou YY, Coffey M, Mansur D, Wasman J, Asa SL, Couce M. Images in endocrine pathology: progressive loss of sustentacular cells in a case of recurrent jugulotympanic paraganglioma over a span of 5 years. Endocr Pathol 2020;31:310-4.
48. Nonaka D, Chiriboga L, Rubin BP. Sox10: a pan-schwannian and melanocytic marker. Am J Surg Pathol 2008;32:1291-8.
49. Othman NH, Ezzat S, Kovacs K, et al. Growth hormone-releasing hormone (GHRH) and GHRH receptor (GHRH-R) isoform expression in ectopic acromegaly. Clin Endocrinol (Oxf) 2001;55:135-40.
50. Tsuta K, Raso MG, Kalhor N, Liu DC, Wistuba II, Moran CA. Sox10-positive sustentacular cells in neuroendocrine carcinoma of the lung. Histopathology 2011;58:276-85.
51. Farhat NA, Powers JF, Shepard-Barry A, Dahia P, Pacak K, Tischler AS. A previously unrecognized monocytic component of pheochromocytoma and paraganglioma. Endocr Pathol 2019;30:90-5.
52. Colombo-Benkmann M, Klimaschewski L, Heym C. Immunohistochemical heterogeneity of nerve cells in the human adrenal gland with special reference to substance P. J Histochem Cytochem 1996;44:369-75.
53. Brindley RL, Bauer MB, Walker LA, et al. Adrenal serotonin derives from accumulation by the antidepressant-sensitive serotonin transporter. Pharmacol Res 2019;140:56-66.
54. Schroeder JO, Asa SL, Kovacs K, Killinger D, Hadley GL, Volpé R. Report of a case of pheochromocytoma producing immunoreactive ACTH and beta-endorphin. J Endocrinol Invest 1984; 7:117-21.
55. Asa SL, Kovacs K, Vale W, Petrusz P, Vecsei P. Immunohistologic localization of corticotrophin-releasing hormone in human tumors. Am J Clin Pathol 1987;87:327-33.
56. Sano T, Asa SL, Kovacs K. Growth hormone-releasing hormone-producing tumors: clinical, biochemical, and morphological manifestations. Endocr Rev 1988;9:357-73.
57. Leibowitz-Amit R, Mete O, Asa SL, Ezzat S, Joshua AM. Malignant pheochromocytoma secreting vasoactive intestinal peptide and response to sunitinib: a case report and literature review. Endocr Pract 2014;20:e145-50.

58. Elder EE, Xu D, Hoog A, et al. KI-67 AND hTERT expression can aid in the distinction between malignant and benign pheochromocytoma and paraganglioma. Mod Pathol 2003;16:246-55.
59. Kimura N, Takayanagi R, Takizawa N, et al. Pathological grading for predicting metastasis in phaeochromocytoma and paraganglioma. Endocr Relat Cancer 2014;21:405-14.
60. Fishbein L, Leshchiner I, Walter V, et al. Comprehensive molecular characterization of pheochromocytoma and paraganglioma. Cancer Cell 2017;31:181-93.
61. Pierre C, Agopiantz M, Brunaud L, et al. COPPS, a composite score integrating pathological features, PS100 and SDHB losses, predicts the risk of metastasis and progression-free survival in pheochromocytomas/paragangliomas. Virchows Arch 2019;474:721-34.
62. Dahia PL, Ross KN, Wright ME, et al. A HIF1α regulatory loop links hypoxia and mitochondrial signals in pheochromocytomas. PLoS Genet 2005;1:72-80.
63. van Nederveen FH, Gaal J, Favier J, et al. An immunohistochemical procedure to detect patients with paraganglioma and phaeochromocytoma with germline SDHB, SDHC, or SDHD gene mutations: a retrospective and prospective analysis. Lancet Oncol 2009;10:764-71.
64. Gill AJ, Benn DE, Chou A, et al. Immunohistochemistry for SDHB triages genetic testing of SDHB, SDHC, and SDHD in paraganglioma-pheochromocytoma syndromes. Hum Pathol 2010;41:805-14.
65. Gill AJ. Succinate dehydrogenase (SDH)-deficient neoplasia. Histopathology 2018;72:106-16.
66. Korpershoek E, Favier J, Gaal J, et al. SDHA immunohistochemistry detects germline SDHA gene mutations in apparently sporadic paragangliomas and pheochromocytomas. J Clin Endocrinol Metab 2011;96:E1472-6.
67. Mete O, Pakbaz S, Lerario AM, Giordano TJ, Asa SL. Significance of alpha-inhibin expression in pheochromocytomas and paragangliomas. Am J Surg Pathol 2021;45:1264-73.
68. Pinato DJ, Ramachandran R, Toussi ST, et al. Immunohistochemical markers of the hypoxic response can identify malignancy in phaeochromocytomas and paragangliomas and optimize the detection of tumours with VHL germline mutations. Br J Cancer 2013;108:429-37.
69. Juhlin CC, Stenman A, Haglund F, et al. Whole-exome sequencing defines the mutational landscape of pheochromocytoma and identifies KMT2D as a recurrently mutated gene. Genes Chromosomes Cancer 2015;54:542-54.
70. Dahia PL. Pheochromocytoma and paraganglioma pathogenesis: learning from genetic heterogeneity. Nat Rev Cancer 2014;14:108-19.
71. Burnichon N, Vescovo L, Amar L, et al. Integrative genomic analysis reveals somatic mutations in pheochromocytoma and paraganglioma. Hum Mol Genet 2011;20:3974-85.
72. Agarwal G, Rajan S, Valiveru RC, et al. Genetic profile of Indian pheochromocytoma and paraganglioma patients—a single institutional study. Indian J Endocrinol Metab 2019;23:486-90.
73. Jiang J, Zhang J, Pang Y, et al. Sino-European differences in the genetic landscape and clinical presentation of pheochromocytoma and paraganglioma. J Clin Endocrinol Metab 2020;105:dgaa502.
74. Pacak K, Jochmanova I, Prodanov T, et al. New syndrome of paraganglioma and somatostatinoma associated with polycythemia. J Clin Oncol 2013;31:1690-8.
75. Darr R, Nambuba J, Del RJ, et al. Novel insights into the polycythemia-paraganglioma-somatostatinoma syndrome. Endocr Relat Cancer 2016;23:899-908.
76. Mete O, Asa SL, Giordano TJ, Papotti M, Sasano H, Volante M. Immunohistochemical biomarkers of adrenal cortical neoplasms. Endocr Pathol 2018;29:137-49.
77. Thompson LD. Pheochromocytoma of the adrenal gland scaled score (PASS) to separate benign from malignant neoplasms: a clinicopathologic and immunophenotypic study of 100 cases. Am J Surg Pathol 2002;26:551-66.
78. Wu D, Tischler AS, Lloyd RV, et al. Observer variation in the application of the Pheochromocytoma of the Adrenal Gland Scaled Score. Am J Surg Pathol 2009;33:599-608.
79. Koh JM, Ahn SH, Kim H, et al. Validation of pathological grading systems for predicting metastatic potential in pheochromocytoma and paraganglioma. PLoS One 2017;12:e0187398.
80. Stenman A, Zedenius J, Juhlin CC. The value of histological algorithms to predict the malignancy potential of pheochromocytomas and abdominal paragangliomas-A meta-analysis and systematic review of the literature. Cancers (Basel) 2019;11:225.
81. Assadipour Y, Sadowski SM, Alimchandani M, et al. SDHB mutation status and tumor size but not tumor grade are important predictors of clinical outcome in pheochromocytoma and abdominal paraganglioma. Surgery 2017;161:230-9.
82. Fishbein L, Khare S, Wubbenhorst B, et al. Whole-exome sequencing identifies somatic ATRX mutations in pheochromocytomas and paragangliomas. Nat Commun 2015;6:6140.
83. Job S, Draskovic I, Burnichon N, et al. Telomerase activation and ATRX mutations are independent risk factors for metastatic pheochromocytoma and paraganglioma. Clin Cancer Res 2019;25:760-70.

84. Hamidi O, Young WF Jr, Gruber L, et al. Outcomes of patients with metastatic phaeochromocytoma and paraganglioma: a systematic review and meta-analysis. Clin Endocrinol (Oxf) 2017;87:440-50.
85. Ayala-Ramirez M, Feng L, Johnson MM, et al. Clinical risk factors for malignancy and overall survival in patients with pheochromocytomas and sympathetic paragangliomas: primary tumor size and primary tumor location as prognostic indicators. J Clin Endocrinol Metab 2011;96:717-25.
86. Douwes Dekker PB, Corver WE, Hogendoorn PC, van der Mey AG, Cornelisse CJ. Multiparameter DNA flow-sorting demonstrates diploidy and SDHD wild-type gene retention in the sustentacular cell compartment of head and neck paragangliomas: chief cells are the only neoplastic component. J Pathol 2004;202:456-62.
87. Mete O, Tischler AS, de Krijger R, et al. Protocol for the examination of specimens from patients with pheochromocytomas and extra-adrenal paragangliomas. Arch Pathol Lab Med 2014;138: 182-8.
88. Thompson LD, Gill AJ, Asa SL, et al. Data set for the reporting of pheochromocytoma and paraganglioma: explanations and recommendations of the guidelines from the International Collaboration on Cancer Reporting. Hum Pathol 2021;110:83-97.
89. Stenman A, Zedenius J, Juhlin CC. Retrospective application of the pathologic tumor-node-metastasis classification system for pheochromocytoma and abdominal paraganglioma in a well characterized cohort with long-term follow-up. Surgery 2019;166:901-6.
90. Kragel PJ, Johnston CA. Pheochromocytoma-ganglioneuroma of the adrenal. Arch Pathol Lab Med 1985;109:470-2.
91. Nigawara K, Suzuki T, Onodera T, et al. [Watery diarrhea, hypokalemia, achlorhydria syndrome due to recurrent malignant pheochromocytoma]. Nihon Naibunpi Gakkai Zasshi 1987;63: 923-33. [Japanese]
92. Min KW, Clemens A, Bell J, Dick H. Malignant peripheral nerve sheath tumor and pheochromocytoma. A composite tumor of the adrenal. Arch Pathol Lab Med 1988;112:266-70.
93. Miettinen M, Saari A. Pheochromocytoma combined with malignant schwannoma: unusual neoplasm of the adrenal medulla. Ultrastruct Pathol 1988;12:513-27.
94. Brady S, Lechan RM, Schwaitzberg SD, Dayal Y, Ziar J, Tischler AS. Composite pheochromocytoma/ganglioneuroma of the adrenal gland associated with multiple endocrine neoplasia 2A: case report with immunohistochemical analysis. Am J Surg Pathol 1997;21:102-8.
95. Comstock JM, Willmore-Payne C, Holden JA, Coffin CM. Composite pheochromocytoma: a clinicopathologic and molecular comparison with ordinary pheochromocytoma and neuroblastoma. Am J Clin Pathol 2009;132:69-73.
96. Tatekawa Y, Muraji T, Nishijima E, Yoshida M, Tsugawa C. Composite pheochromocytoma associated with adrenal neuroblastoma in an infant: a case report. J Pediatr Surg 2006;41:443-5.
97. Tran L, Fitzpatrick C, Cohn SL, Pytel P. Composite tumor with pheochromocytoma and immature neuroblastoma: report of two cases with cytogenetic analysis and discussion of current terminology. Virchows Arch 2017;471:553-7.
98. Matias-Guiu X, Garrastazu MT. Composite phaeochromocytoma-ganglioneuroblastoma in a patient with multiple endocrine neoplasia type IIA. Histopathology 1998;32:281-2.
99. Charfi S, Ayadi L, Ellouze S, et al. [Composite pheochromocytoma associated with multiple endocrine neoplasia type 2B]. Ann Pathol 2008; 28:225-8. [French]
100. Pozza C, Sesti F, Di Dato C, et al. A novel MAX gene mutation variant in a patient with multiple and "composite" neuroendocrine-neuroblastic tumors. Front Endocrinol (Lausanne) 2020;11:234.
101. Gupta S, Zhang J, Erickson LA. Composite pheochromocytoma/paraganglioma-ganglioneuroma: a clinicopathologic study of eight cases with analysis of succinate dehydrogenase. Endocr Pathol 2017;28:269-75.
102. Comino-Mendez I, Tejera AM, Curras-Freixes M, et al. ATRX driver mutation in a composite malignant pheochromocytoma. Cancer Genet 2016;209:272-7.
103. George DJ, Watermeyer GA, Levin D, et al. Composite adrenal phaeochromocytoma-ganglioneuroma causing watery diarrhoea, hypokalaemia and achlorhydria syndrome. Eur J Gastroenterol Hepatol 2010;22:632-4.
104. Ende K, Henkel B, Brodhun M, et al. A 45-year-old female with hypokalemic rhabdomyolysis due to VIP-producing composite pheochromocytoma. Z Gastroenterol 2012;50:589-94.
105. Miettinen M, Chatten J, Paetau A, Stevenson A. Monoclonal antibody NB84 in the differential diagnosis of neuroblastoma and other small round cell tumors. Am J Surg Pathol 1998;22: 327-32.
106. Thomas JO, Nijjar J, Turley H, Micklem K, Gatter KC. NB84: a new monoclonal antibody for the recognition of neuroblastoma in routinely processed material. J Pathol 1991;163:69-75.

107. Gupta R, Sharma A, Arora R, Vijayaraghavan M. Composite phaeochromocytoma with malignant peripheral nerve sheath tumour and rhabdomyosarcomatous differentiation in a patient without von Recklinghausen disease. J Clin Pathol 2009;62:659-61.
108. Fujiwara T, Kawamura M, Sasou S, Hiramori K. Results of surgery for a compound adrenal tumor consisting of pheochromocytoma and ganglioneuroblastoma in an adult: 5-year follow-up. Intern Med 2000;39:58-62.
109. Nakagawara A, Ikeda K, Tsuneyoshi M, Daimaru Y, Enjoji M. Malignant pheochromocytoma with ganglioneuroblastoma elements in a patient with von Recklinghausen's disease. Cancer 1985;55:2794-8.
110. Oudijk L, de Krijger RR, Pacak K, Tischler AS. Adrenal medulla and extra-adrenal paraganglia. In: Mete O, Asa SL, eds. Endocrine pathology. Cambridge: Cambridge University Press; 2016:628-76.
111. Aiba M, Hirayama A, Ito Y, et al. A compound adrenal medullary tumor (pheochromocytoma and ganglioneuroma) and a cortical adenoma in the ipsilateral adrenal gland. A case report with enzyme histochemical and immunohistochemical studies. Am J Surg Pathol 1988;12:559-66.
112. Lau SK, Chu PG, Weiss LM. Mixed cortical adenoma and composite pheochromocytoma-ganglioneuroma: an unusual corticomedullary tumor of the adrenal gland. Ann Diagn Pathol 2011;15:185-9.
113. Juarez D, Brown RW, Ostrowski M, Reardon MJ, Lechago J, Truong LD. Pheochromocytoma associated with neuroendocrine carcinoma. A new type of composite pheochromocytoma. Arch Pathol Lab Med 1999;123:1274-9.
114. Carney JA, Sizemore GW, Tyce GM. Bilateral adrenal medullary hyperplasia in multiple endocrine neoplasia, type 2: the precursor of bilateral pheochromocytoma. Mayo Clin Proc 1975;50:3-10.
115. DeLellis RA, Wolfe HJ, Gagel RF, et al. Adrenal medullary hyperplasia. A morphometric analysis in patients with familial medullary thyroid carcinoma. Am J Pathol 1976;83:177-96.
116. Kreiner E. Weight and shape of the human adrenal medulla in various age groups. Virchows Arch A Pathol Anat Histol 1982;397:7-15.
117. Tischler AS, Powers JF, Pignatello M, Tsokas P, Downing JC, McClain RM. Vitamin D3-induced proliferative lesions in the rat adrenal medulla. Toxicol Sci 1999;51:9-18.
118. Korpershoek E, Petri BJ, Post E, et al. Adrenal medullary hyperplasia is a precursor lesion for pheochromocytoma in MEN2 syndrome. Neoplasia 2014;16:868-73.
119. Falhammar H, Stenman A, Calissendorff J, Juhlin CC. Presentation, treatment, histology, and outcomes in adrenal medullary hyperplasia compared with pheochromocytoma. J Endocr Soc 2019;3:1518-30.
120. Carney JA, Sizemore GW, Sheps SG. Adrenal medullary disease in multiple endocrine neoplasia, type 2: pheochromocytoma and its precursors. Am J Clin Pathol 1976;66:279-90.
121. Grogan RH, Pacak K, Pasche L, Huynh TT, Greco RS. Bilateral adrenal medullary hyperplasia associated with an SDHB mutation. J Clin Oncol 2011;29:e200-2.
122. Toledo SP, Lourenco DM Jr, Sekiya T, et al. Penetrance and clinical features of pheochromocytoma in a six-generation family carrying a germline TMEM127 mutation. J Clin Endocrinol Metab 2015;100:E308-18.
123. Hernandez KG, Ezzat S, Morel CF, et al. Familial pheochromocytoma and renal cell carcinoma syndrome: TMEM127 as a novel candidate gene for the association. Virchows Arch 2015;466: 727-32.
124. Romanet P, Guerin C, Pedini P, et al. Pathological and genetic characterization of bilateral adrenomedullary hyperplasia in a patient with germline MAX mutation. Endocr Pathol 2017;28:302-7.
125. Tischler AS, Semple J. Adrenal medullary nodules in beckwith-wiedemann syndrome resemble extra-adrenal paraganglia. Endocr Pathol 1996;7:265-72.
126. Bongiovanni AM, Yakovac WC, Steiker DD. Study of adrenal glands in childhood: hormonal content correlated with morphologic characteristics. Lab Invest 1961;10:956-67.
127. Naeye RL. Brain-stem and adrenal abnormalities in the sudden-infant-death syndrome. Am J Clin Pathol 1976;66:526-30.
128. Perez-Platz U, Saeger W, Dhom G, Bajanowski T. The pathology of the adrenal glands in sudden infant death syndrome (SIDS). Int J Legal Med 1994;106:244-8.
129. Gosney JR. Adrenal corticomedullary hyperplasia in hypobaric hypoxia. J Pathol 1985;146:59-64.

9 PARAGANGLIA AND PARAGANGLIOMAS

DEVELOPMENT, ANATOMY, AND FUNCTION OF NORMAL PARAGANGLIA

Paraganglia are neuroendocrine organs associated with paraxial sympathetic nerves and branches of the glossopharyngeal and vagus nerves in the head and neck. In the late 19th century, the anatomist Alfred Kohn proposed that these anatomically dispersed structures consisted of cells developmentally, structurally, and functionally related to each other and to sympathetic ganglia. On the basis of that analogy, he coined the term paraganglia (1).

While Kohn's analogy is still largely viable, it is now recognized that there is more developmental divergence between sympathetic ganglia and paraganglia than previously believed. The 1970s concept of a single neural crest–derived progenitor giving rise to either chromaffin cells or neurons, depending on local environmental influences, began to be eroded by studies showing that the developmental fates of these cells are preprogrammed (2).

Current evidence indicates that while most sympathetic neuron progenitors reach their destination by direct migration from the neural crest, paraganglionic neuroendocrine cells are derived, in large part, from progenitors called Schwann cell precursors (SCPs), which are themselves derived from neural crest cells that form the dorsal root ganglia (3–6). It has been suggested that SCPs consist of a pool of neural crest–like cells employed for expansion and diversification of neural crest–derived tissues after the neural crest no longer exists (7). Diversification is further accomplished by additional direct contributions of migratory neural crest cells to chromaffin cell populations and contributions of SCPs to a subset of sympathetic neurons (4,6). This diversification during development may be partly responsible for phenotype differences between paragangliomas in different anatomic locations and hereditary syndromes (7,8).

Familiarity with the normal distribution of paraganglia is important because paragangliomas occur with the greatest frequency in areas where paraganglionic tissue is most abundant during development or in adult life, and occur occasionally in outlier areas. This association was elegantly illustrated by Rex Coupland, the father of modern chromaffin cell biology, in his now classic 1965 book, *The Natural History of the Chromaffin Cell* (fig. 9-1) (9). Paragangliomas should, therefore, be considered in the differential diagnosis of many types of tumor that occur in different parts of the body (10).

The paraganglia associated with sympathetic nerves are known as *sympathetic* or *sympathoadrenal paraganglia*. These include the adrenal medulla, the organ of Zuckerkandl, and multiple microscopic paraganglia that vary in number and distribution. The adrenal medulla in adults and the organ of Zuckerkandl in early development are the only macroscopic examples.

As mapped during fetal and postnatal development, sympathetic paraganglia are most often associated with nerve fibers and ganglia of the preaortic sympathetic plexuses, including the celiac, mesenteric, hypogastric, renal, and gonadal plexuses (figs. 9-1A, 9-2). The organ of Zuckerkandl, at the origin of the inferior mesenteric artery near the superior hypogastric plexus, is the largest single aggregate of extra-adrenal chromaffin tissue but accounts for less than half the volume of all extra-adrenal chromaffin tissue (9). The microscopic paraganglia comprising the remainder are widely distributed rostrally and caudally, and are often prominent near the celiac ganglion and its branches near the renal, celiac, or superior mesenteric arteries (9,11).

The extra-adrenal sympathetic paraganglia are the major source of circulating catecholamines

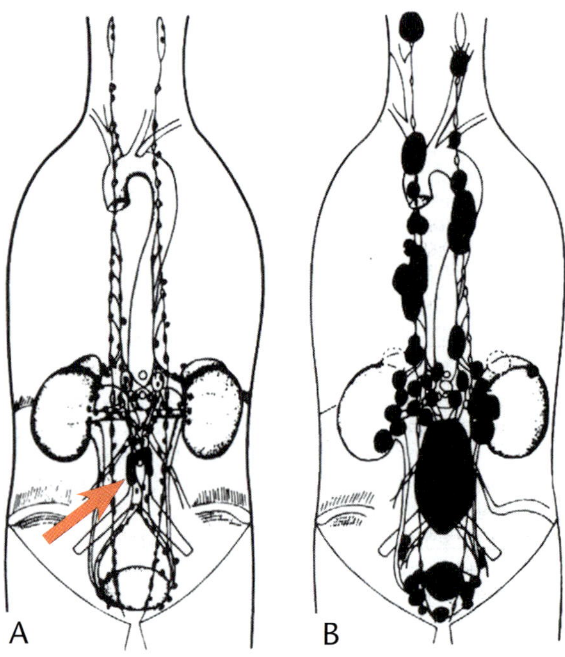

Figure 9-1

SITES OF PARAGANGLIOMA DEVELOPMENT

The relationship between the distribution of extra-adrenal chromaffin tissue in neonates (A) and prevalent sites of sympathetic paragangliomas reported up to 1965 (B) in a classic illustration. Arrow added in A indicates the organ of Zuckerkandl. The two hot spots for paragangliomas in the neck are based in part on carotid body paragangliomas, which occasionally produce significant amounts of norepinephrine or dopamine. Since its initial publication, this figure has gone through many iterations by other authors, but remains essentially correct. (Figs. 64 and 63 from Coupland RE. The natural history of the chromaffin cell. Longmans, Green, and Co; 1965:192-4.)

in fetal life, and are fully formed in humans long before the adrenal medulla (figs. 9-3–9-5). Both the organ of Zuckerkandl and the microscopic paraganglia mature until approximately age 3 years, then begin to involute (11). However, at least some microscopic paraganglia persist in adults (12) and are often found incidentally in microscopic sections taken in or near abdominal, pelvic, and thoracic organs. In some locations, such as the genitourinary tract, they are diagnostic pitfalls, potentially mistaken for carcinoma (fig. 9-6) (13).

The paraganglia associated with cranial nerves are known as *parasympathetic* or *head and neck paraganglia*; the carotid bodies (fig. 9-7) are the only macroscopic examples. Other microscopic paraganglia are distributed in the floor and wall of the middle ear (*jugulotympanic paraganglia*), in or near the nodose ganglion of the vagus nerve (*vagal paraganglia*), or near the thyroid gland or larynx in association with the terminus of the superior laryngeal nerve. Similar to sympathetic paraganglia, these microscopic paraganglia vary in number and distribution. Some undergo involution or modification during development or aging (14,15), but some persist in adults and are confused with other structures. A particular challenge when identifying normal paraganglia and paragangliomas in the head and neck is the presence of ectopic, hyperplastic, or neoplastic parathyroid tissue (figs. 9-8–9-10).

The term parasympathetic paraganglia is an anatomic designation indicating an association with a classic parasympathetic nerve rather than a physiologic definition implying a role in antagonizing sympathetic nervous system functions. With the exception of the carotid bodies, which serve as systemic chemoreceptors (16), the functions of most paraganglia are largely unknown. These structures can receive innervation from both cranial and sympathetic neurons. As an example, the superior laryngeal nerves, which branch from the vagus nerves, include a contribution of fibers from the superior cervical ganglia (17). It is also not strictly correct that "head and neck" is interchangeable with "parasympathetic," because the paravertebral chains of sympathetic ganglia extend into the neck. During fetal development these ganglia and nerves contain neuroendocrine cells comparable to those in abdominal paraganglia (18), and they occasionally give rise to paragangliomas later in life (19).

An important function of both the carotid bodies and the organ of Zuckerkandl is sensing and responding to changes in oxygen concentration. Both of these organs release catecholamines in response to hypoxia. The organ of Zuckerkandl responds directly as an endocrine organ by secreting catecholamines that may protect against fetal heart failure (20). In contrast, the carotid body acts indirectly by transmitting sensory information to neurons in the central nervous system, which regulate cardiovascular function postnatally (16,20). Hypoxia-inducible transcription factor 2A

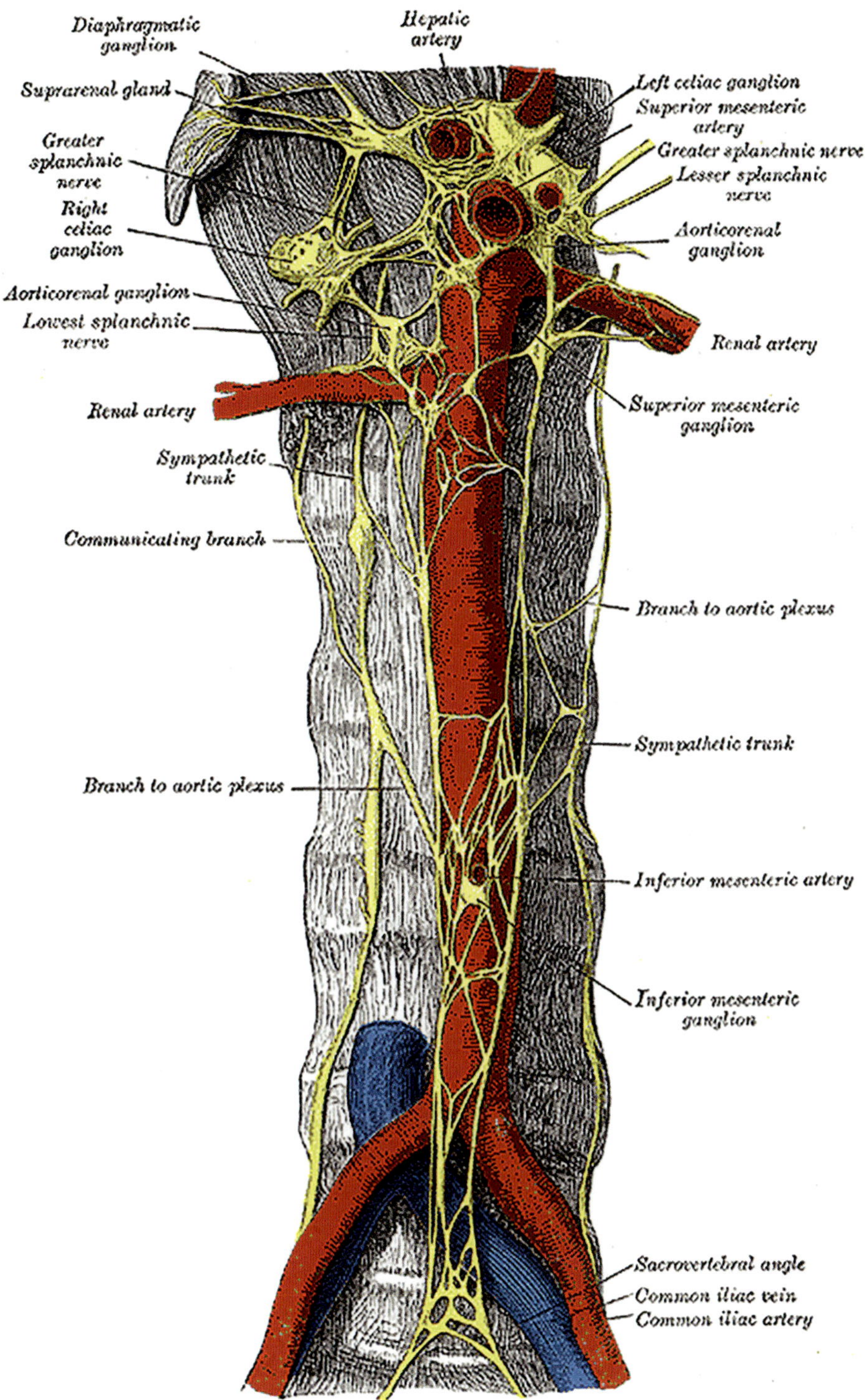

Figure 9-2

ABDOMINAL PORTION OF THE PREAORTIC SYMPATHETIC TRUNK, WITH THE CELIAC AND HYPOGASTRIC PLEXUSES

The major distribution of abdominal paraganglia coincides with the distribution of nerve fibers from these plexuses. (Illustration by Henry Vandyke Carter, fig. 847 from Gray H. Anatomy of the human body. Lea and Febiger; 1918:965.)

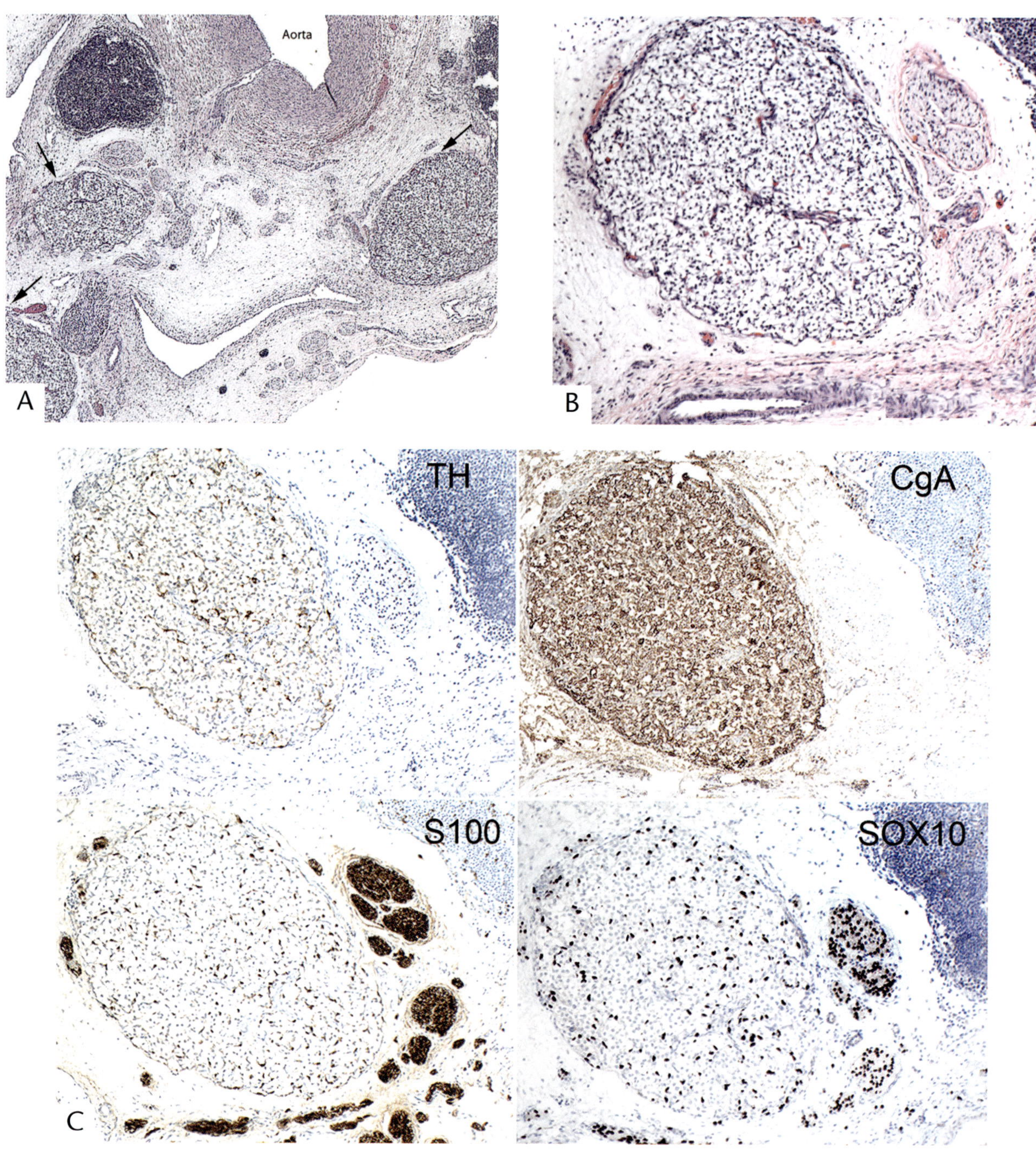

Figure 9-3

NORMAL ABDOMINAL PARAGANGLIA IN AN EARLY SECOND TRIMESTER STILLBORN FETUS

A: Normal paraganglia (arrows) in close proximity to lymph nodes. In a patient with a hereditary paraganglioma syndrome, this anatomic relationship makes the distinction between multiple primary tumors and metastases impossible in imaging studies.

B: Higher magnification of a normal paraganglion shows characteristic clear cells and "zellballen" architecture.

C: Immunohistochemical markers for sustentacular cells and chromaffin cells in a paraganglion in consecutive sections adjacent to A. At this stage chromogranin A (CgA) is expressed in all of the chromaffin cells while expression of tyrosine hydroxylase (TH) is just beginning. S-100 protein and SOX10 are expressed in sustentacular cells within the paraganglion and Schwann cells in an adjacent nerve. The lymph node at top right contain cells positive for S-100 protein but not for SOX10.

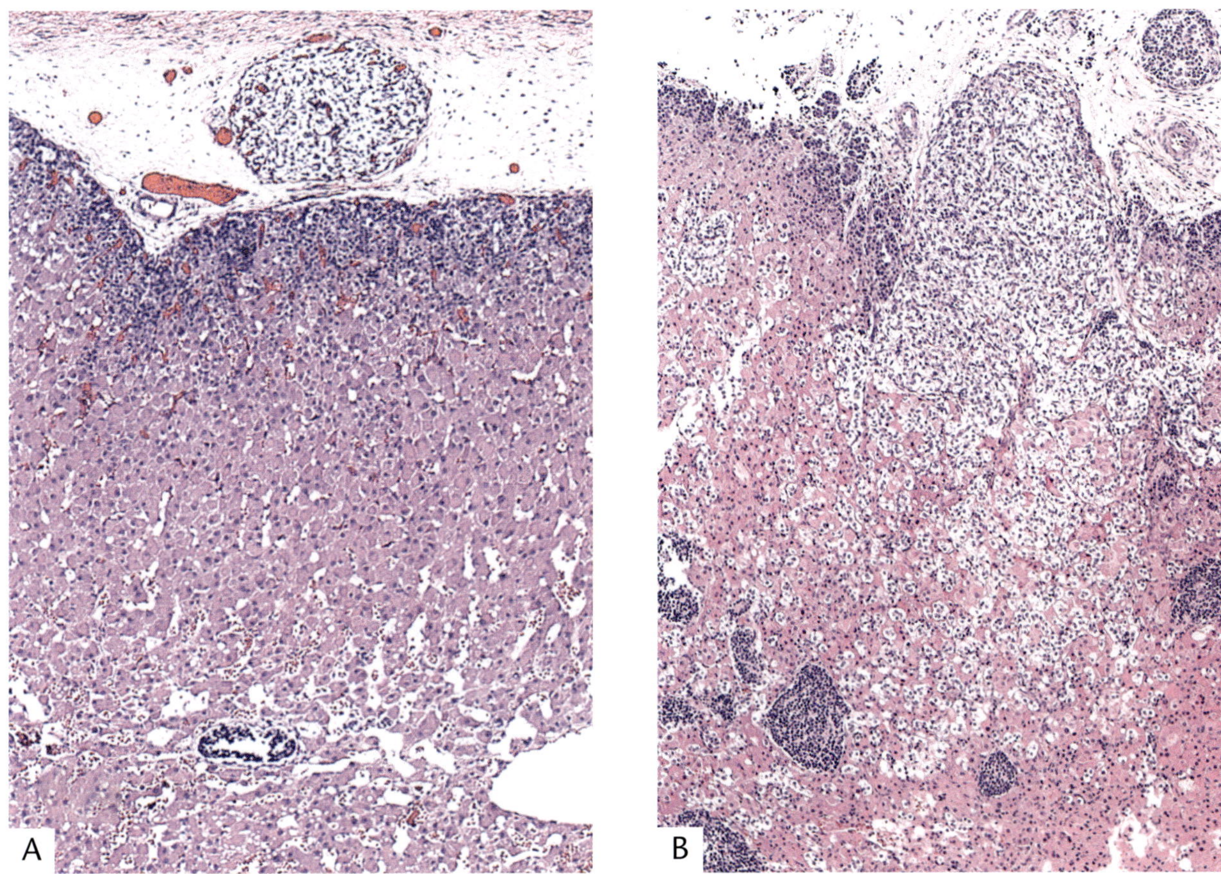

Figure 9-4

ABDOMINAL PARAGANGLIA AND ADJACENT DEVELOPING ADRENAL GLAND

A: Normal paraganglia are fully formed before the adrenal medulla exists. This adrenal gland shows only provisional cortex and a small cluster of primitive adrenal medullary precursors (same specimen as fig. 9-3).

B: In this unusual specimen cells from an extra-adrenal paraganglion are being incorporated into adjacent developing adrenal tissue, which also contains typical nests of neuroblast-like medullary progenitors. A developing ganglion composed of similar-appearing progenitors is present at top right.

(EPAS1), which is critically implicated in the pathobiology of "pseudohypoxic" pheochromocytomas and paragangliomas caused by succinate dehydrogenase mutations (21), is developmentally regulated in a tissue-specific manner and is highly expressed in both the organ of Zuckerkandl and carotid bodies (20).

Normal paraganglia have two characteristic architectural patterns: the well-known nested pattern that Alfred Kohn called "Zellballen" and a trabecular pattern that he called "Zellsträngen." The first term is still widely used, while the second has been dropped. There are also two characteristic cell types: nonepithelial neuroendocrine cells and sustentacular cells.

Prior to the advent of immunohistochemistry, the principal tool for functional characterization of paraganglia was the chromaffin reaction. Paraganglia in head and neck sites were not readily detected by this now obsolete technique because of its low sensitivity, and were therefore referred to as "nonchromaffin paraganglia." The absence of a chromaffin reaction in turn led to multiple names for individual cell types. Hence, the neuroendocrine cells in parasympathetic paraganglia are also known as "type 1" cells, "chief cells," or "glomus cells," and the sustentacular cells are also known as "type 2 cells." "Glomus" is discouraged in pathology literature because the same name

Figure 9-5

NORMAL ABDOMINAL PARAGANGLION

A: Low magnification shows proximity of this paraganglion to adjacent developing ganglia, adrenal gland, and kidney in a mid-second trimester fetus.

B: Higher magnification of paraganglion in a mid-second trimester fetus. This specimen is more mature than the one in figure 9-4, and shows a mixture of chromaffin cells with clear and amphophilic cytoplasm, more fibrous stroma, and partially trabecular architecture.

C: Immunohistochemical stain for tyrosine hydroxylase (TH) in paraganglion and developing sympathetic ganglia from mid-second trimester fetus (section adjacent to that seen in fig. 9-7). This paraganglion is more mature than the one in A, and the neuroendocrine cells all strongly express TH. The TH stain also emphasizes the trabecular architecture.

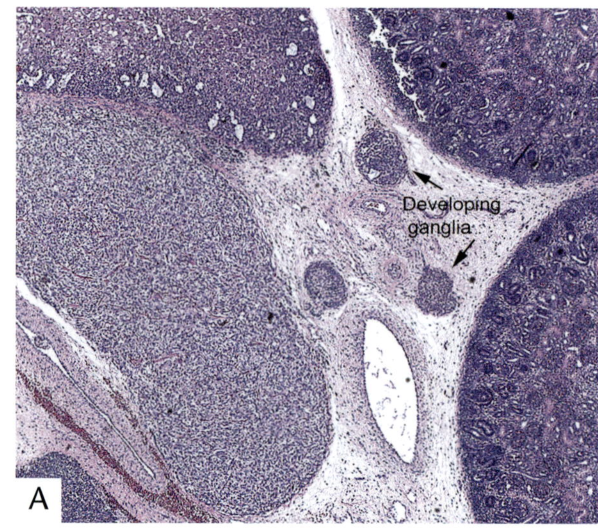

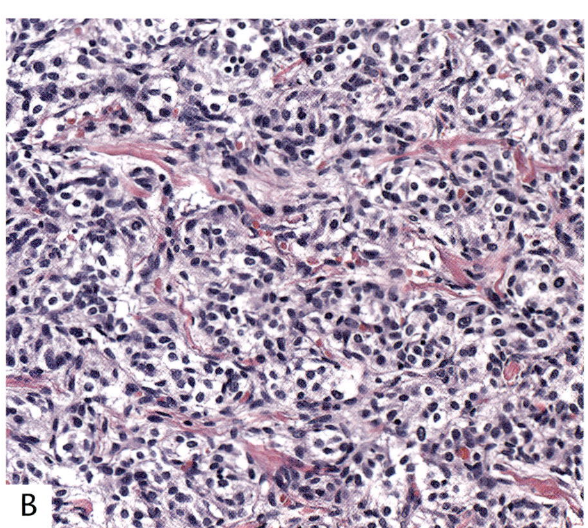

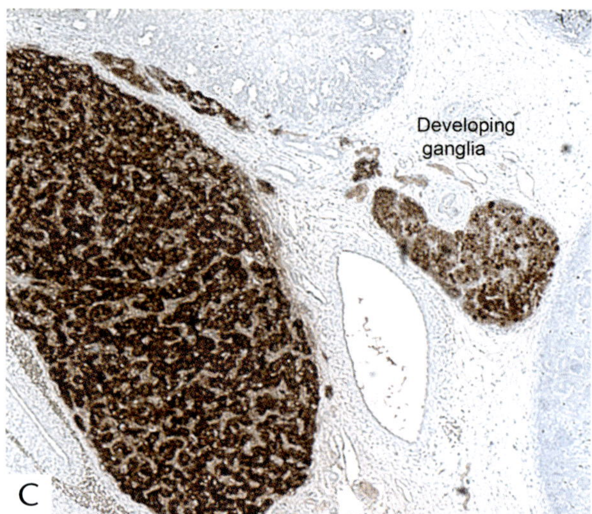

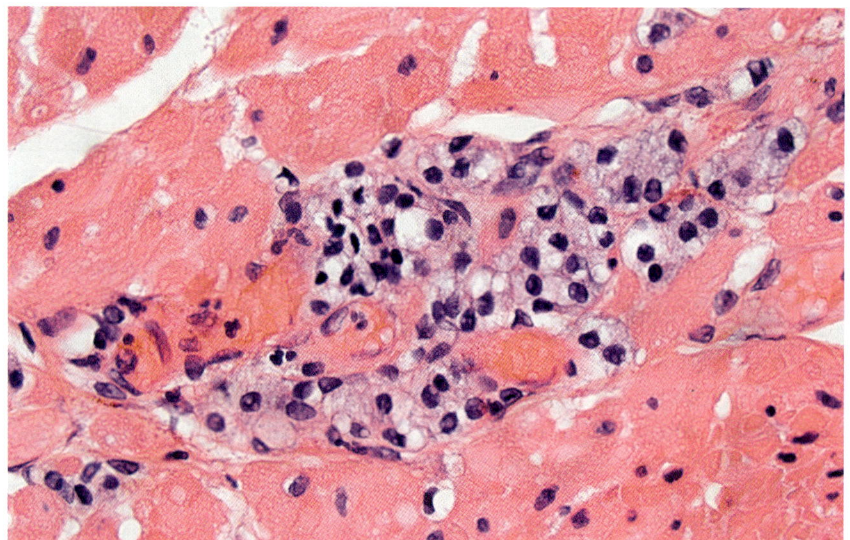

Figure 9-6

NORMAL PARAGANGLION IN THE WALL OF ADULT URINARY BLADDER

Incidentally discovered paraganglia can be mistaken for deposits of prostatic, renal cell, or other carcinomas. As in this example, their identity can be further obscured by bubbly or vacuolated cytoplasm, which is a histologic artefact.

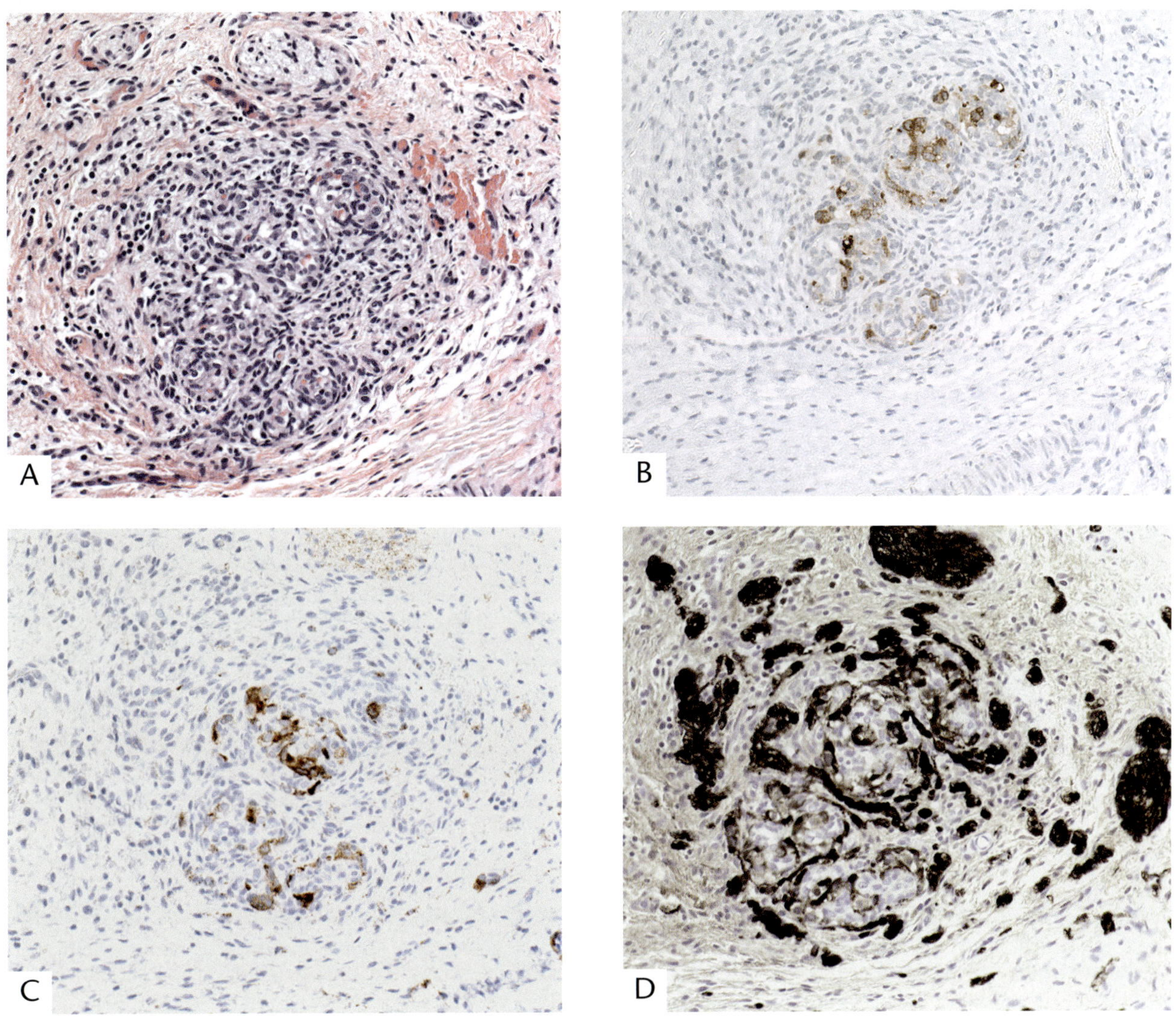

Figure 9-7

NORMAL CAROTID BODY

A: This developing carotid body in an early second trimester stillborn fetus (same case as in figs. 9-3 and 9-4), shows more complex architecture and greater admixture of cell types than the abdominal paraganglia.

B: Tyrosine hydroxylase expression in chief cells of developing carotid body.

C: Chromogranin A in chief cells of developing carotid body.

D: Prominent nerve fibers in developing carotid body highlighted by S-100 protein.

applies to cutaneous temperature sensors and their corresponding tumors, but it still appears frequently in physiology papers.

Extra-adrenal paraganglia are now identified using the same immunohistochemical markers commonly applied to chromaffin cells and sustentacular cells in the adrenal medulla (see chapter 1). Usually these are chromogranin A or synaptophysin as a generic neuroendocrine marker and tyrosine hydroxylase as a functional neuroendocrine marker (see figs. 9-3, 9-5, 9-7) (22). Transcription factors have also recently been employed as lineage markers, with INSM1 used to identify all neuroendocrine cells (23) and GATA3 as a selective, although not entirely specific, marker for paraganglia. GATA3 plays complex roles in sympathoadrenal development, including regulation of downstream and

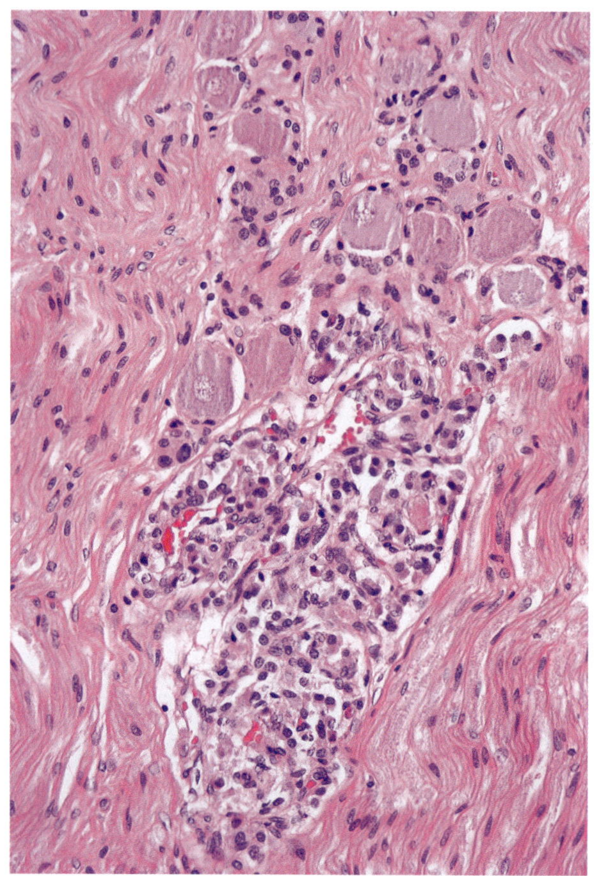

Figure 9-8

NORMAL PARAGANGLION WITHIN A GANGLION OF THE VAGUS NERVE

The paraganglion contrasts with the neurons adjacent to it.

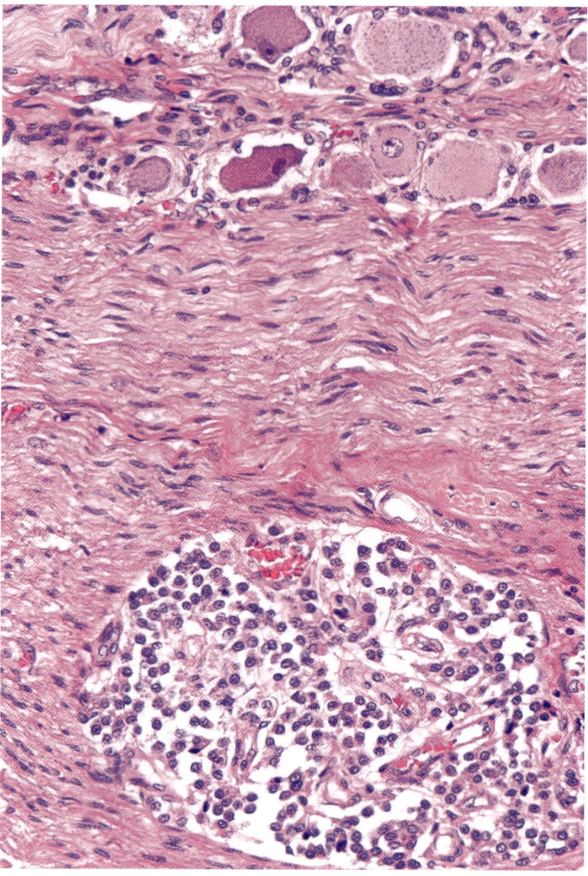

Figure 9-9

ECTOPIC PARATHYROID TISSUE WITHIN VAGUS NERVE

Non-neoplastic ectopic parathyroid tissue and parathyroid adenomas within the vagus, as in this example, are potentially mistaken for vagal paraganglia and paragangliomas.

upstream targets (24). One of these targets, the transcription factor PHOX2B, is a highly specific marker and a potential new tool for the diagnosis of paragangliomas (25,26).

Sustentacular cells are identified by staining for S-100 protein or SOX10 (see figs. 9-3, 9-7) (27). Subsets of these cells may stain for glial fibrillary acidic protein (GFAP). Paraganglia are negative for cytokeratins, which may be an additional helpful marker in cases where the differential diagnosis is between a normal paraganglion and an epithelial tumor deposit. This is a concern in the genitourinary tract where normal paraganglia may be mistaken for prostatic (13) or urothelial cancer because of their often clear cytoplasm (see fig. 9-6).

PARAGANGLIOMAS

Paragangliomas are nonepithelial neuroendocrine tumors originating from paraganglia associated with sympathetic nerves and ganglia anywhere in the body, and with branches of the glossopharyngeal and vagus nerves in the head and neck. On the basis of their anatomic associations, the two groups of tumor are called *sympathetic* or *parasympathetic paragangliomas*.

The 2022 World Health Organization (WHO) classification of endocrine tumors lists a single ICD-0 code for sympathetic paragangliomas, which may occur anywhere in the distribution of normal sympathetic paraganglia. Similarly, there is now a single code for parasympathetic paragangliomas. This contrasts with the 2017

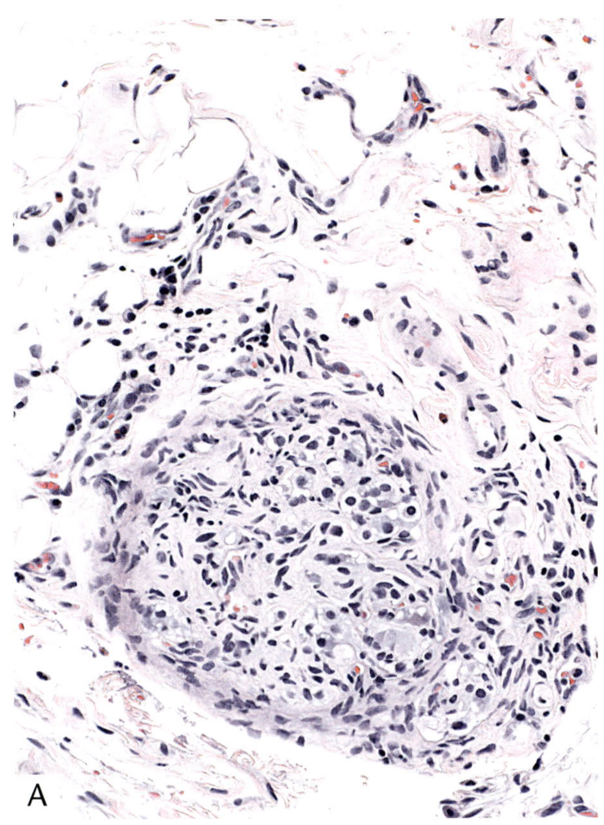

Figure 9-10

NORMAL PERITHYROIDAL PARAGANGLION

A: This paraganglion, discovered incidentally in perithyroidal soft tissue in an adult patient with a multinodular goiter, is architecturally similar to the carotid body in figure 9-7. The intimate association of neuroendocrine cells and nerve fibers is potentially mistaken for perineural infiltration by paraganglioma, medullary thyroid carcinoma, or other tumors. Paragangliomas arising from normal paraganglia in or near the thyroid gland are especially likely to be mistaken for medullary thyroid carcinoma.

B: Normal perithyroidal paraganglion stained for tyrosine hydroxylase.

C: Normal perithyroidal paraganglion stained for chromogranin A.

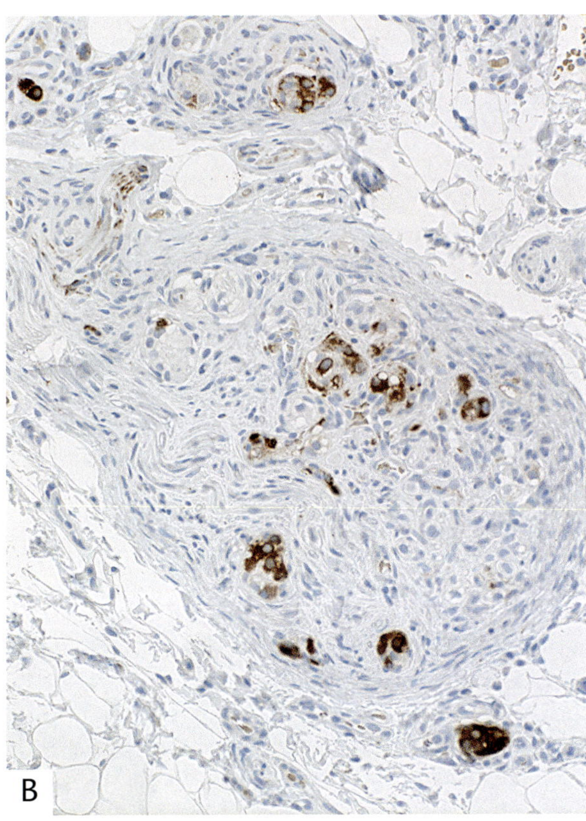

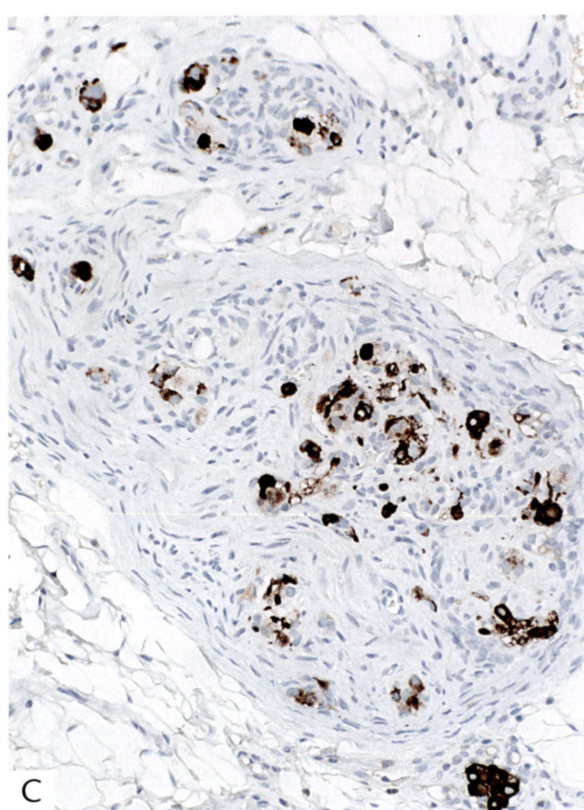

classification (28), in which head and neck paragangliomas were coded according to their four principal anatomic locations: carotid body (40 to 60 percent), jugulotympanic (middle ear; about 30 percent), vagal (about 10 percent), and laryngeal (rare). The change acknowledges that paragangliomas in the head and neck also occur in other locations, including the sella turcica, orbit, paranasal sinuses, parotid gland, thyroid gland, clivus, and mandible (10,29–31).

Although sympathetic paragangliomas are chromaffin cell tumors, the term pheochromocytoma is reserved for tumors in the adrenal medulla. This distinction continues a practice that began as an arbitrary convention in the 1950 Armed Forces Institute of Pathology (AFIP) adrenal gland Fascicle, in response to increasing confusion in the use of terminology (32). However, it may be justified biologically by distinctive characteristics of extra-adrenal versus adrenal tumors. These characteristics include usual inability to synthesize epinephrine and proclivity to occur in patients with hereditary mutations of genes encoding subunits of succinate dehydrogenase (33); they may reflect the different developmental timetables for the adrenal medulla and extra-adrenal paraganglia (see figs. 9-4, 9-5). The WHO definition applies irrespectively of whether or not a tumor produces catecholamines.

Evolving Concepts of Paragangliomas

The WHO 2022 classification of endocrine and neuroendocrine tumors (34) introduces a major change in the conceptual approach to paragangliomas (35). These tumors, like pheochromocytomas, are now considered to have metastatic potential and are assessed in terms of risk stratification. This concept of a continuum of risk is aligned with the current approach to epithelial neuroendocrine tumors and with the 2017 expert consensus proposal by the WHO and International Agency for Research on Cancer (IARC) aiming to achieve a common classification framework (36). Accordingly, the previous binary classification into "benign" and "malignant" paragangliomas is no longer employed. The new approach implicitly recognizes the previous definition of malignancy based on the presence of metastases, but substitutes "metastatic" for malignant in order to avoid the confusion previously caused by a competing definition based solely on invasion. The term "benign paraganglioma" also should not be used. Concurrently with the new concepts and terminology, a staging system for pheochromocytoma and paraganglioma was introduced in the 8th edition of the American Joint Committee on Cancer (AJCC) staging manual (37).

General Features

Epidemiologic data vary widely because of regional and national variations and institutional referral patterns. The regional prevalence of mutations in different hereditary susceptibility genes, tracing back to original patients with founder mutations, may underlie some epidemiologic disparities (see chapter 10).

Approximately 500 to 1,600 new cases of pheochromocytoma and paraganglioma collectively are estimated to occur per year in the United States (38). A 2018 analysis of a comprehensive national pathology database in the Netherlands showed an annual combined incidence of pheochromocytoma and sympathetic paraganglioma of 0.6 per 100,000 person years (39), consistent with the upper end of the estimated US range. Epidemiologic trends in the Dutch study show an increasing incidence and smaller tumor size at diagnosis, most likely reflecting recently increased awareness of these tumors and improved diagnostic techniques. The mean patient age at diagnosis was 51 ± 16 years, and there was a slight male predominance. Paragangliomas comprised 19 percent of the cases: 56 percent abdominal/retroperitoneal, 22 percent paraspinal, 10 percent urinary bladder, 9 percent thoracic, and 3 percent miscellaneous, including mesentery, uterus, ovary, testicle, or spermatic cord (39).

Head and neck paragangliomas account for 3 percent (40) to more than 20 percent (41) of all paragangliomas, and for 0.03 percent of all head and neck tumors (42). As previously noted, "head and neck" and "parasympathetic" are not precisely synonymous. Sympathetic paragangliomas occasionally arise from the cervical sympathetic chains, sometimes even concurrently with carotid body tumors (19). Carotid body paragangliomas predominate in most head and neck series (about 60 percent of cases), followed by jugulotympanic (30 percent) and vagal (10 percent). Laryngeal paragangliomas are rare (43).

The mean patient age at diagnosis is in the 5th to 6th decades but the reported age ranges are wide, especially for laryngeal paragangliomas (5 to 85 years) (44). In contrast to sympathetic paragangliomas, those in the head and neck show a female to male ratio of 2-3 to 1 (43). In addition, carotid body paragangliomas occur with increased frequency in people and animals living at high altitudes (45,46), where the female to male ratio can be as high as 8 to 1. This altitude effect probably reflects the oxygen-sensing functions of the carotid bodies, which experimentally undergo physiologic hyperplasia in response to hypoxia (47). However, an increased incidence of paragangliomas has not been reported in all mountainous regions. At least in some patients, the effect of altitude results from exacerbation of a regionally prevalent hereditary predisposition (48–51).

Both sympathetic and parasympathetic paragangliomas are frequently multifocal, especially in young adults and children. A young patient or the presence of multiple tumors suggests the presence of hereditary disease (52). In a 2017 review of 748 patients with pheochromocytoma or sympathetic paraganglioma, children compared to adults show a higher prevalence of hereditary (80.4 versus 52.6 percent), extra-adrenal (66.3 versus 35.1 percent), multifocal (32.6 versus 13.5 percent), metastatic (49.5 versus 29.1 percent), and recurrent (29.5 versus 14.2 percent) tumors (53). In large head and neck series, about 20 percent of patients with carotid body, 3 percent with middle ear, and 8 to 40 percent with vagal paragangliomas have bilateral or multifocal tumors (30,43).

Clinical Features

Historical clinical data are somewhat confused by ambiguous use of the term "functional," which has variably been based on either clinical signs and symptoms or biochemical testing. Further, results of biochemical testing depend on the metabolites tested and the sensitivity of the assays employed. Even tumors considered biochemically nonfunctional may show immunohistochemical expression of tyrosine hydroxylase, which can be helpful in diagnosis (see Immunohistochemical Findings). Nevertheless, several generalizations apply.

Most patients with sympathetic paragangliomas show signs and symptoms of catecholamine excess (54,55), and almost all show increased levels of normetanephrine and/or methoxytyramine, the metabolites of norepinephrine and dopamine, respectively. These metabolites are generated intratumorally by O-methylation of the parent amines, which leak from secretory vesicles into the tumor cell cytoplasm, and they are therefore more sensitive biomarkers than catecholamines, which are also produced by sympathetic neurons (56). Nonetheless, some abdominal paragangliomas, especially those with mutations of the *SDHB* gene, are biochemically silent. Biochemical testing alone may therefore not be sufficient to detect these tumors even when screening patients who are known mutation carriers (57).

Like normal extra-adrenal paraganglia, paragangliomas typically do not produce epinephrine or metanephrine. This contrasts with most pheochromocytomas (56), exceptions being noradrenergic pheochromocytomas associated with hereditary mutations of *VHL* or *SDHx* genes (58), and adrenergic paragangliomas reported in China associated with somatic *HRAS* and *FGFR1* mutations (59). The secretory profiles obtained from biochemical testing can, therefore, provide clues to tumor location and genotype in most cases.

Paragangliomas that produce substantial proportions of dopamine and methoxytyramine are associated with an increased risk of metastasis and mutations of succinate dehydrogenase genes (56). Patients with tumors that produce exclusively dopamine usually do not exhibit signs of catecholamine excess, and may present with large tumors and mass effects (56). The mass effects can be particularly devastating for patients with paragangliomas in a critical location, such as the heart (60).

In contrast to thoracoabdominal sympathetic paragangliomas, head and neck paragangliomas typically present as mass lesions. Fewer than 4 percent are biochemically functional (55), including those from the cervical sympathetic chain (19). The clinical signs and symptoms depend on the anatomic site, tumor size, and presence or absence of local invasion. Classically, carotid paragangliomas present as asymptomatic neck masses at the angle of the mandible; middle ear paragangliomas typically are associated with pulsatile tinnitus, hearing abnormalities, or aural fullness; and vagal paragangliomas present as neck masses and are more often associated with cranial nerve involvement (43). These tumors are now increasingly discovered as incidentalomas or in the course of screening patients with hereditary tumor syndromes (61). When cranial nerve damage occurs it may result in isolated or multiple palsies of the lower cranial nerves. Signs and symptoms may be etiologically ambiguous, such as hoarseness and dysphagia caused by glossopharyngeal nerve involvement, or isolated facial nerve palsy (62). Horner syndrome may result from damage to the cervical sympathetic chain.

Head and neck paragangliomas are slow-growing tumors, and the risk of cranial nerve damage caused by surgical or other treatment suggests a "wait and scan" strategy may be optimal for some patients (63). This consideration must be balanced against tumor genotype and potentially aggressive behavior associated with mutations of the *SDHB* gene (61,64).

An overall mean growth rate for paragangliomas has minimal meaning because tumor characteristics are dependent on tumor genotype. However, even for the most aggressive tumors, the growth rate is unusually slow compared to many other neoplasms. The median volume doubling times have been calculated as 5.44 years for head and neck paragangliomas, 6.94 years for abdominal paragangliomas, and 11.8 years for thoracic paragangliomas with succinate dehydrogenase mutations in one recent study (65). Another study in which genotype was unknown, head and neck paragangliomas showed average growth of 1.0 mm/year and a median tumor doubling time of 4.2 years (66).

The overall risks of metastasis for sympathetic and parasympathetic paragangliomas vary according to anatomic site. In general, abdominal and retroperitoneal sympathetic paragangliomas pose the highest risk (about 30 versus about 8 percent for pheochromocytoma) (67). Head and neck paragangliomas are considered to be indolent tumors, with a 4 to 6 percent risk of metastasis for those in the carotid body and 2 percent in the middle ear (43). At least 40 percent of paragangliomas, however, are hereditary, and there are strong phenotype correlations with different susceptibility genes (see chapter 10). It is therefore difficult to ascertain the contributions of genotype versus anatomic site to the reported clinical characteristics of sympathetic or parasympathetic paragangliomas in different locations (8). This is especially true because the tumors are rare and much of the clinical literature antedates genetic testing.

Extra-adrenal abdominal paragangliomas often harbor *SDHB* mutations, which may be almost the entire cause of their aggressive behavior (67). In contrast, carotid body paragangliomas are most often associated with *SDHD* mutations (43). In a study of 34 patients with resected carotid body paragangliomas harboring *SDHB* mutations, 6 patients (17.3 percent) developed distant metastases and an equal number developed recurrent tumor at the previous resection site after a median of 126 months (range: 1 to 293 months) (64). In a 2019 multidecade study of paragangliomas from multiple head and neck locations, all but one of 5 tumors that metastasized harbored a mutation of *SDHB* (68). From a molecular standpoint, it is of interest that the typically indolent *SDHD*-mutated head and neck paragangliomas share more features with indolent *VHL* pheochromocytomas than with more aggressive *SDHB*-mutated abdominal paragangliomas (69).

The historical clinical data on the frequency of metastasis may be partly inaccurate because paragangliomas often occur synchronously or metachronously in multiple locations, and multiple primary tumors may have been interpreted as metastases. Consequently, the WHO definition of metastasis now stipulates that tumors diagnosed as metastatic must be in locations where normal paraganglia are not present. The only such locations are bone and histologically confirmed lymph nodes. Paragangliomas in the lung and liver are usually metastases rather than primary tumors, but a caveat is that primary paragangliomas do occasionally occur in those organs (10,70,71). The clinical context, especially the possibility of hereditary susceptibility to multiple tumors, must therefore be considered. While diagnosis of visceral and bone metastases frequently relies on imaging studies, diagnosis of lymph node metastases absolutely requires histologic confirmation because of the overlapping anatomic distributions of lymph nodes and normal paraganglia (see fig. 9-3).

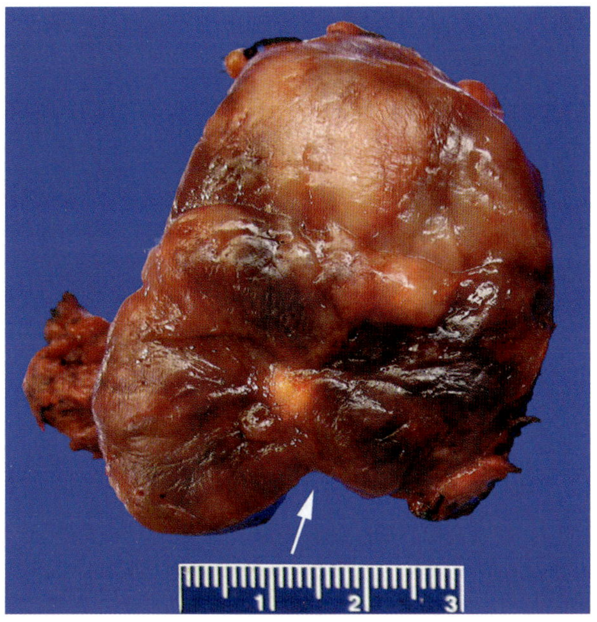

Figure 9-11

ABDOMINAL PARAGANGLIOMA

This paraganglioma from the abdomen at the origin of the superior mesenteric artery shows a groove (arrow) where the specimen abutted the artery. Bulging areas of fleshy pink-gray and pink-tan tissue contrast with slightly retracted areas of congestion and hemorrhage that often develop during surgery.

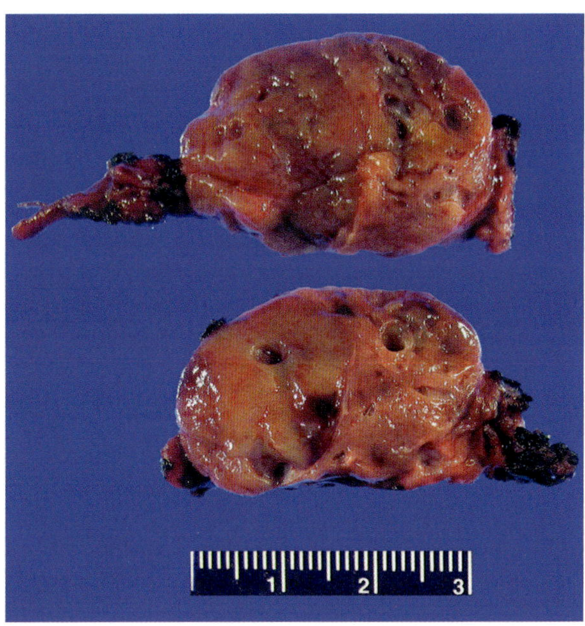

Figure 9-12

CAROTID BODY PARAGANGLIOMA

The tumor is completely encapsulated and has protruding segments of nerve and artery (top left). The cut surface shows focal cystic degeneration and prominent cavernous blood vessels. A distinct large fleshy nodule is present at lower left.

Gross Findings

Paragangliomas are circumscribed, usually encapsulated tumors that grow to diameters greater than 10 cm (72,73). Large tumors are especially likely to occur in the abdomen and retroperitoneum. Capsular surfaces may be grooved by abutting arteries, and these grooves should be noted as landmarks for specimen orientation (fig. 9-11). Relationships with attached segments of nerves and blood vessels should also be documented (fig. 9-12), and these structures should be assessed for vascular and perineural invasion.

The cut surface is pink tan to dark red, corresponding to intratumoral variation in cellularity, congestion, and hemorrhage (figs. 9-11, 9-12). Areas of necrosis may be present, especially in tumors embolized prior to surgery. Coarse nodularity may be observed on the external or cut surfaces, and was associated with risk for local invasion and/or metastases in a series of sympathoadrenal paragangliomas (74). Sections should represent all grossly distinct regions of tumor.

Microscopic Findings

Sympathetic and parasympathetic paragangliomas show a wide range of histologic variation, and are often indistinguishable from each other or from pheochromocytoma. As in pheochromocytoma, the nested zellballen pattern is usually discernible but can be poorly developed or absent. When present, the cell nests can vary markedly in size and shape. Often there is a mixture of nested and trabecular patterns, and some tumors show predominantly trabecular or diffuse growth (fig. 9-13). Areas with varied histoarchitecture and cytology are sometimes present within the same tumor.

The classic nested pattern is often more pronounced in parasympathetic paragangliomas, especially those in the carotid body, than in their sympathetic counterparts. Parasympathetic paragangliomas also tend to show less expression of the functional neuroendocrine markers chromogranin A and tyrosine hydroxylase, although those markers may vary with

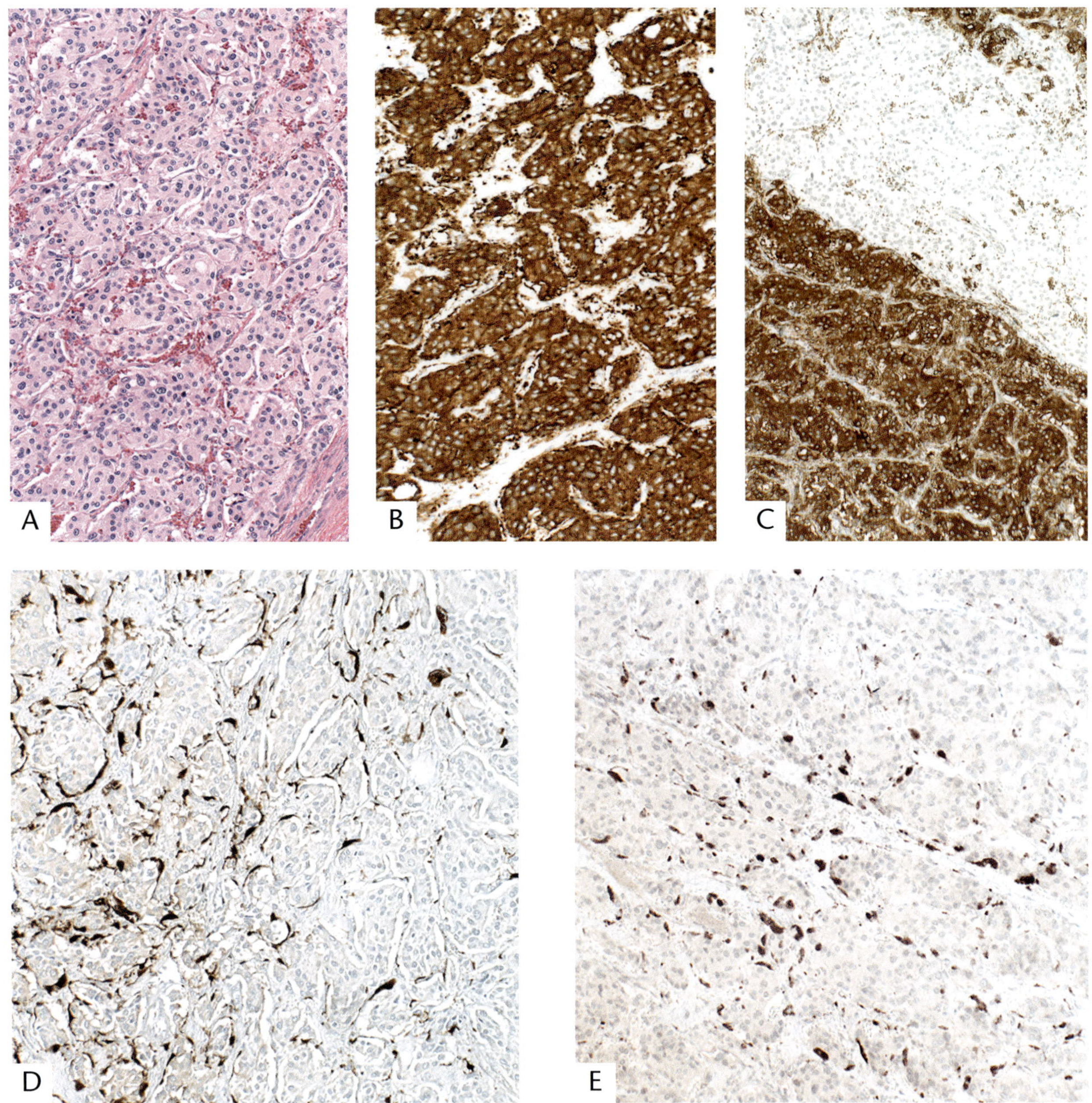

Figure 9-13

ABDOMINAL PARAGANGLIOMA

A: The tumor, from a patient with a hereditary mutation of the *SDHB* gene (same specimen as in fig. 9-11), has mixed nested and trabecular architecture and is composed of monomorphic cells with slightly eosinophilic cytoplasm. This appearance is often associated with *SDHB* mutation.

B: The strong, diffuse expression of synaptophysin highlights a mixed pattern of nested, trabecular, and diffuse histoarchitecture.

C: Immunohistochemical stain for tyrosine hydroxylase shows intratumoral heterogeneity at the functional level, with abrupt transition from strong, diffuse expression to absent expression in a small region of the tumor.

D: Immunohistochemical stain for S-100 protein shows heterogeneous distribution of sustentacular cells in adjacent areas of tumor.

E: Immunohistochemical stain for glial fibrillary acidic protein (GFAP) shows a small focus of GFAP-positive sustentacular cells in a tumor in which S-100 protein-positive cells were much more numerous.

individual tumors and according to patient genotype. Morphologic-functional correlations are illustrated in representative immunohistochemical panels for sympathetic paragangliomas in figures 9-13 and 9-14, and for a carotid body paraganglioma in figure 9-15. In addition to particularly pronounced zellballen, carotid body paragangliomas often display prominent cavernous blood vessels (52) that communicate with extratumoral and capsular vessels (fig. 9-15A). Architectural patterns are highlighted by reticulin stains (fig. 9-15C) or by immunohistochemical markers.

Paragangliomas with unusual architectural patterns can mimic features associated with other neoplasms. For example, a noncanonical arrangement of tumor cells in relation to blood vessels can produce an angioma-like vascular pattern or perivascular pseudorosettes (fig. 9-16). Dramatic intraneural and perineural growth may be seen with vagal paragangliomas, which originate within or near the vagus nerve (fig. 9-17), and intramural spread within the urinary bladder can mimic urothelial carcinoma (fig. 9-18).

A common feature that can markedly alter the histoarchitecture is sclerosis. In one recent study (75), some degree of sclerosis, defined by broad bands of fibrosis between tumor cells and excluding central fibrotic scars, was present in approximately 50 percent of paragangliomas and involved 60 percent or more of tumor area in 13 percent of cases. Sclerosis was found to correlate with hereditary tumor predisposition, and was especially associated with *SDHx* mutations (75).

Sclerosing paraganglioma, characterized by extensive sclerosis resulting in a pseudoinfiltrative pattern of tumor cells (fig. 9-19), is considered by some to be a distinct clinicopathologic entity notable for mimicking malignant epithelial neoplasms (76,77). These tumors occur most frequently in the carotid body but also in other head and neck and thoracoabdominal locations. They are negative for keratins and overall exhibit an immunophenotype consistent with other paragangliomas in the same locations, with comparably low rates of local infiltration, recurrence, or metastasis. Misdiagnosis is most likely to occur when sclerosis is present at the periphery of the tumor. In some cases, fibrous bands extend to or radiate from the adventitia of tumor blood vessels or the tumor capsule, which still remains clearly defined (fig. 9-19B). In cases where the capsule is not clearly delineated microscopically, it may be impossible to discriminate soft tissue invasion from sclerosis. Careful correlation with the gross specimen is therefore essential.

The pathogenesis of sclerosis in paragangliomas is unclear. In a few cases, sclerosis is associated with hemosiderin deposition, suggesting that ischemia or hypoxic signaling contribute to its development (fig. 9-20).

Similar to normal paraganglia, paragangliomas contain two characteristic cell populations, nonepithelial neuroendocrine cells and sustentacular cells. The neuroendocrine cells are typically polygonal, with a range of sizes, and may be smaller or larger than in normal paraganglia (fig. 9-21). The nuclei of the neuroendocrine cells are typically round to oval, with smooth contours, small or absent nucleoli, and stippled chromatin. Nuclear atypia may occur in scattered cells or diffusely. The cytoplasm is most often slightly eosinophilic to amphophilic or clear (fig. 9-21). Granular basophilic cytoplasm is sometimes present, especially in sympathoadrenal paragangliomas (fig. 9-22). Nevertheless, cytologic as well as architectural variation can occur in any location.

Sustentacular cells are sometimes recognized at the periphery of cell nests by their smaller oval or triangular nuclei, but they are readily confused with other cell types, especially perivascular monocytes (78). Their definite identification requires immunohistochemistry.

Cytologic features that may pose particular challenges include nuclear atypia (fig. 9-23), spindle cells (fig. 9-24), cytoplasmic vacuoles (figs. 9-25, 9-26), and pigmented cells (fig. 9-27). In some cases, vacuolization is caused by suboptimal tissue dehydration or use of heat to dry incompletely dehydrated paraffin sections. However, it also occurs in some well-fixed and well-processed tumors, especially those associated with *SDHB* mutations (figs. 9-23, 9-24) (35) (see chapter 10).

The vacuoles in paragangliomas are usually large and irregular, with a scattered distribution, and have sometimes been called "pseudoacini" (fig. 9-25) (79). Small intracytoplasmic vacuoles with flocculent content, a well-known characteristic of SDH-deficient renal cell carcinoma

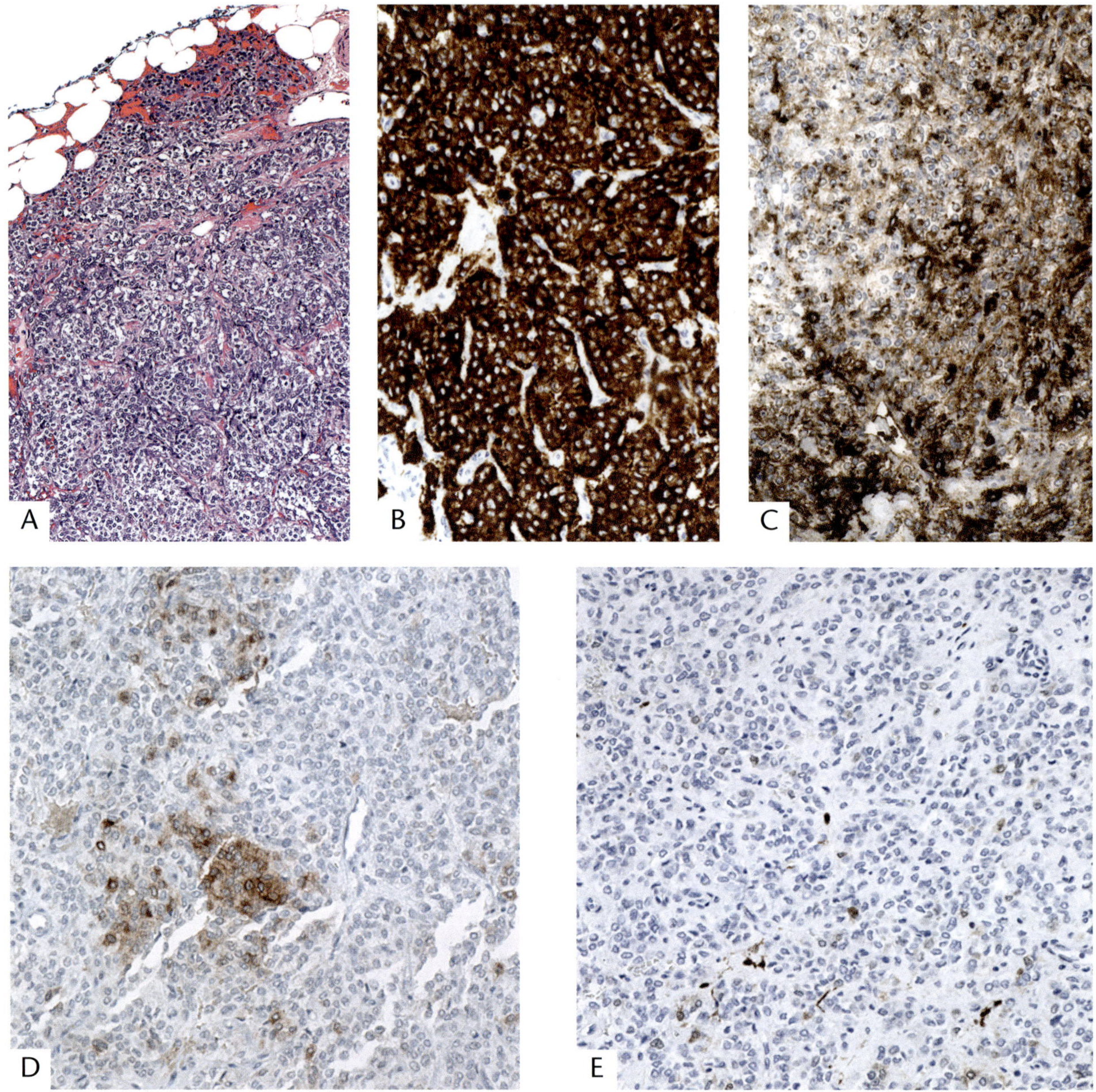

Figure 9-14

MESENTERIC PARAGANGLIOMA

A: This paraganglioma composed of basophilic and clear cells was incidentally discovered in an asymptomatic patient with no other tumors. Fat invasion, poorly defined irregular cell nests, and diffuse growth are putative risk factors for the development of metastases, but no other tumors were detectable. Staining for neuroendocrine markers showed the usual trend for these markers: synaptophysin > chromogranin > tyrosine hydroxylase. Although decreased expression of functional markers in paragangliomas is often associated with mutations of genes encoding subunits of succinate dehydrogenase (SDH), this tumor was SDH intact.

B: Strong diffuse expression of synaptophysin is present throughout this incidental mesenteric paraganglioma.

C: Expression of chromogranin A is lost or greatly reduced in patchy areas, with some cells showing only dot-like perinuclear staining.

D: Expression of tyrosine hydroxylase is weak and focal, consistent with minimal or undetectable elevation of metanephrines or catecholamines.

E: S-100 protein stain shows sparse sustentacular cells in this tumor.

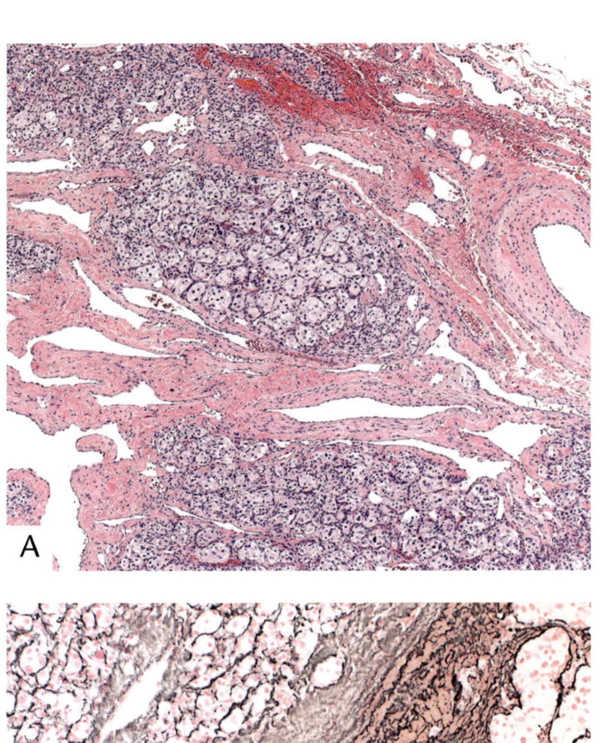

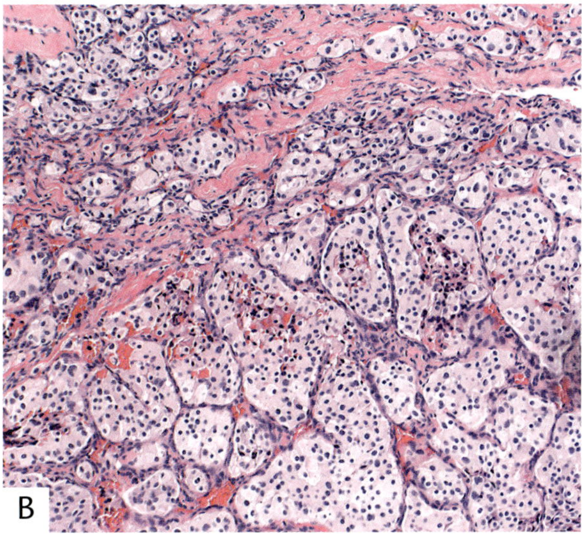

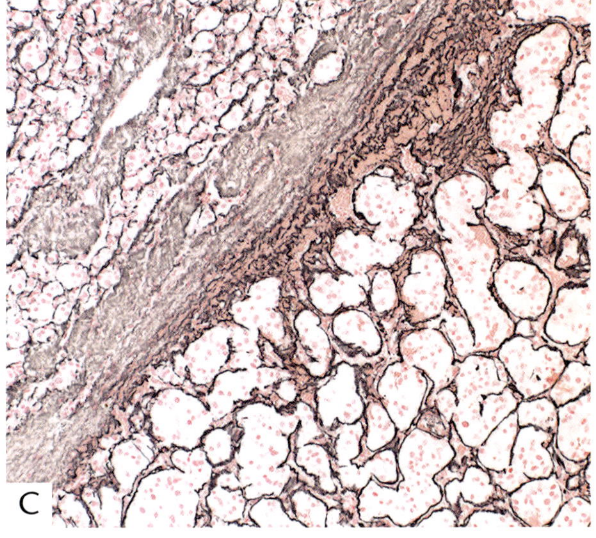

Figure 9-15

CAROTID BODY PARAGANGLIOMA

A: Tumor tissue in most parts of the tumor (same specimen as fig. 9-12) shows uniform round nests of weakly eosinophilic cells and cavernous blood vessels.

B: This section, taken at the interface between the circumscribed fleshy nodule at lower left in the gross photograph (fig. 9-12) and the adjacent tissue, shows a sharp transition between large irregular cell nests in the nodule and adjacent small nests.

C: Reticulin stain highlights different architectural patterns.

D: Immunostain for chromogranin A in this section shows many tumor cells that are negative or express only punctate staining.

E: Immunoreactivity for tyrosine hydroxylase is often focally present in carotid body paragangliomas considered biochemically nonfunctional on the basis of serum or plasma assays for catecholamine metabolites.

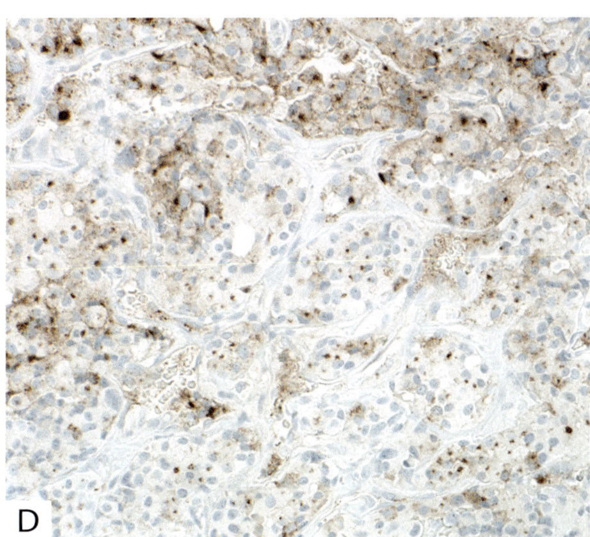

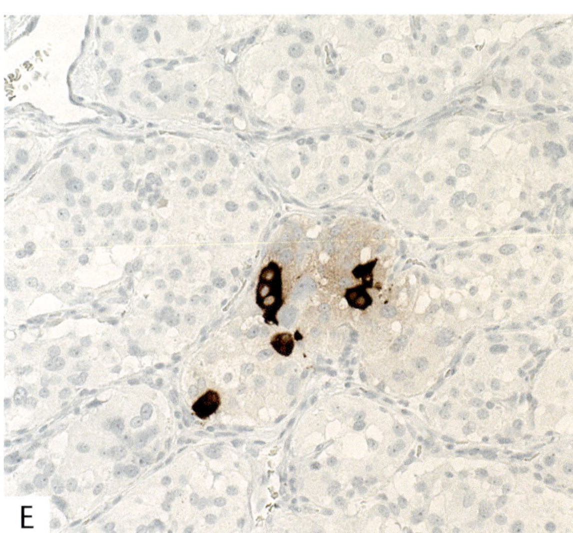

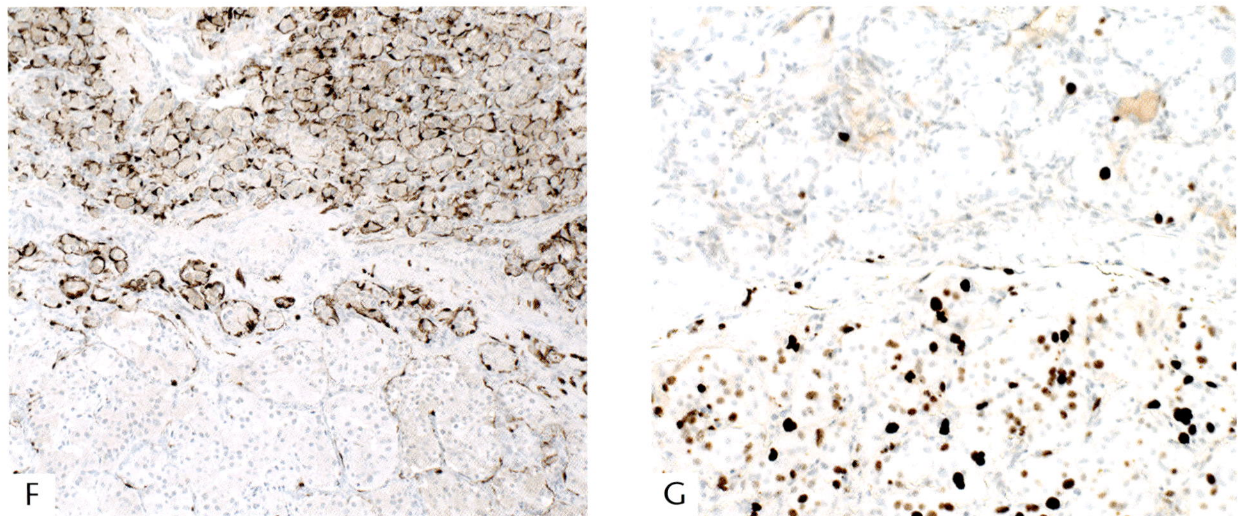

Figure 9-15, continued

F: Immunohistochemical stain for S-100 protein shows varied distribution of sustentacular cells in adjacent areas of tumor.

G: The sparse Ki-67 labeling in the left half of the image is typical of most paragangliomas and is consistent with their slow growth. The abrupt transition to extremely high labeling at right is unusual and is consistent with the fleshy appearance of the tumor-within-tumor nodule seen in figure 9-12.

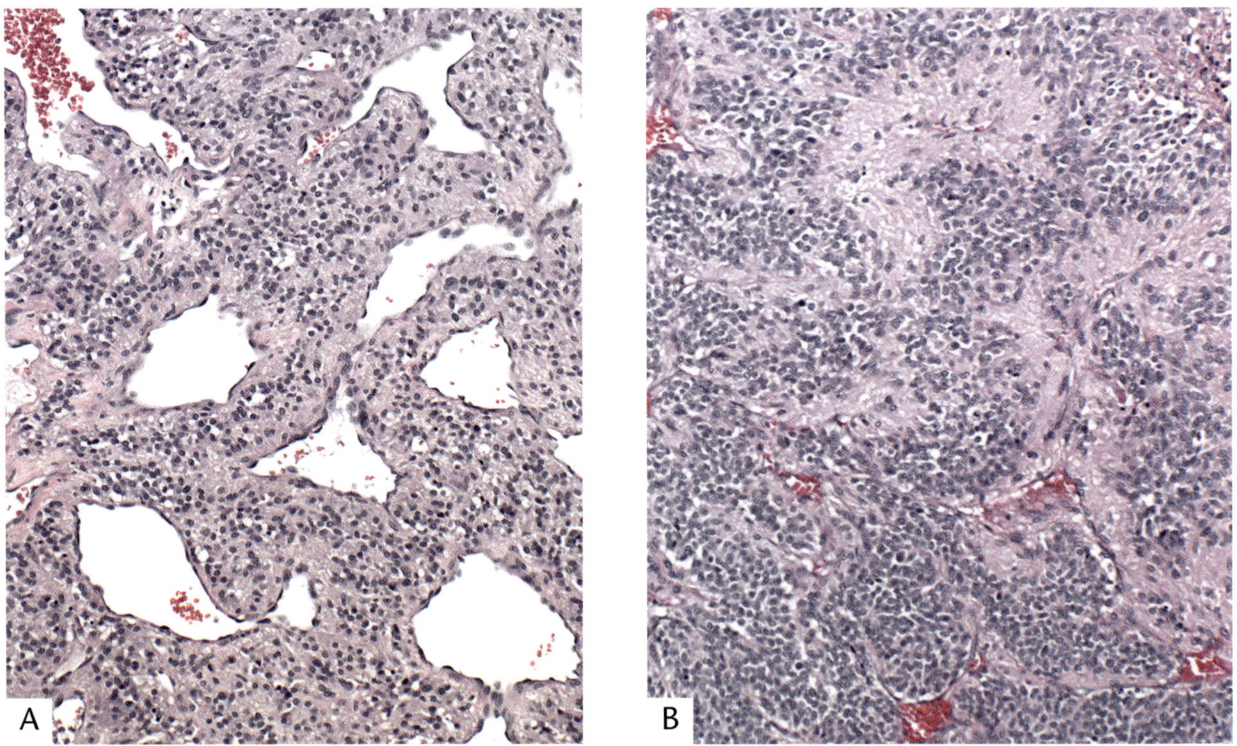

Figure 9-16

VAGAL PARAGANGLIOMA

A: The prominent blood vessels and a pericyte-like arrangement of tumor cells may suggest a vascular tumor in the differential diagnosis.

B: A different area of the same tumor shows a solid growth pattern with prominent perivascular pseudorosettes.

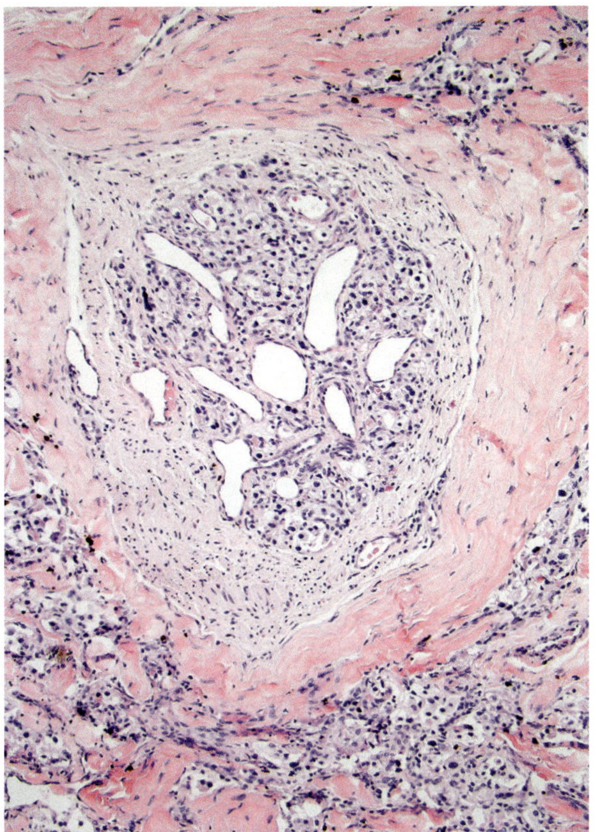

Figure 9-17

VAGAL PARAGANGLIOMA WITH INTRANEURAL GROWTH

Intraneural extension of the tumor at center is accompanied by extensive infiltration of the surrounding tissue.

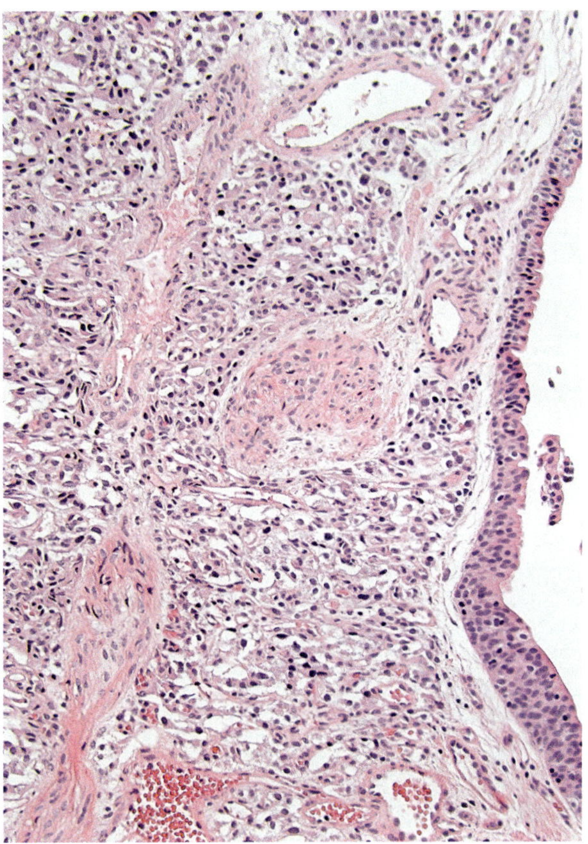

Figure 9-18

URINARY BLADDER PARAGANGLIOMA

This paraganglioma shows extensive superficial growth beneath the mucosa. The tumor is a potential mimic of urothelial carcinoma because of its growth pattern and pale eosinophilic to clear cytoplasm.

(80,81), are extremely rare in SDH-deficient paraganglioma (see fig. 9-26). The dark brown or black granules in pigmented paragangliomas usually consist of lipofuscin, as shown by electron microscopy (see fig. 9-27B). However, cutaneous type melanin is reported occasionally. The differentiation from melanoma is made by expression of neuroendocrine markers. Although melanomas and the sustentacular cells in paragangliomas both express S-100 protein and SOX10, in paragangliomas the distribution of the staining for these markers conforms to the sustentacular cell pattern (82).

The histoarchitecture and cytology of paraganglia and paragangliomas are highly susceptible to artefacts of fixation and processing, which can complicate the diagnosis. These especially include cytoplasmic bubbles and vacuoles that may mimic mucin droplets or signet ring cells (see fig. 9-23). Other potentially confounding morphologic changes are intratumoral congestion and hemorrhage secondary to ligation of blood vessels during surgery (fig. 9-28), or necrosis from preoperative embolization performed to reduce intraoperative bleeding (figs. 9-29, 9-30). New approaches to embolization, including the use of gel beads and shortened intervals between embolization and surgery, may reduce the prevalence of this finding.

Paragangliomas are readily mistaken for other neoplasms, and vice versa. The long list of differential diagnoses varies according to the anatomic site and histologic features of any individual tumor. A sclerosing pattern can

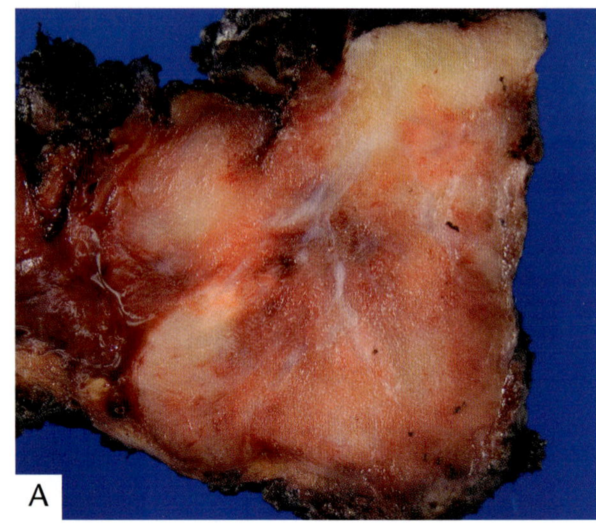

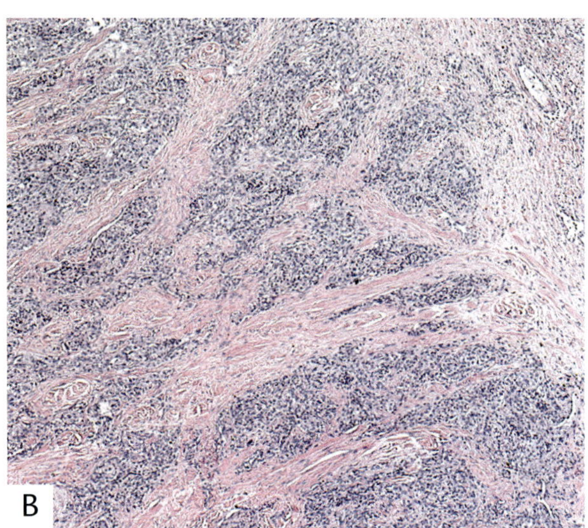

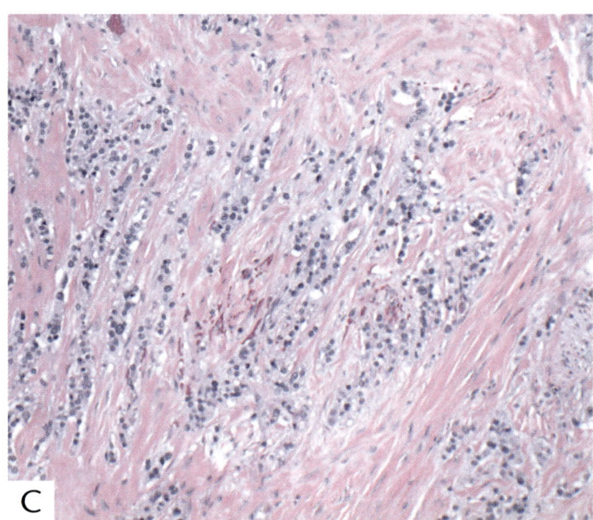

Figure 9-19

CAROTID BODY SCLEROSING PARAGANGLIOMA

A: Sclerosing paragangliomas have been reported as a distinct entity that can mimic infiltrating carcinoma. This sharply circumscribed example shows extensive sclerosis, which is very dense at top right.

B: Section from the lower portion of A shows bands of sclerosis extending to the clearly defined capsule.

C: Section from the densely sclerotic area shows entrapped tumor cells resembling infiltrating carcinoma.

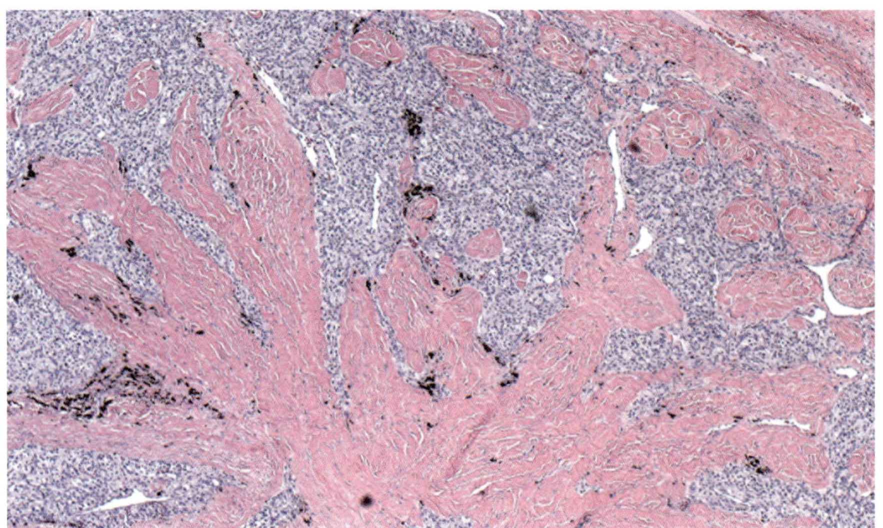

Figure 9-20

CAROTID BODY PARAGANGLIOMA

Areas of sclerosis encroach on the tumor capsule and tumor blood vessels. The presence of hemosiderin suggests repeated ischemia and infarction as possible factors contributing to the development of sclerosis in this case.

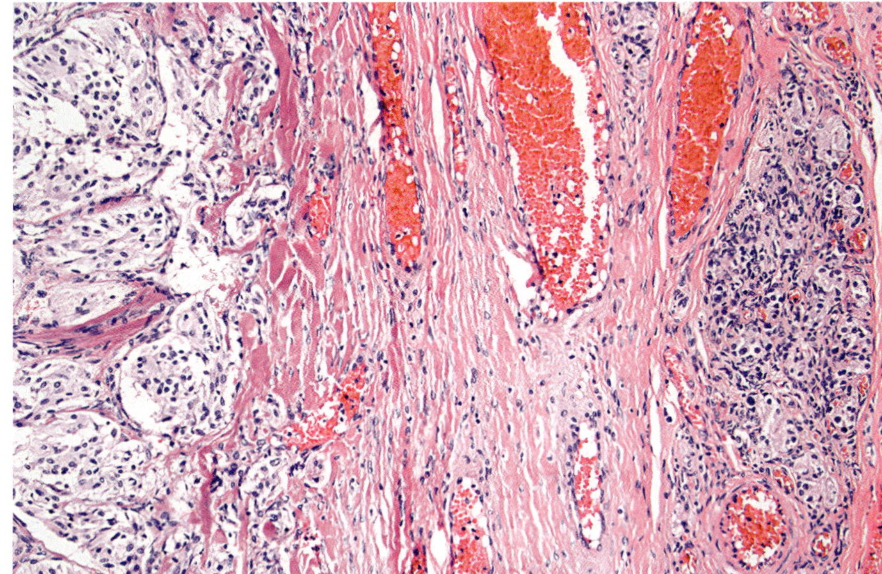

Figure 9-21

CAROTID BODY PARAGANGLIOMA WITH ADJACENT NORMAL CAROTID BODY

The neoplastic cells in this tumor are larger, more spindled, and clearer than the normal carotid body chief cells (at right).

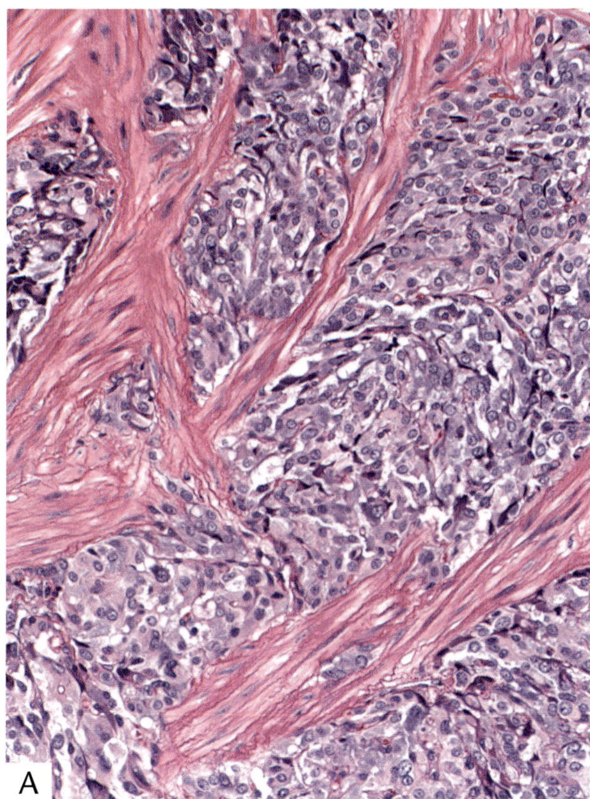

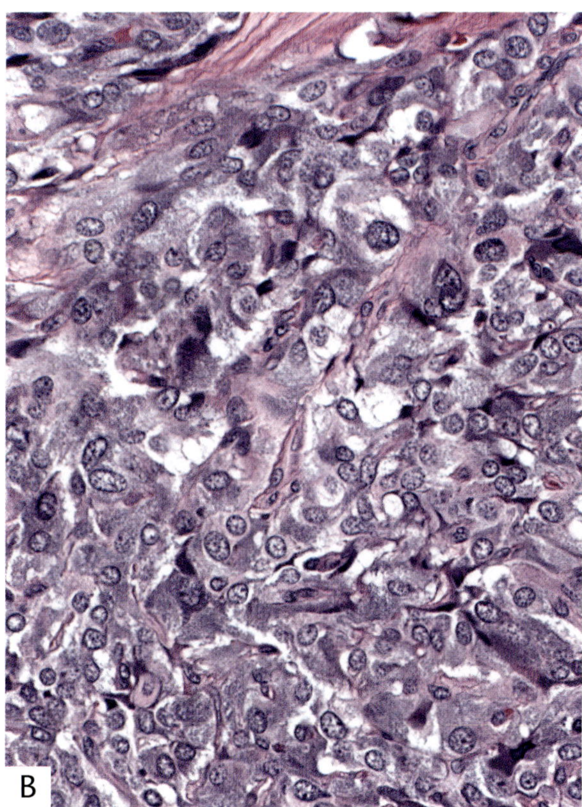

Figure 9-22

URINARY BLADDER PARAGANGLIOMA

A: This paraganglioma extends between muscle bundles in the bladder wall and is a potential mimic of other more aggressive neoplasms because of its growth pattern.

B: Higher magnification of the tumor shows that most neoplastic cells have granular basophilic cytoplasm, which is seen more often in sympathoadrenal than in head and neck paragangliomas.

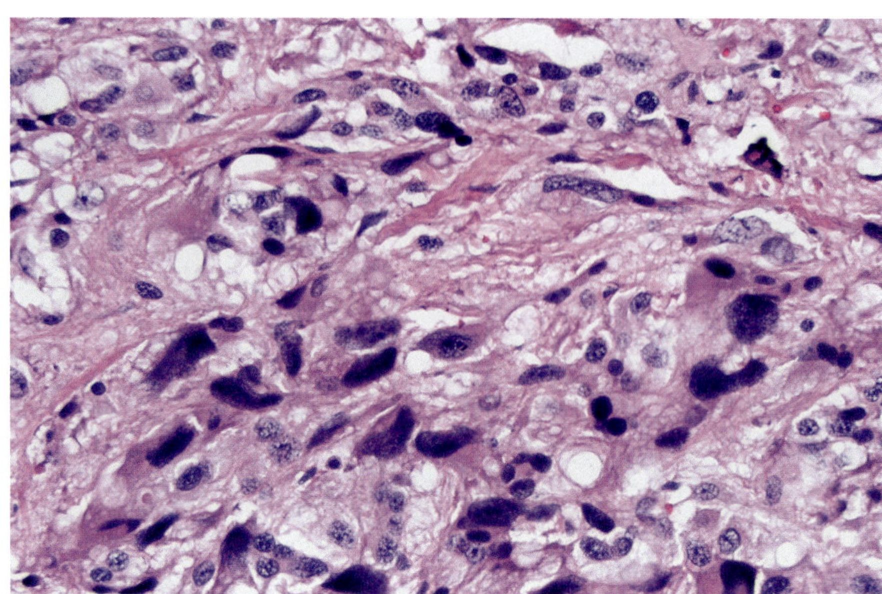

Figure 9-23

MIDDLE EAR (JUGULOTYMPANIC) PARAGANGLIOMA

The presence of nuclear pleomorphism, together with artifactual cytoplasmic vacuolization and clearing in this paraganglioma, may suggest adenocarcinoma as a diagnosis.

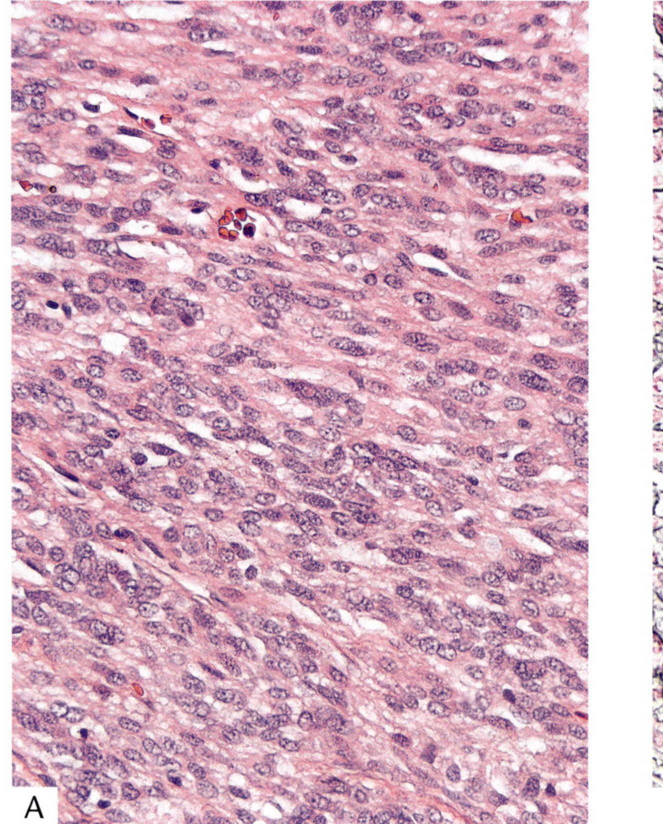

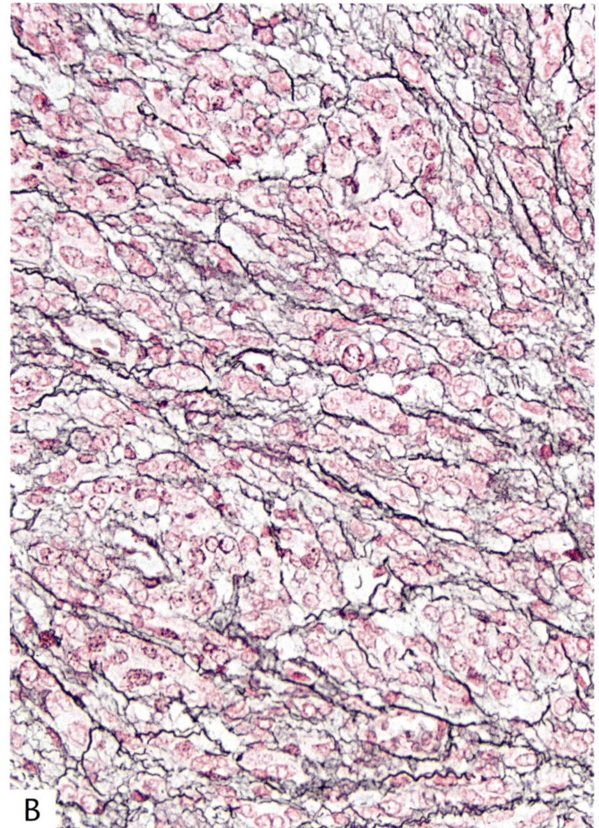

Figure 9-24

CAROTID BODY PARAGANGLIOMA

A: The almost entirely spindle cell composition of this paraganglioma raises the possibility of mesenchymal and other spindle cell neoplasms.

B: Reticulin stain highlights the spindle cell pattern.

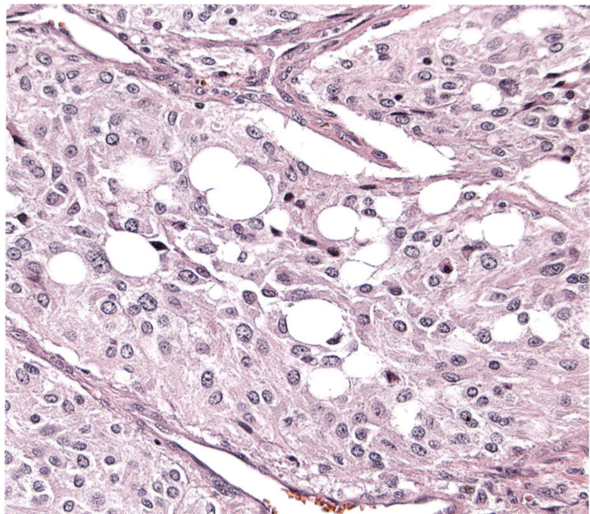

Figure 9-25

ABDOMINAL PARAGANGLIOMA

This tumor is composed of monomorphic cells with slightly eosinophilic cytoplasm and prominent vacuoles creating a pseudoacinar appearance. This appearance is often associated with *SDHB* mutation.

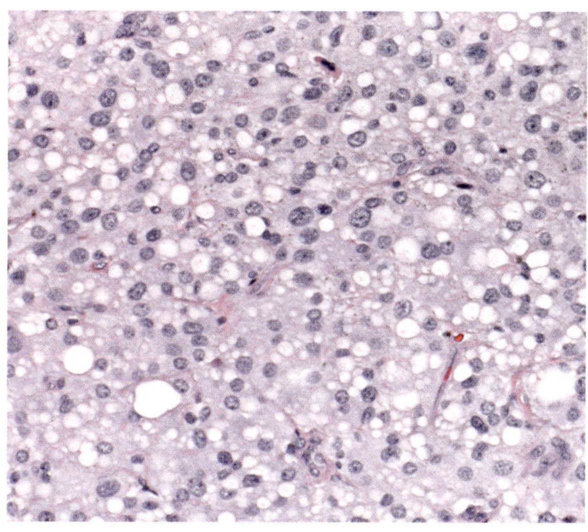

Figure 9-26

SDH-DEFICIENT ABDOMINAL PARAGANGLIOMA

This paraganglioma with a known *SDHB* mutation exhibits small intracytoplasmic vacuoles with flocculent content, which are a well-known characteristic of SDH-deficient renal cell carcinoma but extremely rare in SDH-deficient paraganglioma.

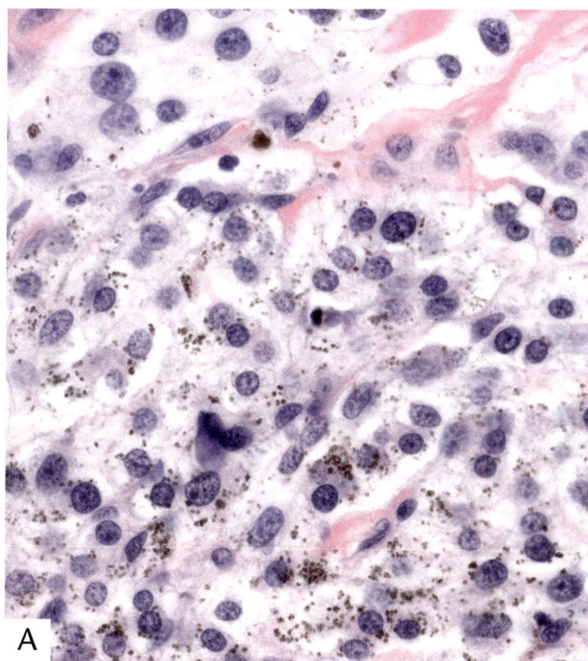

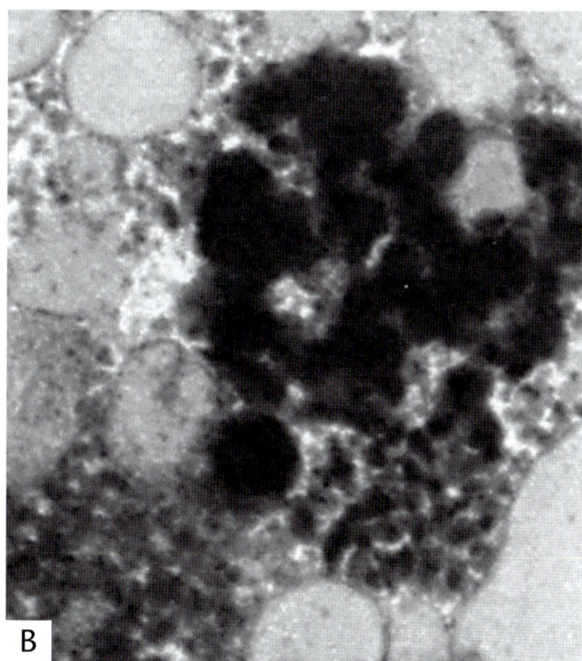

Figure 9-27

PIGMENTED CARDIAC PARAGANGLIOMA

A: Dark brown to black granules are present in the tumor cell cytoplasm.

B: Electron micrograph shows the pigment in A corresponds to amorphous aggregates consistent with lipofuscin/neuromelanin.

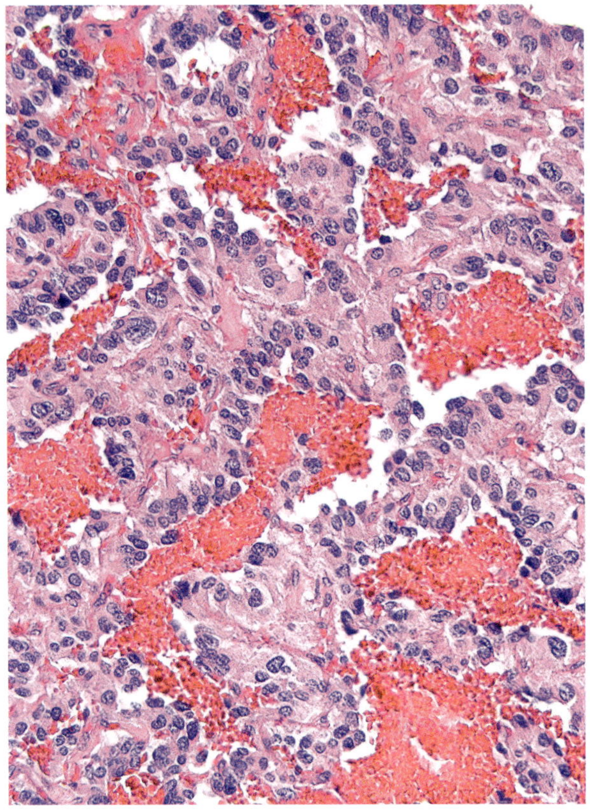

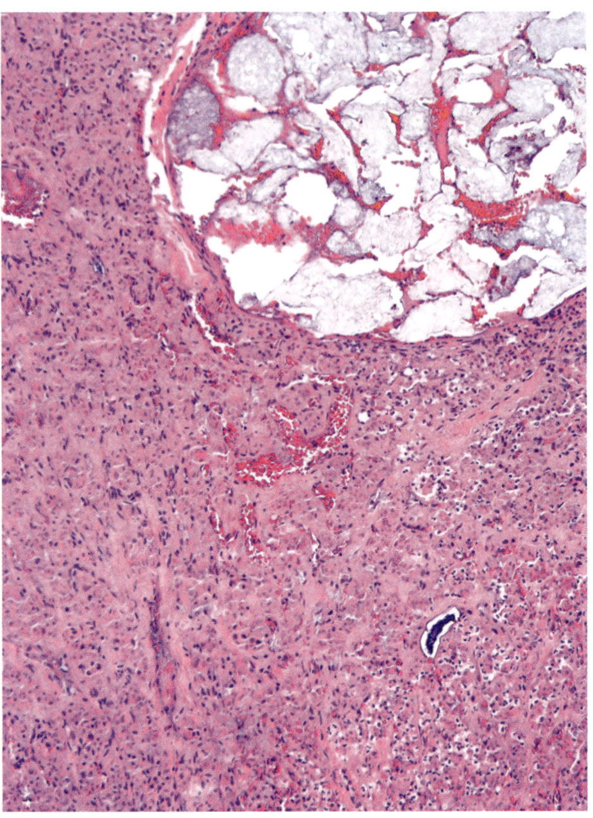

Figure 9-28

VAGAL PARAGANGLIOMA

Pseudopapillary architecture resulting from intraoperative hemorrhage causes this paraganglioma to resemble a papillary carcinoma.

Figure 9-29

CAROTID BODY PARAGANGLIOMA

Extensive necrosis resulted from preoperative embolization performed to reduce bleeding. Embolic foreign material is present in blood vessel at top right.

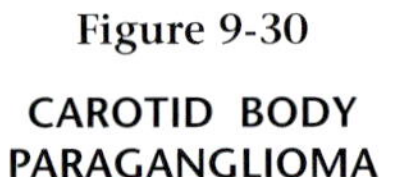

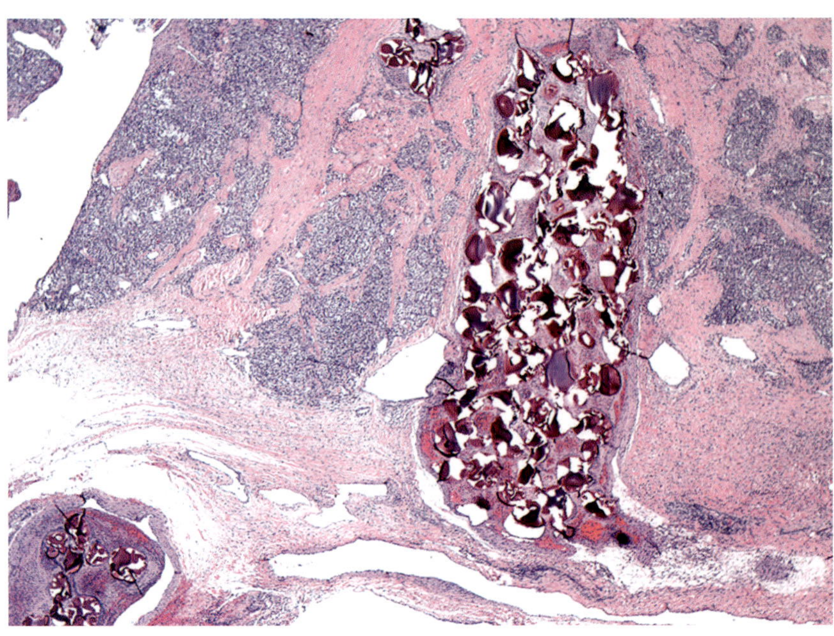

Figure 9-30

CAROTID BODY PARAGANGLIOMA

Preoperative embolization of this tumor was performed with gel beads, which are seen occluding cavernous blood vessels. There is no postembolic necrosis in this case, probably reflecting a shorter interval between embolization and surgery.

mimic infiltrating carcinoma (76), while a vascular pattern can mimic angiomatous tumors. Cytologic variability may further complicate the differential diagnosis. Paragangliomas containing clear cells are especially likely to be mistaken for carcinomas in the urinary bladder (83), prostate gland (84), or kidney (85), while cytoplasmic pigment can suggest melanoma (86,87). Pleomorphism and necrosis (88) can suggest a variety of malignancies.

Diagnosis is especially difficult for paragangliomas in unusual locations, such as the heart (fig. 9-31), where the first step toward arriving at a correct diagnosis is "to think of it." The most challenging differential diagnoses often involve other neuroendocrine tumors, including neuroendocrine tumors of lung (fig. 9-32), small bowel (fig. 9-33) (10), thyroid gland (fig. 9-34) (89), and larynx (43). In general, morphologic features of paragangliomas that favor them over more malignant mimics include sparse mitoses and absence of cell polarity.

Not surprisingly, paragangliomas in the least common locations are the most likely to be misdiagnosed. Even in somewhat usual locations, they may not come to mind if not accompanied by suggestive signs and symptoms. An illustrative analysis of urinary bladder paragangliomas highlights reasons for frequent misdiagnosis of these tumors as urothelial carcinomas (83). Paragangliomas comprise less than 0.5 percent of bladder tumors (83). "Classic" micturition-associated hypertension is present in only a minority of cases, and some patients are asymptomatic. The tumors often arise deeply within the bladder wall and may permeate between muscle fibers as they enlarge. The most common features associated with misdiagnosis are diffuse architecture, focal clear cells, necrosis, and involvement of the muscularis propria (see figs. 9-18, 9-22A). The presence of significant cautery artefact further confounds interpretation of some specimens (83).

Composite Paragangliomas. Composite paragangliomas are histologically similar to composite pheochromocytomas, which are described in chapter 8. They are extraordinarily rare tumors, even compared to their adrenal counterparts. Fewer than 20 well-documented cases have been reported, all located in the abdomen or retroperitoneum and consisting of paraganglioma plus ganglioneuroma. The single

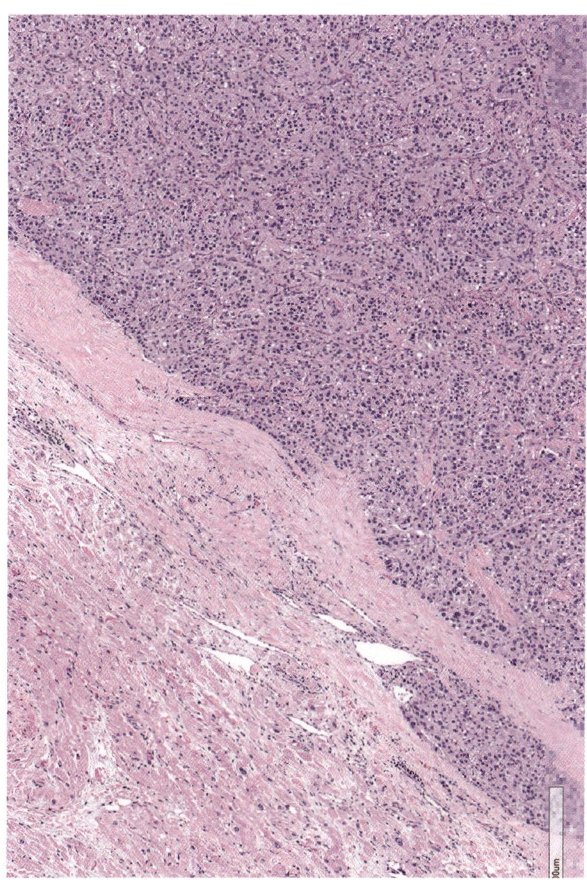

Figure 9-31

CARDIAC PARAGANGLIOMA

This primary cardiac paraganglioma infiltrates myocardium at lower right.

most common location is the urinary bladder (90). There are no unequivocal cases in the head and neck. The tumors previously called "cauda equina paraganglioma" in the cauda equina/filum terminale region and "gangliocytic paraganglioma" in the duodenum are no longer considered to be true paragangliomas, and are classified, respectively, as "cauda equina neuroendocrine tumor" and "composite gangliocytoma/neuroma and neuroendocrine tumor (CoGNET)" in the WHO series 5 bluebooks. True paragangliomas histologically the same as in other locations and without the epithelial component characteristic of CoGNET do occasionally arise in the duodenum (see fig. 9-33), but no composite paragangliomas as currently defined have been reported in that location.

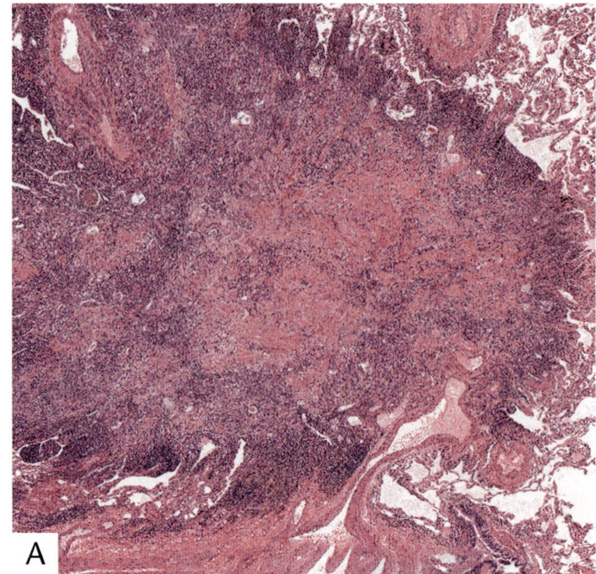

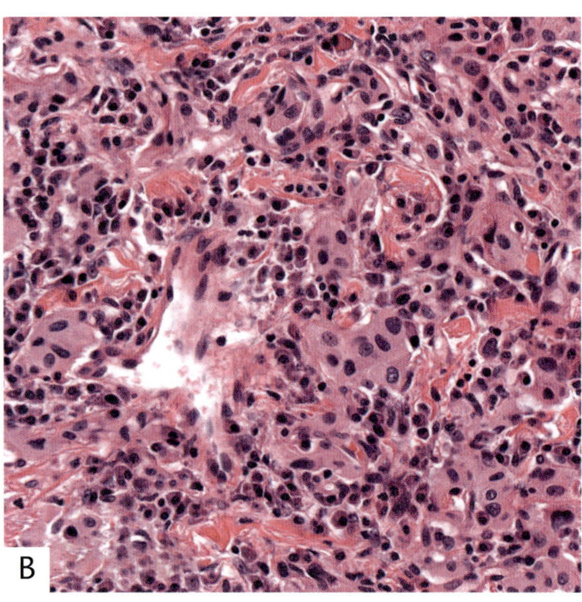

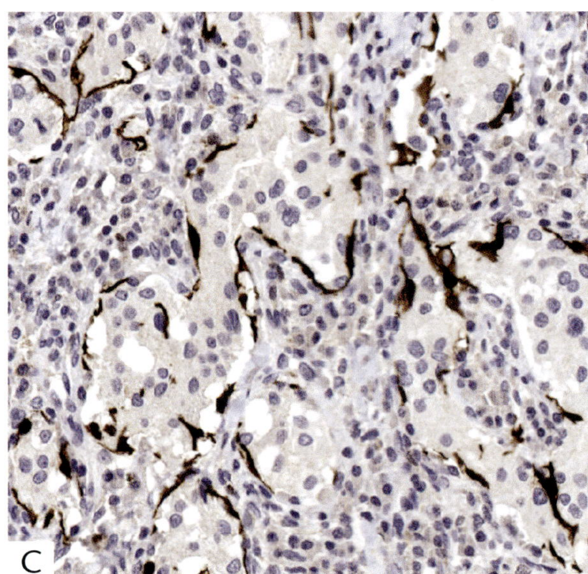

Figure 9-32

PULMONARY PARAGANGLIOMA

A: This paraganglioma was a solitary lesion initially interpreted as an epithelial pulmonary NET. The patient was found to have a hereditary *SDHC* mutation. No other primary tumors were found and this tumor contained abundant sustentacular cells. In view of the clinical and histologic findings and the low metastasis frequency associated with *SDHC* mutation, the tumor was consistent with a primary pulmonary paraganglioma.

B: High magnification shows nested architecture and cells of varied sizes.

C: S-100 protein highlights sustentacular cells.

Genetic data are not available for most composite paragangliomas. Only one case with a germline deletion in *SDHB* has been reported in a Western population (91), while multiple cases with somatic *HRAS* and *BRAF* mutations have been reported in China (92). The latter mutations have a higher prevalence in ordinary paragangliomas in China than in the West.

Immunohistochemical Findings

Applications of Immunohistochemistry in Differential Diagnosis. The differential diagnosis entails consideration of the anatomic site and morphology in conjunction with judicious use of immunohistochemistry. A systematic general approach relies on three components: positive staining for generic neuroendocrine markers (usually chromogranin A or synaptophysin) to establish a neuroendocrine phenotype; absent staining for cytokeratins or other epithelial markers to rule out epithelial neuroendocrine tumors; and tissue-selective functional markers or transcription factors to further focus on the tissue of origin (93).

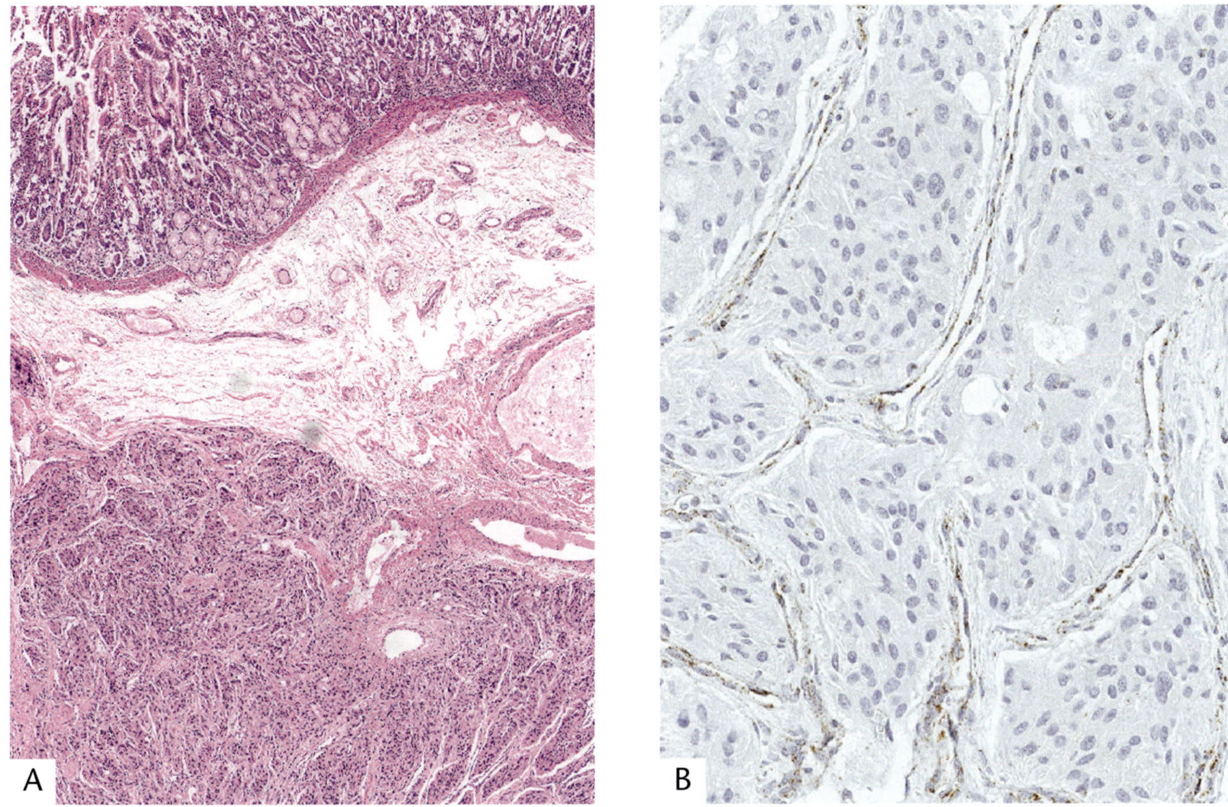

Figure 9-33

DUODENAL PARAGANGLIOMA

A: This tumor was initially interpreted as an epithelial intestinal NET. However, it was negative for keratins and expressed paraganglioma markers. The absence of keratins also weighed against a diagnosis of composite gangliocytoma/neuroma (CogNET, formerly called "gangliocytic paraganglioma"), which would have been a consideration in this area.

B: Immunohistochemical stain shows loss of SDHB in tumor cells and retention in endothelial cells within the tumor.

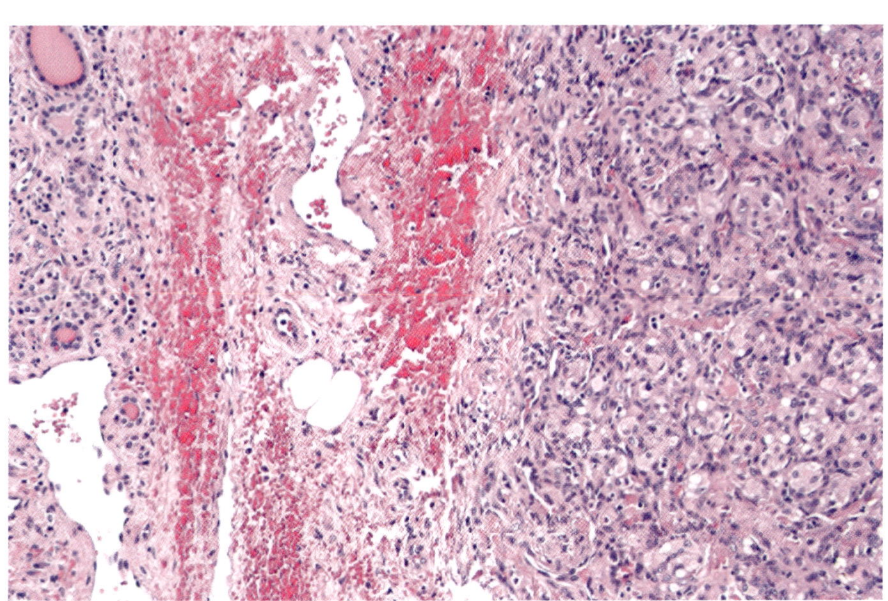

Figure 9-34

THYROID GLAND PARAGANGLIOMA

Thyroid tissue is present on the left.

Transcription factors have become an increasingly valuable tool for the differential diagnosis of challenging cases. Insulinoma-associated protein 1 (INSM1) is reported to have higher specificity but lower sensitivity than conventional functional markers for discriminating both epithelial neuroendocrine tumors and paragangliomas from tumors without neuroendocrine differentiation (94).

Immunoreactivity for GATA3 is useful to distinguish paraganglioma from most epithelial neuroendocrine tumors but is negative in occasional cases. While GATA3 is expressed in many nonendocrine tissues, the only neuroendocrine cells that express this biomarker are normal and neoplastic parathyroid and pituitary tissues (10,93,95). Parathyroid adenomas are of particular concern because they can arise from supernumerary parathyroid glands or ectopic parathyroid tissue occurring in much the same distribution as head and neck paraganglia, including within the vagus nerve (see fig. 9-9) (96,97). Similar to GATA3-negative neuroendocrine tumors (NETs) in the gastroenteropancreatic axis, parathyroid and pituitary adenomas are usually positive for keratins, which should be assessed in conjunction with GATA3 (10). GATA3 is also expressed in urothelium and urothelial carcinoma, and should be avoided for diagnosis of urinary bladder paraganglioma versus urothelial carcinoma (fig. 9-35). The transcription factor PHOX2B has recently been reported to be a highly specific and sensitive marker that may be especially useful for the differential diagnosis of paragangliomas in the head and neck and other challenging locations (25,26).

An important but underutilized functional marker for paragangliomas is tyrosine hydroxylase, which is required for catecholamine biosynthesis (see chapter 1). The expression of tyrosine hydroxylase may not be entirely specific for tumors of neural derivation, as evident from immunoreactivity in a variety of neoplasms (25) and dopamine production in occasional pancreatic NETs (98). Nonetheless, in the appropriate context, tyrosine hydroxylase immunoreactivity is extremely reliable for discriminating paragangliomas from close mimics. An example is medullary thyroid carcinoma, although even in that setting it has been reported (99). Immunoreactivity for synaptophysin is often stronger and more diffusely expressed than staining for chromogranin A or tyrosine hydroxylase, making it the most sensitive of these three markers, but it is also the least specific, as discussed in chapter 1. Tyrosine hydroxylase, which is the most specific, is also the most readily lost and may show only patchy expression even in tumors that are biochemically functional (see figs. 9-13, 9-14). Although most sympathetic paragangliomas are biochemically functional, relative downregulation of neuroendocrine markers has been reported in paragangliomas caused by mutations of genes encoding subunits of succinate dehydrogenase (100).

Paragangliomas in the head and neck express functional neuroendocrine markers variably (see fig. 9-15E–H). Also, in rare cases, they are focally immunoreactive for keratins (fig. 9-36) (101,102). An expanded battery of markers is therefore often required. Chromogranin A, which is the most widely used granin protein in diagnostic pathology, is often negative or only focally positive, and chromogranin B may be preferentially expressed (103,104). Synaptophysin or, less specifically, CD56, may still be present in these cases. In a 2015 single-institution study, 32 percent of head and neck paragangliomas expressed immunoreactivity for tyrosine hydroxylase, even though only 1 percent of patients showed biochemically detectable catecholamine hypersecretion. The remaining 68 percent were negative (105). Despite its low yield, staining for tyrosine hydroxylase can provide useful information even if a diagnosis of paraganglioma is already established independently: if a patient with a tyrosine hydroxylase-negative paraganglioma has persistent or recurrent elevation of catecholamine metabolites, the most likely explanation is a second primary tumor rather than a metastasis.

Demonstration of sustentacular cells by staining for S-100 protein or SOX10 (fig. 9-37) is useful in some cases. Sustentacular cells, however, can be irregularly distributed (see fig. 9-13D) and sparse (see fig. 9-14E) in some primary paragangliomas (78), and are often absent in metastases (106). They can also be numerous in bronchopulmonary (107) and other NETs (93). The prevailing current opinion concerning the nature of sustentacular cells is that they are non-neoplastic (108), and their physiologic roles

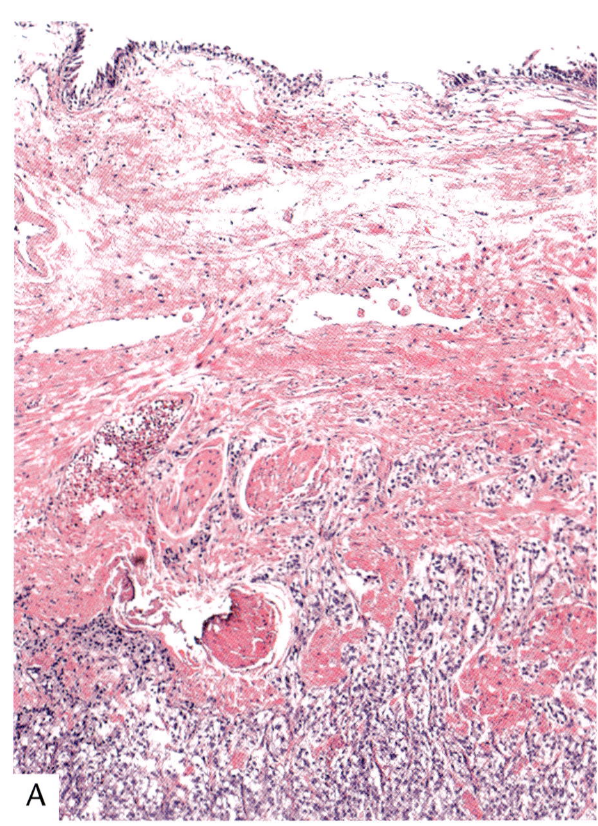

Figure 9-35

URINARY BLADDER PARAGANGLIOMA

A: This extensively invasive urinary bladder paraganglioma deep within the bladder wall is a potential mimic of prostatic or urothelial carcinoma.

B: Diffuse expression of tyrosine hydroxylase confirms the identity of the tumor as paraganglioma.

C: Immunohistochemical staining for GATA3 is positive in bladder paraganglioma but also in the overlying urothelium.

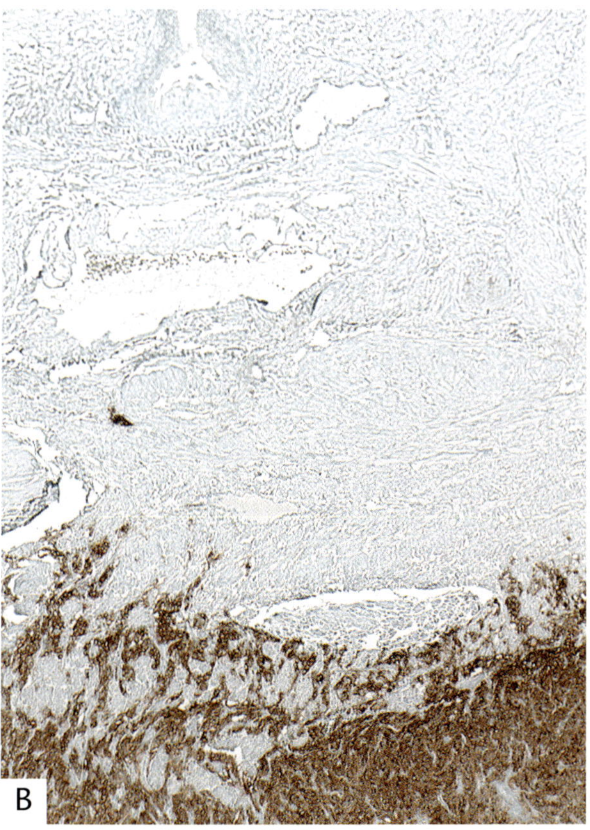

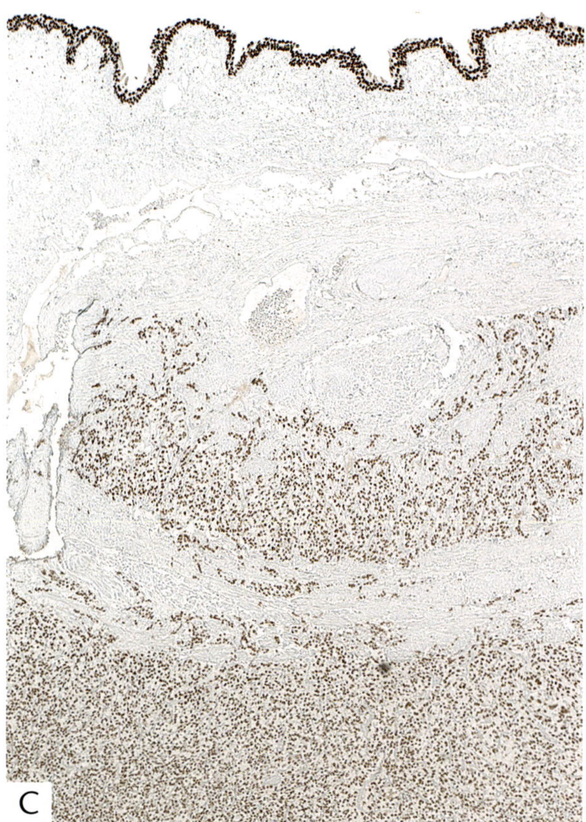

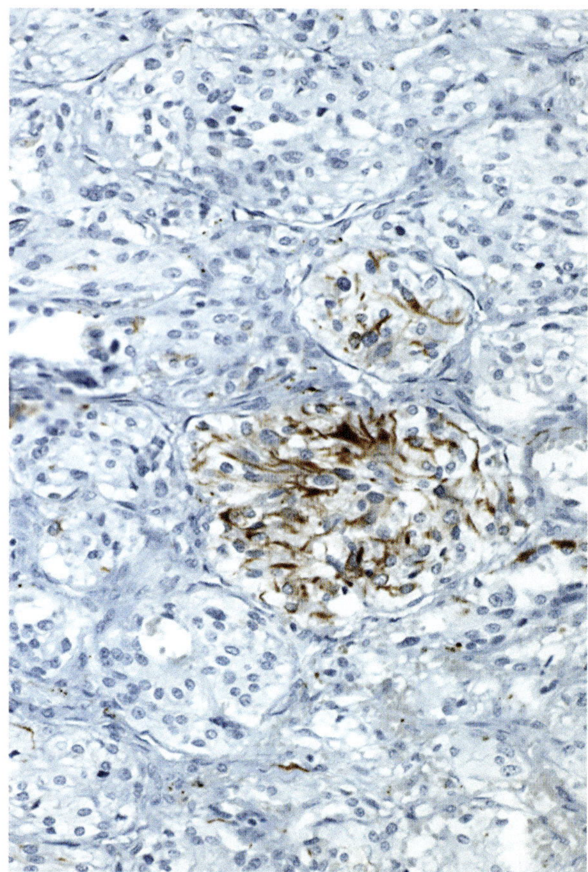

Figure 9-36

CAROTID BODY PARAGANGLIOMA WITH ANOMALOUS KERATIN EXPRESSION

Focal anomalous immunoreactivity for keratins is present in this paraganglioma stained with antibody cocktail AE1/AE3. The tumor was diffusely positive for synaptophysin and chromogranin A and contained numerous sustentacular cells positive for S-100 protein. The tumor was negative for GFAP.

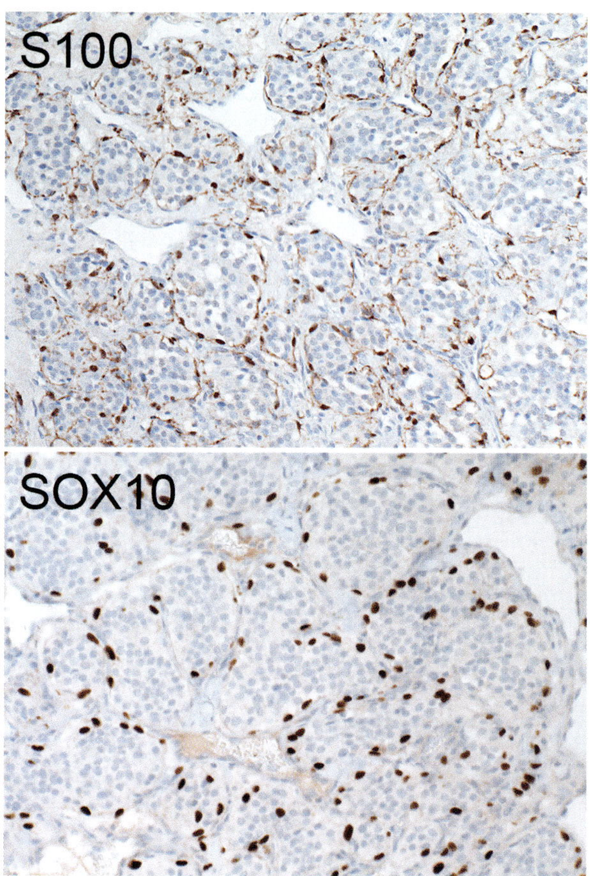

Figure 9-37

CAROTID BODY PARAGANGLIOMA

Immunohistochemical stains for S-100 protein and SOX10 outline the nested architecture in carotid body paraganglioma.

in paragangliomas are unknown. Some paragangliomas contain a small subset of sustentacular cells that stain for GFAP (106); it is not clear how these differ from the majority of sustentacular cells (see fig. 9-13E). It has recently been reported that paragangliomas often contain a population of monocytes that can mimic the distribution of sustentacular cells (fig. 9-38) (78).

Emerging Applications of Immunohistochemistry. Current and emerging immunohistochemical markers have prognostic or predictive applications (93,109,110). Staining for somatostatin receptors helps triage patients for peptide radioreceptor imaging and radionuclide therapy (fig. 9-39) (111–113). The Ki-67 proliferative fraction has been proposed as a parameter for stratifying risk of metastasis (see below) (114), and loss of SDHB (115), ATRX (116), and other markers may also supplement putative histologic criteria for risk assessment.

Increasingly, loss or gain of staining for immunohistochemical markers helps stratify patients for genetic testing for hereditary paraganglioma syndromes or serves as a surrogate marker where testing is not available (see chapter 10). Such markers include loss of SDHB (117), fumarate hydratase (118), or MAX (119). Gain of staining is seen for 2 succinylcysteine, which accumulates as a result of fumarate hydratase loss (120). Alpha-inhibin, which has previously been

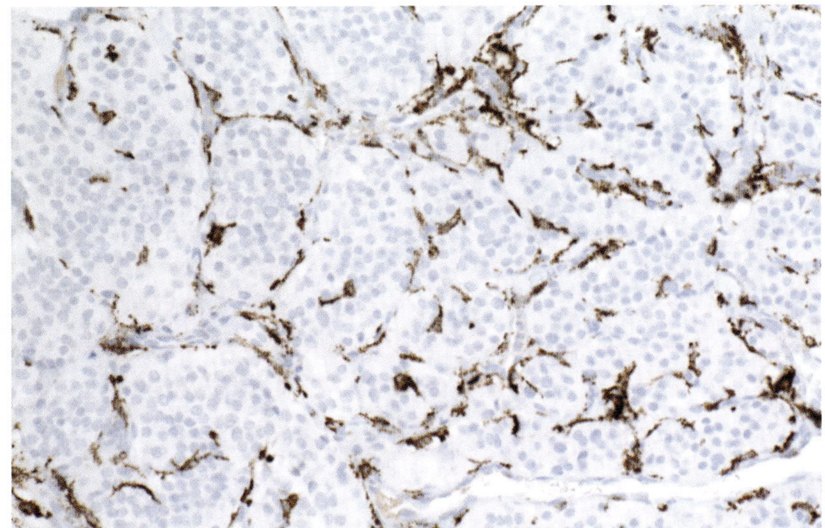

Figure 9-38

CAROTID BODY PARAGANGLIOMA

Immunohistochemical stain for CD163 shows macrophages in a paraganglioma that closely mimic the distribution of sustentacular cells.

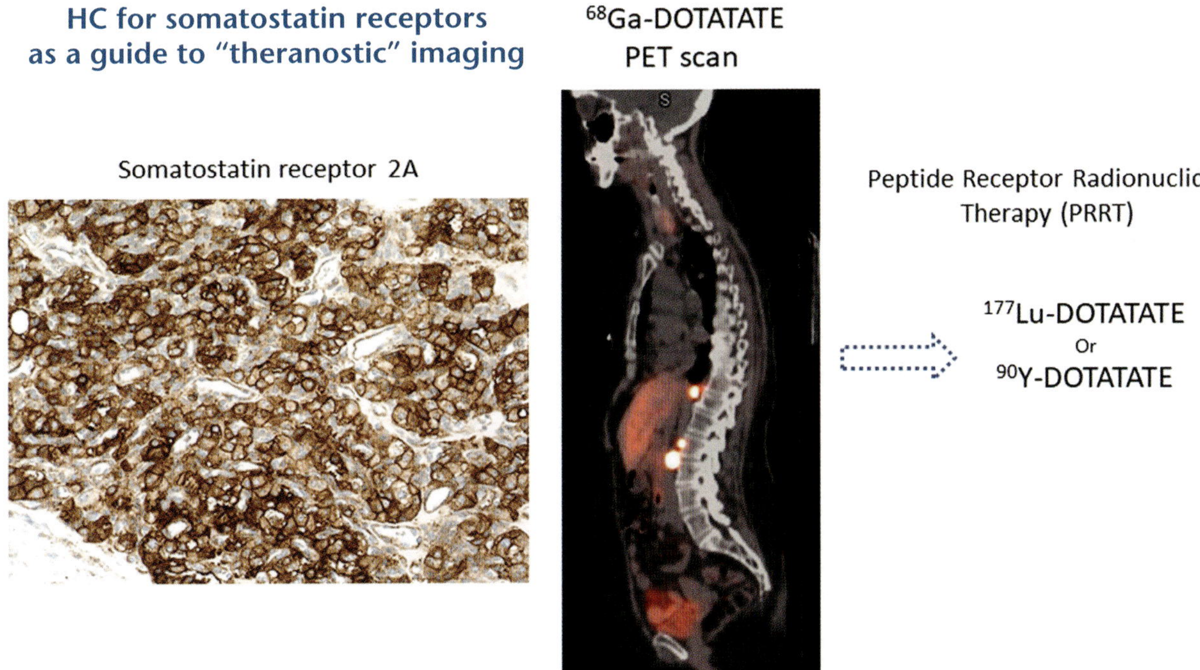

Figure 9-39

ABDOMINAL PARAGANGLIOMA

Immunohistochemical staining for somatostatin receptor 2A, an inexpensive stain, can triage patients for somatostatin receptor imaging and peptide receptor radionuclide therapy (PRRT). This very sensitive positron emission tomography (PET)/computerized tomography (CT) imaging uses ^{68}Ga, a positron emitter, chelated to a bifunctional "DOTA" compound that is also linked to a somatostatin analog. The most widely used analog is octreotate, which has high affinity for SSTR type 2, hence, the term "DOTAtate" scan. This modality is more sensitive and faster than octreotide scans, which have been available for decades and use ^{111}in, a single photon emitter, as the radioactive tracer. For PRRT, ^{68}Ga is replaced by ^{177}Lu, which emits cytotoxic medium energy beta particles, or by ^{90}Y, a high energy beta emitter that may be preferable for larger tumors [ref. 111]. The immunohistochemical stain for SSTR2A in this figure corresponds to one of the lesions in the PET/CT image.

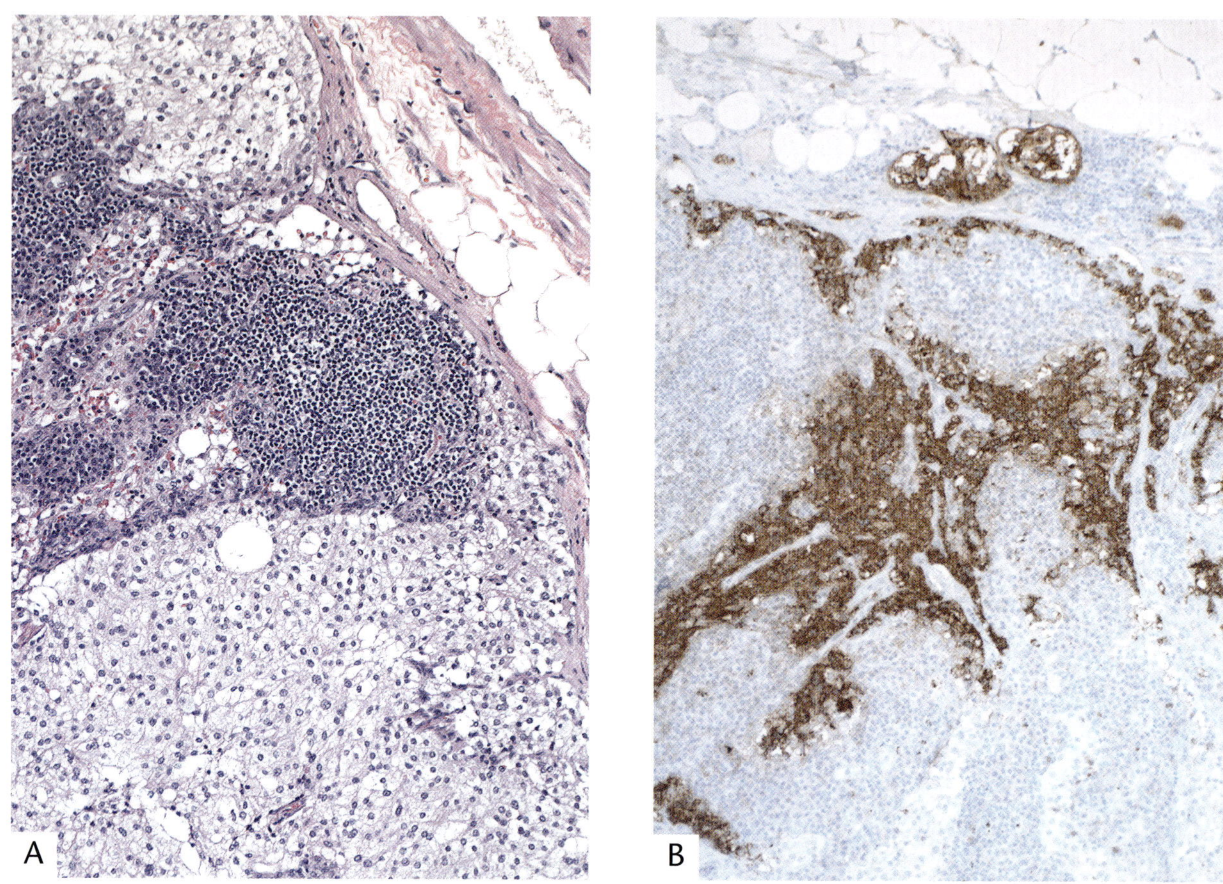

Figure 9-40

LYMPH NODE METASTASIS OF ABDOMINAL PARAGANGLIOMA

A: This metastatic tumor could be mistaken for a carcinoma with clear cell features, but was known to correspond to one of the lesions in fig. 9-39. The diagnosis was confirmed by an immunohistochemical stain for chromogranin A.

B: Metastatic paraganglioma in lymph node, stained for chromogranin A.

used to discriminate adrenal cortical tumors from pheochromocytoma, has recently been reported to be expressed in a high percentage of paragangliomas, especially those with a SDHx/VHL-driven pseudohypoxic signature (121). Membrane staining for carbonic anhydrase IX (CAIX) has been reported as a highly selective, although relatively insensitive, marker for VHL-associated disease (121,122). Even in patients for whom germline genetic testing has been performed, the loss of staining for SDHB, fumarate hydratase, or MAX in tumor tissue can still provide evidence that a gene variant of unknown significance (VUS) is in fact pathogenic.

Diagnosis of Metastases. Metastatic paragangliomas are associated with two diagnostic pitfalls. The first is misdiagnosis of an unequivocal lymph node or bone metastasis that is not already known to be paraganglioma for some other type of tumor. This is easily resolved by immunohistochemistry if the possibility of paraganglioma is kept in mind (figs. 9-40, 9-41).

The second pitfall is the diagnosis of metastatic versus primary paraganglioma. This poses more difficult challenges that have not been fully resolved even for lymph node metastases, because in some cases, residual lymph node tissue is obliterated by tumor. This is especially true in retroperitoneal locations where lymph nodes and paraganglia exist in close proximity (see fig. 9-3A). It may also be difficult to determine whether paragangliomas, for example,

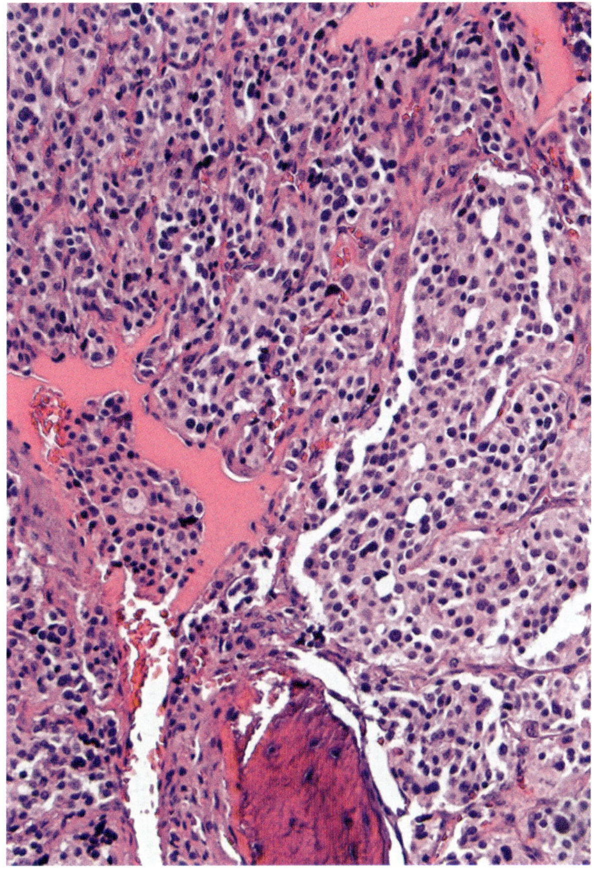

Figure 9-41

PARAGANGLIOMA METASTATIC TO VERTEBRAL BONE

Residual bone is present at bottom center. Since normal paraganglia do not occur in bone, the diagnosis of metastasis is unequivocal.

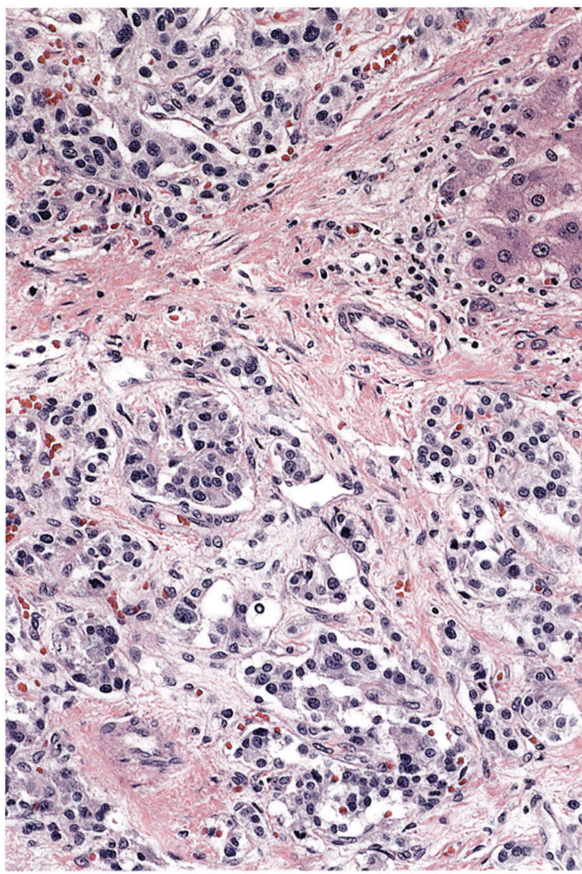

Figure 9-42

PARAGANGLIOMA INVOLVING LIVER BY DIRECT EXTENSION

The liver was infiltrated by tumor cells from a 10-cm partly intrahepatic and partly extrahepatic mass. The mass was a known recurrence of an extrahepatic tumor, and the intrahepatic component contained numerous sustentacular cells. The slightly eosinophilic polygonal tumor cells may suggest hepatocellular carcinoma, but were strongly positive for tyrosine hydroxylase and chromogranin A.

in liver (fig. 9-42), lung, or kidney, result from direct invasion or metastasis, or to distinguish metastases from new primary paragangliomas. Primary paragangliomas can occur, albeit rarely, in these organs (even in the liver, which lacks parenchymal innervation, but has nerves accompany entering blood vessels [123]). This differentiation determines whether there is local surgical treatment and cure versus systemic therapy and ultimately incurable disease.

When a tumor is suspected of being a metastasis, the primary should be examined for morphologic and immunohistochemical comparison. This is often not possible, however. The absence of sustentacular cells is helpful in some cases (106), but rare sustentacular cells are found in bona fide metastases (124). In addition, other cell types positive for S-100 protein, such as in lymphoid tissue (see fig. 9-3B), may mimic sustentacular cells. Immunohistochemical staining for SOX10 may help assess the presence of sustentacular cells in questionable cases (see fig. 9-37). Ultimately, morphology must be considered together with clinical context.

Risk Stratification

Current algorithms for stratifying the risk that a primary paraganglioma or pheochromocytoma will metastasize are currently based predominantly on clinically determined features.

For sympathetic paragangliomas, extra-adrenal location, large tumor size, production of dopamine or methoxytyramine, and hereditary *SDHB* mutation are all associated with increased risk (72,125). Tumors with *SDHB* mutations pose the highest risk in any location (67,126–128). However, they also tend to develop in thoracoabdominal locations where they can become large before diagnosis, and they frequently produce methoxytyramine (128).

Independent risk factors have been difficult to identify. In a 2019 meta-analysis of 21 studies and 703 patients, multivariate analyses showed significant correlation of metastasis with *SDHB* mutation and dopamine but not location (67). Nonetheless, despite the importance of *SDHB*, it is clear that not all paragangliomas that metastasize are associated with this mutation or are hereditary (67), and for some, the driver mutations are unknown. Even tumors with *SDHB* mutations or other high-risk hereditary mutations progress at different rates. At the molecular level this may be associated with secondary somatic mutations in *ATRX* or *TERT*, and loss of ATRX can be demonstrated by immunohistochemistry in some cases (129).

The role of histology in risk assessment is evolving. It is generally accepted that no single histologic feature is a robust predictor of metastases. Various features proposed as risk factors have been combined into scoring systems. The two major scoring systems are Pheochromocytoma of Adrenal Medulla Scaled Score (PASS) (88) and Grading of Adrenal Pheochromocytoma and Paraganglioma (GAPP) (114). GAPP was designed for both pheochromocytomas and sympathetic paragangliomas, while PASS was based on pheochromocytomas but has also been tested on paragangliomas with some success (116).

The parameters in these scoring systems partially overlap but are assessed differently and assigned different weights. The overlapping histologic parameters are large irregular cell nests, comedo necrosis, high cellularity, and diffuse growth. Both systems have been validated in several papers since their inception (116,130), although poor concordance between expert pathologists was noted in a PASS study (131). Comedo necrosis and growth pattern are the most readily recognized parameters, and possibly the most predictive, while cellularity is potentially more subjective. A 2019 meta-analysis of multiple papers concludes that a low score with either PASS or GAPP is a strong predictor of low metastatic risk but that high scores have little predictive value in the absence of adjunctive markers (116). These may include a combination of immunohistochemical markers such as Ki-67 (already included in GAPP), SDHB, and emerging markers such as ATRX (116), together with hormonal profile, genotype, metabolome, and other molecular markers.

Another scoring system, the Composite Pheochromocytoma Paraganglioma Prognostic Score (COPPS), which incorporates staining for SDHB and S-100 protein and uses MCM6 as a proliferation marker, has been proposed (130). No scoring system is currently endorsed by the WHO. However, the WHO does recommend determination of Ki-67 labeling index in paragangliomas as in other neuroendocrine neoplasms. There is evidence that a high labeling index portends a worse prognosis in terms of progression-free survival or metastasis, but the data are limited and standardized thresholds have not been established (130). In contrast to epithelial NETs, paragangliomas usually exhibit very low proliferation, leaving a compressed range in which to discriminate between low and high risk. The proposed high-risk threshold is 3 percent or more in GAPP (114) and 4 percent or more in COPPS (130).

No scoring system currently applies to head and neck paragangliomas. Studies of small series of carotid body paragangliomas have found no correlation of metastasis with vascular invasion, perineural invasion, mitotic activity, or pleomorphism (132) but *SDHB* mutation poses increased risk (64). Ki-67 labeling has been assessed in only a few cases, and may be either low or high in tumors that metastasize.

Cytologic Findings

Biopsy of abdominal/retroperitoneal paragangliomas is generally not advised because of potential bleeding or catecholamine release (133), but sometimes occurs in the course of routine workup for other tumors. Head and neck, especially carotid body, paragangliomas are sometimes sampled by fine-needle aspiration (FNA) in selected cases. FNA cytology in cases biopsied either inadvertently or deliberately shows bloody background and clusters of predominantly globoid to polygonal plasmacytoid cells with indistinct borders and abundant cytoplasm. Nuclei are round to ovoid, with absent to prominent nucleoli. Nuclear spindling, grooves, inclusions, and pleomorphism may be present (134).

Particularly challenging cases are those in which other endocrine or neuroendocrine tumors are a consideration. These include follicular, papillary, and medullary thyroid carcinoma and gastroenteropancreatic or bronchopulmonary neuroendocrine tumors (135). Medullary thyroid carcinoma is the most common misdiagnosis of paraganglioma among thyroid tumors (136). Although subtle distinctions, such as the presence of amyloid in medullary carcinoma or naked nuclei in paraganglioma, aid in distinguishing these entities (89,136), the first step is to be mindful of the possibility of paraganglioma in uncommon locations.

REFERENCES

1. Kohn A. Die paraganglien. Arch Mikr Anat 1903; 52:262-365.
2. Unsicker K, Huber K, Schober A, Kalcheim C. Resolved and open issues in chromaffin cell development. Mech Dev 2013;130:324-9.
3. Furlan A, Dyachuk V, Kastriti ME, et al. Multipotent peripheral glial cells generate neuroendocrine cells of the adrenal medulla. Science 2017;357:eaal3753.
4. Kastriti ME, Kameneva P, Kamenev D, et al. Schwann cell precursors generate the majority of chromaffin cells in Zuckerkandl organ and some sympathetic neurons in paraganglia. Front Mol Neurosci 2019;12:6.
5. Hockman D, Adameyko I, Kaucka M, et al. Striking parallels between carotid body glomus cell and adrenal chromaffin cell development. Dev Biol 2018;444(Suppl 1):S308-24.
6. Kameneva P, Artemov AV, Kastriti ME, et al. Single-cell transcriptomics of human embryos identifies multiple sympathoblast lineages with potential implications for neuroblastoma origin. Nat Genet 2021;53:694-706.
7. Furlan A, Adameyko I. Schwann cell precursor: a neural crest cell in disguise? Dev Biol 2018; 444(Suppl 1):S25-35.
8. Fliedner SM, Brabant G, Lehnert H. Pheochromocytoma and paraganglioma: genotype versus anatomic location as determinants of tumor phenotype. Cell Tissue Res 2018;372:347-65.
9. Coupland RE. The natural history of the chromaffin cell. London: Longmans, Green, and Co; 1965.
10. Asa SL, Ezzat S, Mete O. The diagnosis and clinical significance of paragangliomas in unusual locations. J Clin Med 2018;7:280.
11. Coupland RE. Post-natal fate of the abdominal para-aortic bodies in man. J Anat 1954;88:455-64.
12. Hervonen A, Vaalasti A, Partanen M, Kanerva L, Vaalasti T. The paraganglia, a persisting endocrine system in man. Am J Anat 1976;146:207-10.
13. Rode J, Bentley A, Parkinson C. Paraganglial cells of urinary bladder and prostate: potential diagnostic problem. J Clin Pathol 1990;43:13-6.
14. Guild SR. The glomus jugulare, a nonchromaffin paraganglion, in man. Ann Otol Rhinol Laryngol 1953;62:1045-71.
15. Kjaergaard J. Anatomy of the carotid glomus and carotid glomus-like bodies (non-chromaffin paraganglia). Copenhagen: F.A.D.L.'s Forlag; 1973.
16. Leonard EM, Salman S, Nurse CA. Sensory processing and integration at the carotid body tripartite synapse: neurotransmitter functions and effects of chronic hypoxia. Front Physiol 2018;9:225.
17. Monfared A, Gorti G, Kim D. Microsurgical anatomy of the laryngeal nerves as related to thyroid surgery. Laryngoscope 2002;112:386-92.
18. Coupland RE. The natural history of the chromaffin cell—twenty-five years on the beginning. Arch Histol Cytol 1989;52(Suppl):331-41.
19. Sharif KF, Sims JR, Khorsandi AS, Urken ML. Bilateral carotid body and cervical sympathetic chain paragangliomas: a case report and review of the literature. Otolaryngol Case Rep 2019;10:8-9.

20. Tian H, Hammer RE, Matsumoto AM, Russell DW, McKnight SL. The hypoxia-responsive transcription factor EPAS1 is essential for catecholamine homeostasis and protection against heart failure during embryonic development. Genes Dev 1998;12:3320-4.
21. Tella SH, Taieb D, Pacak K. HIF-2alpha: Achilles' heel of pseudohypoxic subtype paraganglioma and other related conditions. Eur J Cancer 2017; 86:1-4.
22. Tischler AS, Asa SL. Paraganglia. In: Mills SE, ed. Histology for pathologists, 5th ed. Philadelphia: Wolters Kluwer; 2018:102-23.
23. Mahalakshmi B, Baskaran R, Shanmugavadivu M, Nguyen NT, Velmurugan BK. Insulinoma-associated protein 1 (INSM1): a potential biomarker and therapeutic target for neuroendocrine tumors. Cell Oncol (Dordr) 2020;43:367-76.
24. Moriguchi T, Takako N, Hamada M, et al. Gata3 participates in a complex transcriptional feedback network to regulate sympathoadrenal differentiation. Development 2006;133:3871-81.
25. Miyauchi M, Akashi T, Furukawa AK, et al. PHOX2B is a sensitive and specific marker for the histopathological diagnosis of pheochromocytoma and paraganglioma. Endocr Pathol 2022;33:506-18.
26. Nonaka D, Wang BY, Edmondson D, Beckett E, Sun CC. A study of gata3 and phox2b expression in tumors of the autonomic nervous system. Am J Surg Pathol 2013;37:1236-41.
27. Miettinen M, McCue PA, Sarlomo-Rikala M, et al. Sox10—a marker for not only schwannian and melanocytic neoplasms but also myoepithelial cell tumors of soft tissue: a systematic analysis of 5134 tumors. Am J Surg Pathol 2015;39:826-35.
28. Lloyd R, Osamura R, Klöppel G, Rosai J, eds. WHO classification of tumours of endocrine organs, 4th ed. Lyon: IARC Press; 2017.
29. Lyne SB, Polster SP, Fidai S, Pytel P, Yamini B. Primary sellar paraganglioma: case report with literature review and immunohistochemistry resource. World Neurosurg 2019;125:32-6.
30. Papaspyrou K, Mewes T, Rossmann H, et al. Head and neck paragangliomas: report of 175 patients (1989-2010). Head Neck 2012;34:632-7.
31. Huang N, Rayess HM, Svider PF, et al. Orbital paraganglioma: a systematic review. J Neurol Surg B Skull Base 2018;79:407-12.
32. Karsner HT. Tumors of the adrenal. Atlas of Tumor Pathology, 1st Series, Fascicle 29. Washington DC: Armed Forces Institute of Pathology; 1950.
33. Turchini J, Cheung VK, Tischler AS, De Krijger RR, Gill AJ. Pathology and genetics of phaeochromocytoma and paraganglioma. Histopathology 2018;72:97-105.
34. WHO Classification of Tumours Editorial Board. WHO classification of tumours: endocrinene and neuroendocrine tumours, 5th ed. Lyon: IARC; 2022.
35. Mete O, Asa SL, Gill AJ, Kimura N, de Krijger RR, Tischler A. Overview of the 2022 WHO classification of paragangliomas and pheochromocytomas. Endocr Pathol 2022;33:90-114.
36. Rindi G, Klimstra DS, Abedi-Ardekani B, et al. A common classification framework for neuroendocrine neoplasms: an International Agency for Research on Cancer (IARC) and World Health Organization (WHO) expert consensus proposal. Mod Pathol 2018;31:1770-86.
37. Amin MB, Edge SB, American Joint Committee on Cancer. AJCC cancer staging manual, 8th ed. Switzerland: Springer; 2017.
38. Chen H, Sippel RS, O'Dorisio MS, et al. The North American Neuroendocrine Tumor Society consensus guideline for the diagnosis and management of neuroendocrine tumors: pheochromocytoma, paraganglioma, and medullary thyroid cancer. Pancreas 2010;39:775-83.
39. Berends AM, Buitenwerf E, de Krijger RR, et al. Incidence of pheochromocytoma and sympathetic paraganglioma in the Netherlands: a nationwide study and systematic review. Eur J Intern Med 2018;51:68-73.
40. Offergeld C, Brase C, Yaremchuk S, et al. Head and neck paragangliomas: clinical and molecular genetic classification. Clinics (Sao Paulo) 2012; 67(Suppl 1):19-28.
41. Mannelli M, Castellano M, Schiavi F, et al. Clinically guided genetic screening in a large cohort of italian patients with pheochromocytomas and/ or functional or nonfunctional paragangliomas. J Clin Endocrinol Metab 2009;94:1541-7.
42. Sykes JM, Ossoff RH. Paragangliomas of the head and neck. Otolaryngol Clin North Am 1986;19: 755-67.
43. Williams MD. Paragangliomas of the head and neck: an overview from diagnosis to genetics. Head Neck Pathol 2017;11:278-87.
44. Myssiorek D, Rinaldo A, Barnes L, Ferlito A. Laryngeal paraganglioma: an updated critical review. Acta Otolaryngol 2004;124:995-9.
45. Saldana MJ, Salem LE, Travezan R. High altitude hypoxia and chemodectomas. Hum Pathol 1973; 4:251-63.
46. Arias-Stella J, Bustos F. Chronic hypoxia and chemodectomas in bovines at high altitudes. Arch Pathol Lab Med 1976;100:636-9.
47. Lopez-Barneo J. Oxygen sensing and stem cell activation in the hypoxic carotid body. Cell Tissue Res 2018;372:417-25.
48. Enriquez-Vega ME, Munoz-Paredes JG, Cossio-Zazueta A, Ontiveros-Carlos Y, Pacheco-Pittaluga E, Bizueto-Rosas H. [SDHD gene mutation in Mexican population whit carotid body tumor.] Cir Cir 2018;86:38-42. [Spanish]

49. Cerecer-Gil NY, Figuera LE, Llamas FJ, et al. Mutation of SDHB is a cause of hypoxia-related high-altitude paraganglioma. Clin Cancer Res 2010;16:4148-54.
50. Jech M, Alvarado-Cabrero I, Albores-Saavedra J, Dahia PL, Tischler AS. Genetic analysis of high altitude paragangliomas. Endocr Pathol 2006;17:201-2.
51. Astrom K, Cohen JE, Willett-Brozick JE, Aston CE, Baysal BE. Altitude is a phenotypic modifier in hereditary paraganglioma type 1: evidence for an oxygen-sensing defect. Hum Genet 2003;113: 228-37.
52. Tischler AS, deKrijger RR. 15 years of paraganglioma: pathology of pheochromocytoma and paraganglioma. Endocr Relat Cancer 2015;22:T123-33.
53. Pamporaki C, Hamplova B, Peitzsch M, et al. Characteristics of pediatric vs adult pheochromocytomas and paragangliomas. J Clin Endocrinol Metab 2017;102:1122-32.
54. Wen J, Li HZ, Ji ZG, Mao QZ, Shi BB, Yan WG. A decade of clinical experience with extra-adrenal paragangliomas of retroperitoneum: report of 67 cases and a literature review. Urol Ann 2010;2:12-6.
55. Erickson D, Kudva YC, Ebersold MJ, et al. Benign paragangliomas: clinical presentation and treatment outcomes in 236 patients. J Clin Endocrinol Metab 2001;86:5210-6.
56. Eisenhofer G, Klink B, Richter S, Lenders JW, Robledo M. Metabologenomics of phaeochromocytoma and paraganglioma: an integrated approach for personalised biochemical and genetic testing. Clin Biochem Rev 2017;38:69-100.
57. Timmers HJ, Pacak K, Huynh TT, et al. Biochemically silent abdominal paragangliomas in patients with mutations in the succinate dehydrogenase subunit B gene. J Clin Endocrinol Metab 2008;93:4826-32.
58. Timmers HJ, Kozupa A, Eisenhofer G, et al. Clinical presentations, biochemical phenotypes, and genotype-phenotype correlations in patients with succinate dehydrogenase subunit B-associated pheochromocytomas and paragangliomas. J Clin Endocrinol Metab 2007;92:779-86.
59. Jiang J, Zhang J, Pang Y, et al. Sino-European differences in the genetic landscape and clinical presentation of pheochromocytoma and paraganglioma. J Clin Endocrinol Metab 2020;105:dgaa502.
60. Tella SH, Jha A, Taieb D, Horvath KA, Pacak K. Comprehensive review of evaluation and management of cardiac paragangliomas. Heart 2020;106:1202-10.
61. Varoquaux A, Kebebew E, Sebag F, et al. Endocrine tumors associated with the vagus nerve. Endocr Relat Cancer 2016;23:R371-9.
62. Nunez AA, Ramos-Duran LR, Cuetter AC. Glomus jugulare presenting with isolated facial nerve palsy. Surg Res Pract 2014;2014:514086.
63. Jansen TT, Timmers H, Marres HA, Kunst HP. Feasibility of a wait-and-scan period as initial management strategy for head and neck paraganglioma. Head Neck 2017;39:2088-94.
64. Ellis RJ, Patel D, Prodanov T, Nilubol N, Pacak K, Kebebew E. The presence of SDHB mutations should modify surgical indications for carotid body paragangliomas. Ann Surg 2014;260:158-62.
65. Michalowska I, Cwikla JB, Michalski W, et al. Growth rate of paragangliomas related to germline mutations of the SDHX genes. Endocr Pract 2017;23:342-52.
66. Jansen JC, van den Berg R, Kuiper A, van der Mey AG, Zwinderman AH, Cornelisse CJ. Estimation of growth rate in patients with head and neck paragangliomas influences the treatment proposal. Cancer 2000;88:2811-6.
67. Crona J, Lamarca A, Ghosal S, Welin S, Skogseid B, Pacak K. Genotype-phenotype correlations in pheochromocytoma and paraganglioma: a systemic review and individual patient meta-analysis. Endocr Relat Cancer 2019;26:539-50.
68. McCrary HC, Babajanian E, Calquin M, et al. Characterization of malignant head and neck paragangliomas at a single institution across multiple decades. JAMA Otolaryngol Head Neck Surg 2019;145:641-6.
69. Shankavaram U, Fliedner SM, Elkahloun AG, et al. Genotype and tumor locus determine expression profile of pseudohypoxic pheochromocytomas and paragangliomas. Neoplasia 2013;15: 435-47.
70. Aubertine CL, Flieder DB. Primary paraganglioma of the lung. Ann Diagn Pathol 2004;8:237-41.
71. Liao W, Ding ZY, Zhang B, et al. Primary functioning hepatic paraganglioma mimicking hepatocellular carcinoma: a case report and literature review. Medicine (Baltimore) 2018;97:e0293.
72. Eisenhofer G, Lenders JW, Siegert G, et al. Plasma methoxytyramine: a novel biomarker of metastatic pheochromocytoma and paraganglioma in relation to established risk factors of tumour size, location and SDHB mutation status. Eur J Cancer 2012;48:1739-49.
73. Moslemi MK, Abolhasani M, Vafaeimanesh J. Malignant abdominal paraganglioma presenting as a giant intra-peritoneal mass. Int J Surg Case Rep 2012;3:537-40.
74. Linnoila RI, Keiser HR, Steinberg SM, Lack EE. Histopathology of benign versus malignant sympathoadrenal paragangliomas: clinicopathologic study of 120 cases including unusual histologic features. Hum Pathol 1990;21:1168-80.
75. Pucci A, Bacca A, Barravecchia I, et al. Sclerosing paragangliomas: correlations of histological features with patients' genotype and vesicular monoamine transporter expression. Head Neck Pathol 2022;16:998-1011.

76. Santi R, Franchi A, Saladino V, et al. Sclerosing paraganglioma of the carotid body: a potential pitfall of malignancy. Head Neck Pathol 2015; 9:300-4.
77. Plaza JA, Wakely PE Jr, Moran C, Fletcher CD, Suster S. Sclerosing paraganglioma: report of 19 cases of an unusual variant of neuroendocrine tumor that may be mistaken for an aggressive malignant neoplasm. Am J Surg Pathol 2006;30:7-12.
78. Farhat NA, Powers JF, Shepard-Barry A, Dahia P, Pacak K, Tischler AS. A previously unrecognized monocytic component of pheochromocytoma and paraganglioma. Endocr Pathol 2019;30:90-5.
79. Lack EE. Tumors of the adrenal gland and extraadrenal paraganglia. Atlas of Tumor Pathology, 4th Series, Fascicle 8. Silver Spring: ARP Press; 2007.
80. Williamson SR, Eble JN, Amin MB, et al. Succinate dehydrogenase-deficient renal cell carcinoma: detailed characterization of 11 tumors defining a unique subtype of renal cell carcinoma. Mod Pathol 2015;28:80-94.
81. Gill AJ, Hes O, Papathomas T, et al. Succinate dehydrogenase (SDH)-deficient renal carcinoma: a morphologically distinct entity: a clinicopathologic series of 36 tumors from 27 patients. Am J Surg Pathol 2014;38:1588-602.
82. Dong YJ, Zhang ZW, Wang Z, Wang XY, Tian ZZ, Zhang XS. Primary melanotic paraganglioma of thyroid gland: report of a rare case with clinicopathologic and immunohistochemical analysis and a literature review. Clin Med Insights Pathol 2017;10:1179555716684670.
83. Zhou M, Epstein JI, Young RH. Paraganglioma of the urinary bladder: a lesion that may be misdiagnosed as urothelial carcinoma in transurethral resection specimens. Am J Surg Pathol 2004;28:94-100.
84. Wang HH, Chen YL, Kao HL, et al. Extra-adrenal paraganglioma of the prostate. Can Urol Assoc J 2013;7:E370-2.
85. Hempenstall LE, Siriwardana AR, Desai DJ. Investigation of a renal mass: diagnosing renal paraganglioma. Urol Case Rep 2018;21:8-9.
86. Zhao L, Luo J, Zhang H, Da J. Pigmented paraganglioma of the kidney: a case report. Diagn Pathol 2012;7:77.
87. Lack EE, Kim H, Reed K. Pigmented ("black") extraadrenal paraganglioma. Am J Surg Pathol 1998;22:265-9.
88. Thompson LD. Pheochromocytoma of the adrenal gland scaled score (PASS) to separate benign from malignant neoplasms: a clinicopathologic and immunophenotypic study of 100 cases. Am J Surg Pathol 2002;26:551-66.
89. Satturwar SP, Rossi ED, Maleki Z, Cantley RL, Faquin WC, Pantanowitz L. Thyroid paraganglioma: a diagnostic pitfall in thyroid FNA. Cancer Cytopathol 2021;129:439-49.
90. Tzikos G, Menni A, Cheva A, et al. Composite paraganglioma of the celiac trunk: a case report and a comprehensive review of the literature. Front Surg 2022;9:824076.
91. Armstrong R, Greenhalgh KL, Rattenberry E, et al. Succinate dehydrogenase subunit B (SDHB) gene deletion associated with a composite paraganglioma/neuroblastoma. J Med Genet 2009;46:215-6.
92. Chen J, Wu Y, Wang P, Wu H, Tong A, Chang X. Composite pheochromocytoma/paraganglioma-ganglioneuroma: analysis of SDH and ATRX status, and identification of frequent HRAS and BRAF mutations. Endocr Connect 2021;10:926-34.
93. Duan K, Mete O. Algorithmic approach to neuroendocrine tumors in targeted biopsies: Practical applications of immunohistochemical markers. Cancer Cytopathol 2016;124:871-84.
94. Kriegsmann K, Zgorzelski C, Kazdal D, et al. Insulinoma-associated protein 1 (INSM1) in thoracic tumors is less sensitive but more specific compared with synaptophysin, chromogranin A, and CD56. Appl Immunohistochem Mol Morphol 2020;28:237-42.
95. Miettinen M, McCue PA, Sarlomo-Rikala M, et al. GATA3: a multispecific but potentially useful marker in surgical pathology: a systematic analysis of 2500 epithelial and nonepithelial tumors. Am J Surg Pathol 2014;38:13-22.
96. Daruwalla J, Sachithanandan N, Andrews D, Miller JA. Ectopic intravagal parathyroid adenoma. Head Neck 2015;37:E200-4.
97. Lack EE, Delay S, Linnoila RI. Ectopic parathyroid tissue within the vagus nerve. Incidence and possible clinical significance. Arch Pathol Lab Med 1988;112:304-6.
98. Nilubol N, Freedman EM, Quezado MM, Patel D, Kebebew E. Pancreatic neuroendocrine tumor secreting vasoactive intestinal peptide and dopamine with pulmonary emboli: a case report. J Clin Endocrinol Metab 2016;101:3564-7.
99. Mete O, Essa A, Bramdev A, Govender N, Chetty R. MEN2 syndrome-related medullary thyroid carcinoma with focal tyrosine hydroxylase expression: does it represent a hybrid cellular phenotype or functional state of tumor cells? Endocr Pathol 2017;28:362-6.
100. Letouze E, Martinelli C, Loriot C, et al. SDH mutations establish a hypermethylator phenotype in paraganglioma. Cancer Cell 2013;23:739-52.
101. Chetty R, Pillay P, Jaichand V. Cytokeratin expression in adrenal phaeochromocytomas and extra-adrenal paragangliomas. J Clin Pathol 1998;51:477-8.
102. Dermawan JK, Mukhopadhyay S, Shah AA. Frequency and extent of cytokeratin expression in paraganglioma: an immunohistochemical study of 60 cases from 5 anatomic sites and review of the literature. Hum Pathol 2019;93:16-22.

103. Schmid KW, Schroder S, Dockhorn-Dworniczak B, et al. Immunohistochemical demonstration of chromogranin A, chromogranin B, and secretogranin II in extra-adrenal paragangliomas. Mod Pathol 1994;7:347-53.
104. Tischler AS. Pheochromocytoma and extra-adrenal paraganglioma: updates. Arch Pathol Lab Med 2008;132:1272-84.
105. Osinga TE, Korpershoek E, de Krijger RR, et al. Catecholamine-synthesizing enzymes are expressed in parasympathetic head and neck paraganglioma tissue. Neuroendocrinology 2015;101:289-95.
106. Achilles E, Padberg BC, Holl K, Kloppel G, Schroder S. Immunocytochemistry of paragangliomas—value of staining for S-100 protein and glial fibrillary acid protein in diagnosis and prognosis. Histopathology 1991;18:453-8.
107. Tsuta K, Raso MG, Kalhor N, Liu DC, Wistuba, II, Moran CA. Sox10-positive sustentacular cells in neuroendocrine carcinoma of the lung. Histopathology 2011;58:276-85.
108. Douwes Dekker PB, Corver WE, Hogendoorn PC, van der Mey AG, Cornelisse CJ. Multiparameter DNA flow-sorting demonstrates diploidy and SDHD wild-type gene retention in the sustentacular cell compartment of head and neck paragangliomas: chief cells are the only neoplastic component. J Pathol 2004;202:456-62.
109. Cheung VK, Gill AJ, Chou A. Old, new, and emerging immunohistochemical markers in pheochromocytoma and paraganglioma. Endocr Pathol 2018;29:169-75.
110. Juhlin CC. Challenges in paragangliomas and pheochromocytomas: from histology to molecular immunohistochemistry. Endocr Pathol 2021;32:228-44.
111. Iori M, Capponi PC, Rubagotti S, et al. Labelling of ^{90}Y- and ^{177}Lu-DOTA-bioconjugates for targeted radionuclide therapy: a comparison among manual, semiautomated, and fully automated synthesis. Contrast Media Mol Imaging 2017;2017:8160134.
112. Korner M, Waser B, Schonbrunn A, Perren A, Reubi JC. Somatostatin receptor subtype 2A immunohistochemistry using a new monoclonal antibody selects tumors suitable for in vivo somatostatin receptor targeting. Am J Surg Pathol 2012;36:242-52.
113. Janssen I, Blanchet EM, Adams K, et al. Superiority of [68Ga]-DOTATATE PET/CT to other functional imaging modalities in the localization of SDHB-associated metastatic pheochromocytoma and paraganglioma. Clin Cancer Res 2015;21:3888-95.
114. Kimura N, Takayanagi R, Takizawa N, et al. Pathological grading for predicting metastasis in phaeochromocytoma and paraganglioma. Endocr Relat Cancer 2014;21:405-14.
115. Kimura N, Takekoshi K, Horii A, et al. Clinicopathological study of SDHB mutation-related pheochromocytoma and sympathetic paraganglioma. Endocr Relat Cancer 2014;21:L13-6.
116. Stenman A, Zedenius J, Juhlin CC. The value of histological algorithms to predict the malignancy potential of pheochromocytomas and abdominal paragangliomas—a meta-analysis and systematic review of the literature. Cancers (Basel) 2019;11:225.
117. Papathomas TG, Oudijk L, Persu A, et al. SDHB/SDHA immunohistochemistry in pheochromocytomas and paragangliomas: a multicenter interobserver variation analysis using virtual microscopy: a multinational study of the European Network for the Study of Adrenal Tumors (ENS@T). Mod Pathol 2015;28:807-21.
118. Joseph NM, Solomon DA, Frizzell N, Rabban JT, Zaloudek C, Garg K. Morphology and immunohistochemistry for 2SC and FH aid in detection of fumarate hydratase gene aberrations in uterine leiomyomas from young patients. Am J Surg Pathol 2015;39:1529-39.
119. Roszko KL, Blouch E, Blake M, et al. Case report of a prolactinoma in a patient with a novel MAX mutation and bilateral pheochromocytomas. J Endocr Soc 2017;1:1401-7.
120. Maxwell PH. Seeing the smoking gun: a sensitive and specific method to visualize loss of the tumour suppressor, fumarate hydratase, in human tissues. J Pathol 2011;225:1-3.
121. Mete O, Pakbaz S, Lerario AM, Giordano TJ, Asa SL. Significance of alpha-inhibin expression in pheochromocytomas and paragangliomas. Am J Surg Pathol 2021;45:1264-73.
122. Favier J, Meatchi T, Robidel E, et al. Carbonic anhydrase 9 immunohistochemistry as a tool to predict or validate germline and somatic VHL mutations in pheochromocytoma and paraganglioma-a retrospective and prospective study. Mod Pathol 2020;33:57-64.
123. Mizuno K, Ueno Y. Autonomic nervous system and the liver. Hepatol Res 2017;47:160-5.
124. Unger P, Hoffman K, Pertsemlidis D, Thung S, Wolfe D, Kaneko M. S100 protein-positive sustentacular cells in malignant and locally aggressive adrenal pheochromocytomas. Arch Pathol Lab Med 1991;115:484-7.
125. Parasiliti-Caprino M, Lucatello B, Lopez C, et al. Predictors of recurrence of pheochromocytoma and paraganglioma: a multicenter study in Piedmont, Italy. Hypertens Res 2020;43:500-10.
126. Kimura N, Takekoshi K, Naruse M. Risk stratification on pheochromocytoma and paraganglioma from laboratory and clinical medicine. J Clin Med 2018;7:242.

127. Eisenhofer G, Tischler AS. Neuroendocrine cancer. Closing the GAPP on predicting metastases. Nat Rev Endocrinol 2014;10:315-6.
128. Eisenhofer G, Tischler AS, de Krijger RR. Diagnostic tests and biomarkers for pheochromocytoma and extra-adrenal paraganglioma: from routine laboratory methods to disease stratification. Endocr Pathol 2012;23:4-14.
129. Job S, Draskovic I, Burnichon N, et al. Telomerase activation and ATRX mutations are independent risk factors for metastatic pheochromocytoma and paraganglioma. Clin Cancer Res 2019;25:760-70.
130. Desport JC, Jesus P, Fayemendy P, Linut R. [Food and malnutrition in the elderly.] Rev Prat 2018;68:312-8. [French]
131. Wu D, Tischler AS, Lloyd RV, et al. Observer variation in the application of the Pheochromocytoma of the Adrenal Gland Scaled Score. Am J Surg Pathol 2009;33:599-608.
132. Barnes L, Taylor SR. Carotid body paragangliomas. A clinicopathologic and DNA analysis of 13 tumors. Arch Otolaryngol Head Neck Surg 1990;116:447-53.
133. Vanderveen KA, Thompson SM, Callstrom MR, et al. Biopsy of pheochromocytomas and paragangliomas: potential for disaster. Surgery 2009;146:1158-66.
134. Handa U, Kundu R, Mohan H. Cytomorphologic spectrum in aspirates of extra-adrenal paraganglioma. J Cytol 2014;31:79-82.
135. Zeng J, Simsir A, Oweity T, Hajdu C, Cohen S, Shi Y. Peripancreatic paraganglioma mimics pancreatic/gastrointestinal neuroendocrine tumor on fine needle aspiration: report of two cases and review of the literature. Diagn Cytopathol 2017;45:947-52.
136. Taweevisit M, Bunyayothin W, Thorner PS. Thyroid paraganglioma: "naked" nuclei as a clue to diagnosis on imprint cytology. Endocr Pathol 2015;26:232-8.

10 FAMILIAL AND SYNDROME-RELATED PHEOCHROMOCYTOMA AND PARAGANGLIOMA

Pheochromocytomas and paragangliomas (PPGLs) are a group of tumors with an extraordinarily high genetic predisposition. Approximately 40 percent of patients with these tumors harbor a germline mutation conveying susceptibility to the development of these and other neoplasms (1). The genetic alterations are most often hereditary, giving rise to familial disease that can involve only PPGL formation, but usually is associated with other disorders. Some are nonhereditary disorders that cause syndromic manifestations.

At least 20 genes have been implicated in familial disorders that manifest with PPGLs (2-4). Table 10-1 provides a summary of the genes, their chromosomal locations, the syndromes to which they give rise, the most common PPGL locations, and other associated lesions. Genotype-phenotype correlations have been identified that allow the classification of these tumors into clusters based on the transcriptome and signaling pathways involved in their development, which generally correlate with the biochemical profile of the catecholamines they produce (5).

In its initial formulation, the clustering concept defined only two groups: cluster 1, characterized by pseudohypoxic signaling and represented mostly by mutations of *VHL* and Krebs cycle genes encoding members of the succinate dehydrogenase (SDH) complex, and cluster 2, characterized by kinase signaling and represented mostly by mutations of *RET* and *NF1* (6). Additional hereditary susceptibility genes including *EPAS1* (*HIF2A*) and its regulators were subsequently discovered and assigned to cluster 1, while *TMEM127* and *MAX* were added to cluster 2. A third cluster, consisting of pheochromocytomas with somatic *CSDE1* mutations and *MAML3* fusions and characterized by Wnt signaling, was later described (7). This latter group is sporadic and not discussed in this chapter.

It is now recognized that somatic mutations of *RET, NF1, MAX,* and *VHL* also lead to the development of PPGL. In addition, an early postzygotic mutation resulting in somatic mosaicism of genes including *VHL* and *SDHB* has been reported (4), potentially resulting in discordance between germline genetic testing and markers expressed in tumor tissue. Recent research based on single cell sequencing of the neoplastic cells in both hereditary and sporadic PPGLs resolves multiple subclusters under each of the initial headings (8).

The biochemical phenotype of cluster 1 tumors is characterized by the production of dopamine, with or without norepinephrine. Mutations of Krebs cycle-related genes, including members of the SDH complex or fumarate hydratase (FH), are implicated primarily in dopamine production, often referred to as "immature biochemical phenotype." These tumors typically do not give rise to florid symptoms and are often nonfunctional or produce only the dopamine metabolite methoxytyramine; mutations in *VHL* or *EPAS1* are more often associated with noradrenergic function, which is referred to as "immature secretory phenotype."

Cluster 2 (and cluster 3) disease is characterized by a "mature secretory phenotype" in which the production of epinephrine and norepinephrine (adrenaline and noradrenaline) often gives rise to clinical signs and symptoms. The description of biochemical function as "immature" or "mature" does not necessarily indicate the degree of cellular maturity or differentiation because normal paraganglia are fully mature organs that may produce dopamine (9) or norepinephrine (10).

The nonhereditary syndromes frequently involve somatic mosaicism and include Carney triad (paraganglioma, pulmonary chondroma,

Table 10-1

FAMILIAL PARAGANGLIOMA SYNDROMES

Gene	Chromosome Location	Syndrome	Most Common Paraganglioma Locations	Cluster	Other Associated Lesions
RET	10q11.2	MEN2 (multiple endocrine neoplasia type 2)	Adrenal	2	Medullary thyroid carcinoma; parathyroid hyperplasia or neoplasia Mucocutaneous ganglioneuromas (MEN2B)
NF1	17q11.2	NF1 (neurofibromatosis type 1)	Adrenal (rarely extra-adrenal)	2	Neurofibroma and MPNST[a] Ocular manifestations including Lisch nodules and optic gliomas Café-au-lait spots Duodenal neuroendocrine tumor Renal, liver, and epididymal cysts
VHL	3p25.5	VHL (von Hippel-Lindau disease)	Adrenal and extra-adrenal	1	Clear cell renal cell carcinoma Hemangioblastomas Neuroendocrine tumors Pancreatic serous cystadenomas Endolymphatic sac tumors
SDHA	5p15.33	PGL5 (paraganglioma type 5)	Adrenal and extra-adrenal	1	SDH deficient renal cell carcinoma SDH deficient gastrointestinal stromal tumor
SDHB	1p36.13	PGL4 (paraganglioma type 4)	Abdominal and thoracic; head and neck	1	SDH deficient renal cell carcinoma SDH deficient gastrointestinal stromal tumor SDH deficient pituitary neuroendocrine tumor
SDHC	1q.23.3	PGL3 (paraganglioma type 3)	Head and neck, thoracic	1	SDH deficient renal cell carcinoma SDH deficient gastrointestinal stromal tumor SDH deficient pituitary neuroendocrine tumor
SDHD	11q23	PGL1 (paraganglioma type 1)	Adrenal and extra-adrenal	1	SDH deficient renal cell carcinoma SDH deficient gastrointestinal stromal tumor SDH deficient pituitary neuroendocrine tumor
SDHAF2	11q12.2	PGL2 (paraganglioma type 2)	Head and neck	1	Unknown
FH	1q42.1	HLRCC (hereditary leiomyomatosis and renal cell carcinoma)	Adrenal and extra-adrenal	1	Cutaneous and uterine leiomyomas Renal cell carcinoma
MDH2	7q11.23		Abdominal and thoracic	1	Unknown
EGLN1	1q42.1		Adrenal, abdominal, and thoracic	1	Polycythemia
EGLN2	19q13.2		Adrenal, abdominal, and thoracic	1	Polycythemia
KIF1B	1p36.22		Adrenal	2	Ganglioneuroma/neuroblastoma Leiomyosarcoma Lung adenocarcinoma
TMEM127	2q11.2		Adrenal and extra-adrenal	2	Renal cell carcinoma Pituitary neuroendocrine tumor
MAX	14q23.3		Adrenal and extra-adrenal	2	Renal oncocytoma
MEN1	11q13.1		Adrenal and head and neck	?	Parathyroid hyperplasia or neoplasia Pituitary neuroendocrine tumor Neuroendocrine tumors

[a]MPNST = malignant peripheral nerve sheath tumor.

and SDH-deficient gastrointestinal stromal tumor [GIST]) caused by promoter hypermethylation of *SDHC* (11), Pacak-Zhuang syndrome (PZS) caused by early postzygotic mutations of *EPAS1* (12), and Beckwith-Wiedemann syndrome (13) due to abnormal methylation the imprinting centers that regulate several growth-promoting genes, including *CDKN1C, H19, IGF2,* and *KCNQ1OT1*. The hereditary combination of paraganglioma solely with GIST, known as Carney-Stratakis dyad or Carney-Stratakis syndrome, may not be a distinct entity and can be associated with any of the germline *SDHx* gene mutations. A more unusual incompletely understood hereditary association is "3PAS" (pheochromocytoma, paraganglioma, pituitary adenoma), defined by the presence of a pituitary neuroendocrine tumor (PitNET) along with other components of hereditary syndromes (11,14), including SDH-deficient PitNET with conventional paraganglioma syndromes (11).

Pathologists have several new roles with respect to syndromic tumors. One is to alert clinicians to the possibility of a hereditary tumor based on morphologic or immunohistochemical characteristics of the pheochromocytoma or paraganglioma itself, or on recognition of a syndromically associated tumor that may be the initial presentation of hereditary disease (15). Another role is to help determine whether a variant of unknown significance in germline genetic testing is actually pathogenic, as is demonstrated by loss of SDHB, FH, or MAX protein.

As with any immunohistochemical stain, the technical aspects of tissue fixation and stain optimization can result in false positive or false negative results, and the use of biomarkers is not intended to replace molecular genetic testing, but can be helpful to interpret clinical findings and guide such testing. In labs with well-established immunohistochemical protocols, loss of protein expression can validate the significance of a molecular variant of uncertain significance (VUS).

Pathology is also important in determining whether a second type of tumor in a patient with a pheochromocytoma or paraganglioma is actually part of a syndrome or a coincidence. A prime example is immunohistochemical staining for SDHB showing that PitNET caused by loss of SDHB is rare (16). However, a hypothetical biologic consideration in these cases is whether a secondary effect of the hereditary mutation indirectly contributes to tumorigenesis. For example, high levels of circulating succinate resulting from *SDHx* mutations may activate aspects of hypoxic signaling reported in SDH-intact PitNETs (14).

HEREDITARY PPGLS

Clinical Features

Due to the very high likelihood of genetic predisposition, current recommendations suggest genetic testing for all patients with PPGLs (5,17). If this is not possible, investigation should be prompted by young age at presentation, identification of multiple tumors, and identification of extra-adrenal tumors.

Biochemistry is an important component of the assessment of any patient with a PPGL. As indicated above, the profile of catecholamines often predicts specific mutations.

The biochemical features that reflect specific mutations also guide radiologic assessment. In addition, imaging is subject to variability based on the technique used. Cluster 1 tumors tend to express somatostatin receptors at high levels and therefore are well visualized with ^{68}Ga-DOTATATE positron emission tomography (PET)/computerized tomography (CT) (18,19), whereas cluster 2 tumors are of the more traditional type and can be imaged with Iobeguane (metaiodobenzylguanidine; ^{123}I-MIBG) or ^{18}F-DOPA PET/CT. Imaging allows the identification of multifocal disease (fig. 10-1) and may be the initial clue to the diagnosis of a hereditary disorder. The distribution of paraganglia is similar to that of para-aortic lymph nodes, and multifocal paragangliomas are sometimes confused with metastatic malignancy. This also occurs in liver and lung where these lesions may be multifocal primary tumors (20).

The hereditary syndromes usually give rise to multifocal neoplasms, including multiple paragangliomas, but also, as documented in Table 10-1, tumors involving other tissues and organs. Most of these syndromes are autosomal dominant with variable penetrance, as they are due to transmission of an inactivated tumor suppressor gene that presumably requires a "second hit" to inactivate the normal allele. A single exception to this rule is multiple endocrine

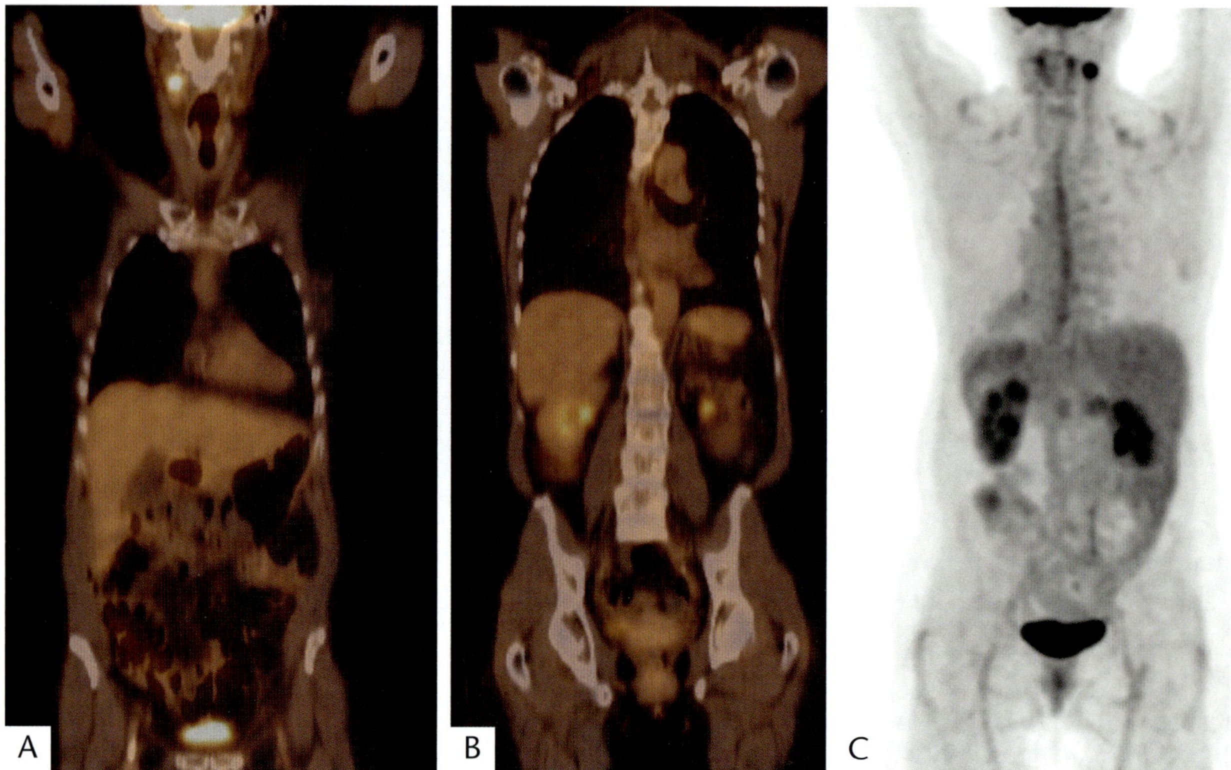

Figure 10-1

MULTIFOCAL PARAGANGLIOMAS IN SDHX DISEASE

A: This patient, with a documented pathogenic germline *SDHB* variant, was diagnosed with a right carotid body tumor. B,C: ^{68}Ga-DOTATATE PET/CT also showed multiple lesions in his retroperitoneum. (A–C: courtesy of Dr. S. Ezzat, Toronto, Canada)

neoplasia type 2 (MEN2), which is associated with an activated oncogene, *RET,* that provides almost 100 percent certainty of developing a medullary thyroid carcinoma. The development of pheochromocytoma in these families is more variable, however, and appears to be related to both the specific genetic alteration and other genetic and tissue modifiers (21).

Multiple Endocrine Neoplasia Type 2

Multiple endocrine neoplasia type 2, first described by Sipple in 1961 (22) and Cushman in 1962 (23) and then subsequently as a variant with mucosal neuromas by Williams in 1966 (24), was recognized to be attributed to germline mutations in the *RET* proto-oncogene in 1993 (25,26). *RET* is a transmembrane tyrosine kinase receptor for glial cell line-derived neurotropic factors (GDNFs) that plays a role in migration of cells from the neural crest; inactivating mutations of *RET* result in Hirschsprung disease (27, 28) and failure of renal development in mice (29).

In patients with MEN2, a number of pathogenic activating mutations, duplications, insertions, or deletions have been identified in different regions of the *RET* coding sequence, including those that code for the extracellular domains, the transmembrane region, and the intracellular split kinase. About 95 percent of *RET* germline mutations are in codons 609, 611, 618, or 620 of exon 10 or codon 634 of exon 11. These genetic alterations result in ligand-independent activation of kinase signaling that results in medullary thyroid carcinoma, pheochromocytoma, and parathyroid hyperplasia/tumors (21). Most of the mutations result in MEN2A but a germline mutation in codon 918 at the site of tyrosine phosphorylation results in MEN2B, a more aggressive form of the disease with the additional features of a marfanoid body habitus, mucosal neuromas, and gastrointestinal ganglioneuromas; MEN2B is usually attributed to a de novo germline event.

Since MEN2 is a unique example of hereditary activation of an oncogene, there is almost 100 percent development of disease in carriers of a mutation, but only medullary thyroid carcinoma has close to 100 percent penetrance. This is the basis for prophylactic thyroidectomy in patients with this disorder; the timing depends on the mutation associated with the most aggressive disease recommended for surgery in infancy, including MEN2B (21). The development of pheochromocytoma and parathyroid disease is more variable; for example, *RET* codon 634 mutations are associated with a high penetrance of pheochromocytoma but mutations in exon 10 have a much lower risk of pheochromocytoma (21,30). Most patients with MEN2B and codon 918 mutations develop pheochromocytoma. Extra-adrenal paragangliomas are not a feature of this disease.

The morphology of the adrenal disease is highly characteristic. It is frequently bilateral and almost always associated with some degree of adrenal medullary hyperplasia, which is diffuse and nodular (fig. 10-2) (30–34). The nodules are usually appreciated grossly and on microscopy: they have variable degrees of cytologic atypia that allow their distinction from the intervening parenchyma. Bilateral disease can be synchronous or asynchronous, but this is the one disorder in which cortical-sparing adrenalectomy is indicated due to the frequency of involvement of both adrenal glands (35,36). It has been suggested that hyaline globules are more common in this disorder, but this was only in comparison to patients with von Hippel-Lindau (VHL) disease and this relationship remains to be proven (37). Metastatic disease is rare. There are currently no specific biomarkers for MEN2-associated pheochromocytoma apart from the associated medullary hyperplasia. Hyperplastic nodules or areas contain identical genetic abnormalities as fully developed pheochromocytomas, regardless of their size (30).

Neurofibromatosis Type 1

Neurofibromatosis type 1 (NF1), originally called von Recklinghausen disease, is an autosomal dominant disorder caused by germline mutation of the *NF1* gene that encodes neurofibromin, a tumor suppressor that regulates cell differentiation and proliferation by signaling through the mTOR pathway (38). Fewer than 10 percent of patients with NF1 develop pheochromocytomas (24,39), but this may reflect a lack of appreciation of hypertension in these patients who may have subclinical disease (38).

Pheochromocytoma is the main paraganglioma associated with NF1 (fig. 10-3), but occasional extra-adrenal paragangliomas have been reported (40). The pheochromocytomas are not usually associated with adrenal medullary hyperplasia, however, they may be more frequently composite tumors containing a benign or malignant neuroblastic component (40). An unusual case report identified a malignant Triton tumor composed of peripheral nerve sheath tumor and skeletal muscle differentiation in a patient with NF1 (41).

Von Hippel-Lindau Disease

Von Hippel-Lindau (VHL) disease is a hereditary autosomal dominant disorder associated with deletions or mutations in the *VHL* tumor suppressor gene (42). The name derives from the description of two individuals, the German ophthalmologist Eugen von Hippel, who described angiomas of the eye in 1904, and Arvid Lindau, who described angiomas of the cerebellum and spine in 1927. The association of these two disorders led to the term von Hippel-Lindau disease.

VHL is typically characterized by retinal angiomas; central nervous system hemangioblastomas; hemangiomas of the adrenal gland, liver, and lung; and cysts in the kidney, pancreas, epididymis, and liver. Additional manifestations include endolymphatic sac tumors, renal cell carcinoma, and neuroendocrine neoplasms (NENs). Pheochromocytomas and paragangliomas are the most common NENs but patients with this disorder also develop pancreatic neuroendocrine tumors.

The protein encoded by VHL is an E3 ubiquitin ligase that targets HIF1α for proteasomal degradation. In low oxygen conditions, downregulation of pVHL allows HIF1α to dimerize with HIF1β, stimulating transcription of VEGF, PDGFβ, and erythropoietin; these are also activated in VHL-associated tumors that exhibit the hypoxia-mediated signaling pathway (43).

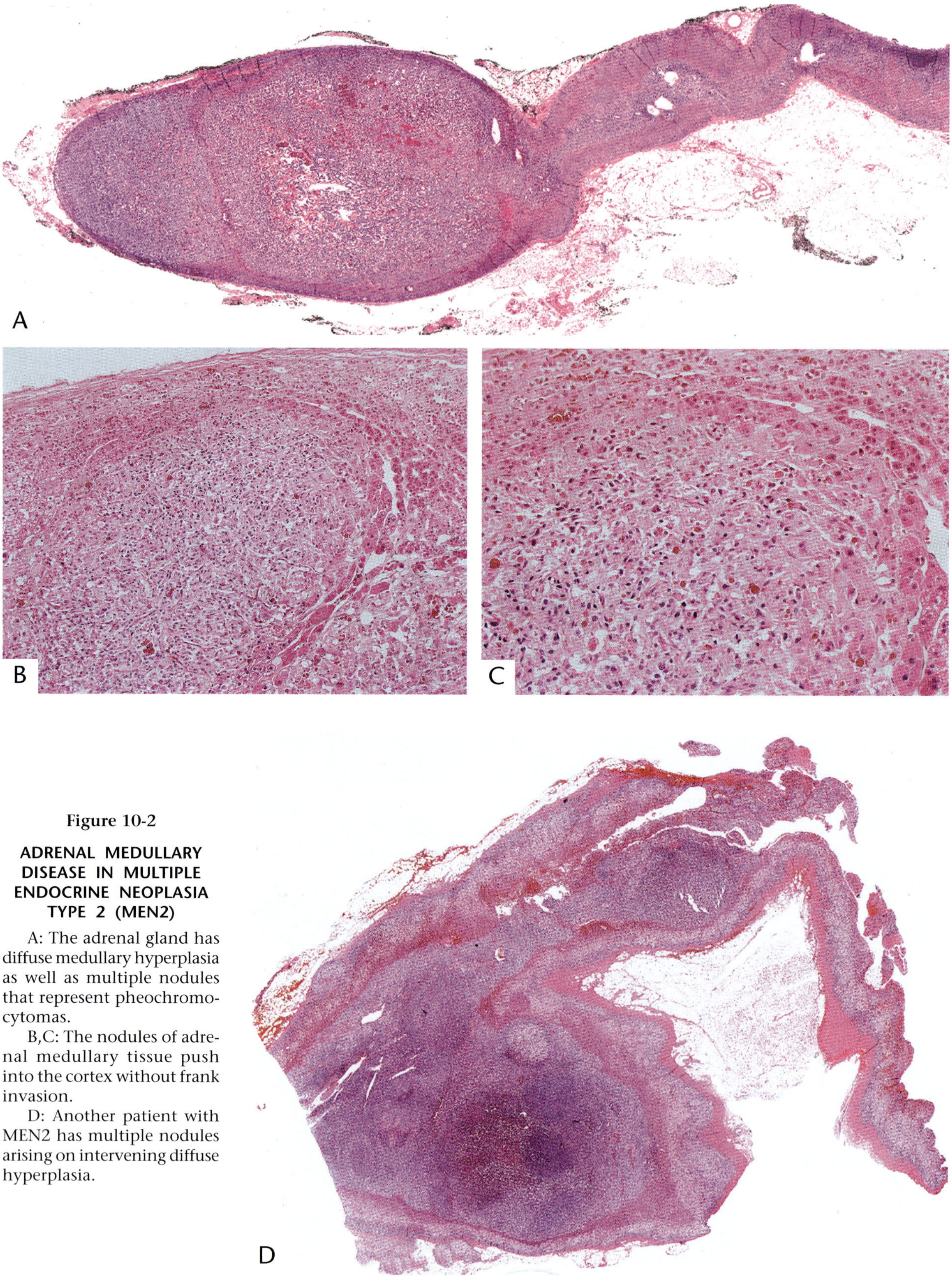

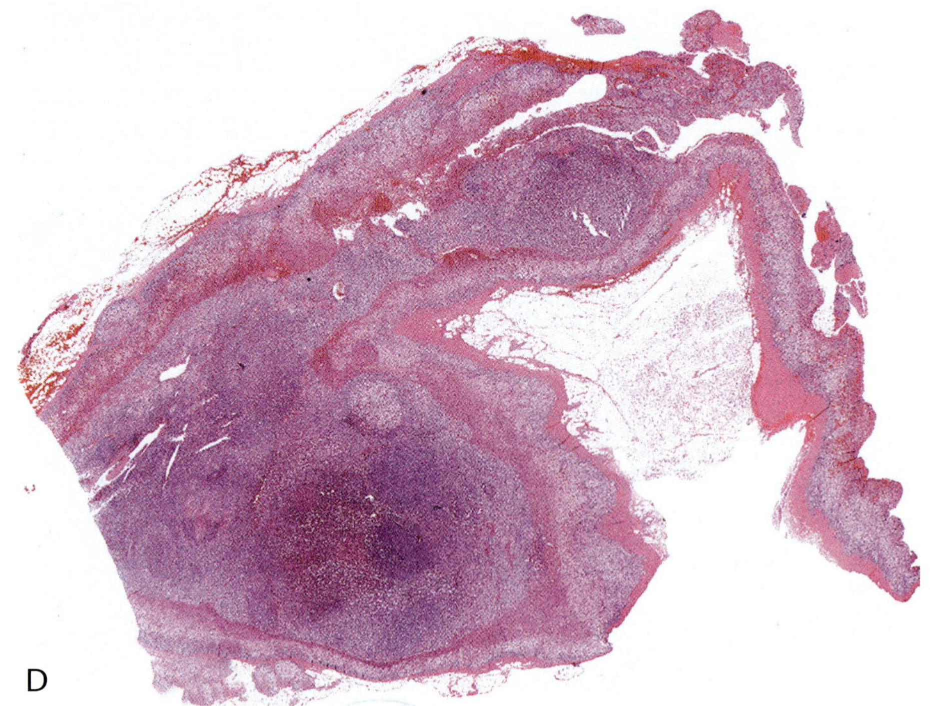

Figure 10-2

ADRENAL MEDULLARY DISEASE IN MULTIPLE ENDOCRINE NEOPLASIA TYPE 2 (MEN2)

A: The adrenal gland has diffuse medullary hyperplasia as well as multiple nodules that represent pheochromocytomas.

B,C: The nodules of adrenal medullary tissue push into the cortex without frank invasion.

D: Another patient with MEN2 has multiple nodules arising on intervening diffuse hyperplasia.

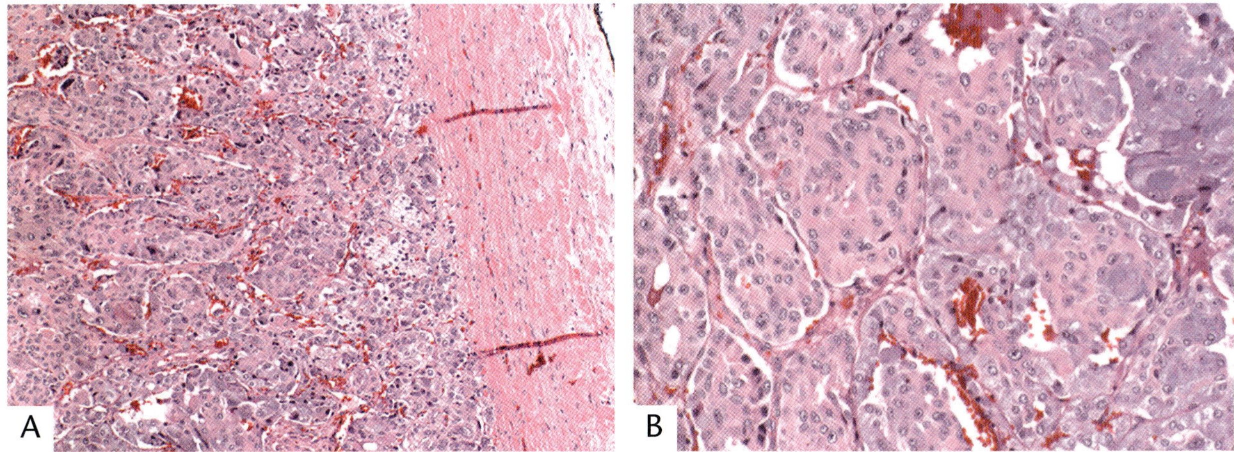

Figure 10-3

PHEOCHROMOCYTOMA IN NEUROFIBROMATOSES TYPE 1 (NF1)

A: The adrenal gland contains a pheochromocytoma.
B: The morphology of the tumor is not distinct from sporadic pheochromocytomas.

VHL has a worldwide incidence of 1 in 36,000 live births (44). The various molecular alterations in the *VHL* gene give rise to two clinical types of VHL (45). Type 1, associated with deletions, frameshifts, and nonsense mutations, is characterized by a low incidence of pheochromocytoma; type 2, associated with missense variants, has a high incidence of paragangliomas including pheochromocytomas.

The morphology of VHL-associated PPGLs is striking for the clear cell morphology and stromal changes (37,46). The clear cell morphology is a feature of other neuroendocrine neoplasms that are seen in VHL including duodenal and pancreatic neuroendocrine tumors (47,48) as well as renal cell carcinomas; this can lead to misdiagnosis. Grossly, the accumulation of lipid that is the source of the clear cytoplasm gives a yellow appearance that is readily mistaken for an adrenal cortical lesion (see fig. 8-12). On microscopy, the vacuolated cytoplasm can mimic clear cell renal cell carcinoma or adrenal cortical cytology (fig. 10-4). Another pitfall is that these tumors can express inhibin, a feature that is often thought to be a biomarker of adrenal cortex (49); synaptophysin is also expressed in both cortex and medulla, and therefore, is not helpful in this distinction. However, these tumors are positive for INSM1, chromogranin, GATA3, and tyrosine hydroxylase, and they are negative for the more specific biomarker of adrenal cortex, steroidogenic factor 1 (SF1). Also, VHL-associated PPGLs can have reduced expression of SDHB, potentially leading to misdiagnosis as an *SDHx*-associated lesion.

On electron microscopy, the tumors have all the features of neuroendocrine tumors with secretory granules. The extensive lipid degeneration can be associated with hemosiderin deposition and focal fibrosis. VHL-associated tumors have strong membranous positivity for CAIX, an important stain for diagnosis (49,50), although the reactivity is often less diffuse than in other *VHL*-mutant tumors and positivity may reflect a somatic mutation rather than germline disease.

Succinate Dehydrogenase-Related Disease

The identification of mutations in the SDH complex in paragangliomas was based on clinical data showing familial clustering of carotid body tumors in a group of Dutch patients (51) with the pattern of a maternally-imprinted paternally-expressed gene (52,53). In 2000, the implicated gene locus was shown to be *SDHD* (54), and the story of the mitochondrial respiratory chain protein complex in neoplasia became evident. Subsequent studies showed germline mutations of *SDHD* in familial pheochromocytoma (55) and in apparently sporadic pheochromocytomas (56), followed by the identification of mutations in *SDHB* and *SDHC* in pheochromocytomas that were not thought

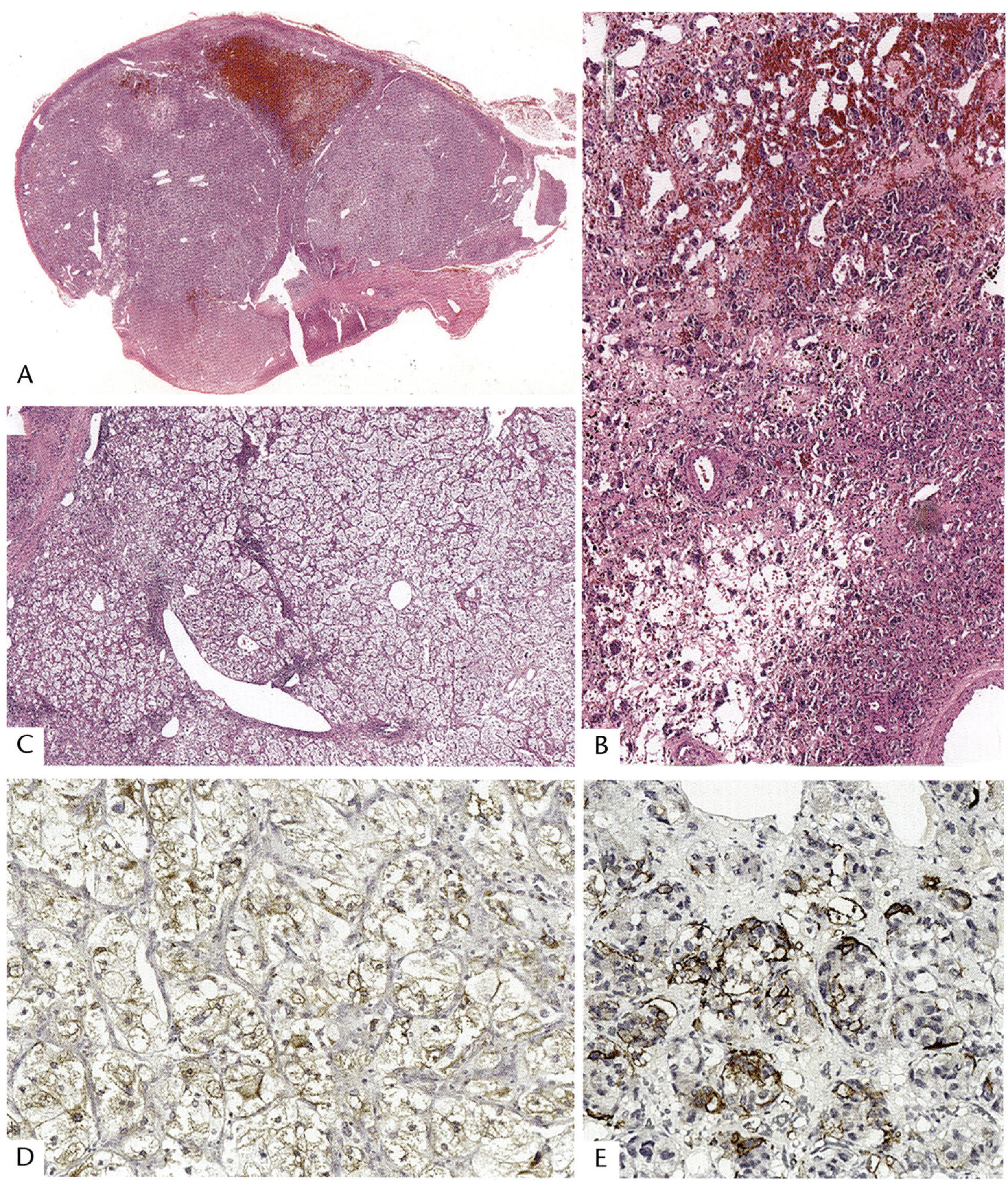

Figure 10-4

ADRENAL PATHOLOGY IN VON HIPPEL-LINDAU (VHL) DISEASE

A: The adrenal gland has multiple nodules that represent pheochromocytomas.
B: The tumors have degenerate stroma with striking edema, focal hemorrhage, and hemosiderin deposition.
C,D: The abundant clear cytoplasm represents lipid, mimicking adrenal cortex (C), but stains for chromogranin (D).
E: Positivity for CAIX is a feature of VHL-associated disease.

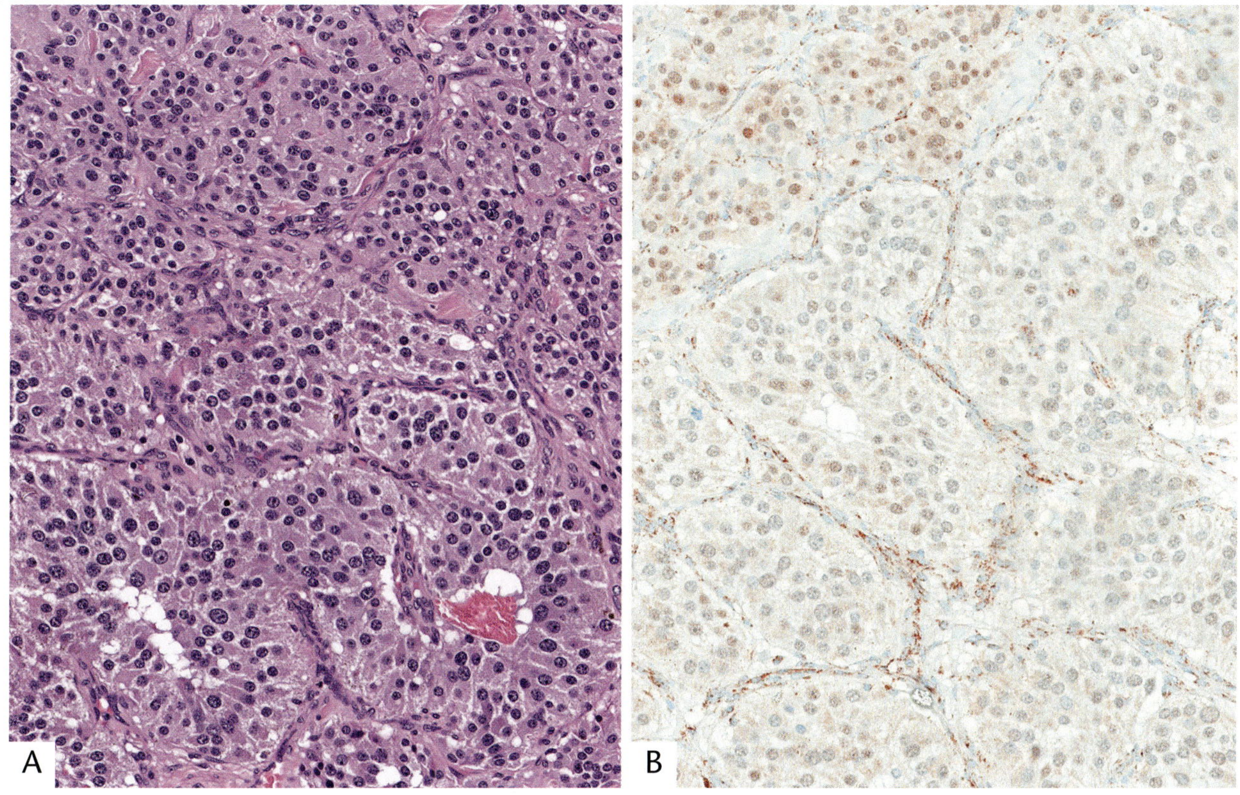

Figure 10-5

SDH-DEFICIENT PARAGANGLIOMA

A: This cardiac paraganglioma from a patient with a pathogenic *SDHB* variant has round nesting architecture, oncocytic cytology with abundant eosinophilic cytoplasm, and focal tendency to form pseudorosettes.

B: SDHB immunohistochemistry is negative, with retained positivity in vascular endothelium serving as an internal positive control.

to be hereditary (57). Thus many tumors that were thought to be sporadic were suddenly recognized to be hereditary.

The mitochondrial respiratory complex 2, known as succinate dehydrogenase because it catalyzes the oxidation of succinate to fumarate on the inner mitochondrial membrane, is composed of four subunits, SDHA, SDHB, SDHC, and SDHD, and two components known as assembly factors, SDHAF1 and SDHAF2. These are each encoded by a distinct nuclear gene, and mutations in each of these genes are associated with paragangliomas; most also give rise to other tumors, most commonly renal cell carcinomas and GISTs. Occasional PitNETs are also reported in association with these disorders (58,59). The clinical syndromes are known as PGL 1 through 5 based on the timing of the initial reports, which do not coincide with the alphabetical nomenclature of the *SDH* genes: PGL1 is caused by *SDHD* mutations since that was the initial candidate gene, PGL2 is caused by *SDHAF*, PGL3 by *SDHC*, PGL4 by *SDHB,* and PGL5 by *SDHA* mutations.

The morphology of these tumors is sometimes distinct (fig. 10-5) but not all *SDHx*-related tumors have the described features that include a prominent circular nested architecture, oncocytic cytology (60), and a peculiar pseudorosette pattern (61). Unlike sporadic paragangliomas, they rarely assume a spindle cell morphology. Immunohistochemistry is helpful in proposing an *SDHx*-related pathogenesis and also in confirming the pathogenic nature of variants of uncertain significance. The use of SDHB immunoreactivity is based on the role of SDHB in the complex; loss of any of the five members results in destabilization of SDHB, which is then negative in tumor cells, with positivity in stromal cells, serving as an internal control (62–64). While occasional reports have identified false negative immunoreactivity in some cases (65), this may be attributed to poor technique and inexperienced interpretation, and SDHB immunoreactivity remains a highly valuable tool for molecular diagnosis (49,62–64). In tumors that lack SDHB immunoreactivity, loss of SDHA staining identifies patients with *SDHA* mutations (66). A potential pitfall is weak staining for SDHB. Inhibin-alpha is another biomarker expressed in paragangliomas of the pseudohypoxic signaling pathway (49).

Mutations in *SDHB* are associated with a higher risk of metastatic behavior. The exact incidence of this is unclear, since many apparent metastases reported in the literature represent multifocal primary tumors in patients with germline disease.

Hereditary Leiomyomatosis and Renal Cell Carcinoma

Mutations in the fumarate hydratase (*FH*) gene encoding fumarate hydratase are the cause of *Reed syndrome*, a familial autosomal dominant condition characterized by numerous cutaneous leiomyomas, uterine leiomyomas in women, and renal cell carcinoma (49,67), giving rise to the noneponymic term *hereditary leiomyomatosis and renal cell carcinoma* (HLRCC). Patients with this disorder have a higher than expected incidence of adrenal and extra-adrenal paragangliomas (7).

There is little documented about the morphology of these tumors. In an example illustrated in figure 10-6, there is a striking angiomatous proliferation of the cortical vasculature, with both smooth muscle and endothelial components. *FH* mutations can be screened by immunohistochemistry for FH, which shows loss in tumor cell cytoplasm and retained staining in stroma (7,65); there is staining for 2-succinocysteine (2SC), a target of fumarate. The incidence of metastatic behavior in these paragangliomas is said to be high.

Malate Dehydrogenase *(MDH2)* Mutations

Malate dehydrogenase is an enzyme that catalyzes the reversible oxidation of malate to oxaloacetate in the mitochondria. The protein is encoded by the *MDH2* gene, which is mutated in rare familial cases of PPGL (68). The metabolic effects are not unlike those of *SDHx* disease and these patients also have a cluster 1 biochemical phenotype. Few cases have been reported and no specific morphologic features are known. No associated disorders are reported.

KIF1 Mutations

Inherited loss-of-function *KIF1β* missense mutations have been reported in neuroblastoma and pheochromocytoma (69), leiomyosarcoma, and lung adenocarcinoma. It is unclear, however, if *KIF1* is actually responsible for the pheochromocytoma, as subsequent reports identified a germline *MAX* mutation in these patients (70).

TMEM-Related Disease

Germline mutations of *TMEM127* have been identified in patients with familial pheochromocytoma (71,72). The same gene was implicated in renal cell carcinoma, and coexistence of these entities has been reported (73). Recently, a patient with this disorder was reported to have acromegaly due to a PitNET (74). The patients may develop multiple tumors and the biochemical profile fits into cluster 2 disease, with secretion of epinephrine and norepinephrine. The pheochromocytomas are characterized by tumor multicentricity and are associated with nodular adrenal medullary hyperplasia (72). Metastatic disease is rare (75).

MAX Mutations

Patients with germline mutations in the gene known as *MAX* (MYC-associated factor X) develop pheochromocytomas that may be bilateral or multifocal within the same gland; paragangliomas also occur in other abdominal sites and in the thorax (76). Germline mutation predisposes to early onset disease. Although these tumors are classified in cluster 2, the profile is perhaps more intermediate, with metanephrine levels that tend to be higher than in patients with *VHL/SDHx*-related disease but lower than in patients with *RET/NF1* mutations (76). They are associated with renal oncocytoma. Metastatic disease occurs occasionally.

MEN1-Associated Paragangliomas

The development of a paraganglioma in patients with MEN1 has been reported in isolated

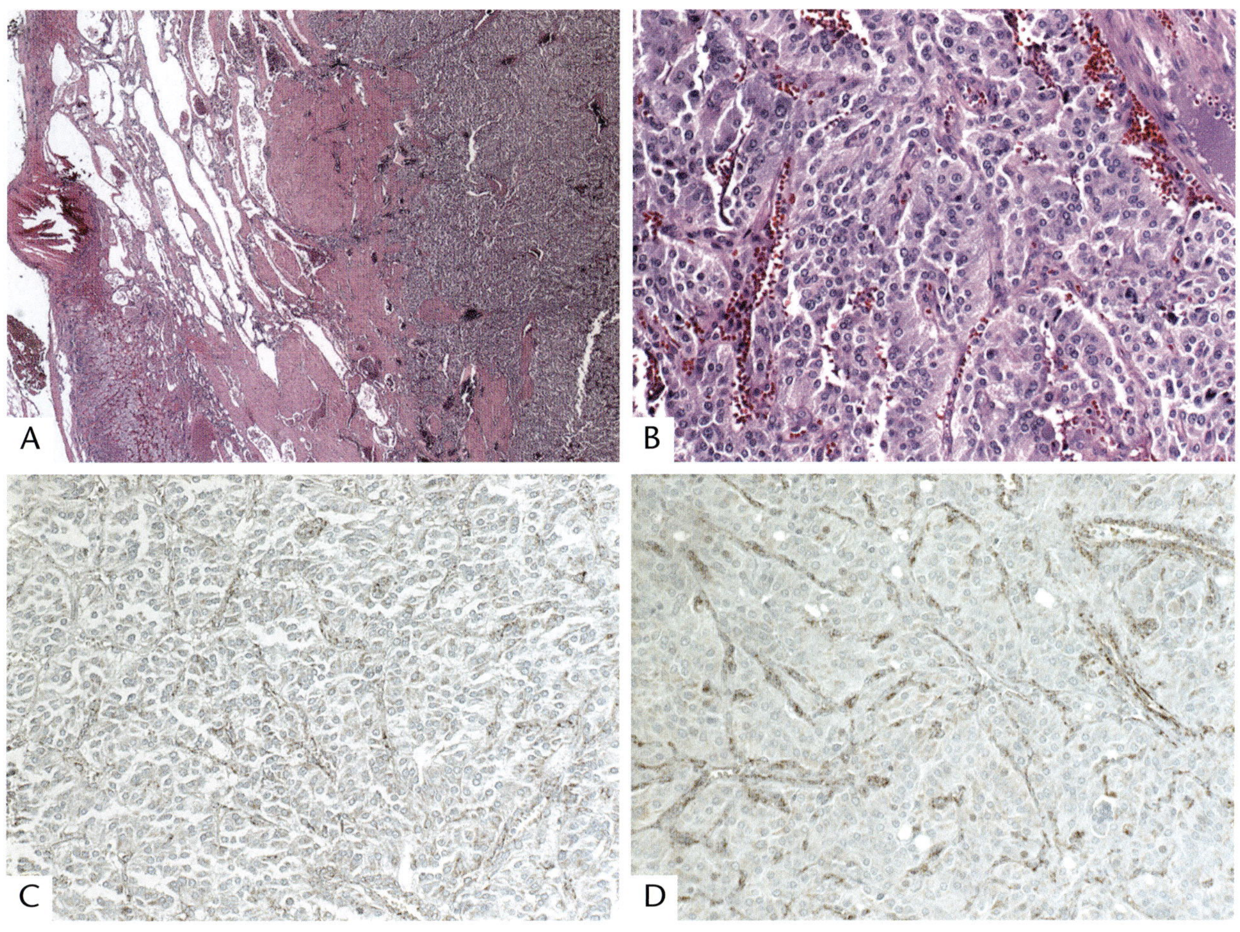

Figure 10-6

PHEOCHROMOCYTOMA AND PARAGANGLIOMA ASSOCIATED WITH GERMLINE FUMARATE HYDRATASE (FH) MUTATION

A: The pheochromocytoma is associated with an unusual angiomatous proliferation of the cortical vasculature, with both smooth muscle and endothelial components.

B: The paraganglioma, diagnosed 6 years later, has no unusual features.

C,D: Staining for FH shows loss in the pheochromocytoma (C) and the paraganglioma (D).

cases (59,77); it remains unclear what the role of MEN1 loss is in these tumors and how frequent this association is. These lesions have no reported specific morphologic features but show loss of menin staining on immunohistochemistry (fig. 10-7) (78).

NONHEREDITARY SYNDROMIC PPGLS

Paraganglioma and Polycythemia Syndromes

Patients with erythrocytosis are known to have germline mutations in the *EGLN1* gene that encodes prolyl hydroxylase domain 2 (PHD2) and the protein that regulates hypoxia-inducible factor (HIF). *EGLN1* mutation with erythrocytosis is also associated with paraganglioma in which there is loss of heterozygosity, indicating its role as a tumor suppressor (79,80). Mutation of *EGLN2*, which encodes the related PHD1, has also been reported (81). Mutations of the *EGLN* genes appear to be a rare cause of familial paraganglioma (82).

The syndrome known as *Pacak-Zhuang syndrome (PZS)* was reported in 2012 when somatic gain-of-function mutations in the gene encoding hypoxia-inducible factor 2 alpha (HIF2A) were identified in two patients presenting with polycythemia and paraganglioma, and one

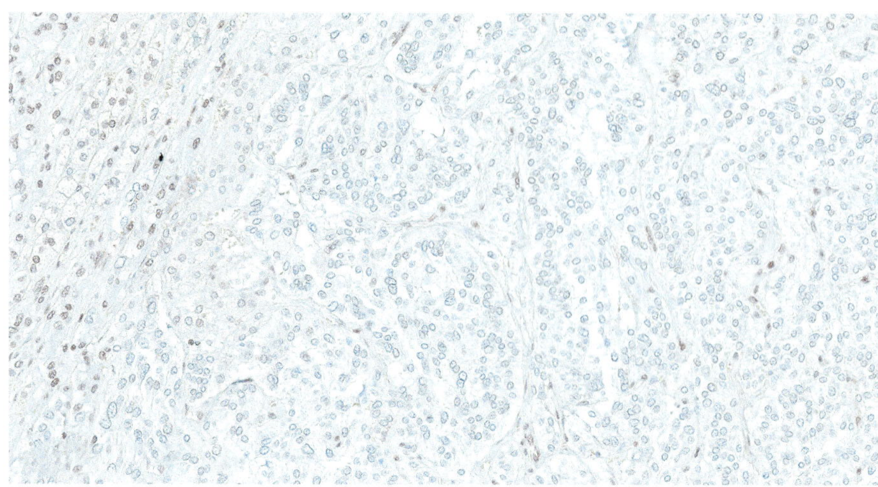

Figure 10-7

PHEOCHROMOCYTOMA IN A PATIENT WITH MEN1

This adrenal pheochromocytoma was identified at the time of surgery for a pancreatic neuroendocrine tumor in a patient with a history of multiglandular parathyroid disease. While the tumor has no specific morphologic features that distinguish it as MEN1-associated disease, loss of menin on immunohistochemistry confirms this diagnosis.

also with a somatostatinoma (83). Additional patients were reported (84,85) and while initially thought to occur only in females, a male patient was also identified (86). This exceptionally rare entity is not known to be familial, since no relatives of affected patients have been reported with the disorder. A genetic predisposition, currently thought to be due to early postzygotic mutations of *EPAS1* (12), is supported by the fact that all reported patients developed secondary polycythemia before the age of 8 years (12). Paragangliomas developed at a median age of 17 years and all were multiple; metastatic spread has been reported (12).

The paragangliomas of PZS associate with cluster 1 disease, as expected with involvement of the pseudohypoxic pathway involvement, and are primarily norepinephrine producing. The duodenal lesions develop later, at an average age of 29 years. Patients have other features resembling VHL disease, including ophthalmologic findings, cystic lesions at multiple sites, and hemangiomas. The long-term prognosis remains unknown. Similar somatic mutations were also identified in the unusual duodenal lesion known as gangliocytic paraganglioma (87), not as part of the syndrome; these are not considered to be true paragangliomas but rather are triphasic tumors composed of epithelial neuroendocrine cells, ganglion cells, and neural stroma (88).

Carney Triad

Carney triad is characterized by the development of paraganglioma, pulmonary chondroma, and SDH-deficient GIST. It shows a slight female predilection and is caused by methylation of the *SDHC* gene; there may be germline mosaicism of the responsible genetic defect (11). The features of these lesions are similar to those of other *SDHx*-related tumors.

Beckwith-Wiedemann Syndrome

Beckwith-Wiedemann syndrome is a disorder characterized by early childhood onset of clinical abnormalities, including omphalocele, umbilical hernia, hypoglycemia, and macrosomia including macroglossia and visceromegaly. The development of malignant tumors, especially renal Wilms tumors, is well documented. There have been several case reports of bilateral adrenal pheochromocytoma (13).

This disorder is attributed to alterations involving the region of chromosome 11p15 that contains several imprinted genes including *IGF2*; there may be loss of methylation on the maternal chromosome at imprinting center 2 (IC2), paternal uniparental disomy, or gains in methylation on the maternal chromosome at imprinting center 1 (IC1). No specific features have been reported in the pheochromocytomas of the few cases reported.

REFERENCES

1. Mete O, Asa SL, Gill AJ, Kimuta N, de Krijger R, Tischler A. Overview of the 2022 WHO classification of paragangliomas and pheochromocytomas. Endocr Pathol 2022;33:115-54.
2. Toledo RA, Burnichon N, Cascon A, et al. Consensus statement on next-generation-sequencing-based diagnostic testing of hereditary phaeochromocytomas and paragangliomas. Nat Rev Endocrinol 2017;13:233-47.
3. Turchini J, Cheung VK, Tischler AS, de Krijger RR, Gill AJ. Pathology and genetics of phaeochromocytoma and paraganglioma. Histopathology 2018;72:97-105.
4. Flores SK, Estrada-Zuniga CM, Thallapureddy K, Armaiz-Pena G, Dahia PL. Insights into mechanisms of pheochromocytomas and paragangliomas driven by known or new genetic drivers. Cancers (Basel) 2021;13:1259-61.
5. Lenders JW, Kerstens MN, Amar L, et al. Genetics, diagnosis, management and future directions of research of phaeochromocytoma and paraganglioma: a position statement and consensus of the Working Group on Endocrine Hypertension of the European Society of Hypertension. J Hypertens 2020;38:1443-56.
6. Dahia PL, Ross KN, Wright ME, et al. A HIF1α regulatory loop links hypoxia and mitochondrial signals in pheochromocytomas. PLoS Genet 2005;1:72-80.
7. Castro-Vega LJ, Buffet A, de Cubas AA, et al. Germline mutations in FH confer predisposition to malignant pheochromocytomas and paragangliomas. Hum Mol Genet 2014;23:2440-6.
8. Zethoven M, Martelotto L, Pattison A, et al. Single-nuclei and bulk-tissue gene-expression analysis of pheochromocytoma and paraganglioma links disease subtypes with tumor microenvironment. Nat Commun 2022;13:6262.
9. Leonard EM, Salman S, Nurse CA. Sensory processing and integration at the carotid body tripartite synapse: neurotransmitter functions and effects of chronic hypoxia. Front Physiol 2018;9:225.
10. Hervonen A, Korkala O. The effect of hypoxia on the catecholamine content of human fetal abdominal paraganglia and adrenal medulla. Acta Obstet Gynecol Scand 1972;51:17-24.
11. Pitsava G, Settas N, Faucz FR, Stratakis CA. Carney triad, Carney-Stratakis syndrome, 3PAS and other tumors due to SDH deficiency. Front Endocrinol (Lausanne) 2021;12:680609.
12. Darr R, Nambuba J, Del RJ, et al. Novel insights into the polycythemia-paraganglioma-somatostatinoma syndrome. Endocr Relat Cancer 2016;23:899-908.
13. Caza T, Manwaring J, Riddell J. Recurrent, bilateral, and metastatic pheochromocytoma in a young patient with Beckwith-Wiedemann syndrome: a genetic link? Can Urol Assoc J 2017; 11:E240-3.
14. Xekouki P, Brennand A, Whitelaw B, Pacak K, Stratakis CA. The 3PAs: an update on the association of pheochromocytomas, paragangliomas, and pituitary tumors. Horm Metab Res 2019;51:419-36.
15. Papathomas TG, Suurd DP, Pacak K, et al. What have we learned from molecular biology of paragangliomas and pheochromocytomas? Endocr Pathol 2021;32:134-53.
16. Gill AJ, Toon CW, Clarkson A, et al. Succinate dehydrogenase deficiency is rare in pituitary adenomas. Am J Surg Pathol 2014;38:560-6.
17. Lenders JW, Duh QY, Eisenhofer G, et al. Pheochromocytoma and paraganglioma: an endocrine society clinical practice guideline. J Clin Endocrinol Metab 2014;99:1915-42.
18. Jha A, Ling A, Millo C, et al. Superiority of ^{68}Ga-DOTATATE over ^{18}F-FDG and anatomic imaging in the detection of succinate dehydrogenase mutation (SDHx)-related pheochromocytoma and paraganglioma in the pediatric population. Eur J Nucl Med Mol Imaging 2018;45:787-97.
19. Janssen I, Blanchet EM, Adams K, et al. Superiority of [68Ga]-DOTATATE PET/CT to other functional imaging modalities in the localization of SDHB-associated metastatic pheochromocytoma and paraganglioma. Clin Cancer Res 2015;21: 3888-95.
20. Asa SL, Ezzat S, Mete O. The diagnosis and clinical significance of paragangliomas in unusual locations. J Clin Med 2018;7:280.
21. Wells SA Jr, Asa SL, Dralle H, et al. Revised American Thyroid Association guidelines for the management of medullary thyroid carcinoma. Thyroid 2015;25:567-610.
22. Sipple JH. The association of phenochromocytoma with carcinoma of the thyroid gland. Am J Med 1961;31:163-6.
23. Cushman P Jr. Familial endocrine tumors; report of two unrelated kindred affected with pheochromocytomas, one also with multiple thyroid carcinomas. Am J Med 1962;32:352-60.
24. Williams ED, Pollock DJ. Multiple mucosal neuromata with endocrine tumors-a syndrome allied to Von Recklinghausen's disease. J Pathol Bacteriol 1966;91:71-80.

25. Mulligan LM, Kwok JB, Healey CS, et al. Germline mutations of the RET proto-oncogene in multiple endocrine neoplasia type 2A. Nature 1993;363:458-60.
26. Donis-Keller H, Dou S, Chi D, et al. Mutations in the RET proto-onocogene are associated with MEN 2A and FMTC. Hum Mol Genet 1993;2:851-6.
27. Edery P, Lyonnet S, Mulligan LM, et al. Mutations of the RET proto-oncogene in Hirschsprung's disease. Nature 1994;367:378-80.
28. Romeo G, Ronchetto P, Luo Y, et al. Point mutations affecting the tyrosine kinase domain of the RET proto-oncogene in Hirschsprung's disease. Nature 1994;367:377-8.
29. Schuchardt A, D'Agati V, Larsson-Blomberg L, Costantini F, Pachnis V. Defects in the kidney and enteric nervous system of mice lacking the tyrosine kinase receptor RET. Nature 1993;367:380-3.
30. Korpershoek E, Petri BJ, Post E, et al. Adrenal medullary hyperplasia is a precursor lesion for pheochromocytoma in MEN2 syndrome. Neoplasia 2014;16:868-73.
31. Carney JA, Sizemore GW, Tyce GM. Bilateral adrenal medullary hyperplasia in multiple endocrine neoplasia, type 2: the precursor of bilateral pheochromocytoma. Mayo Clin Proc 1975;50:3-10.
32. Carney JA, Sizemore GW, Sheps SG. Adrenal medullary disease in multiple endocrine neoplasia, type 2: pheochromocytoma and its precursors. Am J Clin Pathol 1976;66:279-90.
33. DeLellis RA, Wolfe HJ, Gagel RF, et al. Adrenal medullary hyperplasia. A morphometric analysis in patients with familial medullary thyroid carcinoma. Am J Pathol 1976;83:177-96.
34. Mete O, Asa SL. Precursor lesions of endocrine system neoplasms. Pathology 2013;45:316-30.
35. Yip L, Lee JE, Shapiro SE, et al. Surgical management of hereditary pheochromocytoma. J Am Coll Surg 2004;198:525-34.
36. Castinetti F, Taieb D, Henry JF, et al. Management of endocrine disease: outcome of adrenal sparing surgery in heritable pheochromocytoma. Eur J Endocrinol 2016;174:R9-18.
37. Koch CA, Mauro D, Walther MM, et al. Pheochromocytoma in von hippel-lindau disease: distinct histopathologic phenotype compared to pheochromocytoma in multiple endocrine neoplasia type 2. Endocr Pathol 2002;13:17-27.
38. Zinnamosca L, Petramala L, Cotesta D, et al. Neurofibromatosis type 1 (NF1) and pheochromocytoma: prevalence, clinical and cardiovascular aspects. Arch Dermatol Res 2011;303:317-25.
39. Zografos GN, Vasiliadis GK, Zagouri F, et al. Pheochromocytoma associated with neurofibromatosis type 1: concepts and current trends. World J Surg Oncol 2010;8:14.
40. Gupta S, Zhang J, Erickson LA. Composite pheochromocytoma/paraganglioma-ganglioneuroma: a clinicopathologic study of eight cases with analysis of succinate dehydrogenase. Endocr Pathol 2017;28:269-75.
41. Gupta R, Sharma A, Arora R, Vijayaraghavan M. Composite phaeochromocytoma with malignant peripheral nerve sheath tumour and rhabdomyosarcomatous differentiation in a patient without von Recklinghausen disease. J Clin Pathol 2009;62:659-61.
42. Latif F, Tory K, Gnarra J, et al. Identification of the von Hippel-Lindau disease tumor suppressor gene. Science 1993;260:1317-20.
43. Kaelin WG Jr. Molecular basis of the VHL hereditary cancer syndrome. Nat Rev Cancer 2002;2: 673-82.
44. Maher PA. Identification and characterization of a novel, intracelllular isoform of fibroblast growth factor receptor-1(FGFR-1). J Cell Physiol 1995;169:380-90.
45. Nielsen SM, Rhodes L, Blanco I, et al. Von Hippel-Lindau disease: genetics and role of genetic counseling in a multiple neoplasia syndrome. J Clin Oncol 2016;34:2172-81.
46. Ramsay JA, Asa SL, van Nostrand AW, Hassaram ST, de Harven EP. Lipid degeneration in pheochromocytomas mimicking adrenal cortical tumors. Am J Surg Pathol 1987;11:480-6.
47. Chetty R, Kennedy M, Ezzat S, Asa SL. Pancreatic endocrine pathology in von Hippel-Lindau disease: an expanding spectrum of lesions. Endocr Pathol 2004;15:141-8.
48. Gucer H, Szentgyorgyi E, Ezzat S, Asa SL, Mete O. Inhibin-expressing clear cell neuroendocrine tumor of the ampulla: an unusual presentation of von Hippel-Lindau disease. Virchows Arch 2013;463:593-7.
49. Mete O, Pakbaz S, Lerario AM, Giordano TJ, Asa SL. Significance of alpha-inhibin expression in pheochromocytomas and paragangliomas. Am J Surg Pathol 2021;45:1264-73.
50. Pinato DJ, Ramachandran R, Toussi ST, et al. Immunohistochemical markers of the hypoxic response can identify malignancy in phaeochromocytomas and paragangliomas and optimize the detection of tumours with VHL germline mutations. Br J Cancer 2013;108:429-37.
51. Kroll AJ, Alexander B, Cochios F, Pechet L. Hereditary deficiencies of clotting factors VII and X associated with carotid body tumors. N Engl J Med 1964;270:6-13.
52. van der Mey AG, Maaswinkel-Mooy PD, Cornelisse CJ, Schmidt PH, van de Kamp JJ. Genomic imprinting in hereditary glomus tumours: evidence for new genetic theory. Lancet 1989;2: 1291-4.

53. Heutink P, van der Mey AG, Sandkuijl LA, et al. A gene subject to genomic imprinting and responsible for hereditary paragangliomas maps to chromosome 11q23-qter. Hum Mol Genet 1992; 1:7-10.
54. Baysal BE, Ferrell RE, Willett-Brozick JE, et al. Mutations in SDHD, a mitochondrial complex II gene, in hereditary paraganglioma. Science 2000;287:848-51.
55. Astuti D, Douglas F, Lennard TW, et al. Germline SDHD mutation in familial phaeochromocytoma. Lancet 2001;357:1181-2.
56. Gimm O, Armanios M, Dziema H, Neumann HP, Eng C. Somatic and occult germ-line mutations in SDHD, a mitochondrial complex II gene, in nonfamilial pheochromocytoma. Cancer Res 2000;60:6822-5.
57. Neumann HP, Bausch B, McWhinney SR, et al. Germ-line mutations in nonsyndromic pheochromocytoma. N Engl J Med 2002;346:1459-66.
58. Papathomas TG, Gaal J, Corssmit EP, et al. Non-pheochromocytoma (PCC)/paraganglioma (PGL) tumors in patients with succinate dehydrogenase-related PCC-PGL syndromes: a clinicopathological and molecular analysis. Eur J Endocrinol 2014;170:1-12.
59. Denes J, Swords F, Rattenberry E, et al. Heterogeneous genetic background of the association of pheochromocytoma/paraganglioma and pituitary adenoma: results from a large patient cohort. J Clin Endocrinol Metab 2015;100:E531-41.
60. Turchini J, Gill AJ. Morphologic clues to succinate dehydrogenase (SDH) deficiency in pheochromocytomas and paragangliomas. Am J Surg Pathol 2020;44:422-4.
61. Kimura N, Takekoshi K, Horii A, et al. Clinicopathological study of SDHB mutation-related pheochromocytoma and sympathetic paraganglioma. Endocr Relat Cancer 2014;21:L13-6.
62. van Nederveen FH, Gaal J, Favier J, et al. An immunohistochemical procedure to detect patients with paraganglioma and phaeochromocytoma with germline SDHB, SDHC, or SDHD gene mutations: a retrospective and prospective analysis. Lancet Oncol 2009;10:764-71.
63. Gill AJ, Benn DE, Chou A, et al. Immunohistochemistry for SDHB triages genetic testing of SDHB, SDHC, and SDHD in paraganglioma-pheochromocytoma syndromes. Hum Pathol 2010;41:805-14.
64. Oudijk L, Gaal J, de Krijger RR. The role of immunohistochemistry and molecular analysis of succinate dehydrogenase in the diagnosis of endocrine and non-endocrine tumors and related syndromes. Endocr Pathol 2019;30:64-73.
65. Udager AM, Magers MJ, Goerke DM, et al. The utility of SDHB and FH immunohistochemistry in patients evaluated for hereditary paraganglioma-pheochromocytoma syndromes. Hum Pathol 2018;71:47-54.
66. Korpershoek E, Favier J, Gaal J, et al. SDHA immunohistochemistry detects germline SDHA gene mutations in apparently sporadic paragangliomas and pheochromocytomas. J Clin Endocrinol Metab 2011;96:E1472-6.
67. Toro JR, Nickerson ML, Wei MH, et al. Mutations in the fumarate hydratase gene cause hereditary leiomyomatosis and renal cell cancer in families in North America. Am J Hum Genet 2003;73:95-106.
68. Calsina B, Curras-Freixes M, Buffet A, et al. Role of MDH2 pathogenic variant in pheochromocytoma and paraganglioma patients. Genet Med 2018;20:1652-62.
69. Schlisio S, Kenchappa RS, Vredeveld LC, et al. The kinesin KIF1Bβ acts downstream from EglN3 to induce apoptosis and is a potential 1p36 tumor suppressor. Genes Dev 2008;22:884-93.
70. Cardot Bauters C, Leteurtre E, Carnaille B, et al. Genetic predisposition to neural crest-derived tumors: revisiting the role of KIF1B. Endocr Connect 2020;9:1042-50.
71. Yao L, Schiavi F, Cascon A, et al. Spectrum and prevalence of FP/TMEM127 gene mutations in pheochromocytomas and paragangliomas. JAMA 2010;304:2611-9.
72. Toledo SP, Lourenco DM Jr, Sekiya T, et al. Penetrance and clinical features of pheochromocytoma in a six-generation family carrying a germline TMEM127 mutation. J Clin Endocrinol Metab 2015;100:E308-18.
73. Hernandez KG, Ezzat S, Morel CF, et al. Familial pheochromocytoma and renal cell carcinoma syndrome: TMEM127 as a novel candidate gene for the association. Virchows Arch 2015;466:727-32.
74. Stutz B, Korbonits M, Kothbauer K, Muller W, Fischli S. Identification of a TMEM127 variant in a patient with paraganglioma and acromegaly. Endocrinol Diabetes Metab Case Rep 2020; 2020:20-0019.
75. Armaiz-Pena G, Flores SK, Cheng ZM, et al. Genotype-phenotype features of germline variants of the TMEM127 pheochromocytoma susceptibility gene: a 10-year update. J Clin Endocrinol Metab 2021;106:e350-64.
76. Burnichon N, Cascon A, Schiavi F, et al. MAX mutations cause hereditary and sporadic pheochromocytoma and paraganglioma. Clin Cancer Res 2012;18:2828-37.
77. Jamilloux Y, Favier J, Pertuit M, et al. A MEN1 syndrome with a paraganglioma. Eur J Hum Genet 2014;22:283-5.
78. Asa SL, Mohamed A. Menin loss in pheochromocytoma of multiple endocrine neoplasia type 1. Endocr Pathol 2023;34:156-8.

79. Ladroue C, Carcenac R, Leporrier M, et al. PHD2 mutation and congenital erythrocytosis with paraganglioma. N Engl J Med 2008;359:2685-92.
80. Fishbein L, Leshchiner I, Walter V, et al. Comprehensive molecular characterization of pheochromocytoma and paraganglioma. Cancer Cell 2017;31:181-93.
81. Yang C, Zhuang Z, Fliedner SM, et al. Germ-line PHD1 and PHD2 mutations detected in patients with pheochromocytoma/paraganglioma-polycythemia. J Mol Med (Berl) 2015;93:93-104.
82. Astuti D, Ricketts CJ, Chowdhury R, et al. Mutation analysis of HIF prolyl hydroxylases (PHD/EGLN) in individuals with features of phaeochromocytoma and renal cell carcinoma susceptibility. Endocr Relat Cancer 2011;18:73-83.
83. Zhuang Z, Yang C, Lorenzo F, et al. Somatic HIF2A gain-of-function mutations in paraganglioma with polycythemia. N Engl J Med 2012;367:922-30.
84. Favier J, Buffet A, Gimenez-Roqueplo AP. HIF2A mutations in paraganglioma with polycythemia. N Engl J Med 2012;367:2161-2.
85. Pacak K, Jochmanova I, Prodanov T, et al. New syndrome of paraganglioma and somatostatinoma associated with polycythemia. J Clin Oncol 2013;31:1690-8.
86. Toyoda H, Hirayama J, Sugimoto Y, et al. Polycythemia and paraganglioma with a novel somatic HIF2A mutation in a male. Pediatrics 2014;133: e1787-91.
87. Zhuang Z, Yang C, Ryska A, et al. HIF2A gain-of-function mutations detected in duodenal gangliocytic paraganglioma. Endocr Relat Cancer 2016;23:L13-6.
88. Okubo Y, Yoshioka E, Suzuki M, et al. Diagnosis, Pathological findings, and clinical management of gangliocytic paraganglioma: a systematic review. Front Oncol 2018;8:291.

11 NEUROBLASTOMA, GANGLIONEUROBLASTOMA, AND GANGLIONEUROMA

NEUROBLASTOMA AND GANGLIONEUROBLASTOMA

Neuroblastoma (NBL) is a primitive neoplasm of neuroectodermal origin that arises in anatomic sites corresponding to the distribution of the sympathoadrenal autonomic nervous system. It is subdivided into three subtypes: undifferentiated, poorly differentiated, and differentiating, on the basis of specific histopathologic criteria (see below in Microscopic Findings). *Ganglioneuroblastoma (GNBL)* is a closely related neoplasm showing variable cytodifferentiation into ganglion cells, accompanied by a spindle cell schwannian stroma. There are two forms of GNBL: intermixed and nodular, which are clinically relevant to distinguish, since the latter carries a poorer prognosis (1,2). The term *neuroblastic* or *peripheral neuroblastic tumor* encompasses the full morphologic range of NBL and GNBL, including the histologically fully mature and clinically benign *ganglioneuroma (GN)*.

The origin of the adrenal medulla, containing chromaffin cells, and the sympathetic autonomic nervous system, composed of sympathetic neurons, has long been studied and debated. Chromaffin cells were thought to develop from a neural crest precursor cell that migrated to the dorsal aortic area and also gave rise to sympathetic neurons (3,4). However, this concept has now been challenged by several studies that have shown that progenitor cells have partly overlapping but distinct expression profiles (5–8). It appears now that chromaffin cells derive from peripheral glial stem cells, known as Schwann cell progenitor cells, which have limited capability for expansion and migrate along preganglionic nerves to reach their adrenal medulla destination. In contrast, sympathetic neurons are still thought to originate from neural crest progenitor cells (9,10).

While these new findings are supported by cell lineage analysis and by single cell RNA sequencing experiments, it has been shown recently that the divergence is not absolute and that Schwann cell precursors appear to be responsible for a small, but relevant, fraction of sympathetic neurons; additionally, a proportion of chromaffin cells may derive from migrating neural crest cells (10–12). Given the fact that NBLs (and by inference GNBL and GN as well) develop both in the adrenal gland (medulla) and along the cervical, thoracic, and abdominal sympathetic nervous system, both cell types, i.e., Schwann cell progenitor cells and neural crest progenitor cells, are potentially implicated as cells of origin.

Epidemiology

NBL and GNBL are the most common extracranial solid tumors in infancy and childhood, with an incidence of 9.7 cases per million in the entire pediatric age group in White populations and a slightly lower incidence in Black individuals (6.8 cases per million). There is a slight male predominance (13). In the first year of life, they together comprise the most common malignancy, with an incidence of 25 to 50 per million individuals (14).

Most (approximately 90 percent) of these tumors arise in patients under the age of 10 years and there is a gradual decline in incidence with increasing age (fig. 11-1). Nevertheless, NBL has been described in adolescents and even in young adults, accounting for less than 5 percent of all NBL cases (15). Fewer than 100 adult-onset cases have been published, and most of these have a poor outcome, usually due to late detection of the disease (16).

The incidence has remained stable over the past decades (with the exception of countries that introduced and subsequently abandoned

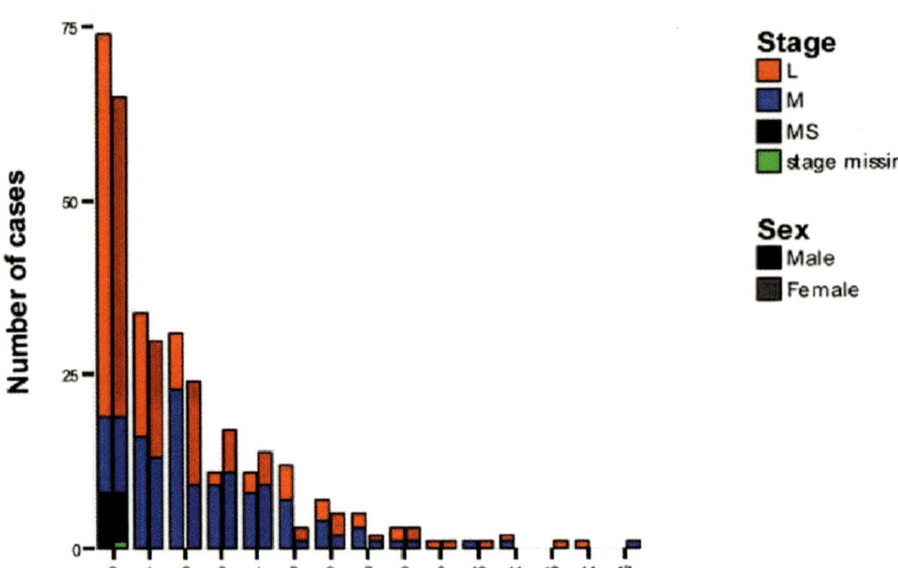

Figure 11-1

DISTRIBUTION OF NEUROBLASTOMA BY SEX AND STAGE OVER AGE

Nearly 90 percent of patients are below 5 years of age at diagnosis. The left side of each pair of columns represents male patients and the right side of each pair of columns represents female patients. (Fig. 5 from Träger C. Neuroblastoma incidence, biology and outcome: with special emphasis on quantitative analysis of tyrosine hydroxylase mRNA in blood and bone marrow. Stockholm: Karolinska Institute; 2009:27.)

Table 11-1

SIGNS AND SYMPTOMS OF NEUROBLASTOMA

General
Abdominal distension/mass effect
(Bone) pain
Hypertension
Failure to thrive
Fever
Weight loss
Anemia/thrombocytopenia
Specific
Horner syndrome
Watery diarrhea due to VIP[a] overproduction
Opsoclonus/myoclonus syndrome
Heterochromia iridis
Central hypoventilation syndrome (Ondine curse)
Blueberry muffin baby

[a]VIP = vasoactive intestinal protein.

national screening programs (see under Spontaneous Regression and Maturation) (7–19). Some ethnic and racial minority groups in the United States have a lower NBL incidence than the White population, in parallel with the lower incidence in certain areas in Africa (20,21). NBL accounts for up to 8 percent of childhood malignancies, but is responsible for 15 percent of pediatric cancer-related mortality, according to Surveillance, Epidemiology, and End Result (SEER) data (22).

Familial cases exist in NBL but are rare (1 to 2 percent of cases) (23). There are two genes that predispose for familial NBL: anaplastic lymphoma kinase (*ALK*), in which germline gain-of-function mutations occur, and paired-like homeobox 2B (*PHOX2B*), in which germline loss-of-function mutations have been described and in which patients may also have central congenital hypoventilation syndrome (Ondine curse) and/or Hirschsprung disease (24–27). Apart from germline mutations, somatic *ALK* mutations and amplifications have been found in up to 15 percent of sporadically occurring NBLs (28–30). Genome-wide association studies have shown that NBL is a very complex genetic disease with a large number of germline-based genetic abnormalities, of which at least 12 associations have currently been validated (31). Each of these has a modest effect on NBL development, but they may cooperate to yield a higher tumor risk. Thus far, no algorithms with regard to screening or counseling for tumor risk have been developed.

Clinical Features

Clinical signs and symptoms are linked to the primary site of the tumor, to size and extent of disease, including metastatic sites, and to a limited number of special situations that will be described below (Table 11-1). Localized disease frequently presents as an incidental finding, but large tumors may become apparent due to mass effect (abdominal distension) or pain. Hypertension may develop due to hormonal

Site	Age at Diagnosis		Total (%)
	≤1 Year (%)	>1 Year (%)	
Cervical	4	0.5	1
Thoracic	29	14	19
Abdominal			
Adrenal	25	40	35
Nonadrenal	26	32	30
Pelvic	3	2.5	2
Other	13	9	12
Unknown	0	2	1

Figure 11-2

ANATOMIC DISTRIBUTION OF NEUROBLASTOMA

The adrenal gland is the most frequent primary location of neuroblastoma (NB), followed by other locations in the abdomen. (Table 56-1 from Brodeur GM, Castleberry RP. Neuroblastoma. In: Pizzo PA, Poplack DG, eds. Principles and practice of pediatric oncology, 3rd ed. Philadelphia: Lippincott-Raven; 1997:761-97.)

overproduction or local compression of the renal artery through tumor growth, but appears limited to a minority (4 percent) of patients (32). In case of (extensive) metastasis, infants and children may show constitutional symptoms such as failure to thrive, fever, or weight loss. When there is extensive skeletal localization, bone pain and the consequences of impaired bone marrow function, such as anemia and thrombocytopenia, may occur.

Patients with a primary tumor or tumor localization in the head and neck area may present with Horner syndrome due to involvement of the superior cervical ganglion. This syndrome consists of a combination of one or more of the following: ptosis, miosis, enophthalmos, and anhidrosis.

Another presentation of NBL and GNBL is through a watery diarrhea syndrome, due to overproduction of vasoactive intestinal peptide (VIP). These symptoms may present before the diagnosis of NBL has been made, and in such cases, it mostly concerns tumors with a favorable prognosis. However, a second group of cases has also been described, in which watery diarrhea occurs after chemotherapeutic treatment of NBL, so-called secondary "VIP-omas," which, in contrast to the first group, represent high-stage disease with a worse prognosis (32).

Opsoclonus/myoclonus ("dancing eyes, dancing feet") syndrome (OMS) is a rare presentation in NBL and GNBL patients, usually associated with a favorable prognosis and regarded as a paraneoplastic syndrome related to an autoimmune response to the cerebellum. Only 2 to 3 percent of NBL patients are affected, but conversely, NBL or GNBL is diagnosed in 50 to 80 percent of children with OMS (33). Finally, heterochromia iridis is a rare, but potentially important sign associated with cervical or mediastinal NBL or GNBL. The heterochromia results from prenatal or postnatal interruption of sympathetic tracts that apparently mediate pigmentation of the iris (34).

Anatomic Distribution

Primary NBL and GNBL develop from sympathetic ganglia that are located on either side of the spinal cord and from the adrenal medulla. This implies a wide range of locations, from the cervical region, through the thorax (posterior mediastinum) and abdomen (including retroperitoneum), down to the pelvic area. Approximately 40 to 50 percent of primary tumors arise from the adrenal medulla, and they may be bilateral in rare instances (less than 1 percent).

The primary site of NBL and GNBL (fig. 11-2) is related to survival, with adrenal gland primaries, particularly, carrying a worse prognosis (35). When tumors arise in the paraspinal sympathetic ganglia, they may extend intraspinally and compress the spinal cord, resulting in neurologic impairment (fig. 11-3). Such so-called dumbbell NBLs, as assessed radiologically, occur

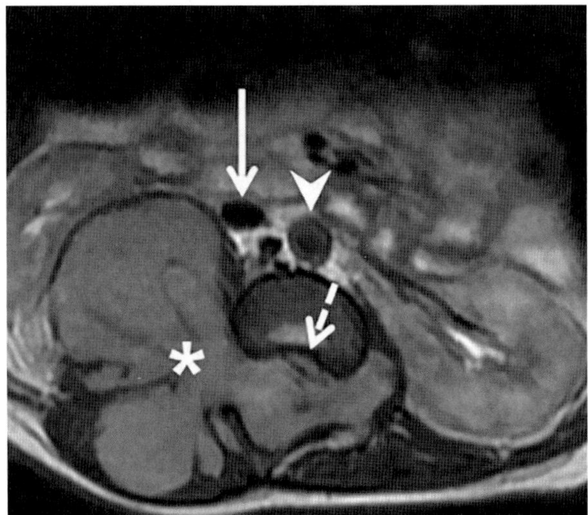

Figure 11-3

PARAVERTEBRAL DUMB-BELL NEUROBLASTOMA CAUSING SPINAL CORD COMPRESSION AND NEUROLOGIC IMPAIRMENT

Axial T2-weighted magnetic resonance image (MRI) at diagnosis. The primary tumor (*) is centered on the right paravertebral chain. It invades the psoas and spinal muscles and fills the spinal canal, compressing the spinal cord (dotted arrow). The tumor is totally separated from the inferior vena cava (arrow) and the aorta (arrowhead). (Fig. 3C from Brisse HJ, Blanc T, Schleiermacher G, et al. Radiogenomics of neuroblastomas: relationships between imaging phenotypes, tumor genomic profile and survival. PLoS One 2017;12:e0185190.)

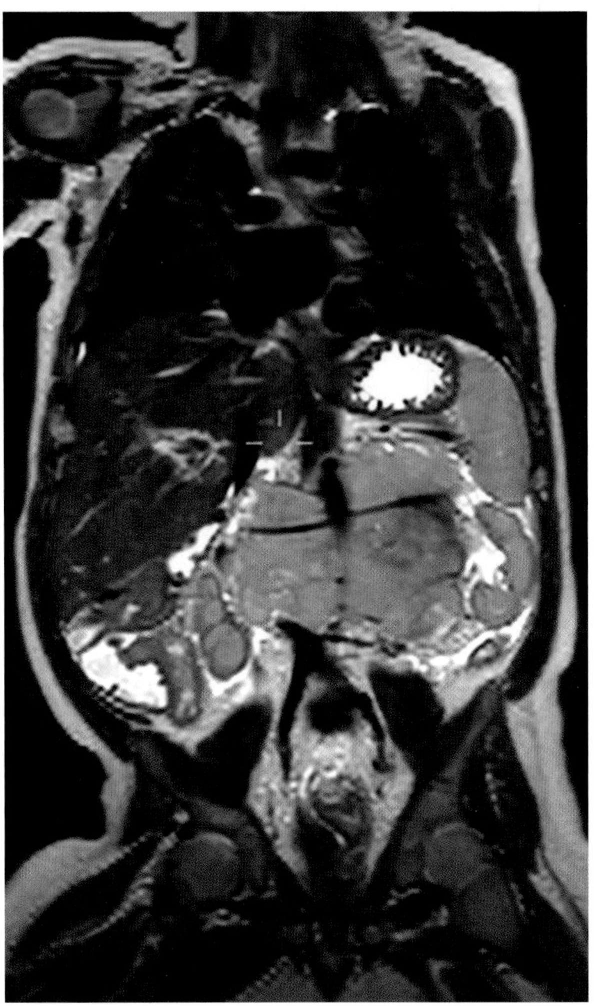

Figure 11-4

MRI SCAN OF METASTASIZED ABDOMINAL NEUROBLASTOMA

in 10 percent of abdominal NBLs, 28 percent of thoracic tumors, and 25 percent of cervical tumors (36).

Metastatic locations of NBL and GNBL have a wide range, but are frequently seen in bone marrow, bone, lymph nodes, and liver (figs. 11-4, 11-5). Multiple skin manifestations of NBL may result in the clinical picture of a "blueberry muffin baby." Less frequent locations are skin, gonads, and cranial locations such leptomeninges and dura (37). Metastases to the brain parenchyma are rare but do occur, especially during progression or relapse of stage 4 NBL (38). Placental localization has been reported in cases with congenital manifestation of NBL, but tumor cells are limited to the vascular spaces and do not form metastatic deposits (39).

Staging and Risk Classification

Various staging systems for NBL have been developed. The International Neuroblastoma Staging System (INSS) was used between 1990 and 2010, and was based on the extent of surgical excision at diagnosis and the extent of disease (Table 11-2) (40). This staging system was problematic because it did not allow pretreatment classification.

The International Neuroblastoma Risk Group Staging System (INRGSS), based on image-defined risk factors (IDRF), is now widely used (Table 11-3) (41). There are four INRGSS stages: L1, L2, M, and MS. L1 is reserved for local disease, without the presence of any IDRFs, whereas in L2 one or more IDRFs are present. M represents all instances of metastatic disease with the exclusion of MS, which overlaps with the former stage 4S. Stage MS is for children under 18 months of age that have limited metastatic disease. It is now recognized that these patients have a favorable prognosis when they can be grouped in the low- or intermediate-risk groups on the basis of age and molecular characteristics, although they may also fall into the high-risk group, with a consequently worse prognosis (42).

The INRG task force used data from over 8,000 children with NBL, with up to 35 risk factors, to come up with a pretreatment risk group stratification (42b). The most significant of these risk factors are age, histology, differentiation grade, and results of molecular analysis (including MYCN status, 11q status, and ploidy) (Table 11-3). The risk grouping seen in this Table forms the basis for treatment plans and clinical trials (see Treatment). The rapid developments with respect to molecular techniques and data will probably lead to further refinement of risk groups in the near future. One example of this concerns a recent study that suggests that low- and high-grade NBLs may be distinguished on the basis of their transcriptomes, a finding that may be introduced in clinical practice, if confirmed (43).

Risk groups are designed according to event-free survival (EFS). Thus, very low risk corresponds to EFS over 85 percent, low risk to EFS of 75 to 85 percent, intermediate risk to EFS of 50 to 75 percent, and high risk has EFS below 50 percent. For each combination of risk factors, EFS and overall survival (OS) have been calculated in a regression tree in the original publication, yielding insight into the relative contribution of each risk factor (43,44).

Gross Findings

NBLs and GNBLs generally have a nonspecific gross morphology. Some tumors that are localized and not attached to vital (vascular) structures can be resected without preceding chemotherapy (fig. 11-6). However, NBLs have a tendency to envelop vessels and nerves, making complete resection difficult (fig. 11-7). In larger tumors it may be difficult or even impossible to determine the primary site of origin, for instance, when they arise in or near the adrenal gland.

NBLs and GNBLs are usually round to ovoid and of varying diameter. If they are located in

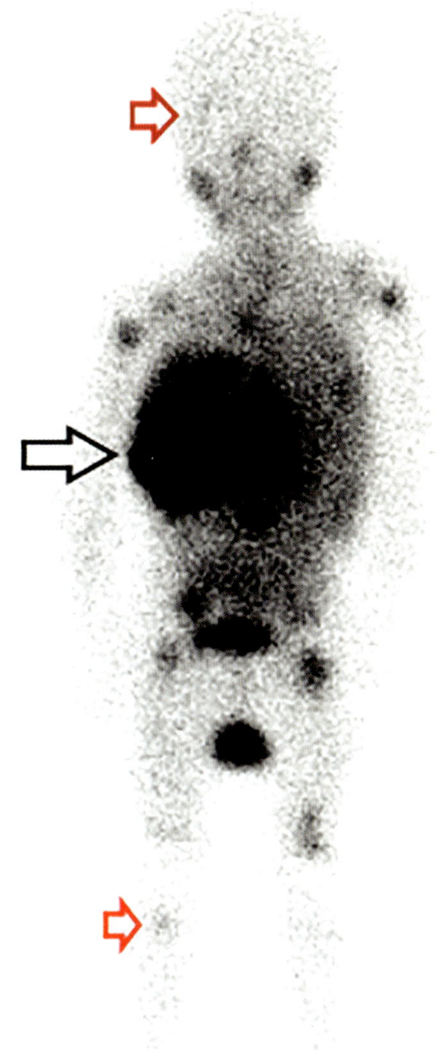

Figure 11-5

METAIODOBENZYLGUANIDINE (MIBG) SCAN OF WIDELY METASTASIZED NEUROBLASTOMA

A 2-year old girl presented with fever and a painless swelling in the abdomen. Blood and urine cultures were negative and initial ultrasound was suspicious of a renal tumor. After referral to an expert center, clinical examination (slight hypertension), laboratory testing (increased urinary catecholamines), and abdominal MRI imaging showed that a diagnosis of metastatic NB was more likely. The MRI depicted a right-sided adrenal mass with retroperitoneal lymph node metastases. Additional MIBG scintigraphy shows increased uptake in the primary adrenal NB and locoregional lymph nodes (black arrow), but also extensive bone marrow metastases throughout. At the top, is a classic orbital metastasis and at the bottom, a tibial metastasis (red arrows). Histopathology showed an undifferentiated NB, with *MYCN* amplification, indicative of a high-risk NB patient. (Courtesy of Dr. A. J. Braat, University Medical Center Utrecht, The Netherlands)

Table 11-2

THE INTERNATIONAL NEUROBLASTOMA STAGING SYSTEM (INSS)[a]

Stage 1	Localized tumor with complete gross excision, with or without microscopic residual disease; representative ipsilateral lymph nodes negative for tumor microscopically (nodes atttached and removed with the primary tumor may be positive)
Stage 2A	Localized tumor with incomplete gross excision, representative ipsilateral nonadherent lymph nodes negative for tumor microscopically
Stage 2B	Localized tumor with or without complete gross excision, with ipsilateral nonadherent lymph nodes positive for tumor. Enlarged contralateral lymph nodes must be negative microscopically
Stage 3	Unresectable unilateral tumor infiltrating across the midline, with or without regional lymph node involvement; or localized unilateral tumor with contralateral regional lymph node involvement; or midline tumor with bilateral extension by infiltration (unresectable) or by lymph node involvement
Stage 4	Any primary tymor with dissemination to distant lymph nodes, bone, bone marrow, liver, skin, and/or other organs (except as defined for stage 4S)
Stage 4S	Localized primary tumor (as defined for stage 1, 2A, or 2B), with dissemination limited to skin, liver, and/or bone marrow (limited to infants <1 year of age)

[a]Modified table from Brodeur GM, Pritchard J, Berthold F, et al. Revisions of the international criteria for neuroblastoma diagnosis, staging, and response to treatment. J Clin Oncol 1993;11:1466-77.

Table 11-3

THE INTERNATIONAL NEUROBLASTOMA RISK GROUP STAGING SYSTEM (INRGSS)[a]

INRG Stage	Age (months)	Histologic Category	Grade of Tumor Differentiation	MYCN	11q Aberration	Ploidy	Pretreatment Risk Group
L1/L2		GN maturing; GNB intermixed					A very low
L1		Any, except GN maturing of GNB intermixed		NA			B very low
				Amp			K high
L2	<18	Any, except GN maturing of GNB intermixed		NA	No		D low
					Yes		G intermediate
	>18	GNB nodular; neuroblastoma	Differentiating	NA	No		E low
					Yes		H intermediate
			Poorly differentiated or undifferentiated	NA			
				Amp		Hyperploid	N high
M	<18			NA		Diploid	F low
	<12			NA		Diploid	I Intermediate
	12 to <18			NA			J Intermediate
	<18			Amp			O High
	≥18						P High
MS	<18			NA	No		C Very low
					Yes		Q High
				Amp			R High

[a]Figure 2 from Cohn SL, Pearson AD, London WB, et al. The International Neuroblastoma Risk Group (INRG) classification system: an INRG task force report. J Clin Oncol 2008;27:295.

the adrenal gland, preexistent adrenal cortex may be recognized by the bright yellow color. Otherwise, they are usually pale white to brown on the outside. On cut surface, they may be gray-white to tan and mostly firm, sometimes with a vaguely nodular aspect. Consistency depends on the composition of the tumor, with mature schwannian stroma of GNBL intermixed yielding a firm texture (fig. 11-8).

Hemorrhagic areas, necrosis, and calcification are frequently present in NBLs, even in the absence of preceding chemotherapy (fig. 11-9).

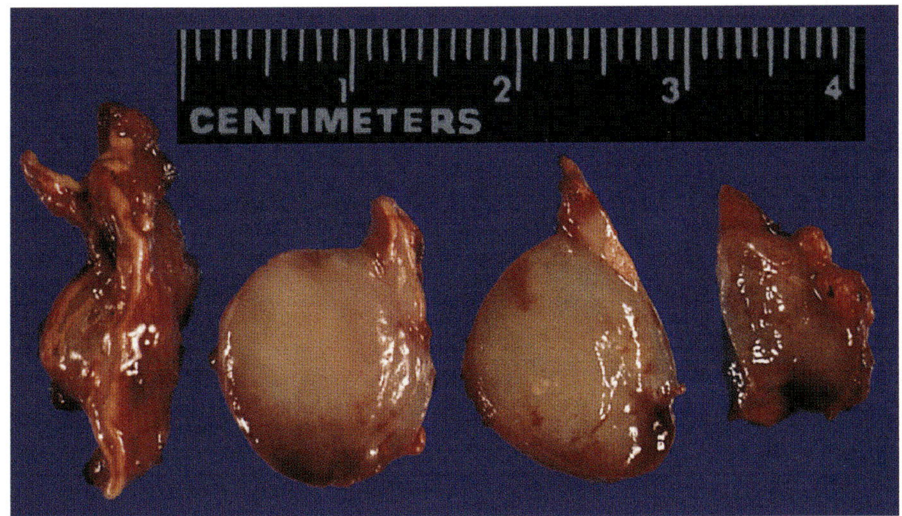

Figure 11-6

RESECTED NEUROBLASTOMA

Gross image of sliced resection specimen of NB without previous chemotherapy treatment. Preexistent adrenal gland in left slice and possible calcifications in second slice from the right.

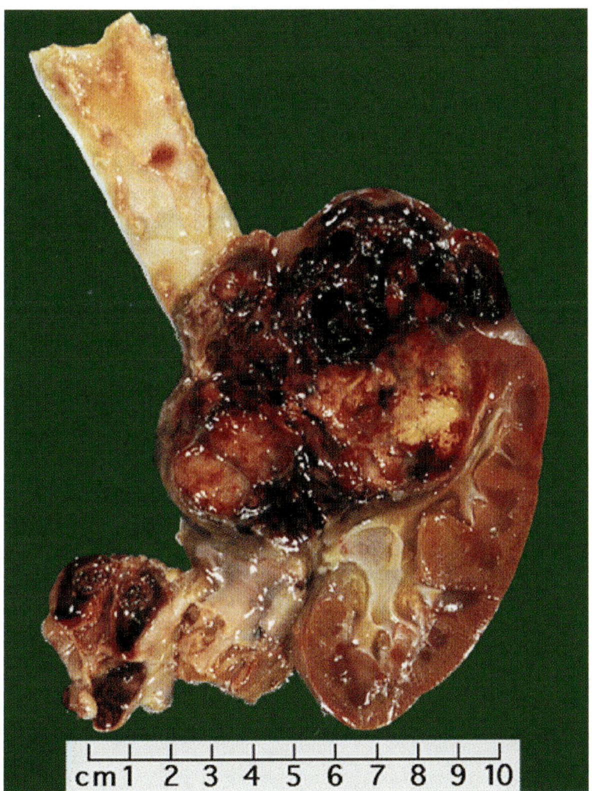

Figure 11-7

HEMORRHAGIC NEUROBLASTOMA

NB of the left adrenal gland at autopsy. The tumor replaced the entire adrenal gland and invaded the renal parenchyma. Coarse lobulations are seen on cross section, with bulging nodules showing irregular areas of hemorrhage. The tumor had a soft, almost encephaloid, texture. Several regional lymph nodes are involved by metastases.

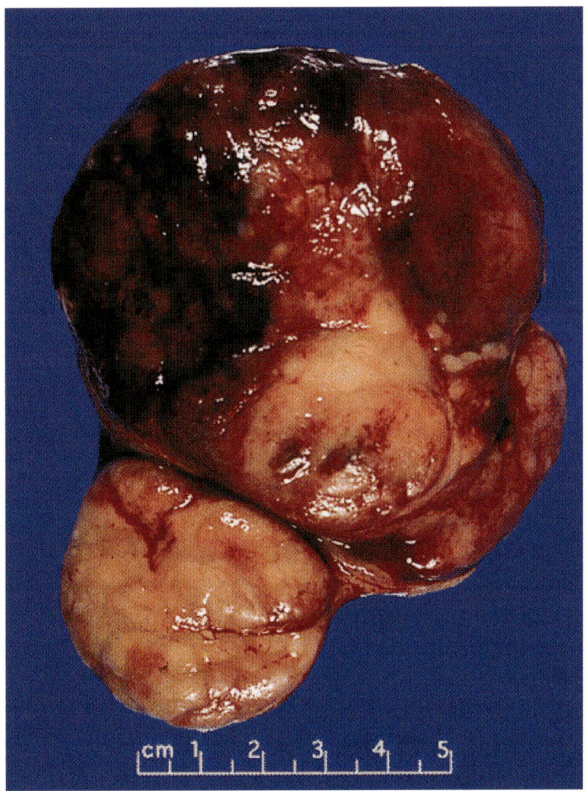

Figure 11-8

STROMA-POOR NEUROBLASTOMA

On cross section, a variation from hemorrhagic, bulging, irregular lobules to more pale, homogeneous foci is seen. Because of ganglion cell differentiation, this tumor in older terminology would be classified as a ganglioneuroblastoma (GNB).

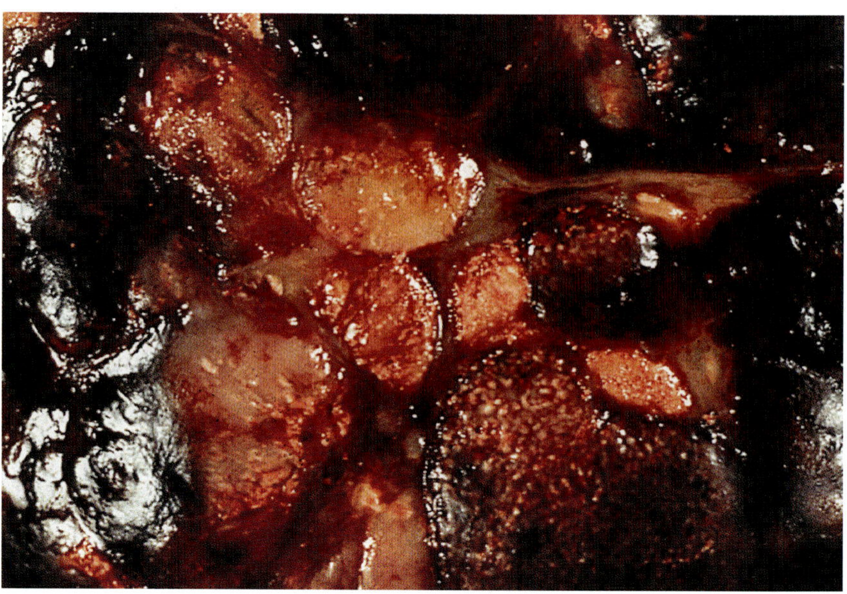

Figure 11-9

STROMA-POOR NEUROBLASTOMA

In the upper abdomen in a teenage patient, there are bulging nodules of congested and hemorrhagic tumor. Punctate yellowish areas represent dystrophic calcification.

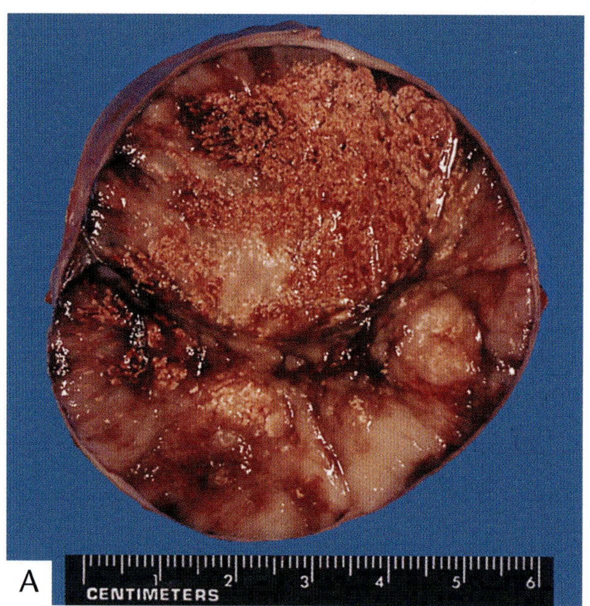

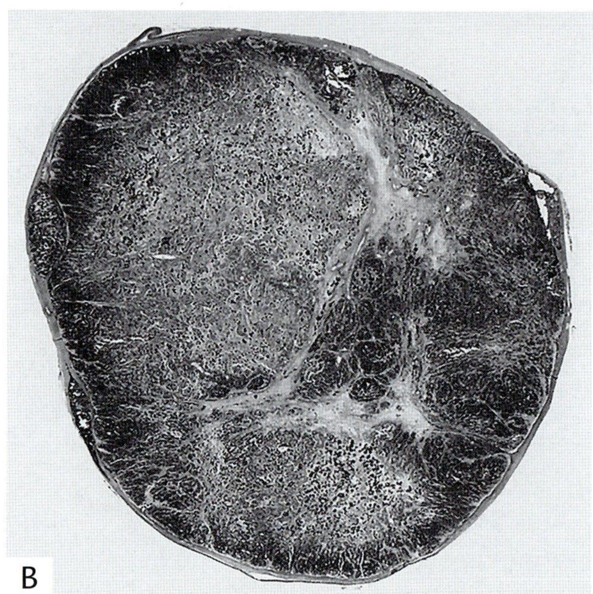

Figure 11-10

ADRENAL NEUROBLASTOMA

A: Cross-section of a 174-g adrenal NB from a 5-month-old boy. There is extensive calcification within the tumor, which appears as yellowish coarse granular areas.

B: Specimen radiograph with identical orientation shows extensive foci of dystrophic calcification.

Calcifications may render some tumors difficult to section or there may be a gritty sensation upon slicing (fig. 11-10).

In many instances, primary resection cannot be achieved, and resection or debulking is performed following chemotherapy. Such resections frequently arrive piecemeal in the pathology laboratory and resection margins are not an issue with regard to pathology reporting. Instead, the amount of chemotherapy effect should be assessed upon gross morphology, taking into account areas of hemorrhage, cystic

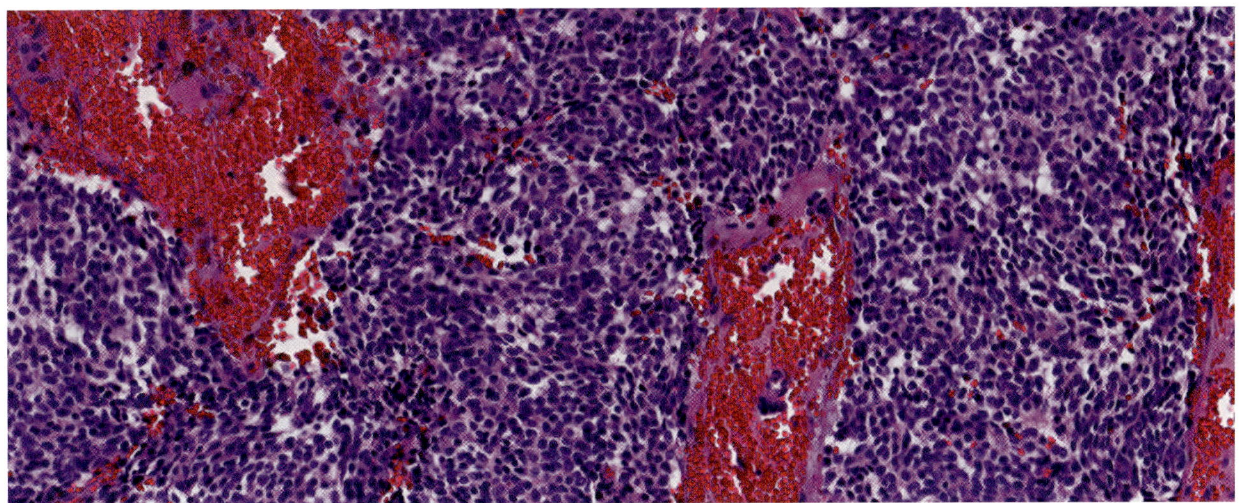

Figure 11-11

SMALL BLUE ROUND CELL TUMOR

Immunohistochemically, this tumor was shown to be undifferentiated NB. The cellular tumor is composed of undifferentiated cells with atypical nuclei and scant cytoplasm. No stroma or ganglion cell differentiation is present.

degeneration, calcification, and necrosis. Apart from these secondary changes, the features are similar to untreated NBL or GNBL.

Sampling needs to be generous, to allow comparison between the gross and microscopic findings and to allow sufficient representation of the different components of NBL and GNBL (see below). When there are grossly recognizable nodules, all of these need to be sampled, to allow the detection of the GNBL nodular category, which carries a worse prognosis than its GNBL intermixed counterpart. In general, at least 1 block per centimeter tumor diameter needs to be examined microscopically.

An unusual gross appearance may be encountered with congenital NBLs, which may present as unilocular adrenal cysts that must be distinguished from benign cystic lesions, including hemorrhagic pseudocysts (45). These cystic tumors may be filled with serous or hemorrhagic content and have only a thin patchy layer of neoplastic cells on the cyst wall.

Microscopic Findings

NBLs and GNBLs consist of two components: immature or maturing neurons and stroma. The former compartment consists of neuroblasts or neuroblastic cells that may exhibit varying degrees of differentiation. The least differentiated cell, the neuroblast, has a relatively small, hyperchromatic, round nucleus with little pleomorphism and scant cytoplasm. Sometimes, the dispersed chromatin gives the impression of stippling or speckling of the nucleus, and rarely, nuclear inclusions are described. If the tumor is entirely composed of such cells, it falls into the category of a small blue round cell tumor for which additional immunohistochemistry is necessary to allow a definitive diagnosis (figs. 11-11, 11-12).

Frequently, however, part of the tumor cell population shows signs of differentiation in the direction of ganglion cells. These cells have ample eosinophilic cytoplasm and an eccentric nucleus with a prominent nucleolus. Many tumor cells in neuroblastic tumors show varying degrees of differentiation toward ganglion cells, exhibiting an increase in the amount of cytoplasm and nuclear size, and completely mature ganglion cells may be found next to less mature cells (figs. 11-13, 11-14).

Some NBLs show large cell histology, anaplasia, or foci of spindle-shaped cells, which may resemble those of other childhood tumors, such as rhabdomyosarcoma (figs. 11-15–11-17) (46,47). No clinical or prognostic correlations of such alternative histology are known, although two large cell NBLs were described in combination with *MYC* amplification and MYC protein

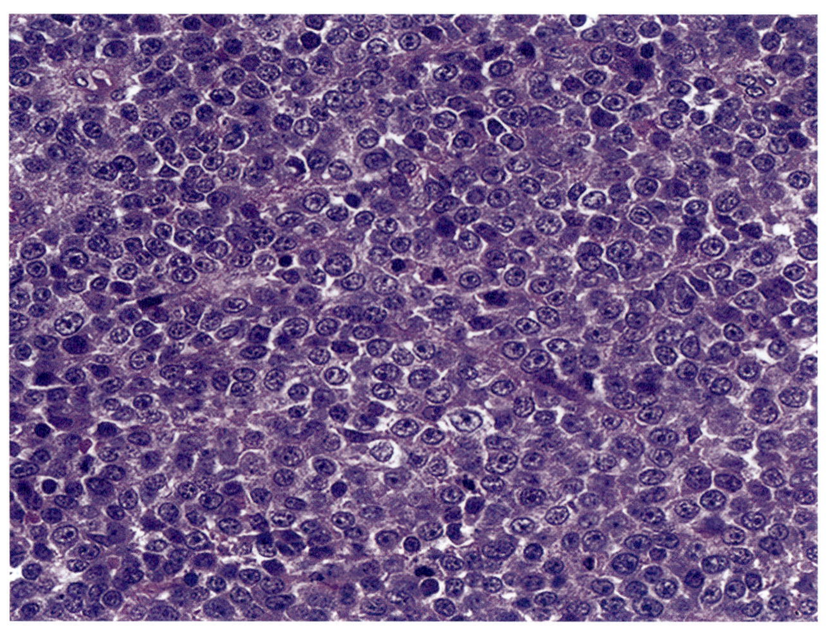

Figure 11-12

UNDIFFERENTIATED NEUROBLASTOMA

Closely packed, poorly differentiated neuroblasts form solid sheets. Patterns such as this may cause confusion with other childhood neoplasms such as malignant lymphoma. Most nuclei have a single, small nucleolus. The tumor had been fixed in B-5 fixative.

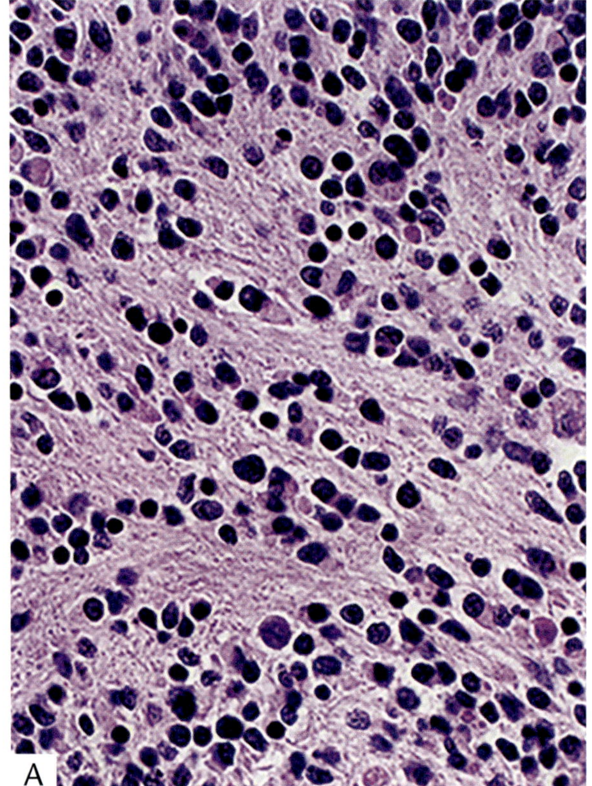

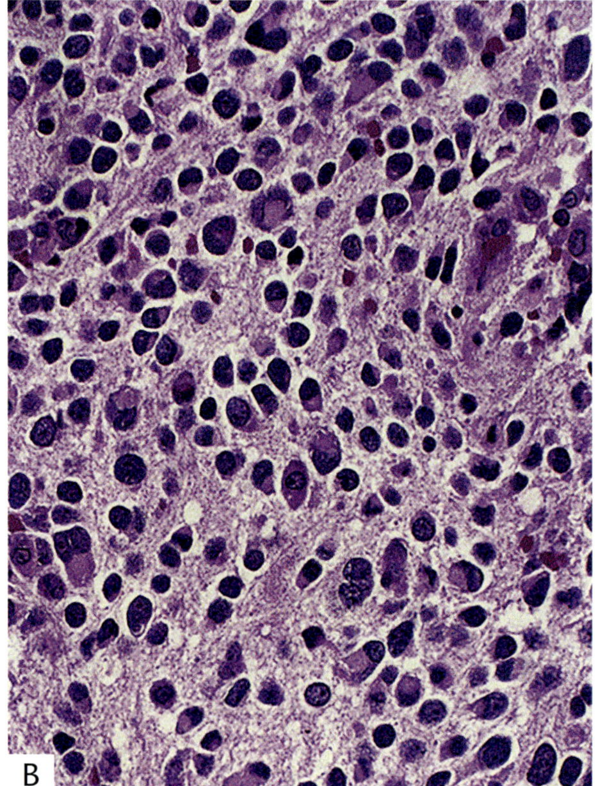

Figure 11-13

STROMA-POOR NEUROBLASTOMA

A: The tumor cells are separated by pale pink fibrillary material representing neuritic cell processes. The tumor was largely undifferentiated, with only rare cells showing early ganglion cell differentiation.

B: A different tumor with abundant, fairly immature ganglion cells.

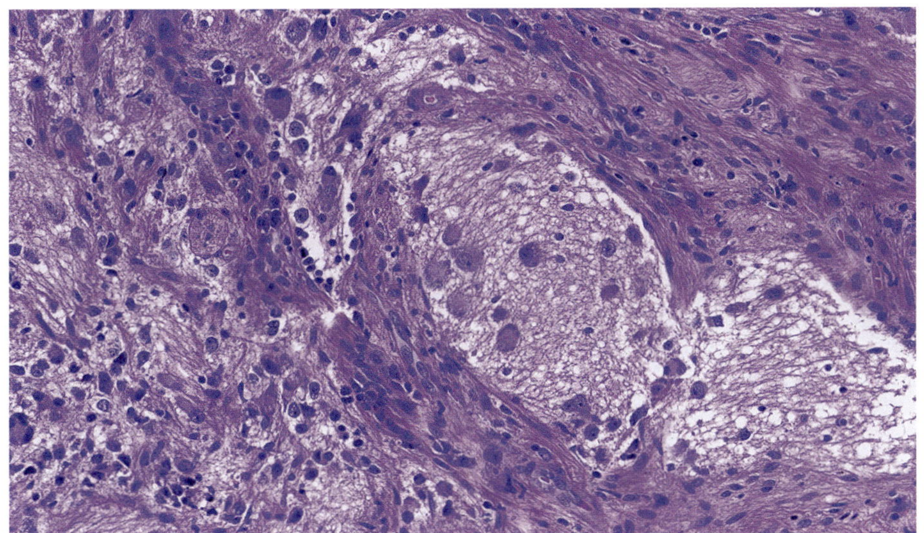

Figure 11-14

DIFFERENTIATING NEUROBLASTOMA

This tumor shows areas of ganglion cell differentiation, which are still in an immature neuropil matrix, flanked by areas of schwannian stroma (upper right), but also by areas of immature neuroblasts (lower left). When there is more than 5 percent ganglion cell differentiation and less than 50 percent schwannian stroma, tumors are classified as differentiating NB.

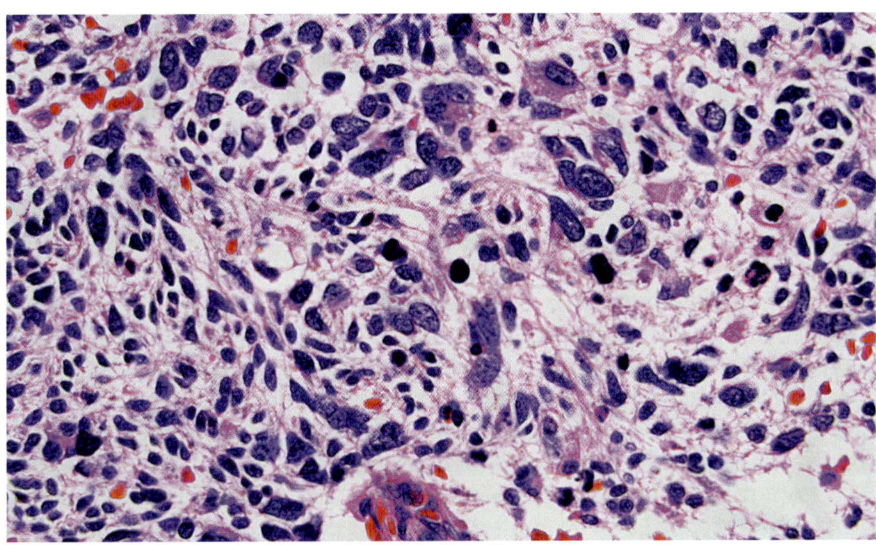

Figure 11-15

ANAPLASTIC NEUROBLASTOMA

The tumor cells have pleomorphic, hyperchromatic nuclei.

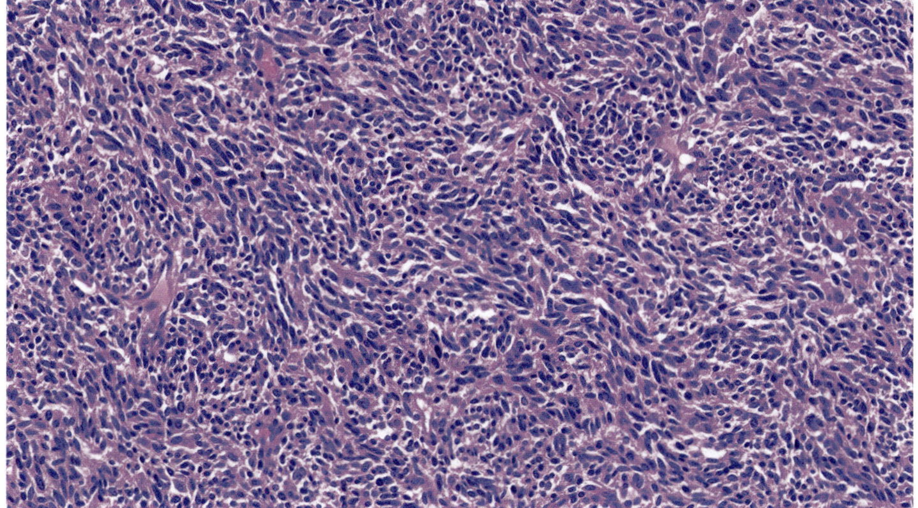

Figure 11-16

UNDIFFERENTIATED NEUROBLASTOMA WITH SPINDLE CELL DIFFERENTIATION

This stage 3 NB in a 5-year-old girl was not pretreated and showed this peculiar undifferentiated spindle cell morphology without presence of neuropil, ganglion cell differentiation, or schwannian stroma. Chromogranin A and synaptophysin were diffusely and strongly positive.

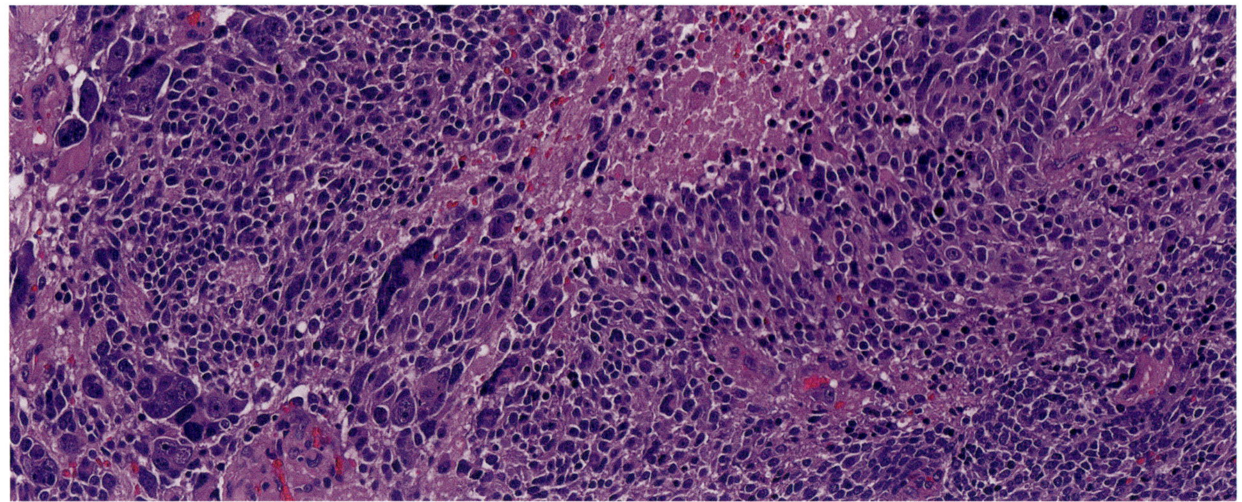

Figure 11-17

UNDIFFERENTIATED NEUROBLASTOMA WITH LARGE CELL HISTOLOGY

This tumor was originally diagnosed as malignant rhabdoid tumor on the basis of morphology and loss of BRG1 immunostaining, until resection showed foci of ganglion cell differentiation next to remaining undifferentiated areas and extensive necrosis. (Images are from the initial biopsy specimen.)

overexpression (but not *MYCN* amplification or protein overexpression), which is also rare in NBL (47).

Neuropil is recognized as slightly eosinophilic fibrillary material representing naked neuritic cell processes, and is present as smaller or larger areas between neuroblastic cells. Schwannian stroma is composed of fascicles of spindle cells (Schwann cells) that envelop neuritic processes of ganglion cells (fig. 11-18). In GNBL intermixed there may still be naked neuritic processes of neuroblasts, but in GN such processes from ganglion cells are covered by Schwann cell cytoplasm. Although Schwann cells are regarded as an integral part of NBL or GNBL, they are not neoplastic and may play a role in regulating the differentiation of the neoplastic neuronal component (48). The schwannian stroma usually forms paucicellular bundles, in which ganglion cells may be present (figs. 11-19, 11-20).

Architecturally, NBL and GNBL frequently show a nodular pattern of immature, maturing, or mature cells, intermingled with varying amounts of stroma. These nodules are separated by delicate fibrovascular septa. Within the nodules, the tumor cells may form Homer Wright rosettes, which are in fact pseudorosettes, because the central core is composed of neuropil, a vessel, or collagen, whereas true rosettes, so-called Flexner-Wintersteiner rosettes, have an empty central core (figs. 11-21, 11-22).

NBL and GNBL may present with calcification, necrosis, and hemorrhage, even in the absence of previous treatment, and sometimes inflammatory infiltrates with lymphocytes and plasma cells (figs. 11-23–11-26). The latter are usually easily distinguished from tumor cells, since lymphocyte nuclei are clearly smaller than those of neuroblasts. In case of doubt or when there is heavy inflammatory infiltration, immunohistochemical stains for both lymphoid cells (CD3, CD20) and for neuroblastic cells (see Immunohistochemical Findings) may be used.

The classification of neuroblastic tumors, the International Neuroblastoma Pathology Classification (INPC), was proposed almost two decades ago, and is, with minor modifications, still valid (49,50). The INPC is based on the relative amounts of mature and immature cellular and stromal components, and essentially distinguishes NBL, further subdivided into undifferentiated, poorly differentiated, and differentiating; GNBL nodular; GNBL intermixed; and GNBL, maturing and mature (Table 11-4).

In undifferentiated NBL there are only neuroblasts, without any ganglion cell differentiation,

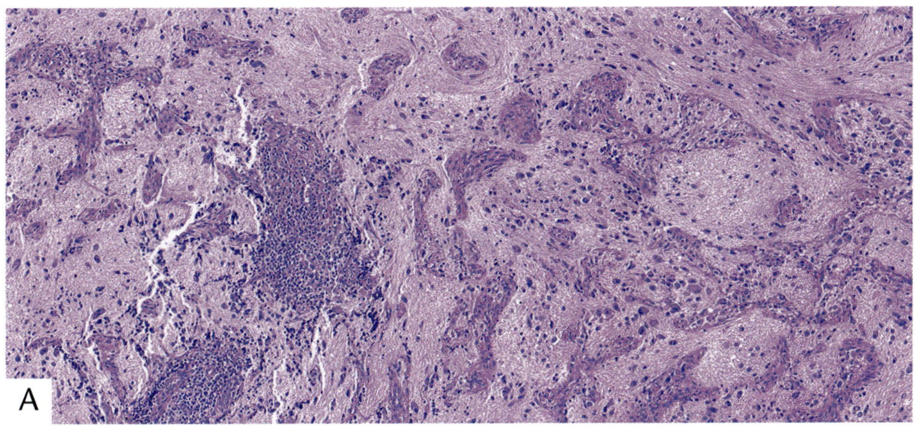

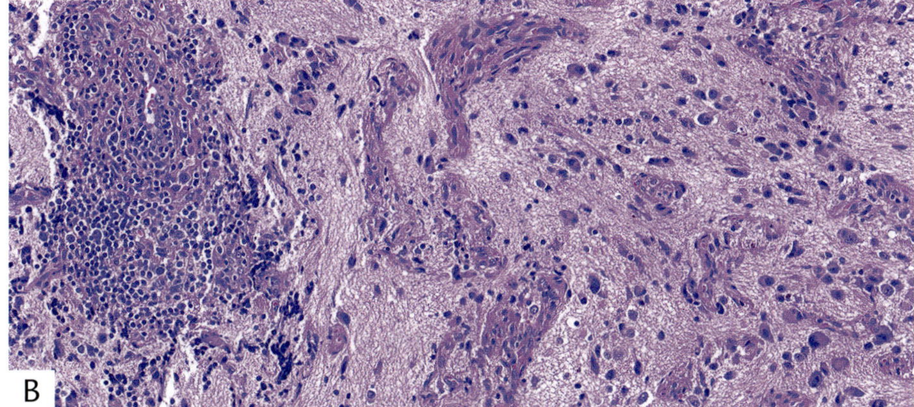

Figure 11-18

DIFFERENTIATING NEUROBLASTOMA

A: This 18-month-old boy with opsoclonus myoclonus syndrome had a paravertebral tumor, which was shown to be a neuroblastic tumor with more than 5 percent ganglion cell differentiation, classifying it as differentiating NB. Note the extensive amount of immature stroma (neuropil).

B: Higher magnification shows neuroblasts and ganglion cells of varying degrees of differentiation embedded in a fibrillary immature matrix (neuropil).

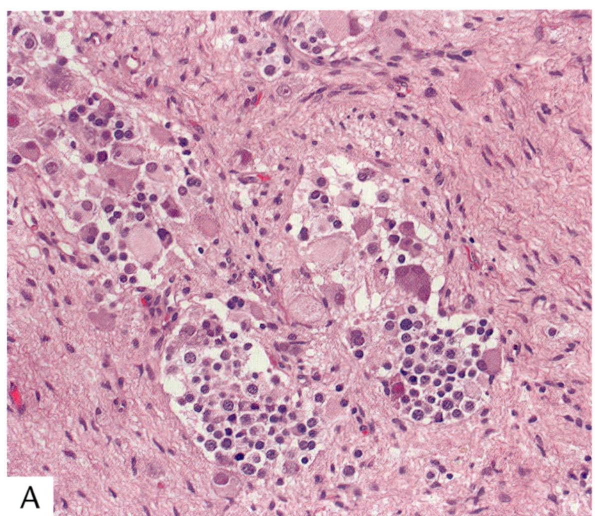

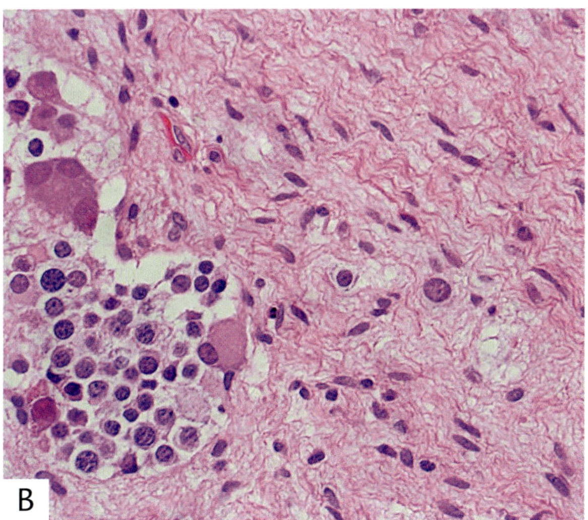

Figure 11-19

GANGLIONEUROBLASTOMA INTERMIXED

This tumor had more than 50 percent of schwannian stroma and may be called stroma rich.

A: There is a mixture of mature and immature elements with schwannian stroma to the lower left and upper right.

B: Focus of neuroblasts to the left, with some cells showing complete or partial ganglion cell differentiation. Schwann cell component seen in the upper right.

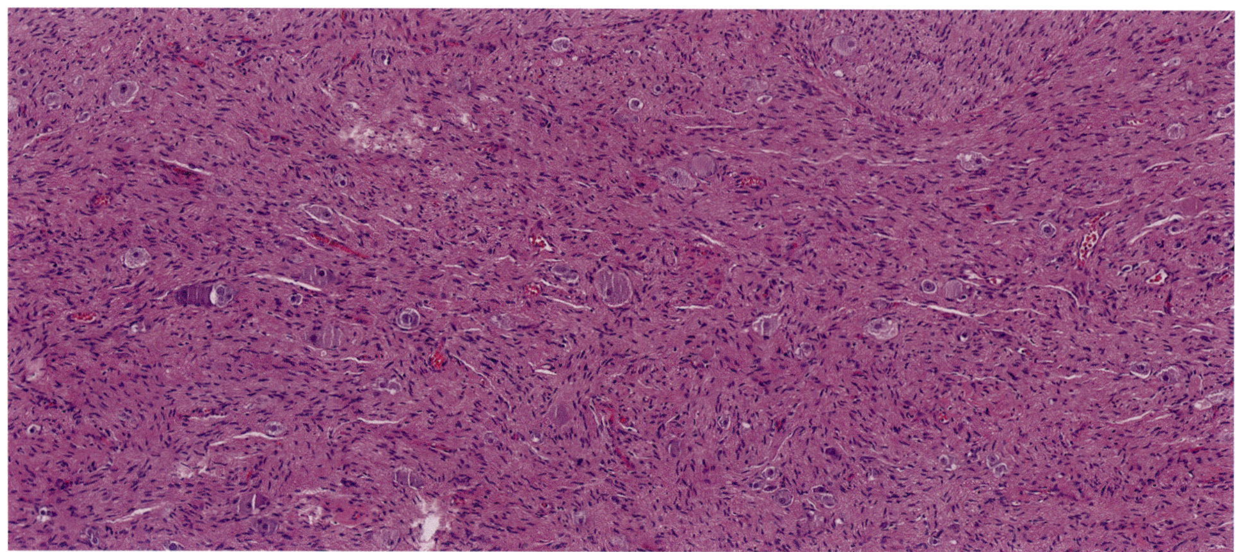

Figure 11-20

NEUROBLASTOMA FOLLOWING CHEMOTHERAPY

This resection specimen showed the typical heterogeneity of neuroblastic tumors with areas of immature cells and stroma (not shown), calcification, but also completely matured areas as shown in this image, with mature ganglion cells embedded in schwannian stroma.

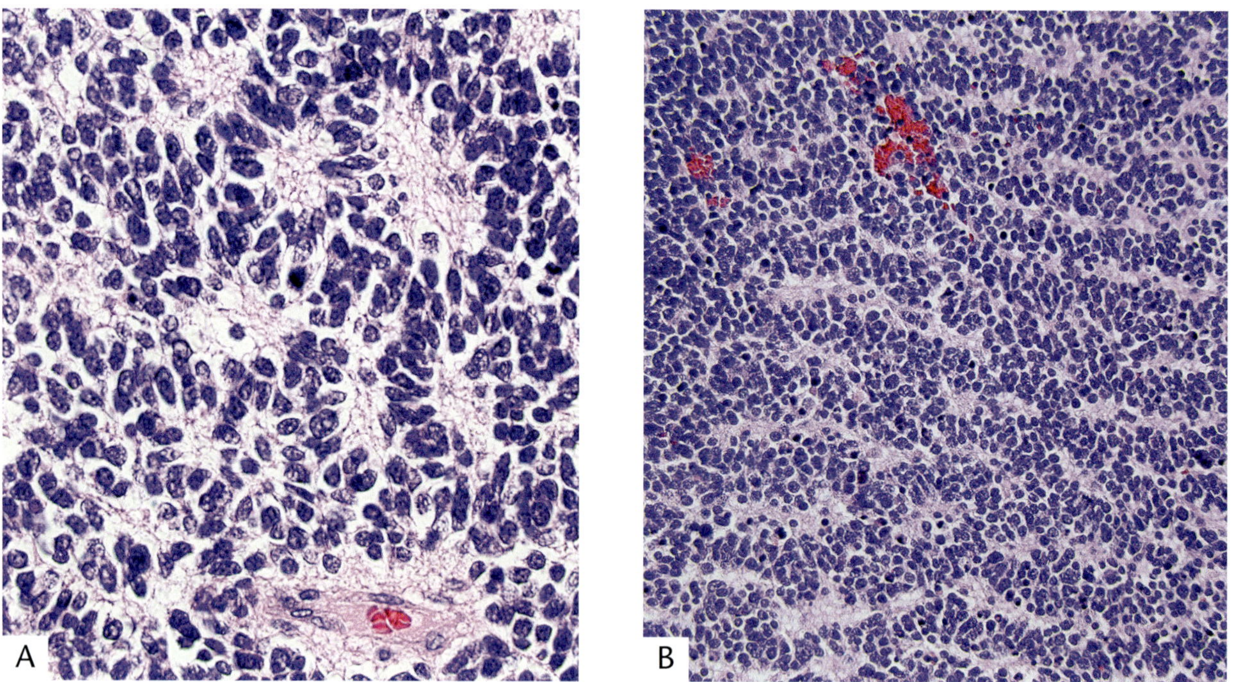

Figure 11-21

HOMER WRIGHT PSEUDOROSETTES IN NEUROBLASTOMA

A: Neuroblasts are grouped around a core of slightly eosinophilic fibrillary material, so-called neuropil, which are neuritic processes.

B: This kind of palisading of Homer Wright pseudorosettes is unusual in NB.

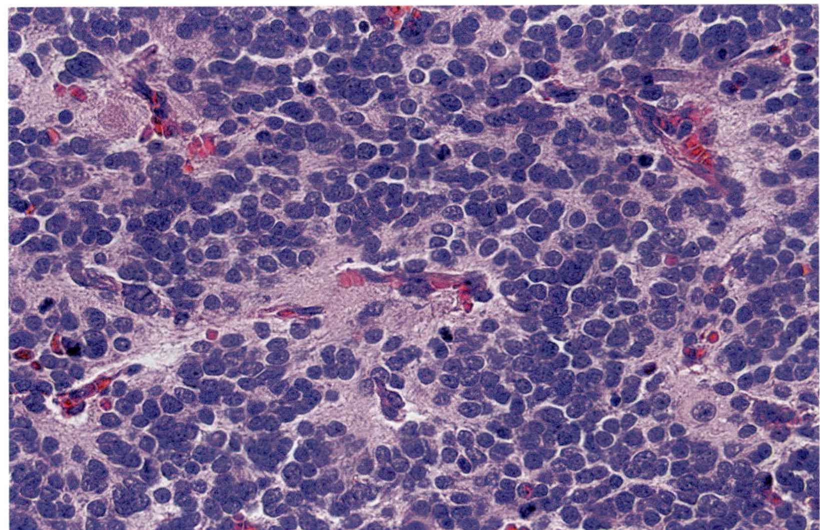

Figure 11-22

VASCULAR PSEUDOROSETTES IN NEUROBLASTOMA

Apart from Homer Wright pseudorosettes, vascular pseudorosettes are commonly found, sometimes as a result of palisading of neuroblastic tumor cells around vessels.

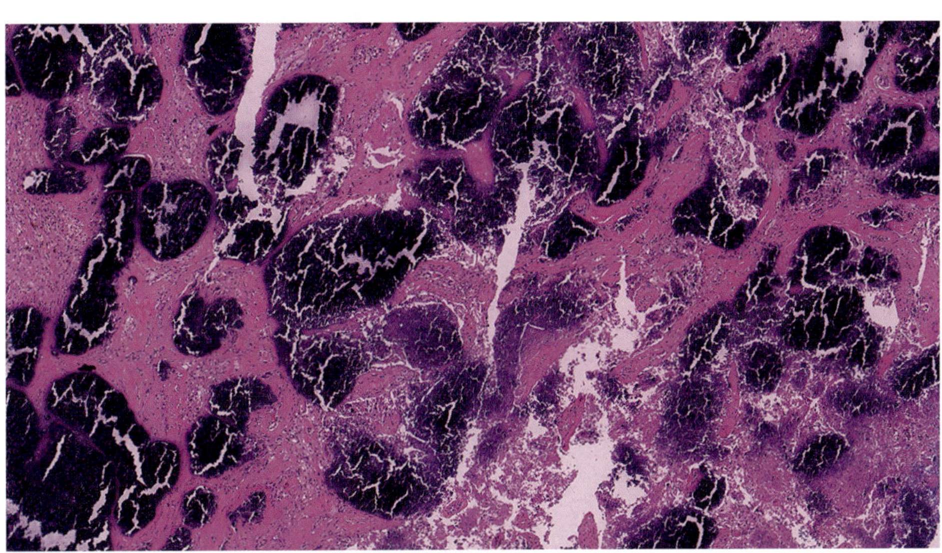

Figure 11-23

POST-TREATMENT CHANGE IN NEUROBLASTOMA

There is extensive calcification and fibrosis, without any vital tumor tissue present.

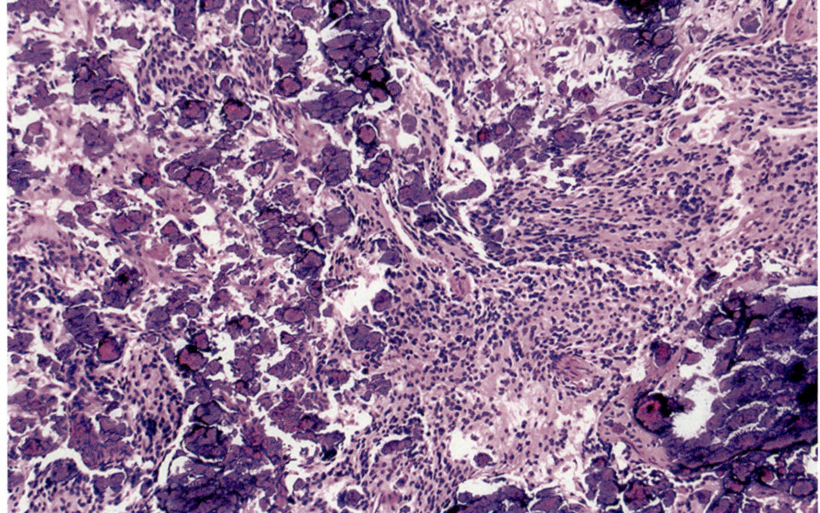

Figure 11-24

POST-TREATMENT CHANGE IN NEUROBLASTOMA

Vital neuroblastic tumor cells are seen between the areas of extensive calcification.

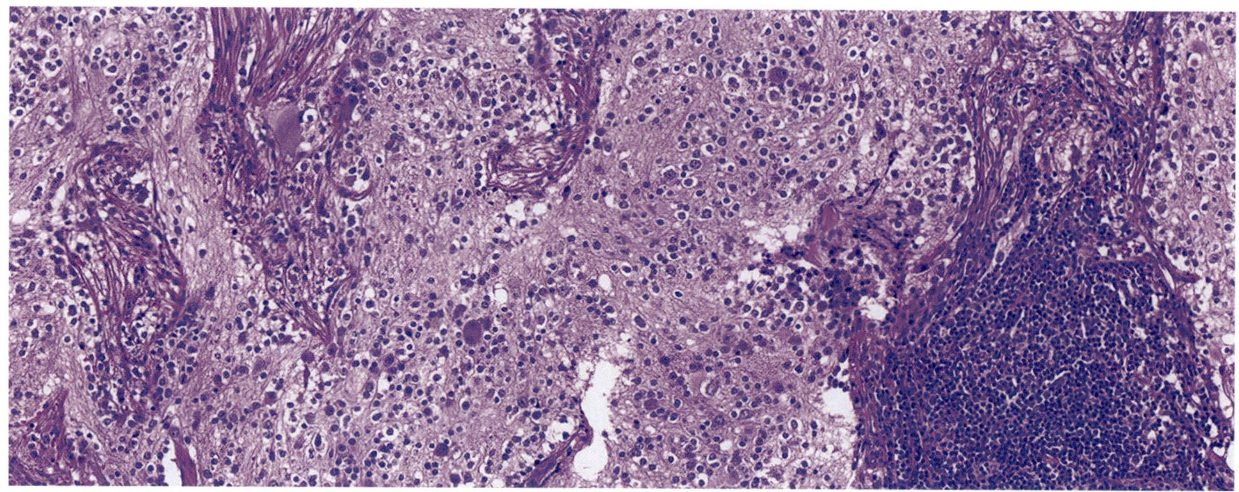

Figure 11-25

POST-TREATMENT CHANGE IN NEUROBLASTOMA

Resection of left adrenal gland in 7-year-old boy with pretreated NB shows limited chemotherapy effect, but frequent lymphocytic infiltrates, in addition to vital tumor tissue with immature stroma and neuroblasts.

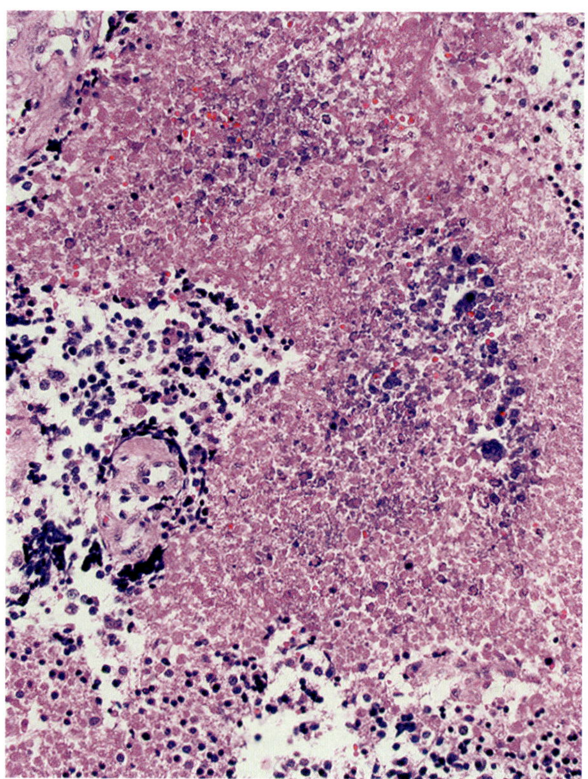

Figure 11-26

POST-TREATMENT CHANGE IN NEUROBLASTOMA

Largely necrotic NB, with focal calcification. The biologic potential of the remaining nucleated tumor cells is frequently difficult to estimate.

Table 11-4

NEUROBLASTOMA SUBCLASSIFICATION

Tumor Type	Subtype
Neuroblastoma	Undifferentiated Poorly differentiated Differentiating
Ganglioneuroblastoma intermixed	
Ganglioneuroblastoma nodular	
Ganglioneuroma	Maturing Mature

making this subtype the most difficult to diagnose, and often requiring immunohistochemistry (fig. 11-27). The tiniest amount of stroma may give a clue to the diagnosis, and although the definition of NBL allows up to 50 percent of schwannian stroma, frequently no mature stroma can be found. This is also the case for the other two subtypes of NBL, poorly differentiated and differentiating (figs. 11-28, 11-29). Both show some degree of ganglion cell differentiation, by definition less than 5 percent in the former and more than 5 percent in the latter. In addition, both contain neuropil, the stroma that characterizes NBL. GNBL contains more than 50 percent of schwannian stroma and can therefore be recognized easily on the

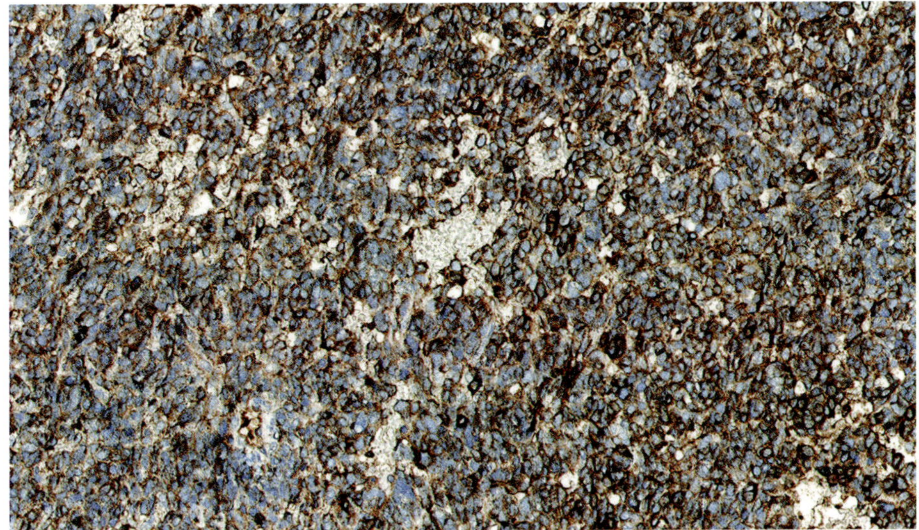

Figure 11-27

UNDIFFERENTIATED NEUROBLASTOMA

This tumor could not be diagnosed by morphology only, but was shown to be a NB on the basis of positive cytoplasmic staining of chromogranin A (shown here), synaptophysin, and tyrosine hydroxylase, and nuclear staining of PHOX2B (not shown).

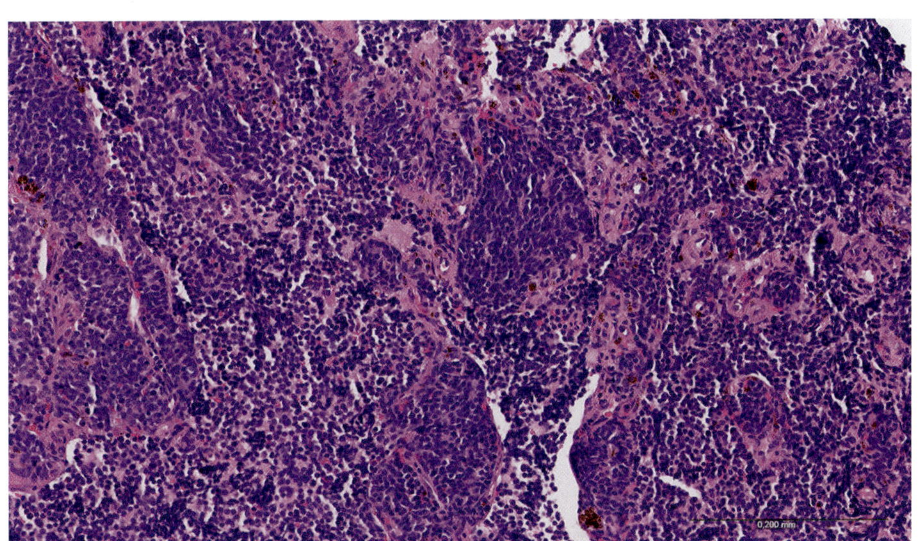

Figure 11-28

POORLY DIFFERENTIATED NEUROBLASTOMA

Very cellular tumor, composed of small blue round cells. The scattered small amount of neuropil indicates the neuroblastic nature of the lesion.

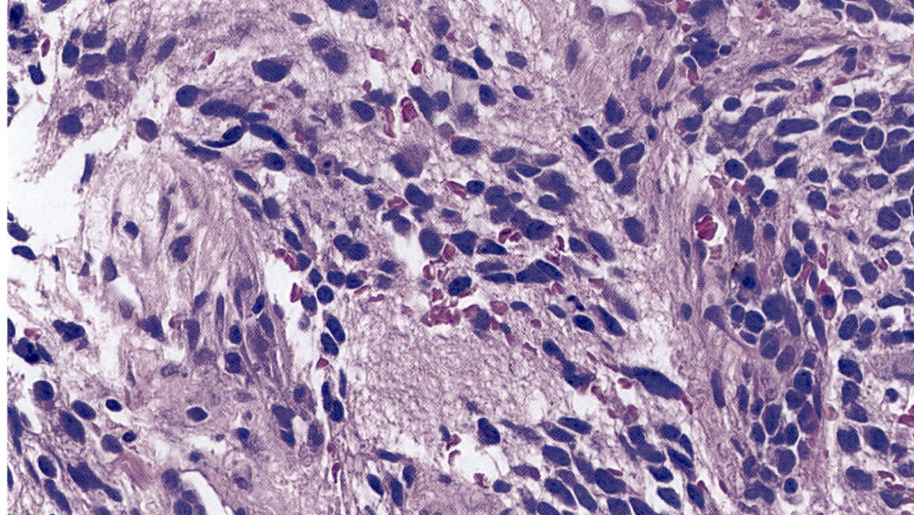

Figure 11-29

DIFFERENTIATING NEUROBLASTOMA

Left adrenal biopsy of 13-month-old boy shows morphologically heterogeneous tumor with varying ganglion cell differentiation, next to areas with neuroblasts, separated by neuropil.

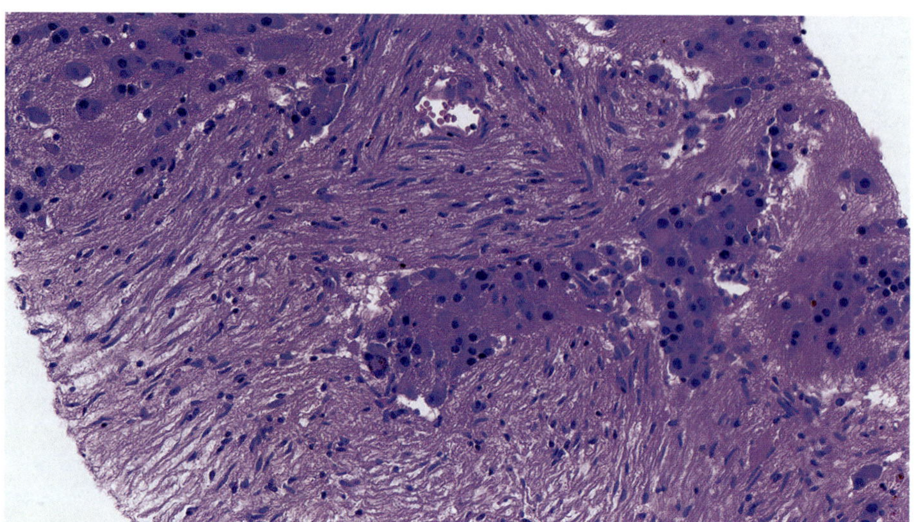

Figure 11-30

GANGLIONEUROBLASTOMA INTERMIXED

Biopsy of paravertebral mass in 11-year-old girl shows areas of schwannian stroma, intermingled with areas of immature stroma containing cells with varying differentiation toward ganglion cells, but also including neuroblasts.

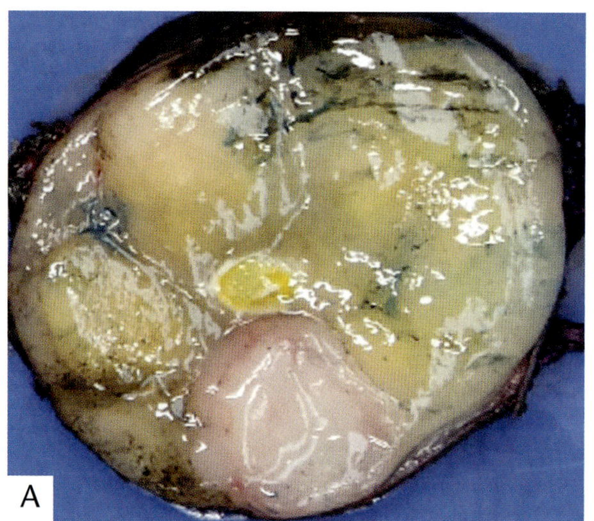

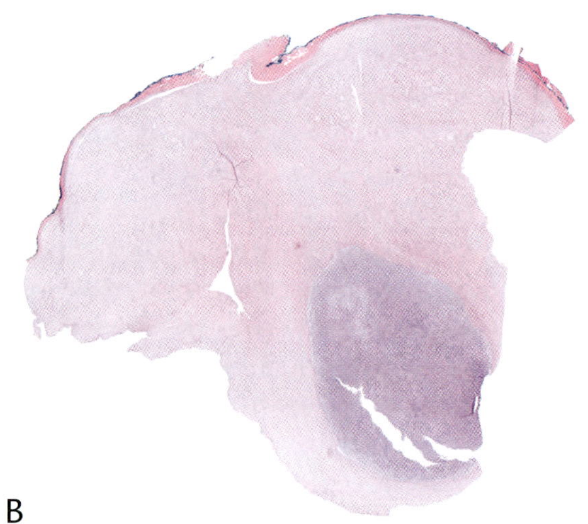

Figure 11-31

GANGLIONEUROBLASTOMA NODULAR

A: The pink nodule corresponds to a neuroblastic nodule (see B), whereas the yellow parts represent GN.
B: There is a cellular neuroblastic area on the bottom in a background of GN.

basis of histology alone. In the intermixed subtype there are areas of stroma intermingled with cellular areas, including at least one area with neuroblasts and/or neuropil (fig. 11-30).

In contrast, the nodular subtype of GNBL contains at least one (expansile) nodule with neuroblasts, which may already be apparent macroscopically (fig. 11-31). Care should be taken at the time of gross examination to sample tissue from such a nodule for microscopic examination to allow this diagnosis to be made correctly. If no neuroblasts are present, tumors are classified as GN. Maturing or mature subtype is distinguished by the ganglion cell status, with completely mature ganglion cells (and thus mature subtype) defined as completely covered with satellite cells (fig. 11-32). Consequently, GN maturing subtype is defined as a tumor with maturing ganglion cells that are not or not completely covered by satellite cells.

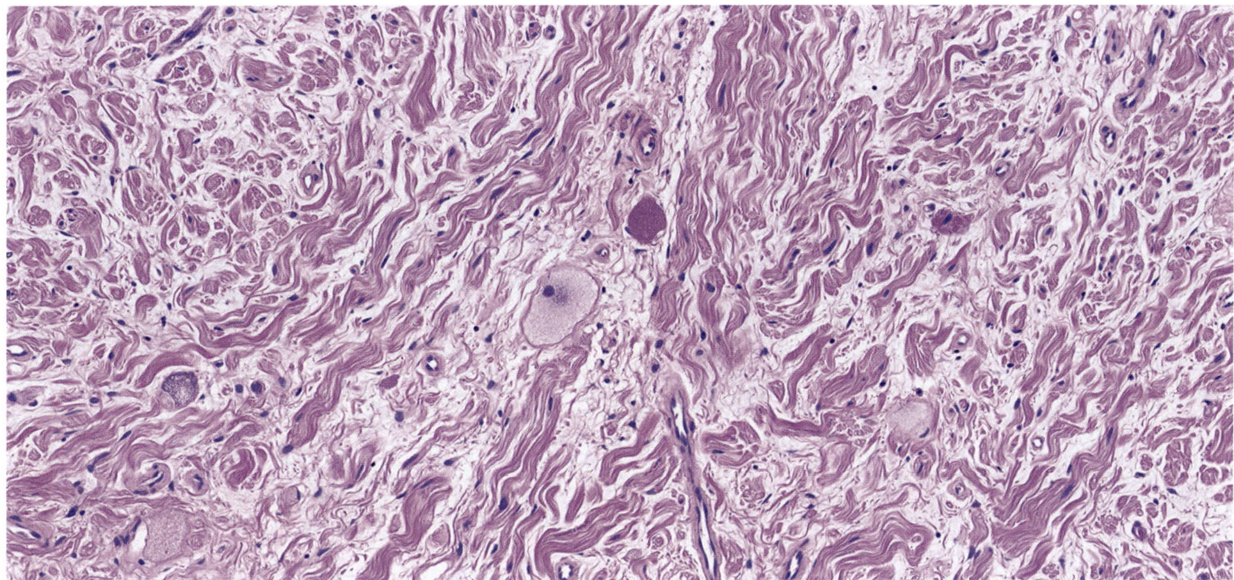

Figure 11-32

GANGLIONEUROMA

This groin lesion in a 15-year-old boy consisted entirely of mature elements, both the stroma, which is paucicellular schwannian stroma, and the cellular compartment, which was composed of scattered mature ganglion cells, with the typical ample amount of cytoplasm, eccentric nuclei, and a single prominent nucleolus.

With the advent of the various forms of chemotherapy for the treatment of NBL and GNBL, pathologists frequently see these tumors in their native form only in pretreatment biopsy specimens, for diagnostic purposes. When primary surgery is feasible, a resection specimen is obtained without secondary changes. In those cases, and for biopsies, the INPC classification can be applied, with the cautionary note that biopsies may be subject to sample bias. In the majority of cases, however, surgical resection specimens become available only after extensive chemotherapeutic treatment, yielding a specimen with varying degrees of related changes. While it is generally relevant to describe the extent and nature of chemotherapy-induced change, as well as the degree of differentiation of the remaining vital tumor tissue, there are currently no reliable guidelines for prognostic correlation of chemotherapy-induced changes; they cannot serve as an indicator of patient prognosis, especially in relation to biologically unfavorable tumors. Nevertheless, quantification of such components, such as vital tumor and therapy-induced change, should be attempted.

Therapy-induced change may be in the form of calcification, hemorrhage, necrosis, or fibrosis, with collections of foamy macrophages. In addition, areas of lymphocytic infiltrate may be found. An interesting finding is the occurrence of therapy-induced maturation, with areas containing hyperplastic nerves and large conglomerates of ganglion cells, with or without schwannian stroma. Such components may be present in untreated tumors as well (see Spontaneous Regression and Maturation), but they are particularly prominent in treated tumors. Frequently, the amount of remaining vital tumor is small, requiring extensive sampling and meticulous searching. Calcified areas may obscure small areas of immature neuroblastic cells, or ganglioneuroblastic areas may be found in the middle of mature stroma (fig. 11-33).

One additional important element that needs to be mentioned in the pathology report is the mitosis karyorrhexis index (MKI). This is a combination of all mitoses and all karyorrhectic figures in 5,000 tumor cells and has been described in the INPC (50). This combination has been chosen on purpose to reduce the inter- and

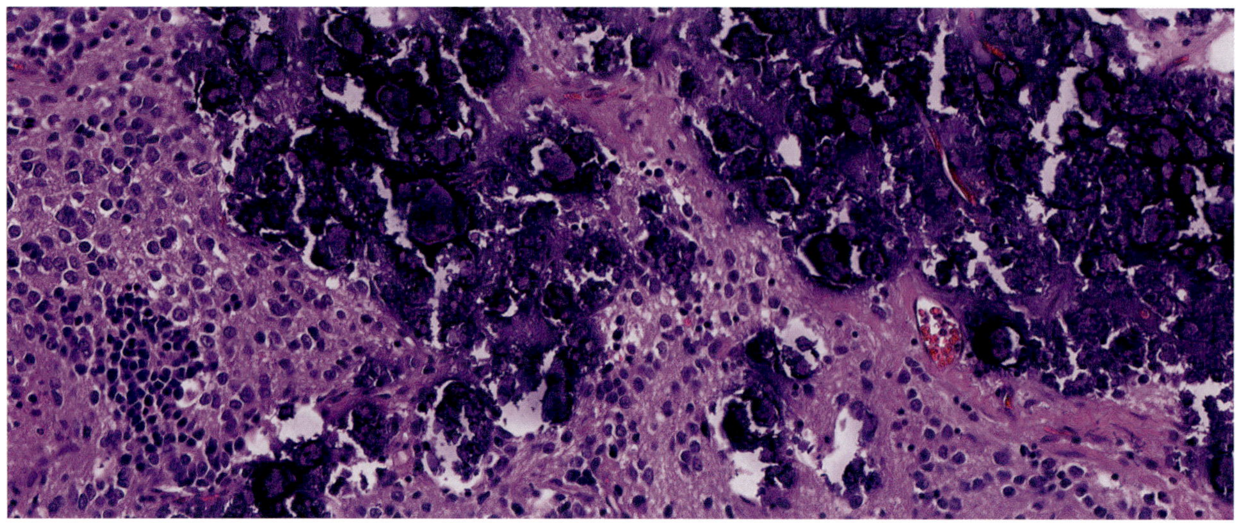

Figure 11-33

NEUROBLASTOMA WITH EXTENSIVE DYSTROPHIC CALCIFICATION

NBs may show varying degrees of calcification, both in primary tumors and following chemotherapy. This can already be appreciated at gross inspection or sectioning, and may obscure the vital tumor component in some instances. In this case, vital neuroblasts are easily seen.

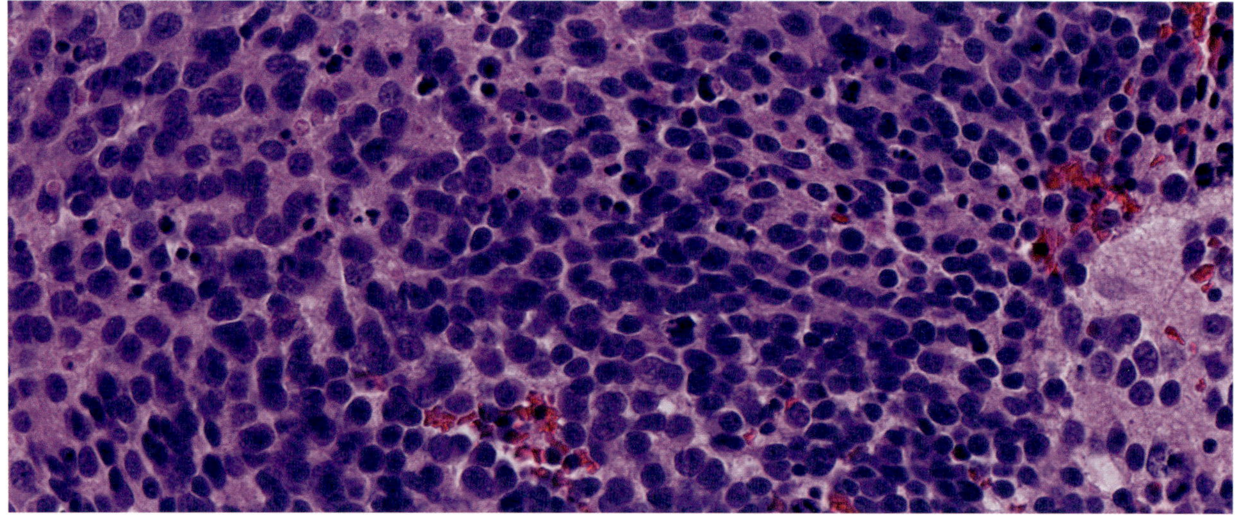

Figure 11-34

NEUROBLASTOMA WITH NUMEROUS MITOTIC AND APOPTOTIC FIGURES

Area of undifferentiated NB with numerous apoptotic bodies and a few potential mitotic figures, illustrating the approach to mitosis karyorrhexis counting.

intraobserver variation in the identification of mitotic figures (fig. 11-34). The MKI carries prognostic relevance when divided into three categories: <100/5000 cells (L-MKI) is favorable in children less than 60 months of age and unfavorable in those over 60 months; <200/5000 cells (I-MKI) is favorable in children less than 18 months of age and unfavorable in those over 18 months; >200/5000 cells (H-MKI) is unfavorable without any further age cut-off.

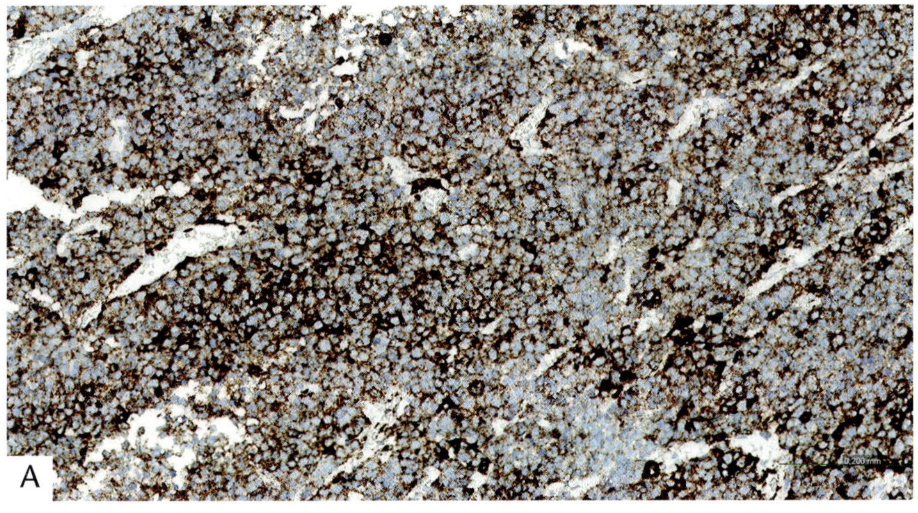

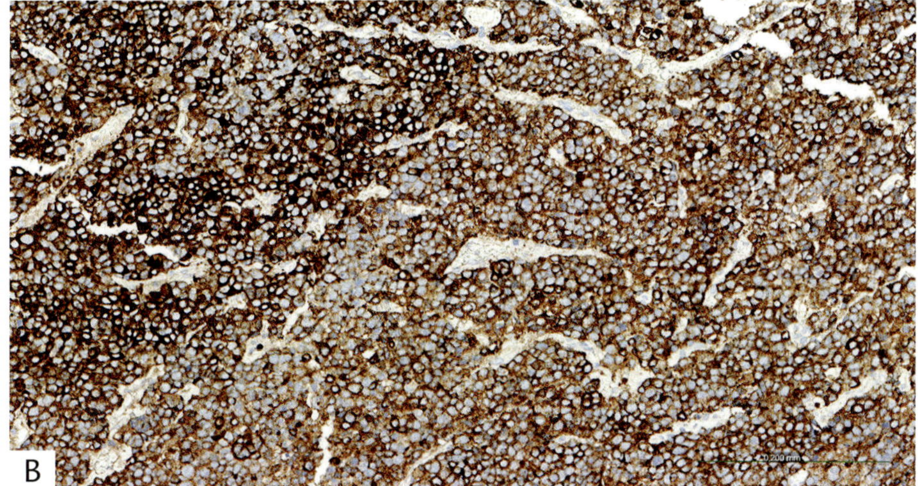

Figure 11-35

UNDIFFERENTIATED NEUROBLASTOMA: IMMUNOHISTOCHEMISTRY

A: Undifferentiated NB for which immunohistochemistry was needed for classification shows extensive cytoplasmic staining for synaptophysin with a slightly granular aspect.

B: Strong cytoplasmic staining for chromogranin A.

Immunohistochemical Findings and Differential Diagnosis

The use of immunohistochemistry in NBL and GNBL is limited to a number of specific indications. First and foremost is the diagnosis of NBL in the context of a so-called small blue round cell tumor, where the differential diagnosis includes lymphoblastic lymphoma, rhabdomyosarcoma, Ewing sarcoma and related sarcoma types, and in rare instances, Wilms tumor, desmoplastic round cell tumor, medulloblastoma, synovial sarcoma, retinoblastoma, small cell osteosarcoma, and mesenchymal chondrosarcoma. Patient age, tumor location, and additional clinical information usually narrow this differential diagnosis and a well-chosen panel of immunohistochemical markers almost always results in complete resolution of differential diagnostic uncertainty. NBL and GNBL show positive staining for a number of markers. Some of these, including PGP9.5, CD56, CD57, and neuron-specific enolase (NSE), have high sensitivity but lack specificity. Those with higher specificity, however, including synaptophysin and chromogranin A, have lower sensitivity (fig. 11-35).

NBL84 sensitivity may be up to 95 percent, although it lacks specificity for some of the differential diagnostic considerations, including Ewing sarcoma, desmoplastic round cell tumor, small cell osteosarcoma, and medulloblastoma (51,52). The recently described neuroepithelial transcription factor, INSM1, analyzed immunohistochemically in neuroblastic tumors in comparison to other embryonal neoplasms,

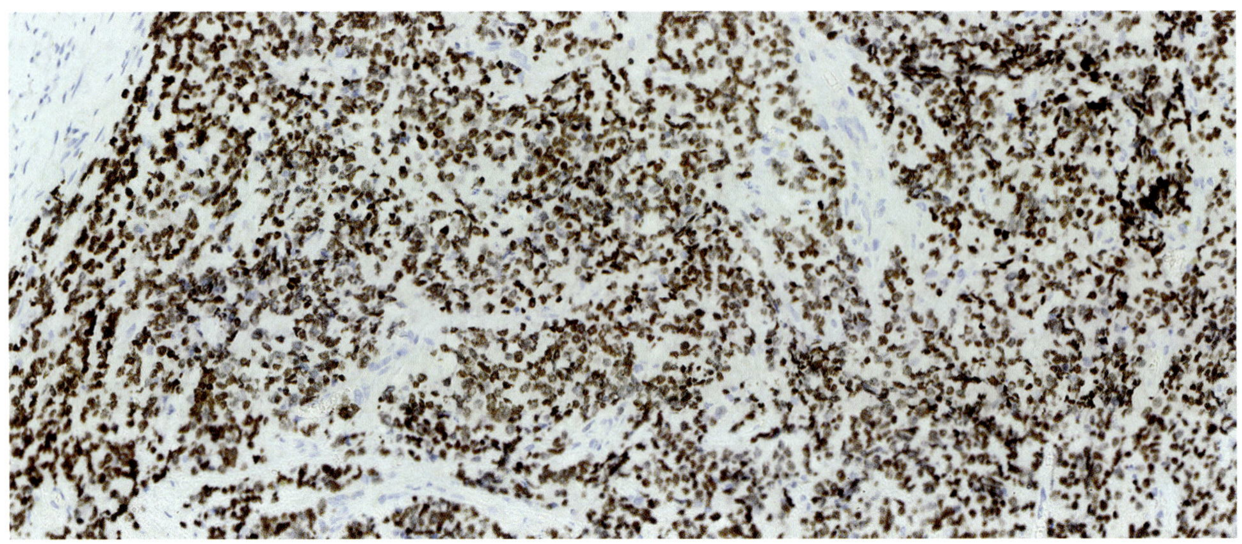

Figure 11-36

IMMUNOHISTOCHEMISTRY WITH PHOX2B

The same tumor as figure 11-35 shows nuclear reactivity for PHOX2B, a nuclear transcription factor specific for sympathetic neurons and chromaffin cells.

was expressed in 78 percent of a small series of NBLs, GNBLs, and GNs, in at least 20 percent of nuclei, but focal or patchy staining was also reported in a proportion of Ewing sarcomas, rhabdomyosarcomas, and Wilms tumors (53).

Catecholamine biosynthetic pathway enzymes, including tyrosine hydroxylase and dopamine beta-hydroxylase, are fairly specific, but may also be found in scattered cells in the gastrointestinal tract, neural tissues, and ganglia. Sensitivity depends on the degree of tumor differentiation. Another relatively new nuclear marker is PHOX2B, directed to the paired-like homeobox 2B protein, a nuclear transcription factor in sympathetic neurons and chromaffin cells, which has 100 percent sensitivity and 100 percent specificity for tumors of the autonomic nervous system (fig. 11-36) (53b).

These positive markers should be complemented with markers that are usually negative in NBL and GNBL, but show reactivity in tumors that feature in the differential diagnosis. Such markers include desmin and myogenin for the exclusion or demonstration of rhabdomyosarcoma, CD99 for Ewing sarcoma, terminal deoxynucleotidyl transferase (TdT) and lymphoid markers for lymphoblastic lymphoma, WT1 for Wilms tumor, and possibly epithelial membrane antigen (EMA), keratins, and WT1 (C-terminus) for desmoplastic small round cell tumor.

The second application for immunohistochemistry in the workup of NBL and GNBL is in the detection of minimal residual disease in the bone marrow. For this purpose a panel of at least two of three antibodies (chromogranin A, synaptophysin, PHOX2B, and tyrosine hydroxylase) should be used (figs. 11-37, 11-38). This advice is one of a series of recommendations done by the International Neuroblastoma Response Criteria Bone Marrow Working Group, which includes consensus criteria for the collection, analysis, and reporting of the percentage area of infiltration of bone marrow by NBL tumor cells in trephine biopsies (54). In addition, the quantitative analysis of tumor in bone marrow aspirates by bone marrow cytology and reverse transcription polymerase chain reaction (RT-PCR) for tyrosine hydroxylase or PHOX2B is discussed by this group.

Molecular Genetic Findings

Although molecular diagnostics in NBL risk stratification play a prognostic role, molecular analysis also has an important role in the differential diagnosis of NBL, distinguishing it from the entities mentioned in the section above

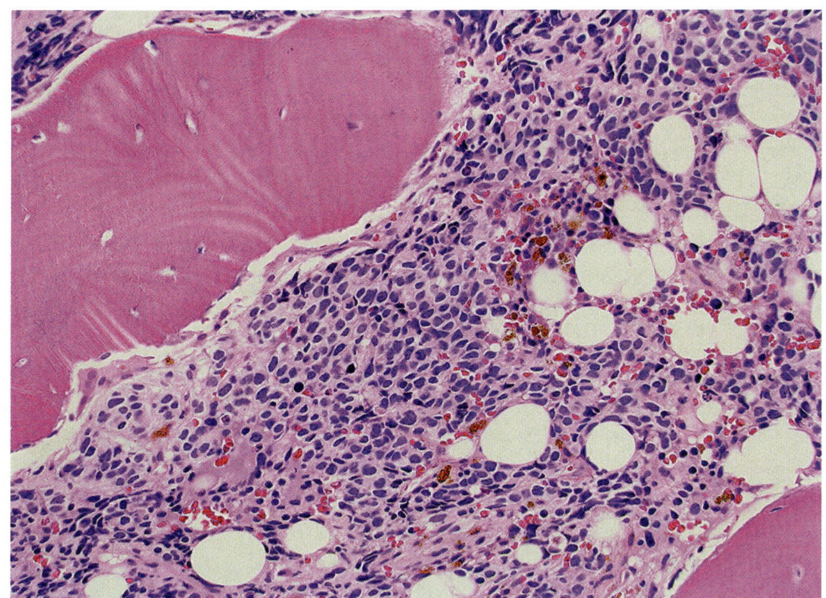

Figure 11-37

NEUROBLASTOMA METASTATIC TO BONE AND BONE MARROW

This trephine biopsy shows extensive involvement by NB, which is even appreciated without immunohistochemistry.

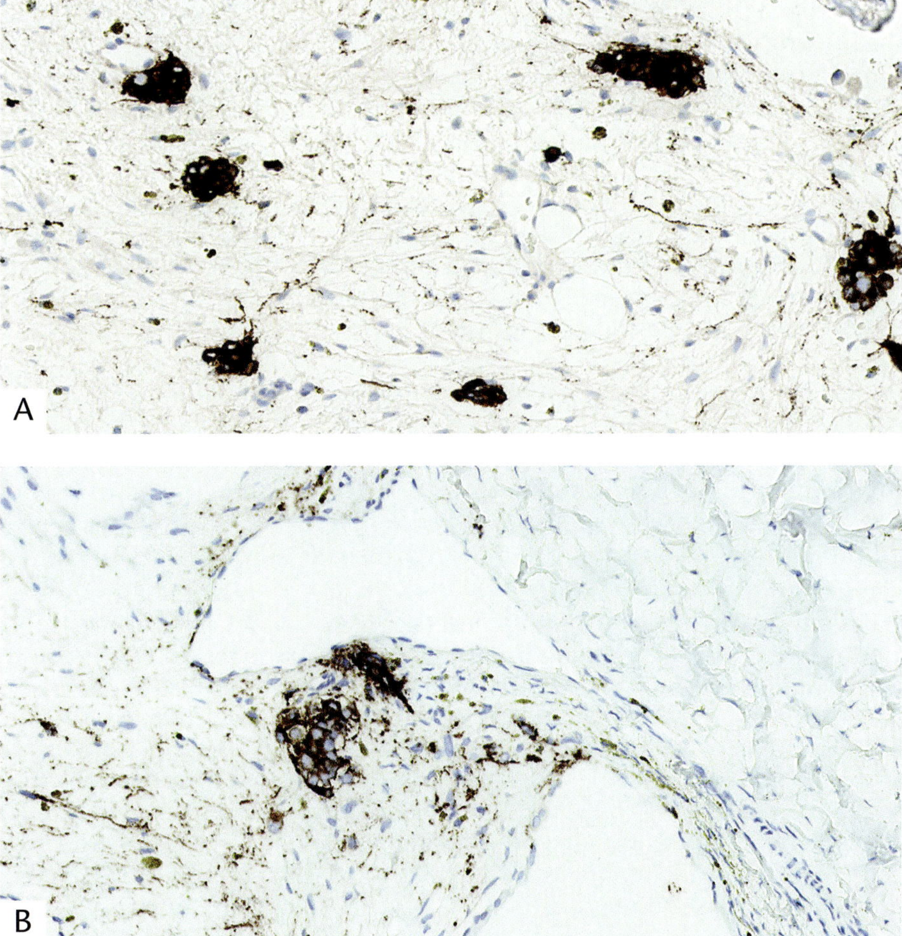

Figure 11-38

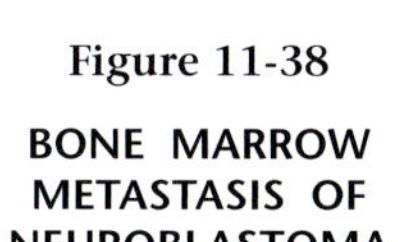

BONE MARROW METASTASIS OF NEUROBLASTOMA

A: Chromogranin A highlights the presence of isolated fields of tumor cells in paucicellular bone marrow following chemotherapy.

B: Synaptophysin highlights the presence of isolated fields of tumor cells in paucicellular bone marrow following chemotherapy.

Table 11-5

RECURRENT MOLECULAR ABNORMALITIES IN SMALL BLUE ROUND CELL TUMORS

Tumor Type	Molecular Abnormality
Ewing sarcoma	*FET::ETS* fusion
Alveolar rhabdomyosarcoma	*PAX3::FOXO1* or *PAX7::FOXO1* fusion[a]
Synovial sarcoma	*SS18::SSX* fusion
Wilms tumor	No pathognomonic molecular abnormalities
Desmoplastic round cell tumor	*EWSR1::WT1* fusion

[a]Alternative *PAX3* fusions have been described.

(Table 11-5). Many of these entities have highly specific chromosomal abnormalities.

An increasing number of molecular tests has been incorporated in the diagnostic workup of NBL, because of their prognostic value and their use in risk stratification. The most important of these is the status of the v-myc avian myelocytomatosis viral oncogene NB-derived homolog (*MYCN*). Amplification, defined as more than 4 copies of *MYCN* per cell in relation to a centromeric reference probe of chromosome 2, performed by fluorescent in situ hybridization (FISH) analysis, and occurring in about 20 percent of NBL patients, is associated with clinically aggressive advanced-stage disease and treatment failure, and has a profound negative prognostic effect (55–58). *MYCN*-amplified tumors frequently have unfavorable histologic characteristics, belonging to the undifferentiated or poorly differentiated subgroups, displaying a high MKI, and having prominent nucleoli. Interestingly, in high-risk patients without *MYCN* amplification, other molecular mechanisms leading to MYC pathway activation are operational, suggesting that this pathway is pivotal for the high-risk phenotype (59). An example of this is the presence of MYC protein expression in NBL without *MYCN* amplification, which is also correlated with worse prognosis, older age, and higher stage (60). MYC expression may be a characteristic of the so-called large cell NBL, a peculiar histologic NBL subtype showing large cells with open nuclei, nucleolar hypertrophy, and aberrant desmin expression (47).

Cell ploidy is another important prognostic variable. Counter intuitively, diploid tumors have an inferior prognosis than triploid or hyperdiploid tumors (55). This is probably because hyperdiploid tumors have a better response to chemotherapy (61). The favorable prognostic association of ploidy is mostly evident in infants under 1 year of age, whereas it is absent in children over 2 years, probably because tumors in the latter age group also contain structural chromosomal rearrangements (61). Ploidy can easily be studied by flow cytometry, which is available in most laboratories. However, this technique is not able to separate hyperdiploid cases from diploid tumors with segmental chromosomal rearrangements, which should be studied with cytogenetics or single nucleotide polymorphism (SNP) analysis.

A third and very relevant molecular parameter is the analysis of segmental chromosomal copy number changes, which is currently performed by SNP array. While the mechanism of segmental chromosomal loss is unknown, and no tumor suppressor genes or oncogenes have so far been identified in regions of loss or gain, respectively, it has been shown that the number of chromosome breakpoints is related to prognosis, with seven or more of these associated with worse outcome. The most frequent losses are of 1p (23 to 35 percent), 3p, 4p, and 11q (30 to 40 percent) and the most frequent gains are 1q, 2p, and 17q (over 50 percent); these are frequently correlated with other unfavorable genetic and clinical features (62–64). Segmental chromosomal alterations predict a worse outcome in patients without *MYCN* amplification and in low-stage disease, and are the most reliable indicator of relapse of NBL (65,66).

As in other types of childhood cancer, few mutations in driver genes have been found, whether germline or somatic. The two genes for which germline mutations have been described, *PHOX2B* and *ALK*, have already been mentioned above. For the *ALK* gene, but not for *PHOX2B*, gene amplifications (in 2 to 4 percent of NBL patients), copy number gains (in 20 to 30 percent), and somatic mutations (in up to 15 percent) have been described (28–30,67–69). All of these lead to overexpression of receptor protein kinase, detectable by ALK immunohistochemistry, which may also show ALK protein expression without concomitant genetic abnormalities (70). Although *ALK* amplifications or mutations are associated with poor outcome, the role of ALK overexpression in the prognosis is subject to debate (71–73). Clinical trials with ALK inhibitors are currently under way to show the effect on survival (74,75).

The increasing availability of next-generation sequencing techniques has allowed detection of other germline or somatic mutations, including somatic mutations in the RNA-helicase alpha-thalassemia/mental retardation syndrome X-linked gene (*ATRX*), and death-associated protein 6 (*DAXX*) in less than 10 percent of patients and promoter rearrangements in *TERT* in approximately 20 percent of patients, all of which may confer a survival advantage to tumor cells by elongation of telomeres (76–79). *ATRX* mutations are associated with chemoresistance in longstanding NBLs that show slow growth, and are mutually exclusive with *TERT* rearrangements.

Treatment

An extensive description of treatment protocols for NBL patients is beyond the scope of this chapter. In this paragraph, a brief account of the extremely wide range of therapeutic strategies is given. In patients with adrenal tumors less than 3.1 cm (when solid) and less than 5.0 cm (when cystic) and younger than 6 months of age, a watchful waiting strategy has been employed in North America by the Children's Oncology Group, with 100 percent 3-year overall survival rate (80). Low-risk patients with INRG L1 (see Table 11-3) have excellent survival rates with surgery only (81). Chemotherapy in this patient category is reserved for those with incomplete surgical resection or with recurrent or progressive disease (82). Intermediate-risk patients (INRG L2 or INRG M with favorable biology) are also treated with chemotherapy (82). Stage MS, the peculiar stage that presents in children less than 18 months old, with small L1 or L2 primary tumors and metastases limited to skin, liver, and bone marrow (less than 10 percent involvement), has a high rate of spontaneous regression/maturation (see next section) and thus these children are treated with supportive care or chemotherapy (83).

In contrast to the relatively mild treatment and generally favorable outcome for each of the above groups, high-risk NBL patients (INRG L2 and M with unfavorable biology) have a dismal outcome, with a 5-year survival rate below 50 percent despite extensive multimodal therapy. Typically, these children start on chemotherapy, followed by cytoreductive (debulking) surgery, and myeloablative chemotherapy with autologous stem cell transplantation (84). Further treatment options have become available through the use of ^{131}I-metaiodobenzylguanidine for targeted radiotherapy in relapsed and refractory patients, resulting in increased survival, and through the use of immunotherapy with monoclonal antibodies to the disialoganglioside GD2, which is present in NBL tumor cells (85,86). Based on the latter, chimeric antigen receptor (CAR) T lymphocytes recognizing the GD2 antigen have been constructed and shown to have an antitumor effect, leading to tumor remission and increased survival, although such therapy cannot currently be considered as standard therapy (87,88).

Spontaneous Regression and Maturation

One peculiar aspect of NBL is the ability of some tumors to undergo regression or maturation to GN without the use of any form of therapy. There are multiple patients described in the literature that originally presented with a histologically confirmed NBL or GNBL, who went on to have only GN or no detectable tumor at all. This is most evident for patients with stage MS (the former stage 4S) disease (83,89–92).

Mass screening programs, as have been performed in Japan, the United States, and Europe, have provided further evidence for the presence and prevalence of spontaneous regression, since they show much higher rates of NBL than at baseline (1 to 2,000 versus 1 to 8,000), although mortality from disease remains unchanged (93–96). Presumably, many lesions detected in this way would not have gone on to a clinically manifest tumor. One may speculate that at least in some of these cases spontaneous regression or maturation has occurred. Evidence for this phenomenon is occasionally seen in patients with congenital NBLs that are detected incidentally before birth and subsequently followed postnatally.

The mechanisms behind spontaneous regression have been and still are of great interest to both the research community as well as pediatric oncologists treating infants and children

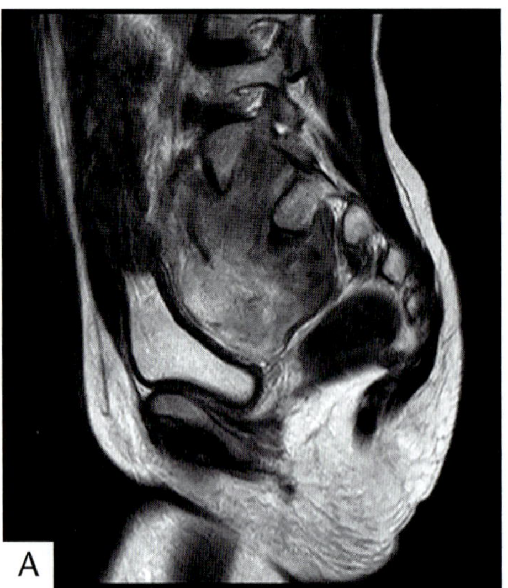

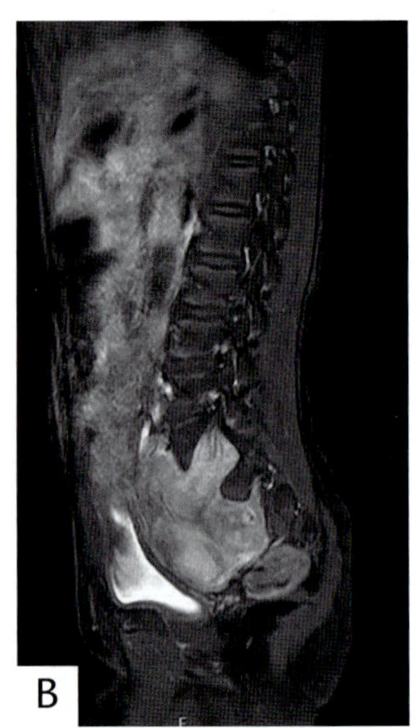

Figure 11-39

PRESACRAL GANGLIONEUROMA EXTENDING INTO THE SACRAL NEUROFORAMINA

MRI from a 7-year-old girl. Sagittal T2 (A) and sagittal T1 with fat suppression after intravenous gadolinium contrast (B). (Courtesy of Dr. A. S. Littooij, University Medical Center Utrecht, The Netherlands)

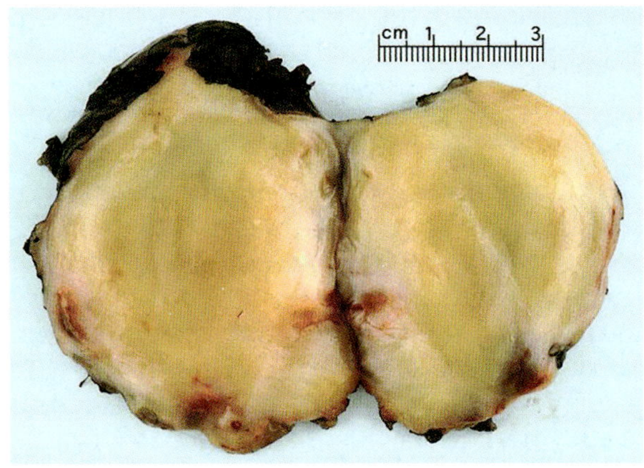

Figure 11-40

GANGLIONEUROMA

Gross image of bisected tumor 9 cm in diameter, with a smooth, glistening, tan-yellow surface. The tumor is well-circumscribed on the outer aspect.

with NBL, as they may provide clues for the treatment, not only of low-stage, but also of high-stage patients. Potential mechanisms for which evidence has suggested that they may play a role include neurotrophin deprivation, loss of telomerase activity, immunologic mechanisms, and epigenetic regulation (89–92,97). The first of these mechanisms, especially regarding the presence of neurotrophin receptors such as TrkA and the dependence on nerve growth factor (NGF) for survival of neuronal (precursor) cells, holds promise, now that pan-Trk inhibitors are being tested in clinical trials (89,98–100).

GANGLIONEUROMA

Ganglioneuromas (GN) are fully mature, benign neoplasms that usually occur in an older age group than NBL. Many patients are over 7 years of age (89–91). Most GNs arise in the posterior mediastinum, followed by the retroperitoneum, particularly the presacral space, but they can occur in a variety of other locations (fig. 11-39) (101–112).

Symmetric dumbbell-shaped GNs of the cervical spine that extend intradurally have been recently reported in a patient with von Recklinghausen disease (113). Only a small proportion of GNs are adrenal in origin (103,114). Calcification has been detected by radiographic study in 41 percent of GNs and this may be detected on gross examination as well (115,116). Rare instances of so-called masculinizing adrenal GN are reported, where an admixture of typical GN with Leydig cells is found (117,118).

GNs are characteristically well-circumscribed or even encapsulated (101,104,119), but microscopic studies of sections taken from the periphery of some tumors do not entirely support this claim. GNs vary in size and may measure up to 15 cm and weigh 5,000 g (104,120). The tumors are firm and may have a rubbery texture. On cross-section they are gray-white to tan-yellow (fig. 11-40).

Occasionally, there is a trabecular or whorled appearance, reminiscent of a leiomyoma. In some cases, particularly in children, GN should be carefully examined to exclude stroma-rich

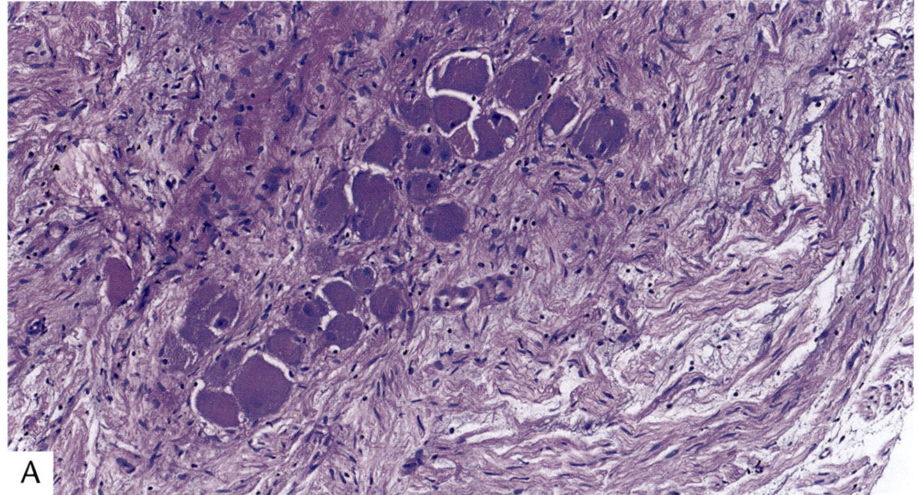

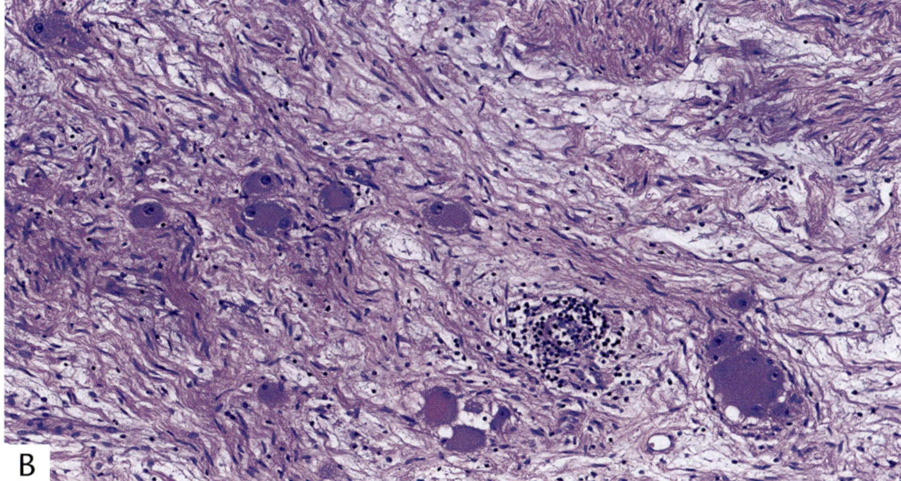

Figure 11-41

GANGLIONEUROMA

A: This tumor is composed of multiple ganglion cells (which may vary considerably in distribution and number) and abundant Schwann cells. Some of the mature ganglion cells have Nissl substance in the peripheral cytoplasm.

B: Some ganglion cells are binucleated or multinucleated, and the background schwannian stroma has a slightly myxoid character, but remains paucicellular.

GNBL, which may grossly and microscopically resemble GN. Adrenal GN may be sharply circumscribed and surrounded by preexistent adrenal cortex and capsule, but the tumor may extend focally beyond the confines of the capsule. GNs contain a variable admixture of mature and maturing ganglion cells and an abundance of mature Schwann cells (fig. 11-41).

The Schwann cells are usually arranged in small fascicles which intersect at various angles, and the bundles are separated by loose myxoid stroma. There is considerable variation in the number of ganglion cells. When they are rare, this may give the mistaken impression of a neurofibroma, while a chance section through a neurofibroma may show a secondarily incorporated ganglion, which can lead to a mistaken diagnosis of GN (104). Rare GNs have one or more ganglia or myelinated nerve bundles that are secondarily incorporated into the tumor. It is possible that such foci represent an unusually high degree of differentiation within a GN (fig. 11-42). Otherwise, well-differentiated ganglion cells in GN have compact eosinophilic cytoplasm with distinct cell borders and a single eccentric nucleus with a prominent nucleolus, although binucleated or multinucleated ganglion cells also occur in GN and may be used for the distinction from preexistent structures. Small amounts of mature fat are seen in some tumors, especially at the periphery.

Adrenal GNs are usually sharply demarcated from the adrenal cortex, although some admixture may be present. Cellular atypia, mitotic activity, and necrosis are not features of GN. Mast cells may be seen in some GNs, similar to

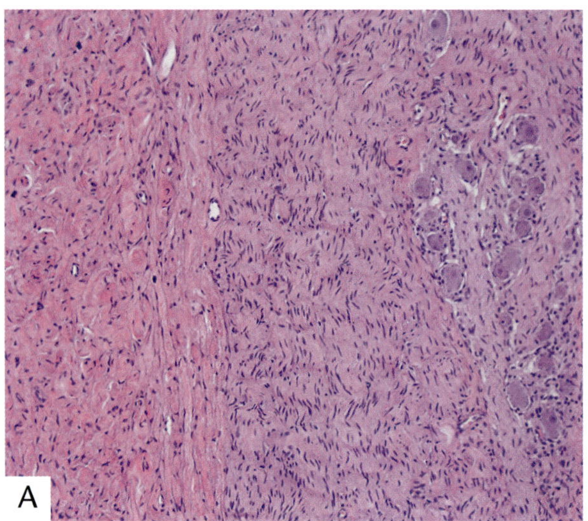

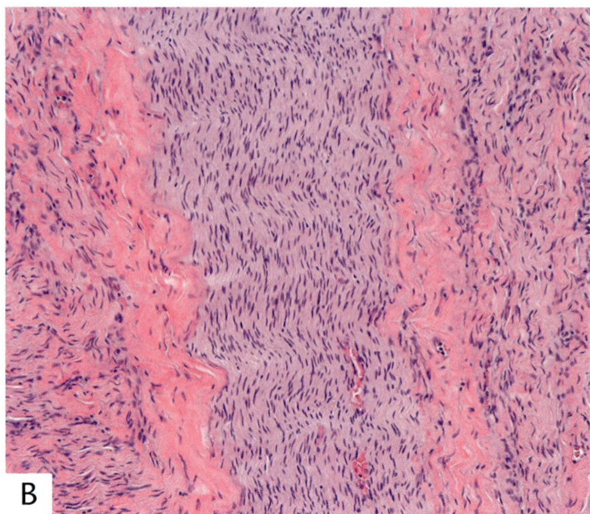

Figure 11-42

GANGLIONEUROMA

A: This tumor contained a few small ganglia. A small myelinated nerve is present at one pole of a ganglion. The ganglion is probably incorporated secondarily in the tumor.

B: A large bundle of myelinated nerve is present in this GN and is most likely secondarily incorporated into the tumor.

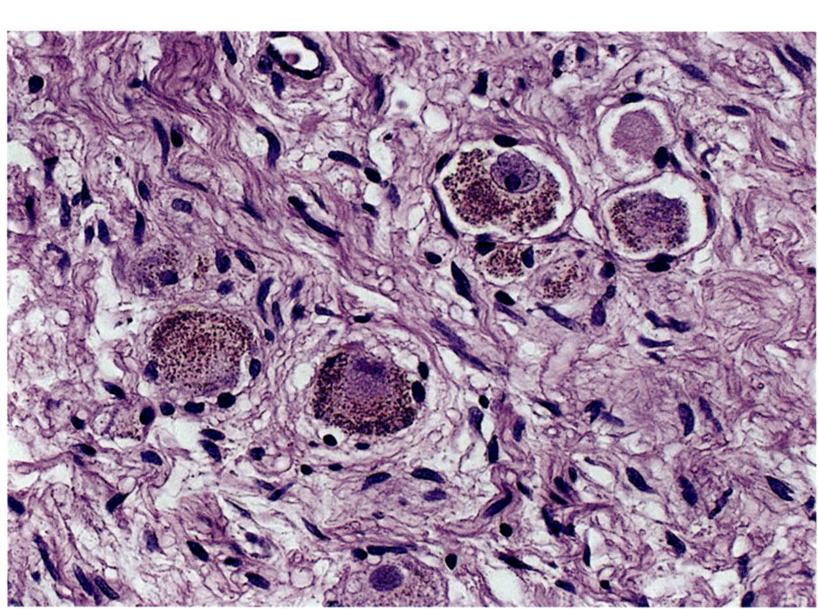

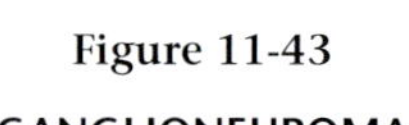

Figure 11-43

GANGLIONEUROMA

The ganglion cells contain abundant granular pigment consistent with lipofuscin or neuromelanin. Mature ganglion cells are encircled by satellite cells.

neurofibromas. Some ganglion cells contain pigment consistent with lipofuscin or neuromelanin (fig. 11-43). Occasional GNs have areas of dense collagenous stroma, and, as in other neuroblastic tumors, areas of chronic inflammation may be present, which should not be mistaken for primitive neuroblasts. There is still some controversy whether GNs arise de novo or by maturation or differentiation of preexisting NBLs or GNBLs.

Although rare, examples of malignant transformation of GN into malignant peripheral nerve sheath tumor (MPNST), either de novo or following radiation therapy for NBL or GNBL, or into NBL have been reported (fig. 11-44) (114,121–127). In addition, a case of rhabdomyosarcoma arising from a dormant dumbbell-shaped GN of the lumbar spine has been described (128).

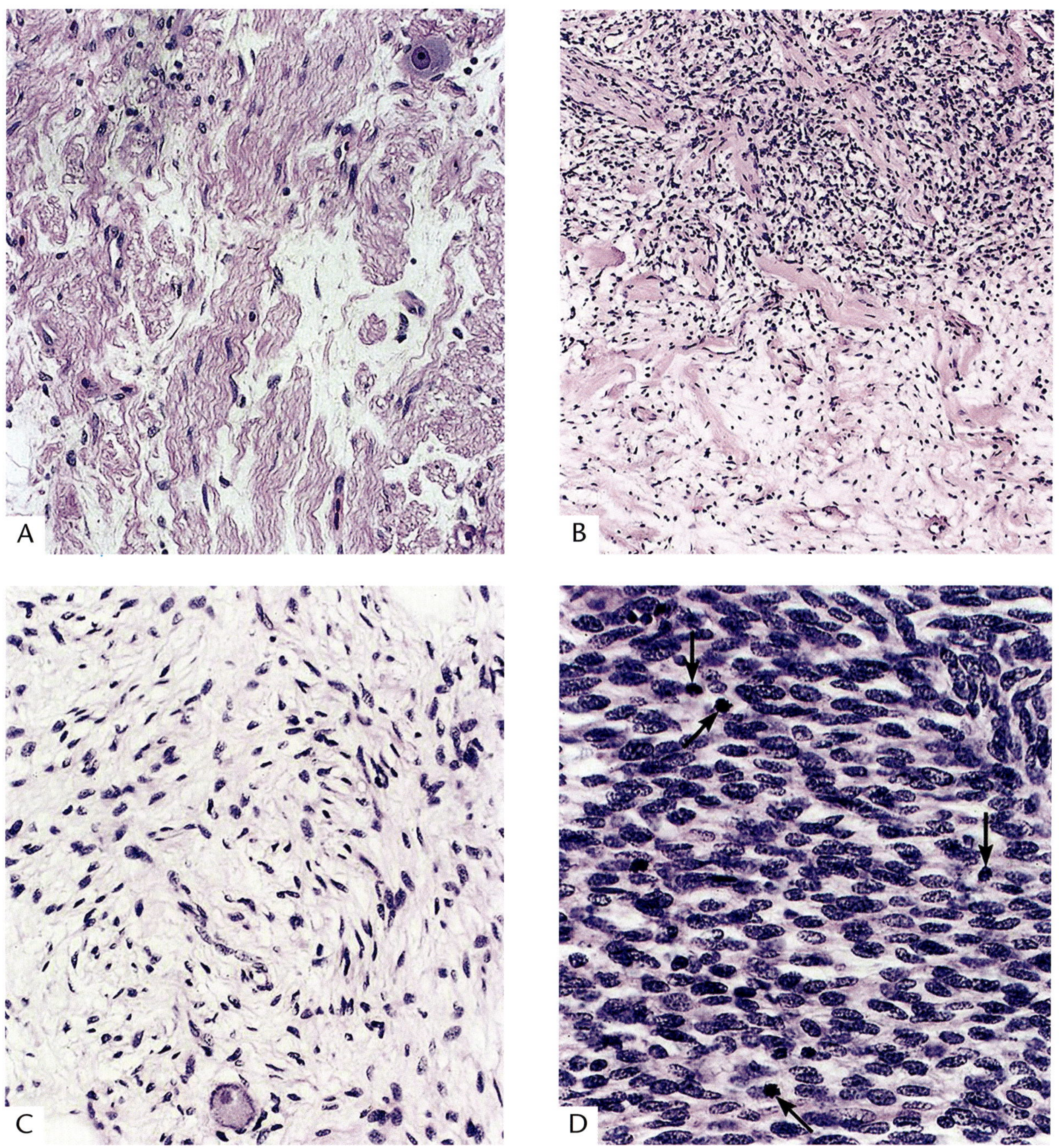

Figure 11-44

MALIGNANT TRANSFORMATION OF A GANGLIONEUROMA

This tumor first presented at 13 months as an unresectable abdominal NB and was treated with abdominal mega-voltage irradiation. At age 17 years, the patient presented again with an abdominal tumor.

A: Mature GN was seen in much of the resected tumor when the patient was a little over 18 years of age during a debulking procedure.

B: Other areas of the same tumor show transition to a malignant peripheral nerve sheath tumor.

C: Transition zone of a malignant peripheral nerve sheath tumor.

D: Some areas show pure malignant peripheral nerve sheath tumor with numerous mitotic figures (arrows) while other areas are necrotic.

REFERENCES

1. Peuchmaur M, d'Amore ES, Joshi VV, et al. Revision of the International Neuroblastoma Pathology Classification: confirmation of favorable and unfavorable prognostic subsets in ganglioneuroblastoma, nodular. Cancer 2003;98:2274-81.
2. Angelini P, London WB, Cohn SL, et al. Characteristics and outcome of patients with ganglioneuroblastoma, nodular subtype: a report from the INRG project. Eur J Cancer 2012;48:1185-91.
3. Huber K, Kalcheim C, Unsicker K. The development of the chromaffin cell lineage from the neural crest. Auton Neurosci 2009;151:10-6.
4. Saito D, Takase Y, Murai H, Takahashi Y. The dorsal aorta initiates a molecular cascade that instructs sympatho-adrenal specification. Science 2012;336:1578-81.
5. Chan WH, Gonsalvez DG, Young HM, Southard-Smith EM, Cane KN, Anderson CR. Differences in CART expression and cell cycle behavior discriminate sympathetic neuroblast from chromaffin cell lineages in mouse sympathoadrenal cells. Dev Neurobiol 2016;76:137-49.
6. Ernsberger U, Esposito L, Partimo S, et al. Expression of neuronal markers suggests heterogeneity of chick sympathoadrenal cells prior to invasion of the adrenal anlagen. Cell Tissue Res 2005;319:1-13.
7. Furlan A, Dyachuk V, Kastriti ME, et al. Multipotent peripheral glial cells generate neuroendocrine cells of the adrenal medulla. Science 2017; 357:eaal3753.
8. Lumb R, Tata M, Xu X, et al. Neuropilins guide preganglionic sympathetic axons and chromaffin cell precursors to establish the adrenal medulla. Development 2018;145:dev162552.
9. Furlan A, Adameyko I. Schwann cell precursor: a neural crest cell in disguise? Dev Biol 2018; 444(Suppl 1):S25-35.
10. Kameneva P, Artemov AV, Kastriti ME, et al. Single-cell transcriptomics of human embryos identifies multiple sympathoblast lineages with potential implications for neuroblastoma origin. Nat Genet 2021;53:694-706.
11. Huber K, Janoueix-Lerosey I, Kummer W, Rohrer H, Tischler AS. The sympathetic nervous system: malignancy, disease, and novel functions. Cell Tissue Res 2018;372:163-70.
12. Kastriti ME, Kameneva P, Kamenev D, et al. Schwann cell precursors generate the majority of chromaffin cells in zuckerkandl organ and some sympathetic neurons in paraganglia. Front Mol Neurosci 2019;12:6.
13. SEER cancer statistics review, 1975-2017. https://seer.cancer. gov/csr/1975_2017.
14. Stiller CA, Parkin DM. International variations in the incidence of neuroblastoma. Int J Cancer 1992;52:538-43.
15. Mossé YP, Deyell RJ, Berthold F, et al. Neuroblastoma in older children, adolescents and young adults: a report from the International Neuroblastoma Risk Group project. Pediatr Blood Cancer 2014;61:627-35.
16. Jrebi NY, Iqbal CW, Joliat GR, Sebo TJ, Farley DR. Review of our experience with neuroblastoma and ganglioneuroblastoma in adults. World J Surg 2014;38:2871-4.
17. Nakata K, Ito Y, Magadi W, et al. Childhood cancer incidence and survival in Japan and England: a population-based study (1993-2010). Cancer Science 2018;109:422-34.
18. Berthold F, Spix C, Kaatsch P, Lampert F. Incidence, survival, and treatment of localized and metastatic neuroblastoma in Germany 1979-2015. Pediatr Drugs 2017;19:577-93.
19. Isaevska E, Manasievska M, Alessi D, et al. Cancer incidence rates and trends among children and adolescents in Piedmont, 1967-2011. PLoS One 2018;12:e0181805.
20. Friedrich P, Itriago E, Rodriguez-Calindo C, Ribeiro K. Racial and ethnic disparities in the incidence of pediatric extracranial embryonal tumors. J Natl Cancer Inst 2017;109:djx050.
21. Miller RW. Ethnic differences in cancer occurrence: genetic and environmental influences with particular reference to neuroblastoma. In: Mulvihall JJ, Miller RW, Fraumeni JF Jr, eds. Genetics of human cancer. New York: Raven Press; 1977:1-14.
22. Miller RW, Young JL Jr, Novakovic B. Childhood cancer. Cancer 1995;75:395-405.
23. Matthay KK, Maris JM, Schleiermacher G, et al. Neuroblastoma. Nat Rev Dis Primers 2016;2: 16078.
24. Mossé YP, Laudenslager M, Longo L, et al. Identification of ALK as a major familial neuroblastoma predisposition gene. Nature 2008;455:930-5.
25. Janoueix-Lerosey I, Lequin D, Brugières L, et al. Somatic and germline activating mutations of the ALK kinase receptor in neuroblastoma. Nature 2008;455:967-70.
26. Trochet D, Bourdeaut F, Janoueix-Lerosey I, et al. Germline mutations of the paired-like homeobox 2B (PHOX2B) gene in neuroblastoma. Am J Hum Genet 2004;74:761-4.

27. Amiel J, Laudier B, Attié-Bitach T, et al. Polyalanine expansion and frameshift mutations of the paired-like homeobox gene PHOX2B in congenital central hypoventilation syndrome. Nat Genet 2003;33:459-61.
28. Bresler SC, Weiser DA, Huwe PJ, et al. ALK mutations confer differential oncogenic activation and sensitivity to ALK inhibition therapy in neuroblastoma. Cancer Cell 2014;26:682-94.
29. George RE, Sanda T, Hanna M, et al. Activating mutations in ALK provide a therapeutic target in neuroblastoma. Nature 2008;455:975-8.
30. Janoueix-Lerosey I, Lopez-Delisle L, Delattre O, Rohrer H. The ALK receptor in sympathetic neuron development and neuroblastoma. Nature 2008;455:967-70.
31. Manolio TA, Collins FS, Cox NJ, et al. Finding the missing heritability of complex diseases. Nature 2009;461:747-53.
32. Bourdeaut F, de Carli E, Timsit S, et al. VIP hypersecretion as primary or secondary syndrome in neuroblastoma: a retrospective study by the Société Française des Cancers de l'Enfant (SFCE). Pediatr Blood Cancer 2009;52:585-90.
33. Hero B, Schleiermacher G. Update on pediatric opsoclonus myoclonus syndrome. Neuropediatrics 2013;44:324-9.
34. Sayed AK, Miller BA, Lack EE, Sallan SE, Levey RH. Heterochromia iridis and Horner's syndrome due to paravertebral neurilemmoma. J Surg Oncol 1983;22:15-6.
35. Vo KT, Matthay KK, Neuhaus J, et al. Clinical, biologic, and prognostic differences on the basis of primary tumor site in neuroblastoma: a report from the international neuroblastoma risk group project. J Clin Oncol 2014;32:3169-76.
36. King D, Goodman J, Hawk T, Boles ET Jr, Sayers MP. Dumbbell neuroblastomas in children. Arch Surg 1975;110:888-91.
37. Kumar M, Batra G, Saun A, Singh R. Blueberry muffin baby: an unusual presentation of infantile neuroblastoma. Ind J Med Paed Oncol 2018; 39:263-5.
38. Fleck S, Marx S, Bobak C, et al. Neuroblastoma with intracerebral metastases and the need for neurosurgery: a single-center experience. J Neurosurg Pediatr 2019:1-6.
39. Kume A, Morikawa T, Ogawa M, Yamashita A, Yamaguchi S, Fukayama M. Congenital neuroblastoma with placental involvement. Int J Clin Exp Pathol 2014;7:8198-204.
40. Brodeur GM, Pritchard J, Berthold F, et al. Revisions of the international criteria for neuroblastoma diagnosis, staging, and response to treatment. J Clin Oncol 1993;11:1466-77.
41. Monclair T, Brodeur GM, Ambros PF, et al. The International Neuroblastoma Risk Group (INRG) staging system: an INRG Task Force report. J Clin Oncol 2009;27:298-303.
42. Iehara T, Hiyama E, Tajiri T, et al. Is the prognosis of stage 4s neuroblastoma in patients 12 months of age and older really excellent? Eur J Cancer 2012;48:1707-12.
42b. Cohn SL, Pearson AD, London WB, et al. The International Neuroblastoma Risk Group (INRG) classification system: an INRG Task Force report. J Clin Oncol 2009;27:289-97.
43. Bedoya-Reina OC, Li W, Arceo M, et al. Single-nuclei transcriptomes from human adrenal gland reveal distinct cellular identities of low and high-risk neuroblastoma tumors. Nat Commun 2021;12:5309.
44. Irwin MS, Naranjo A, Zhang FF, et al. Revised Neuroblastoma Risk Classification System: a report from the Children's Oncology Group. J Clin Oncol 2021;39:3229-41.
45. Eo H, Kim JH, Jang KM, et al. Comparison of clinico-radiological features between congenital cystic neuroblastoma and neonatal adrenal hemorrhagic pseudocyst. Korean J Radiol 2011;12:52-8.
46. Navarro S, Noguera R, Pellin A, Mejia C, Ruiz A, Llombart-Bosch A. Pleomorphic anaplastic neuroblastoma. Med Pediatr Oncol 2000;35:498-502.
47. Matsuno R, Gifford AJ, Fang J, et al. Rare MYC-amplified neuroblastoma with large cell histology. Pediatr Dev Pathol 2018;21:461-6.
48. Weiss T, Taschner-Mandl S, Janker L, et al. Schwann cell plasticity regulates neuroblastic tumor cell differentiation via epidermal growth factor-like protein 8. Nat Commun 2021;12:1624.
49. Shimada H, Ambros IM, Dehner LP, Hata J, Joshi VV, Roald B. Terminology and morphologic criteria of neuroblastic tumors: recommendations by the International Neuroblastoma Pathology Committee. Cancer 1999;86:349-63.
50. Peuchmaur M, d'Amore ES, Joshi VV, et al. Revision of the International Neuroblastoma Pathology Classification: confirmation of favorable and unfavorable prognostic subsets in ganglioneuroblastoma, nodular. Cancer 2003;98:2274-81.
51. Miettinen M, Chatten J, Paetau A, Stevenson A. Monoclonal antibody NB84 in the differential diagnosis of neuroblastoma and other small round cell tumors.Am J Surg Pathol 1998;22:327-32.
52. Sebire NJ, Gibson S, Rampling D, Williams S, Malone M, Ramsay AD. Immunohistochemical findings in embryonal small round cell tumors with molecular diagnostic confirmation. Appl Immunohistochem Mol Morphol 2005;13:1-5.
53. Wang H, Krishnan C, Charville GW. INSM1 expression in peripheral neuroblastic tumors and other embryonal neoplasms. Pediatr Dev Pathol 2019;22:440-8.

53b. Bielle F, Fréneaux P, Jeanne-Pasquier C, et al. PHOX2B immunolabeling: a novel tool for the diagnosis of undifferentiated neuroblastomas among childhood small round blue-cell tumors. Am J Surg Pathol 2012;36:1141-9.
54. Burchill SA, Beiske K, Shimada H, et al. Recommendations for the standardization of bone marrow disease assessment and reporting in children with neuroblastoma on behalf of the International Neuroblastoma Response Criteria Bone Marrow Working Group. Cancer 2017;123: 1095-105.
55. Ambros PF, Ambros IM, Brodeur GM, et al. International consensus for neuroblastoma molecular diagnostics: report from the International Neuroblastoma Risk Group (INRG) Biology Committee. Br J Cancer 2009;100:1471-82.
56. Bagatell R, Beck-Popovic M, London WB, et al. Significance of MYCN amplification in international neuroblastoma staging system stage 1 and 2 neuroblastoma: a report from the International Neuroblastoma Risk Group database. J Clin Oncol 2009;27:365-70.
57. Brodeur GM, Seeger RC, Schwab M, Varmus HE, Bishop JM. Amplification of N-myc in untreated human neuroblastomas correlates with advanced disease stage. Science 1984;224:1121-4.
58. Campbell K, Gastier-Foster JM, Mann M, et al. Association of MYCN copy number with clinical features, tumor biology, and outcomes in neuroblastoma: a report from the Children's Oncology Group. Cancer 2017;123:4224-35.
59. Liu X, Mazanek P, Dam V, et al. Deregulated Wnt/beta-catenin program in high-risk neuroblastomas without MYCN amplification. Oncogene 2008;27:1478-88.
60. Wang LL, Teshiba R, Ikegaki N, et al. Augmented expression of MYC and/or MYCN protein defines highly aggressive MYC-driven neuroblastoma: a Children's Oncology Group study. Br J Cancer 2015;113:57-63.
61. George RE, London WB, Cohn SL, et al. Hyperdiploidy plus nonamplified MYCN confers a favorable prognosis in children 12 to 18 months old with disseminated neuroblastoma: a Pediatric Oncology Group study. J Clin Oncol 2005;23:6466-73.
62. Attiyeh EF, London WB, Mossé YP, et al. Chromosome 1p and 11q deletions and outcome in neuroblastoma. N Engl J Med 2005;353:2243-53.
63. Bown N, Cotterill S, Lastowska M, et al. Gain of chromosome arm 17q and adverse outcome in patients with neuroblastoma. N Engl J Med 1999;340:1954-61.
64. Schleiermacher G, Mosseri V, London WB, et al. Segmental chromosomal alterations have prognostic impact in neuroblastoma: a report from the INRG project. Br J Cancer 2012;107:1418-22.
65. Caron H, van Sluis P, de Kraker J, et al. Allelic loss of chromosome 1p as a predictor of unfavorable outcome in patients with neuroblastoma. N Engl J Med 1996;334:225-30.
66. Lastowska M, Cotterill S, Pearson AD, et al. Gain of chromosome arm 17q predicts unfavourable outcome in neuroblastoma patients. U.K. Children's Cancer Study Group and the U.K. Cancer Cytogenetics Group. Eur J Cancer 1997;33:1627-33.
67. Carén H, Abel F, Kogner P, Martinsson T. High incidence of DNA mutations and gene amplifications of the ALK gene in advanced sporadic neuroblastoma tumours. Biochem J 2008;416:153-9.
68. Subramaniam MM, Piqueras M, Navarro S, Noguera R. Aberrant copy numbers of ALK gene is a frequent genetic alteration in neuroblastomas. Hum Pathol 2009;40:1638-42.
69. Wang M, Zhou C, Sun Q, et al. ALK amplification and protein expression predict inferior prognosis in neuroblastomas. Exp Mol Pathol 2013;95:124-30.
70. Duijkers FA, Gaal J, Meijerink JP, et al. High anaplastic lymphoma kinase immunohistochemical staining in neuroblastoma and ganglioneuroblastoma is an independent predictor of poor outcome. Am J Pathol 2012;180:1223-31.
71. Carpenter EL, Mossé YP. Targeting ALK in neuroblastoma—preclinical and clinical advancements. Nat Rev Clin Oncol 2012;9:391-9.
72. Mossé YP. Anaplastic lymphoma kinase as a cancer target in pediatric malignancies. Clin Cancer Res 2016;22:546-52.
73. Kamihara J, Bourdeaut F, Foulkes WD, et al. Retinoblastoma and neuroblastoma predisposition and surveillance. Clin Cancer Res 2017;23:e98-106.
74. Schulte JH, Schulte S, Heukamp LC, et al. Targeted therapy for neuroblastoma: ALK inhibitors. Klin Padiatr 2013;225:303-8.
75. Javanmardi N, Fransson S, Djos A, et al. Low frequency ALK hotspots mutations in neuroblastoma tumours detected by ultra-deep sequencing: implications for ALK inhibitor treatment. Sci Rep 2019;9:2199.
76. Cheung NK, Zhang J, Lu C, et al. Association of age at diagnosis and genetic mutations in patients with neuroblastoma. JAMA 2012;307: 1062-71.
77. Molenaar JJ, Koster J, Zwijnenburg DA, et al. Sequencing of neuroblastoma identifies chromothripsis and defects in neuritogenesis genes. Nature 2012;483:589-93.
78. Peifer M, Hertwig F, Roels F, et al. Telomerase activation by genomic rearrangements in high-risk neuroblastoma. Nature 2015;526:700-4.
79. Pugh TJ, Morozova O, Attiyeh EF, et al. The genetic landscape of high-risk neuroblastoma. Nat Genetics 2013;45:279-84.

80. Nuchtern JG, London WB, Barnewolt CE, et al. A prospective study of expectant observation as primary therapy for neuroblastoma in young infants: a Children's Oncology Group study. Ann Surg 2012;256:573-80.
81. Strother DR, London WB, Schmidt ML, et al. Outcome after surgery alone or with restricted use of chemotherapy for patients with low-risk neuroblastoma: results of Children's Oncology Group study P9641. J Clin Oncol 2012;30:1842-8.
82. Baker DL, Schmidt ML, Cohn SL, et al. Outcome after reduced chemotherapy for intermediate-risk neuroblastoma. N Engl J Med 2010;363:1313-23.
83. Nickerson HJ, Matthay KK, Seeger RC, et al. Favorable biology and outcome of stage IV-S neuroblastoma with supportive care or minimal therapy: a Children's Cancer Group study. J Clin Oncol 2000;18:477-86.
84. Matthay KK, Villablanca JG, Seeger RC, et al. Treatment of high-risk neuroblastoma with intensive chemotherapy, radiotherapy, autologous bone marrow transplantation, and 13-cis-retinoic acid. Children's Cancer Group. N Engl J Med 1999;341:1165-73.
85. Zhou MJ, Doral MY, DuBois SG, Villablanca JG, Yanik GA, Matthay KK. Different outcomes for relapsed versus refractory neuroblastoma after therapy with ^{131}I-metaiodobenzylguanidine (^{131}I-MIBG). Eur J Cancer 2015;51:2465-72.
86. Yu AL, Gilman AL, Ozkaynak MF, et al. Anti-GD2 antibody with GM-CSF, interleukin-2, and isotretinoin for neuroblastoma. N Engl J Med 2010;363:1324-34.
87. Pule MA, Savoldo B, Myers GD, et al. Virus-specific T cells engineered to coexpress tumor-specific receptors: persistence and antitumor activity in individuals with neuroblastoma. Nat Med 2008; 14:1264-70.
88. Louis CU, Savoldo B, Dotti G, et al. Antitumor activity and long-term fate of chimeric antigen receptor-positive T cells in patients with neuroblastoma. Blood 2011;118:6050-6.
89. Diede SJ. Spontaneous regression of metastatic cancer: learning from neuroblastoma. Nat Rev Cancer 2014;14:71-2.
90. Matthay KK. Stage 4S neuroblastoma: what makes it special. J Clin Oncol 1998;16:2003-6.
91. Nakagawara A. Molecular basis of spontaneous regression of neuroblastoma: role of neurotrophic signals and genetic abnormalities. Hum Cell 1998;11:115-24.
92. Pritchard J, Hickman JA. Why does stage 4s neuroblastoma regress spontaneously? Lancet 1994; 344:869-70.
93. Bessho F. Comparison of the incidences of neuroblastoma for screened and unscreened cohorts. Acta Paediatr 1999;88:404-6.
94. Schilling FH, Spix C, Berthold F, et al. Neuroblastoma screening at one year of age. N Engl J Med 2002;346:1047-53.
95. Woods WG, Gao RN, Shuster JJ, et al. Screening of infants and mortality due to neuroblastoma. N Engl J Med 2002;346:1041-6.
96. Yamamoto K, Ohta S, Ito E, et al. Marginal decrease in mortality and marked increase in incidence as a result of neuroblastoma screening at 6 months of age: cohort study in seven prefectures in Japan. J Clin Oncol 2002;20:1209-14.
97. Brodeur GM, Bagatell R. Mechanisms of neuroblastoma regression. Nat Rev Clin Oncol 2014; 11:704-13.
98. Croucher JL, Iyer R, Li N, et al. TrkB inhibition by GNF-4256 slows growth and enhances chemotherapeutic efficacy in neuroblastoma xenografts. Cancer Chemother Pharmacol 2015;75:131-41.
99. Iyer R, Varela CR, Minturn JE, et al. AZ64 inhibits TrkB and enhances the efficacy of chemotherapy and local radiation in neuroblastoma xenografts. Cancer Chemother Pharmacol 2012;70:477-86.
100. Iyer R, Wehrmann L, Golden RL, et al. Entrectinib is a potent inhibitor of Trk-driven neuroblastomas in a xenograft mouse model. Cancer Lett 2016;372:179-86.
101. Bove KE, McAdams AJ. Composite ganglioneuroblastoma. An assessment of the signficance of histological maturation in neuroblastoma diagnosed beyond infancy. Arch Pathol Lab Med 1981;105:325-30.
102. Stout AP. Ganglioneuroma of the sympathetic nervous system. Surg Gynecol Obstet 1947;84: 101-10.
103. Weiss SW, Goldblum JR, eds. Enzinger and Weiss's soft tissue tumors, 4th ed. St. Louis: Mosby; 2001:1284-8.
104. Carpenter WB, Kernohan JW. Retroperitoneal ganglioneuromas and neurofibromas. A clinicopathological study. Cancer 1963;16:788-97.
105. Wyman HE, Chappell BS, Jones WR Jr. Ganglioneuroma of bladder: report of a case. J Urol 1950;63:526-32.
106. Nassiri M, Ghazi C, Stivers JR, Nadji M. Ganglioneuroma of the prostate. A novel finding of neurofibromatosis. Arch Pathol Lab Med 1994; 118:938-9.
107. Mithofer K, Grabowski EF, Rosenberg AE, Ryan DP, Mankin HJ. Symptomatic ganglioneuroma of bone. A case report. J Bone Joint Surg 1999; 81:1589-95.
108. Christein JD, Kim AW, Jakate S, Deziel DJ. Central pancreatectomy for a pancreatic ganglioneuroma in a patient with previous neuroblastoma. Pancreatology 2002;2:557-60.

109. Wallace CA, Hallman JR, Sangueza OP. Primary cutaneous ganglioneuroma: a report of two cases and literature review. Am J Dermatopathol 2003;25:239-42.
110. Cannon TC, Brown HH, Hughes BM, Wenger AN, Flynn SB, Westfall CT. Orbital ganglioneuroma in a patient with chronic progressive proptosis. Arch Opthalmol 2004;122:1712-4.
111. Pardalidis NP, Grigoriadis K, Papatsoris AG, Kosmaoglu EV, Horti M. Primary paratesticular adult ganglioneuroma. Urology 2004;63:584-5.
112. Shekitka KM, Sobin LH. Ganglioneuroma of the gastrointestinal tract. Relation to von Recklinghausen disease and other multiple tumor syndromes. Am J Surg Pathol 1994;18:250-7.
113. Kyoshima K, Sakai K, Kanaji M, et al. Symmetric dumbbell ganglioneuromas of bilateral C2 and C3 roots with intradural extension associated with von Recklinghausen's disease: case report. Surg Neurol 2004;61:468-73.
114. Shawa H, Elsayes KM, Javadi S, et al. Adrenal ganglioneuroma: features and outcomes of 27 cases at a referral cancer centre. Clin Endocrinol 2014;80:342-7.
115. Hamilton JP, Koop CE. Ganglioneuromas in children. Surg Gynecol Obstet 1965;121:803-12.
116. Pachter MR, Lattes R. Neurogenous tumors of the mediastinum: a clinicopathologic study bases on 50 cases. Dis Chest 1963;44:79-87.
117. Mack E, Sarto GE, Crummy AB, Carlson IH, Curet LB, Wu J. Virilizing adrenal ganglioneuroma. JAMA 1978;239:2273-4.
118. Aguirre P, Scully RE. Testosterone-secreting adrenal ganglioneuroma containing Leydig cells. Am J Surg Pathol 1983;7:699-705.
119. Abell MR, Hart WR, Olson JR. Tumors of the peripheral nervous system. Hum Pathol 1970; 1:503-51.
120. Ackerman LV, Taylor FH. Neurogenous tumor within the thorax: a clinicopathologic evaluation of forty-eight cases. Cancer 1951;4:669-91.
121. Fletcher CD, Fernando IN, Braimbridge MV, McKee PH, Lyall JR. Malignant nerve sheath tumour arising in a ganglioneuroma. Histopathology 1988;12:445-8.
122. Banks E, Yum M, Brodhecker C, Goheen M. A malignant peripheral nerve sheath tumor in association with a paratesticular ganglioneuroma. Cancer 1989;64:1738-42.
123. Ghali VS, Gold JE, Vincent RA, Cosgrove JM. Malignant peripheral nerve sheath tumor arising spontaneously from retroperitoneal ganglioneuroma: a case report, review of the literature, and immunohistochemical study. Hum Pathol 1992;23:72-5.
124. de Chadarévian JP, MaePascasio J, Halligan GE, et al. Malignant peripheral nerve sheath tumor arising from an adrenal ganglioneuroma in a 6-year-old boy. Pediatr Dev Pathol 2004;7:277-84.
125. Ricci A Jr, Parham DM, Woodruff JM, Callihan T, Green A, Erlandson RA. Malignant peripheral nerve sheath tumors arising from ganglioneuromas. Am J Surg Pathol 1984;8:19-29.
126. Keller SM, Papazoglou S, McKeever P, Baker A, Roth JA. Late occurrence of malignancy in a ganglioneuroma 19 years following radiation therapy to a neuroblastoma. J Surg Oncol 1984; 25:227-31.
127. Chandrasoma P, Shibata D, Radin R, Brown LP, Koss M. Malignant peripheral nerve sheath tumor arising in an adrenal ganglioneuroma in an adult male homosexual. Cancer 1986;57:2022-5.
128. Kimura S, Kawaguchi S, Wada T, Nagoya S, Yamashita T, Kikuchi K. Rhabdomyosarcoma arising from a dormant dumbbell ganglioneuroma of the lumbar spine: a case report. Spine (Phila Pa 1976) 2002;27:E513-7.

12 OTHER ADRENAL NEOPLASMS AND TUMOR-LIKE LESIONS

Various neoplastic and non-neoplastic lesions cause enlargement of the adrenal glands. Aside from cortical and chromaffin cell tumors, the neoplasms most often encountered are primary benign or malignant tumors arising from nonendocrine cell types that reside in the normal adrenal gland, or metastases to the adrenal gland from distant sites. There are also a few case reports of both endocrine and nonendocrine tumors potentially related to the adrenal gland developmentally, either by steroidogenic lineage or by proximity to the adrenogenital/gonadal ridge (see chapter 1). Non-neoplastic causes of adrenal enlargement include hemorrhages, cysts, amyloidosis, infections, and noninfectious inflammatory conditions. Each of these entities can be confusing both clinically and radiologically. In some cases, the clinical context and imaging characteristics allow for a presumptive diagnosis (1–3), but in many instances diagnosis is by fine needle aspiration biopsy or examination of surgically resected specimens.

Nonendocrine adrenal lesions are now most often discovered as incidentalomas, defined as masses larger than 1 cm detected during imaging for unrelated conditions. The strictest definition excludes lesions found during the workup for staging of known cancers (1). With current imaging technology, the prevalence of incidentalomas is estimated at approximately 4 percent. In a compilation of 25 autopsy studies it was 6 percent (1).

Adrenal incidentalomas are uncommon below age 30, and the prevalence increases with age. Approximately 10 to 15 percent are bilateral. In developed countries, the most frequent diagnoses are adrenal cortical adenoma (up to 80 percent), followed by myelolipomas, adrenal cortical carcinomas, pheochromocytomas, cysts, and unexpected metastatic tumors (each less than 10 percent, with varying prevalence in different series). In developing countries, inflammatory lesions and cysts comprise a higher proportion of cases (4). The prevalence and nature of adrenal incidentalomas also vary between Western and Asian patient populations, and with the type of imaging equipment available (2).

While the imaging studies leading to the discovery of incidentalomas often result from nonspecific abdominal, genitourinary, or systemic complaints, an occasional additional route to detection of nonendocrine adrenal masses is through the clinical workup of patients who are ultimately diagnosed with *pseudopheochromocytoma* or *pseudo-Cushing syndrome*. Pseudopheochromocytoma is a syndrome defined by transient episodes of paroxysmal hypertension, elevated epinephrine and metanephrine (5), and other signs and symptoms suggesting pheochromocytoma, with no pheochromocytoma detected. It differs from panic attacks and related conditions in that the episodes are not overtly triggered by fear or anxiety (6). Pseudo-Cushing is defined by mild to moderate adrenocorticotropic hormone (ACTH)-dependent hypersecretion of cortisol that can be caused by various neural and metabolic mechanisms (7). Regardless of the route to discovery, once an adrenal mass is found, integrated radiologic, hormonal, and clinical evaluation is required to guide patient management.

Several algorithms have been developed to help distinguish functional from nonfunctional and benign from malignant lesions (8,9). The diagnostic workup of adrenal masses frequently begins with anatomic imaging modalities that may include computerized tomography (CT), magnetic resonance imaging (MRI), or ultrasound. These may be followed by functional imaging, including "theranostic" molecular imaging. An example of theranostic imaging is the use of meta-iodobenzylguanidine (MIBG) in the workup of suspected pheochromocytoma (10). For some types of adrenal lesions, a helpful diagnostic feature in imaging studies is the presence and pattern of calcification, which while not specific, are part of an integrated imaging approach that also assesses size, heterogeneity, contrast enhancement, and washout, and other parameters depending on the modality employed.

Adrenal calcification is associated with primary adrenal tumors including neuroblastoma, cortical adenoma and carcinoma, pheochromocytoma, and various nonendocrine tumors (11). Adrenal cysts, some infectious diseases such as histoplasmosis (12), and metastases to the adrenal gland are also often calcified (11). Occasionally, the adrenal gland contains metaplastic bone, presumably as a remote consequence of necrosis or hemorrhage. Bilateral adrenal calcifications have also been seen in metabolic diseases, including Wolman disease, a rare autosomal recessive disorder of lipid metabolism caused by deficiency of lysosomal acid lipase (12b,12c), and adenosine deaminase-deficient severe combined immunodeficiency.

PRIMARY NONENDOCRINE TUMORS

The morphologic and immunohistochemical features of primary nonendocrine adrenal tumors are basically the same as for their counterparts in other locations. For some tumors, the diagnosis is particularly difficult because of because of the histologic resemblance of various types of nonendocrine tumors to each other and to adrenal cortical neoplasms.

Mesenchymal Tumors

Myelolipoma. Myelolipomas are benign tumor-like lesions consisting of mature adipose tissue admixed in varying proportions with trilineage hematopoietic elements. These tumors are classified as mesenchymal neoplasms in the 2017 World Health Organization (WHO) classification of tumors of endocrine origin (13). They are now the second most common adrenal incidentaloma, after cortical adenomas in some series (14). While most myelolipomas arise in the adrenal gland, extra-adrenal locations have been reported, including liver, kidney, spleen, mediastinum, lung, nasal cavity, meninges, and eyes (14). Myelolipoma has also been reported in ectopic adrenal cortex (15).

In a 2018 review of all available case reports (14), the average age of patients with adrenal myelolipoma at diagnosis was about 51 years, with a range from 1 to 83 years and an equal sex distribution. Twelve percent of the tumors were bilateral. The estimated incidence at autopsy varies from 0.08 to 0.2 percent; in one study it was 0.01 percent for the age group between 36 and 65 years (16). A female predominance has been reported in a series of Chinese patients (17).

Patients undergoing surgical resection of adrenal myelolipomas are typically asymptomatic or have a variety of nonspecific complaints or findings, such as abdominal pain or discomfort, flank pain, or loin pain. Approximately 5 percent have a palpable abdominal mass (14). A potential complication of large tumors is spontaneous rupture, noted in 4.5 percent of patients, which can result in chronic retroperitoneal hemorrhage or hemorrhagic shock (14,18). In addition, adrenal myelolipomas are frequently associated with hypertension and type 2 diabetes (14).

Several theories have been advanced for the pathogenesis of myelolipomas, with none conclusive (14). However, the tumors often appear to be monoclonal, as judged from nonrandom X-chromosome analysis in a single small study of female patients (19). A contributory role of intrinsic adrenal cortical disorders and/or excess ACTH is suggested by the association with congenital adrenal hyperplasia in 10 percent of all patients analyzed, while adrenal hypersecretory disorders, including hypercortisolism, primary hyperaldosteronism, and hyperandrogenism were found in an additional 7.5 percent of cases (14). These associations are consistent with experimental findings by Hans Selye in 1950 showing that prolonged, combined treatment with testosterone and crude anterior pituitary preparations converted the zona reticularis and inner fasciculata of rat adrenal glands into typical bone marrow-containing trilineage hemopoietic tissue and fat cells (20).

Endocrine disorders affecting steroidogenesis are also associated with adrenal myelolipoma, including Carney complex, polycystic ovary syndrome, and Nelson syndrome, in addition to a variety of possibly coincidental endocrine and nonendocrine tumors in isolated cases as recently reviewed (14). A patient with alimentary tract ganglioneuromatosis-lipomatosis, adrenal myelolipomas, pancreatic telangiectasia, and

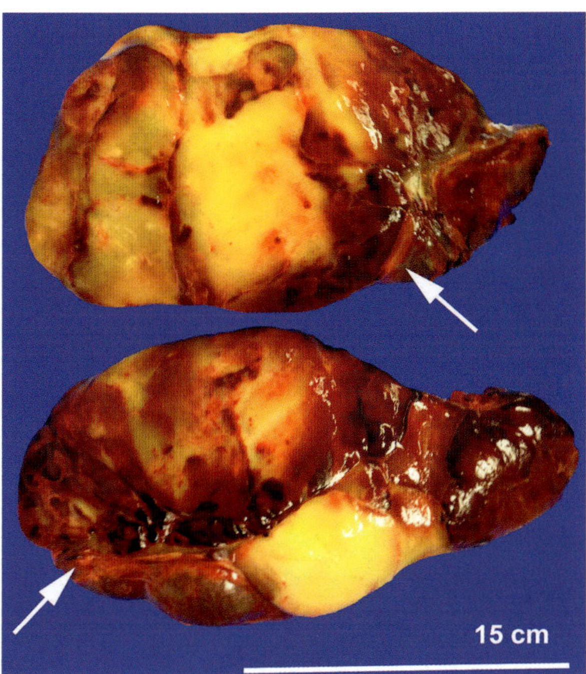

Figure 12-1

BILATERAL GIANT ADRENAL MYELOLIPOMAS

The myelolipomas are from an adult man with poorly controlled congenital adrenal hyperplasia. The red-brown areas contain hematopoietic cells and blood and the yellow patches consist of fat. Yellow-orange bands traversing the specimens are adrenal cortex (arrows).

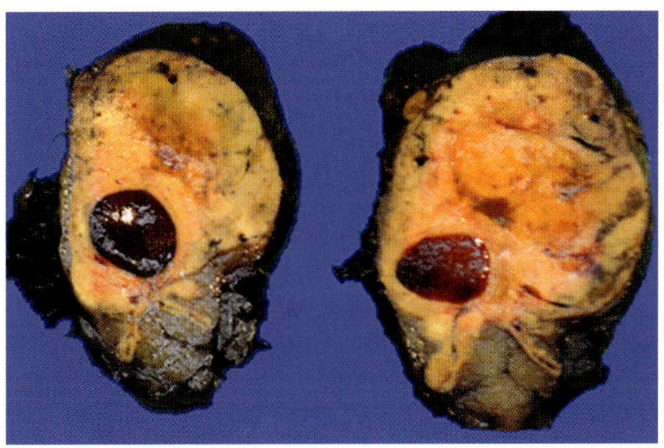

Figure 12-2

ADRENAL MYELOLIPOMA

The tumor consists almost entirely of fat. A somewhat unusual circumscribed focus of hematopoietic cells resembles a tumor within a tumor.

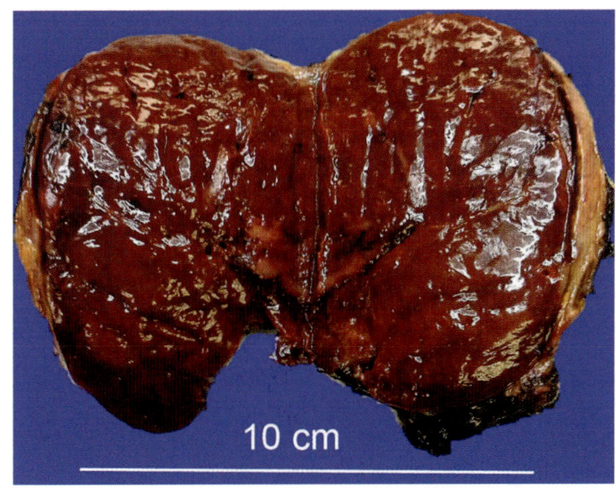

Figure 12-3

ADRENAL MYELOLIPOMA

This tumor consists almost entirely of hematopoietic cells. A peripheral rim of normal adrenal cortex is present at right.

multinodular goiter has been reported as having a possible neuroendocrine syndrome (21). There is usually no relationship to anemia or other hematopoietic abnormalities (16), although some patients with thalassemia are reported (14).

Adrenal myelolipomas vary from small lesions discovered incidentally at autopsy to immense masses, the largest to date weighing 11.4 kg (22). Giant myelolipomas, defined by diameter over 10 cm, are sometimes discovered incidentally but are especially associated with congenital adrenal hyperplasia, where more than 50 percent are bilateral (fig. 12-1) (14). The tumors are well circumscribed on gross inspection, but rarely encapsulated. The color varies from pale yellow to deep red or red-brown, depending upon the proportions of fat and hematopoietic elements (figs. 12-2, 12-3).

The contour of adrenal myelolipoma is smooth, wavy, or irregular, and may show intermingling of cortical cells with elements of the myelolipoma. Microscopically, the surrounding cortical cells can appear relatively normal or be compressed. There is a variable mixture of mature fat with hematopoietic elements (fig. 12-4).

Occasionally, infarction or hemorrhage occurs along with secondary hematoma formation or fibrosis. Foci of ossification are occasionally present. The diagnosis is sometimes established by fine needle aspiration biopsy (fig. 12-5) in conjunction with a typical heterogeneous appearance on imaging studies. The differential diagnosis principally includes other fatty tumors.

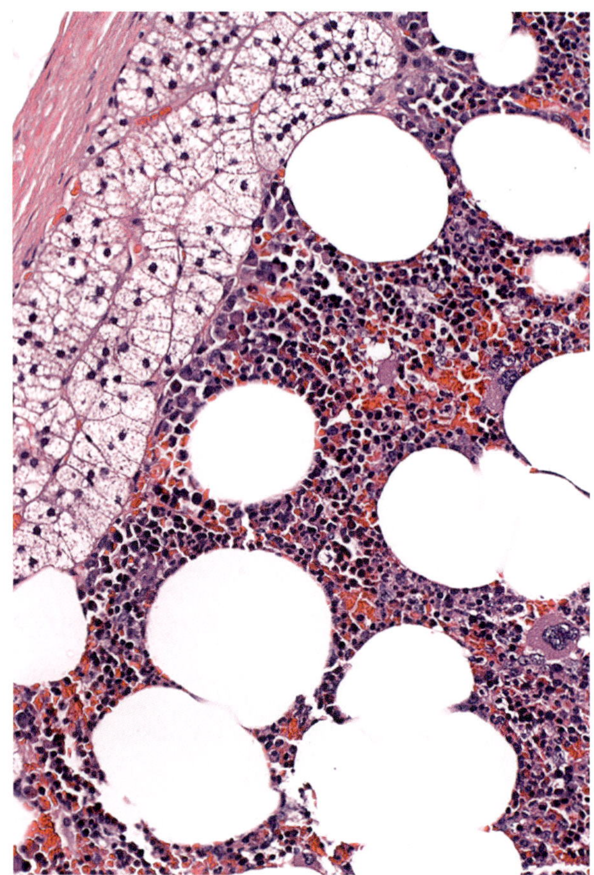

Figure 12-4

ADRENAL MYELOLIPOMA

The tumor consists of tri-lineage hematopoietic cells with a moderate amount of admixed fat. A peripheral rim of normal adrenal cortex is present at left.

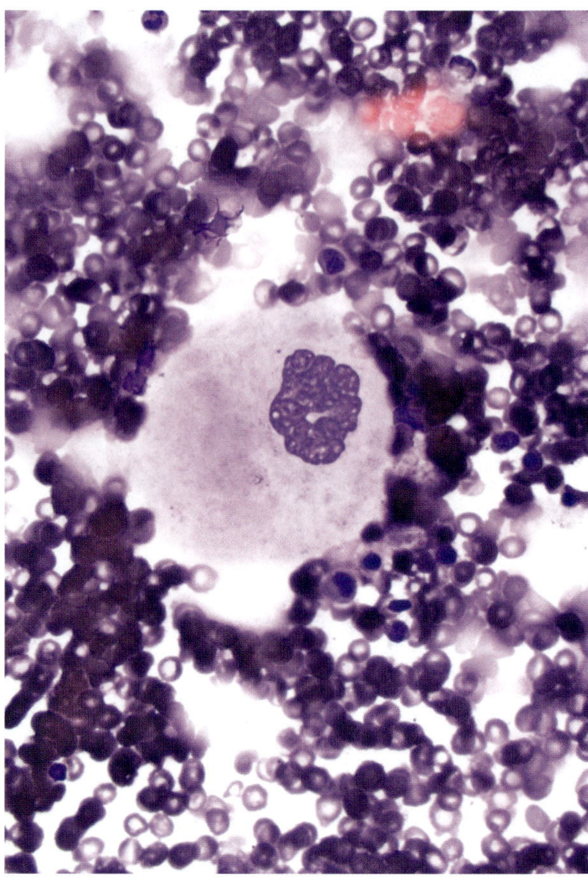

Figure 12-5

ADRENAL MYELOLIPOMA

Although the fine needle aspirate is sparse, the prominent megakaryocyte is a helpful diagnostic feature.

Fatty Tumors. The most frequently encountered primary fatty tumors in the adrenal gland are lipomas and cortical tumors with lipomatous change. Other rare examples include angiomyolipomas (see PEComas), liposarcomas (17,23), hibernomas (17), and teratomas (17).

Vascular Tumors. The adrenal glands occasionally give rise to benign or malignant vascular tumors, most of which are derived from blood vessels. An important diagnostic consideration is the need to distinguish these tumors from a reparative vascular proliferation, which can occur in recanalization of hemorrhage in other types of tumor or in non-neoplastic conditions. Adrenal hemorrhage in infarcted cortical adenomas can be robustly recanalized by the numerous endothelial cells that reside in the adrenal cortex, producing both large and small vascular channels with occasional hobnail cells. In some cases it is difficult to determine whether a vascular proliferative lesion is a true neoplasm or an end result of this process (fig. 12-6).

A massive hemorrhagic adrenal cortical adenoma has been reported mimicking an angiosarcoma (24). Conversely, adrenal angiosarcomas can be extremely bloody lesions, and careful search may be required to identify tumor cells within pools of blood.

Adrenal hemangiomas are extremely rare and most are discovered incidentally at autopsy. The files of the Armed Forces Institute of Pathology (AFIP) listed one adrenal angioma for every 10,000 autopsy protocols accessioned between 1943 and 1958 (frequency of 0.01 percent) (25).

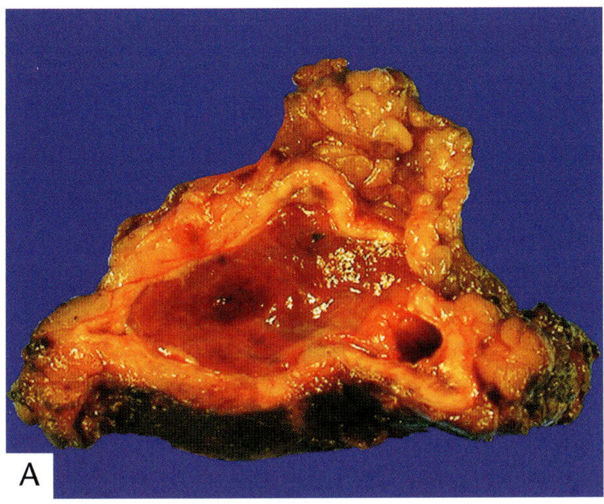

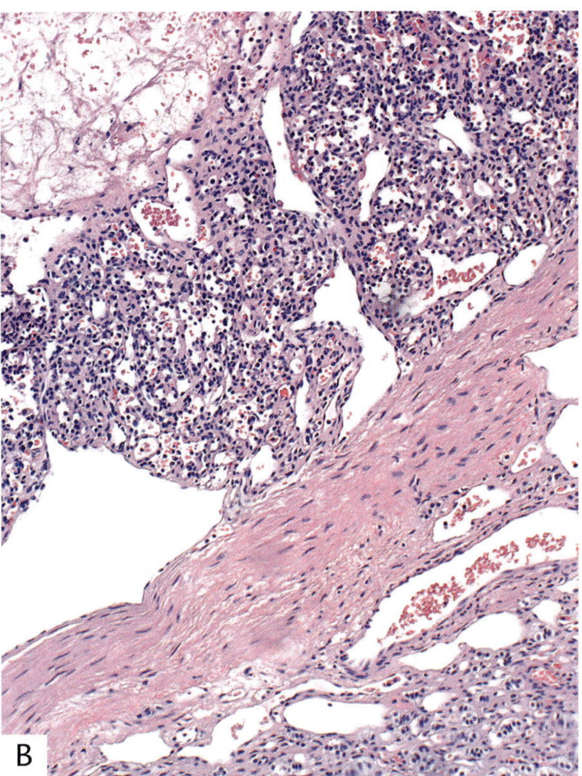

Figure 12-6

PRIMARY ADRENAL HEMANGIOMA GROSSLY MIMICKING A RESOLVING HEMATOMA

A: The gross appearance of this retracted fibrinous and hemorrhagic lesion is suggestive of a resolving hematoma.

B: Histologic sections show resolving hemorrhage (top left) associated with a capillary hemangioma. The tumor is traversed by a nerve within the adrenal medulla.

The patients are usually in the 3rd to the 8th decades of life, and most are women. Only rarely do patients present with symptoms related to a mass effect by the adrenal tumor. The primary adrenal hemangioma in figure 12-7 occurred in a patient with symptoms suggesting pheochromocytoma.

Adrenal hemangiomas discovered at autopsy are usually less than 2 cm in diameter, while those removed surgically typically measure up to 22 cm. Most are unilateral and solitary, but occasional bilateral examples have been reported (1). Capillary (fig. 12-8) or cavernous (fig. 12-9) features may predominate, and the architecture is sometimes mixed (see fig. 12-7). Intravascular extension can be present (see fig. 12-7). Extramedullary hematopoiesis is occasionally seen.

An unusual type of hemangioma reported as anastomosing hemangioma (26) has a predilection for the kidney but rare cases arising in the adrenal gland have been reported (27). These benign lesions have compact architecture, with tightly packed anastomosing capillary channels and areas of hobnail cells, but are distinguished from angiosarcomas by the lack of multilayered endothelium, little or no atypia, and only rare or absent mitoses. They are frequently associated with vascular thrombi and occasionally show extramedullary hematopoiesis (26,28). Variable findings are a lobular architecture and a prominent framework of nonendothelial supporting cells creating a "spleen-like" pattern of vascular channels (26) lined by hobnail endothelial cells (see fig. 12-8). Features of these benign lesions can be seen in hemangiomas that do not exhibit the full diagnostic spectrum (see fig. 12-7).

Adrenal hemangiomas sometimes occur in patients with other vascular lesions. A small adrenal hemangioma (benign hemangioendothelioma of infancy) from a newborn infant who also had a much larger hepatic hemangioma with identical histology is illustrated in figure 12-9. The child also had the Kasabach-Merritt syndrome with thrombocytopenic purpura and died of high output cardiac failure.

Approximately 40 examples of primary *adrenal angiosarcoma* have been reported in English publications (29), one coexisting with an adrenal hemangioma (30). One example apparently arose in a cortical adenoma (31). A few have been reported in patients with occupational exposure to vinyl chloride, arsenicals, or other

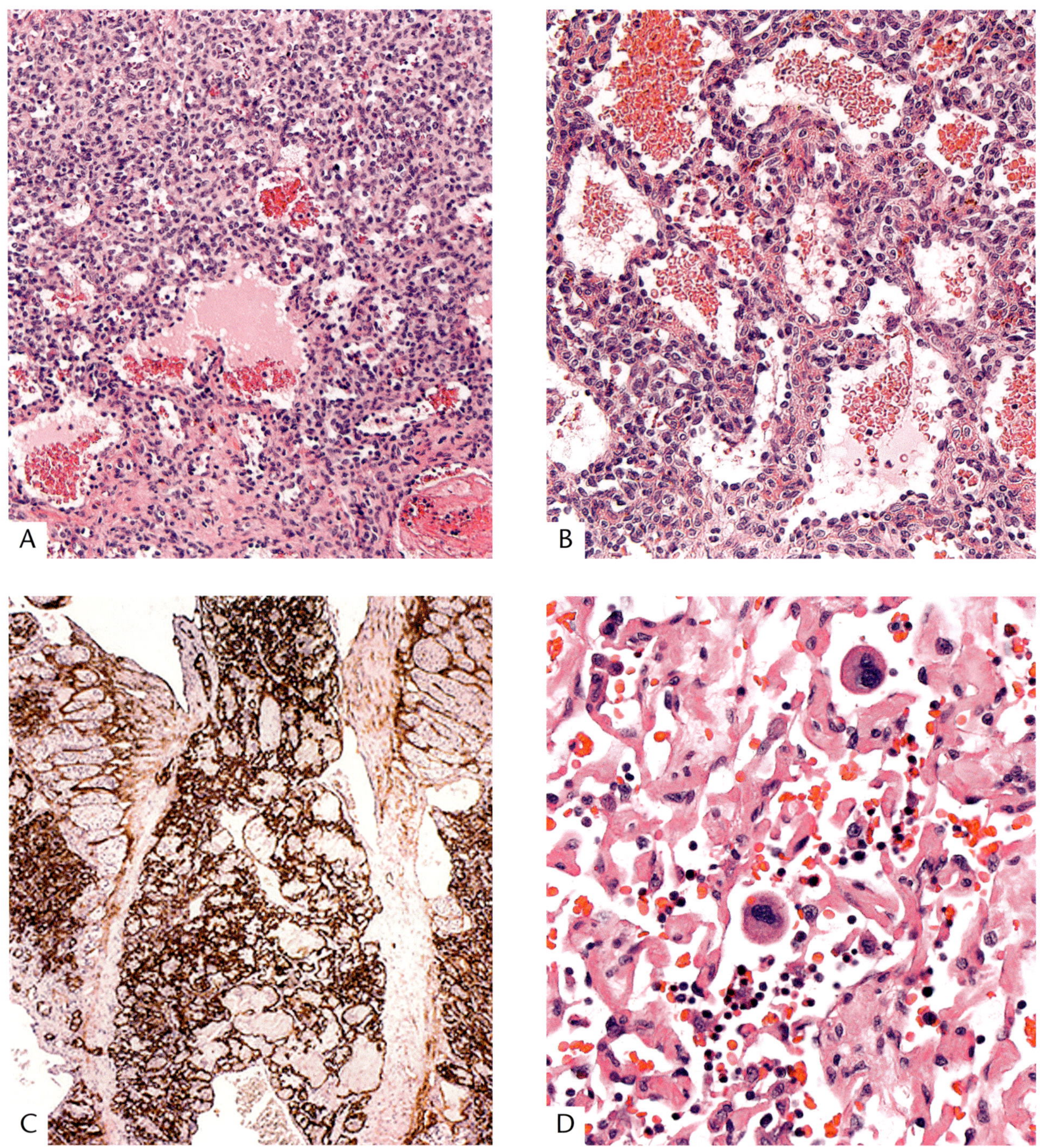

Figure 12-7

PRIMARY ADRENAL HEMANGIOMA

A: The subcortical area has a delicate capillary pattern, while more central areas are cavernous and have nonendothelial supporting cells. Fibrin thrombi are also present.

B: The more cavernous pattern, with dilated vascular spaces and delicate septa, shows prominent cytologically bland hobnail cells, which are not indicative of malignancy.

C: A portion of the adrenal hemangioma is located within the lumen of a large venous tributary of the central adrenal vein. Endothelial cells immunostain for CD34.

D: Adrenal hemangioma with extramedullary hematopoiesis (different case from A–C).

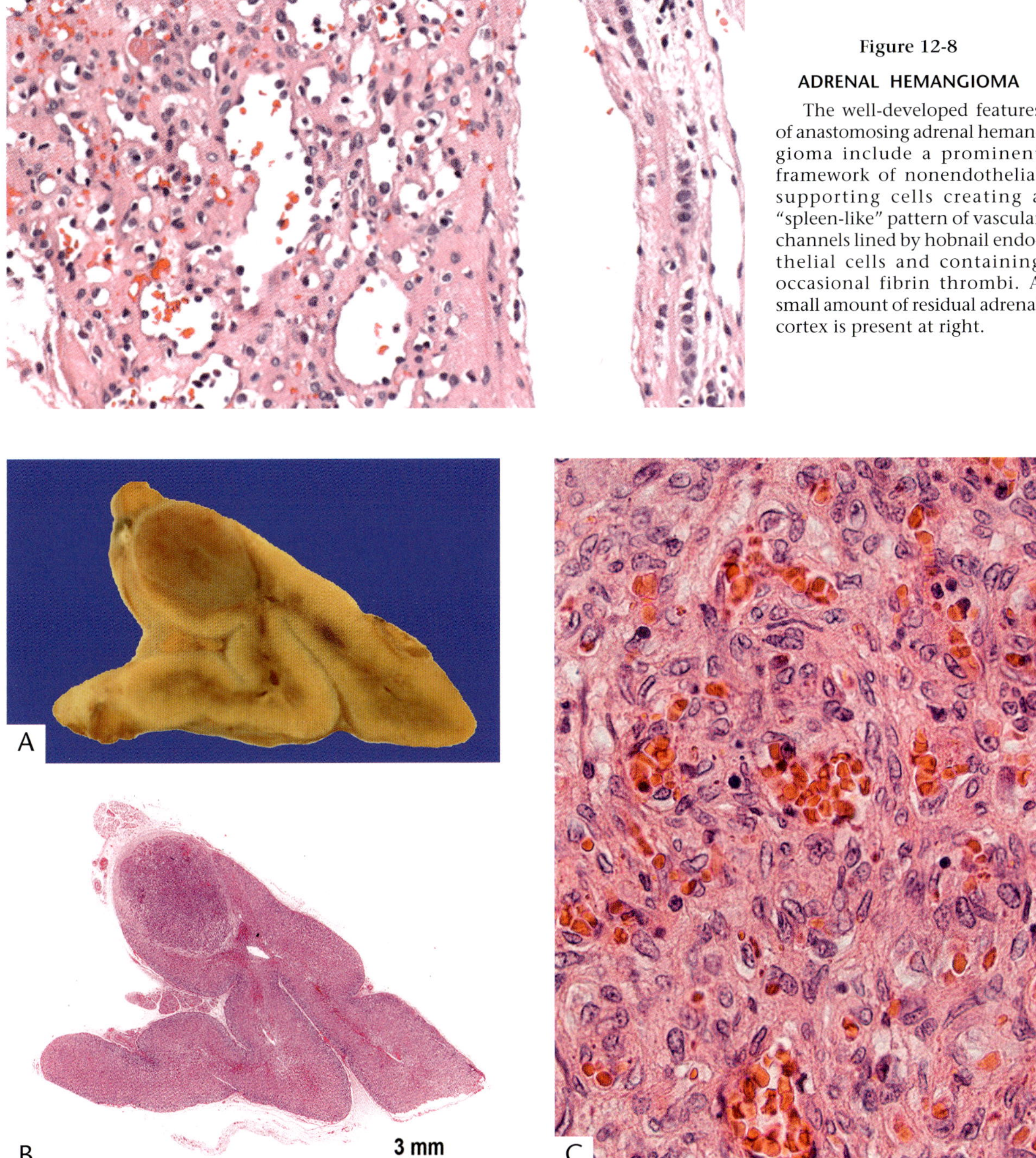

Figure 12-8

ADRENAL HEMANGIOMA

The well-developed features of anastomosing adrenal hemangioma include a prominent framework of nonendothelial supporting cells creating a "spleen-like" pattern of vascular channels lined by hobnail endothelial cells and containing occasional fibrin thrombi. A small amount of residual adrenal cortex is present at right.

Figure 12-9

JUVENILE (CAPILLARY) ADRENAL HEMANGIOMA

A,B: Small juvenile hemangioma of the adrenal gland in a newborn. There were numerous cutaneous hemangiomas and a large vascular tumor of the liver with identical histology. The adrenal tumor is uniformly tan, about 5 mm in diameter, and located in the ridge or crista of the gland. The infant also had the Kasabach-Merritt syndrome with thrombocytopenia.

C: The growth pattern in this juvenile hemangioma is more solid, with plump endothelial cells and relatively inconspicuous vascular lumens.

putative carcinogens (32). One case was reported in association with mesenteric fibromatosis (32).

Adrenal angiosarcoma usually presents little difficulty in diagnosis when there is a clear vasoformative pattern (fig. 12-10A). Problems arise with tumors having a solid pattern composed of epithelioid or spindle cells (seen focally in fig. 12-10B and to a greater extent in fig. 12-10C). For these tumors, the distinction from benign epithelioid vascular tumors is based on malignant cytologic features, and distinction from other epithelioid or spindle cell malignancies usually requires immunohistochemistry. Diagnostic pitfalls can be expression of cytokeratins, which is reported in 20 to 50 percent of cases, and occasional anomalous expression of neuroendocrine markers, recently reviewed (33). The transcription factor ERG is positive in most cases (33), and can be especially helpful in discriminating angiosarcoma from extraordinarily rare primary adrenal epithelioid sarcomas (34).

Lymphangioma of the adrenal gland is exceedingly rare. A review of 53 reported cases showed a mean patient age of 39.5 years, and a female predominance (35). The lining can be highlighted by immunohistochemical staining for the generic endothelial markers CD31 or CD34. Lymphangiomas are distinguished from hemangiomas by staining for podoplanin (D2-40), a marker that also stains adenomatoid tumors, and some reported cases are best classified as adenomatoid tumor. A cavernous lymphangioma of the adrenal gland is illustrated in figure 12-11.

Smooth Muscle Tumors. Primary adrenal leiomyomas and leiomyosarcomas are very rare tumors associated with the adrenal vein or its tributaries. Fewer than 20 *adrenal leiomyomas* had been reported up to about 2013, occurring in both children and adults (36). The tumors were bilateral in 2 of 4 pediatric patients and 3 of 12 adults.

Approximately 30 cases of primary adrenal leiomyosarcomas have been reported (37). The tumors all arose in adults, and were bilateral in two cases. Men and women were affected equally. Four patients had acquired immunodeficiency syndrome (AIDS), and in some cases, latent Epstein-Barr (EB) virus was also demonstrated in the tumor cells (37).

Histologically, adrenal leiomyomas (fig. 12-12) and leiomyosarcomas (fig. 12-13) are identical to their counterparts in other sites. Leiomyosarcomas can have large areas of hemorrhage and necrosis, an infiltrative pattern, and a high mitotic rate. Immunohistochemistry confirms the smooth muscle nature of the tumors in questionable cases.

Adenomatoid Tumor. Primary adenomatoid tumor of the adrenal gland is a rare benign tumor of mesothelial origin that typically occurs in adult men. Approximately 30 cases have been reported (38). On gross examination, the tumors are smooth, white, and homogeneous. Microscopically, they are not truly encapsulated, hence may give a false impression of local invasion (39). The architectural pattern is a variable combination of canalicular (angiomatoid) areas lined by flattened mesothelium and cord-like (gland-like) areas of cuboidal cells, usually with a small amount of intervening connective tissue (fig. 12-14). Cystic areas and inflammatory cells are sometimes present. The nuclei have a vesicular quality, often with a single, small, dot-like nucleolus; there may be prominent cytoplasmic vacuoles. This classic histomorphology often gives adenomatoid tumors a sieve-like appearance due to cytoplasmic vacuoles and gland-like spaces.

A mesothelial origin is confirmed by positive immunohistochemical staining for calretinin, WT1, podoplanin, cytokeratins, and HBME1. Retention of BAP1 distinguishes adenomatoid tumor (fig. 12-14) from malignant mesothelioma, especially the adenomatoid variant, which can closely mimic adenomatoid tumor but is distinguished by loss of BAP1 (40). Composite tumors consisting of adenomatoid tumor and cortical adenoma (41) or myelolipoma (42) have been reported. A major entity in the differential diagnosis is lymphangioma (38).

Angiomyolipomas and Other Perivascular Epithelioid Cell Tumors (PEComas). Angiomyolipomas and other PEComas involving the adrenal gland most often arise in the kidney and involve the adrenal gland secondarily, but occasional primary adrenal examples have been reported (23,33,43). The tumors occur predominantly in women and are frequently associated with tuberous sclerosis complex.

Classic angiomyolipomas are triphasic tumors consisting of perivascular epithelioid cells clustered around irregular thick-walled blood

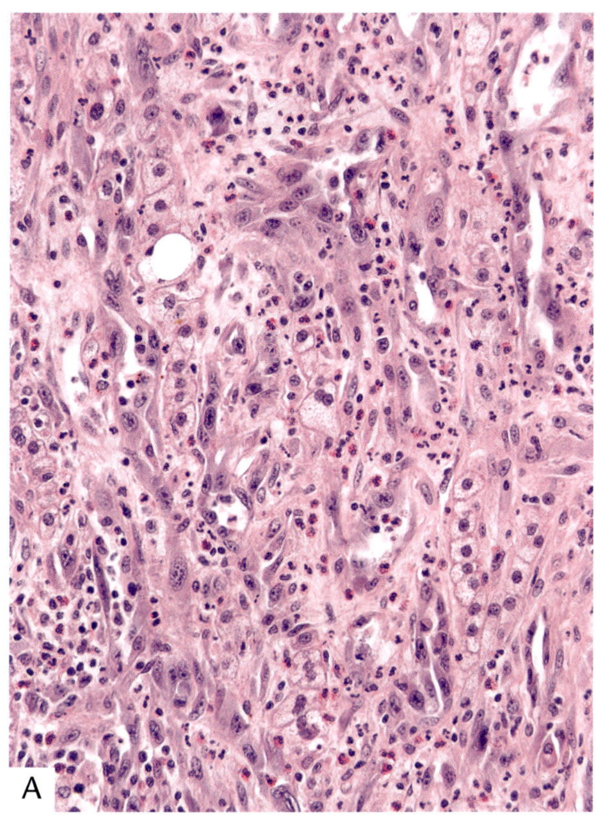

Figure 12-10

PRIMARY ADRENAL ANGIOSARCOMA

A: There is well-developed vasoformative architecture with sinusoidal growth between cords of vacuolated cortical cells. Acute inflammatory cells and eosinophils are also present.

B: Primary adrenal angiosarcoma (same case as A) growing in part as a high-grade tumor with epithelioid features. Residual adrenal cortical cells are also present.

C: High-grade epithelioid angiosarcoma of adrenal gland surrounds small nests of residual cortical cells. Tumor cells have prominent nucleoli and small clear spaces, suggesting primitive vascular lumen formation.

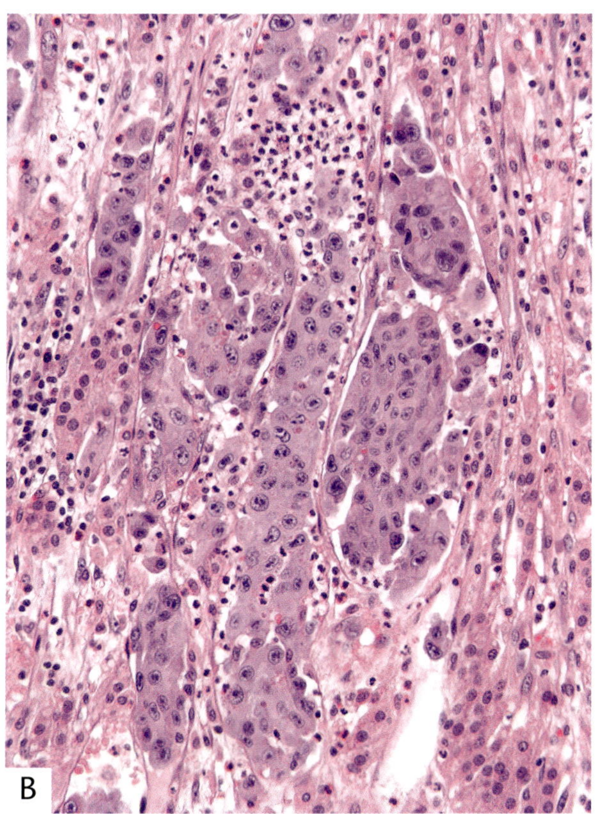

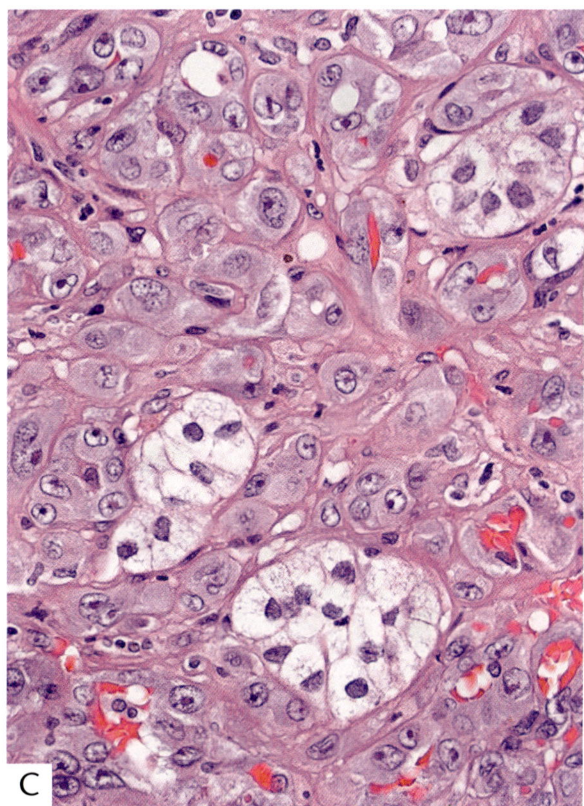

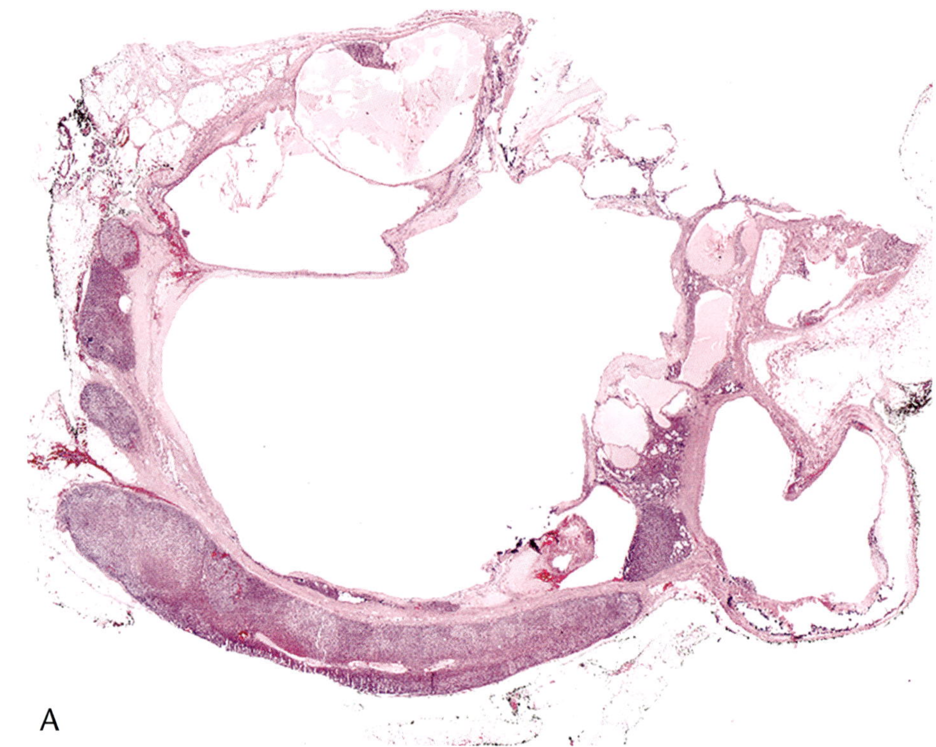

Figure 12-11

ADRENAL LYMPHANGIOMA

A: The tumor appears as a multilocular cyst filled in part by proteinaceous fluid. The cyst had clear watery contents. Portions of adrenal gland are splayed over the tumor surface.

B: Lymphangioma has a cavernous (or cystic) architecture with proteinaceous fluid and some collections of lymphocytes. Lymphoid aggregates were also present in the interstitium elsewhere in this lymphangioma. Mildly compressed adrenal cortex is present (left upper corner).

C: The flattened lymphatic endothelium of adrenal lymphangioma is strongly immunoreactive for CD31.

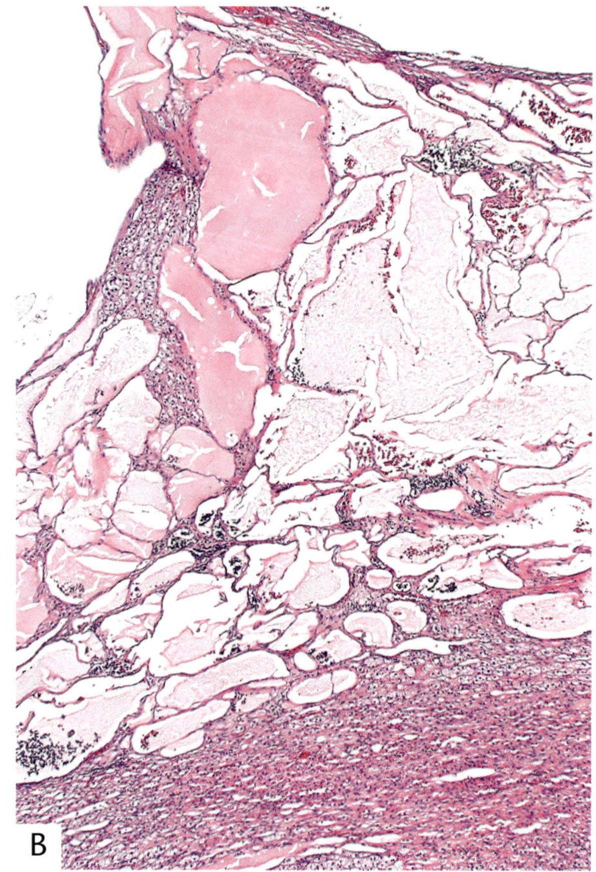

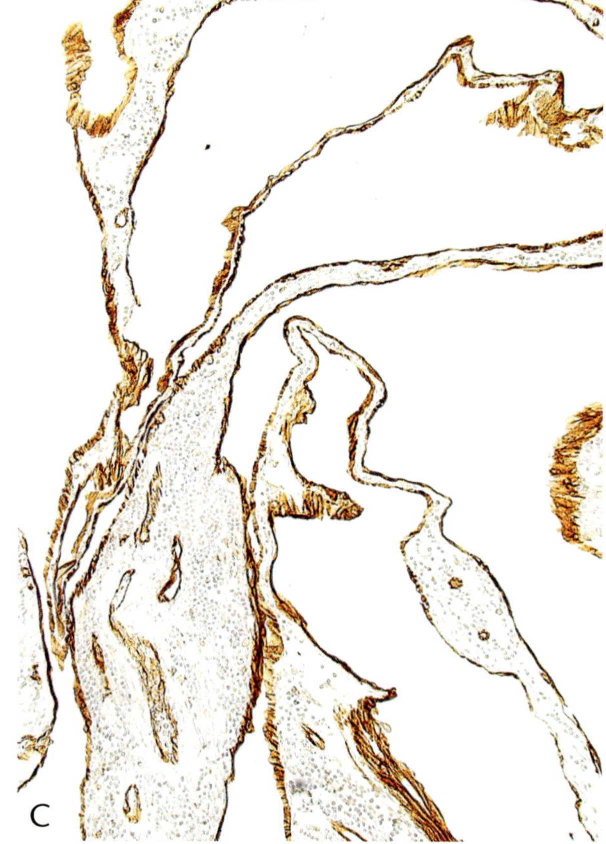

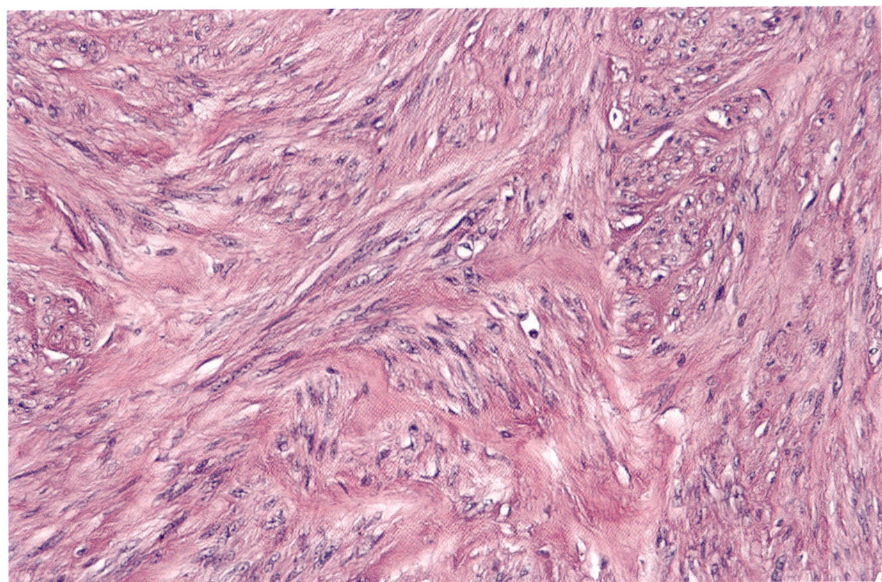

Figure 12-12

PRIMARY ADRENAL LEIOMYOMA

Primary adrenal leiomyoma is composed of interlacing fascicles of spindle cells in various planes of section. No mitotic figures were identified and there is no necrosis.

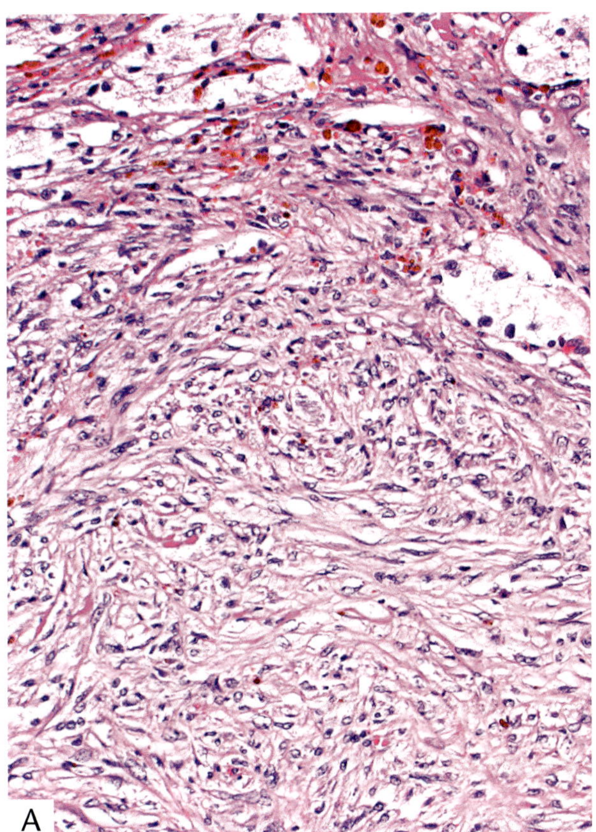

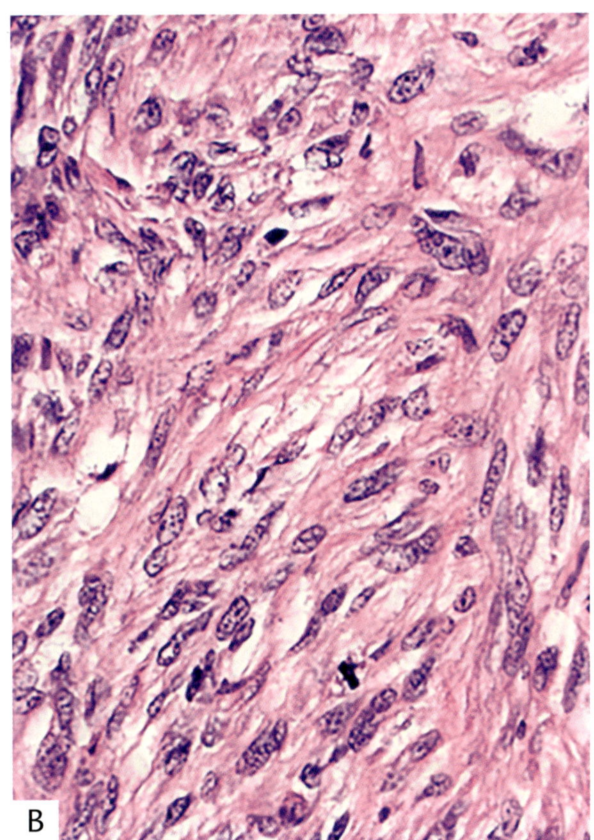

Figure 12-13

PRIMARY ADRENAL LEIOMYOSARCOMA

A: The leiomyosarcoma infiltrates the adrenal gland. Nests of residual cortical cells are present at the top.

B: This adrenal leiomyosarcoma had areas of confluent tumor necrosis. The mitotic rate averaged 15 per 10 high-power fields in some of the more active areas.

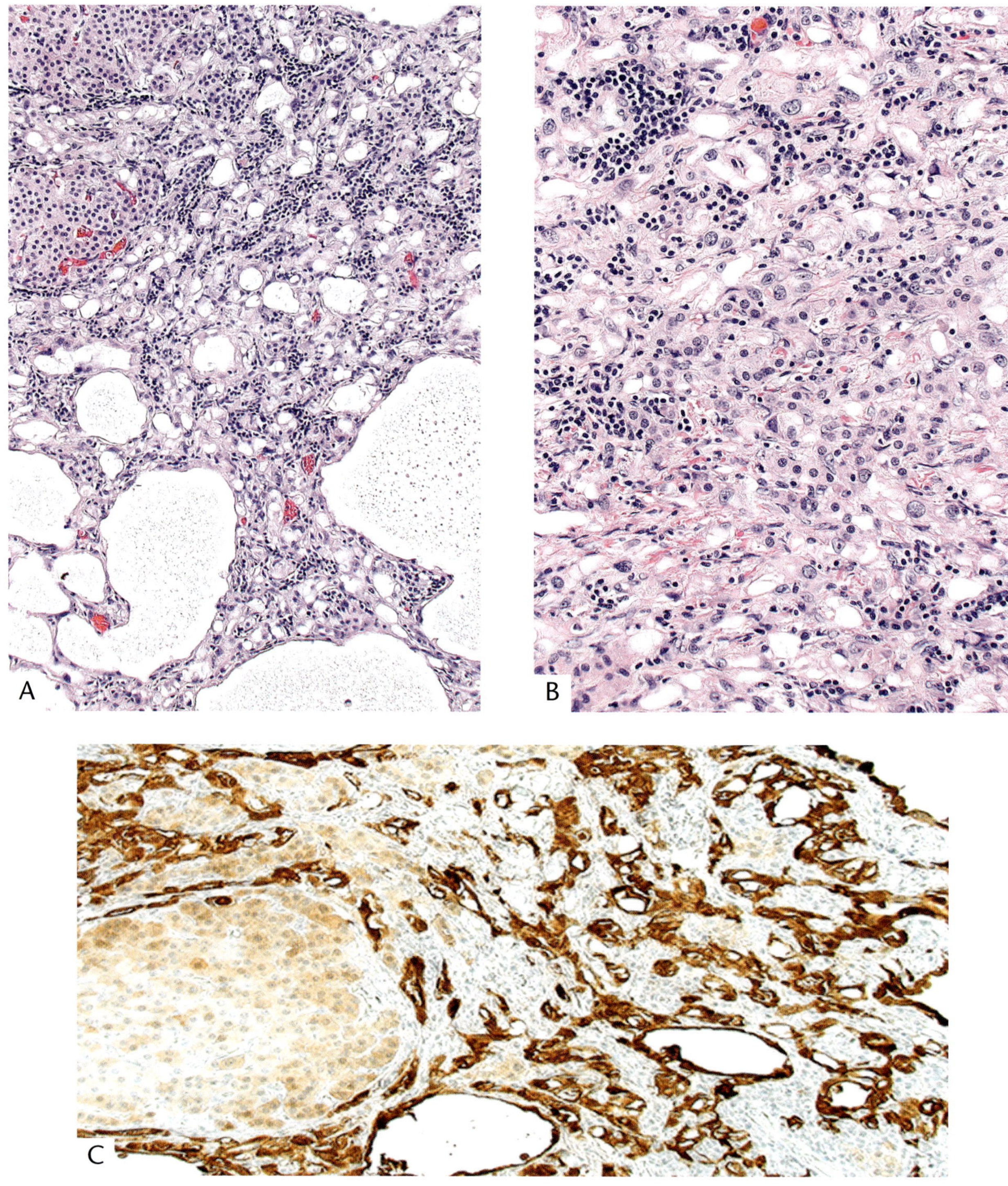

Figure 12-14

PRIMARY ADRENAL ADENOMATOID TUMOR

A: Numerous small gland-like spaces transition into cysts with a flattened lining. At top left, the tumor insinuates between and envelops small nests of adrenal cortical cells.

B: Higher magnification of a more solid area.

C: Immunohistochemical staining of adenomatoid tumors: calretinin.

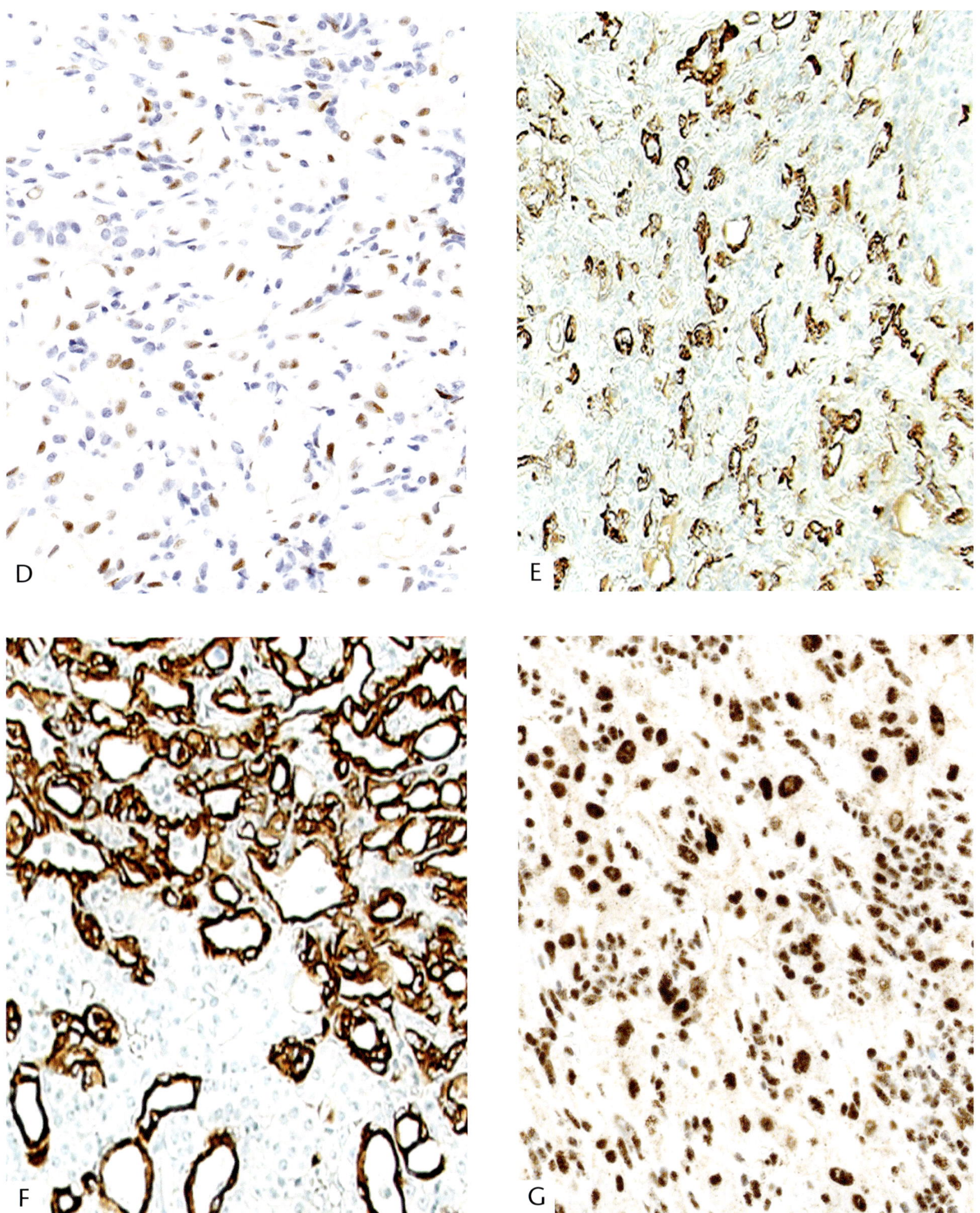

Figure 12-14, continued

D–G: Immunohistochemical staining of adenomatoid tumors: D, WT1; E, podoplanin; F, pancytokeratin; G, BAP1. The infiltrative pattern of adenomatoid tumors can mimic carcinoma in some cases (four different tumors shown).

vessels in a background of mature adipose tissue. They are usually benign but may be locally invasive and occasionally give rise to sarcomatous metastases. Other PEComas with little or no fat, varying proportions of epithelioid and spindle cells, clear or eosinophilic cytoplasm, and varied architectural patterns occasionally originate in the adrenal gland (33,43,43b) and have a higher malignant potential (figs. 12-15, 12-16). Adverse histologic features include marked cytologic atypia, high mitotic activity, and tumor necrosis (33). Because of their diverse features, the list of differential diagnoses is large, both histologically and radiologically (44). Considerations include renal cell carcinoma, adrenal cortical carcinoma, other mesenchymal tumors, and metastatic carcinomas.

PEComas are characterized immunohistochemically by the expression of both smooth muscle markers and melanocytic markers such as HMB-45, Melan-A/MART-1, MITF, and cathepsin K. Marker expression is variable, and expression of at least one smooth muscle and one melanocytic marker is often required for a confident diagnosis (33,43b). Melanocytic markers and cathepsin K, however, are not entirely specific for PEComa and can be expressed in bona fide smooth muscle tumors (45). The expression of melanocytic markers in PEComas is associated with *TSC2* alterations or *TFE3* fusions, which are also the main drivers of tumor development, but a one to one correlation is not always present (45). Nuclear immunoreactivity for TFE3 protein and fluorescent in situ hybridization (FISH) for the detection of chromosome translocations involving the *TFE3* gene at Xp11.2 may serve as ancillary techniques (46,47).

An additional diagnostic pitfall is that malignant epithelioid PEComas can closely resemble sarcomatoid adrenal cortical carcinoma, which can also stain for Melan-A/MART-1 in some cases. Exclusion of other adrenal cortical markers may therefore be required for diagnosis (14,43).

Neural Tumors

Primary neural tumors, excluding neuroblastomas, in the adrenal gland are rare, although their presence is not unexpected given the glands' extensive innervation (see chapter 1). Examples include *schwannomas* (48), including one classified as a microcystic/reticular schwannoma (49), *neurofibromas* (50), and *malignant peripheral nerve sheath tumors (MPNSTs)*. Apart from their varying histology, adrenal and periadrenal schwannomas appear to have a different underlying molecular etiology compared to schwannomas elsewhere (50b). One example of MPNST occurred in association with an ipsilateral pheochromocytoma, an example of a rare composite tumor (51), while another arose in an adrenal ganglioneuroma in an adult man who was human immunodeficiency virus (HIV) positive (52).

MPNSTs can exhibit spindle cell, epithelioid, or pleomorphic morphology, mimicking a variety of other malignant tumors (fig. 12-17). High-grade MPNSTs may show limited or absent expression of S-100 protein and SOX10. A feature recently reported to immunohistochemically identify most cases is loss of histone H3 trimethylated at lysine 27 (H3K27me3) (53).

Adrenal ganglioneuromas containing Leydig cells with pathognomonic Reinke crystalloids (fig. 12-18) have been reported in two virilized women (54,55), one of whom also had ovarian stromal hyperthecosis and hilus cell hyperplasia (55). Although these cases are reported as virilizing ganglioneuromas, the actual virilizing hormone was derived from a steroidogenic cell type and not a neuron. It has been suggested that secretory products of the ganglioneuromas may have played a role in the development of Leydig cells in these tumors, based on the analogous observation that hilus cells in the normal ovary are also associated with nerve fibers (56). Leydig cells are rare within the adrenal gland, but the existence of these cells and tumors derived from them is explained, in part, by the close embryologic relationship between the developing gonad and the adrenal primordium.

Primary Adrenal Melanoma

Only about 20 *primary adrenal melanomas* have been reported. When considering this diagnosis, and in assessing reported cases, it should be remembered that melanomas frequently metastasize to the adrenal gland, and that metastatic melanoma from occult primary sites can first present in the adrenal gland (57). The validity of the diagnosis may, therefore, always be in doubt. Stringent diagnostic criteria proposed include: presence of malignant melanoma in only one

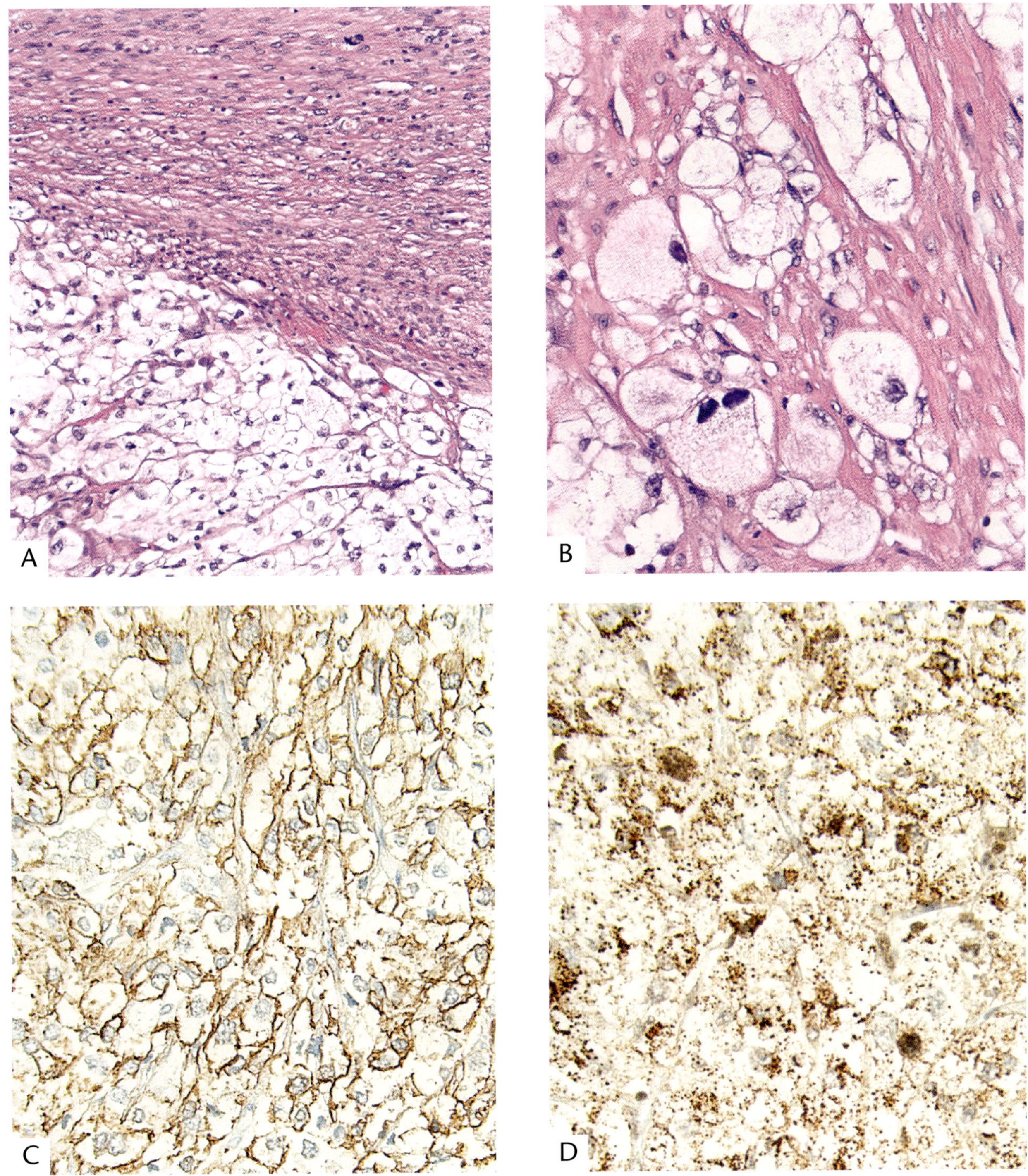

Figure 12-15

VARIED APPEARANCES OF ADRENAL PECOMA

A: Sharply circumscribed areas of eosinophilic spindle cells and uniform clear cells resemble renal cell carcinoma.

B: Different area of the same tumor shows a markedly varied population of small and ballooned clear cells interspersed with spindle cells.

C: Clear cells and spindle cells stain for smooth muscle actin.

D: Clear cells and spindle cells stain for HMB-45.

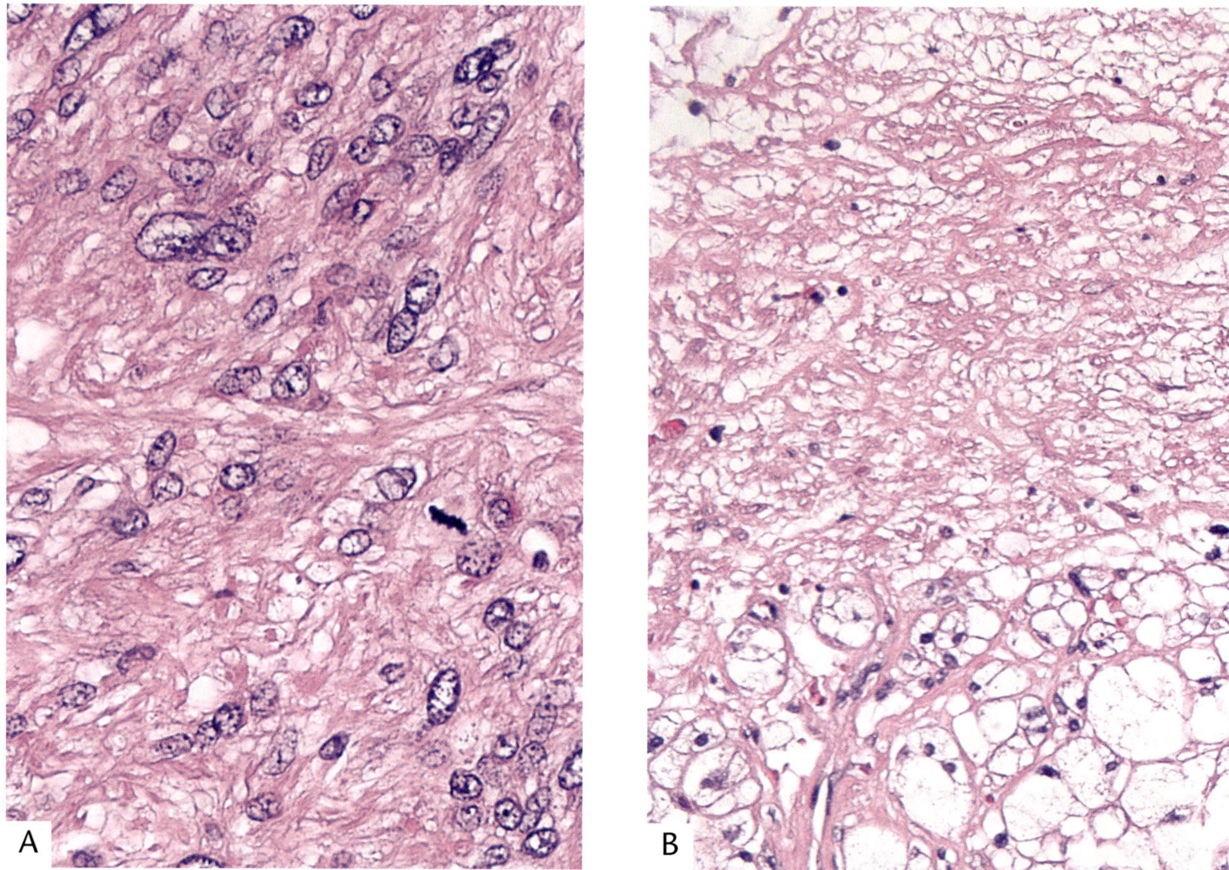

Figure 12-16

MALIGNANT ADRENAL PECOMA

A: Spindle cell area with cytologic atypia and a mitosis.
B: Area of extensive necrosis (top of field).

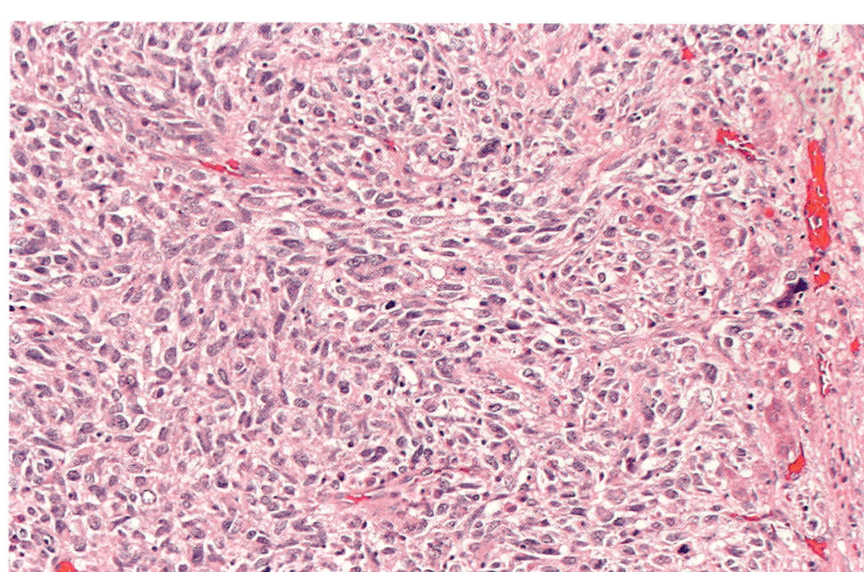

Figure 12-17

MALIGNANT PERIPHERAL NERVE SHEATH TUMOR

Interlacing fascicles of spindle cells and epithelioid cells are present beneath a small area of residual adrenal cortex.

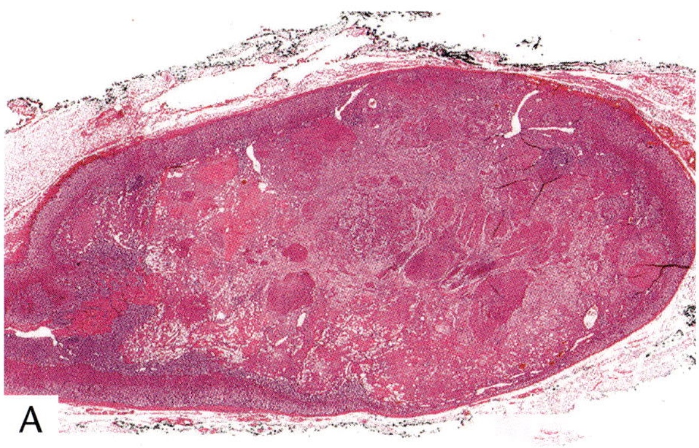

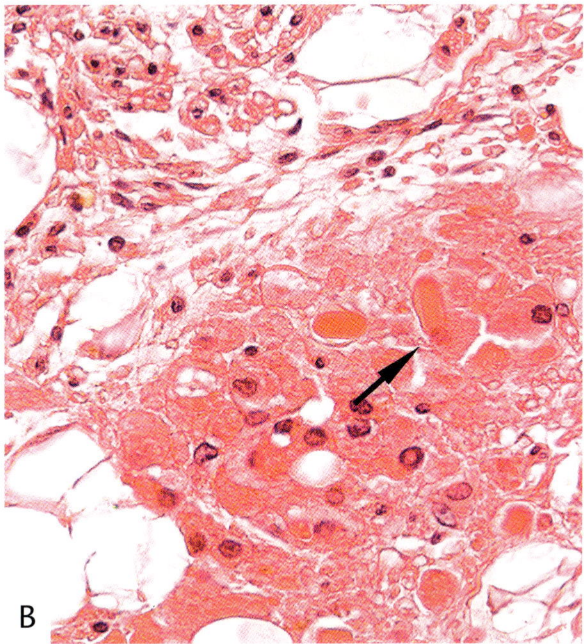

Figure 12-18

MASCULINIZING ADRENAL GANGLIONEUROMA

A: The right adrenal tumor measured 4 cm in size and was completely enveloped by a thin rim of adrenal cortical tissue. The histology was typical for ganglioneuroma, but scattered throughout were small collections of Leydig cells.

B: Leydig cells contain prominent lipochrome pigment and pathognomonic Reinke crystalloid (arrow).

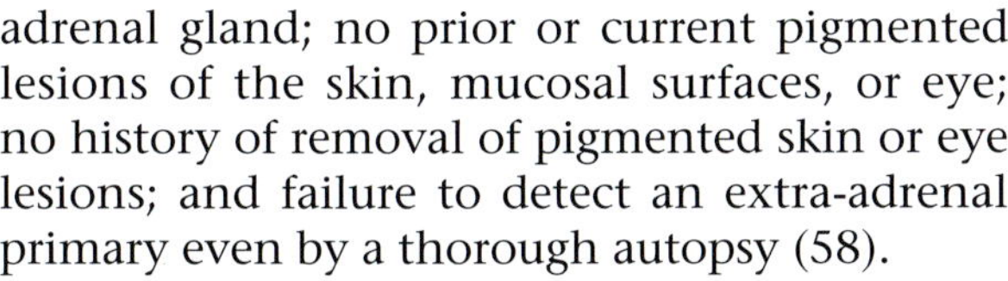
adrenal gland; no prior or current pigmented lesions of the skin, mucosal surfaces, or eye; no history of removal of pigmented skin or eye lesions; and failure to detect an extra-adrenal primary even by a thorough autopsy (58).

An important entity in the differential diagnosis is pigmented pheochromocytoma, which typically contains neuromelanin/lipofuscin pigment but occasionally contains melanosomes or premelanosomes (59) and shows focal expression of HMB-45, a melanosome marker (60). Other pigmented adrenal lesions, such as adrenal hematoma with hemosiderin-laden macrophages and pigmented (black) cortical adenomas, are also part of the differential diagnosis.

The histology and cytology are generally the same as for melanoma arising in more conventional sites, and fine needle aspiration may provide the initial diagnosis. Some primary adrenal melanomas are amelanotic or sparsely pigmented, and may resemble another tumor, such as pheochromocytoma. In those cases immunohistochemistry may be needed to confirm the diagnosis. The approach may require demonstration of multiple melanocytic markers (e.g., MART1, S-100 protein, and MITF) and absence of pheochromocytoma markers (e.g., chromogranin A, synaptophysin, and tyrosine hydroxylase). In pheochromocytomas, S-100 protein is usually expressed in a pattern corresponding to the distribution of sustentacular cells, in contrast to the diffuse expression in melanomas. In some cases, however, chromaffin cells also stain for S-100 protein, although usually less intense than sustentacular cells.

Primary melanoma of the adrenal gland is highly malignant and usually fatal within 2 years of diagnosis (61). A tumor meeting the criteria for primary adrenal melanoma in a patient who had no evidence of melanoma in any other site following a rigorous investigation using a multidisciplinary approach is illustrated in figure 12-19.

The existence of primary adrenal melanomas and the ability of some pheochromocytomas to express cutaneous type melanin have long been hypothesized to be based on the neural crest origin of melanocytes and chromaffin cells. However, melanocytes delaminate early from the neural crest and migrate dorsolaterally to the skin, a route different from the ventral migration of sympathoadrenal precursors. A possibly deeper understanding comes from modern lineage tracing techniques showing that a subset of melanocytes is derived from pluripotent neural crest-derived stem cells known as Schwann cell precursors (SCPs) that migrate along nerve fibers from dorsal root

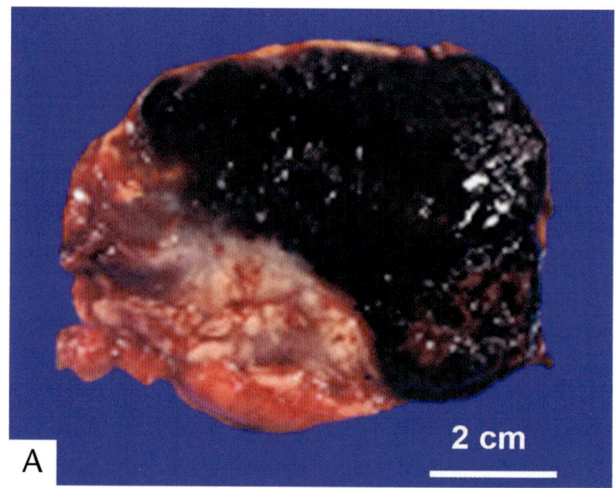

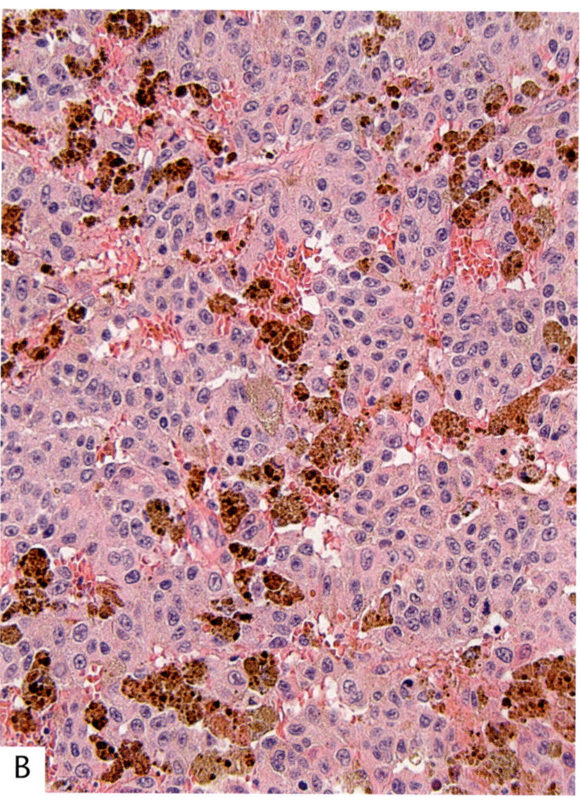

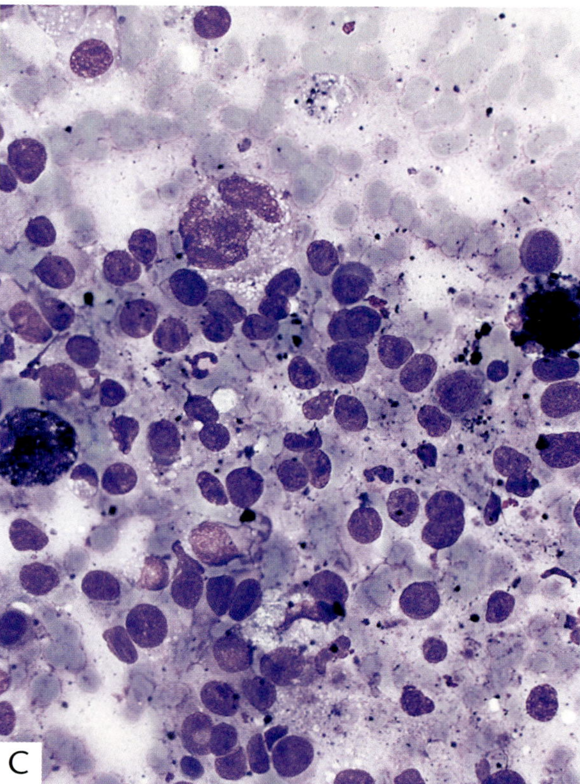

Figure 12-19

PRIMARY ADRENAL MELANOMA

A: A 48-year-old woman complained of abdominal pain; imaging studies revealed a right adrenal mass; the patient had no evidence of melanoma in any other site following rigorous investigation using a multidisciplinary approach. Part of this unusual appearing gross specimen is deeply pigmented and the remainder is amelanotic, with fibrosis and necrosis.

B: Melanin pigment is abundant and the tumor cells are moderately pleomorphic with prominent nucleoli.

C: Fine needle aspiration of an adrenal mass shows numerous malignant cells, with many containing coarse pigment granules. Many tumor cells had disrupted cytoplasm with dispersed pigment granules (Diff Quik stain).

ganglia (62). These precursors, which also give rise to chromaffin cells in the adrenal gland and paraganglia, are discussed in chapter 1.

Other Rare Primary Tumors

There are a few isolated case reports of both endocrine and nonendocrine tumors potentially related to the adrenal gland developmentally, either by lineage or by proximity to the adrenogenital/gonadal ridge (see chapter 1). The former include granulosa cell tumor (63) and a fibrous tumor of the adrenal gland resembling ovarian fibroma (fig. 12-20). This lesion is possibly related to "ovarian thecal metaplasia" (covered in chapter 1), and may have a sex-cord stromal derivation but has also been suggested to have a myofibroblastic origin (64). Tumors possibly related by proximity of their developmental precursors include teratoma (65,66), a malignant mixed germ cell tumor (67), and a

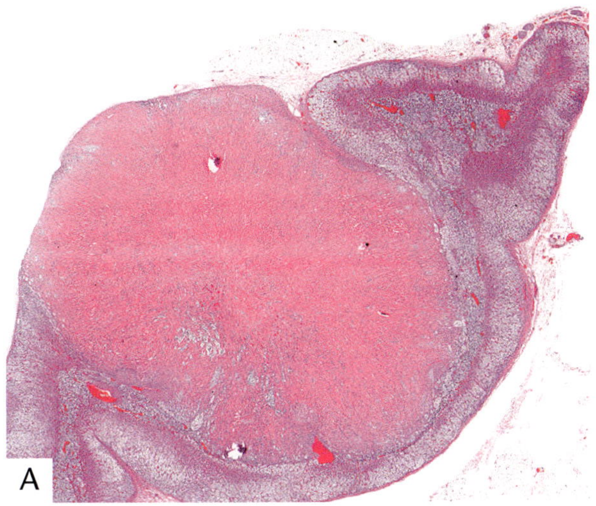

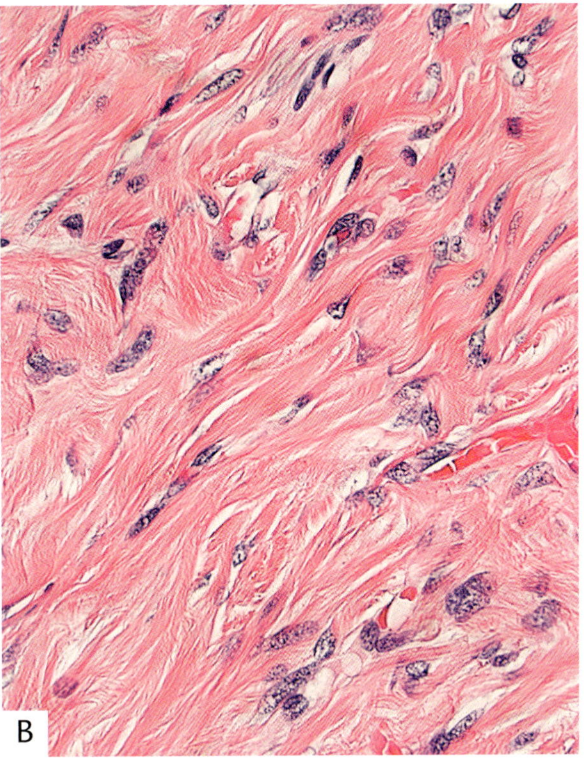

Figure 12-20

ADRENAL CORTICAL "FIBROMA"

A: This tumor measured about 7 mm and was an incidental finding in a patient undergoing radical nephrectomy for a clear cell renal cell carcinoma.

B: The tumor is densely fibrotic with bland spindle cells and resembles an ovarian fibroma.

primary tumor with the morphologic features of an extrarenal Wilms tumor reported in a 4-year-old boy (68).

Some primary adrenal tumors in isolated reports have no apparent relationship to the normal adrenal gland. Examples include a Ewing sarcoma (64,65), Castleman disease (69), non-Hodgkin lymphomas (70), and, extremely rarely, Hodgkin lymphoma (71). Primary adrenal lymphomas represent less than 1 percent of non-Hodgkin lymphomas and less than 3 percent of all extranodal lymphomas (70). In some cases, adrenal involvement is accompanied or followed by more widespread disease, and the categorization as primary versus metastasis is uncertain even with strict exclusion criteria.

ADRENAL ENLARGEMENT AND INSUFFICIENCY CAUSED BY INFECTION

The most common causes of adrenal insufficiency are autoimmune adrenalitis, which accounts for most cases in Western countries, and granulomatous infections, which are responsible for most cases in the developing world (72). While autoimmune disease typically causes adrenal shrinkage, granulomatous infections commonly cause adrenal enlargement, which can be unilateral or bilateral and can occur in immunocompetent or immunocompromised patients (73,74).

Like autoimmune adrenalitis, adrenal infections are frequently missed clinically and can result in potentially life-threatening adrenal insufficiency (Addison disease). *Tuberculosis* remains the most common infectious agent causing adrenal gland destruction. However, adrenal involvement in patients with disseminated fungal infections, especially histoplasmosis, is increasingly recognized in previously nonendemic developing countries (73). Other fungal infections that can affect the adrenal gland include paracoccidioidomycosis, and less commonly, disseminated blastomycosis, coccidioidomycosis, and cryptococcosis (72). The destructive effects of different fungi, however, are modulated by the interplay between host and microbial factors (72).

More than 40 percent of patients with disseminated *histoplasmosis* have adrenal hypofunction (73). The extensive adrenal destruction caused by histoplasmosis appears to be

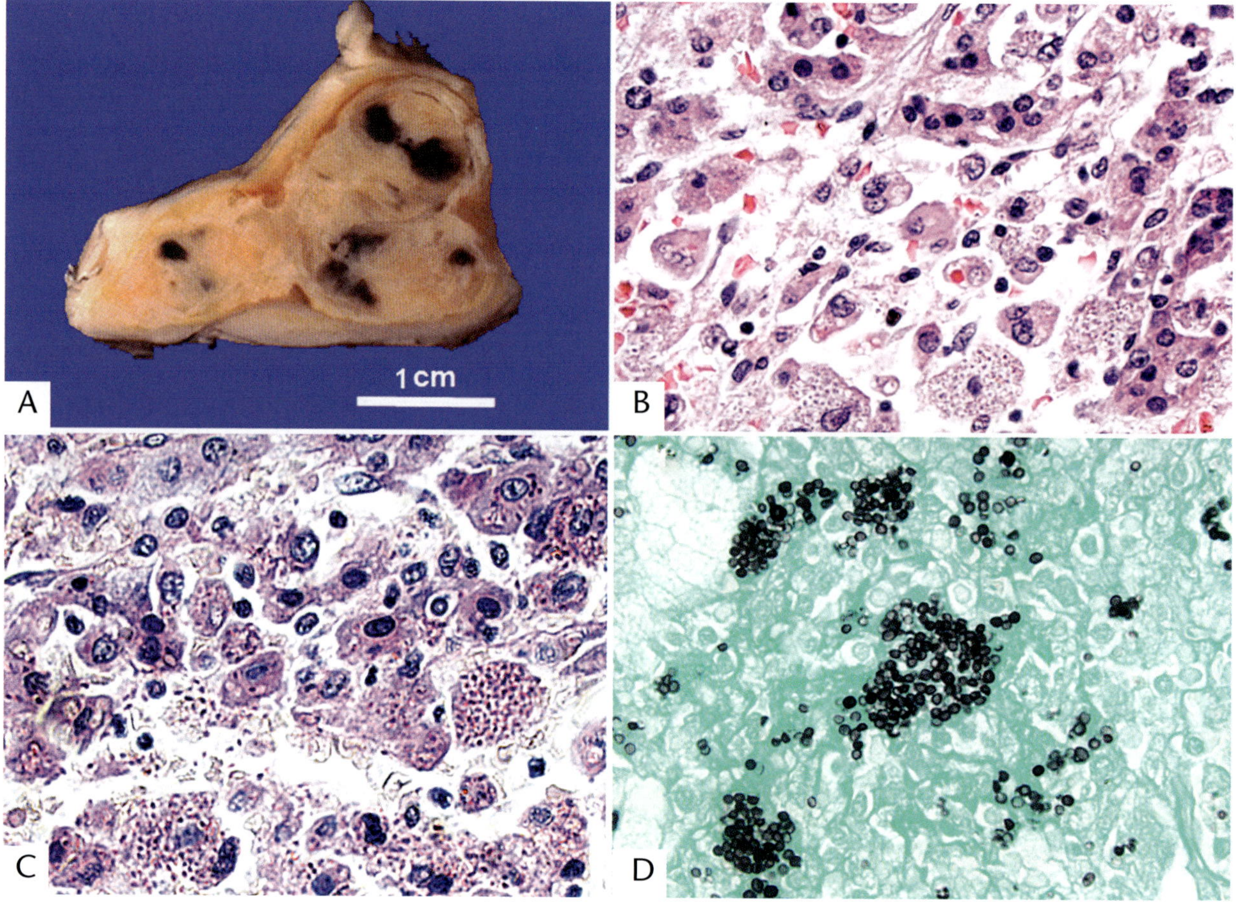

Figure 12-21

ADRENAL HISTOPLASMOSIS

A: Transverse section of an enlarged adrenal gland from a patient with disseminated histoplasmosis. There are broad zones of caseous necrosis and small areas of hemorrhage.

B: Numerous intracellular organisms typical for *Histoplasma capsulatum* are present in macrophages within sinusoids of the adrenal cortex (hematoxylin and eosin [H&E] stain).

C: Organisms typical for *H. capsulatum* are sharply delineated with periodic acid–Schiff (PAS) stain.

D: Gomori methenamine silver stain of a cell block prepared from fine needle aspiration biopsy of enlarged adrenal gland shows numerous round to ovoid yeast forms of *H. capsulatum*. Some organisms are isolated and single while others have a budding configuration.

associated with extracapsular adrenal vasculitis, which leads to adrenal infarction (72) and caseation necrosis. Numerous organisms are found within macrophages (fig. 12-21). Some of the parasitized cells are adrenal cortical cells.

Adrenal abscess formation resulting from bacterial infection is a rare cause of adrenal enlargement. It is usually unilateral, but occasionally bilateral. Radiographically, it may be difficult to differentiate an adrenal abscess from a primary or secondary adrenal neoplasm. Adrenal abscess formation has been reported more often in children where the suppurative process may complicate neonatal adrenal hemorrhage. In most cases, involvement by bacteria, usually *Escherichia coli*, followed by group B *Streptococcus* and *Bacteroides*, is attributable to hematogenous dissemination. Malakoplakia of the adrenal gland has also been reported in association with *E. coli* infection (75). Tuberculosis can cause tumefactive enlargement and necrotizing granulomatous inflammation of the adrenal glands, similar to histoplasmosis (73). However, fewer than 2 percent of patients with disseminated

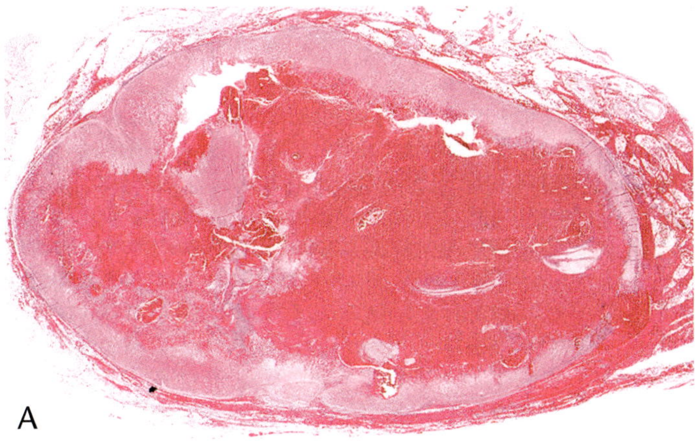

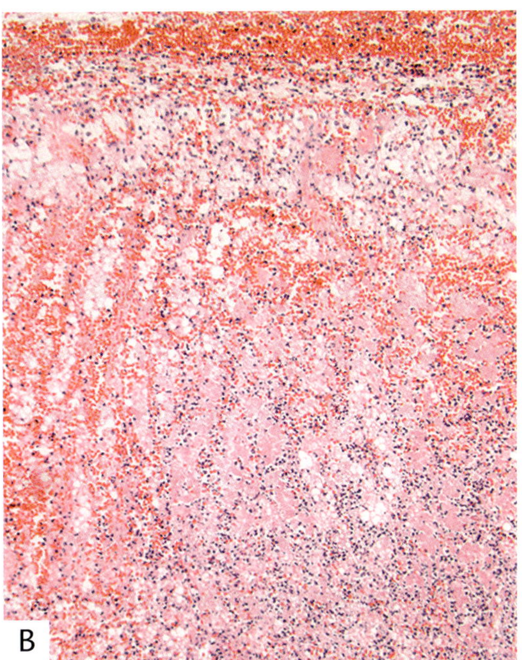

Figure 12-22

ADRENAL HEMORRHAGE IN WATERHOUSE-FRIDERICHSEN SYNDROME

A: Low magnification image of a fresh hematoma.
B: Hemorrhagic necrosis of overlying adrenal cortex.

tuberculosis present with signs or symptoms of adrenal cortical insufficiency (76).

Some adrenal infections cause both necrosis and hemorrhage. These include viral diseases such as varicella-zoster virus or cytomegalovirus, fungal infection, especially disseminated mucormycosis, and opportunistic infection by *Pneumocystis jiroveci*. With these infections there may be a mild degree of adrenal gland enlargement. There is fresh or clotted blood in various stages of organization, and the adjacent adrenal cortex or medulla may have areas of necrosis.

ADRENAL ENLARGEMENT WITH HEMORRHAGE AND HEMATOMA FORMATION

Adrenal hemorrhage can present in neonates, children, or adult patients as unilateral or bilateral adrenal enlargement. It can be focal or diffuse, ranging in size from only a few millimeters to 15 cm, and involve up to 2 L of blood. Etiologies include mechanical trauma, sepsis with or without Waterhouse-Friderichsen syndrome (fig. 12-22), coagulopathies, steroid or ACTH administration, primary or metastatic tumors, adrenal vein thrombosis, and complications of adrenal venography (77). Adrenal hemorrhage is also associated with anticoagulant drugs and with the phospholipid antibody syndrome, where it may present with bilateral hemorrhagic infarction (78).

The susceptibility of the adrenal gland to massive hemorrhage is attributed to its vascular supply and pivotal role in physiologic response to stress; likened to a "vascular dam," any rise in adrenal venous pressure (e.g., vasoconstriction during shock) can lead to hemorrhage into the gland. In patients with neonatal and other traumatic adrenal hemorrhage, the right adrenal gland is more commonly affected (79). This is attributed to compression of the right adrenal gland between the liver and spine, and transmission of elevated venous pressure from the inferior vena cava through the short draining adrenal vein on the right side (3). Adrenal hemorrhage can be complicated by adrenal insufficiency, secondary infection with abscess formation, or retroperitoneal hemorrhage, and may be life-threatening in some patients who have associated sepsis, malignancy, history of anticoagulant or steroid treatment, or major surgery (80).

Neonatal adrenal hemorrhage is often associated with a complicated delivery or asphyxia, and poses a specific diagnostic challenge because it may simulate the radiographic features of congenital hemorrhagic or cystic neuroblastoma. These two entities, which are the major diagnostic considerations for neonatal adrenal masses (79), can frequently be differentiated by a combination of clinical and imaging characteristics: prenatal detection, tiny flecks of calcification, vascularity on color Doppler ultrasound, and evolving complex architecture all favor neuroblastoma (79). Initial clinical management is usually conservative, but many patients eventually undergo interval adrenalectomy either because of persistent symptoms or concern for possible neoplasia (77). Over time, benign adrenal hematomas may undergo regression, sometimes with residual dystrophic calcification and possibly cyst formation (77,79).

ADRENAL CYSTS

The current classification of *adrenal cysts* continues to be based on a 1966 iteration of a schema dating to 1906 in which these lesions are categorized according to the presence or absence of a lining and the type of lining identified. The categories are epithelial cysts, endothelial cysts, pseudocysts, and parasitic cysts. Endothelial cysts are further divided into angiomatous and lymphangiomatous types (81). While this system may have use as a memory device, it provides limited clinically useful diagnostic information, in part because the categories overlap and are applied inconsistently across series.

An important confounder is the category of epithelial cysts, which has been variably interpreted to include cystic neoplasms (82). If cystic neoplasms are excluded, "true" epithelial cysts are rare and most cysts that appear to fit this category have mesothelial characteristics by immunohistochemistry (83,84). Another major source of confusion is the category of pseudocysts, which do not have a lining, and can result from hematomas in non-neoplastic conditions or in hemorrhagic tumors (82,85). Although more precise classifications of adrenal cysts have been proposed (84), there is presently no consensus on their adoption. In short, the main consideration in diagnosing an adrenal cyst is usually to distinguish a non-neoplastic cystic mass from a cystic neoplasm.

Predominantly cystic adrenal masses are rare. In a 2013 study representing the largest series to date (82), they were found only 31 of 4,500 combined surgical and autopsy adrenal gland specimens (0.7 percent) seen over a 20-year period. Thirty-nine percent of the lesions were non-neoplastic pseudocysts, 6 percent endothelial cysts, and 55 percent epithelial cysts which included multiple pheochromocytomas and one adrenal cortical carcinoma. In the previously largest series (85), involving 41 surgical cases and excluding cystic neoplasms from the epithelial cyst category, 32 (78 percent) were pseudocysts, including cortical carcinomas, cortical adenomas, and pheochromocytomas; 8 (20 percent) were endothelial cysts including an angiosarcoma; and only 1 was an epithelial cyst. Rare causes of a cystic adrenal mass not encountered in these two large series include arteriovenous malformation (86), endometriosis (87), polycystic kidney disease (88), Beckwith-Wiedemann syndrome (8), neuroblastoma (79), and teratoma (65).

Adrenal cysts can occur at any age, but are more commonly seen in the 5th and 6th decades of life, with a female predominance (82, 85). Fewer than 5 percent occur in children, and they are very rare in the neonatal period. Approximately 6 to 15 percent are bilateral (82). Signs and symptoms, when present, are usually nonspecific, and include dull flank pain, gastrointestinal complaints such as epigastric distress or indigestion, and a palpable abdominal mass. There may be constitutional symptoms such as malaise and weakness, or anemia resulting from intracystic hemorrhage.

Adrenal pseudocysts can be very large, sometimes up to 50 cm in diameter. They are typically unilocular, The wall is 1- to 5-mm thick, can be focally hyalinized, and may show prominent dystrophic calcification that can be diagnostically helpful when seen as a curvilinear rim calcification in CT scans. Microscopically, some cysts have smooth muscle in their wall that is continuous with the smooth muscle of the adrenal vein. The lining is smooth or shaggy, and may have a partially organized, fibrinoid appearance without an apparent cellular lining. Elevated nodules or plaques may be present and, on close inspection, yellow mottled areas of residual adrenal cortex may be seen within the pseudocyst wall.

Pseudocyst contents typically consist of yellow-brown or bloody amorphous material with hemosiderin deposition, foamy macrophages, and cholesterol clefts, consistent with fresh or altered blood from recent and old hemorrhage.

In some cases, there is an amorphous fibrin-rich exudate showing recanalization and organization, which may be aided by endothelial outgrowth from the microvasculature of the attached adrenal cortex (fig. 12-23). The proliferative endothelial component can be very florid and become complicated by papillary endothelial hyperplasia.

Unusual variant features of adrenal pseudocysts include lipomatous and myelolipomatous metaplasia (89), and metaplastic bone formation. A rare example of mammary carcinoma manifesting as an intracystic metastasis has been reported (89). In contrast to pseudocysts, endothelial cysts usually measure less than 2 cm and have a smooth lining. Lymphangiomatous cysts are often multilocular and contain milky or clear fluid. Their classification as cysts versus lymphangiomas (fig. 12-23) is somewhat arbitrary. The linings of both lymphangiomatous and capillary endothelial cysts stain for CD31, while lymphangiomatous cysts also stain for podoplanin. Parasitic cysts may contain ova or parasites in their walls or lumens. They are rarely encountered in nonendemic regions and are usually associated with systemic disease. Most cases are caused by Echinococcus (90). One incidentally detected echinococcal cyst of the adrenal gland (90) and one case of visceral leishmaniasis presenting as an isolated adrenal cystic mass have been reported (91).

TUMORS METASTATIC TO ADRENAL GLANDS

Incidence and Common Primary Sites

While primary adrenal tumors are of great interest in many respects, metastases to the adrenal glands from other primary sites are much more common. The adrenal glands are reported to be the fifth most common anatomic site involved by metastases overall, surpassed only by lymph nodes, lung, liver, and bone (92). In an autopsy study of 1,000 carcinomas, the adrenal glands were secondarily involved in 27 percent of cases (93). The leading sources of adrenal metastases were breast (53.9 percent), lung (35.6 percent), kidney (24 percent), stomach (21 percent), pancreas (19 percent), ovary (17 percent), and colon (14.4 percent). Endometrial carcinomas rarely give rise to adrenal metastases (94).

In a more recent study of 464 patients, lung primaries predominated, with adenocarcinomas accounting for about two thirds of these cases, followed by small cell, squamous cell, and large cell cancers (95). Approximately 50 percent of the adrenal metastases were bilateral, emphasizing that bilaterality itself can be a clue that an adrenal mass is a metastasis. Adrenal metastases were most frequent in elderly patients (mean age, 62 +/- 13 years), and were often detected a short time after diagnosis of a primary tumor (mean latent period, 7 months). Four percent of adrenal metastases (n = 20) were symptomatic, and those tended to be bigger than asymptomatic lesions and to occur in younger patients. Five patients presented with adrenal insufficiency (Addison disease) and one with massive peritoneal hemorrhage. Ninety percent (n = 421) of all adrenal metastases were carcinomas and 56 percent of these were adenocarcinoma (95). Autopsy studies have shown that in patients with breast carcinoma the frequency of adrenal metastases decreases with age and increases in proportion to total number of metastases in other sites (96).

It is believed that the metastatic seeding of the adrenal gland occurs by a hematogenous route and that the high incidence of metastases to the adrenal gland, despite the gland's relatively small size, can be explained by high blood flow resulting from the triple arterial blood supply (superior suprarenal artery branching from the inferior phrenic artery, middle suprarenal artery from the abdominal aorta, and inferior suprarenal artery from the renal artery) and sinusoidal vasculature.

It has recently been hypothesized that lymphatic spread occurs, in some cases, by retrograde flow through the thoracic duct (97). If correct, this could explain occasional reports that patients with solitary adrenal metastases have a higher likelihood of an ipsilateral than a contralateral primary tumor, and that those with an ipsilateral primary who underwent adrenalectomy had a greater 5-year survival rate than patients with a contralateral primary (97).

Although no current guidelines recommend treatment of adrenal metastasis based on laterality, it is accepted that adrenalectomy for a clinically isolated adrenal metastasis can prolong survival in selected patients (98), and approaches to this type of therapy are continuing to

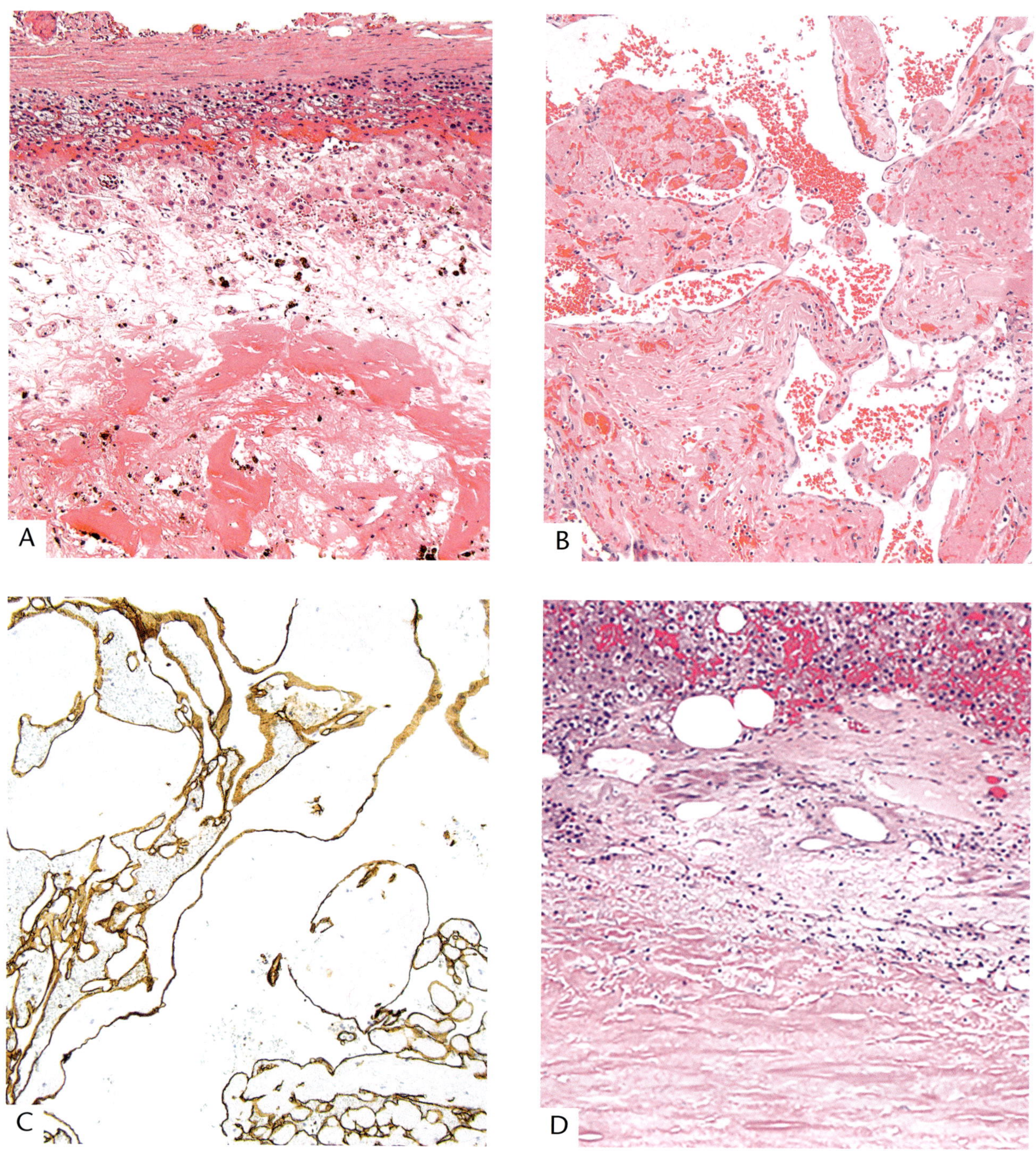

Figure 12-23

ADRENAL PSEUDOCYST

A: Adrenal pseudocyst shows degenerating recent and old hemorrhage.

B: Fibrin-rich matter within an adrenal pseudocyst undergoing organization by endothelial cells emanating from the rich capillary plexus of attached cortex.

C: CD34 immunostain shows organizing fibrin-rich matter with complex maze of endothelial cells. This pattern can be mistaken for a vascular neoplasm.

D: The thick fibrous wall of a longstanding adrenal pseudocyst (different case) is at the bottom of field. There is no endothelial or other lining. Congested adrenal cortex is at top. Foci of dystrophic calcification were present in other areas.

evolve. In a retrospective analysis of patients with non-small cell lung cancer, adrenalectomy together with surgical excision of the primary tumor resulted in significantly improved survival, particularly for patients with metachronous adrenal metastasis and those who were negative for lymph node metastasis (98).

Imaging Characteristics of Metastatic Disease

Adrenal glands involved by metastatic tumor vary in size and may have a smooth or lobulated contour. Involvement frequently is bilateral. Most tumor deposits are of soft tissue density unless large and necrotic (3). Variable enhancement is evident on CT scan. On MRI, there is usually a heterogeneous signal intensity on T2-weighted images and sometimes T1-weighted images as well, in which the gland appears isointense or hypointense to liver. If the metastatic tumor is hemorrhagic, there is usually a high signal on T1- and T2-weighted images. An adrenal metastasis may resemble an adrenal cortical adenoma on CT scan, and various techniques and mathematical models have been proposed for differentiating these lesions (99).

Pathology Pitfalls and Common Differential Diagnoses

As with primary adrenal tumors, adrenal metastases are increasingly recognized during life with the aid of high-resolution imaging studies such as CT scan and MRI. During the staging workup for patients with a known malignancy elsewhere, the finding of an adrenal mass can be very significant in terms of both prognosis and management. In this setting, CT or ultrasound-guided fine needle aspiration biopsy of the mass is a definitive method for distinguishing between a metastasis and a primary adrenal tumor. In cases of isolated adrenal metastases, an adrenalectomy may be required. There is usually no problem in making the histologic or cytologic diagnosis of metastatic carcinoma, although some metastases mimic adrenal cortical carcinomas or other primary adrenal tumors, and some primary adrenal tumors are mistaken for metastases. Immunohistochemical stains tailored to the specific diagnostic considerations are important in these cases.

Grossly, metastatic tumors usually lack the yellow-orange hue of an adrenal cortical primary. They may appear gray-white, with or without zones of necrosis. Gross black to brown pigmentation may arouse suspicion for a metastatic melanoma, while a gelatinous appearance may suggest adenocarcinoma. Metastatic deposits may become massive, often with extensive necrosis and obliteration of much of the gland.

Microscopically, metastatic carcinoma may retain well-developed structural characteristics, such as gland formation, or have a predominantly solid growth pattern within the rich microvasculature of the adrenal gland (figs. 12-24–12-27). Spread of tumor within vascular channels may be evident adjacent to the main site of metastasis, and may extend into small capsular extrusions of cortex. Occasionally, there is direct invasion of tributaries of the central adrenal vein, and with overgrowth of the adrenal vascular system, there may be retrograde extension into the inferior vena cava. In cases with extensive necrosis and inflammation, it may be difficult to identify adrenal remnants and to discriminate tumor cells from other cell types, especially in biopsy specimens.

Immunohistochemistry can be valuable both for confirming the adrenal location and for highlighting the different cell populations. In fine needle aspiration biopsies, particularly those in which the specimen is sparse or artifactually distorted, immunohistochemistry can help avoid diagnostic pitfalls, including metastatic adenocarcinoma mimicking normal adrenal cortical cells (100), and cells from a benign adrenal cortical nodule mimicking a small cell malignancy (101). Specific types of metastatic tumors are particularly likely to present diagnostic challenges by simulating primary adrenal neoplasms. These include poorly differentiated lung cancers, hepatocellular carcinoma, and renal cell carcinoma. Metastatic poorly differentiated non-small cell lung carcinomas can simulate adrenal cortical carcinoma both histologically (see fig. 12-25) and clinically, especially in cases where a combination of unilateral adrenal enlargement and multiple pulmonary nodules suggests pulmonary metastases from a primary adrenal tumor.

Adrenal involvement by SMARCA4-deficient thoracic tumors may be a potential pitfall in the differential diagnosis with adrenal cortical carcinoma due to the epithelioid undifferentiated morphology and frequent expression of

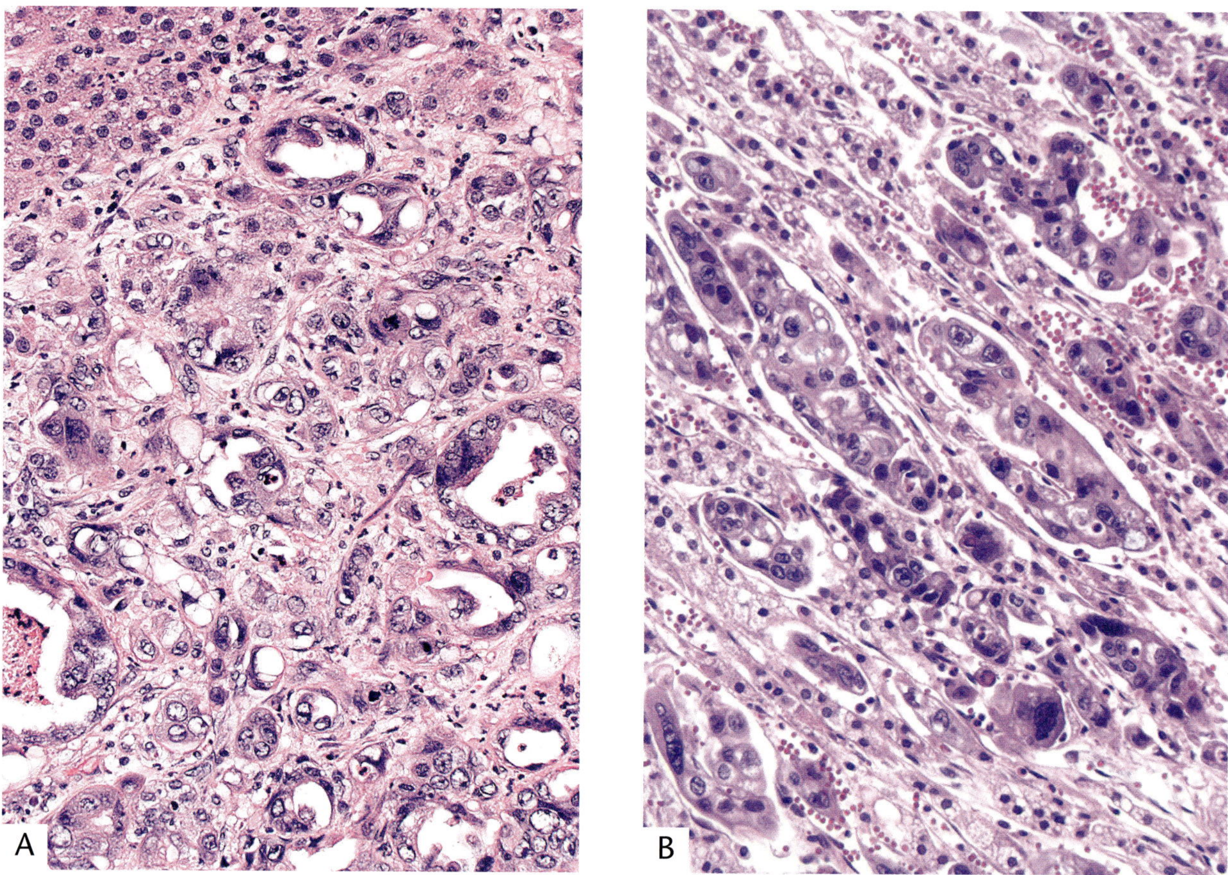

Figure 12-24

ADENOCARCINOMA OF BREAST METASTATIC TO ADRENAL GLAND

A: The pattern of growth shows a well-developed glandular arrangement of tumor cells.
B: Metastatic poorly differentiated adenocarcinoma has a sinusoidal pattern, with growth as loose plugs of tumor within delicate vascular spaces.

synaptophysin. With metastatic small cell lung tumors or other neuroendocrine neoplasms producing ectopic ACTH, there is also hypersecretion of cortisol. In that unusual situation there may be recognizable hyperplasia of residual adrenal cortex (102).

The adrenal glands are the second most common site of metastasis from hepatocellular carcinoma (HCC) after the lungs. Adrenal metastases from HCC are found in 8.4 percent of autopsy cases, and can be the initial presentation of metastatic disease (103) or may be clinically identified years after diagnosis of the liver primary (see fig. 12-26). An unusual example of adrenal seeding of HCC via adrenohepatic fusion has been reported (104). In a comparative study of immunohistochemical markers, arginase-1 and HepPar1 had the highest sensitivity for well-differentiated HCC, while arginase-1 and glypican-3 had the highest sensitivity for poorly differentiated HCC (see fig. 12-26) (105).

Adrenal metastases from renal cell carcinomas pose particular challenges because of both their histologic features and the proximity of the kidneys to the adrenal glands. Somewhat paradoxically, the annual detection rate of these lesions during life has decreased with improved imaging technology, probably because of earlier detection of primary renal cell carcinomas (106). In recent series, up to 25 percent of patients with renal cell carcinoma presented with metastases and 3.5 to 5.0 percent have adrenal metastases (106). Although metastasis is usually to the ipsilateral adrenal gland, renal cell carcinoma rarely

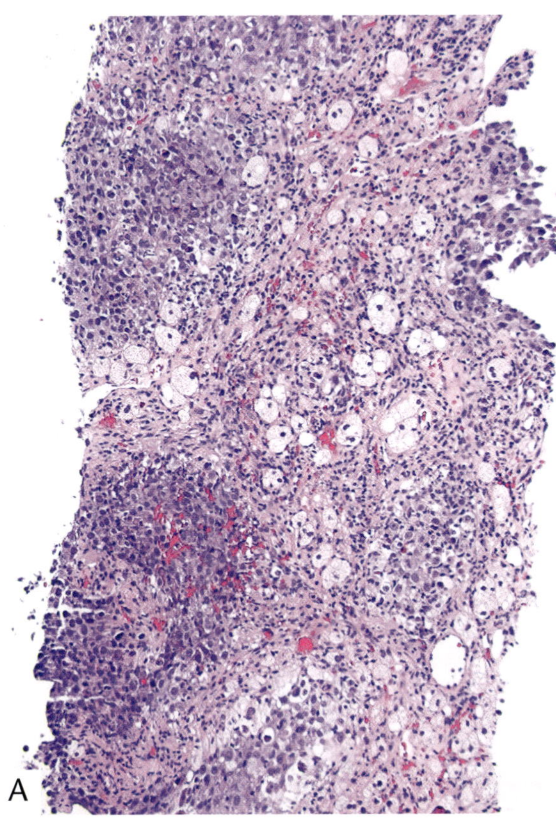

Figure 12-25

ADENOCARCINOMA METASTATIC TO ADRENAL GLAND

A: Core needle biopsy of an adrenal mass in a patient with non-small cell lung carcinoma. The tumor replaces much of adrenal tissue, with only small nests of residual pale-staining cortical cells visible.

B: Immunostain for cytokeratin 8/18 vividly highlights metastatic carcinoma.

C: Residual cortical cells are more readily apparent with immunostain for Melan-A.

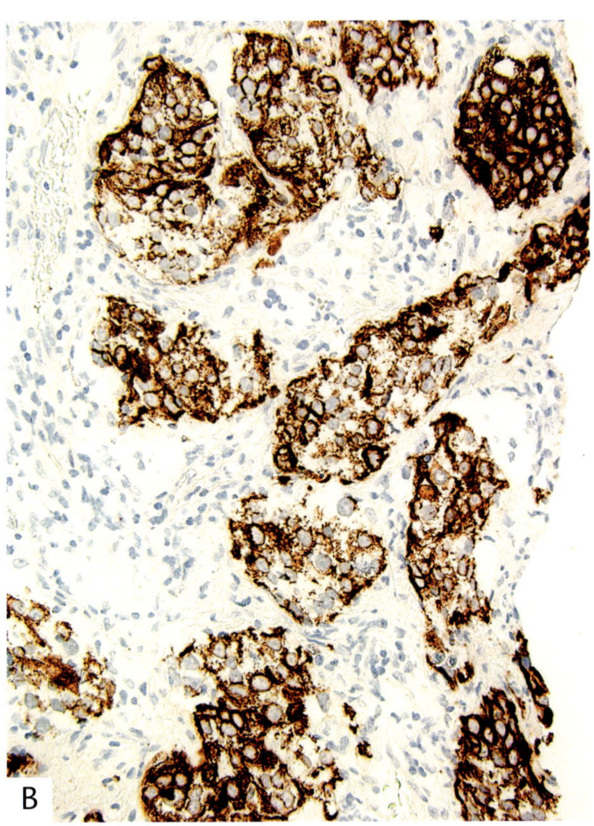

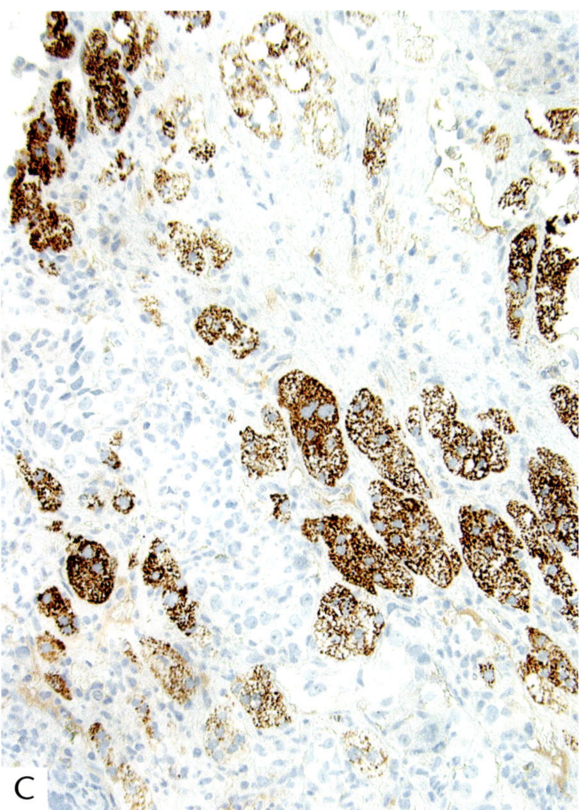

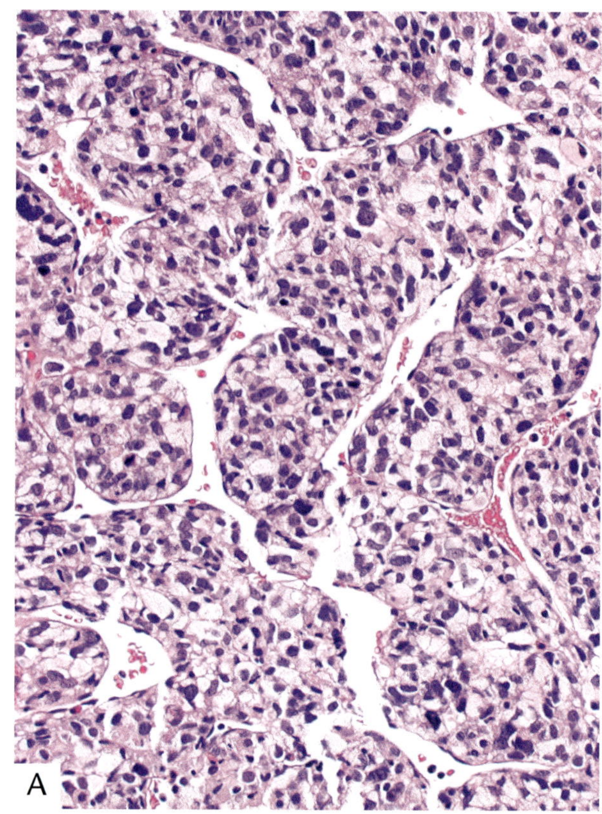

Figure 12-26

HEPATOCELLULAR CARCINOMA METASTATIC TO ADRENAL GLAND

A: Hepatocellular carcinoma metastatic to the adrenal gland can simulate a primary adrenal cortical carcinoma. This metastasis has a prominent trabecular pattern with delicate sinusoidal lining.

B: This poorly differentiated hepatocellular carcinoma metastatic to adrenal gland (cortex at top of field) has abundant eosinophilic cytoplasm, prominent nucleoli, and numerous intracytoplasmic hyaline globules. The features suggest a high-grade adrenal cortical carcinoma.

C: Immunostain for arginase-1 (in a different case) shows both nuclear and cytoplasmic immunoreactivity in a metastatic hepatocellular carcinoma.

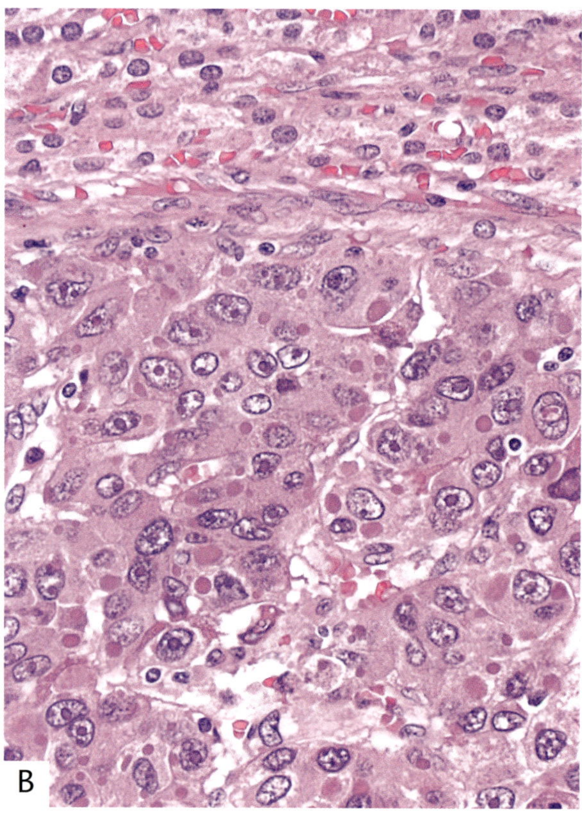

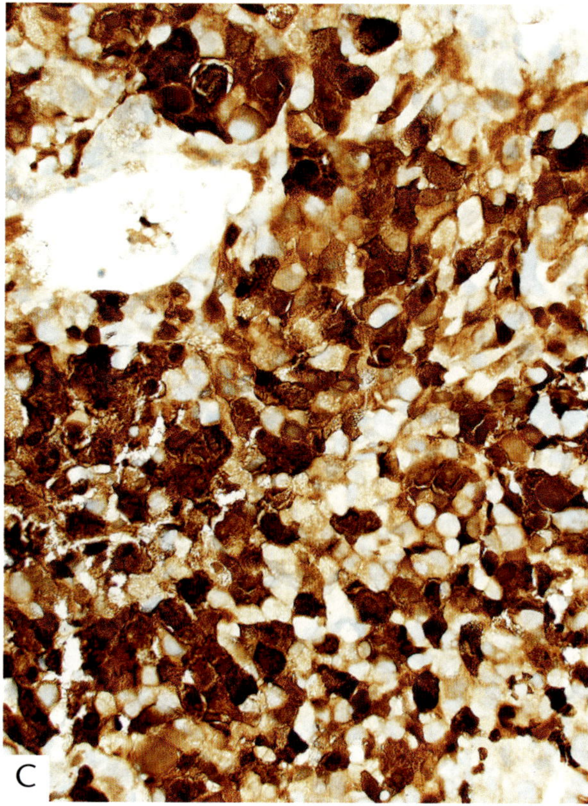

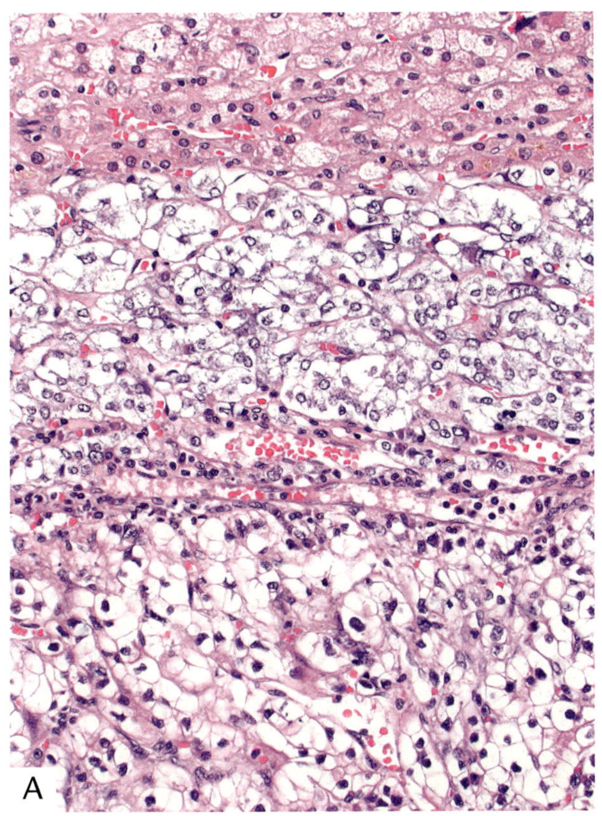

Figure 12-27

RENAL CELL CARCINOMA METASTATIC TO ADRENAL GLAND

A: Adrenal metastasis from an ipsilateral renal cell carcinoma. The patient had undergone radical nephrectomy for a large renal cell carcinoma. Metastatic clear cell renal carcinoma is present in bottom third of field, adrenal medulla is in middle third, and adrenal cortex is at top of field with zona reticularis abutting on medulla.

B: Adrenal metastasis of clear cell renal carcinoma in a different case has a nesting pattern with rich microvasculature. Optically clear cytoplasm is seen.

C: Fine needle aspiration of renal cell carcinoma metastatic to adrenal gland in a different case. The loosely dispersed cells have relatively uniform nuclei (Diff Quik stain).

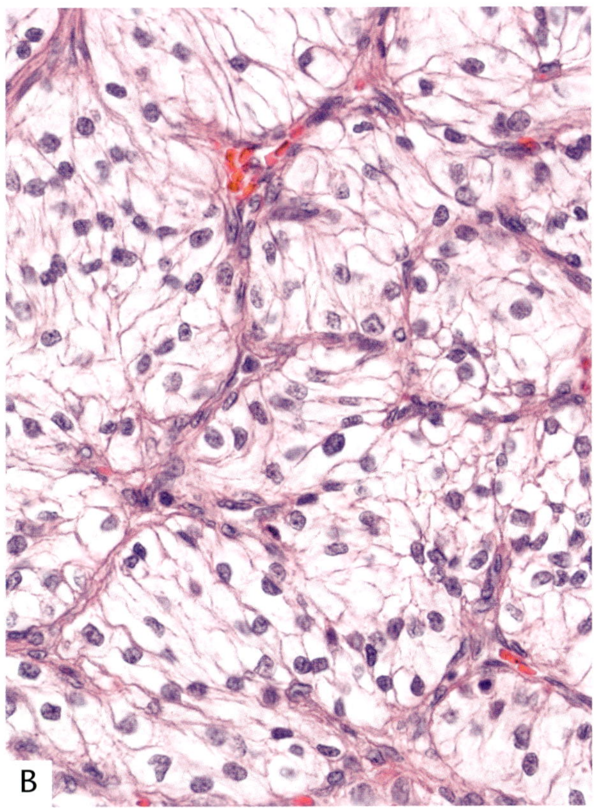

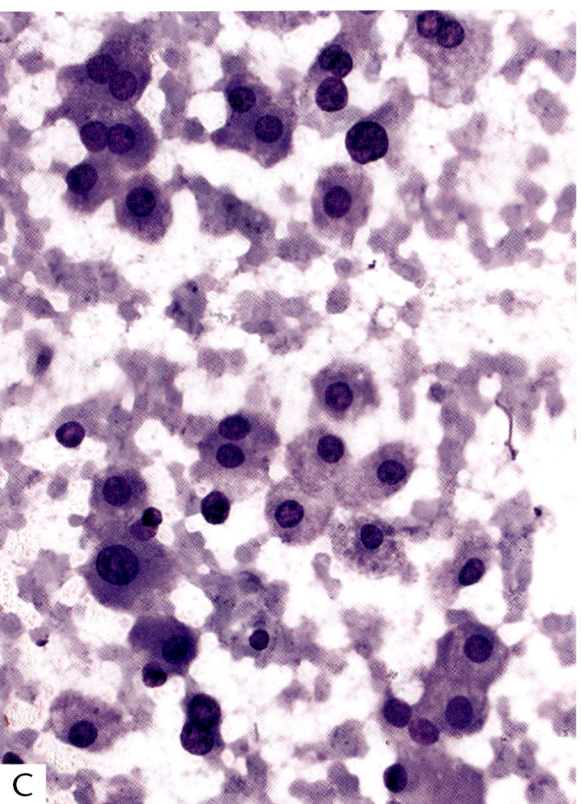

presents with contralateral or bilateral adrenal metastases (106). Contralateral metastasis in particular can potentially be confused with an adrenal cortical neoplasm. Renal cell carcinoma can also involve the ipsilateral adrenal gland by direct extension. Histologically, renal cell carcinoma may have an alveolar or nesting pattern similar to that of an adrenal cortical neoplasm, but the optical clarity of the cytoplasm differs from the finely vacuolated cytoplasm seen in many adrenal cortical nodules and tumors (fig. 12-27). Correlation of clinical, radiologic, and endocrinologic data and intraoperative findings are important for the correct diagnosis.

Other Origins of Adrenal Metastases

Lymphoma. Secondary involvement of the adrenal gland is seen in approximately 25 percent of patients with proven systemic lymphoma at autopsy and about 4 percent antemortem as detected by CT scan (107). Approximately 90 percent are non-Hodgkin lymphomas of B-cell phenotype, and most of the remainder are T-cell lymphomas. In contrast to the distribution of epithelial tumors in the adrenal glands, most secondary adrenal lymphomas are unilateral (107) while up to 70 percent of primary adrenal lymphomas are bilateral (70). Massive bilateral adrenal involvement can result in adrenal insufficiency (Addison disease) (70), which may be a presenting manifestation (70). Spontaneous adrenal hemorrhage is a rare complication (108).

Intravascular lymphomatosis, a peculiar type of non-Hodgkin lymphoma with destructive angiotropism, can involve the adrenal glands and cause gross enlargement (109). The disorder was originally believed to be a primary endothelial malignancy, hence, the term malignant angioendotheliomatosis, but was subsequently shown to be a lymphoma of large B-cell or immunoblastic type (109). The malignant cells proliferate within small vessels in most organs, but have an affinity for the central nervous system, with varied neurologic manifestations. There may be massive involvement of the adrenal glands, which can result in Addison disease, and may account for the fever, hypotension, and electrolyte abnormalities frequently seen in some patients.

The diagnosis of adrenal involvement by lymphoma is confirmed by fine needle aspiration biopsy. Problems in the differential diagnosis (e.g., metastatic undifferentiated carcinoma) are usually resolved by positive immunostaining for lymphoid markers and negative staining for epithelial and other markers (fig. 12-28).

Mesenchymal Neoplasms. The adrenal glands are only rarely involved secondarily by mesenchymal neoplasms. Angiosarcoma involves the adrenal gland secondarily. There may be incidental involvement of the adrenal glands by Kaposi sarcoma (fig. 12-29) as a manifestation of more widely disseminated tumor in patients with AIDS. Both angiosarcoma and Kaposi sarcoma may have a sinusoidal pattern, with extension between residual columns and cords of cortical cells, and there may be extension outside the gland into periadrenal fat. Meningeal hemangiopericytoma has been reported metastatic to the adrenal gland and is associated with multiple metastases to bones and lungs (110).

Tumor-to-Tumor Metastases. Tumor-to-tumor metastases from extra-adrenal primary malignancies to primary adrenal neoplasms including pheochromocytoma (111) and adrenal cortical adenoma (112) have occasionally been reported.

Secondary Adrenal Cortical Insufficiency (Addison Disease) Caused by Adrenal Metastases

Acute adrenal insufficiency can result from the stress of illness superimposed on smoldering borderline insufficiency or from sudden massive adrenal damage. Acute insufficiency caused by adrenal metastases is a life-threatening crisis, and can be the initial presentation (113,114), while borderline insufficiency may contribute to morbidity and mortality in undiagnosed patients (109). Patients with bilateral adrenal metastases should be evaluated for the presence of adrenal insufficiency and receive careful follow-up, even if baseline testing is normal (113). However, in the literature, it is not clear how often this is actually done (70).

It has been estimated that 80 to 90 percent of the adrenal gland must be replaced or destroyed before adrenal insufficiency is detected (113), which attests to the remarkable functional reserve of the adrenal gland with regard to cortical function. The true incidence of subclinical adrenal insufficiency due to adrenal metastases

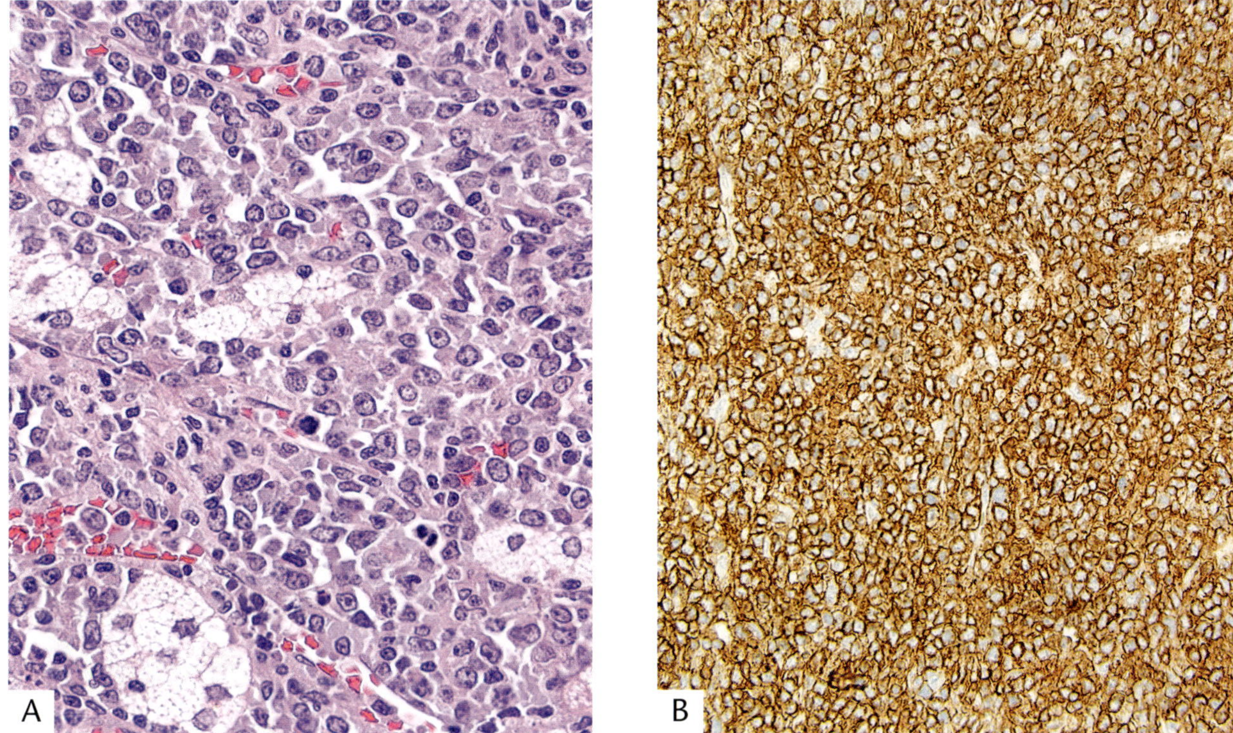

Figure 12-28

ADRENAL INVOLVEMENT BY LYMPHOMA

A: The tumor, a large B-cell lymphoma, replaces much of the adrenal gland, with only a few small nests of pale-staining cortical cells present in this field.

B: Immunostain for CD20 is diffusely positive.

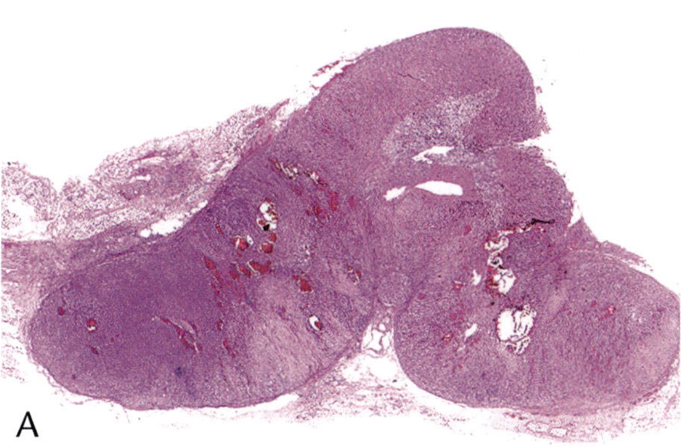

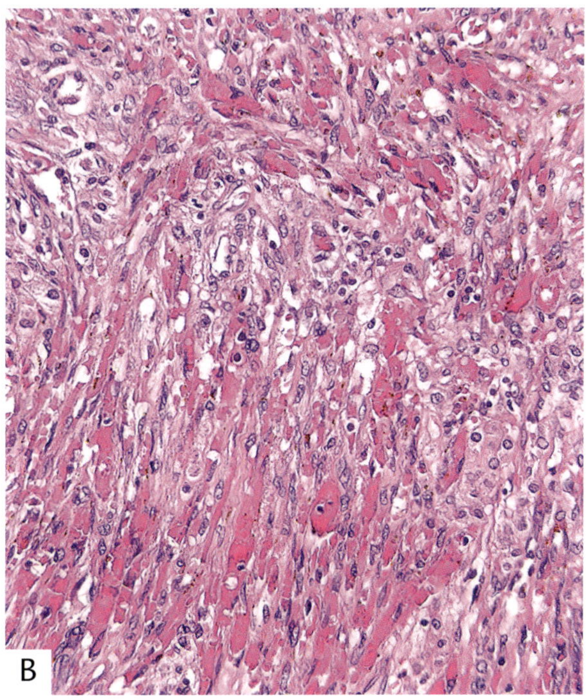

Figure 12-29

KAPOSI SARCOMA EXTENSIVELY INVOLVING ADRENAL GLAND

A: Kaposi sarcoma involving the adrenal gland in an adult male with acquired immunodeficiency syndrome (AIDS). The tumor obscures much of normal histology and architecture of the gland and extends into periadrenal adipose tissue.

B: Slit-like spaces are suffused with blood. Only few compressed adrenal cortical cells are recognized between spindle cells. Rare hyaline globules were present in other areas.

is difficult to determine because the relatively nonspecific symptoms may be confused with cachexia or the fluid and electrolyte imbalance associated with cancer. However, in a prospective study, provocative testing showed adrenal insufficiency in 33 percent of patients with bilateral adrenal metastases (115).

A wide variety of metastatic malignancies cause Addison disease by massive adrenal replacement, most commonly, lung and breast carcinomas. Others include renal cell carcinoma, gastric carcinoma, colonic adenocarcinoma, pancreatic carcinoma, transitional cell carcinoma, and melanoma.

REFERENCES

1. Ioachimescu AG, Remer EM, Hamrahian AH. Adrenal incidentalomas: a disease of modern technology offering opportunities for improved patient care. Endocrinol Metab Clin North Am 2015;44:335-54.
2. Ahn SH, Kim JH, Baek SH, et al. Characteristics of adrenal incidentalomas in a large, prospective computed tomography-based multicenter study: the COAR study in Korea. Yonsei Med J 2018; 59:501-10.
3. Lattin GE Jr, Sturgill ED, Tujo CA, et al. From the radiologic pathology archives: adrenal tumors and tumor-like conditions in the adult: radiologic-pathologic correlation. Radiographics 2014;34:805-29.
4. Bhargav PR, Mishra A, Agarwal G, Agarwal A, Verma AK, Mishra SK. Adrenal incidentalomas: experience in a developing country. World J Surg 2008;32:1802-8.
5. Sharabi Y, Goldstein DS, Bentho O, et al. Sympathoadrenal function in patients with paroxysmal hypertension: pseudopheochromocytoma. J Hypertens 2007;25:2286-95.
6. Mann SJ. Labile and paroxysmal hypertension: common clinical dilemmas in need of treatment studies. Curr Cardiol Rep 2015;17:99.
7. Chabre O. The difficulties of pseudo-Cushing's syndrome (or "non-neoplastic hypercortisolism"). Ann Endocrinol (Paris) 2018;79:138-45.
8. MacFarland SP, Mostoufi-Moab S, Zelley K, et al. Management of adrenal masses in patients with Beckwith-Wiedemann syndrome. Pediatr Blood Cancer 2017;64:10.1002/pbc.26432.
9. Mayo-Smith WW, Song JH, Boland GL, et al. Management of incidental adrenal masses: a white paper of the ACR Incidental Findings Committee. J Am Coll Radiol 2017;14:1038-44.
10. Taieb D, Pacak K. Molecular imaging and theranostic approaches in pheochromocytoma and paraganglioma. Cell Tissue Res 2018;372:393-401.
11. Johnson PT, Horton KM, Fishman EK. Adrenal mass imaging with multidetector CT: pathologic conditions, pearls, and pitfalls. Radiographics 2009;29:1333-51.
12. Kenney PJ, Stanley RJ. Calcified adrenal masses. Urol Radiol 1987;9:9-15.

12b. Wolman M, Sterk VV, Gatt S, Frenkel M. Primary familial xanthomatosis with involvement and calcification of the adrenals. Report of two more cases in siblings of a previously described infant. Pediatrics 1961;28:742-57.

12c. Marshall WC, Ockenden BG, Fosbrooke AS, Cumings JN. Wolman's disease. A rare lipidosis with adrenal calcification. Arch Dis Child 1969; 44:331-41.

13. Lam AK. Update on adrenal tumours in 2017 World Health Organization (WHO) of endocrine tumours. Endocr Pathol 2017;28:213-27.
14. Decmann A, Perge P, Toth M, Igaz P. Adrenal myelolipoma: a comprehensive review. Endocrine 2018;59:7-15.
15. Stein SH, Latour F, Frost SS. Myelolipoma arising from ectopic adrenal cortex: case report and review of the literature. Am J Gastroenterol 1986; 81:999-1001.
16. Plaut A. Myelolipoma in the adrenal cortex; myeloadipose structures. Am J Pathol 1958;34:487-515.
17. Lam AK. Lipomatous tumours in adrenal gland: WHO updates and clinical implications. Endocr Relat Cancer 2017;24:R65-79.
18. Liu HP, Chang WY, Chien ST, et al. Intra-abdominal bleeding with hemorrhagic shock: a case of adrenal myelolipoma and review of literature. BMC Surg 2017;17:74.
19. Bishop E, Eble JN, Cheng L, et al. Adrenal myelolipomas show nonrandom X-chromosome inactivation in hematopoietic elements and fat: support for a clonal origin of myelolipomas. Am J Surg Pathol 2006;30:838-43.

20. Selye H, Stone H. Hormonally induced transformation of adrenal into myeloid tissue. Am J Pathol 1950;26:211-33.
21. Hegstrom JL, Kircher T. Alimentary tract ganglioneuromatosis-lipomatosis, adrenal myelolipomas, pancreatic telangiectasias, and multinodular thyroid goiter. A possible neuroendocrine syndrome. Am J Clin Pathol 1985;83:744-7.
22. Allison KH, Mann GN, Norwood TH, Rubin BP. An unusual case of multiple giant myelolipomas: clinical and pathogenetic implications. Endocr Pathol 2003;14:93-100.
23. Duralska M, Podkowinska-Polak R, Kuzaka P, Kuzaka B, Dzwonkowski J, Otto M. Adrenal angiomyolipoma: a rare localization. Pol Arch Intern Med 2019;129:52-3.
24. Granger JK, Houn HY, Collins C. Massive hemorrhagic functional adrenal adenoma histologically mimicking angiosarcoma. Report of a case with immunohistochemical study. Am J Surg Pathol 1991;15:699-704.
25. Plaut A. Hemangiomas and related lesions of the adrenal gland. Virchows Arch Pathol Anat Physiol Klin Med 1962;335:345-55.
26. Montgomery E, Epstein JI. Anastomosing hemangioma of the genitourinary tract: a lesion mimicking angiosarcoma. Am J Surg Pathol 2009;33:1364-9.
27. Ross M, Polcari A, Picken M, Sankary H, Milner J. Anastomosing hemangioma arising from the adrenal gland. Urology 2012;80:e27-8.
28. Lappa E, Drakos E. Anastomosing hemangioma: short review of a benign mimicker of angiosarcoma. Arch Pathol Lab Med 2020;144:240-4.
29. Cornejo KM, Hutchinson L, Cyr MS, et al. MYC Analysis by fluorescent in situ hybridization and immunohistochemistry in primary adrenal angiosarcoma (PAA): a series of four cases. Endocr Pathol 2015;26:334-41.
30. Nakagawa N, Takahashi M, Maeda K, Fujimura N, Yufu M. Case report: adrenal haemangioma coexisting with malignant haemangioendothelioma. Clin Radiol 1986;37:97-9.
31. Hendry S, Forrest C. Epithelioid angiosarcoma arising in an adrenal cortical adenoma: a case report and review of the literature. Int J Surg Pathol 2014;22:744-8.
32. Criscuolo M, Valerio J, Gianicolo ME, Gianicolo EA, Portaluri M. A vinyl chloride-exposed worker with an adrenal gland angiosarcoma: a case report. Ind Health 2014;52:66-70.
33. Hung YP, Hornick JL. Immunohistochemical biomarkers of mesenchymal neoplasms in endocrine organs: diagnostic pitfalls and recent discoveries. Endocr Pathol 2018;29:189-98.
34. Huang X, Nayar R, Zhou H. Primary adrenal gland epithelioid sarcoma: a case report and literature review. Diagn Cytopathol 2019;47:918-21.
35. Michalopoulos N, Laskou S, Karayannopoulou G, Pavlidis L, Kanellos I. Adrenal gland lymphangiomas. Indian J Surg 2015;77:1334-42.
36. Parelkar SV, Sampat NP, Sanghvi BV, et al. Case report of bilateral adrenal leiomyoma with review of literature. Pediatr Surg Int 2013;29:655-8.
37. Zhou Y, Tang Y, Tang J, Deng F, Gong G, Dai Y. Primary adrenal leiomyosarcoma: a case report and review of literature. Int J Clin Exp Pathol 2015;8:4258-63.
38. Zhao M, Li C, Zheng J, Yan M, Sun K, Wang Z. Cystic lymphangioma-like adenomatoid tumor of the adrenal gland: report of a rare case and review of the literature. Int J Clin Exp Pathol 2013;6:943-50.
39. El-Daly H, Rao P, Palazzo F, Gudi M. A rare entity of an unusual site: adenomatoid tumour of the adrenal gland: a case report and review of the literature. Patholog Res Int 2010;2010:702472.
40. Erber R, Warth A, Muley T, Hartmann A, Herpel E, Agaimy A. BAP1 loss is a useful adjunct to distinguish malignant mesothelioma including the adenomatoid-like variant from benign adenomatoid tumors. Appl Immunohistochem Mol Morphol 2020;28:67-73.
41. Taskin OC, Gucer H, Mete O. An unusual adrenal cortical nodule: composite adrenal cortical adenoma and adenomatoid tumor. Endocr Pathol 2015;26:370-3.
42. Timonera ER, Paiva ME, Lopes JM, Eloy C, van der Kwast T, Asa SL. Composite adenomatoid tumor and myelolipoma of adrenal gland: report of 2 cases. Arch Pathol Lab Med 2008;132:265-7.
43. Pant L, Kalita D, Chopra R, Das A, Jain G. Malignant perivascular epithelioid cell tumor (PEcoma) of the adrenal gland: report of a rare case posing diagnostic challenge with the role of immunohistochemistry in the diagnosis. Endocr Pathol 2015;26:129-34.

43b. Wakefield CB, Sadow PM, Hornick JL, Fletcher CD, Barletta JA, Anderson WJ. PEComa of the adrenal gland: a clinicopathologic series of 7 cases. Am J Surg Pathol 2023;47:1316-24.

44. Schieda N, Kielar AZ, Al Dandan O, McInnes MD, Flood TA. Ten uncommon and unusual variants of renal angiomyolipoma (AML): radiologic-pathologic correlation. Clin Radiol 2015;70:206-20.
45. Bennett JA, Ordulu Z, Pinto A, et al. Uterine PEComas: correlation between melanocytic marker expression and TSC alterations/TFE3 fusions. Mod Pathol 2022;35:515-23.

46. Argani P, Lal P, Hutchinson B, Lui MY, Reuter VE, Ladanyi M. Aberrant nuclear immunoreactivity for TFE3 in neoplasms with TFE3 gene fusions: a sensitive and specific immunohistochemical assay. Am J Surg Pathol 2003;27:750-61.
47. Argani P, Zhong M, Reuter VE, et al. TFE3-fusion variant analysis defines specific clinicopathologic associations among Xp11 translocation cancers. Am J Surg Pathol 2016;40:723-37.
48. Lau SK, Spagnolo DV, Weiss LM. Schwannoma of the adrenal gland: report of two cases. Am J Surg Pathol 2006;30:630-4.
49. Zhou J, Zhang D, Wang G, et al. Primary adrenal microcystic/reticular schwannoma: clinicopathological and immunohistochemical studies of an extremely rare case. Int J Clin Exp Pathol 2015;8:5808-11.
50. Gupta P, Aggarwal R, Sarangi R. Solitary neurofibroma of the adrenal gland not associated with type-1 neurofibromatosis. Urol Ann 2015;7:124-6.
50b. Stenman A, Falhammar H, Zedenius J, Juhlin CC. Adrenal and periadrenal schwannoma: histological, molecular and clinical characterization of an institutional case series. Endocrine 2023;82:631-7.
51. Min KW, Clemens A, Bell J, Dick H. Malignant peripheral nerve sheath tumor and pheochromocytoma. A composite tumor of the adrenal. Arch Pathol Lab Med 1988;112:266-70.
52. Chandrasoma P, Shibata D, Radin R, Brown LP, Koss M. Malignant peripheral nerve sheath tumor arising in an adrenal ganglioneuroma in an adult male homosexual. Cancer 1986;57:2022-5.
53. Prieto-Granada CN, Wiesner T, Messina JL, Jungbluth AA, Chi P, Antonescu CR. Loss of H3K27me3 expression is a highly sensitive marker for sporadic and radiation-induced MPNST. Am J Surg Pathol 2016;40:479-89.
54. Mack E, Sarto GE, Crummy AB, Carlson IH, Curet LB, Wu J. Virilizing adrenal ganglioneuroma. JAMA 1978;239:2273-4.
55. Aguirre P, Scully RE. Testosterone-secreting adrenal ganglioneuroma containing Leydig cells. Am J Surg Pathol 1983;7:699-705.
56. Horvath E, Chalvardjian A, Kovacs K, Singer W. Leydig-like cells in the adrenals of a woman with ectopic ACTH syndrome. Hum Pathol 1980;11:284-7.
57. Ejaz S, Shawa H, Henderson SA, Habra MA. Melanoma of unknown primary origin presenting as a rapidly enlarging adrenal mass. BMJ Case Rep 2013;2013:bcr2013009727.
58. Carstens PH, Kuhns JG, Ghazi C. Primary malignant melanomas of the lung and adrenal. Hum Pathol 1984;15:910-4.
59. Landas SK, Leigh C, Bonsib SM, Layne K. Occurrence of melanin in pheochromocytoma. Mod Pathol 1993;6:175-8.
60. Maison N, Korpershoek E, Eisenhofer G, Robledo M, de Krijger R, Beuschlein F. Somatic RET mutation in a patient with pigmented adrenal pheochromocytoma. Endocrinol Diabetes Metab Case Rep 2016;2016:150117.
61. Gonzalez-Saez L, Pita-Fernandez S, Lorenzo-Patino MJ, Arnal-Monreal F, Machuca-Santacruz J, Romero-Gonzalez J. Primary melanoma of the adrenal gland: a case report and review of the literature. J Med Case Rep 2011;5:273.
62. Adameyko I, Lallemend F, Aquino JB, et al. Schwann cell precursors from nerve innervation are a cellular origin of melanocytes in skin. Cell 2009;139:366-79.
63. Orselli RC, Bassler TJ. Theca granuloma cell tumor arising in adrenal. Cancer 1973;31:474-7.
64. Mete O, Raphael S, Pirzada A, Asa SL. Is adrenal ovarian thecal metaplasia a misnomer? Report of three cases of radial scar-like spindle cell myofibroblastic nodule of the adrenal gland. Endocr Pathol 2011;22:222-5.
65. Bedri S, Erfanian K, Schwaitzberg S, Tischler AS. Mature cystic teratoma involving adrenal gland. Endocr Pathol 2002;13:59-64.
66. Garg A, Pollak-Christian E, Unnikrishnan N. A rare adrenal mass in a 3-month-old: a case report and literature review. Case Rep Pediatr 2017;2017:4542321.
67. Chen L, Fang L, Liu Z, et al. Giant adrenal germ cell tumour in a 59-year-old woman. Can Urol Assoc J 2016;10:E201-3.
68. Santonja C, Diaz MA, Dehner LP. A unique dysembryonic neoplasm of the adrenal gland composed of nephrogenic rests in a child. Am J Surg Pathol 1996;20:118-24.
69. Debatin JF, Spritzer CE, Dunnick NR. Castleman disease of the adrenal gland: MR imaging features. AJR Am J Roentgenol 1991;157:781-3.
70. Laurent C, Casasnovas O, Martin L, et al. Adrenal lymphoma: presentation, management and prognosis. QJM 2017;110:103-9.
71. Haykal T, Zayed Y. Primary adrenal Hodgkin lymphoma. Clin Case Rep 2019;7:568-9.
72. Paolo WF Jr, Nosanchuk JD. Adrenal infections. Int J Infect Dis 2006;10:343-53.
73. Koene RJ, Catanese J, Sarosi GA. Adrenal hypofunction from histoplasmosis: a literature review from 1971 to 2012. Infection 2013;41:757-9.
74. Gajendra S, Sharma R, Goel S, et al. Adrenal histoplasmosis in immunocompetent patients presenting as adrenal insufficiency. Turk Patoloji Derg 2016;32:105-11.
75. Benjamin E, Fox H. Malakoplakia of the adrenal gland. J Clin Pathol 1981;34:606-11.

76. Slavin RE, Walsh TJ, Pollack AD. Late generalized tuberculosis: a clinical pathologic analysis and comparison of 100 cases in the preantibiotic and antibiotic eras. Medicine (Baltimore) 1980; 59:352-66.
77. Ali A, Singh G, Balasubramanian SP. Acute non-traumatic adrenal haemorrhage-management, pathology and clinical outcomes. Gland Surg 2018;7:428-32.
78. Berneis K, Buitrago-Tellez C, Muller B, Keller U, Tsakiris DA. Antiphospholipid syndrome and endocrine damage: why bilateral adrenal thrombosis? Eur J Haematol 2003;71:299-302.
79. Eo H, Kim JH, Jang KM, et al. Comparison of clinico-radiological features between congenital cystic neuroblastoma and neonatal adrenal hemorrhagic pseudocyst. Korean J Radiol 2011; 12:52-8.
80. Vella A, Nippoldt TB, Morris JC 3rd. Adrenal hemorrhage: a 25-year experience at the Mayo Clinic. Mayo Clin Proc 2001;76:161-8.
81. Foster DG. Adrenal cysts. Review of literature and report of case. Arch Surg 1966;92:131-43.
82. Sebastiano C, Zhao X, Deng FM, Das K. Cystic lesions of the adrenal gland: our experience over the last 20 years. Hum Pathol 2013;44:1797-803.
83. Suh J, Heimann A, Cohen H. True adrenal mesothelial cyst in a patient with flank pain and hematuria: a case report. Endocr Pathol 2008; 19:203-5.
84. Koperski L, Pihowicz P, Szczepankiewicz B, et al. Clinicopathological and immunohistochemical analysis of epithelial-lined (true) cysts of the adrenal gland with proposal of a new histogenetic categorization. Pathol Res Pract 2017;213:1089-96.
85. Erickson LA, Lloyd RV, Hartman R, Thompson G. Cystic adrenal neoplasms. Cancer 2004;101: 1537-44.
86. Fernandez-Vega I, Camacho-Urkaray E, Guerra-Merino I. Huge adrenal hemorrhagic endothelial cyst secondary to an adrenal arteriovenous malformation and mimicking a malignant lesion. Endocr Pathol 2014;25:443-5.
87. Lee HJ, Park YM, Jee BC, Kim YB, Suh CS. Various anatomic locations of surgically proven endometriosis: a single-center experience. Obstet Gynecol Sci 2015;58:53-8.
88. Bastide C, Boyer L, Djellouli N, Baguet JC, Viallet JF. [Bilateral adrenal cysts and hepatorenal polycystic disease]. Presse Med 1997;26:711-2. [French]
89. Gaffey MJ, Mills SE, Medeiros LJ, Weiss LM. Unusual variants of adrenal pseudocysts with intracystic fat, myelolipomatous metaplasia, and metastatic carcinoma. Am J Clin Pathol 1990;94:706-13.
90. Akbulut S. Incidentally detected hydatid cyst of the adrenal gland: a case report. World J Clin Cases 2016;4:269-72.
91. Brenner DS, Jacobs SC, Drachenberg CB, Papadimitriou JC. Isolated visceral leishmaniasis presenting as an adrenal cystic mass. Arch Pathol Lab Med 2000;124:1553-6.
92. Disibio G, French SW. Metastatic patterns of cancers: results from a large autopsy study. Arch Pathol Lab Med 2008;132:931-9.
93. Abrams HL, Spiro R, Goldstein N. Metastases in carcinoma; analysis of 1000 autopsied cases. Cancer 1950;3:74-85.
94. Zaidi SS, Lakhani VT, Fadare O, Khabele D. Adrenal gland metastasis is an unusual manifestation of endometrial cancer. Case Rep Surg 2013;2013:428456.
95. Lam KY, Lo CY. Metastatic tumours of the adrenal glands: a 30-year experience in a teaching hospital. Clin Endocrinol (Oxf) 2002;56:95-101.
96. Cummings MC, Simpson PT, Reid LE, et al. Metastatic progression of breast cancer: insights from 50 years of autopsies. J Pathol 2014;232: 23-31.
97. Bazhenova L, Newton P, Mason J, Bethel K, Nieva J, Kuhn P. Adrenal metastases in lung cancer: clinical implications of a mathematical model. J Thorac Oncol 2014;9:442-6.
98. Gao XL, Zhang KW, Tang MB, Zhang KJ, Fang LN, Liu W. Pooled analysis for surgical treatment for isolated adrenal metastasis and non-small cell lung cancer. Interact Cardiovasc Thorac Surg 2017;24:1-7.
99. Ng CS, Altinmakas E, Wei W, et al. Utility of intermediate-delay washout CT Images for differentiation of malignant and benign adrenal lesions: a multivariate analysis. AJR Am J Roentgenol 2018;211:W109-15.
100. Mitchell ML, Ryan FP Jr, Shermer RW. Pulmonary adenocarcinoma metastatic to the adrenal gland mimicking normal adrenal cortical epithelium on fine needle aspiration. Acta Cytol 1985;29:994-8.
101. Min KW, Song J, Boesenberg M, Acebey J. Adrenal cortical nodule mimicking small round cell malignancy on fine needle aspiration. Acta Cytol 1988;32:543-6.
102. Satoh H, Saito R, Hisata S, et al. An ectopic ACTH-producing small cell lung carcinoma associated with enhanced corticosteroid biosynthesis in the peritumoral areas of adrenal metastasis. Lung Cancer 2012;76:486-90.
103. Tsalis K, Zacharakis E, Sapidis N, Lambrou I, Zacharakis E, Betsis D. Adrenal metastasis as first presentation of hepatocellular carcinoma. World J Surg Oncol 2005;3:50.

104. Okano K, Usuki H, Maeta H. Adrenal metastasis from hepatocellular carcinoma through an adrenohepatic fusion. J Clin Gastroenterol 2004;38:912.
105. Nguyen T, Phillips D, Jain D, et al. Comparison of 5 immunohistochemical markers of hepatocellular differentiation for the diagnosis of hepatocellular carcinoma. Arch Pathol Lab Med 2015;139:1028-34.
106. Siemer S, Lehmann J, Kamradt J, et al. Adrenal metastases in 1635 patients with renal cell carcinoma: outcome and indication for adrenalectomy. J Urol 2004;171:2155-9.
107. Altinmakas E, Ucisik-Keser FE, Medeiros LJ, Ng CS. CT and ^{18}F- FDG-PET-CT findings in secondary adrenal lymphoma with pathologic correlation. Acad Radiol 2019;26:e108-14.
108. Kartsios C, Kaloyannidis P, Yannaki E, et al. Spontaneous adrenal haemorrhage as a manifestation of isolated relapse of non-Hodgkin's lymphoma. Acta Haematol 2003;110:197-9.
109. Askarian F, Xu D. Adrenal enlargement and insufficiency: a common presentation of intravascular large B-cell lymphoma. Am J Hematol 2006;81:411-3.
110. Pistolesi S, Fontanini G, Barellini L, et al. Meningeal hemangiopericytoma metastatic to the adrenal gland with multiple metastases to bones and lungs: a case report. Tumori 2004; 90:147-50.
111. Seitz G, Schüder G. Neoplasm to neoplasm metastasis pheochromocytoma harboring a metastasis of breast cancer. Pathol Res Pract 1987;182:228-32.
112. Martin JT, Alkhoury F, Helton S, Fiedler P, Sakharova O, Yood S. Metastatic adenocarcinoma within a functioning adrenal adenoma: a case report. Cases J 2009;2:7965.
113. Carvalho F, Louro F, Zakout R. Adrenal insufficiency in metastatic lung cancer. World J Oncol 2015;6:375-7.
114. Akcay MN, Tekin SB, Akcay G. Addisonian crisis due to adrenal gland metastasis in Hodgkin's disease. Int J Clin Pract 2003;57:840-1.
115. Redman BG, Pazdur R, Zingas AP, Loredo R. Prospective evaluation of adrenal insufficiency in patients with adrenal metastasis. Cancer 1987;60:103-7.

INDEX*

A

*In a series of numbers, those in boldface indicate the main discussion of the entity.